Power Electronic Systems

POWER ELECTRONIC SYSTEMS

THEORY AND DESIGN

JAI P. AGRAWAL

Prentice Hall

Upper Saddle River, New Jersey
Columbus, Ohio

Library of Congress Cataloging-in-Publication Data

Agrawal, Jai P.
Power electronic systems : theory and design / Jai P. Agrawal.
p. cm.
Includes index.
ISBN 0-13-442880-3
1. Power electronics. I. Title.

TK7881.15 .A38 2001
00-059845

Editor in Chief: Stephen Helba
Assistant Vice President and Publisher: Charles E. Stewart, Jr.
Production Editor: Alexandrina Benedicto Wolf
Production Coordination: Lisa Garboski, bookworks
Design Coordinator: Robin G. Chukes
Cover Designer: Thomas Borah
Cover Image: Marjory Dressler
Production Manager: Matthew Ottenweller
Marketing Manager: Barbara Rose

This book was set in Times Roman by Carlisle Communications, Ltd. It was printed and bound by R.R. Donnelley & Sons Company. The cover was printed by Phoenix Color Corp.

Prentice Hall

10 9 8 7 6 5 4 3 2 1
ISBN 0-13-442880-3

My Parents:
Shri Babu Ram Samaraya
Shrimati Janak Kishori
My Gurus and My Motherland Bharat

Preface

The field of power electronics encompasses the application of fundamental concepts in several disciplines: electronic devices and circuits, signals and systems, motor drives, and control systems. A course in power electronics should tie together all these diverse fundamental concepts into a consolidated core and add the considerations of power utilization, quality, interfacing, and design issues, which are special to the field of power electronics. This book is written with the above focus in mind with a conscious effort to simplify mathematics by use of MATLAB™ and to concentrate on developing a robust understanding of the subject. This book is intended as a textbook for a course on power electronics for junior and senior undergraduate students in electrical engineering programs. It could also serve as a self-learning book for practicing engineers. The prerequisites for this book are courses on electrical circuits and systems, electronic devices and circuits, and mathematics courses in calculus, differential equations, Fourier series and transformations, and linear algebra.

This book, intended for a one- or two-semester course, is divided into four parts. Part I presents an overview of the field of power electronics and the review of important mathematical concepts such as determining the average and rms values and the harmonic profile of waveforms, which are essential for understanding the rest of the chapters.

Part II provides an understanding of components used in the design of power electronics circuits in the generic categories of power diodes, transistors, and thyristors. Design of current sources, inductors, and transformers illustrated using design examples. Concepts are introduced for power losses during switching transitions, and on-state and off-state of semiconductor devices. This part also introduces the driver and protection circuits for each device discussed.

Part III discusses the classes of switch-mode converters: dc-dc, dc-ac, ac-dc, ac-ac, and the resonant converters. The focus is on topologies, performance measures, and performance characteristics.

Part IV covers the application systems such as power factor correction, electric utility interfacing, converter control, power supply, electronic ballast, and motor drives. This part also presents practical design issues such as temperature control, selection of heat sinks, protection, packaging, shielding, and layout.

Each chapter contains several, design examples to reinforce the concept learned, which illustrate the decision choices and selection of components. MATLAB® has been used extensively in these examples and also in the elaboration of the converter operation. PSPICE® simulation examples are included wherever possible.

Chapters 1 and 2 provide review and focus on the characteristics of components used in power electronic circuits. Chapters 3 to 5, on semiconductor switching components, may be covered at a faster pace if students have a strong prerequisite in semiconductor devices. Chapters 6 to 11 form the core of the course. Chapters 12, 14, and 15 discuss the design requirements of some specific illustrations and, therefore, can be covered at a faster rate. Chapter 16 on thermal and other design issues, in my opinion, must not be excluded. Chapter 13 presents the development of average and state-space average models of power converter systems followed by the derivation of the transfer function. It is left to instructor's discretion whether to cover it in the undergraduate course or not.

Several reviewers provided valuable assistance during the development of this text, and I am grateful for their input. They are Charles L. Bachman, Southern Polytechnic State University; Shamala Chickamenahalli, Wayne State University; Alexander E. Emanuel, Worcester Polytechnic Institute; Michael L. Holcombe, Purdue University; Rickie L. Miller, Ferris State University; Medhat M. Morcos, Kansas State University; and Shekhar Pradhan, Bluefield State College. As a final note, I am grateful to my wife Vaidehi, my children Sanjay, Vivek and Kshama for their patience and encouragement during the time I devoted to writing and revising this book.

Jai P. Agrawal

CONTENTS

POWER ELECTRONIC SYSTEMS

PART ONE
INTRODUCTION

CHAPTER 1

Bird's-Eye View of Power Electronics Systems

OBJECTIVES

1. The overall systems view of power electronic converters
2. The desired performance characteristics, which include power efficiency, power factor, undesirable harmonic content of voltages and currents, input–output relationships, parameters of control, etc.
3. Introduction to modeling of elements and systems of power electronic converters
4. Introduction to MATLAB methods of solution and graphic techniques

1.1 INTRODUCTION

Electric power is the muscle of modern industry and power electronics makes its utilization smarter. Managing power is an essential gradient in system design in all fields of electrical engineering but not necessarily the main concern except in power electronics and power generating and distribution systems. For example, in communication engineering, the premium is placed on the accuracy and quality of signal transmission and reception. Power electronics is solely concerned with the processing of electric energy. The electric energy by itself is not useful to mankind. It must be converted to other energy forms such as heat, light, sound and mechanical energies, which are directly useful to human beings and society. The focus in power electronics is on conversion, efficiency of conversion and control of energy.

Electric power is used in almost every part of modern home and society. An efficient use of power, therefore, is not only tremendously advantageous but has become essential. The objective of power electronics is to improve the quality and utilization of electric power. Generation, transmission, distribution and utilization of electric power takes place at different levels to suit the available components, processes and technologies involved. Interfacing of different technologies needs intelligent control and conversion techniques. Power electronics addresses these issues.

Recent advances in semiconductor switching devices have contributed very significantly to research and reengineering in the field of modern power electronics. Power electronics applications span a wide range on the power scale: milliwatts in

TABLE 1-1 POWER ELECTRONICS APPLICATIONS

Aerospace	
Aircraft power systems	Space vehicle power systems
Satellite power systems	

Automotive	
Alarms and security systems	Electric vehicles
Audio and RF amplifiers	Regulators
Battery chargers	

Commercial	
Advertising neon signs and displays	Electronic ballasts
Battery chargers	Hand power tools
Blenders	Light dimmers
Computers	Mercury-arc lamps
Electric blankets	Photocopiers
Electric fans	Vending machines
	Washing machines

Games and entertainment	
Games and toys	Televisions
Movie projectors	

Home appliances	
Audio and RF amplifiers	Photography
Food warming trays	Refrigerators
Garage door openers	Sewing machines
Phonographic equipment	

Industrial	
Air conditioning	Conveyors
Battery chargers	Cranes and hoists
Blowers	Dryers
Boilers	Electric furnaces and ovens
Chemical processing equipment	Electric vehicles
Contactor and circuit breakers	Electromagnets
	Electroplating

Industrial *(continued)*	
Electronic ignitions	Oil-well drilling equipment
Electrostatic precipitators	Paper mill machinery
Elevators	Power supplies
Flashers	Printing press machinery
Gas turbine starters	Pumps and compressors
Generator exciters	Relays
Grinders and mixers	Servo systems
High-voltage DC (HVDC)	Steel mill instrumentation
Induction heating	Temperature controls
Laser power supplies	Timers
Linear induction motion control	Traffic signal controls
Machine tools	Ultrasonic generators
Mining power equipments	Uninterruptible power supplies (UPS)
Motor drives and starters	Vacuum cleaners
Movers	Welding equipment
Nuclear reactor control	

Medical	
Fitness machines	Medical instrumentation
Laser power supplies	

Security systems	
Alarms and security systems	Radar/Sonar

Telecommunications	
Solar power supplies	VLF transmitters
Uninterruptible power supplies (UPS)	Wireless communication power supplies

Transportation	
Magnetic levitation	Trains and locomotives
Motor drives	

Utility systems
Power factor correction and VAR compensators

wireless personal communication sets or cordless screwdrivers to megawatts in High Voltage DC transmission systems and huge industrial motor drives.

1.2 A SIMPLE VOLTAGE CONVERTER

Consider a simple example of designing a 5-volts dc power supply for TTL ICs from a 12-volts battery. A voltage divider circuit, shown in Figure 1.1(a), is the first thought that comes to mind. Selecting resistor $\mathbf{R_2}$ arbitrarily to be 1 kohm, $\mathbf{R_1}$ is determined to be 1.4 kohms.

$$\frac{R_2}{R_1 + R_2} = \frac{5}{12}, \quad Rs = \frac{R_1 R_2}{R_1 + R_2} = 583.3 \; ohms$$

The Thevenin equivalent circuit of voltage divider, given in Figure 1.1(b), consists of a 5 volts dc source in series with a source resistance R_S of 583.3 ohms. This circuit has two problems:

1. The voltage at the terminals of voltage divider, V_O, is 5 volts only when no current is drawn. It reduces rapidly as more current is drawn from it.

   $$V_O = 5 - 583.3\, I_O$$

   I_O is the current drawn from the output terminals, also called the load current. V_O will drop by 0.5 volts (maximum permitted in standard TTL ICs) if a current of 0.86 mA is drawn from it. This current is hardly sufficient for a single IC. The problem of varying output voltage with increasing load current is referred to as the voltage regulation problem. Ideally, V_O should not vary for any amount of current drawn; that is, the output voltage should have 100% regulation, or exhibits zero regulation error.

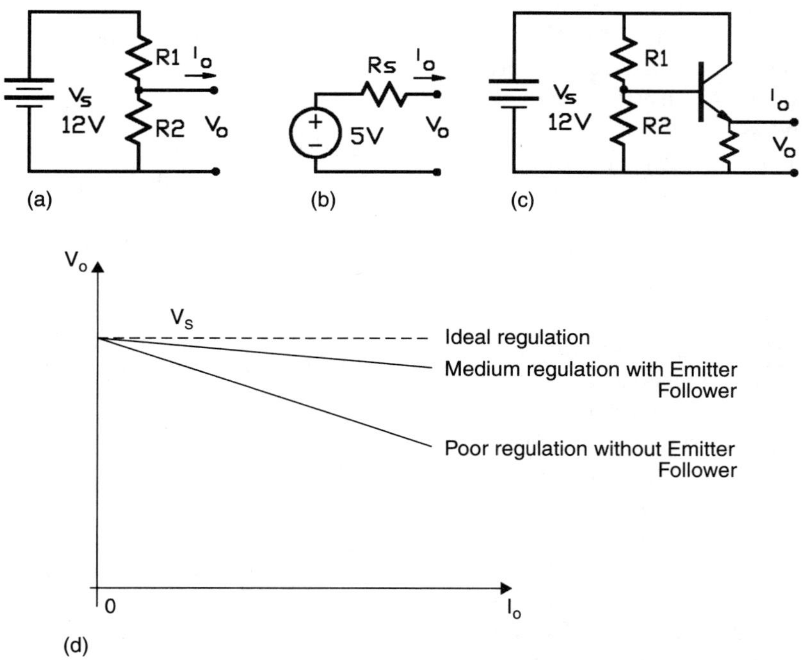

Figure 1.1
(a) A voltage divider (b) Thevenin's equivalent circuit
(c) voltage divider with emitter follower and
(d) voltage regulation characteristics

2. Power is continuously drained from the source even when load current is zero, 60 milliwatts. This power drain is converted to heat in resistors R_1 and R_2. For a load current of 1 mA, the actual power delivered is 4.4 milliwatts, a power efficiency of 6.8%.

Heat generated in resistors must be removed from the circuit for a reliable operation. The power is removed by using large size components or by use of heat sinks.

The voltage regulation can be improved by reducing the value of source resistance $\mathbf{R_S}$ using an emitter follower as shown in Figure 1.1(c). The transistor is biased in the linear region of the characteristics, hence, it dissipates an additional amount of power:

$$P_{loss} = (V_S - V_O)\, I_O$$

High power is dissipated at high load current. Again, this power is converted to heat, and consequently the temperature of the transistor package raises. A raise in temperature reduces the performance and reliability of the transistor. The dissipated heat is removed by using heat sink.

A better design uses a switch in place of the transistor, as shown in Figure 1.2(a). The switch may be a bipolar junction transistor operated either in the cut-off or the saturation region. The switch is turned on and off alternately in a periodic fashion. The output voltage, after the switch, has a pulse waveform, see Figure 1.2(b). The voltage waveform contains a dc component and an ac component. The ac component is removed by passing it through a filter. The voltage V_O at the output terminals is the dc component, the value of which is determined by the on-time of the switch. Assuming the transistor switch to be ideal, no power is lost in the transistor.

The filter circuit consists of inductors and capacitors, so no power is lost in the filter. Overall, no power is lost in the voltage converter. The practical switch-mode converters, however, have some power losses owing to parasitic elements such as the winding resistance of inductors, the leakage resistance in capacitors, and nonideal switching char-

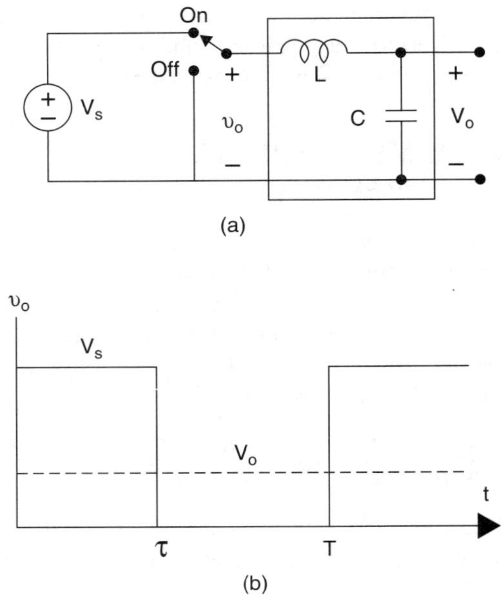

Figure 1.2
(a) The switch-mode voltage divider and (b) output voltage waveform

acteristics of semiconductor devices. The switch-mode converter exhibits significantly higher power efficiency than the nonswitch-mode converter.

The power electronic circuit operation requires temporary storage of energy in inductors and capacitors. Sizes of inductors and capacitors are dictated by the amount of energy to be stored. Furthermore, the ac components of voltage or currents cannot be removed completely by filters. These are smaller priced items when high power applications are involved.

1.3 SYSTEMS VIEW OF POWER ELECTRONIC CONVERTERS

Figure 1.3(a) shows a single-input single-output power converter system. The source provides power, the converter converts it into a usable form to suit the sink where the power is utilized, that is, removed out of the system. The source is either a dc

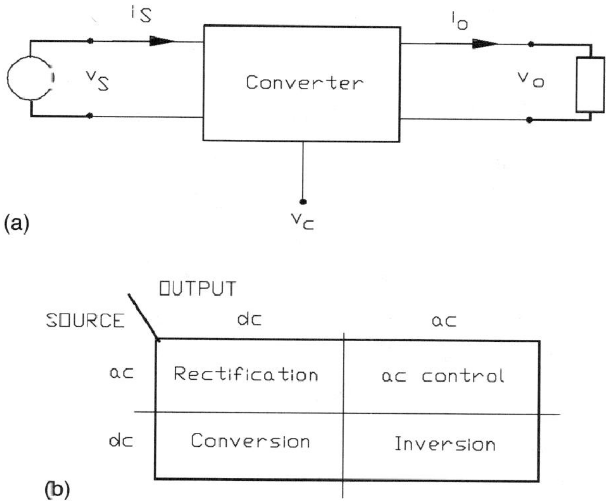

Figure 1.3
(a) A power electronic converter system and
(b) converter classification

or a periodic ac voltage/current source. The sink is either an electric load R, L or C, or another source, or a device where the electrical energy is converted into nonelectrical form such as a motor. The $\mathbf{v_c}$ is the signal that controls a specific characteristic of the output variable. The voltage and current waveforms on the source and the sink sides are either the single-phase or a multiphase (usually three-phase) type. The converter system is comprised of switches, reactive components L, C, and transformers Switches include two-terminal devices such as diodes and three-terminal devices such as transistors or thyristors (SCRs).

A more general power electronic converter system may contain more than one input source and one or more output variables.

For simplifying the analysis of converter operation, the switches and other components are assumed to be ideal (ideality to be described later), linear and invariant over the time of observation. The switches are assumed to have the unrestricted voltage or current carrying capacities. Switching devices must be selected with these characteristics in mind.

Power electronic converters may be classified into four categories on the basis of the type of the source and the type of desired output characteristics:

1. dc to ac converter
2. ac to dc converter
3. dc to dc converter
4. ac to ac converter

DC to AC Converter (Inverter)

The dc to ac converter (inverter) takes power from a dc source (voltage or current) and delivers to a load. The output variable of the converter is a low-distortion ac voltage or ac current of single-phase or multi-phase. A practical inverter has either a battery, a solar powered dc voltage source or a line frequency (50-60 Hertz) derived dc voltage source (often unregulated). The dc-ac converters are widely used from very low-power portable electronic systems such as the flashlight discharge system in a photography camera to very high-power industrial systems. The frequency of output variable may vary from line frequency 50-60 Hz in emergency lighting systems to high frequency 1MHz in induction heating systems.

A basic inverter topology is shown in Figure 1.4(a). The input dc is converted by using a single switch into an ac of high frequency or low frequency depending on the application. When the switch is connected to the on-position, the source provides full positive voltage to the load. After a time interval, which can be controlled, the switch is connected to the off-position, providing the zero voltage to the load. The output voltage waveform is a pulsating dc waveform, as shown in Figure 1.4(b). A filter may be used at the output to shape the waveform as desired. Usually, the desired waveform is a sine wave in low-frequency applications such as the battery powered ac sources for emergency applications.

The three-phase ac waveform is generated by performing three parallel switching. The switches are turned on in such a manner that no two switches are on simultaneously. This is done by turning on switch S2 after a delay of $\frac{1}{3}$ of the switching period

T/3 with respect to switch S1 and delay the turning on of S3 further by T/3 with respect to S2. See Figure 1.5.

A more complex topology consists of a network of switches. The amplitude and frequency of the ac waveform is controlled by controlling the on-time of switches, the frequency of switching and the sequence of switching. The waveform shaping filter consists of reactive components, inductors and capacitors. The filter removes the undesired content of the ac waveform. The instantaneous difference in energy between the input and output sides is temporarily stored in the filter.

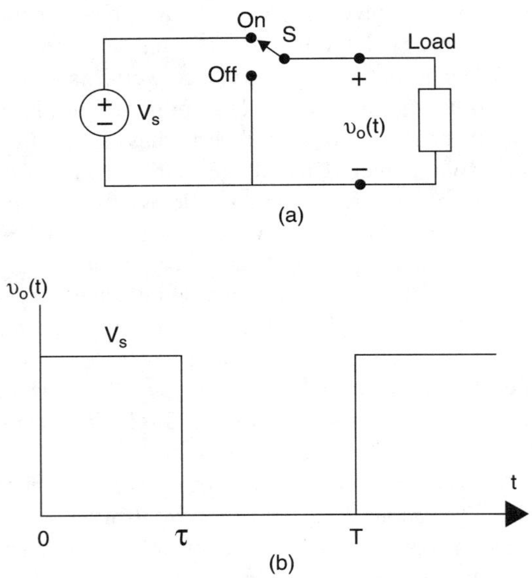

Figure 1.4
(a) A basic dc-ac converter (inverter) topology and (b) output waveform

Applications:

Aircraft and space power supplies

Uninterruptible power supplies

Variable-frequency ac motor drives

Aircraft variable-speed constant frequency supplies

Induction heating supplies

AC to DC Converter

The ac to dc converter (rectifier) takes power from one or more ac voltage/current sources of single or multiple phases and delivers to a load. The output variable is a low-ripple dc voltage or dc current. Many practical ac-dc converters use a voltage source with line frequency of 50 or 60 Hertz, single-phase or three-phase. The ac-dc converters are widely used from very low-power battery charger and dc power supply systems to very high-power dc motor drive systems.

A basic ac-dc converter topology is shown in Figure 1.6(a) and (b). In Figure 1.6(a), polarity of ac source drives the diode on and off. The diode D1 is on during the first half period where the ac source voltage is positive. The diode becomes off when the ac source voltage is negative, thus, giving a zero voltage in the second half period. The diode is basically an uncontrolled or a polarity controlled switch.

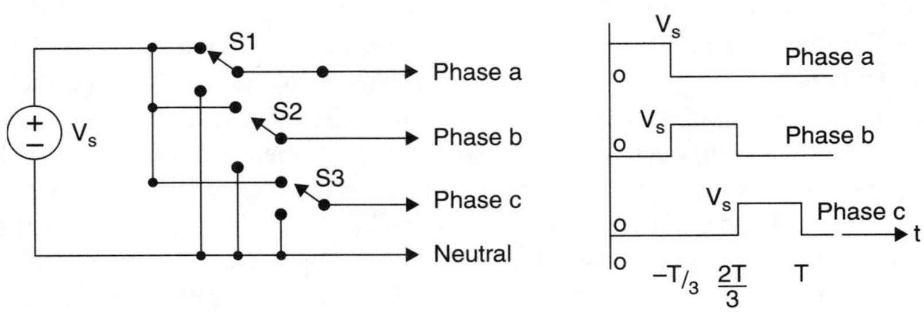

Figure 1.5
A basic 3-phase dc-ac converter

The voltage waveform after the diode contains a dc component along with ac component. The ac component is called *voltage ripple*. The L and C after the diode constitute a filter, which removes the ac component and the dc component is provided to the load.

In Figure 1.6(b), the instant of turning on and off switch relative to the ac source waveform, determines the shape of the output voltage waveform. Consequently the dc and ac components of the waveform can be controlled as desired. The L-C filter removes the ac component and the dc component is provided to the load. The reader may notice that if we ignore the placement of the filter, the ac-dc topology can be obtained by replacing the dc source with the ac source in the dc-ac inverter topology. Hence, the dc-ac and ac-dc converters are functionally similar except for the reverse direction of flow of energy.

The frequency of switching does not have to be restricted by the source frequency. It may be made several times higher, which would enable reduction in the value of filtering components and the consequent reduction in size and weight and increase in portability.

A more complex topology consists of a network of switches. The dc component of the waveform before the filter is controlled by controlling the instant of turning on switches relative to the ac source waveform, the on-time of switches and the sequence of switching. The instantaneous difference in energy between the input and output sides is temporarily stored in the filter components.

Applications:

DC motor drives

Regulated dc power supplies

High voltage dc transmission

Wind generator converters

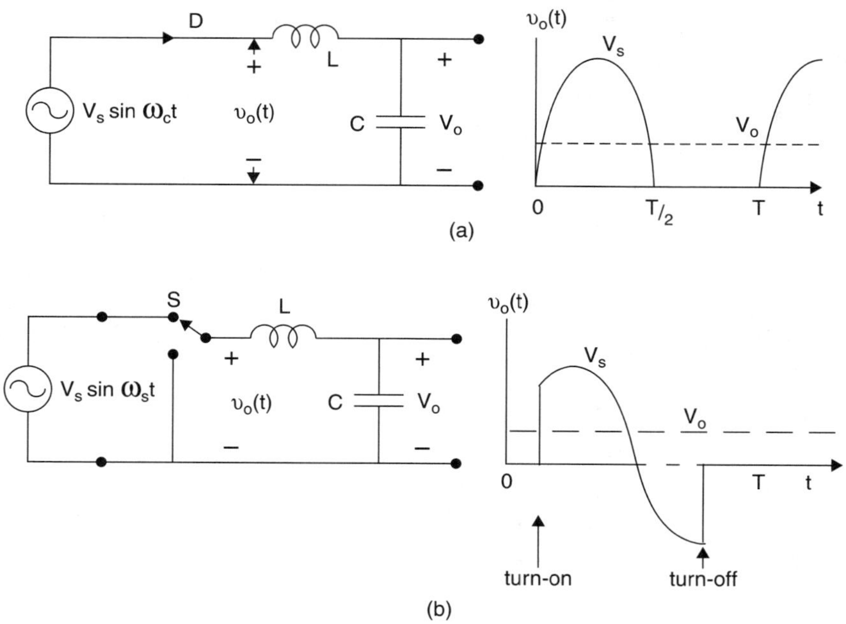

Figure 1.6
A basic ac-dc converter topology and output waveform (a) using diode and (b) using a switch

DC to DC Converter

The dc to dc converter takes power from a dc source (voltage or current) and delivers to a load. The output variable of the converter is one or more low-ripple dc voltages or dc currents (often regulated). The dc output voltage may be different in amplitude than the input source voltage. A practical dc-dc converter has either a battery, a solar powered dc voltage source or a line frequency (50-60 Hertz) derived dc voltage source (often unregulated). The dc-dc converters are widely used from very low-power portable electronic systems to very high-power industrial systems. The very high-power dc-dc converter, such as in electric traction, is also referred as the *chopper*.

The dc-dc converter is a combination of an dc-ac converter followed by an ac-dc converter. The voltage converter of Figure 1.2 is actually a basic dc-dc converter. The input dc is first converted to ac, usually of high frequency, and then the ac is converted to dc. When the switch is connected to the on position, the source provides full positive voltage to the load through the filter. After a time interval, which can be controlled, the switch is connected to the off position, providing the zero voltage to the load. The output voltage waveform after the switch is an ac waveform, as shown in Figure 1.2(b). The L and C after the switches constitute a filter, which removes the ac content of the voltage waveform, the dc content is available to the load.

A more complex topology consists of a network of switches. The ac waveform is controlled by the on-time of the switch and the sequence of switching. The ac is converted to dc by rectification and filtering or filtering alone. The rectification circuit contains a network of diodes or switches. The rectification with switches is referred to as *synchronous rectification*. The filter consists of reactive components, inductors and capacitors. The reactive components are required to provide a relative *short* to the ac contents of the waveform after the switches and the dc component is given to the load. The instantaneous difference in energy between the input and output sides is temporarily stored in the filter. Thus, the input dc voltage or current is transformed to a dc voltage or current of different amplitude. This converter can also be thought of as a dc transformer.

The switching frequency plays a very important part in design. High switching frequency enables reduction in the value of filtering components and the consequent reduction in size and weight and the increase in portability. The dc-dc converter with electrical isolation between the input and output is shown in Figure 1.7. The transformer in the middle may help in obtaining wide difference in voltages or currents on the input and output sides.

Applications:

Electric transportation

High performance regulated power supplies

Electronic ballasts

DC motor drive systems

AC to AC Converter

The ac to ac controller takes power from an ac voltage or current source and delivers to a load. The output variable is a low-distortion ac voltage or ac cur-

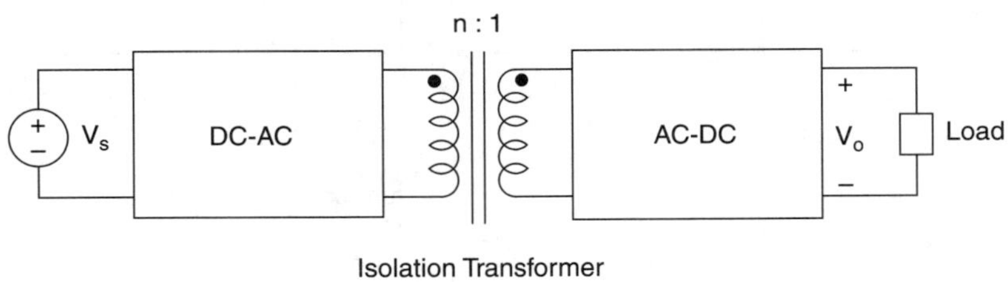

Figure 1.7
An isolated dc-dc converter

rent of equal or different frequency than the input ac source. The conversion may also involve single-phase to multiphase or vice versa. The practical ac controllers usually use line frequency 50-60 Hertz voltage sources, single or three phase. The ac-ac converter where the output frequency is lower than the frequency of the source is called the *cyclo-converter.* The output frequency in cyclo-converter is a simple fraction such as $\frac{1}{3}$, $\frac{1}{5}$ and so on of the source frequency. They are used in very high-power industrial applications. The ac to ac converter with no change of frequency is referred to as the *ac controller.*

A basic ac-ac converter topology is shown in Figure 1.8 in which the output frequency is the same as the source frequency. The switch is turned off at the end of each half period of the source waveform. The instant of turning on the switch relative to the ac source waveform determines the shape of the output voltage waveform, which in turn determines the rms voltage at the output terminals.

A more complex ac-ac conversion is achieved by first converting the input ac to the dc and the dc to the ac of desired frequency, amplitude and number of phases. Such a converter is called the *dc-link ac-ac converter,* shown in Figure 1.9. In this type of conversion, there is no restriction on the output frequency relative to the source frequency.

Applications:

Light dimmers	AC motor speed controls
Voltage regulators	Electronic tap changers
VAR regulators	Solid-state relays

1.4 ELEMENTS OF THE CONVERTER SYSTEMS: MODELING AND ASSUMPTIONS

Real-life elements of power electronic converter systems, the components and subsystems, are very complex, nonlinear in general and time-variant in

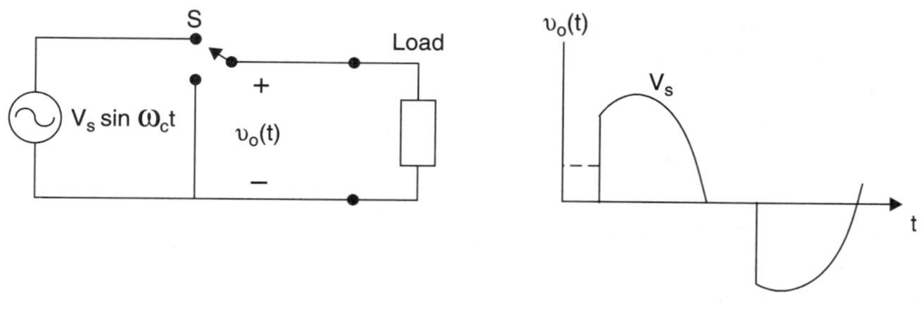

Figure 1.8
A basic ac-ac converter (controller) topology

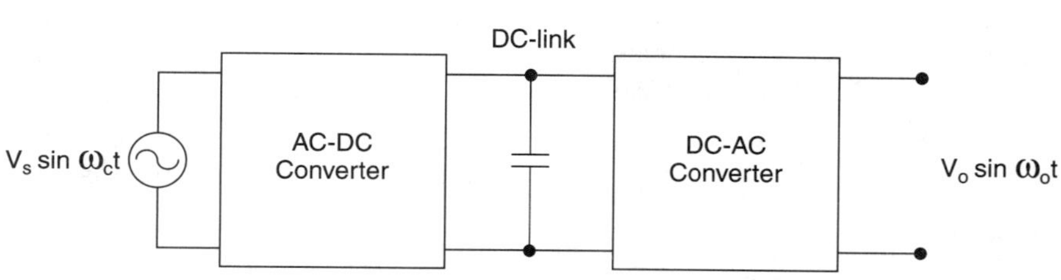

Figure 1.9
A dc-link ac-ac converter

nature. However, if these elements are used over a smaller time duration, within a limited range of its characteristics, fortunately they may be approximated (or say modeled) by linear time-invariant elements. We must keep in mind that the system analysis with such models is only as accurate as the models themselves are. No model is perfect and, hence, no analysis is all-inclusive. At best they provide us with the guidelines to enable us to perform a satisfactory engineering task. Engineers, nevertheless, always indulge in modeling the unknown and the not-so-well-known elements and systems.

1.4.1 Resistance

The v-I relation of the resistance element R is the well-known Ohm's law:

$$v = R\,i \tag{1-1}$$

R is the only energy dissipating element in power electronic systems. It is generally absent as a stand-alone element but exists as the load and as a parasitic element. The parasitic elements are the source resistance of battery and other dc sources, resistance of conductors, winding resistance in inductors, transformers and motors and the leakage resistance in capacitors.

In general, resistance is independent of frequency except for that in the conductors. The resistance of a solid conductor increases with the increase in frequency of current flowing through it. The high frequency current tends to concentrate near the surface of the conductor. The phenomenon is called the *skin effect.* The problem is solved by increasing the surface of conductors. A solid conductor with large diameter is changed to a bunch of conductors with smaller diameters for increasing the surface area. The bunched conductor is called the multistrand conductor.

1.4.2 Inductance

Inductance in power electronic converters is found as a stand-alone element, a load and a parasitic element such as self-inductance of conductors in the power distribution systems, lead inductance in the components and leakage inductance in transformers and motors.

The v-i relation of an inductor L is

$$v_L = L\frac{di_L}{dt} \text{ and } i_L = I_L(0) + \frac{1}{L}\int_0^t v_L\,dt \tag{1-2}$$

$I_L(0)$ is the initial current on the inductor. A large inductor, when observed over a small time segment **dt,** may be considered to be carrying a constant current because

$$\frac{di_L}{dt} = \frac{v_L}{L} \approx 0$$

Hence a large inductor may be modeled by a current source over the segment **dt.**

1.4.3 Capacitance

Like the inductance, the capacitance in power electronic converters is found as a stand-alone element, a load and a parasitic element. The parasitic capacitance is the interwinding and interlayer capacitance in transformers and the inherent capacitance in diodes, transistors and thyristors.

The v-i relation of a capacitor C is

$$i_C = C\frac{dv_C}{dt} \text{ and } v_C = Vc\,(0) + \frac{1}{C}\int_0^t i_C\,dt \tag{1-3}$$

$I_L(0)$ is the initial current on the inductor. A large capacitor, when observed over a small time segment **dt,** may be considered to have a constant voltage at its terminals because

$$\frac{dv_C}{dt} = \frac{i_C}{C} \approx 0$$

Hence a large capacitor may be modeled by a voltage source over the segment **dt.** These models of large inductor and capacitors work very satisfactorily when the circuits including them are excited by pulsating sources of high frequency.

1.4.4 Sources

A source is an energy giving element. The sources may be categorized in several ways: **voltage source** or **current source,** and **dc source** or **ac source.** The ac sources may be further classified on the basis of the number of phases: 1-phase, 3-phase or multi-phase. The equivalent circuit of a voltage source and a current source are shown in Figure 1.10.

The $v_S(t)$ and $i_S(t)$ are respectively the internally generated voltage and current waveforms. Z_S is the source impedance, constituted of the resistance and inductance of conductors. In the voltage source, the voltage at the source terminals is a function of the current flowing:

$$v_t(t) = v_S(t) - Z_S\, i_S(t) \tag{1-4}$$

The source current normally flows out of the terminals, but in some conditions the current flow may be in the reverse direction. The terminal voltage, thus, may be different in amplitude and waveform shape than the internally generated voltage source. An ideal voltage source has zero source impedance; therefore, its terminal voltage waveform is independent of current flow.

In the current source, the current from the source terminals is a function of the compliance voltage across its terminals:

$$i_t(t) = i_S(t) - \frac{v_S(t)}{Z_S} \tag{1-5}$$

Z_S is the source impedance, constituted of the resistance and capacitance between the source terminals and other parasitic elements. The compliance voltage may be positive or negative. The current from the source terminals, thus, may be different in amplitude and waveform shape than the internally generated current source. An ideal current source has infinite source impedance; therefore, the current waveform flowing out of its terminals is independent of the compliance voltage.

DC Source

A source that provides a voltage of constant amplitude at all times, is called the ***dc voltage source.***

A battery is closest to an ideal voltage source. A precharged large value capacitor is also a voltage source. The larger the value the more charge it can store and for longer time. The battery can also be considered as a very large value capacitor.

The current sources do not exist in nature. However, a preenergized large value inductor can be treated as a current source. Remember that a current source must be stored with a closed circuit at its terminals unlike an open circuit battery. A practical current source can be synthesized from a dc or ac voltage source in series with a large value inductor (see Figure 1.11).

Example 1-1

A precharged capacitor is to be used as the dc voltage source of 10 volts. The voltage is held to within 1% of 10 volts over a time interval of 50 S while supplying a current of 1 amp. Find the capacitor value.

$V_S = 10 \quad I_S = 1\ amp \quad \Delta V_S = 0.01\ V_S \quad \Delta t = 50\ \mu S$

From equation (1-2)

$$I_S = C\frac{\Delta V_S}{\Delta t} \Rightarrow C = \frac{I_S \Delta t}{\Delta V_S}$$

$$C = \frac{1x50x10^{-6}}{.01x10} = 500\ \mu F$$

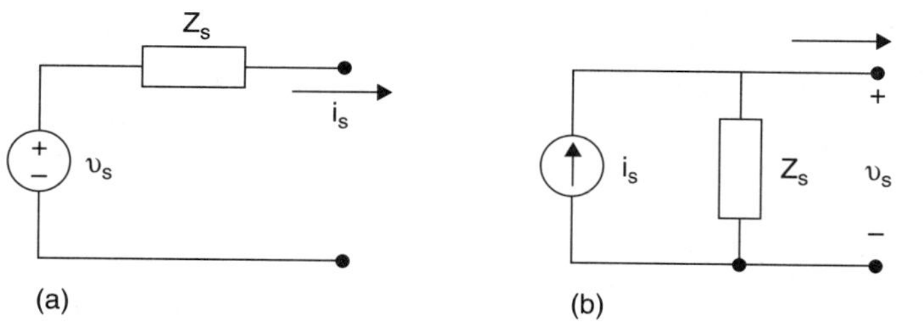

Figure 1.10
(a) A voltage source and (b) a current source

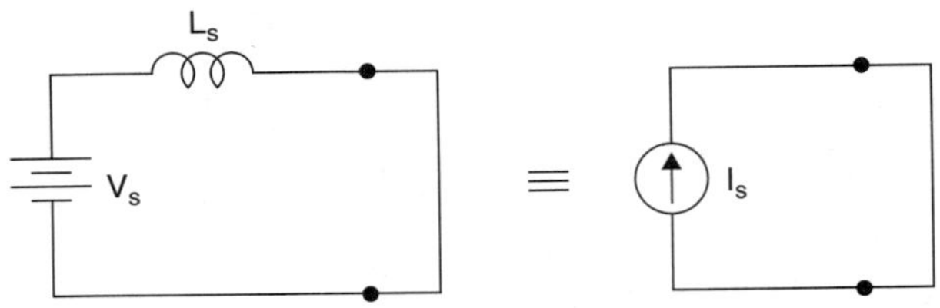

Figure 1.11
A synthesized dc current source

Example 1-2

Design a dc current source of 10 amps from a 12-volts battery for use in a converter operating at the switching frequency of 20 KHz. The current should not vary by more than 1% during a switching period. Use the circuit given in Figure 1.11.

The equivalent circuit of a practical large inductor in series with a battery is given in Figure 1.11. The dc current is determined by the total resistance in the circuit **R,** which is the sum of the winding resistance of the inductor and the source resistance of the battery (assumed zero in this example). The inductor current rises exponentially and settles to a steady state value given by

$$I_S = \frac{V_S}{R} = 10 \; amps$$

$$\Delta I_S = 0.01\, I_S \qquad T_S = \Delta t = 50 \; \mu S$$

The inductor should be wound by the copper wire of a suitable gauge so that the winding resistance is not more than 1.2 ohms. Furthermore, from equation (1-1),

$$V_S = L\frac{\Delta I_S}{\Delta t} \Rightarrow L = V_S \frac{\Delta t}{\Delta I_S}$$

The inductance is calculated from

$$L = \frac{V_S \, \Delta t}{\Delta I_S} = \frac{12 \times 50 \times 10^{-6}}{0.01 \times 10} = 6 \; mH$$

AC Source

A source, which provides a voltage waveform with periodic variation of amplitude over time, is called the **ac voltage source.**

The ac current source does not exist in nature. A practical ac current source can be synthesized from an ac voltage source in series with a large value inductor (see Figure 1.12).

Sinusoidal waveform is a basic ac waveform. A periodic waveform of any shape can be expressed as the sum of several basic sinusoidal waveforms such as the following:

$$v_S(t) = \bar{V}_S + \sum_{n=1}^{\infty} V_{Sn} \sin(n\omega_S t + \theta_n)$$

$$\omega_S = \frac{2\pi}{T_S} \qquad (1\text{-}6)$$

ω_s is the frequency of the ac source in radians per second and **t** the time. $\mathbf{T_S}$ is the period. The n ω_S t is called the conduction angle. $\mathbf{\bar{V}_S}$ is the dc component of the waveform, it is also the average voltage of a period. The sinusoid of n ω_S is called the n-th harmonic. $\mathbf{V_{Sn}}$ is the peak amplitude of the n-th harmonic. The harmonic with n = 1 is referred to as the fundamental harmonic. The root-mean-square voltage of the ac waveform can be calculated from the harmonics as (see Appendix at the end of this book),

$$\tilde{V}_S = \bar{V}_S + \sqrt{\sum_{n=1}^{\infty} \tilde{V}_{Sn}^{2}} \qquad (1\text{-}7)$$

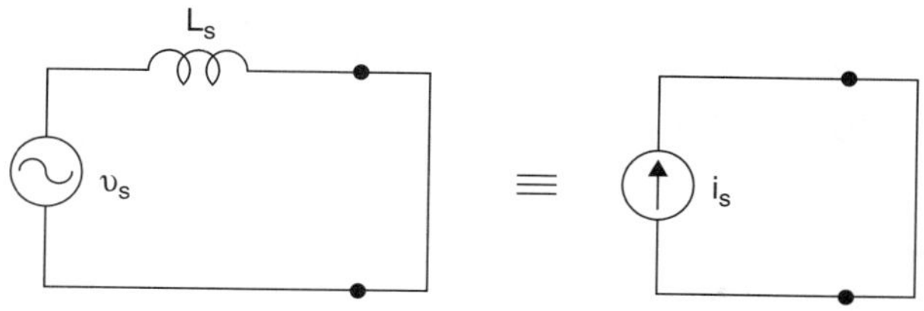

Figure 1.12
A synthesized ac current source

The average power delivered by the ac waveform to a resistor R is equal to the sum of power provided by all harmonics together,

$$P_S = \frac{\tilde{V}_S^2}{R} = \frac{\overline{V}_S^2}{R} + \sum_{n=1}^{\infty} \frac{\tilde{V}_{Sn}^2}{R} \tag{1-8}$$

A multiple-phase ac source consists of two or more identical waveforms that are phase shifted with respect to each other by 2π/P where P is the number of phases. A 3-phase sinusoidal ac voltage source is constituted of three identical sinusoidal voltage waveforms that are phase shifted by 2π/3 = 120° with respect to each other. The multiple-phase source increases the amount of power transfer from the source to the load.

A P-phase voltage source is expressed as follows:

$$v_S(t) = \sum_{n=1}^{\infty} V_{Sn}\left[\sin n\omega_S t + \sin n\left(\omega_S t + \frac{2\pi}{p}\right) + \ldots + \sin n\left(\omega_S t + \frac{(p-1)\pi}{p}\right)\right] \tag{1-9}$$

The 3-phase, 4-wire ac power provides the three phase voltages and a connection to the neutral (the common) node on the source side. This type of connection is referred to as Y configuration or Star configuration. Another configuration of 3-phase and 3-wire is called Δ-configuration. See Figure 1.13 for a Y-connected and a Δ-connected 3-phase ac electric utility.

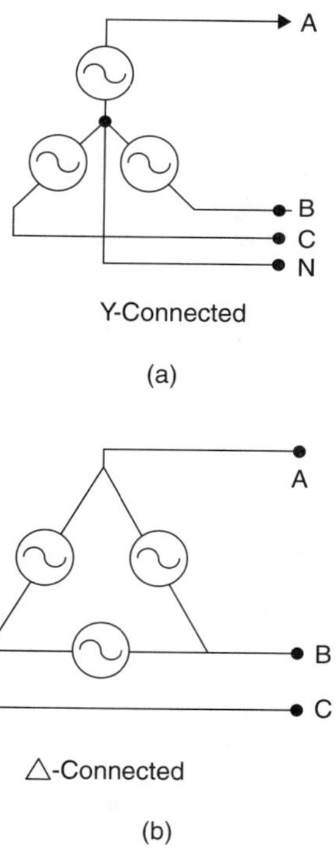

Figure 1.13
(a) Star-connected and (b) Δ-connected 3-phase ac electric utility

The single-phase and 3-phase ac voltages in the United States are respectively 120 volts rms and 220 volts rms at 60 Hz. The 3-phase supply in Europe is 220 volts rms at 50 Hz.

Example 1-3

A square-wave voltage source, 150 V peak, 50 Hz, is connected to a load of 10 mH inductor. The initial current in the inductor is zero. Plot the voltage, the current and the instantaneous power waveforms using MATLAB. Plot the harmonic profile of the current and the instantaneous power up to five harmonics.

v_S (t) : square-wave voltage source, L = 10 mH, $I_L(0) = 0$,

From equation (1-1)

$$i_L = I_L(0) + \frac{1}{L}\int_0^t v_S(t)\,dt$$

Instantaneous power is given by

$$p(t) = v_S(t)\,i_L(t)$$

```
L=10*10^(-3); IL0=0; Ts=1/50; dt=Ts/ 512;
  % 512 points in one time period
t1=0:dt:Ts/2-dt; t2=Ts/2:dt:Ts-dt; t=[t1 t2];
  % one time period
vL=[150*ones(size(t1)) -150*ones(size(t2))];
  % the source voltage applied to the inductor
iL=IL0+(1/L)*integ(vL, t); % inductor current
plot(t,vl, t,iL)
axis([0, Ts,-200, 200])
xlabel('time')
ylabel('amplitude')
p=vL.*iL; %instantaneous power
pause %press enter to get the next plot
[f,iLh,ph]=harmonic(iL,5); % harmonic profile
xlabel('harmonic number')
ylabel('amplitude')
title('Harmonic Profile')
[f,ph,ph]=harmonic(p,5);
xlabel('harmonic number')
ylabel('amplitude')
title('Harmonic Profile of instantaneous power')
```

See Figure 1.14 on pages 17–18.

Example 1-4

Repeat example 1-3 for the following voltage source waveform. Calculate the average power.

The MATLAB program is similar to the program in Example 1-3. The differences are given below:

```
t1=0:dt:Ts/4-dt; t2=Ts/4:dt:Ts*3/4-dt; t3=Ts*3/4:dt:Ts;
  t=[t1 t2 t3];
vl=[10*ones(size(t1)), -5*ones(size(t2)), zeros(size(t3)) ];
Pavg=mean(p)
```

Pavg = 0.0731 which is approximately zero.

See Figure 1.15 on pages 19–20.

Example 1-5

A triangular-wave current source, 100 A peak 50 Hz, charges a 10 μF capacitor. The capacitor is initially completely discharged. Draw the voltage and current waveforms using MATLAB.

The source current waveform, which is also the current into the capacitor, is

$$i_s = i_C = -100 + \left[\frac{200}{T_s/2}\right]t \qquad 0 \le t < T_s/2$$

$$= 100 - \left[\frac{200}{T_s/2}\right](t - T_s/2) \qquad T_s/2 \le t < T_s$$

The capacitor voltage is determined from equation (1-16).

```
C=10*10^(-6); VC0=0; Ts=1/50; dt=Ts/512;
  % 512 points in one period
t1=0:dt:Ts/2-dt; t2=Ts/2:dt:Ts−dt; t=[t1 t2]; % one period
ic=[−100+(200/(Ts/2))*t1, 100−(200/(Ts/2))*(t2−Ts/2)];
  % capacitor current
vc=VC0+(1/C)*integ(ic, t); %capacitor voltage
plot(t,ic, t,vc/100)
grid
xlabel('time')
ylabel('amplitude')
gtext('ic')
gtext('vc')
```

See Figure 1.16 on page 21.

1.4.5 Electrical switch

An ideal electrical switch (see Figure 1.12) may be defined as following:

In the on-state, $\mathbf{v_{sw}=0}$ at all amplitudes and waveforms of the current $\mathbf{i_{sw}}$

In the off-state, $\mathbf{i_{sw}=0}$ at all amplitudes and waveforms of the voltage $\mathbf{v_{sw}}$

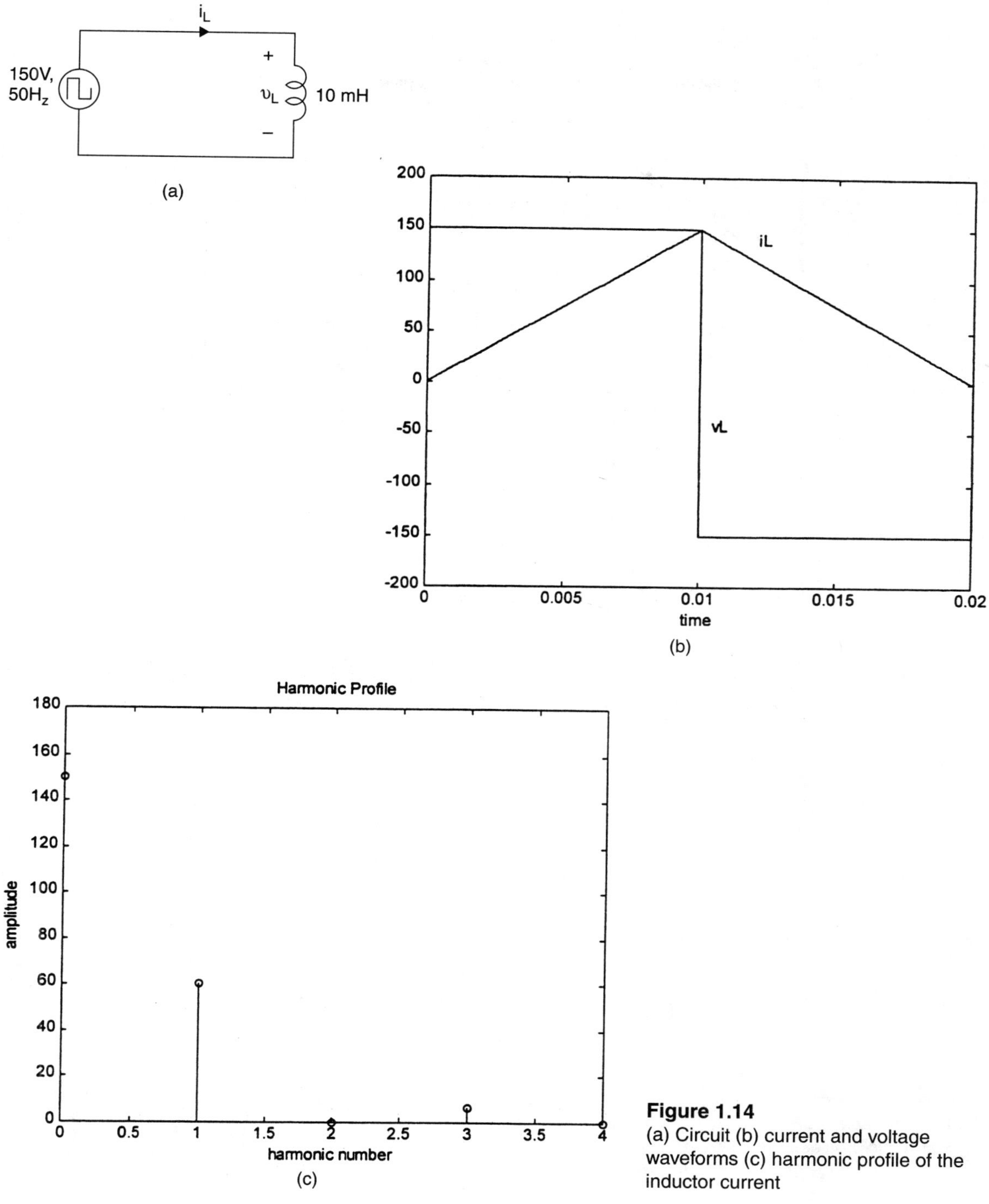

Figure 1.14
(a) Circuit (b) current and voltage waveforms (c) harmonic profile of the inductor current

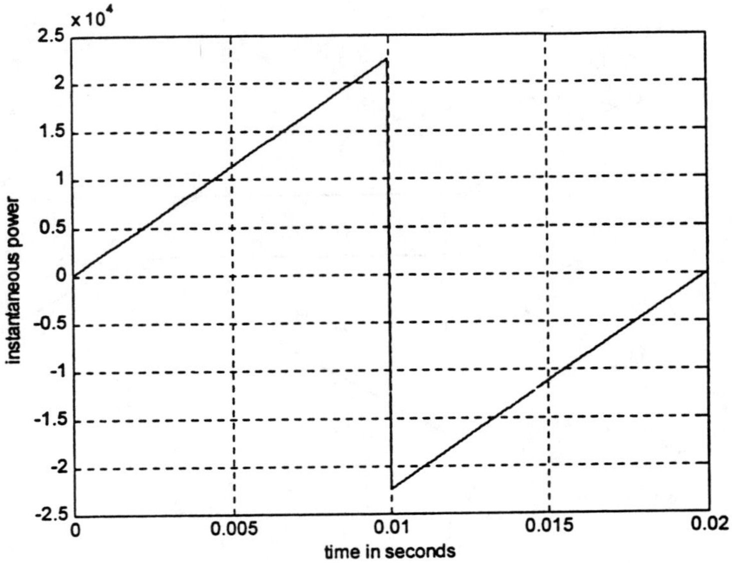

(d)

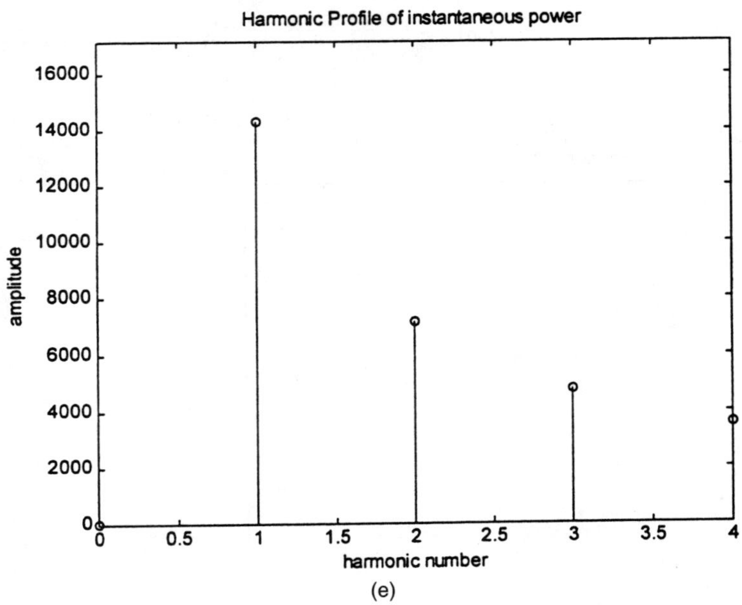

(e)

Figure 1.14 ***(continued)***
(d) Instantaneous power and
(e) Harmonic profile of the instantaneous power

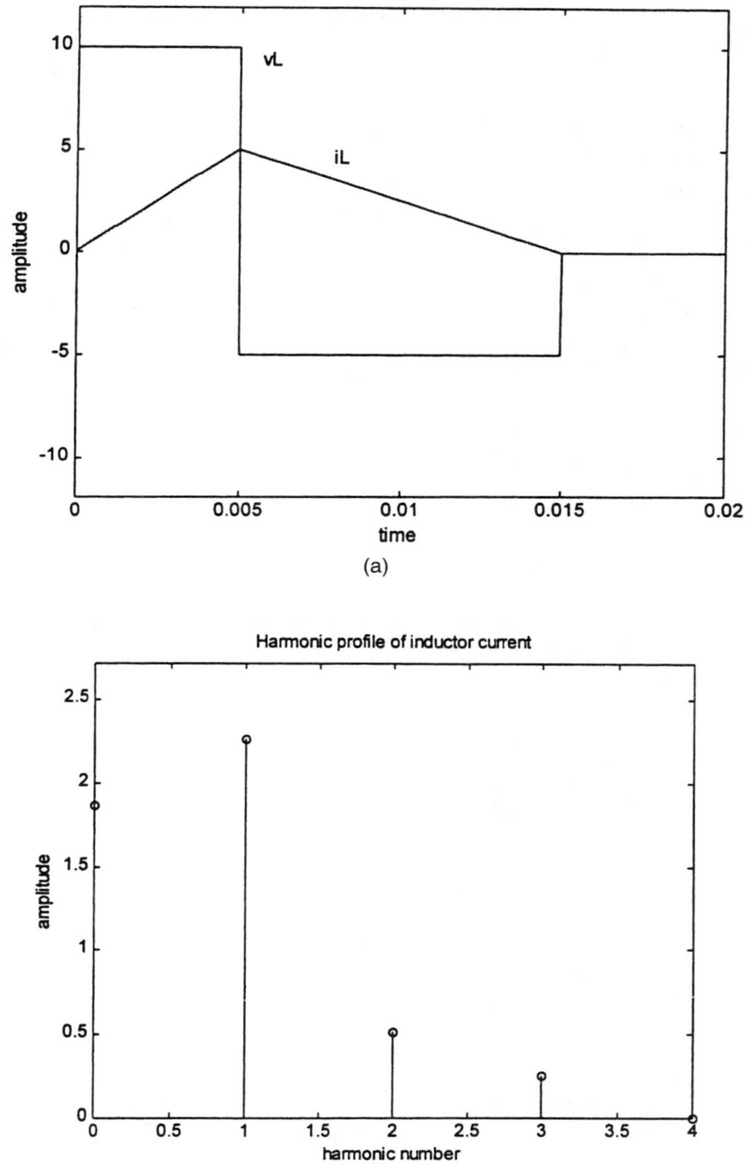

Figure 1.15
(a) current and voltage waveforms
(b) Harmonic profile of the inductor current

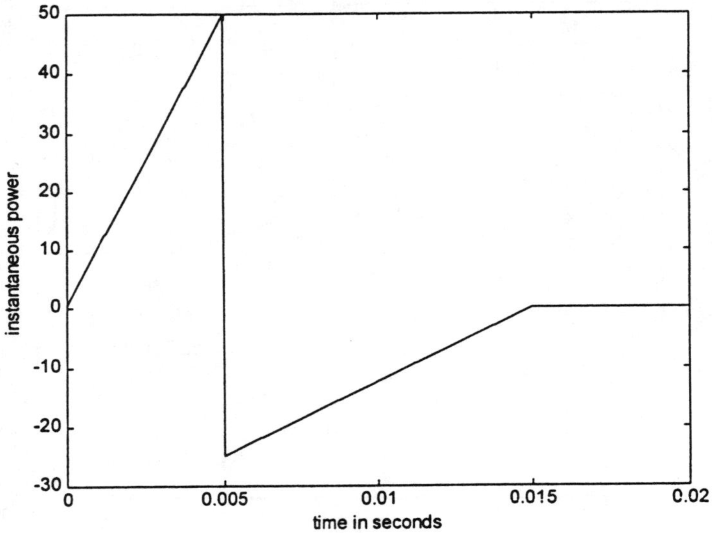

(c)

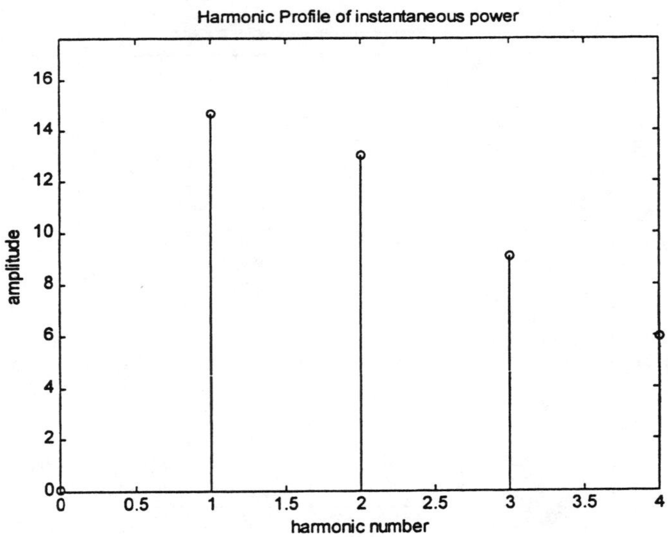

(d)

Figure 1.15 ***(continued)***
(c) Instantaneous power and
(d) Harmonic profile of the instantaneous power

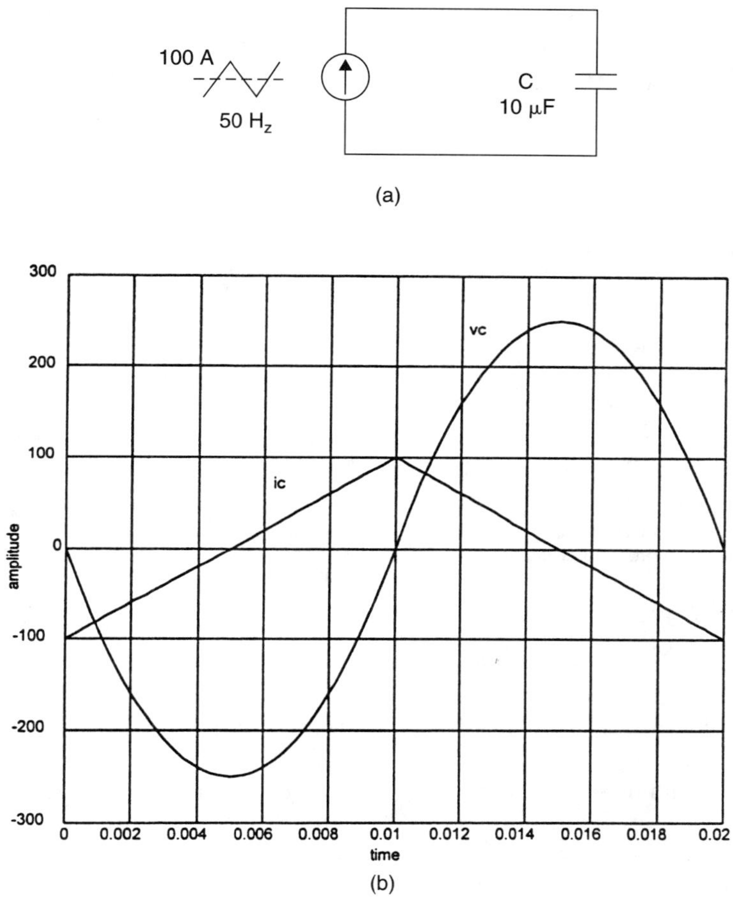

Figure 1.16
(a) Circuit and (b) current and voltage waveforms

The switch is turned on and off by a control signal $\mathbf{v}_c$, which may be a voltage, a current signal, a light signal or a temperature signal.

The switches may be categorized on the basis of (a) the direction of the current flow in the on-state, (b) the polarity of voltage in the off-state and (c) the triggering mechanism.

A bidirectional switch, shown without any arrow on the switch arm in Figure 1.17, permits the current in both directions in the on-state. A bipolar switch can withstand both positive and negative voltages in the off-state.

In practice, only the unidirectional and unipolar switch is realizable electrically. Such a switch (shown by an arrow on the switch arm) can be synthesized from a two-terminal device like the diode or the three-terminal device like the transistor and thyristor. The bidirectional switch can be synthesized from connecting two unidirectional switches in antiparallel manner as shown in Figure 1.17. A transistor is closest to a unidirectional switch. Silicon Controlled Rectifier (SCR) is a unidirectional bipolar switch. The ideal switch is also supposed to be lossless, that is, it has zero on-state

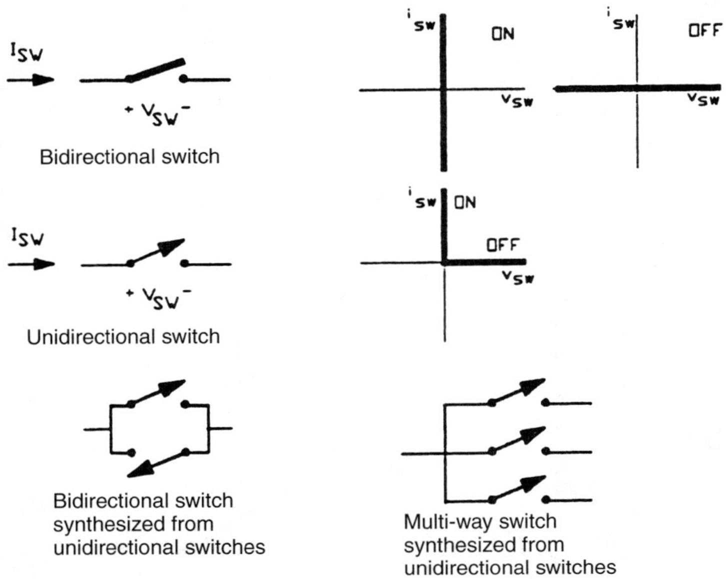

Figure 1.17
Ideal switch and characteristics

losses, zero off-state losses and zero switching losses. Switching devices in a practical design example must be selected with these desired characteristics in mind.

The control signal $\mathbf{v_c}$ may be a threshold voltage (or current) level above which the switch is turned on. The control signal may also be a voltage (or current) pulse above the threshold amplitude.

Two terminal devices like diodes are used as the self-triggered or self-controlled switches. The voltage on the switch $\mathbf{v_{sw}}$ is itself the control signal. The control level is fixed and not controllable. The diode is called the uncontrolled switch.

Examples of three-terminal power electronic switching devices are power Bipolar Junction Transistor BJT, Silicon Controlled Rectifier SCR, Gate-Turn-OFF Thyristor GTO, Metal-Oxide Silicon Field Effect Transistor MOSFET, Insulated Gate Bipolar Transistor IGBT, Mos Controlled Thyristor MCT, etc. All switches have non-zero but small voltage across them in the on-state. SCRs, GTOs and MCTs require narrow control pulses, hence called the pulse triggered devices. BJTs, MOSFETs and IGBTs require dc voltage or current level for triggering, hence they are called the level sensitive (triggered) devices (more on the switches in the next chapter).

The above ideal switch is also a one-position (or one-way) switch. A multi-way switch is synthesized from several one-way switches as shown in Figure 1.17.

1.4.6 Transformer

A transformer in power electronic converters is used for transforming one set of ac V and I in one system to a different set of V and I in other systems with or without electrical isolation. The frequency of V or I waveforms does not change during the transformation. DC voltages do not transform. The law of conservation of energy holds good in the transformation.

A transformer consists of two or more magnetically coupled windings, wound on a core structure that has a closed magnetic path or paths. A trans-

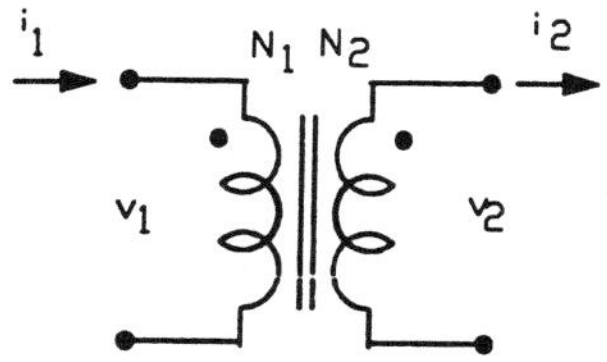

Figure 1.18
A two-winding transformer

former consisting of two windings is shown in Figure 1.18. Dots on the windings indicate that the voltages on the nearest nodes have similar polarity. The voltage on each winding, imposed or induced, is given by the time derivative of the flux linkage of the concerned winding. If a magnetic flux links $\mathbf{N_i}$ turns of the i-th winding, the induced voltage is given by

$$v_i = N_i \frac{d\phi}{dt} : i = 1, 2, \ldots \qquad (1\text{-}10)$$

Therefore,

$$\frac{v_1}{N_1} = \frac{v_2}{N_2} = \ldots \qquad (1\text{-}11)$$

If the current in each winding is assumed to enter the winding marked by a dot, the law of conservation of energy (power) dictates

$$v_1 i_1 = v_2 i_2 = \ldots \qquad (1\text{-}12)$$

An interesting meaning of the above relation is that the voltage and current on a winding in the transformer appears as scaled on the other windings. Likewise, the impedance Z across a winding appears as scaled on the other windings.

$$v_j = \frac{N_j}{N_i} v_i, \quad Z_j = \left(\frac{N_j}{N_i}\right)^2 Z_i : i \neq j \quad (1\text{-}13)$$

N_j/N_I is called the turn ratio of the transformer. A transformer in power electronic converter is strictly treated as a scalar. It is often eliminated in analytical models by merging the scale factors with the transfer and reflective parameters, to be discussed in a later section. Because of the property of scaling, the transformer is also used for impedance matching of the converter to the load.

Example 1-6

Find the input capacitance of an ideal 10:1 transformer, which has a 0.47 F capacitor in the secondary winding.

Using Laplace transforms

$$Z_{in} = \left(\frac{N_1}{N_2}\right)^2 \frac{1}{sC} = 10^2 \frac{1}{s\,0.47\ \mu F} = \frac{1}{s\,0.0047\ \mu F}$$

The input capacitance as seen on the primary side of the transformer is 0.0047 F.

1.4.7 Loads and sinks

Electrical loads can be classified in the following categories:

Resistive:	heaters, ovens and furnaces
Inductive:	electromagnetic relays, cranes, pickups and hoists
Capacitive:	displays, gas discharge lamps such as the halogen and mercury vapor lamps and lasers
DC current loads:	dc motor in constant torque applications
DC voltage loads:	dc motor in constant speed applications, batteries during charging
AC voltage loads:	ac voltage BUS, induction and synchronous motors in constant speed applications
AC current loads:	induction and synchronous motors in constant torque applications

Resistive loads are self-explanatory and are modeled by pure resistances. Converters with resistive loads are generally operated at low frequencies.

An inductive load consists mainly of large inductance. Electrical energy stored in the inductor is converted to mechanical energy such as in the relays, electromagnetic pickups and motors. In the electromagnetic pickups and dc motors in constant torque applications, load demands constant dc current. The **dc current load** may be modeled by a **dc current sink.** A current sink has similar terminal characteristics as those of a dc current source except that the sink removes the energy out of the system whereas the source generates and brings energy into the system. A current sink is represented by a rectangular element with an arrow indicating the direction of the current flow. The dc current sink may also be used to represent a large inductor in series with a low value load resistance.

Examples of **dc voltage load** are dc motors at constant speed and batteries under charging condition. This type of load may be considered as being a large value capacitor in parallel with a resistance to account for the sinking of energy out of the system in nonelectrical form. A model for the dc voltage load is a **dc voltage sink,** represented by a rectangular element with polarity markings.

1.5 CONVERTER TOPOLOGY

The converter topology is a converter schematic diagram without component details. The intention is to show the generic interconnection of elements of the circuit rather than the details.

1.5.1 Inductor in converter topology

A sudden change in the inductor current involves infinitely high di/dt, see equation (1-2), causing infinitely high voltage across the inductor and consequently on other components associated with it in the circuit. Alternately, application of a finite voltage across an inductor, cannot cause an instantaneous jump in inductor current.

Power electronic converter topologies with an inductor and a unidirectional switch in series is not tenable. Whenever such a combination is needed, a path for continuity of inductor current is provided in the event of the opening of the switch. A diode is generally employed across the switch to provide the current continuity through the series inductor. This diode is referred as the ***free-wheeling diode.*** The same is true when the inductor is replaced by a current sink.

A current source cannot be switched in series to an inductor or a current sink if the current ampli-

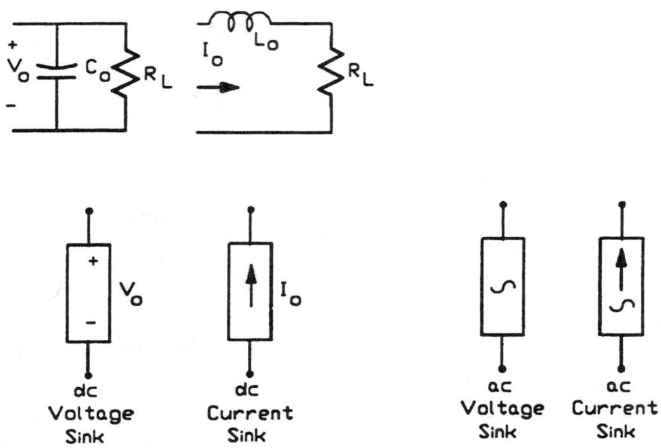

Figure 1.19
Voltage sink and current sink

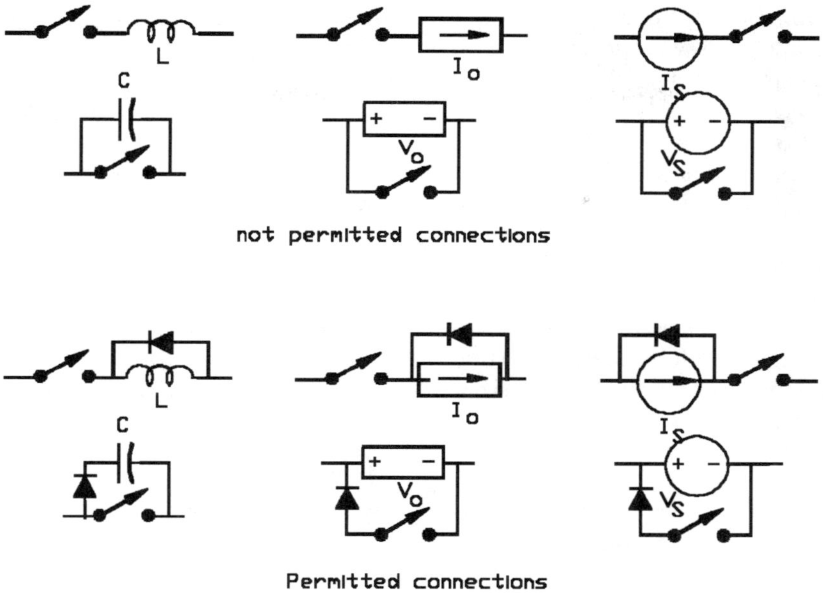

Figure 1.20
Permissible and nonpermissible topological connections

tudes of the two are different. Alternately, a current source can be switched in series to only a noninductive load. The current sink, however, can be switched in series with a capacitor or a voltage sink or a voltage source. See Figure 1.20 for the permissible and nonpermissible connections.

If a circuit that includes an inductor is subjected to periodic excitation, the average voltage across the inductor in a period must be equal to zero in the steady-state conditions. A nonzero average voltage (dc) across inductor means the inductor current will eventually increase to infinity.

$$\int_{t_0}^{t_0+T} v_L \, dt = 0 \tag{1-14}$$

Furthermore, under the periodic excitation in steady state, the inductor current at the end of a period must be equal to the inductor current at the beginning of the period because

$$i_L(t_0 + T) - i_L(t_0) = \frac{1}{L}\int_{t_0}^{t_0+T} v_L \, dt = 0 \tag{1-15}$$

The energy stored in an inductor is given by

$$\frac{dE}{dt} = i_L \frac{d\phi}{dt} = i_L \frac{d}{dt}(Li_L) \tag{1-16}$$

In general, the inductance **L** of a coil, with moving core material such as in the motors, is a function of time, hence, it cannot be taken out of the integral. For a fixed inductance, the above equation leads to the well-known energy equation:

$$E = \int_{i_L(0)}^{i_L} Li_L di_L = \frac{1}{2}Li_L^2 - \frac{1}{2}Li_L^2(0) \tag{1-17}$$

The $\mathbf{i_L(0)}$ and $\mathbf{i_L}$ are respectively the initial and the present inductor current.

1.5.2 Capacitor in converter topology

A sudden change in the voltage across a capacitor involves infinitely high dv/dt, causing infinitely high current in or out of it. Alternately, a finite current in or out of a capacitor cannot cause an instantaneous jump in the capacitor voltage.

Power electronic converter topologies with a capacitor and a unidirectional switch in parallel in any branch of a circuit are not tenable. Whenever such a combination is needed, a blocking diode is added in series with the capacitor to prevent the shorting in the event of the closing of the switch.

A voltage source cannot be switched in parallel with a capacitor or a voltage sink if the terminal voltages of the two are different. Alternately, a voltage source can be switched in parallel with only a noncapacitive load. The voltage sink, however, can be switched in parallel with an inductor, current sink or a current source. See Figure 1.20 for the permissible and nonpermissible connections.

If a circuit that includes a capacitor is subjected to periodic excitation, the average current in or out of the capacitor in a period must equal zero in the steady-state conditions. A nonzero average current (dc) in or out of the capacitor means the capacitor voltage will ramp to infinity.

$$\int_{t_0}^{t_0+T} i_C \, dt = 0 \qquad (1\text{-}18)$$

Like the case of inductors, under the periodic excitation in the steady-state, the capacitor voltage at the end of a period must equal the capacitor voltage at the beginning of the period.

$$v_C(t_0 + T) = v_C(t_0) \qquad (1\text{-}19)$$

The energy stored in (or given out by) a capacitor is given by

$$dE = v_C dQ = v_C \, d(Cv_C) \qquad (1\text{-}20)$$

In general, the capacitance of a capacitor with a moving dielectric material is a function of time, hence it cannot be taken out of the integral. For a fixed capacitor, the above equation leads to the well-known energy relation:

$$E = \int_{v_C(0)}^{v_C} v_C \, dv_C = \frac{1}{2} C v_C^2 - \frac{1}{2} C v_C^2(0) \qquad (1\text{-}21)$$

The $\mathbf{v_C(0)}$ and $\mathbf{v_C}$ are respectively the initial and the present capacitor voltage.

Example 1-7

The circuit shown in Figure 1.21, also called the Buck converter, is switched at 100 KHz. The voltage source is 12-V dc battery and the voltage sink V_o is the model of a large value capacitor in parallel with a 10 ohms resistor. The inductor is large enough to maintain a zero ripple current and the capacitor is large enough to have a constant voltage across it. The switch is turned on for a 75% duration of the switching period. The smoothing inductor is assumed to be infinitely large. Find analytically the average output voltage, the average output current, the average current from the source and the power drawn from the source. Sketch the voltage waveform after the switch using MATLAB and confirm the average voltage and average source current.

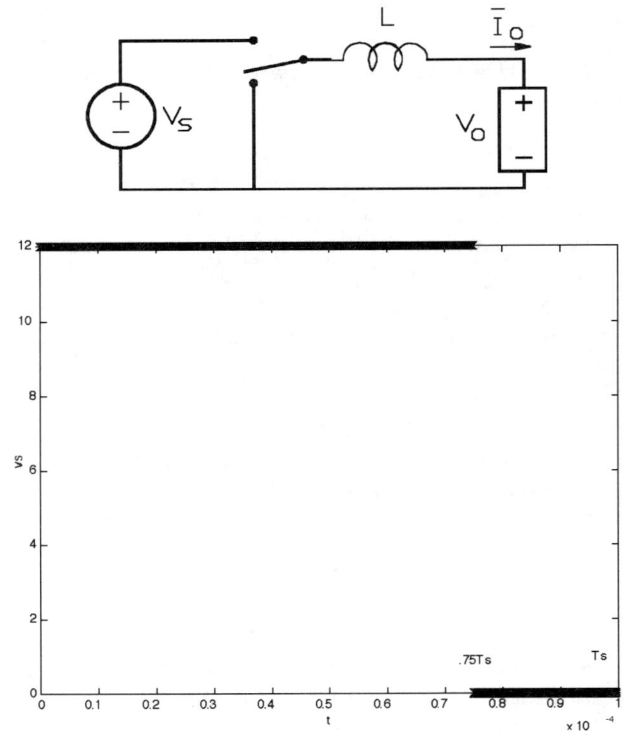

Figure 1.21
Buck converter

The voltage waveform after the switch is a pulse wave alternating between 12 v and zero volts. The duration of 12-volt level is $0.75T_s$ (the on-period). In steady state, the average voltage across the infinite inductor is zero, hence the average voltage (dc voltage) on the resistor is the same as the average voltage after the switch:

$$\bar{V}_o = 12 \times \frac{.75T_s}{T_s} = 9\ volts$$

The dc current in the resistor is

$$\bar{I}_o = \frac{\bar{V}_o}{R} = 0.9A$$

The current through the switch, which is also the current from the source, is also a pulse wave with 75% duty cycle, the peak value being equal to the dc current through the inductor. The average value of the source current is thus calculated as follows:

$$\bar{I}_s = \bar{I}_o \times \frac{.75T_s}{T_s} = 0.675A$$

```
Ts=10e-5; R=10; VS=12; dt=Ts/500;
  %incremental step in the time array
t1=0:dt:.75*Ts-dt; t2=.75*Ts:dt:Ts-dt; t=[t1 t2];
  %time array
vs=[VS*ones(size(t1)) zeros(size(t2))];
  %dc source voltage array
Vodc=mean(vs) % average voltage after the switch
Iodc=Vodc/R % dc current in the load resistance
is=[Iodc*ones(size(t1)) zeros(size(t2))];
  % source current array
Isdc=mean(is) % dc average of the source current
P=VS*Isdc %power supplied by the dc source
plot(t,vs, '*') %voltage waveform after the switch
```

Vodc= 9.0000 Iodc= .9000 Isdc= .6750 P= 8.1000

Example 1-8

Consider a converter circuit, shown in Figure 1.22, is switched at 100 KHz. The ideal current source is 10 A dc and the current sink load I_o is the model of a large value inductor in series with a 10 ohms resistor. The capacitor C is large enough to maintain a zero ripple voltage across the current sink I_o. The switch is turned on for a 75% duration of the switching period. Find analytically the average output voltage, average output current, the average voltage across the current source and the power drawn from the source. Sketch the waveforms of the output voltage and current and the voltage across the current source using MATLAB and confirm the average values.

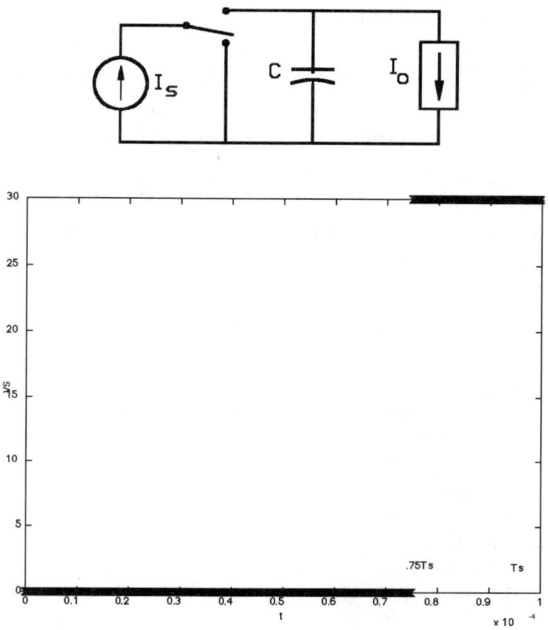

Figure 1.22
A dc-dc converter

The current waveform after the switch is a pulse wave alternating between 10 A and zero current. The duration of 10 A level is $0.25T_s$ (the off period). In steady state, the average current into the infinite capacitor is zero, hence the average current (dc current) in the resistor is the same as the average current after the switch:

$$\bar{I}_o = 10 \times \frac{.25T_s}{T_s} = 2.5\ A$$

The dc voltage across the resistor is

$$\bar{V}_o = \bar{I}_o R = 25V$$

The waveform of the output voltage is identical in shape to the current waveform in the resistor, which in turn is also the current waveform after the switch. The voltage across the current source is the same as the output voltage, that is a pulse wave with 25% duty cycle, the peak value being equal to the dc output voltage across the resistor. Average value of the voltage across the current source is thus calculated as follows:

$$\overline{V}_s = \overline{V}_o \times \frac{.25T_s}{T_s} = 6.25V$$

```
Ts=10e-5; R=10; IS=10; dt=Ts/500;
t1=0:dt:.75*Ts-dt; t2=.75*Ts:dt:Ts-dt; t=[t1 t2];
is=[zeros(size(t1)) IS*ones(size(t2))];
  % current after the switch
Iodc=mean(is) % average dc current
Vodc=Iodc*R % dc output voltage
vs=[zeros(size(t1)) Vodc*ones(size(t2))];
  %voltage across the current source
Vsdc=mean(vs) %dc average dc voltage across the
  current source
P=Vsdc*IS % power supplied by the source
plot(t,vs,'*')
```

Iodc = 3.0000 Vodc = 30.0000 Vsdc = 7.5000 P = 90

1.6 CONVERTER OPERATION AND DESIRED CHARACTERISTICS

Converter operation involves the turning on and off of switches in a periodic sequence. The timing control is determined in many ways: externally, internally by the states of circuit variables (voltages or currents) or by the transitions or polarity inversions in the ac source waveform.

Converter operation consists of two phases: the transient phase and the steady-state phase. The voltage and current waveforms in transient phase are unstable and vary from period to period. In steady state, all waveforms are stabilized.

Some of the desirable characteristics of the converter operation are as follows:

1. The output voltage (current) should be set to the desired value and waveform.
2. The output voltage (or current) should be insensitive to variations in load, input voltage (or current) and component values.
3. The converter should cause minimum distortion and loading on the source side.
4. The output voltage or current should have minimum distortion and minimum number of undesired components.
5. A small range of control parameter should be able to control a large variation in output voltage (or current),
6. The control parameter should have a large range.
7. The converter should exhibit minimum power loss and maximum resistance to transients.
8. The converter elements should store minimum energy, even temporarily.

1.7 CONVERTER PERFORMANCE MEASURES

The converter performance consists of the transient and steady state parts. The important transient performance measures are:

Transient current from the source:

The amount of initial current that converter draws from the source when the power is turned on to it. This current is called the *Inrush current.*

Short circuit current from the source

The current drawn from the source when the output terminals are shorted.

Surge protection

1. Difference between the maximum transient voltage and the nominal output voltage when the power is turned on and turned off to the converter
2. The amount of transient voltage when a large load is switched in or out at the output terminals of the converter

Switch stress

1. Highest voltage and the highest rate of rise or fall of voltage across the switching device
2. Highest current and the highest rate of rise or fall of current through the switching device

The steady-state converter performance measures are:

Forward Transfer characteristics
Reflective characteristics
Control characteristics
Harmonic profile

1.7.1 Forward transfer characteristics

The forward transfer characteristics refer to the relation between a variable at the output terminals to a variable at the input terminals of the converter. Two important characteristics refer to forward voltage transfer and forward current transfer, as defined below:

Voltage transfer ratio

$$T_{vv} = \frac{v_o}{v_s} \qquad (1\text{-}22)$$

Current transfer ratio

$$T_{ii} = \frac{i_o}{i_s} \qquad (1\text{-}23)$$

The subscript 'o' refers to the output variable and the subscript 's' refers to the source side variables.

For example, in the basic dc-dc converter, as shown in Figure 1.2, the ratio of the output dc voltage to the input dc voltage of the source is an important characteristic, called *Converter gain M,* as defined below:

$$M = \frac{\overline{V}_O}{V_s} \qquad (1\text{-}24)$$

where $\overline{V}_O$ and V_s are respectively the average dc output voltage and the dc source voltage. The converter gain M may also be defined as the ratio of the average dc current to the average dc current from the dc voltage source.

Another example is the forward characteristic in an ac-dc converter, which is defined as the ratio of the output dc voltage to the input rms voltage of the source:

$$M = \frac{\overline{V}_O}{\tilde{V}_s} \qquad (1\text{-}25)$$

where $\overline{V}_s$ is the rms voltage of the ac voltage source.

	Output variable	**Input variable**
ac-dc converter:	average value	rms value
dc-ac converter:	dc value	rms of the fundamental harmonic
dc-dc converter:	average value	dc value
ac-ac converter:	rms value	rms of the fundamental harmonic

1.7.2 Reflective characteristics

The reflective characteristics give the effect of output conditions on the voltage and current variables on the source side of the converter.

Voltage reflection ratio

$$R_{vv} = \frac{v_s}{v_o} \qquad (1\text{-}26)$$

Current reflection ratio

$$R_{ii} = \frac{i_s}{i_o} \qquad (1\text{-}27)$$

For example, the power electronic designer of the dc-dc converter of Figure 1.2 would have interest in finding the effect on the dc current drawn from the source due to the average dc current into the load at the output terminals:

$$R_{ii} = \frac{\overline{I}_S}{\overline{I}_O} \qquad (1\text{-}28)$$

where $\overline{I}_S$ and $\overline{I}_O$ are respectively the average dc source current and the average dc current to the load. In an ac-dc converter, the numerator is the rms current from the ac source.

1.7.3 Control characteristics

The power converter systems are often controlled for achieving desired performance characteristics. For example, in the dc-dc converter, the desired converter gain is achieved by controlling the on-time of the switch. The control characteristics refer to the plot of the forward transfer characteristic such as the converter gain as a function of the on-time of the switch, as shown in Figure 1.23.

A feedback control system may be used for actual control. The error between the actual and the desired output characteristic is used to vary the control parameter. It is desirable to have a small variation in the control parameter that would control a large variation in the output variable; that is, the converter operation is preferred in the high slope region of the control characteristic curve.

1.7.4 Harmonic profile

The voltage and the current at the input and output terminals of a converter, in general, have non-sinusoidal waveforms. Harmonic profile is the plot of amplitude versus frequency for the ac waveform of voltage, current or instantaneous power. Following measures are used for the dissemination of information from the harmonic profile:

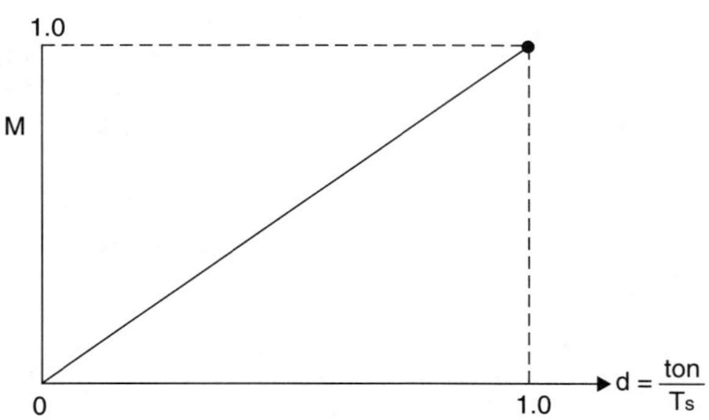

Figure 1.23
Control characteristic of dc-dc converter of Figure 1.2

Average

Root-mean-square (rms) Value

Peak-Peak Ripple

Lowest Undesired Harmonic Frequency

Total Harmonic Distortion (THD)

Power Factor

Crest Factor

Average of the waveform is calculated as the mean value of the waveform over a period.

$$\overline{V} = \frac{1}{T}\int_0^T v(t)dt \qquad (1\text{-}29)$$

The average of a voltage waveform is also the dc value of the waveform.

Root-mean-square value of the waveform is calculated as the square-root of the mean of the squared waveform over a period

$$\tilde{V} = \sqrt{\frac{1}{T}\int_0^T v^2\,(t)\,dt} \qquad (1\text{-}30)$$

Consider the pulse waveform of v(t) in Figure 1.24. The average and rms voltage are calculated using MATLAB as follows:

```
T=1; dt=T/512; t1=0:dt:T/3-dt; t2=T/3:dt:T; t=[t1 t2];
v=[5*ones(size(t1)), zeros(size(t2)) ]; %voltage waveform
  in a period
vdc=mean(v) %average voltage
vrms=sqrt(mean(v.^2)) %rma value of pulse wave v(t)
```

Vdc = 1.6602 vrms = 2.8811

Peak-Peak Ripple in the waveform, is calculated as the amount of peak to peak deviation from the average value of the waveform.

```
vripple=v-vdc;
pp_ripple=max(vripple)-min(vripple)
```

pp_ripple = 5.0

The harmonic profile of the voltage waveform can be obtained as follows:

```
[f,v1,ph]=harmonic(v,10); %Harmonic profile of v(t)
xlabel('harmonic number')
ylabel('amplitude')
title('Harmonic Profile')
ph*180/pi %phase angle of the fundamental harmonic
   cf v(t)
v1=abs(vh(2))*sin(2*pi*t/T+ph); %fundamental harmonic
   waveform
plot(t,v, t, v1, '—') % the waveform of v(t) and its
   fundamental harmonic are plotted in Figure 1.18.
axis([0,1,−6,6])
grid
xlabel('time in seconds')
ylabel('amplitude')
title('a pulse wave')
```

Lowest Undesired Harmonic Frequency is an important design criteria. The lowest undesired harmonic in the dc-ac inverter is designed to be as far away as possible from the fundamental frequency. The lowest undesired harmonic in v(t) of Figure 1.23 after the fundamental harmonic is the second harmonic.

Total Harmonic Distortion (THD) of the current from a sinusoidal voltage source is defined as follows:

$$THD = \sqrt{\frac{\tilde{I}_S^2 - \tilde{I}_{S1}^2}{\tilde{I}_S^2}} \tag{1-31}$$

$_S$ and $_{S1}$ are respectively the rms value of the current waveform and the rms value of the fundamental harmonic current. THD is used for the voltage and current waveforms on both the input and output terminals of the converter system.

Power Factor is a very important measure of the power electronic converter performance. Before we define the power factor, we must learn a few other terms:

Apparent power S flowing through a pair of terminals is defined as the product of the rms voltage across the terminals and the rms current flowing through them. S has another name, VA. It is a hypothetical power. S is a complex value and is resolved in real power P and the reactive power Q, as given below:

$$S = \tilde{V}\tilde{I} = P + jQ \tag{1-32}$$

$\tilde{V}$ and $\tilde{I}$ are the rms values of the voltage and current waveforms respectively. The real power and the reactive power are in turn defined as

$$P = \tilde{V}\tilde{I}\cos\phi \qquad Q = \tilde{V}\tilde{I}\sin\phi \tag{1-33}$$

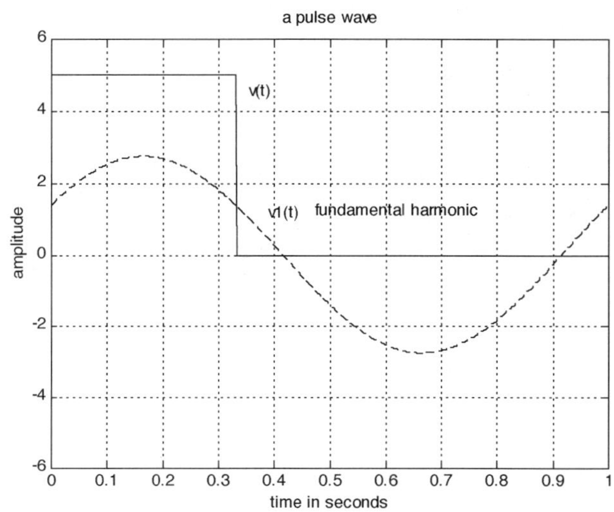

Figure 1.24
A pulse wave v(t) and its fundamental harmonic waveform

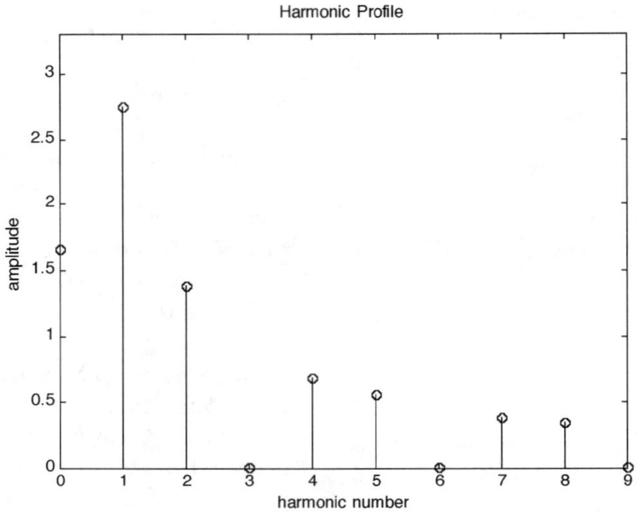

Figure 1.25
The harmonic profile of v(t)

is the phase angle between the voltage and current waveforms. If the source of power is the electric utility in a power distribution system, the reactive power Q is the amplitude of the power oscillation that takes place between the load and the ac source. The customer is able to utilize the real power P but has to pay for S.

Power factor is defined as the ratio of the real power P_1 that is delivered by the fundamental component of the voltage to the apparent power S flowing across a pair of terminals:

$$PF = \frac{P_1}{S} = \frac{\tilde{V}_1 \tilde{I}_1 \cos\phi}{\tilde{V}\tilde{I}} \qquad (1\text{-}34)$$

$\tilde{V_1}$ and $\tilde{I_1}$ are the rms values of the fundamental harmonics of the voltage and current waveforms respectively.

The power factor for negative phase angle is called the *lagging power factor*. The one for the positive phase angle is called the *leading power factor*. Power factor varies from 0 to unity. The unity power factor signifies that all the power flowing through the terminals is real power. A smaller value means that the power flowing through the terminals also consists of the reactive power, which does not contribute to the real power and, therefore, is a penalty.

In the special case of an ac source connected to a resistance, V_1 and I_1 have no phase difference. Then the power factor is calculated as

$$PF = \frac{P_1}{S} = \frac{\tilde{V}_1 \tilde{I}_1}{\tilde{V}\tilde{I}} \qquad (1\text{-}35)$$

MATLAB instructions to determine the power factor, the rms value of the fundamental harmonic of v(t) and the THD are given below:

```
v1rms=vh(2)/sqrt(2) %rms value of pulse wave v(t)
irms=vrms/R; i1rms=v1rms/R;
  %the rms values of current in R
pf=(v1rms*i1rms)/(vrms*irms) %power factor
THD=((sqrt(vrms^2 - v1rms^2))/vrms)* 100
  %percentage Total harmonic distribution in v(t)
```

v1rms = 1.9446 pf = 0.4556 THD = 73.79 %

Crest Factor is the ratio of peak voltage/current to the average or rms voltage/current variables. Low crest factors are desirable.

1.7.5 Component stresses

Voltage and current in the switch may consist of high transients (other than the surges) which may be considerably higher than the nominal operating values. These transient values dictate the selection of switching devices and hence determine the cost. The component stress is measured as the ratio of the peak voltage (current) to the rms value on the component.

1.8 PROTECTION

Electronic switches in the power electronic converters are subjected to high rates of change in voltage and current that may rupture their physical structure and cause permanent damage. The switching devices must be protected from high transition conditions. Furthermore, the converters must be protected from the external surge conditions that may exist in the electric utility supply or may be acquired from other systems in the vicinity through radiation and electromagnetic interference.

The switching converter generates transients of its own due to switching at high power levels. The transients should not be allowed to cause distortions in the electric utility supply.

1.8.1 Surge voltage protection

Selenium diodes and Metal Oxide Varistors (MOV) are commonly used for the protection of power electronic systems from the input voltage surge transients. Figure 1.26 shows the application of the surge protection devices.

In normal operation, the selenium diode is biased to conduct a very low current. At the occurrence of surge voltage, the selenium diode breaks down and conducts heavy current to limit the voltage across it, thus protecting the converter from being subjected to high voltage transients. The designer must ensure that the selenium diode must be capable of dissipating the surge energy, the surge voltage multiplied by the current. For ac line applications, two selenium diodes are connected back-to-back.

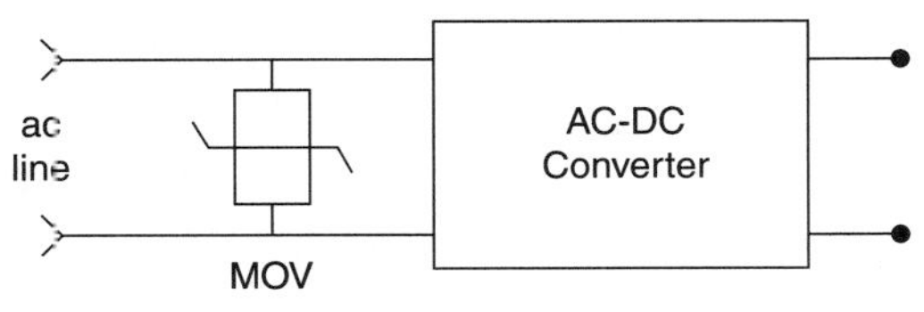

Figure 1.26
Surge voltage protection

Metal-oxide-varister is a device that exhibits nonlinear resistance. It is constituted of metal oxide particles separated by insulation film. The insulation film conducts at high voltages and thus limits the voltage across the converter.

1.8.2 Protection against excessive current

Breaking the connection to the electric utility power supply physically, called fusing, see figure 1.27, is generally used for protection from short circuit or excessive current conditions. The fuse may be a thin wire that burns off at excessive current. It may also be a semiconductor element that exhibits very low resistance in normal operating condition but undergoes in a permanent high resistance mode an excessive current condition. The fuse may be used in series with both the wires of a single-phase utility and on all three lines of the three-phase utility. The equipment manufacturers, however, recommend connecting fuses in series with each device separately. The choice is generally guided by cost constraints.

1.8.3 Protection of switching devices

The switch protector must perform two functions: (1) to limit the stresses on the switch to safe values, and (2) to divert the energy during switching transition from the switch to another circuit. The switch stresses are the maximum transient voltage, the maximum transient current, the rate of change of switch voltage dv/dt and the rate of change of switch current di/dt. The protection is specifically required at the turn-on, turn-off of power to the converter and in over-voltage conditions. The high rate of voltage and current transition give rise to local hot spots in the

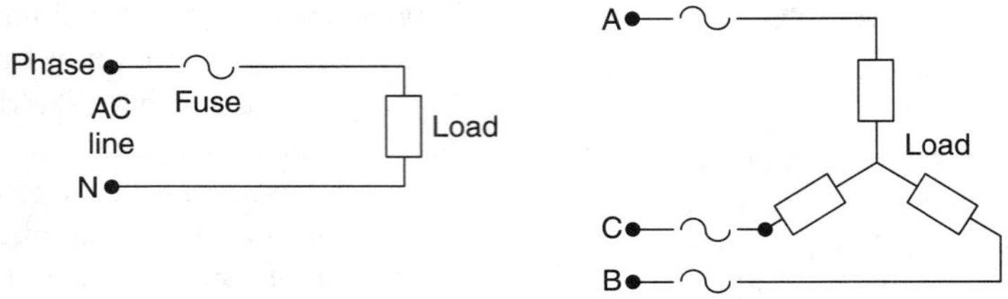

Figure 1.27
Fuse placement in electric utility supply

switching devices that if persist for some time, may permanently damage the devices. Special circuits are used to slow down the high rate of transition. These circuits are called *snubbering* circuits. An example of a snubber circuit is shown in Figure 1.28.

The switching devices are further protected from failure by lowering the losses during switching transitions. Turning on and off of a switch at zero current or zero voltage is the surest way of minimizing switching losses. Special circuit configurations and techniques are employed to achieve them. Soft starting technique is one such technique. The general topic of lowering switching losses is discussed in Chapter 2 and the protection circuits for specific switching devices are discussed in Chapters 3, 4 and 5.

1.8.4 Over-voltage protection

The current carrying conductors in the electric utility system and the conductors for interconnection among subsystems have a very significant amount of self-inductance. This inductance is primarily responsible for long duration, high voltage spikes on the switching devices. Special circuits are required to protect the switching devices from the over voltages. Chapter 2 covers this topic in more detail.

1.9 PACKAGING

The packaging of converter circuits and subsystems plays a very important role in the performance and reliability of power electronic converters. The package design must address several multidiscipline issues:

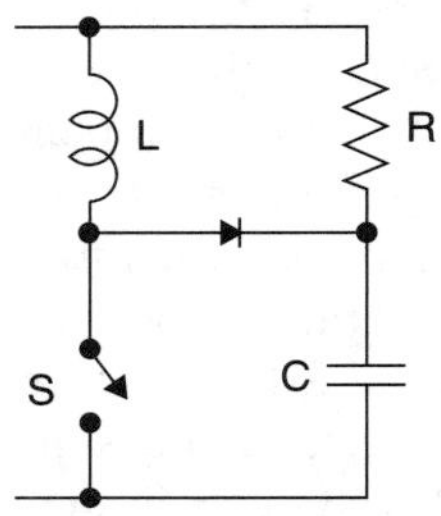

Figure 1.28
Snubber circuit for protection of switches

1. Removing internal heat and control of internal temperature profile of components through heat sinking
2. Shielding of converter from external and intra-component electromagnetic interference through proper layout of components
3. Reducing parasitic elements such as the self-inductance, stray capacitance and induction, etc. through proper layout.

Power losses in electrical components appear in the form of heat. Accumulation of heat energy increases temperature of the internal structure of the components. The heat must be removed from the body of the component by transferring it to the ambient environment. The heat transfer takes place in three ways:

1. Conduction
2. Convection and/or
3. Radiation

Figure 1.29
Heat sinks and bus bars

Heat flows from one stationary interface to another through vibratory motion of atoms or molecules. This mode of heat transfer is called conduction. Convection is the mechanical transport of heat by a moving fluid or gas. The fluid flow may be natural or forced. Fluid flow in natural convection is governed by temperature gradients. Fans and pumps are used for forced convection. In heat transfer by radiation, the heat energy is converted into electromagnetic radiation, which is absorbed by other components in the vicinity. In enclosed packages, the components at elevated temperature radiate heat to the neighboring components at lower temperature. As a consequence, the neighboring components are required to function at elevated temperatures, something they were not planned for.

Fortunately, the heat transfer in power electronic systems is dominated by conduction and convection processes. The thermal design issues and selection of heat sinking systems are discussed in Chapter 16. The power path conductors, such as the bus bars, are required to be wide and with large areas of cross section to minimize self-inductance.

Problems

1-1 Calculate first analytically and then using MATLAB the average and rms voltage for the following waveforms:

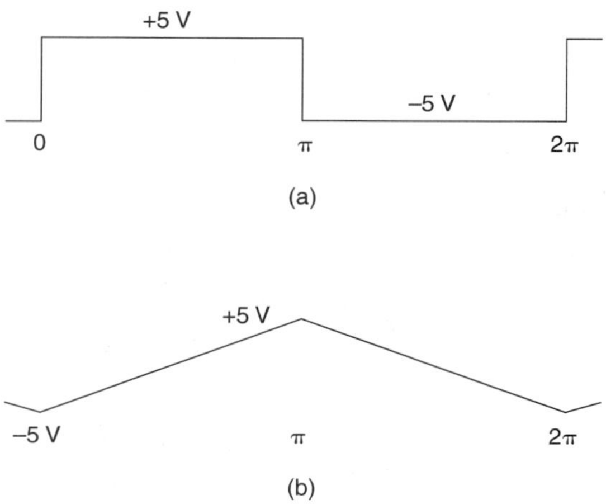

Figure P1.1

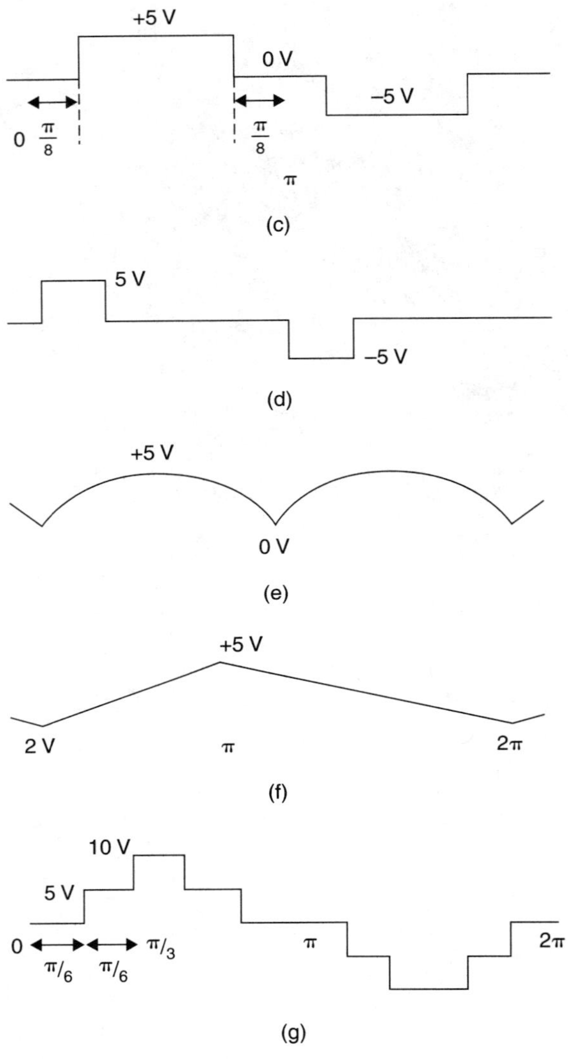

Figure P1.1

1-2 Find the ratios of the rms values of the fundamental component to the overall rms values and the THDs of the waveforms in Problem 1-1.

1-3 A square-wave voltage source, 300 V peak-peak, 400 Hz, is connected to a 25 mH inductor.

a. Analytically, sketch the steady-state current waveform.

b. Calculate the peak value of the current waveform.

c. Draw the voltage and current waveforms using MATLAB.

d. Would you recommend connecting a capacitor in place of the inductor? Support your argument.

1-4 For the circuit in Problem 1-3,

a. calculate the rms values of the voltage and the current.

b. Plot the harmonic profile of the voltage and current waveforms up to nine harmonics.

c. Calculate the average power delivered at the fundamental, the power factor and the total harmonic distortion in the current.

1-5 Plot the instantaneous power waveform and the frequency spectrum of the power for Problem 1-3 using MATLAB.

1-6 A triangular-wave voltage source, 400 V peak-peak, 400 Hz, is connected to a 47 F capacitor.

a. Analytically, sketch the steady-state current waveform.

b. Calculate the peak current.

c. Draw the voltage and current waveforms using MATLAB.

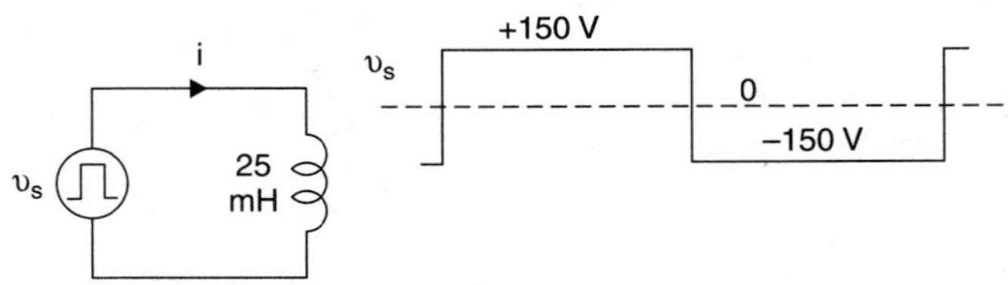

Figure P1.3

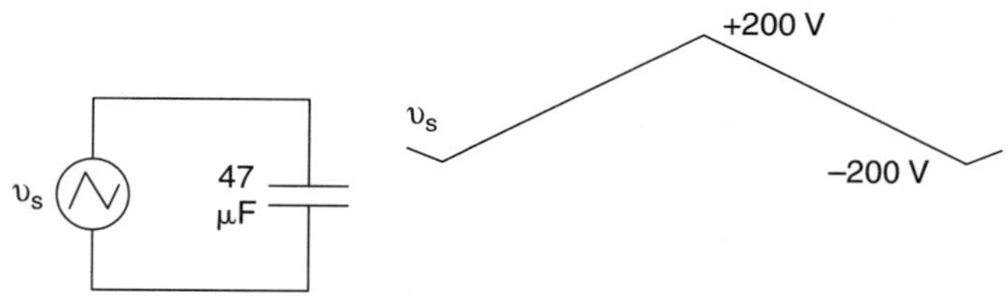

Figure P1.6

1-7 For the circuit in Problem 1-6,

a. plot the harmonic profile of the voltage and current waveforms up to nine harmonics.

b. Calculate the average power delivered at the fundamental, the power factor and the total harmonic distortion in the current.

1-8 Consider a Pulse Width Modulated converter shown in Figure P1.8, also called the Buck converter, is switched at 100 KHz. The switch is turned on for a 75% duration of the switching period. The smoothening inductor is assumed to be infinitely large.

a. Find analytically the average output voltage, the average output current and the average input current in the resistor.

b. Plot the waveforms for the output voltage v_O, the output current i_O and the source current i_s using MATLAB and confirm the average values calculated in part (a).

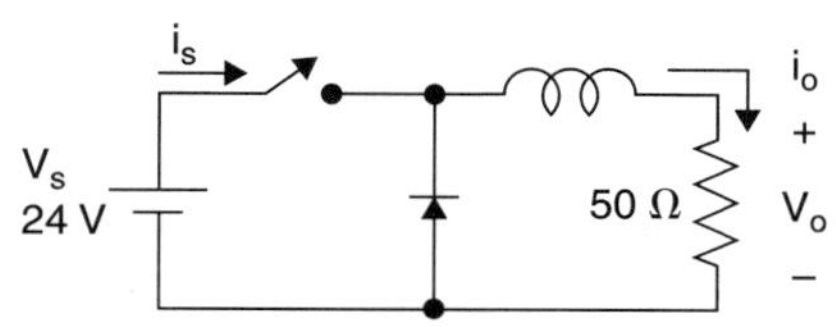

Figure P1.8

1-9 A 100 μF capacitor is connected across the load resistor in the Buck converter in Problem 1-8. Plot waveforms of the output voltage v_O, I_o and the current in the capacitor i_C. Draw the harmonic profile of the capacitor current.

1-10 If the inductor in Problem 1-8 is 1 mH, draw the inductor current waveform using MATLAB.

1-11 A 12-volt battery is delivering a current waveform as shown in Figure P1.11. Find the average current and the power delivered by the battery.

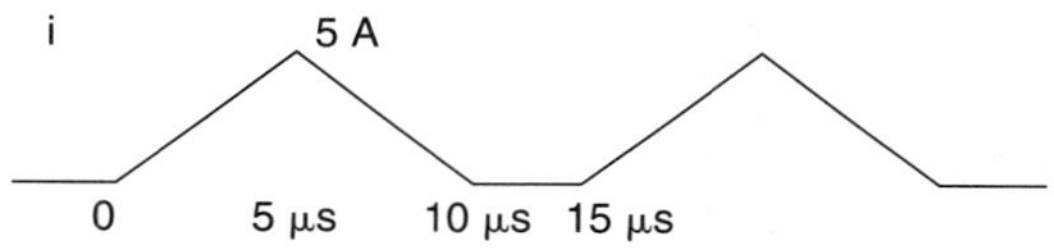

Figure P1.11

1-12 A 120 rms, 60 Hz, source is delivering 10 A rms with a lagging phase of 30°.

a. Is the load inductive or capacitive?

b. Plot the voltage, the current and the instantaneous power as a function of the conduction angle $\theta = 2\pi 60t$.

c. Calculate the apparent power, the real power and the reactive power. Use MATLAB.

1-13 Find the voltage transfer ratio for the following circuit:

a. $T_{vv} = \dfrac{\overline{V_O}}{\overline{V_S}}$

b. $T_{vv} = \dfrac{\overline{V_O}}{\hat{V}_S}$

1-14 A balanced 3-phase load is supplied in steady state by a balanced 3-phase voltage source. The phase voltage is 120 volts rms,

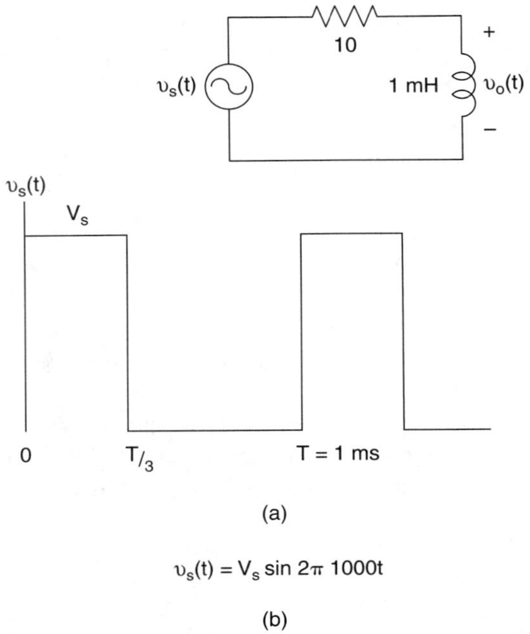

Figure P1.13

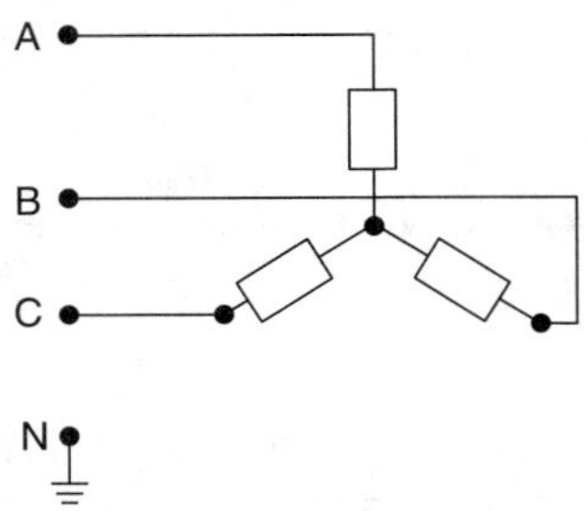

Figure P1.14

60 Hz. The load draws 50 KW at a lagging power factor of 0.75.

a. Calculate the rms values of phase currents and the magnitudes of per phase load impedance.

b. Determine the per phase capacitance to be added to improve the power factor to 0.95 lagging.

1-15 A 120 V rms 60 Hz source supplies a pulse current as shown in Figure P1.15. The fundamental component of the current is in phase with the source voltage waveform. Find the average power delivered.

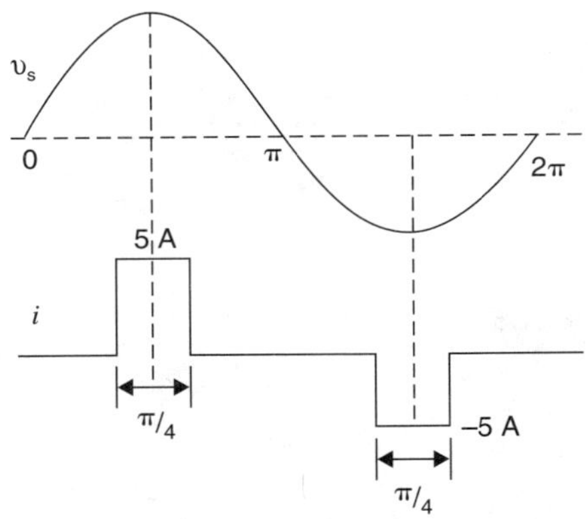

Figure P1.15

1-16 For Problem 1-15, find the pulse width of the current for which the 3rd harmonic is zero.

1-17 Repeat Problem 1-15, with two thin pulses as shown in Figure P1.17.

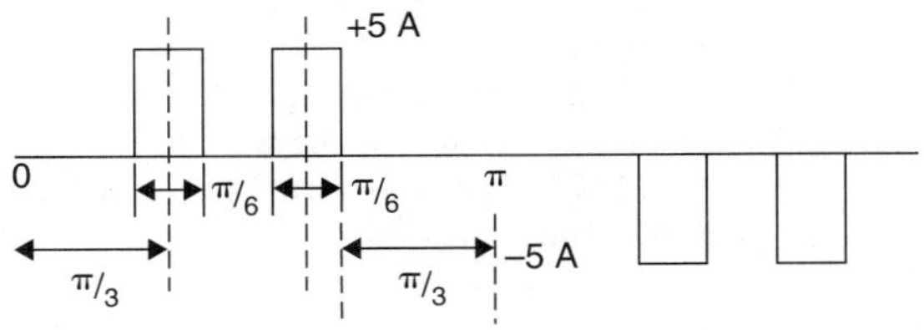

Figure P1.17

1-18 If the sinusoidal source in Problem 1-15 is replaced by a square-wave of a peak voltage equal to that of the sinusoid, find

a. the average power delivered,

b. the power factor, and

c. plot the power frequency spectrum.

1-19 A cycloconverter output voltage of frequency $f_O = 1/T_O$ is shown in Figure P1-19. The half period $T_O/2$ is made of two half-cycles of a high frequency waveform. Plot the harmonic profile up to ten harmonics of f_O.

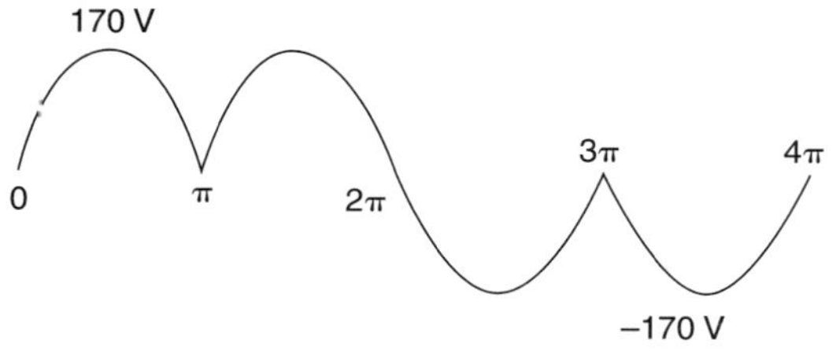

Figure P1.19

1-20 Plot the ratio of the rms value of the fundamental component to the rms value of the complete waveform in the cycloconverter of Figure P1.19 as the function of the number of half cycles of the high frequency waveform per one half period $T_O/2$ (up to a maximum of 15).

1-21 An electrical switch has the switching characteristics shown in Figure P1.21. Find the maximum switching rate that can be achieved if the average power loss in the switch is limited to 100 watts. Plot the instantaneous power dissipated in the switch and find the maximum power.

1-22 A precharged capacitor 1 μF is to be used as the dc voltage source of 10 volts in a power electronic converter operation. The voltage is held to within 1% of 10 volts over a period of the converter operation in steady state while supplying a current of 100 mA. Find the time period.

1-23 Design a dc current source of 12 amps from a 12-volts battery for use in a converter that operates at the switching frequency of 40 KHz. The current should not vary by more than 1% during the switching period when the load resistance varies from 10 ohms to 1 kohms. Give the inductance value and the number of turns.

1-24 Find the equivalent circuit for the circuit in Figure P1-24 as referred to

a. the primary winding and
b. the secondary winding.

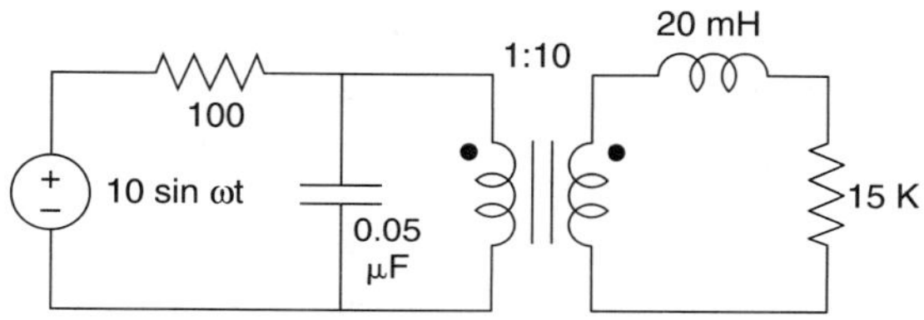

Figure P1.24

1-25 Repeat Problem 1-24 for the circuit in Figure P1.25.

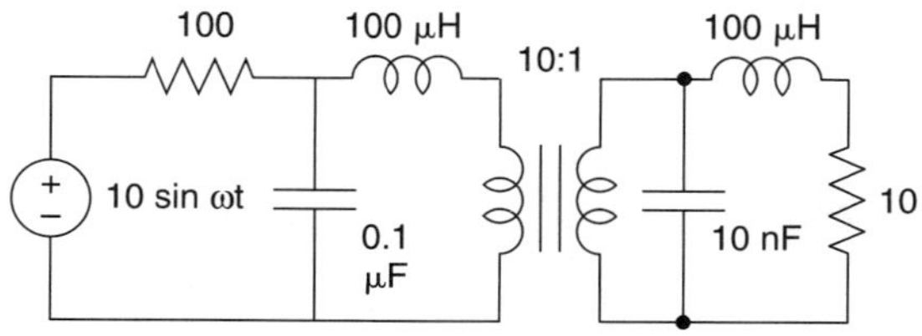

Figure P1.25

1-26 Repeat Problem 1-24 for the circuit in Figure P1.26.

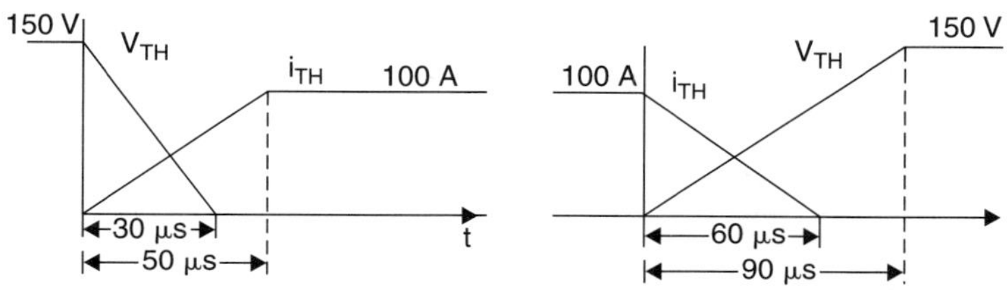

Figure P1.21

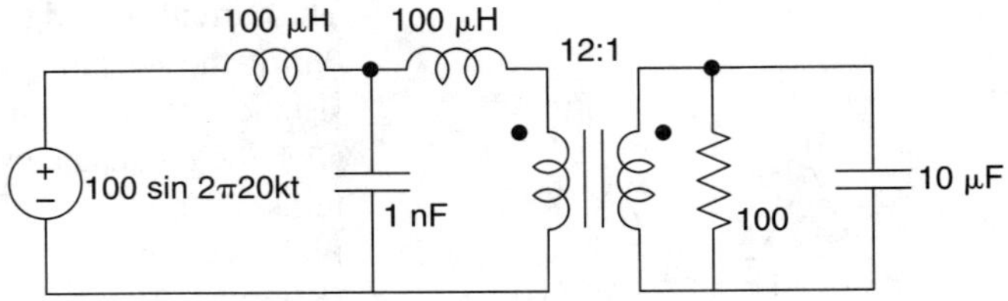

Figure P1.26

Part Two

POWER ELECTRONIC COMPONENTS

CHAPTER 2
Understanding Components

OBJECTIVES

After the completion of this chapter the reader should understand:

1. The fundamentals of magnetism and magnetic materials
2. The basic features of magnetic components like the inductor, transformer and motor
3. The design of inductors
4. The design of transformers
5. The types of capacitors
6. The basic switching with resistive, inductive and capacitive loads: the switching losses
7. The circuits for the protection of switches: snubbering
8. The zero-current and zero-voltage switching

2.1 INTRODUCTION

Power circuits and systems are constituted from inductors, capacitors, transformers, motors and switches. Power electronics is strongly dependent on the component characteristics, which in a real world are not very ideal. The materials from which they are fabricated and the process of fabrication have a lot to do with the nonideal behavior of the components. Hence, it is of paramount importance that we develop an overall understanding of materials and components before discussing the operation of power electronic circuits. In this chapter we will survey the basic concepts and characteristics of magnetic components such as inductors, transformers and motors and electric components such as capacitors and semiconductor switches.

2.2 BASICS OF MAGNETISM

2.2.1 Magnetostatic field

A conductor of length **dl** carrying a current **i** produces a magnetic field around it. The magnetic field intensity **H** is obtained from

$$d\vec{H} = \frac{I}{4\pi r^2}\, \vec{dl}\; x \vec{a} \quad \text{(Biot-Savart Law)} \qquad (2\text{-}1)$$

Vector **a** is the unit vector in the radial direction. Magnetic field **dH** is a vector whose direction is determined from the cross product of unit radial vector and the direction of current flow. Hence, the

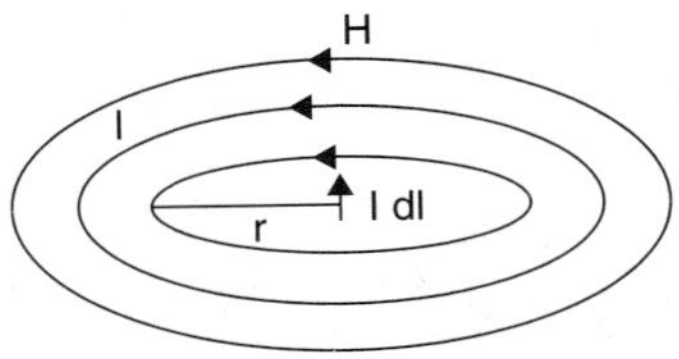

Figure 2.1
Magnetic field around a current element

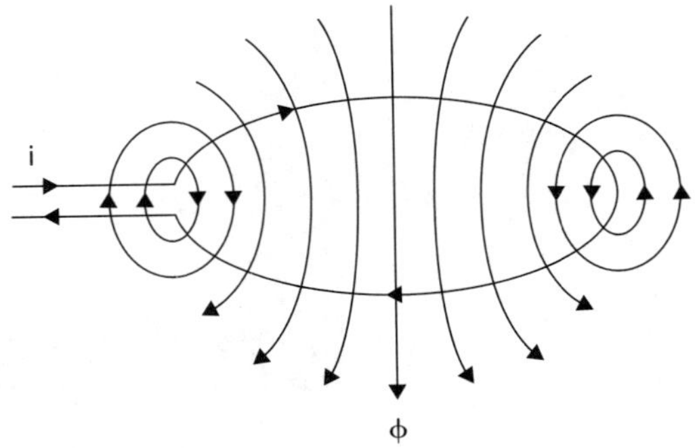

Figure 2.2
Magnetic flux lines of a circular shaped coil

magnetic field intensity **H** due to a conductor of length L is obtained from the following integral:

$$\vec{H} = \frac{1}{4\pi r^2}\int_L I\, d\vec{l} \times \vec{a} \qquad (2\text{-}2)$$

The circular lines around the current carrying conductor, which indicate the direction of magnetic field vectors, are called the magnetic lines of force or simply the magnetic lines of flux. Unlike the radial electric flux lines, the magnetic flux lines are closed curves. If the conductor is shaped into a circle, the magnetic field at the center of the circle is the strongest as it is the sum of the magnetic fields due to all elemental **dl's** of the circular shaped conductor, see Figure 2.2.

The magnetic flux through a bounded surface and the magnetic flux density are denoted by ϕ and **B** respectively and are defined as the following:

$$\phi = \int_A \vec{B} \,.\, d\vec{A}, \ \vec{B} = \mu\vec{H} \qquad (2\text{-}3)$$

A is the surface area. The μ is permeability of the medium in which the magnetic field is established. In a uniform medium (core material), permeability is time invariant and isotropic (uniform in all directions). Above elemental linear relation is not found in practical materials. Recall the famous B-H curves and hysteresis loops of the magnetic materials you have known from the course in electromechanical energy conversion. Nevertheless, the linear relationship may be safely assumed to be valid for small signal operations. There will be more on nonlinearity in a later section of this chapter.

TABLE 2.1

Material	μ_r Relative permeability
Silver	0.9999736
Copper	0.9999906
Water	0.9999912
Hydrogen	1.0000000
Air	1.0000004
Aluminum	1.0000210
Cobalt	250
Nickel	600
Soft iron	5000
Silicon-iron	7000

Permeability is often expressed as the product of permeability in free space μ_0 ($=4\pi 10^{-7}$ weber/m^2) and relative permeability μ_r:

$$\mu = \mu_0\, \mu_r \qquad (2\text{-}4)$$

Relative permeability of some important materials is given in Table 2.1.

Ampere's law states that the line integral of magnetic field **H** along a closed path **l** equals the current flowing through the surface bounded by that path.

$$\int \vec{H} \cdot d\vec{l} = I \quad \text{(Ampere's Law)} \qquad (2\text{-}5)$$

Potential energy density stored in a magnetostatic field (H field alone) is given by

$$W_H = \frac{1}{2}\vec{B}.\vec{H} = \frac{1}{2}\mu H^2 \quad (2\text{-}6)$$

2.2.2 Faraday's law and induced voltage

Rate of flow of charge **dq/dt** is current **i.** A moving charge establishes a magnetic field in the plane that is normal to the line of movement. The magnetic field is also produced by a stationary charge that is varying with time. Likewise, a conductor moving through a magnetic field develops a voltage between its end terminals. The terminal voltage also develops on a stationary conductor in a time varying magnetic flux. The induced voltage is given by

$$v = 0\int \vec{E}.\vec{dl} = -\frac{d\lambda}{dt} \quad \text{(Faraday's Law)} \quad (2\text{-}7)$$

λ is the total magnetic flux (also called the flux linkage) which is defined as

$$\lambda = N\phi \quad (2\text{-}8)$$

where N is the number of times the moving current carrying conductor encounters the magnetic flux. Inductors and transformers can store energy in the magnetic field. Capacitors store energy in the electric field. In the next section, we will focus on the inductors and transformers. Capacitors are discussed in the later sections.

2.3 MAGNETIC MATERIALS

Although all solids contain polarized dipoles, they are randomly oriented. The degree of alignment, measured by relative permeability μ_r, is high in magnetic material such as iron. Permeability of metals such as copper and aluminum are close to that of air (see Table 2.2). Application of an external magnetic field or an electromagnetic field created by a coil wound around a piece of iron force dipoles to align with it. Alignment becomes complete at a certain amplitude of the externally applied field. Flux density thus achieved is called saturation flux density.

Two broad classes of materials are used in making cores: ***ferrous alloys*** and **magnetic ceramics** called ***ferrites.*** Ferrous alloys are used at frequencies 1kHz and below. Ferrites are used in high frequency applications. Ferrous alloys contain mostly iron with small amounts of silicon and other metals such as cobalt, nickel, molybdenum and chromium. These materials are characterized by moderate permeability, high saturation flux densities, high electrical conductivity and high curie temperature, a temperature above which the magnetization is lost.

TABLE 2.2

	Curie temp (C)	B_s (T)	μ_r	ρ (μohm-cm)
Low-Si iron	760	2.2	2700	10
Core iron (1%)	810	2.1	4500	25
Si steel (2.5%)	780	2.0	5000	40
48% Ni alloy	450	1.5	40000	48
80% Ni, 4% Mo alloy	460	0.8	50000	58
50% Co alloy	950	2.3	10000	35
Ferrite (Mn-Zn)	130–225		0.3–0.4	$1\text{–}4\times10^3$
Ferrite (Ni-Zn)	100–500		0.2–0.3	150
Metglass	370	1.6	10000	125

With the addition of materials such as cobalt, nickel, molybdenum and chromium, the permeability can be increased and core losses can be reduced up to about 20 kHz. High conductivity causes high eddy current losses, therefore, use of ferrous metals is limited to low frequencies. More about eddy current losses in a later section.

Ferrites are ceramics made of ferrous oxide, zinc, and either manganese or nickel (in higher frequency applications). Ferrites are characterized by high permeability, low saturation flux densities, typically 0.3 Tesla, low conductivity and low curie temperatures. Low conductivity causes low eddy current losses, hence, ferrite is the most preferred material in high frequency applications.

METGLASS, a trade name for amorphous alloy of iron and other metals, is in between the ferrous alloys and ferrites. It has lower conductivity than ferrous alloys, higher saturation densities than ferrites and can be formed into thin metallic ribbons.

A useful table below gives some of the important units in magnetics:

Magnetic field intensity H:	1 Oersted	= 79.5 Ampere-turns/m
	1 Oersted	= 0.795 ampere-turns/cm
Magnetic flux density B:	1 tesla (weber/m^2)	= 10^4 gausses
Area:	1 inch^2	= 6.452 cm^2
	1 circular mil	= 5.07×10^{-6} cm^2
Power density	1 mW/cm^3	= 0.094 W/lb

2.4 MAGNETIC CIRCUITS

We know from the previous section that the magnetic flux lines are closed paths. A closed path is analogous to an electrical circuit. Magnetic flux is analogous to current in an electrical circuit. To extend the analogy further, we would examine the expression of magnetic flux ϕ in an N-turn coil wound over a magnetic core of homogenous permeability and constant magnetic field intensity H. The flux is given by (from equation (2-10)):

$$\phi = BA = \mu HA = \mu \frac{Ni}{l} A \qquad (2\text{-}9)$$

Magnetic field **H** is uniform over the magnetic path **l** of homogenous permeability. Reorganizing the above equation as follows:

$$\phi = \frac{\mu A}{l} Ni \qquad (2\text{-}10)$$

Ni, the ampere-turns, also called the magnetomotive force **mmf,** may be treated as analogous to the voltage source in electrical circuits. Then we can define a property ***reluctance*** $\mathfrak{R}$ as below:

$$\mathfrak{R} = \frac{Ni}{\phi} = \frac{l}{\mu A} \qquad (2\text{-}11)$$

$$\mathfrak{R} = \frac{(l/\mu_r)}{\mu_0 A} \qquad (2\text{-}12)$$

Term $\mathbf{l/\mu_r}$ is the equivalent magnetic length in free space. Equation (2-11) is used in homogenous permeability medium and Equation (2-12) in heterogeneous permeability medium. Reluctance is analogous to resistance in electrical circuits, and is the property of the core material. Reluctance is what the word means, exhibited by the material to the flow of the magnetic flux. The analogy is complete and it presents us with the power of very well established electrical circuit techniques of analysis.

	Electrical		Magnetic	
across variable	voltage	v	ampere turns	Ni
through variable	current	i	flux	ϕ
Resistive property	Resistance	R	Reluctance	$\mathfrak{R}$

The magnetic flux lines leaking into air in the vicinity of a magnetic circuit is a common phenomenon. Magnetic flux thus lost is called the leakage flux and often must be accounted for in the design of magnetics. On the other hand magnetic components with all air core are found being used quite often in practice.

2.5 INDUCTOR

Inductor is an element that stores energy in the magnetic field. It is comprised of a conductor wound as a several turn coil on a core with a closed magnetic path. The magnetic path may comprise of material of homogenous or heterogeneous permeability. The homogenous paths may be comprised of all ferrous alloys or ferrites or all air. A heterogeneous permeability path is a combination of magnetic core path and air gaps. Figure 2.3(a) shows some of the commonly used core shape: **'C'**, **'E'**, **toroids** and **Pot cores.**

2.5.1 Inductor with homogenous permeability core

Consider an inductor wound on an all magnetic core such as the toroidal core, which has permeability μ and a linear B-H relation, shown in Figure 2.3. The current flow (or more specifically the charge flow) through the coil produces magnetic field intensity **H**

$$\int_l \vec{H}.d\vec{l} = Ni \approx H\,l \quad : l = 2\pi r \tag{2-13}$$

The approximate sign is due to the assumed mean path length **l** as shown in the center of the toroid. Magnetic flux density is expressed by

$$B = \mu\, H = \frac{\mu N i}{l} \tag{2-14}$$

Flux linkage is calculated as the magnetic flux ϕ multiplied by the number of turns in the coil (number of times the current circles around the flux):

$$\lambda = N\phi = N\,B\,A = \frac{\mu\, N^2 A}{l}\, i \tag{2-15}$$

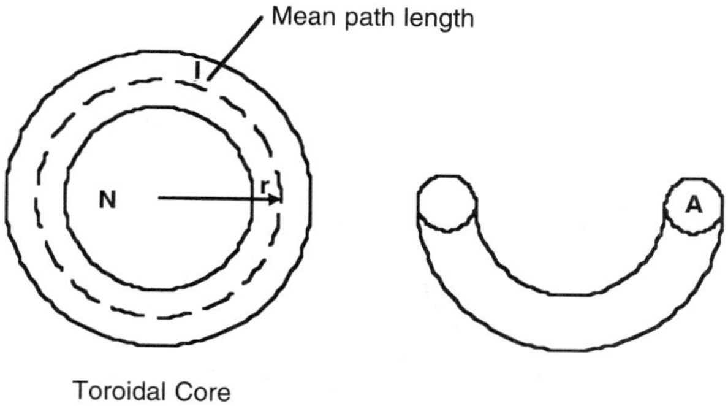

Figure 2.4
A toroidal inductor

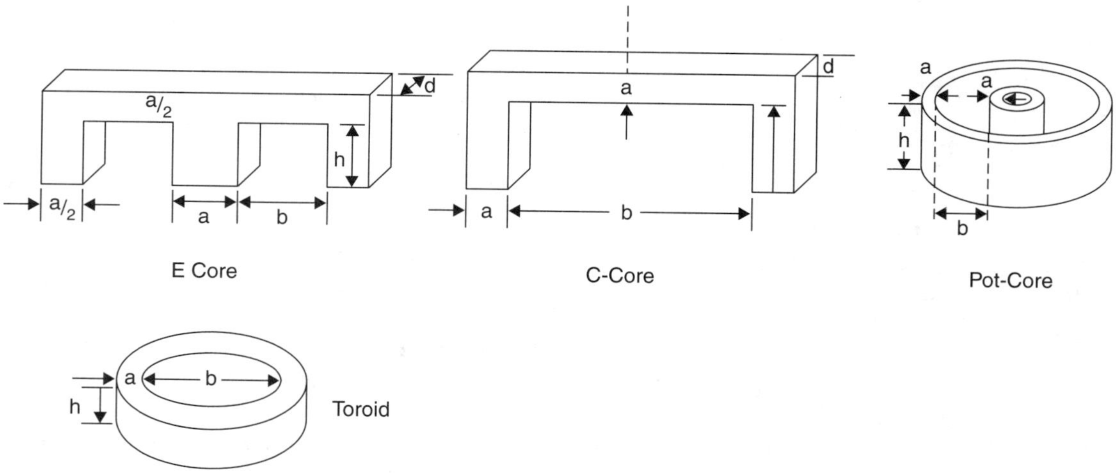

Figure 2.3
Ferrite cores

Hence, we could say

$$\lambda = L\,i \; : \; L = \mu\, N^2 \frac{A}{l} = \mu_0\, N^2 \frac{A}{(l/\mu_r)} \qquad (2\text{-}16)$$

where, **L** is called inductance and is measured in **henry.** The l/μ_r is the equivalent magnetic path length in air.

Example 2-1

Find the number of turns in a 1 mH toroidal inductor. The toroid specifications are given below:

Outer diameter	*Inner diameter*	*permeability μ_r*
= 40 mm.	*= 30 mm.*	*= 1000*

Mean geometric path length of the toroidal core

$$l = \frac{\pi(od + id)}{2} = \pi \frac{(.040 + .030)}{2} \approx 0.11 \text{ meters}$$

The core area

$$A = \pi \frac{(od - id)^2}{4} = 0.785 \times 10^{-4} \text{ sq.m.}$$

The number of turns is calculated from eq. (2-16) as follows:

$$N = \sqrt{\frac{Ll}{\mu_r \mu_0 A}}$$

$$= \sqrt{\frac{1 \times 10^{-3} \times 0.11}{1000 \times 4\pi \times 10^{-7} \times 0.785 \times 10^{-4}}}$$

$$= \approx 101 \text{ turns}$$

Figure 2.5 shows an inductor wound on an E-E core. N turns of a conductor are wound tightly on the center limb of the core. Areas of cross section of all limbs except the center limb are equal and $\mathbf{A_2}$. The area of cross section of the center limb $\mathbf{A_1}$. **Ni,** ampere-turns, applied to the winding, produces a magnetic flux ϕ in the center limb. Magnetic flux flows in two closed paths around the two windows of the core. Each section of the magnetic circuit offers reluctance that may be represented by lumped elements. Reluctance is determined by the area of cross section and the path length. Since the core material is homogenous throughout the structure, we use geometric length instead of magnetic length for our calculations. Geometric path length in any limb differs over the cross section. An approximate reluctance can, however, be calculated using mean path lengths along the dotted lines, as given by the following:

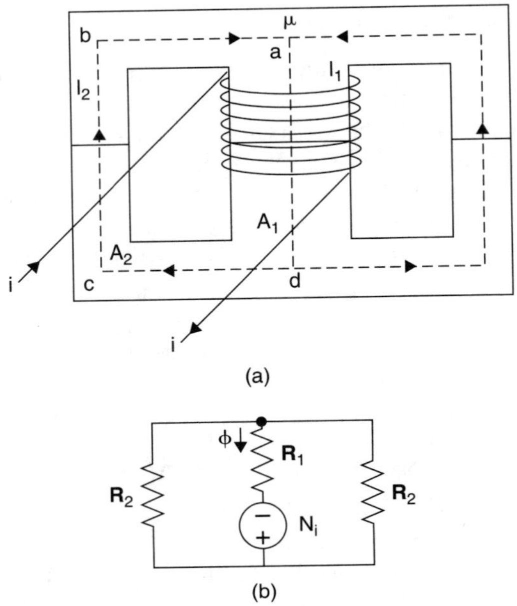

Figure 2.5
(a) An inductor wound on the E-core and (b) magnetic equivalent circuit

$$\Re_1 = \frac{l_1}{\mu\, A_1}, \Re_2 = \frac{l_2}{\mu\, A_2} \qquad (2\text{-}17)$$

$\mathbf{l_1}$ is the mean length of the central path **a-d** and $\mathbf{l_2}$ is the length of the side path denoted by **a-b-c-d.** In practice, the flux lines curve around the corners, hence the dotted lines are again an approximation. We would also assume the leakage flux is zero. Complete magnetic circuit is shown in Figure 2.7(b). Flux ϕ in the center limb is divided equally in the two side sections because both sections have equal reluctance. Using Kirchoff's mesh and nodal equations,

$$Ni = \left(\Re_1 + \frac{\Re_2}{2}\right)\phi \tag{2-18}$$

Flux linkage λ is calculated as following:

$$\lambda = N\phi = \frac{N^2 i}{\Re_1 + \Re_2/2} \tag{2-19}$$

The coil inductance is calculated from

$$L = \frac{\lambda}{i} = \frac{N^2}{\Re_1 + \Re_2/2} = \frac{\mu N^2}{(l_1/A_1) + (l_2/A_2)} \tag{2-20}$$

Leakage inductance may be accounted for by assuming a parallel magnetic circuit with another reluctance.

Permeability in practice is a highly temperature sensitive parameter. As we shall see later, the permeability is also a nonlinear function of higher currents in the vicinity of saturation region on B-H curve. Hence inductance of the above structure is sensitive to the ambient conditions.

Practical inductors have nonzero winding resistances. The winding resistance is also dependent on the frequency of the current flowing through the inductor, as we shall see in later sections of this chapter. The quality factor of inductors that accounts for these losses is called the Q factor:

$$Q = \frac{\omega L}{R} \tag{2-21}$$

The ω is the frequency in radians of the current through the inductor.

2.5.2 Inductor with heterogeneous permeability core

Nonuniform permeability generally means discontinuity in permeability of the magnetic path. The magnetic path of many practical inductors consists of a high permeability material such as the Si-steel or the ferrite and an air gap. Permeability of air is very stable with respect to the ambient conditions. All-air core, therefore, should make an ideal inductor. Size-wise such an inductor for high values, such as that required in power electronics would be unwieldy in size. Intentional air gap in an otherwise high permeability core may be used to stabilize the inductance values. Figure 2.6(a) shows a C-C core structure with two air gaps of length **g/2.** Permeability of the core material and the air gap are $\mu_c = \mu_{rc}\mu_0$ and μ_0 respectively. Magnetically the two air gaps are in series. The ampere-turns of the N-turn coil establishes a magnetic flux ϕ. Combined mean magnetic path of the high permeability part is $\mathbf{l_c}$. The equivalent magnetic circuit with Reluctance is shown in Figure 2.6(b).

Using eq. (2-11) and Kirchoff's mesh law, we get

$$Ni = (\Re_c + \Re_g)\,\phi = \left(\frac{(l_c/\mu_{rc})}{\mu_0 A} + \frac{g}{\mu_0 A}\right)\phi \tag{2-22}$$

$\mathbf{l_c/\mu_{rc}}$ and **g** are the magnetic path lengths of the high permeability material and the air gap.

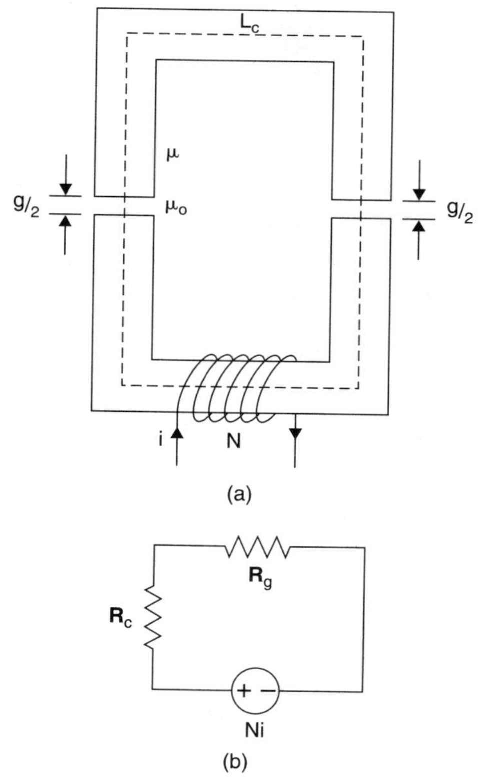

Figure 2.6
(a) Air gaped inductor and (b) magnetic equivalent circuit

Inductance is, therefore, given by

$$L = \frac{\lambda}{i} = \frac{N\phi}{i} = \frac{\mu_{0=} N^2 A}{(l_c/\mu_{rc}) + g} \qquad (2\text{-}23)$$

Inductance value for large air gaps becomes smaller but more independent of the core permeability. Inductance is then determined solely by the air gap geometry, as shown in the following equation:

$$L = \frac{\mu_0 N^2 A}{g} \quad where\ g \ll \frac{l_c}{\mu_{rc}} \qquad (2\text{-}24)$$

Energy Storage Density

Energy storage density in the inductor are governed by eq. (2-5):

$$w = \frac{1}{2}\vec{B}.\vec{H} = \frac{1}{2}\frac{B^2}{\mu} \qquad (2\text{-}25)$$

Total stored energy has two parts: in the core volume and the air gap volume

$$W = \frac{B^2}{2\mu_c} v_c + \frac{B^2}{2\mu_0} v_g \qquad (2\text{-}26)$$

v_c and v_g are the volumes of the high permeability core and the air gap portions of the inductor respectively. Since $\mu_c >> \mu_0$, energy storage density in the air gap is much larger than in the high permeability core. Hence, most of the energy in a gaped inductor is stored in the gap portion. For the iron core with a saturation flux density $\mathbf{B_{max} = 1.4\ T}$, and ferrite core, with $\mathbf{B_{max} = 0.3\ T}$, the maximum energy storage densities in the gap are given below:

$$W_g - iron = \frac{1.4^2}{2\mu_0} = 1\ J/cm^3$$

$$W_g - ferrite = \frac{0.3^2}{2\mu_0} = 0.05\ J/cm^3$$

These are very important parameters in the design of inductors. The inductor specifications are the inductance value and the current capacity, that is, the stored energy value and hence it completely determines the volume of the air gap.

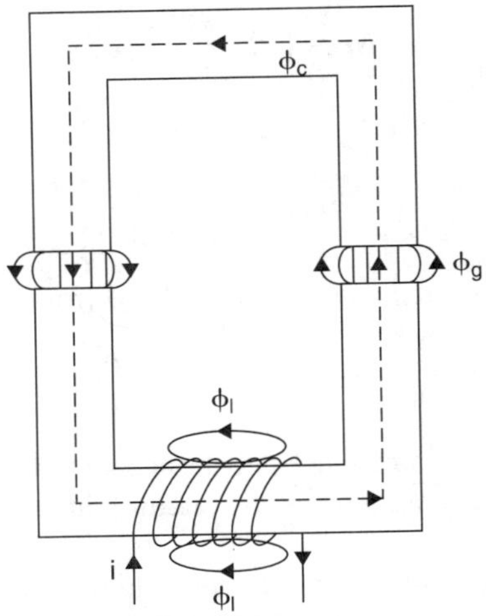

Figure 2.7
Leakage flux

Leakage Flux

In this section, we will discuss leakage fluxes, which are very common in the inductors. Flux may leak from the sides of the winding. The leakage fluxes can be included in the analysis by a reluctance $\Re_l$ in parallel with the ampere-turn source, see Figure 2.7.

Another effect is the bulging out flux lines from the cross sections of the air gap. The bulging field is termed the **fringing field.** Burgeoning of the air gap is accounted for by a larger area of the air gap thus reducing the reluctance of the air gap to $\Re_{g'}$. The new value of inductance is calculated as following:

$$L = N^2\left(\frac{1}{\Re_1} + \frac{1}{\Re_c + \Re_{g'}}\right) \qquad (2\text{-}27)$$

Since $\Re_l > 0$ and $\Re_{g'} < \Re_g$, this calculation leads to a larger inductance value than given by equation (2-23). Fringing flux is larger for larger gap length. Hence for a fixed gap volume, it is advisable to select a small gap length and a wide cross section area.

2.6 NONLINEARITIES IN MAGNETIC COMPONENTS

Magnetization of a core, measured as the flux density **B,** is the process of physically aligning the internal tiny magnetic dipoles throughout the core material. These tiny dipoles are either single atom or groups of atoms, which have the ability of charge polarization. In the absence of any outside forcing agency, the dipoles are randomly aligned in order to neutralize each other. Application of external magnetic field aligns the tiny dipoles physically in a direction so as to minimize their stresses. More alignments take place with the application of higher magnetic field intensity **H.**

Most prominent nonlinearity in magnetic components are (a) saturation, (b) residual flux and (c) hysteresis. Most prominent among losses in the magnetic components are (a) copper losses, (b) core losses and (c) leakage losses. The nonlinearity losses and their repercussions are discussed in the subsections below.

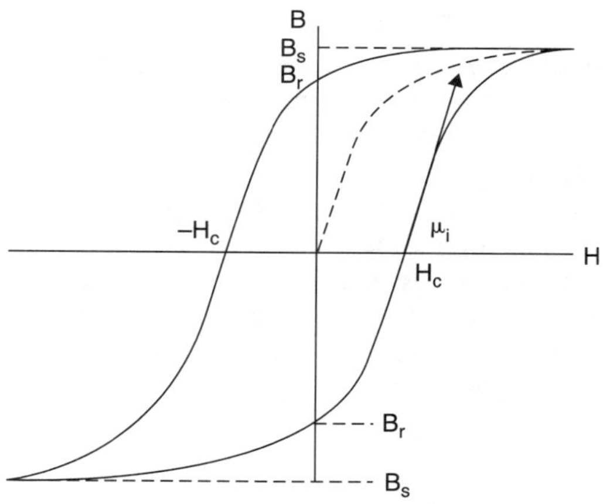

Figure 2.8
B-H curve

2.6.1 Saturation

The relation between the magnetic flux density **B** and magnetic field intensity **H,** as we have assumed to be linear in previous sections, is in reality a very complex relationship. In fact, it is not even homogeneous throughout the volume of the core material. The permeability μ, as we know, is the average value of a tensor parameter, which varies with directions in a core volume. This relationship of average permeability, the slope of the **B-H** curve, is shown in Figure 2.8. The solid curve indicates the steady-state operation in a cyclic magnetization in both directions. The dashed segment is the trace during the first start-up period from $\mathbf{B} = \mathbf{H} = 0$ point. Magnetic flux density **B** saturates in both the forward and backward directions (above and below $B = 0$ axis) when there are no more internal dipoles left to align. Note that the permeability in the saturation region is very low, which practically means a very large change in field intensity is needed to affect a slight change in flux density.

The saturation is avoided in most applications. An electromagnet in saturation will draw heavy current from the magnetizing source. Saturation of cores in inductors and transformers causes the permeability to decrease so much that it becomes comparable to the free air permeability μ_0. Thus more flux is leaked out of the core into the air, causing poor magnetic coupling among the windings of a transformer.

Saturation in the first start-up phase (starting from $\mathbf{B} = \mathbf{H} = 0$) occurs at smaller values of $\mathbf{H}$ (= Ni) than in the steady state in a cyclic operation. This is called the start-up problem. Hence it is advisable to limit the magnetizing current during the first start-up period. The saturation characteristics have been exploited to advantage in some applications such as the magnetic amplifiers, electronic lamp ballast, self-oscillating inverter, etc.

2.6.2 Residual flux

The second nonlinearity refers to the residual magnetism in the core material even after the removal of external magnetic field. The residual flux density is denoted by $\mathbf{B_r}$. A magnetic flux intensity of opposite polarity is required to completely dealign all dipoles. This phenomenon can be explained by the inherent mechanical inertia of dipoles. Residual flux density $\mathbf{B_r}$ differs widely with the core materials. The start-up problem of early arrival of saturation is more troubling than the higher values of $\mathbf{B_r}$.

2.6.3 Hysteresis

Magnetization and demagnetization amounts to physical alignment and dealignment of dipoles inside the bulk of the core. These two processes are not identical, in fact they are very different. They are indicated by different paths of forward magnetization and reverse magnetization curves shown by arrows. This process may be compared to the relative positions of two toothed gears in engagement in forward and backward motions. This phenomenon is called hysteresis.

A piece-wise linear B-Ni relationship for a homogenous core is shown in Figure 2.9(a). The waveforms of magnetizing current and the corresponding magnetic flux density produced in a homogenous core are given in Figure 2.9(b). The Flux density in the first cycle depends on the residual flux density in the core. Three segments in the first cycle of **B** correspond to the positive, zero and negative residual flux densities $\mathbf{B_r}$. Note that in the case of positive $\mathbf{B_r}$, saturation occurs at considerably lower value of **Ni** in the first cycle (start-up problem) than in subsequent cycles.

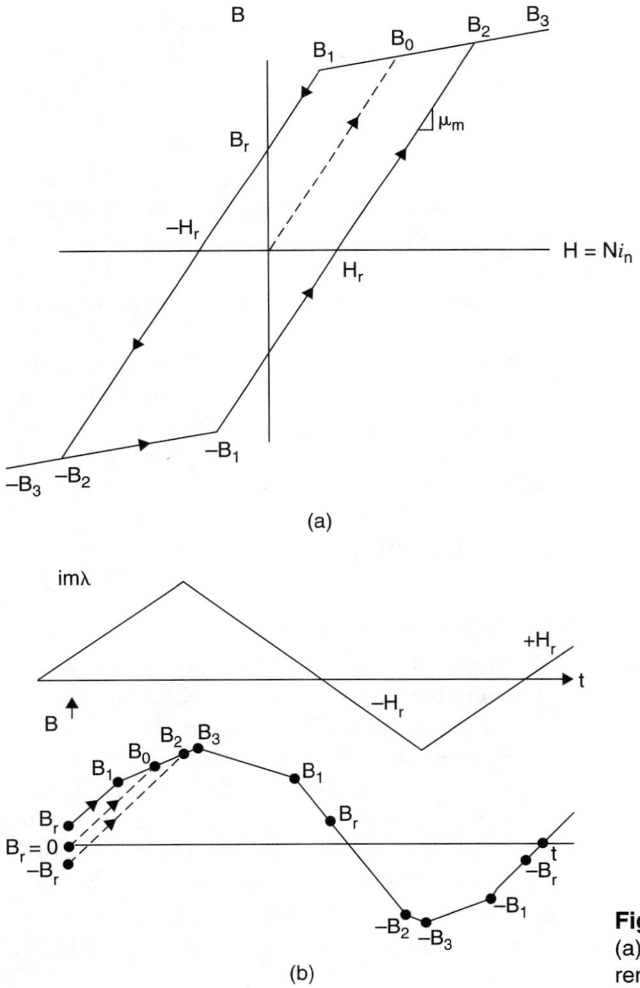

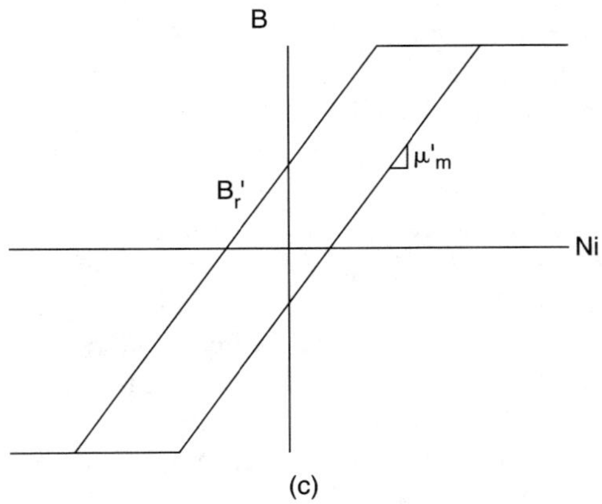

Figure 2.9
(a) Piece-wise linear B-H curve (b) magnetizing current and flux density waveforms and (c) B-Ni loop

Energy stored in the magnetic field is calculated from integrating the energy storage density W_H (equation 2-5) over the core volume as follows:

$$W_H = \frac{1}{2}\int_v \vec{B}.\vec{H}\, dv \qquad (2\text{-}28)$$

Energy stored during the forward and reverse magnetization are not equal. The difference appears as heat. In a cyclic magnetization in forward and reverse directions, such as by a periodic bipolar magnetizing current, energy loss increases with frequency.

Introduction of air gap in the magnetic path effectively reduces the residual flux density. This can be shown by the example of air gaped inductor of Figure 2.8. We will use the piece-wise linear curve of Figure 2.9 for the illustration. In the nonsaturated region, the magnetic flux density is given by the following:

$$B = \mp B_r + \mu H_m = \mu_0 H_g \qquad (2\text{-}29)$$

where $\mathbf{H_m}$ and μ_m are respectively the magnetic flux density and permeability in the core and $\mathbf{H_0}$ and μ_0 are respectively the magnetic flux density and permeability in the air gap. Ampere-turns **Ni** is given by

$$Ni = H_g\, g + H_m l_m \qquad (2\text{-}30)$$

where **g** and $\mathbf{l_m}$ are respectively the air gap length and the magnetic path length in the core.

$$\begin{aligned} Ni &= \frac{B}{\mu_0} g + \frac{B + B_r}{\mu_m} l_m \\ &= \frac{B}{\mu_0}\left[g + l_m \frac{\mu_0}{\mu_m}\right] + \frac{B_r}{\mu_m} l_m \end{aligned} \qquad (2\text{-}31)$$

After some more manipulations, we get

$$B = -B_{r'} + \mu_{m'} \frac{Ni}{g + l_m} \qquad (2\text{-}32)$$

where

$$B_{r'} = B_r \left[\frac{1}{1 + \frac{g}{l_m}\frac{\mu_m}{\mu_0}}\right] \text{ and } \mu_{m'} = \mu_m \left[\frac{l_m + g}{l_m + g\frac{\mu_m}{\mu_0}}\right]$$

$\mathbf{B_{r'}}$ and $\mu_{m'}$ are respectively the effective residual flux density and permeability of the combined structure. In a practical case $\mathbf{l_m >> g}$, hence,

$$\mu_{m'} = \mu_m \left[\frac{1}{1 + \frac{g}{l_m}\frac{\mu_m}{\mu_0}}\right] \qquad (2\text{-}33)$$

Both parameters are lowered by the air gap. The resulting effect **B-Ni** loop is shown in Figure 2.9(c). Note also that the onset of saturation in the start up phase is also delayed.

2.7 LOSSES IN MAGNETIC COMPONENTS

The power losses in magnetic components consist of three parts: copper losses ($\mathbf{i^2R}$), eddy current losses in the windings and hysteresis losses. All components of losses increase with the increase in operating frequency.

2.7.1 Copper losses

Copper losses are due to the ohmic resistance of the windings in the magnetic components. Losses increase as the square of the current. Therefore the current capacity is the limiting factor in determining the wire size. Typical current densities in copper wires are selected to be 100 to 1000 Ampere/cm^2. Lower current density causes smaller rise in temperature of the magnetic component.

Conductor resistance increases with temperature and frequency. The later effect is due to the concentration of current near the surface of the conductor. This action, termed the skin effect, is due to the internal magnetic fields generated by current distribution throughout the cross section of a conductor. The core of a circular conductor is circled by a greater number of flux lines and thus has higher inductance than the area near the surface. The current flows through the lower inductance (reactance) areas near the surface. Current concentration near the conductor surface increases the effective resistivity of the conductor material. The current concentration varies exponentially along the radius, being highest at the surface. The characteristic skin depth of the current concentration below the surface is given by

$$\delta = \frac{1}{\sqrt{\pi f \mu \sigma}} \tag{2-34}$$

f, μ and σ are respectively the frequency, permeability and conductivity of the conductor. Relative permeability of most of the conductor metals is approximately 1.0. The net result is that the ac resistance of a conductor is higher than the dc resistance. Skin depth in copper is about 7.5 mm at 60 Hz. Hence, insignificant current flows through the center of a conductor of radius 10 mm. The skin depth reduces to below 0.5 mm beyond 10 KHz. The central portion of a solid conductor may be removed (use hollow conductors) without any increase in the resistance. The hollow portion can be used for flowing cooling liquid as is done in high frequency induction heating machines. The central hollow portion may also be used for reinforcing strength as is done in aluminum conductors by reinforcing the center with steel tubes.

To find the ratio of ac resistance versus the dc resistance, consider a round solid conductor of radius r and the skin depth δ. The dc and ac resistances per unit conductor length are

$$R_{dc} = \frac{\rho}{\pi r^2}$$

$$R_{ac} = \frac{\rho}{\pi[r^2 - (r - \delta)^2]}$$

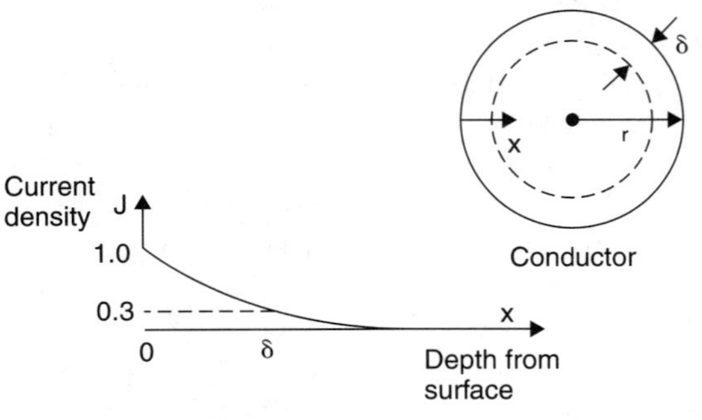

Figure 2.10
Skin effect in metallic conductors

$$= \frac{\rho}{\pi[2r\delta - \delta^2]}$$

$$= \frac{\rho}{\pi r^2\left[\frac{2\delta}{r} - (\frac{\delta}{r})^2\right]}$$

$$\frac{R_{ac}}{R_{dc}} = \frac{1}{\left[\frac{2\delta}{r} - (\frac{\delta}{r})^2\right]} \tag{2-35}$$

For a conductor with r = δ, the ratio equals unity; that is, the skin effect can be ignored if the diameter of a circular conductor is less than 2δ. The above expression is a crude approximation. Even a conductor with a radius smaller than δ has an ac resistance larger than the dc resistance.

Increase in resistance due to skin effect may also be avoided by increasing the surface area of the conductors. A solid large diameter conductor is usually replaced by a bunch of several small diameter conductors insulated from each other, called stranding. The stranded wires are commercially known as ***litz*** wires. Litz wires are very effective at frequencies below 500 kHz. Beyond that inter-strand capacitance make them less effective. Litz wires give higher quality factor **Q** than the solid wire at frequencies below 1–2 MHz.

2.7.2 Eddy current losses

Magnetic core material contains conducting metals in its composition. Hence, a plane in the core structure that is perpendicular to the direction of magnetic flux lines serves as a short-circuited conductor (although lossy). This conductor is in effect a short-circuited secondary winding that has a circulating current flowing that is caused by the transformer action. This current is termed **eddy current.** The ensued losses are purely $\mathbf{i^2R}$ losses, the resistance is however dependent on frequency due to the skin effect. The equivalent series resistance due to the eddy current effect, $\mathbf{R_e}$, **in series with the inductor** is given by

$$R_e \approx 0.04134 \frac{t^2 f^2}{\rho} \mu L \tag{2-36}$$

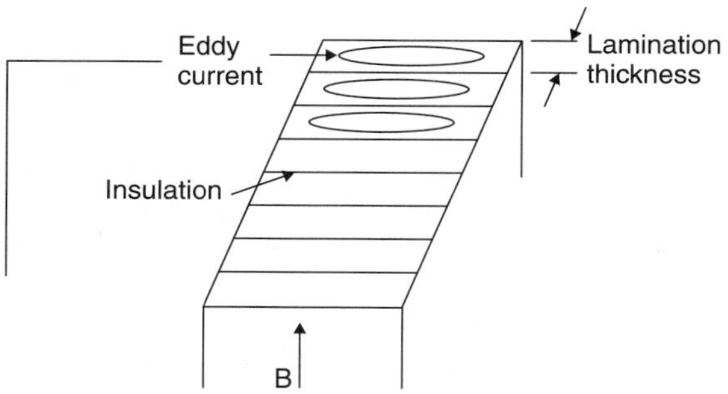

Figure 2.11
Eddy currents in a laminated core

Thickness **t** is in centimeters and resistivity ρ in micro-ohm-cm. L is the inductance in henrys. This formula has ignored secondary effect of skin effect in the metallic core. Eddy current losses increase as the square of core thickness and frequency of operation. Cores made of ferrous metals are laminated and insulated in the plane normal to the flux lines to break the path of eddy currents. In reality, the path is not broken, but broken into shorter loops confined to the thickness of the laminate, see Figure 2.11.

Hence, the laminate thickness is made as small as mechanically possible. Laminated steel cores are not usable beyond 20 KHz. Eddy current losses are reduced by increasing the resistivity of the core material as is done in ferrite cores by adding carbon particles in the core composition. Ferrite cores also have high permeabilities and are therefore preferred at higher frequencies.

Cores are also made from powered iron. These cores contain small iron particles, electrically separated, and thus have high resistivity. The iron particles are dimensionally smaller than one skin depth δ, thus suitable for higher frequency operation than the laminated cores. The powdered cores also have distributed air gap and high permeability and are therefore preferred at high frequencies.

2.7.3 Hysteresis losses

Hysteresis is a form of magnetic friction. Hysteresis loss is the unrecoverable energy that is expended in rotating the magnetic dipoles inside the core material. In a cyclic magnetization in both the forward and reverse directions, the energy lost per cycle per unit volume is obtained from the area under the hysteresis loop. Consider the typical flux density waveform in a choke that sustains an ac current superimposed over a dc component, shown in Figure 2.12(a). The hysteresis loop traversed on the B-H plane is shown in Figure 2.12(b). $\mathbf{W_H}$ is the energy expended given by equation (2-28).

$$W_H = \frac{1}{2} x\,(2B_{ac}.\, 2\,H_{ac})xv = 2\,\frac{B_{ac}^2}{\mu}\,v \qquad (2\text{-}37)$$

Hysteresis losses are obtained as

$$P_H = W_H f = 2\frac{B_{ac}^2}{\mu}\,v\,f \qquad (2\text{-}38)$$

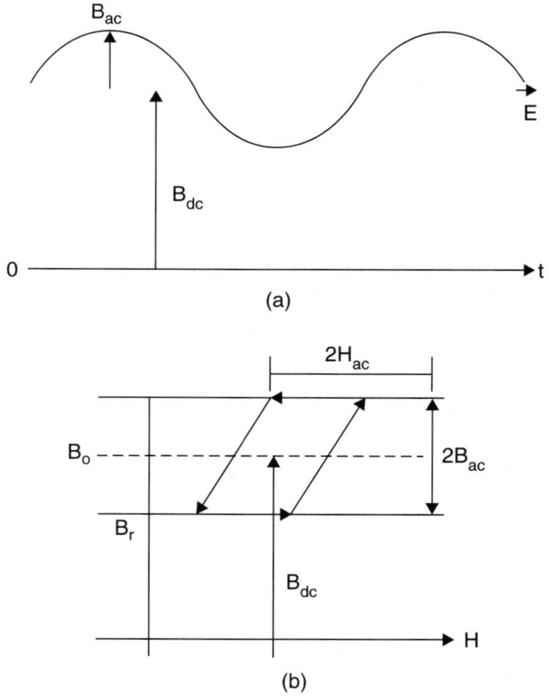

Figure 2.12
(a) Magnetic flux density waveform in a choke and (b) hysteresis loop

where **f** and ν are respectively the frequency of magnetizing current and the core volume. The above relation is an approximate one and on the higher side. Core manufacturers provide curves for these losses.

2.8 WINDINGS

Conductors are wound on a bobbin made of an insulating material. The bobbin with the winding is placed in the core as shown in Figure 2.13. For the reasons of mechanical ease of fabrications, the bobbin does not snugly fit in the window area of the core. Furthermore, the copper conductors are circular in shape and also have interlayer insulation material. Hence the window area is not completely filled by the conductor metal. A fill factor k is commonly used in transformer design that is defined as the ratio of the actual copper area to the window area when the window is packed to full occupancy. The fill factor is typically 0.4 for litz wires and 0.5–0.6 for solid round wires.

The core window area A_w is obtained from the total copper area of the winding divided by the fill factor k

$$A_w = \frac{N A_{cu}}{k} \tag{2-39}$$

A_{cu} is the cross section area of the copper conductor. N is the number of turns in the winding.

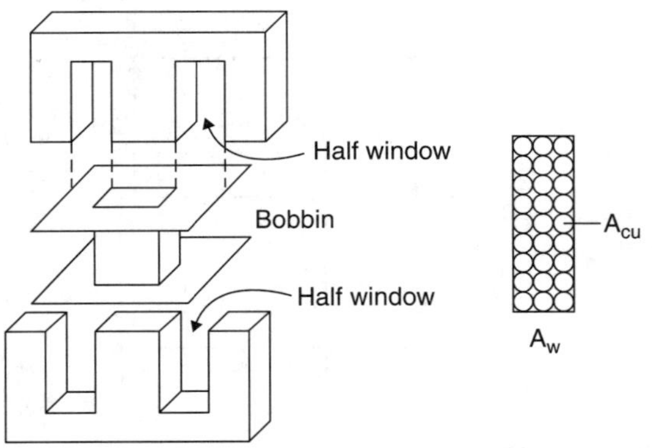

Figure 2.13
Winding bobbin and core window

2.9 TYPES OF INDUCTORS

Inductors are used in mainly two kinds of applications in power electronic systems: (1) filtering current ripple in a dc circuit and (2) used as an inductive component in an ac circuit. Inductors of the first kind are often called chokes. Design considerations are different for the two types of inductors.

2.9.1 Design of a filter inductor (choke)

Chokes carry dc current, store energy in the magnetic field and are designed to minimize the ripple current. Copper losses are mainly due to the dc resistance of the copper conductor. The current ripple is a very small fraction of the dc current, hence, the hysteresis losses and eddy current losses can be ignored.

The selection of core dimensions restricts the number of turns for a given voltage rating. The number of turns, the conductor cross section area, are in turn determined by the current density and the permissible temperature rise of transformer assembly. Power loss in the inductor, $I^2_{max} R_{dc}$, is responsible for temperature rise in the core. If the thermal resistance of the core is R_{th} and the temperature rise over the ambient, the power loss is

$$P = \frac{\Delta T}{R_{th}} = I^2_{\max} R_{dc} = I^2_{\max}\, \rho \frac{N l_t}{A_{cu}} \tag{2-40}$$

where ρ is the resistivity of the copper conductor, l_t is the average length of the conductor per turn and A_{cu} is the area of the cross section of the copper conductor.

Maximum energy stored in the no-air gap inductor is given by equation (2-35)

$$W_{\max} = \frac{1}{2} L I^2_{\max} = \frac{B^2_{\max}}{2\mu_c} l_c A_c \tag{2-41}$$

where μ_c is the core permeability, l_c is the effective core length and A_c the cross sectional area of the core. Furthermore,

$$\frac{1}{2} L I^2_{\max} = \frac{1}{2} A_L (N I_{\max})^2; \; where\ A_L = \frac{L}{N^2}$$

A_L is used as a design parameter and it can be expressed as follows:

$$A_L = \frac{1}{(N\,I_{max})^2}\,L\,I^2_{max}$$

It can be shown from equation (2-40) that

$$(N\,I_{max})^2 = \frac{\Delta T\,N\,A_{cu}}{R_{th}\,\rho\,l_t}\,1$$

Using equation (2-10),

$$(N\,I_{max})^2 = \frac{\Delta T\,k\,A_w}{R_{th}\,\rho\,l_t}$$

$$A_L = \frac{R_{th}\,\rho}{k\,\Delta T}\,\frac{l_t}{A_w}\,L\,I^2_{max} \tag{2-42}$$

The l_t/A_w is purely the geometrical data of the core. Vendors provide the curves of A_L versus LI^2_{max} for various sizes of cores for a safe value of rise in temperature above the ambient. Figure 2.14(a) shows curves for the ferrite E cores (courtesy of Magnetics: A Division of Spang & Company).

The largest value of A_L from the curves gives the smallest size core. The number of turns is then calculated from the A_L value. The maximum flux density is selected from the design curves of core-loss versus the flux density with frequency as the parameter, shown in Figure 2.14(b). Most of the energy of the inductor is stored in the air gap. The air gap in the ferrite core is calculated from the following equation:

$$\frac{1}{2}\,L\,I^2_{max} = \frac{B^2_{max}}{2\,\mu_0}\,g\,A_c \tag{2-43}$$

Example 2-2 Design of a Filter Inductor (Choke)

The following circuit shows the filter circuit of a dc to dc converter. Output voltage is required to be 5 volts dc with zero ripple. Capacitor C has therefore been assumed to be infinitely large. DC current through the inductor is 10 A with a peak-peak current ripple of 1 A and the input voltage waveform from the dc-dc converter is a 10 volts square wave of 20 kHz. Design the filter inductor using Mn-Zn ferrite core with B_s= .3T and μ_r = 150.

Inductor current is a triangle wave with an average of 10 A. Voltage and current waveforms on the filter inductor are shown in Figure 2.15. Period T = 1/f = 50 μS.

$$I_0 = \frac{I_{max} + I_{min}}{2} = 10$$

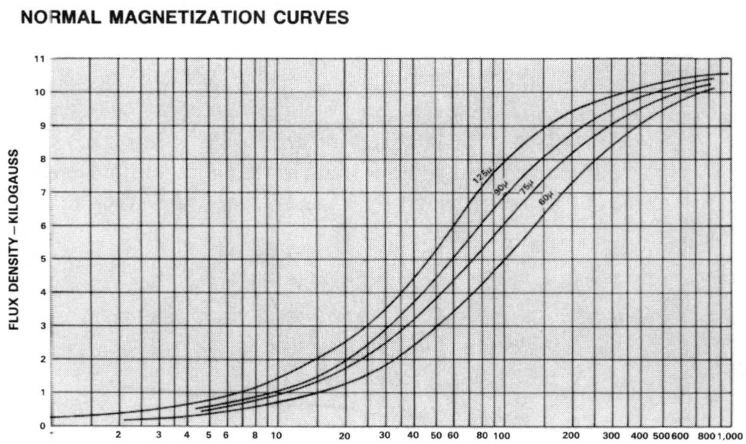

(a)

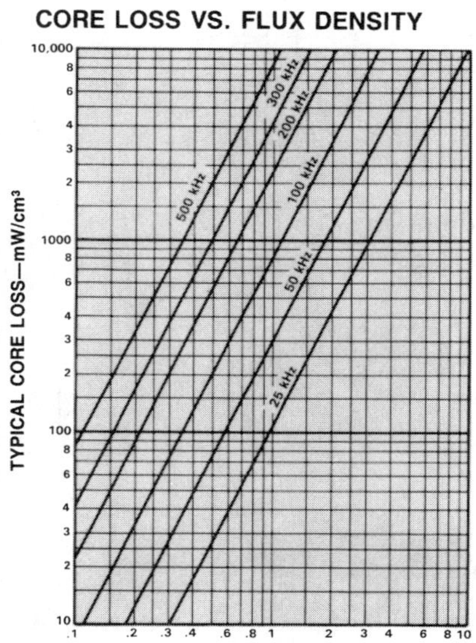

(b)

Figure 2.14
(a) Ferrite core selection curves for a fill factor k = 0.5 and
(b) core-loss as a function of the flux density and the frequency

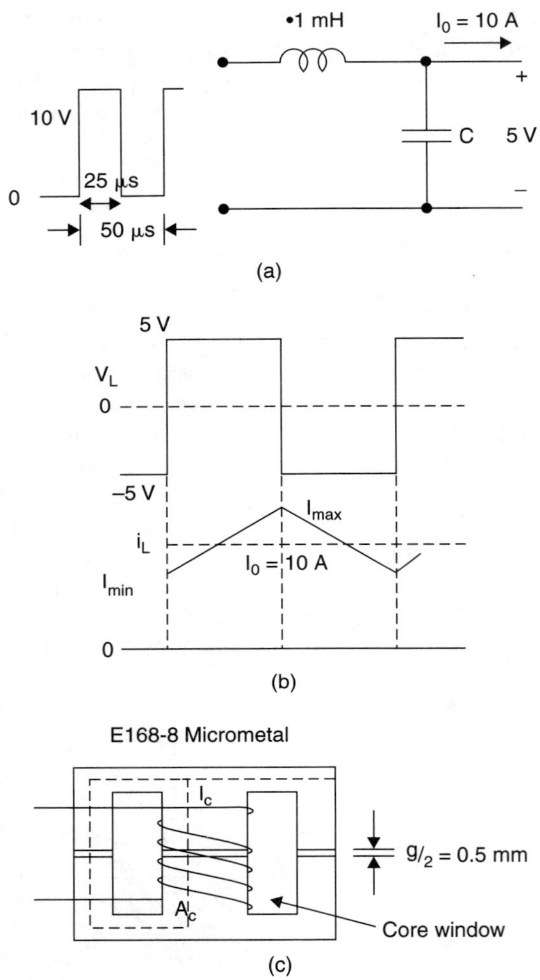

Figure 2.15
Filter inductor in dc to dc converter

and

$$I_{\max} - I_{\min} = 1\,A = \frac{1}{L}\int_0^{T/2} v_L$$

$$dt = \frac{1}{L} 5 \times 25 \times 10^{-6}$$

$L = 125\ \mu H$ and $I_{max} = 10.5$ A. We will select an E-E ferrite core with air-gap type construction as shown in Figure 2.15. Most of the energy is stored in the air gap (see problem at the end of the chapter), hence,

$$W_m \approx \frac{B_s^2}{2\,\mu_0} g$$

$$A_c \frac{1}{2} L I_0^2 = \frac{1}{2} \times 125 \times 10^{-6} \times 10^2 = 0.00625$$

$$g\,A_c = \frac{0.00625 \times 2 \times 4\pi X \times 10^{-7}}{.3^2}$$

$$= 1.745 \times 10^{-7}\,m^3$$

We select a standard Mn-Zn ferrite E-E core, MICROMETAL E168-8, with a core area of the central limb, $A_c = 15 \times 12\ mm^2$ and mean magnetic path length of 109.4 mm. The core selection is discussed in a later section. The air gap for our design is found to be approximately 1 mm. This air gap is divided in two equal parts: 0.5mm in the central limb and 0.5 mm on the side limbs. Equation 2-32 is satisfied where the magnetic path length $\mathbf{l_c}$ is more than the product of air gap g and the relative permeability μ_r. The A_L value for the core is 97.0 nH/N^2, hence, the number of turns required for L = 125 μH are calculated as

$$N = \sqrt{\frac{nH}{A_L}} = \sqrt{\frac{125 \times 10^3}{97}} = 35.89$$

Thus, the number of turns is selected to be N = 36 turns. The copper wire size is selected for a low current density (based on 375 circular mil/amp) from the table provided by Micrometals, Inc. AWG 14 is a good choice.

Example 2-3 Alternative Method of Core Selection for Example 2-2

The LI^2_{max} is calculated from

$$L\,I_{\max}^2 = .125(mH) \times (10.5)^2 \approx 13.8$$

From the curves of Figure 2.14, we find largest value of A_L=720 mH per 1000 turns for a type G ferrite core (Magnetics: A division of Spang & Company).

The number of turns are calculated from

$$N = 1000\sqrt{\frac{L(mH)}{A_L}} = 1000\sqrt{\frac{.125}{720}} = 13.5$$

A rounded number of 14 turns may be used. The wire gauge can be determined from the wire size table provided by the vendor.

Usually the selection of air-gap, type of standard core and copper wire size needs more than one design iteration.

2.9.2 Design of inductor for ac application

Even in a non-sinusoidal ac excitation of an inductor, most of the energy is stored in the fundamental harmonic. Similarly, most of the power losses are attributed to the fundamental harmonic. We will, therefore, discuss the inductor design for the sinusoidal excitations only.

The magnetic field produced by the sinusoidal voltage on the primary winding is

$$v = \sqrt{2}\tilde{V}\sin\omega t$$

Using

$$v = \_N\frac{d\phi}{dt} = \_NA_c\frac{dB}{dt}$$

and

$$B = B_{\max}\cos\omega t$$

$$B_{\max} = \sqrt{2}\tilde{B}; \quad \tilde{B} = \frac{\tilde{V}}{N\,\omega\,A_c} \qquad (2\text{-}44)$$

where A_c is the core area and ω is the frequency in radians per second of the voltage across the inductor. Core size is a function of several parameters such as the number of turns per volt, the frequency and the saturation magnetic flux density. A very useful determinate for core size is

$$\frac{N_p}{\tilde{V}_p}A_c = \frac{1}{2\pi f\tilde{B}} \qquad (2\text{-}45)$$

A higher value of **fB** product gives a fewer number of turns and small core dimensions. Many core vendors provide plots of **fB** product versus frequency that may be used for the selection of appropriate core material. One such plot is shown in Figure 2.16. Selection of core dimensions restricts the number of turns for a given voltage. The number of turns, the conductor cross section area, are in turn determined by the current density and the permissible temperature rise of transformer assembly. Hence, the selection of current density is a recursive process, fine tuned after the design run with an approximate value.

The Volt-Ampere of the inductor provides a very useful design parameter, from equation (2-44),

$$S = \tilde{V}\tilde{I} = N\omega\tilde{B}A_c \cdot A_{cu}\tilde{J}$$

A_{cu} is the copper area. J is the rms current density of the wire at the concerned frequency. Using the fill factor k and the window area instead of A_{cu}, see equation (2-39),

$$S = 2\pi k\,A_c\,A_w f\tilde{B}\tilde{J}$$

The product of core area and the window area $A_{cu}A_w$ is called the area product and it is a very useful design parameter in selecting the core and the number of turns per volt.

$$AP = A_{c\,A_w} = \frac{\tilde{V}\tilde{I}}{2\pi k f\tilde{B}\tilde{J}} \qquad (2\text{-}46)$$

Vendors provide the A_p product for every core size. Power losses in winding and core determine the temperature rise on the inductor assembly,

$$P = \frac{\Delta T}{R_{th}} = \textit{winding loss} + \textit{eddy current loss} + \textit{Hysteresis loss}$$

$$\textit{winding losses} = \tilde{I}^2 R_{ac} = \tilde{J}^2 A_{cu}\frac{R_{ac}}{R_{dc}}\frac{\rho N l_t}{A_{cu}}$$

$$= \tilde{J}^2\rho N l_t\frac{R_{ac}}{R_{dc}} \qquad (2\text{-}47)$$

The eddy current losses are proportional to the square of current density, the frequency and the core volume. The hysteresis losses are calculated from equation (2-38). The eddy current losses combined with the hysteresis losses are termed as the core

Figure 2.16
(a) Frequency-flux density product for ferrite cores, (b) P_o vs. A_P monogram and (c) Ferrite E core data from Magnetics: A Division of Spang & Company

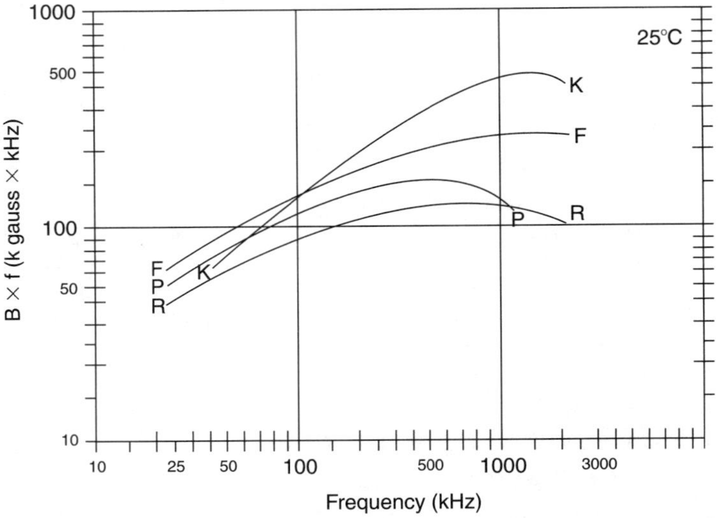

(a)

(b)

MAGNETIC DATA

TO ORDER SEE NOTE ↓	A_L VALUES MH/1000 Turns											
PART NUMBER (Note 1)	COMBI-NATION	MATERIALS μ2300 R (min.)	μ2500 P (min.)	μ3000 F ±25%	μ5000 J (min.)	l_e (cm)	A_e (cm²)	MINIMUM AREA (cm²)	V_e (cm³)	SET NOM. Wt. (gms)	BOBBIN WINDOW AREA (cm²)	WaAc (cm⁴) (Note 2)
45015-EC	E-E	3,615	3,930	6,300	8,200	7.72	1.77	1.42	13.6	70	1.26	2.23
45021-EC	E-E	4,600	5,000	8,000	8,010	9.29	2.25	2.13	20.9	108	1.78	4.00
45528-EC	E-E	4,720	5,130	8,220	9,375	12.3	3.50	3.46	43.1	212	2.83	9.91
45530-EC	E-E	5,640	6,130	9,800	11,190	12.3	4.17	4.13	51.4	255	2.83	11.8
46016-EC	E-E	4,300	4,680	6,590	7,445	11.0	2.48	2.40	27.2	135	2.89	7.16
47228-EC	E-E	4,470	4,860	7,780	8,885	13.7	3.68	3.63	50.3	264	4.02	14.8
48020-EC	E-E	3,505	3,810	6,000	6,940	18.5	3.89	3.82	72.1	357	7.91	30.8

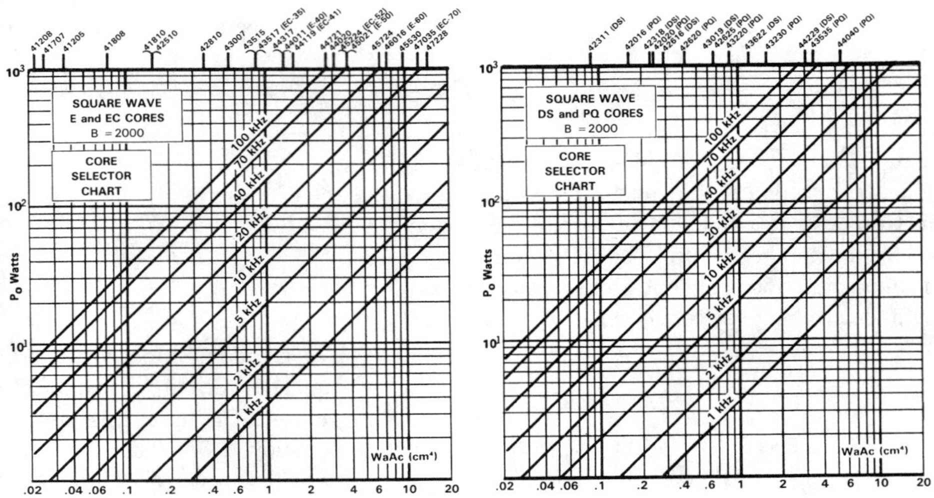

(c)

losses. The core vendors provide curves for determining the core losses for various sizes and types of the cores. Some vendors provide the plot of the output power versus the A_p product.

Example 2-4 Design of Inductor for AC Application

Specifications: $L = 125\ \mu H$, $V_{rms} = 100$ V, $I_{rms} = 6.4$ A, f = 20 kHz, $\Delta T = 50°C$, Sinusoidal waveforms.

Attempt the design with double E ferrite core. Select the core with the highest fB product from the vendor's design curves, Figure 2.16(a) (Magnetics: A Division of Spang & Company). Our selection is 'F' type core that has an fB product of 60 Kgauss-KHz.

$$\text{The Volt Amps, } S = 100 \times 6.4 = 640$$

Approximating S as the power, the A_P product (W_aA_C product in the vendor data of Magnetics on page 4.7) is determined to be 8.5 cm^4 from the plot of P_o versus A_P given in Figure 2.16(b). We have used the data for the square wave design using ferrite E cores. For saturating square wave inverters, the above found A_P product is divided by 1.8.

Select a core that has an A_p product more than the value determined above. A higher value of A_P will facilitate ease in winding. The vendor curve also gives the core type 45530 F, which has the following specifications:

AP = 11.8

permeability $\mu = 3000$
Core volume = 51.4 cm^3
Core Area $A_C = 4.17\ cm^2$
Core losses (@ 4 mW/cm^3) = 204.6 mW
A_L = 9,800 mH per 1000 turns (no air gap)

The number of turns are calculated from

$$N = 1000\sqrt{\frac{L(mH)}{A_L}} = 1000\sqrt{\frac{.125}{9800}} = 3.57 \approx 4$$

The wire gauge can be determined from the wire size table provided by the vendor. This selection at the current density of 375 circular mils per ampere is AWG 15. The equivalent litz wire is 360/44.

2.10 TRANSFORMERS

The transformer consists of electrically isolated but magnetically coupled inductors. Two or more windings share completely or partially a common magnetic flux. The common flux may be produced by the current in a single winding or a number of windings. Figure 2.17(a) shows a transformer consisting of three windings in which all three windings share a common flux completely and Figure 2.17(b) shows the symbol for it. The two solid lines between the left and right side windings indicate magnetic core. The left side winding is generally depicted as the source of ampere-turns and is called arbitrarily the primary winding. The remaining windings are called secondary windings. Winding terminals, marked by dots, have identical polarity of voltages. Figure 2.17(c) shows the equivalent magnetic circuit.

Voltages applied or induced on each winding in an n-winding transformer are related as below:

$$v_i = -\frac{d\lambda_i}{dt} = -N_i\frac{d\phi_i}{dt}\ ; i = 1, 2, 3, \ldots \quad (2\text{-}48)$$

where λ_i and N_i are respectively the flux linkage and the number of turns of the I-th winding. The ϕ_i is the magnetic flux linking the i-th winding part of which may be a common shared flux. The energy stored in the magnetic core is given by equation (2-28). Current in each winding is assumed to enter the winding at the terminal marked by the dot. The law of conservation of energy dictates

$$\int \{v_1 i_1 + v_2 i_2 + \ldots\}\, dt = W_m \quad (2\text{-}49)$$

W_m is the energy stored in the core. Part of the stored energy is lost in the core as eddy current losses and hysteresis losses.

2.10.1 Two-winding transformer

Consider an ideal transformer with two windings wound on a homogenous core with perfect coupling. Both windings share a common magnetic

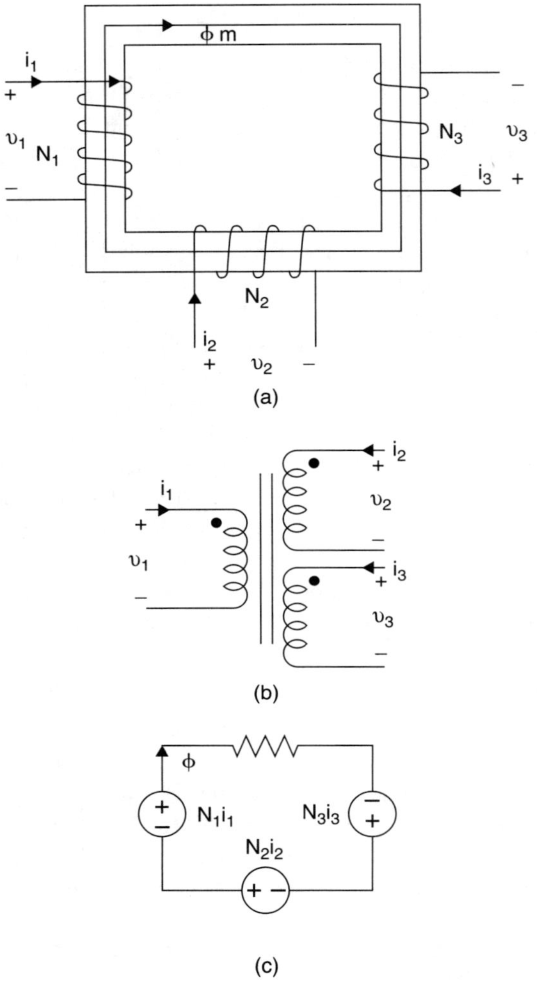

Figure 2.17
(a) A three winding transformer, (b) transformer symbol and (c) magnetic circuit

flux. The transformer does not store any energy, hence, equation (2-48) and (2-49) yield

$$\frac{v_1}{N_1} = \frac{v_2}{N_2} \tag{2-50}$$

$$v_1 i_1 + v_2 i_2 = 0 \tag{2-51}$$

$$i_2 = -\frac{N_1}{N_2} i_1 \tag{2-52}$$

The negative sign means that the current in the second winding is leaving the dotted terminal.

Equation (2-52) implies that the magnetic field H equals zero

$$H \cdot l_m = N_1 i_1 + N_2 i_2 = 0 \tag{2-53}$$

$\mathbf{l_m}$ is the magnetic path length of the core. A finite magnetic flux density **B** with a zero magnetic field intensity **H** shows that the permeability has been assumed to be infinitely high, which in practical terms means the core does not require any current to magnetize the core. An interesting meaning of the above relation is that the voltage and the current on the secondary winding in the transformer appears as scaled on the primary side as v_2' and i_2' respectively. Likewise, an impedance Z across a winding appears as scaled on the primary side as Z_2'.

$$v_2 = \frac{N_1}{N_2} v_2, Z_2 = \left(\frac{N_1}{N_2}\right)^2 Z_2 \tag{2-54}$$

A practical transformer has all the nonlinearity and losses that are associated with inductors. The magnetic field intensity is nonzero. Permeability μ is not only finite but also has hysteresis and consequently some energy is lost in the core. In addition the coupling of the transformer windings is not perfect owing to the presence of leakage fluxes around the core structure.

Next we include the leakage fluxes in the ideal transformer. Let ϕ_m be the common flux and ϕ_{k1} and ϕ_{k2} be the leakage fluxes in winding 1 and winding 2 respectively, see Figure 2.18. Note the direction of the i_2 in this circuit, which is the actual direction it flows. The fluxes in the two windings are

$$\phi_1 = \phi_{k1} + \phi_m \ , \ \phi_2 = \phi_m - \phi_{k2} \tag{2-55}$$

$$\phi_m = \frac{N_1 i_1 - N_2 i_2}{\Re_c} = \frac{N_1}{\Re_c}\left(i_1 - \frac{N_2}{N_1} i_2\right) = \frac{N_1 i_m}{\Re_c} \tag{2-56}$$

$$\phi_m = \frac{L_m i_m}{N_1} \tag{2-57}$$

$$\phi_{k1} = \frac{N_1 i_1}{\Re_{k1}} = \frac{L_{k1} i_1}{N_1}, \ \phi_{k2} = \frac{N_2 i_2}{\Re_{k2}} = \frac{L_{k2} i_2}{N_2}$$

The voltages on the two windings are

$$v_1 = N\frac{d\phi_1}{dt}, \qquad v_2 = N_2\frac{d\phi_2}{dt}$$

$$v_1 = N_1\frac{d\phi_1}{dt} = N_1\frac{d\phi_{k1}}{dt} + N_1\frac{d\phi_m}{dt}$$

$$v_1 = L_{k1}\frac{di_1}{dt} + L_m\frac{di_m}{dt} \tag{2-58}$$

Similarly the voltage on the second winding is expressed by

$$v_2 = N_2\frac{d\phi_2}{dt} = -N_2\frac{d\phi_{k2}}{dt} + N_2\frac{d\phi_m}{dt}$$

$$v_2 = -L_{k2}\frac{di_2}{dt} + \frac{N_2}{N_1}L_m\frac{di_m}{dt} \tag{2-59}$$

$$L_m\frac{di_m}{dt} = v_2 + \left(\frac{N_1}{N_2}\right)^2 L_{k2}\frac{di_2}{dt} \tag{2-60}$$

where $i_2 = \frac{N_2}{N_1}i_2, \qquad v_2 = \frac{N_1}{N_2}v_2$

Equations (2-58) and (2-60) can be shown to represent an equivalent circuit shown below:

i_2 and v_2 are respectively the current and voltage on the secondary winding as reflected on the side of the primary winding.

An alternate expression of primary and secondary winding voltages is often used

$$v_1 = L_{11}\frac{di_1}{dt} + M\frac{di_2}{dt}$$

$$v_2 = M\frac{di_1}{dt} + L_{22}\frac{di_2}{dt} \tag{2-61}$$

where L_{11} and L_{22} are respectively the self-inductances of the primary and secondary windings. M is called the mutual inductance and it is dependent on

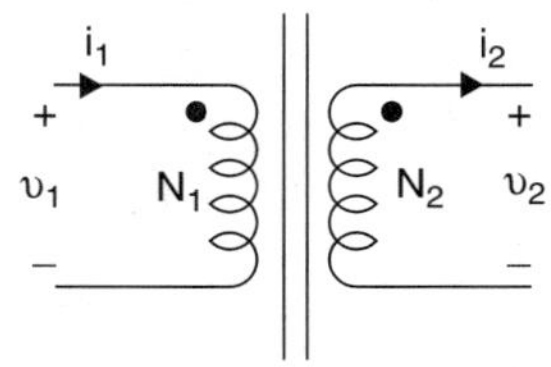

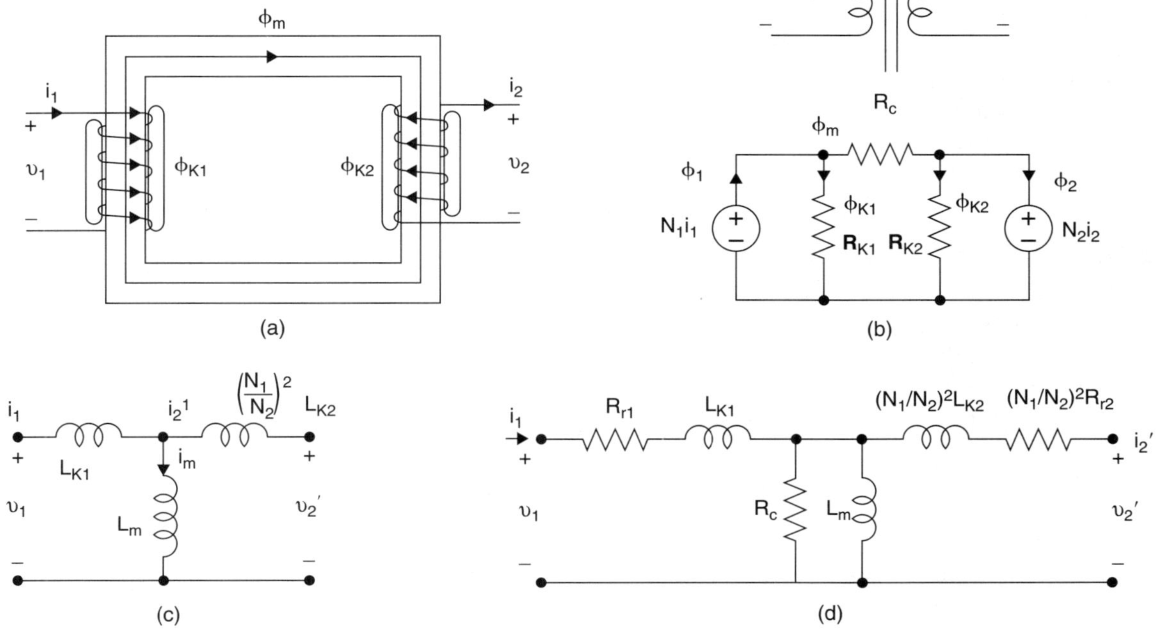

Figure 2.18
(a) A two-winding transformer, (b) magnetic circuit, (c) equivalent circuit and (d) a more general equivalent circuit

the amount of magnetic coupling between the two windings. It can be shown (see problem 2-9) that

$$M = \frac{N_2}{N_1} L_m \qquad (2\text{-}62)$$

$$L_{11} = L_{k1} + L_m, \quad L_{22} = L_{k2} + \frac{N_2}{N_1} L_m$$

A more general equivalent circuit of the two-winding transformer is shown in Figure 2.18(c). The copper losses are due to the resistances in series with each winding and subscripted by r. The resistances are strictly skin-effect dependent. Eddy current and hysteresis losses are accounted for by a resistance $\mathbf{R_c}$ in parallel with the magnetizing inductance $\mathbf{L_m}$, which is connected in parallel with the primary winding.

2.10.2 Transformer design

The transformer design involves selection of core material, core dimensions, turns per volt and wire size. The design procedure for the transformer is similar to that given for the ac inductor in Example 2-4. The procedure gives the design of the primary winding. The number of turns and wire gauge in the secondary winding are determined using the turns ratio and the current in the secondary winding.

2.10.3 Multiwinding Transformer

Multiwinding transformers are popularly used in power supplies. Usually one of the windings is the primary source of ampere-turns. The multiple windings are wound in series or in parallel magnetic circuits. Figure 2.17 shows a series wound three-winding transformer.

Series coupled windings may also be wound on top of one another to increase magnetic coupling in a practical transformer. All windings share a common magnetic flux, hence, from equation (2-48) the voltage per turn is equal in all windings,

$$\frac{v_1}{N_1} = \frac{v_2}{N_2} = \frac{v_3}{N_3} = \ldots \qquad (2\text{-}63)$$

Since energy storage is not the usual objective of most of the practical transformers of this kind, the algebraic sum of ampere-turns is zero:

$$v_1 i_1 + v_2 i_2 + v_3 i_3 + \ldots = 0$$
$$N_1 i_1 + N_2 i_2 + \ldots = 0 \qquad (2\text{-}64)$$

Figure 2.19 shows the magnetic equivalent circuit of the parallel wound three-winding transformer and its equivalent magnetic circuit.

Voltage on the i-th winding is given by:

$$v_i = -N_i \frac{d\phi_i}{dt}; \; i = 1, 2, 3, \ldots$$

where ϕ_i is the flux linking the i-th winding having $\mathbf{N_i}$ turns. Again, energy storage is not the usual objective of most of the practical transformers of this kind, the algebraic sum of all fluxes as shown in the magnetic equivalent circuit is zero:

$$\phi_1 + \phi_2 + \ldots = 0$$

Hence, the sum of voltage per turn for all windings equals zero.

$$\frac{v_1}{N_1} + \frac{v_2}{N_2} + \ldots = 0 \qquad (2\text{-}65)$$

2.10.4 Energy storage transformers

Some transformers are designed to merely store electromagnetic energy in its magnetizing inductance. An example is the transformer used in a flyback converter topology, which will be discussed in Chapter 6. The energy stored, as given by equation (2-41) is proportional to the square of the flux density and the core volume but inversely proportional to the core permeability. Hence energy storage transformers use low permeability cores and higher core volumes. The materials with high saturation density are preferred. Powdered iron cores or the ferrite cores are the proper choices for this application.

2.10.5 Current and potential transformers

The current transformer (**CT**) and potential transformer (**PT**) are special purpose transformers em-

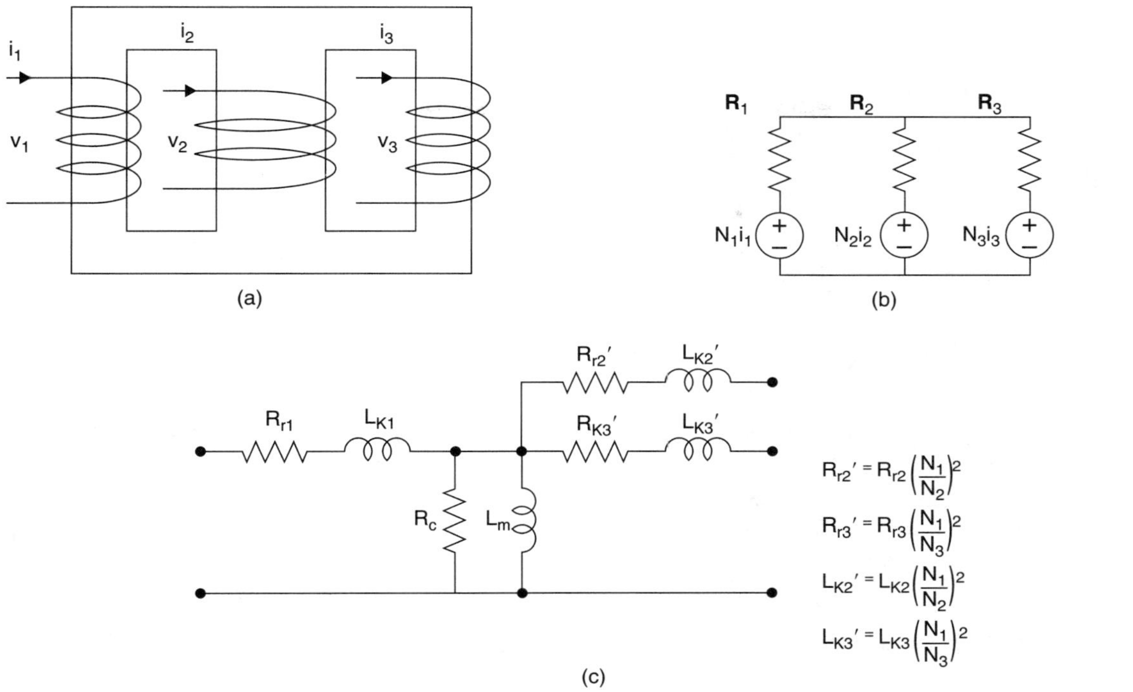

Figure 2.19
(a) A parallel wound three-winding transformer (b) magnetic equivalent circuit and (c) electrical equivalent circuit

ployed for sensing current and voltages respectively in power electronics instrumentation.

The desired characteristics of current transformers are (a) a high current transfer ratio, (b) a low magnetizing current and (c) small size. The primary winding in a current transformer usually consists of a single turn of high current carrying capacity. The secondary winding consists of a greater number of turns so that the secondary current is many times smaller than the primary current. The transformer size is kept small by selecting a high permeability core with no air gaps such as the ferrite toroidal cores. The current transformers are designed to have high values of magnetizing inductance so as to limit the magnetizing current to be a small fraction of the current being sensed.

The desired characteristics of potential transformers are (a) a high voltage transfer ratio, (b) low leakage inductances and (c) small size. Leakage inductance is minimized by the use of all-contained-flux type core structures such as the POT cores or cup cores. Sensing circuits using current and potential transformers are shown in Figure 2.20.

2.11 MOTORS

Electric motors are basically transformers with a two-part magnetic core and air gaps: **stator,** a stationary part, and **rotor,** a rotating part. Primary and secondary windings are wound respectively on the stator and rotor respectively. Misalignment in the directions of the magnetic fields created by current in the stator and rotor windings produce mechanical pull on the rotor. Depending on the type of excitation and geometry of construction, motors can be

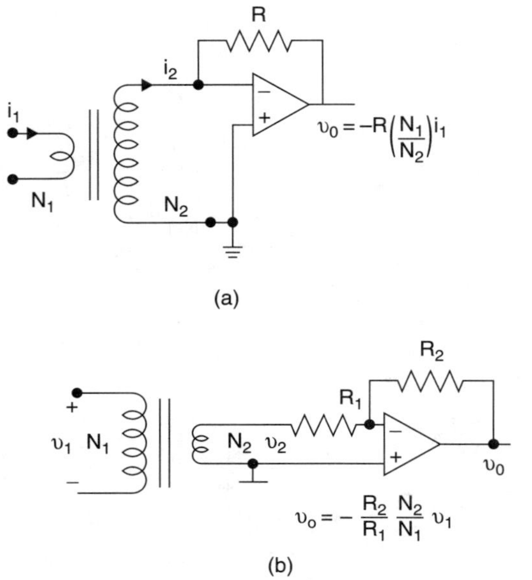

Figure 2.20
(a) Current and (b) potential transformers in sensing circuits

put in four broad classes: dc motors, induction motors, synchronous motors and stepper motors.

2.11.1 DC motors

A stator may be a permanent magnet or an electromagnet in which the energizing winding is excited by a dc voltage. Stator winding is also called field winding. Rotor winding is excited by a pulsating dc to sustain the motor rotation. Pulsating dc is obtained by mechanical chopping as is done by commutator or electrical chopping using diodes and transistors as switches as in brushless motors. Equivalent circuit of a typical dc motor is given in Figure 2.21. Kirchoff's law voltage equation is

$$V_s = R_a i_a + L_a \frac{di_a}{dt} + E_m : E_m = k_e \omega \phi_f$$

$$\text{and} \quad \phi_f = k_f i_f \tag{2-66}$$

where $\mathbf{E_g}$ is called the counter emf of the motor and $\mathbf{i_a}$ is the rotor (armature) current. The $\mathbf{R_a}$ and $\mathbf{L_a}$ are respectively the rotor winding resistance and in-

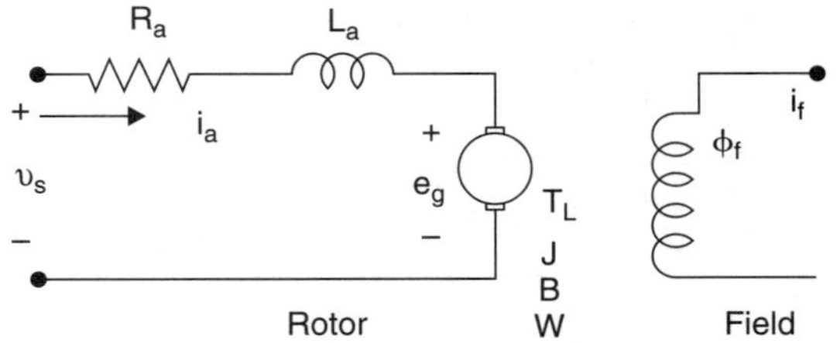

Figure 2.21
A dc motor equivalent circuit

ductance. The $\mathbf{k_e}$, ω and ϕ_f are respectively the motor constant, motor speed in radian/sec and magnetic flux in the stator winding or the stator permanent magnet. The developed mechanical torque is given by

$$T_m = \frac{E_g i_a}{\omega} = K_m \phi_s i_a \tag{2-67}$$

Note that in the above model we have ignored electrical losses such as in the commutator, core losses and mechanical losses such as friction. The following equation governs the mechanics of the motor:

$$T_m = J\frac{d\omega}{dt} + B\omega + T_L \tag{2-68}$$

J and B are respectively the inertia and damping constant of the motor and the load combined. T_L is the load torque.

In self-excited dc motors, the input voltage $\mathbf{V_s}$ also powers the stator field winding. A dc motor with the series connection of the stator and rotor windings is called a series motor. The dc motor with these two windings in parallel is called the shunt motor.

2.11.2 Induction motors

The squirrel-cage induction motor is the workhorse of the industry. The stator winding is excited by an ac voltage single-phase or three-phase. The rotor winding is comprised of shorted conductor bars. The schematic of an induction motor is shown in

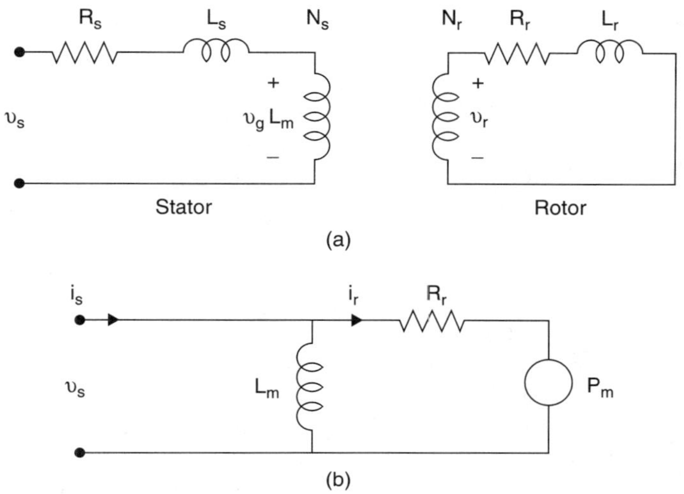

Figure 2.22
(a) Equivalent circuit of a single-phase induction motor and (b) simplified equivalent circuit of a single-phase induction motor

Figure 2.22(a). Winding resistance R_s and leakage resistance L_s on the stator can be ignored in comparison to usually large values of the magnetizing inductance L_m. The rotor resistance R_r is more predominant on the rotor side than the rotor inductance L_r. N_s is the number of turns on the stator winding and N_r is the equivalent number of turns on the rotor side. AC excitation of stator winding by sine wave v_s produces a magnetic field and counter emf v_g which are given by

$$v_g = v_s = V_s \sin \omega_e t$$

$$\phi_g = \frac{V_s}{N_s \omega_e} \cos \omega_e t = \Phi_g \cos \omega_e t \quad (2\text{-}69)$$

The stator winding is wound around several poles. The spatial arrangement of the stator winding and the poles are such that the magnetic field appears to vary with time as well as in space over the circumference in the air gap between the stator and rotor. The speed of rotation of this magnetic field, called the synchronous speed, is given by

$$\omega_s = \frac{2}{p} \omega_e \quad (2\text{-}70)$$

where p is the number of poles wound in the stator. Dependent upon the load torque, the rotor slips in speed with respect to the synchronous speed of the air-gap field. The slip in rotor speed is given by

$$s = \frac{\omega_s - \omega_r}{\omega_s}$$

$$\omega_r = \omega_s(1 - s) \quad (2\text{-}71)$$

ω_r is the rotor speed in radians per second. The shorted rotor conductor experiences the air-gap field varying at a speed that is the difference between the synchronous speed ω_S and the rotor speed ω_S, called the slip speed ω_{sl}. The slip speed and the flux experienced by the rotor conductor are given below:

$$\omega_{sl} = s\, \omega_s = \omega_s - \omega_r \quad (2\text{-}72)$$

$$\phi_r = \Phi_g \cos \omega_{sl} t \quad (2\text{-}73)$$

Note that the flux amplitude is the same as that of the air-gap flux ϕ_g. The voltage induced on the rotor conductors is given by

$$v_r = -N_r \frac{d\phi_r}{dt} = V_r \sin \omega_{sl} t : V_r = N_r \Phi_g \omega_{sl}$$

The rms rotor voltage is then obtained in terms of the rms input voltage as follows:

$$\tilde{V}_r = \tilde{V}_s \frac{N_r}{N_s} \frac{\omega_{sl}}{\omega_e} \tag{2-74}$$

For simplicity, assume $N_s = N_r$ then

$$\tilde{V}_r = \tilde{V}_s \frac{\omega_{sl}}{\omega_e} \tag{2-75}$$

The resultant approximate equivalent circuit as referred to on the stator side is shown in Figure 2.22(b). Part of the power passing through the air gap is expended in the rotor resistance as heat and the rest is the developed electromechanical power available to drive the motor torque. The power passing through the air gap $\mathbf{P_g}$ is calculated as the real part of the product of the counter emf v_g and the rotor current i_r:

$$P_g = \tilde{V}_s \tilde{I}_r : \tilde{I}_r = \frac{\tilde{V}_r}{R_r} \tag{2-76}$$

The developed electromechanical power is given by

$$P_m = P_g - P_r = \tilde{V}_S \frac{\tilde{V}_r}{R_r} - \frac{\tilde{V}_r^2}{R_r}$$

Using equation (2-75),

$$P_m = \frac{\tilde{V}_S^2}{R_r} \frac{\omega_{sl}}{\omega_e} \left[1 - \frac{\omega_{sl}}{\omega_e} \right] \tag{2-77}$$

Since the slip frequency is much smaller than the input electrical frequency, we could approximate

$$P_m \approx \frac{\tilde{V}_S^2}{R_r} \frac{\omega_{sl}}{\omega_e} = \frac{\tilde{V}_S^2}{R_r} \frac{2\,s}{p} \tag{2-78}$$

The motor torque is given by

$$T_m = \frac{P_m}{\omega_r} \approx \frac{\tilde{V}_S^2}{R_r} \frac{\omega_{sl}}{\omega_e \omega_r}$$

$$\approx \left(\frac{\tilde{V}_S}{\omega_e} \right)^2 \frac{p}{2} \frac{s}{(1-s)^2} \omega_r \tag{2-79}$$

Motor torque at a given rotor speed can be controlled by the ratio $\mathbf{V_s/\omega_e}$, input voltage to input frequency.

2.11.3 Synchronous motors

Synchronous motors are used in compressors, computer peripherals and robotics. Stator winding is excited by a single-phase or a three-phase ac voltage. Rotor winding, also called field winding, is excited by a dc voltage. The rotor may also be a permanent magnet such as in brushless dc motors.

Motor rotates at a constant speed called the synchronous speed that is a function of the electrical frequency of the ac voltage applied to the stator winding and is given by equation (2-70) in which p is the number of rotor poles.

Figure 2.23 shows a two-pole salient rotor type synchronous and its equivalent circuit. The R_s and L_{sl} are respectively the stator winding resistance and the leakage inductance. The $\mathbf{v_a}$ is the induced voltage due to the field produced by the ac current in the stator winding. The $\mathbf{v_f}$ is the induced voltage due to the rotating field of the rotor electromagnet.

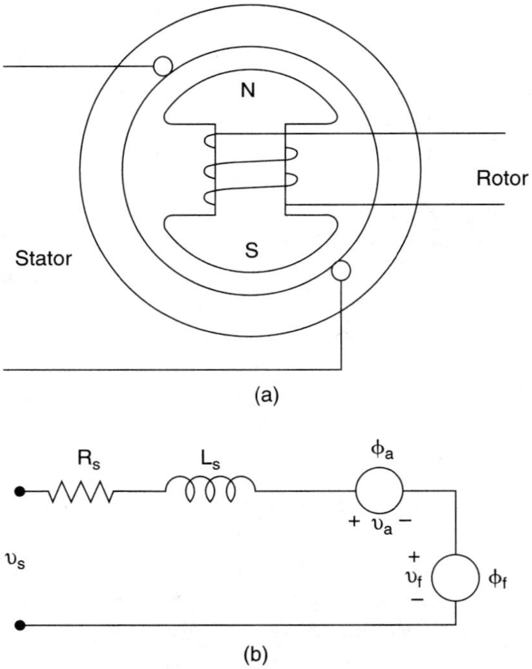

Figure 2.23
A salient 2-pole single-phase synchronous motor

Stator current is a sinusoid of frequency ω_e with a phase difference δ with respect to the input voltage,

$$i_s = I_s \sin(\omega_e t + \delta) \qquad (2\text{-}80)$$

The rotor current is dc, hence, establishes a dc magnetic flux in the rotor,

$$\Phi_f = K_f I_f \qquad (2\text{-}81)$$

Due to the rotating rotor, this magnetic flux appears as a rotating flux at the synchronous speed ω_s in the air gap between the stator and the rotor. However, this field is experienced by the stator winding as a flux that is varying at the applied electrical frequency ω_e.

$$\phi_f = \Phi_f \sin \omega_e t \qquad (2\text{-}82)$$

Rotating flux induces a counter emf in the stator winding given by

$$v_f = -N_s \frac{d\phi_f}{dt} = N_s \omega_e \Phi_f \cos \omega_e t \qquad (2\text{-}83)$$

AC current in the stator winding on the other hand produces another magnetic flux in the magnetizing inductance L_m that is given by

$$\phi_a = \frac{L_m}{N_s} i_s \qquad (2\text{-}84)$$

Voltage on the magnetizing inductance is given by

$$v_a = -N_s \frac{d\phi_a}{dt} = L_m I_s \sin \omega_e t \qquad (2\text{-}85)$$

Electrical power that is converted to the mechanical power is given by

$$\begin{aligned} P_m &= \tilde{V}_f \tilde{I}_s \cos(\delta - \pi/2) \\ &= \frac{1}{2} N_s \omega_e \Phi_f I_s \sin \delta \\ &= K_T \omega_e I_f I_s \sin \delta \qquad (2\text{-}86) \end{aligned}$$

The synchronous motor exhibits a leading power factor. The motor torque is given by

$$T_m = \frac{P_m}{\omega_s} = K_t I_f I_s \sin \delta \qquad (2\text{-}87)$$

The motor torque can be controlled by the dc rotor current or the ac current in the stator winding.

2.12 ELECTROSTATICS AND DIELECTRIC MATERIALS

Recall from section 2.2 that the total electric charge enclosed by a surface is equal to the total current enclosed by that surface. Hence, unlike the magnetic flux lines, the electric flux lines (or electric field vector) are not closed contours (or closes at infinity). The electric flux density **D** and the electric field **E** are related by ϵ (epsilon), called **permittivity** of the medium (or the material). The permittivity is expressed as a product of the permittivity of free space ϵ_0 ($=10^{-9}/36\pi$ F/m) and the relative permittivity ϵ_r also called **dielectric constant:**

$$\epsilon = \epsilon_0 \epsilon_r \qquad (2\text{-}88)$$

In a linear, isotropic (uniform in all directions) and homogenous medium, the relative permittivity ϵ_r is a constant. In general, the relative permittivity of a practical medium is nonhomogeneous and nonisotropic and nonlinear, hence a tensor. However, in power electronic applications, the relative permittivity is assumed to be constant.

Another parameter of importance in capacitors is called **dielectric strength,** the maximum value of electric field that the dielectric material can withstand.

Table 2.3 gives ϵ_r and **dielectric strength** for some materials. Both parameters are a function of temperature and humidity.

2.13 CAPACITORS

A device that stores a charge is called the capacitor. The charge on the capacitor and the voltage across its terminals are related by

$$q = C v \qquad (2\text{-}89)$$

C is the capacitance in farads. A basic capacitor is made of two parallel plates with a dielectric material sandwiched between them. The two plates hold equal and opposite charges. Two opposite polarity charges separated by a distance develop an

TABLE 2.3

	Relative permittivity	Dielectric strength
	ϵ_r	**E**(V/m)
Barium titanate	1200	7.5×10^6
water (distilled)	81	
paper	7	12×10^6
glass	5–10	35×10^6
mica	6	70×10^6
porcelain	6	
quartz (fused)	5	30×10^6
rubber (hard)	3.1	25×10^6
wood	2.5–8.0	
polysterene	2.55	
polypropylene	2.25	
polycarbonate	3.0	
air	1.00059	3×10^6

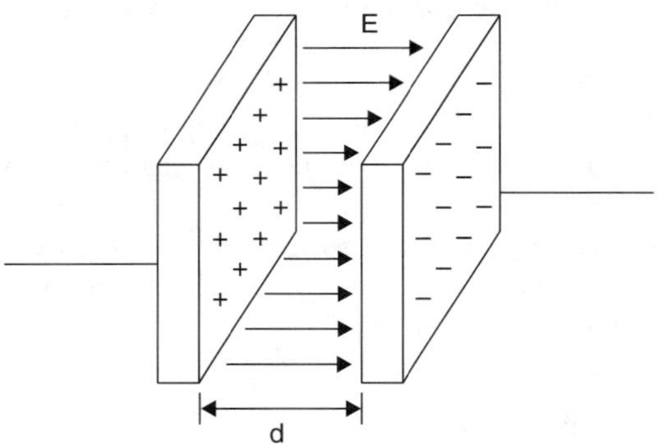

Figure 2.24
A parallel plate capacitor

electrostatic field (or simply the electric field) **E** as shown in Figure 2.24:

$$\int_a^b \vec{E} \,.\, d\vec{l} = E\,d = v_{ab} \tag{2-90}$$

d is the spacing between the plates. The electric field vector is normal to the plate surface area A and points from a to b. Total electric flux (charge **q**) on the plate structure is calculated as the electric flux density **D** multiplied by the area:

$$q = \int d\psi = D\,A = \epsilon\, E\, A = \epsilon \frac{v}{d} A \tag{2-91}$$

Hence, from equations (2-89) and (2-91)

$$C = \epsilon \frac{A}{d} = \epsilon_0 \frac{A}{d/\epsilon_r} \tag{2-92}$$

$\mathbf{d/\epsilon_r}$ is the equivalent electric separation in free space between the plates.

Energy stored in the two-plate capacitor is calculated from

$$W = \int v\, dq = \int C\, v\, dv = \frac{1}{2} C\, v^2 = \frac{1}{2} \epsilon_0 E^2 v_r \tag{2-93}$$

where

$$v_r = \epsilon_r A d = \epsilon_r v_d$$

v_d and v_r are respectively the physical volume of the dielectric medium and its equivalent volume in vacuum. The term multiplied to the relative dielectric volume is the electrostatic energy storage density of the capacitor.

Practical dielectric materials are lossy (conduct current). Figure 2.25 shows two equivalent circuits of a practical capacitor. The $\mathbf{R_p}$ is the **dielectric leakage resistance.** The $\mathbf{R_s}$ is called the **equivalent series resistance ESR** of the capacitor. Impedance of the practical capacitor is

$$Z = R_s + \frac{1}{j\omega C_s} \tag{2-94}$$

where

$$R_s = \frac{R_p}{1 + \omega^2 R_p^{\,2} C^2}, \quad C_s = C\left[1 + \frac{1}{\omega^2 R_p^{\,2} C^2}\right]$$

The impedance is shown on the phasor space as a vector at an angle δ less than −90°. The equivalent series resistance **ESR** is responsible for the power losses in the capacitor. The **tan** δ is called the **Dissipation factor:**

$$DF = \tan\delta = \frac{1}{\omega RC} \tag{2-95}$$

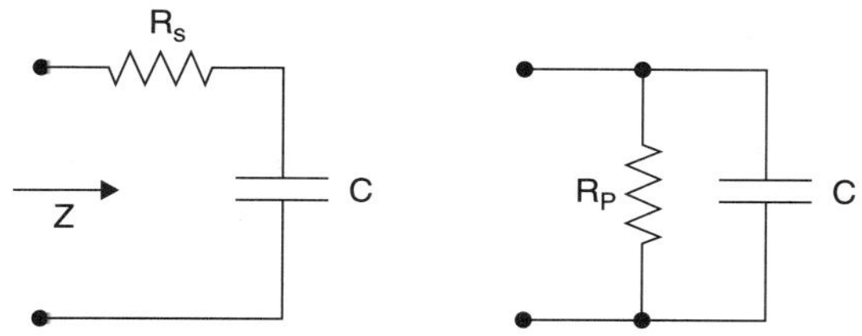

Figure 2.25
Series and parallel equivalent circuits of a practical capacitor

TABLE 2.4

	tan δ (1 KHz)	tan δ (1 MHz)
Porcelain	0.014	0.008
Teflon	0.0003	0.0002

ω is the frequency in radians of the voltage across the capacitor. The dissipation factor is often specified at one frequency. Table 2.4 gives the dissipation factors of various dielectric materials.

2.13.1 Film capacitors

Film capacitors are comprised of two conductor foils separated by an insulating film made of a dielectric material such as polycarbonate, polysterene, polypropylene, mica, etc. Conductor foils and the insulator film are rolled into a cylinder or a cake form. Polypropylene film has a much higher dielectric strength and more temperature stability than polycarbonate film. The dielectric constant (permittivity) is, however, higher for polycarbonate than for polypropylene and polysterene film. Hence the capacitance /volume ratio is higher for the former.

2.13.2 Electrolytic capacitors

Electrolytic capacitors have high capacitance/volume but they are polarized. The aluminum electrodes are coated by aluminum oxide, which serves as the dielectric material. The oxide coated electrodes are placed in an electrically conductive liquid electrolyte. The resistance of the electrolyte adds to the equivalent series resistance ESR. The ESR of electrolytic capacitors is thus poor. The ESR is further aggravated at low temperatures due to crystallization of the electrolyte. For this region the electrolytic capacitors are not suitable for high frequency applications.

2.13.3 Multilayer ceramic capacitors

The dielectric in these capacitors is ceramic. Multilayer construction increases the capacitance/volume ratio but still they are lower than in the film capacitors. The ceramic capacitors have very low series resistance and therefore are most suitable for high frequency applications such as power supply filtering.

2.14 SEMICONDUCTOR SWITCHES

Semiconductor switches are very important and crucial components in power electronic systems. These switches are meant to be the substitutions of the mechanical switches, but they are severely limited by the properties of the semiconductor materials and the process of manufacturing. We will study about these switches in detail in later chapters. In the following sections, we examine the losses incurred during the switching process in a generic switch.

2.14.1 Switching losses

Power losses in the power electronic converters are comprised of (1) the switching losses and (2) the parasitic losses. The parasitic losses account for the losses due to the winding resistances of the inductors and transformers, the dielectric losses of capacitors, the eddy and the hysteresis losses. The switching losses are significant and can be managed. They can be further divided in three components: (a) the on-state losses, (b) the off-state losses and (3) the losses in the transition states.

On-State Losses

The electrical switches conduct heavy current and have nonzero voltage across the switch in the on-state. The on-state power losses are given by

$$P_{on} = v_{son}\, i_f \qquad (2\text{-}96)$$

The $\mathbf{v_{son}}$ and $\mathbf{i_f}$ are respectively the switch voltage in the on-state and the forward current through the switch. For example the typical power diodes and the power transistors have nearly 0.5 to 1 volt across them in the on-state. The forward currents can be hundreds to thousands of amperes. The on-state power losses are very significant.

Off-State Losses

The electrical switches withstand high voltages and have nonzero leakage current through the switch in the off-state. The off-state power losses are given by

$$P_{off} = v_{soff}\, i_r \tag{2-97}$$

The $\mathbf{v_{soff}}$ and $\mathbf{i_r}$ are respectively the reverse bias voltage in the off-state and the reverse current through the switch. For example the typical power diodes and the power transistors have high reverse voltages in hundreds to thousands of volts and microamps to milliamps through them in the off-state.

Transition-State Losses

The practical switching devices have limited capabilities of rate of voltage transition and the rate of current steering. These nonabrupt transition rates give rise to power losses in the switching devices. We will examine these switching losses in two cases separately: the inductive and capacitive loads.

Switching with Inductive Load

The inductor is assumed to be large so that the current through it in steady state is nearly constant $\mathbf{I_o}$. Assume that initially the switch is off. The inductor current is $+\mathbf{I_o}$ and freewheels through diode **D1.** When the switch is turned on, the current through the switch begins to build up linearly (an assumption) to $+\mathbf{I_o}$ while the diode **D1** is still on. The on diode has zero voltage across it (an ideal diode), hence, the voltage on the switch is held constant at $+\mathbf{V_S}$. When the current buildup is over, the diode **D1** ceases to conduct and the voltage on the switch ramps linearly (again an assumption) down to zero.

When the switch is turned off, the voltage begins to build up linearly to $+\mathbf{V_S}$ while the diode **D1** is off. While the diode is off, the current through the switch equals the inductor current, which is constant $\mathbf{I_o}$. After the switch voltage reaches zero, the current through the switch begins to decrease below $\mathbf{I_o}$, as the remaining current is now steered through the diode **D1,** which has now turned on. The current through the switch ramps down to zero ultimately.

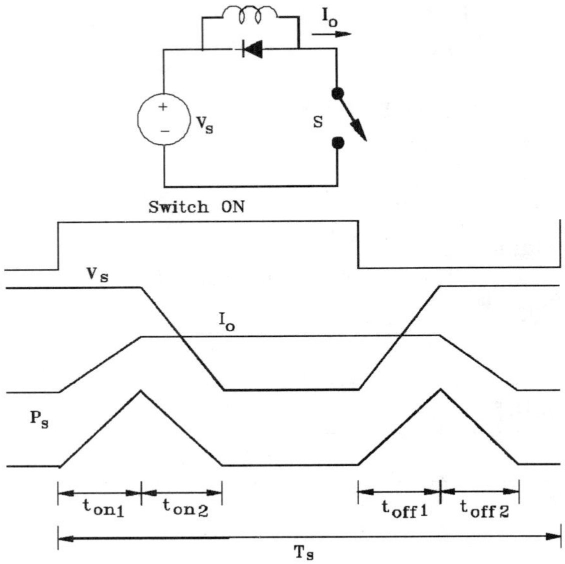

Figure 2.26
Switching waveforms with inductive load

The switching losses are given by

$$P_{sw} = \frac{1}{2} V_S I_o [t_{on\,1} + t_{on\,2} + t_{off1} + t_{off2}] f_s \tag{2-98}$$

The switching power losses increase linearly with the switching frequency like in the resistive case but about six times more. The upper bound on the switching frequency is also about half.

$$f_{smax} = \frac{1}{t_{on1} + t_{on2} + t_{off1} + t_{off2}} \tag{2-99}$$

Switching with Capacitive Load

The capacitor is assumed to be large so that the voltage through it in steady state is nearly constant $\mathbf{V_o}$. Assume that initially the switch is on, hence, the current through the switch is $\mathbf{I_S}$. The capacitor voltage is $\mathbf{V_o}$, the voltage across the switch is zero and the diode **D1** is reverse biased. When the switch is turned off, the switch voltage begins to ramp up to $+V_o$ while the diode **D1** is still off. During this buildup, the current through the switch is held constant at I_S. When the voltage buildup is over, the diode **D1** begins to conduct and the voltage on the switch is clamped at V_o, and the current through the switch ramps linearly (again an assumption) down to zero.

When the switch is closed, the current begins to build up linearly to I_S while the diode **D1** is still on. The voltage on the switch remains clamped at V_o. After the switch current reaches I_S, the diode turns off and the voltage on the switch begins to ramp down to zero.

The switching power losses in the case of capacitive load also have similar dependence as in the case of inductive loads.

The switching losses can be usually minimized in two ways: (a) divert the energy from the switch to a lossy or non-lossy circuit or (b) switch at either zero current or at zero voltage. The first is called snubbering and the later is known as zero-voltage and zero-current switching.

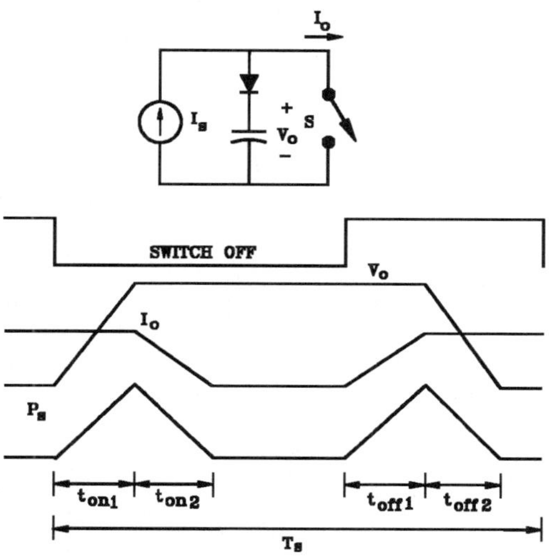

Figure 2.27
Switching waveforms with capacitive load

2.15 SNUBBERING: PROTECTION OF SWITCHING DEVICES

The switch protector must perform two functions: (1) to limit the stresses on the switch to safe values, and (2) to divert the energy during switching transition from the switch to another circuit and thus reduce the power losses in the switch. The switch stresses are the maximum transient voltage, the maximum transient current, the rate of change of switch voltage dv/dt, and the rate of change of switch current di/dt. The protection is specifically required at the turn-on, turn-off of power to the converter and in overvoltage conditions. The high rate of voltage and current transition give rise to local hot spots in the switching devices that if persist for some time, may permanently damage the devices. Special circuits are used to slow down the high rate of transition. These circuits are called *snubbering* circuits.

The following sections will discuss three types of generic snubber circuits: turn-on, turn-off and over-voltage protection snubbers.

2.15.1 Turn-off snubber

The turn-off snubber circuit is shown in Figure 2.28(a). The inductor maintains a dc current I_O. It reduces the switching losses by reducing the voltage across the switch during the transition of current through the switch. In the following discussion, we assume ideal diodes and ideal switch. The switch is also assumed to have limited maximum rates of rise and fall transitions of voltage and current.

Without Snubber Capacitor

Prior to turnoff, the switch current is I_0 at zero switch voltage and the diode D_1 is off. As the switch turnoff is initiated by applying the proper control signal to the switch, the switch voltage begins to rise from 0 to V_S linearly (a fair assumption for a semiconductor switch) in time t_{f1}. The diode D_1 remains off until the rise in switch voltage is complete; hence, the switch would be forced to conduct the full current I_O until time t_{f1}. After time t_{f1}, the

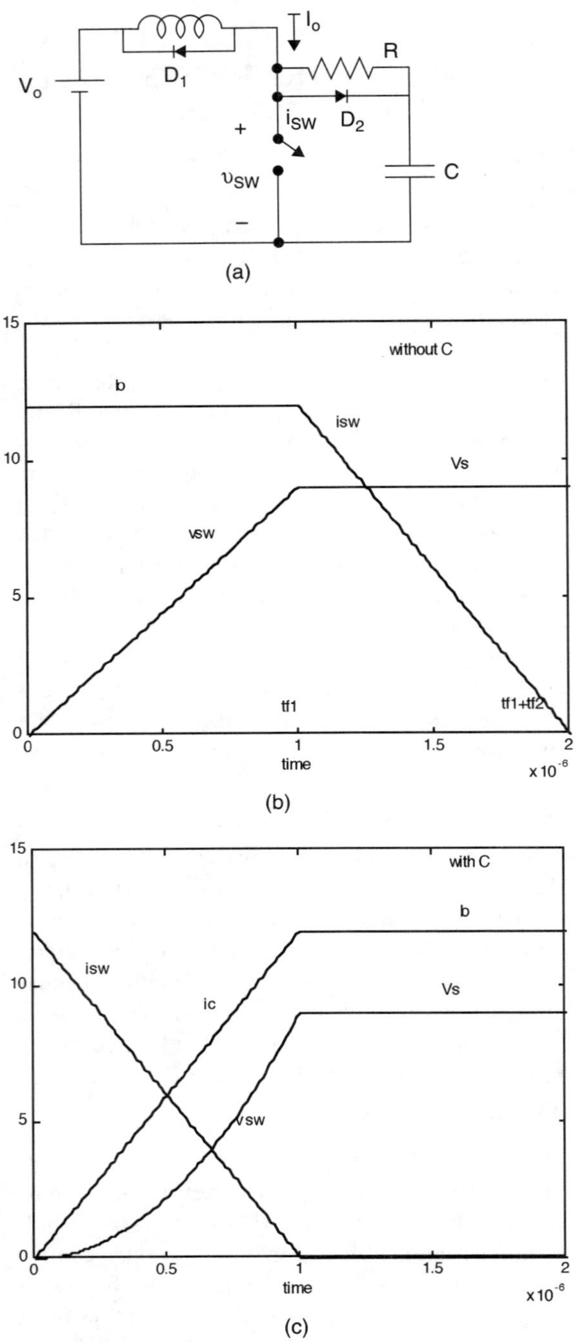

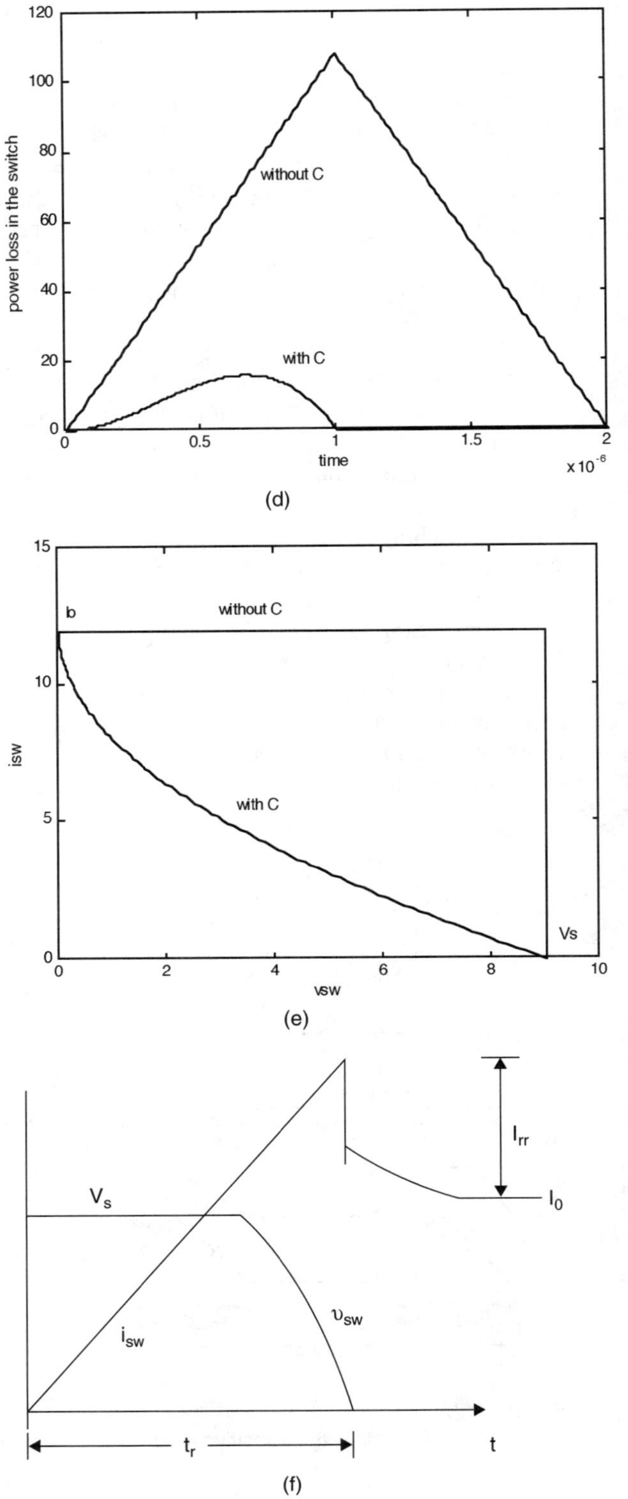

Figure 2.28
(a) Turn-off snubber (b) switch voltage and switch current without capacitor C (c) switch voltage and switch current with capacitor C (d) switching losses
(e) switching trajectory on i_{sw}-v_{sw} plane and (f) turn-on waveforms with turn-off snubber included

switch current decreases linearly (again a reasonable assumption) for duration t_{f2}

$$i_{sw} = I_0\left(1 - \frac{t}{t_{f2}}\right) \quad (2\text{-}100)$$

The balance current I_0-i_{sw} is diverted through diode D_1 for freewheeling through the current source. The waveforms of the voltage and the current through the switch are shown in Figure 2.28(b) for V_s=9 Volts and I_o =12 Amps and $t_{f1} = t_{f2} = 1$ μs. The instantaneous power and the switching trajectory on the I_{sw}-v_{sw} plane are shown in Figures 2.28 (d) and (e) respectively.

With Snubber Capacitor

Prior to turnoff, the switch current is I_0 at zero switch voltage and the diode D_1 is off. As the switch turnoff is initiated by applying the proper control signal to the switch, the switch current decreases linearly as given by equation (2-100). The balance current I_0-i_{sw} is diverted to the capacitor through the snubber diode D_2 and charges the capacitor.

$$v_{sw} = v_c = \frac{1}{C}\int_0^t i_{sw}\,dt = \frac{1}{2}\frac{I_0}{C}\frac{t^2}{t_f} \quad (2\text{-}101)$$

The capacitor finally charges to voltage V_S in a time that is determined by the value of the capacitor. Let us assume a case where the capacitor is just enough to charge the capacitor to V_s in time t_{f2}. The cases with other values of C are also interesting but are left for the readers to work out themselves after following the analysis given below.

The value of the capacitor in this case is determined from

$$C = \frac{I_0 t_{f2}^2}{2V_s} \quad (2\text{-}102)$$

The waveforms of the voltage and the current through the switch are shown in Figure 2.28(c). The switch voltage rise begins with a small slope. The initial holding time t_{f1} for the switch current at the value I_O is absent. The instantaneous power loss in the switch is

$$p_{sw} = v_{sw}i_{sw} = \frac{I_0^2}{2C_{t_{f2}}}\left(1 - \frac{t}{t_f}\right)t^2 \quad (2\text{-}103)$$

The instantaneous power waveform and the switching trajectory on the I_{sw}-v_{sw} plane are shown in Figures 2.28 (d) and (e) respectively. Note the very significant reduction in the power loss through the switch in comparison with the case without the snubber capacitor.

The maximum rate of dv_{sw}/dt occurs when the current has dropped to a low or the zero value.

$$\frac{dv_{sw}}{dt}\Big|_{max} = \frac{I_0}{C} \quad (2\text{-}104)$$

The above equation may be used to calculate the capacitance required for a switch with a specified maximum dv/dt (slew) rating.

Example 2-5

Following MATLAB program gives the plot of the turn-off losses in the switch and the switching trajectory without and with turn-off snubber.

```
Vs=9; tf1=1e-6; tf2=tf1; Io=12;
C=Io*tf1/(2*V0); % the critical value of capacitor
t1=0:tf1/100:tf1; t2=tf1:tf2/100:tf1+tf2; t=[t1, t2];
  % time array
vsw1=[Vs * t1/tf1, Vs *ones(size(t2))];
  % switch voltage without C
isw1=[Io .*ones(size(t1)), Io*(1-(t2-tf1)/(tf2))];
  % switch current without C
psw1=vsw1.* isw1; % instantaneous power in switch
  without C
isw2=[Io*(1-t1/tf1), zeros(size(t2))];
  % switch current with C
ic=I0-isw2; % capacitor current
vsw2 =[Io ./(2*C*tf1) .*t1 .^2, Vs* ones(size(t2))];
  % switch voltage with C
psw2=vsw2 .* isw2; % instantaneous power in the switch
  with C
plot(t,vsw1, t,isw1)
plot(t,vsw2, t,isw2, t,ic)
axis([0,tf1+tf2,0,15])
plot(t,psw1,t,psw2)
plot(vsw1, isw1, vsw2,isw2) % switching trajectory
axis([0,10,0,15])
```

Reduction in turn-off losses is evident. Total turn-off time is also smaller than in the circuit with no turn-off snubber.

The discussion of the turn-off snubber is incomplete without examining its effect on the turn-on process. To understand turnon, first we consider R = 0. Prior to turnon, the current I_O freewheels through diode D_1 and the inductor. The diode D_2 is short-circuited. The snubber capacitor is charged to V_S. As the switch is turned on, the diode D_1 keeps conducting until all its current is diverted to the switch, thus maintaining a constant voltage V_S on the switch. When the switch current builds up to I_O value, the switch voltage begins to decrease but its rate is retarded by the snubber capacitor, which is already charged to V_S. The snubber capacitor dumps its energy on the switch. This is responsible for the continuing linear rise of switch current (an assumption) beyond the I_0 value until the switch voltage drops to zero. The switch current drops back to the I_0 value as soon as the capacitor discharges completely, see Figure 2.28(f). The current overshoot, I_{rr}, is determined by the amount of the charge on the snubber capacitor and the characteristics of diode D_1, which we are not familiar with at this point. Readers are encouraged to refer to the references for detail later.

A nonzero resistance R takes away the capacitor energy in the form of heat out of the system thus retarding the current rise beyond I_0. Diode D_2 isolates the rate of fall of the switch voltage from the snubber capacitor. Designers restrict the amount of current overshoot to 20% of I_O by selecting R from the following equation:

$$R = \frac{V_S}{0.2\, I_O}$$

2.15.2 Turn-on snubber

The turn-on snubber circuit is shown in Figure 2.29(a). It reduces the switching losses by reducing the voltage across the switch during the current transition through the switch. Again, this assumes ideal diodes and ideal switch and the switch has limited maximum rates of rise and fall transitions of voltage and current.

Prior to turnon, the switch voltage is equal to V_S at zero switch current and the diode D_1 is on. As the switch turnon is initiated, the switch current increases linearly (a reasonable assumption).

$$i_{sw} = I_0 \frac{t}{t_r} \qquad (2\text{-}105)$$

The switch current i_{sw}, however, overshoots beyond I_O. The amount of current overshoot, I_{rr}, is determined by the amount of the energy on the snubber inductor and the characteristics of diode D_1, which we are not familiar with at this point. Readers are encouraged to refer to the references for detail later.

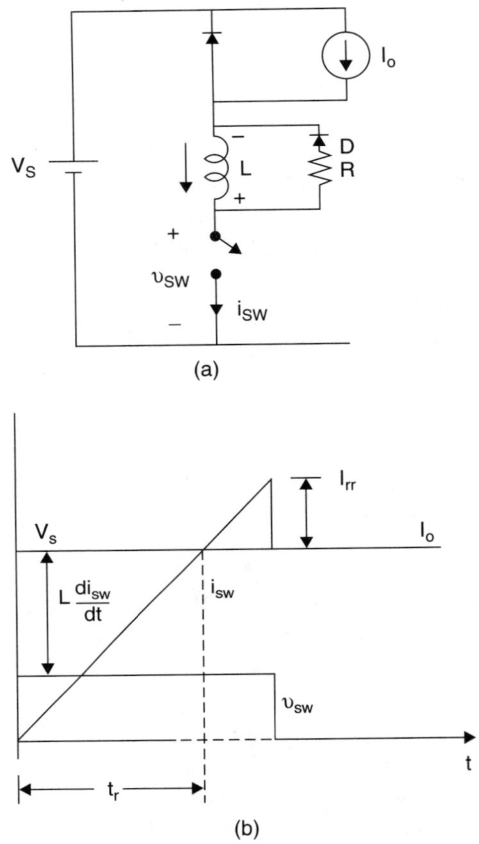

Figure 2.29
(a) Turn-on snubber circuit and (b) waveforms

Switch current also flows through the snubber inductor. Hence, the inductor voltage instantly reduces the voltage across the switch,

$$v_{sw} = V_S - L\frac{di_{sw}}{dt} = V_S - \frac{L\,I_0}{t_r} \quad (2\text{-}106)$$

The instantaneous power loss in the switch is

$$p_{sw} = v_{sw}\,i_{sw} = V_S I_0\left[1 - \frac{L}{t_{f1}}\frac{I_0}{V_S}\right]\frac{t}{t_r} \quad (2\text{-}107)$$

The turn-on snubber would worsen the subsequent turn-off process but for the diode D_2 and resistance R in parallel with the inductor. To understand turnoff, first we consider R to be 'open circuit'. The current through the switch continues to be at I_0 even after the switch is turned off. The voltage on the inductor remains zero. The switch voltage builds up linearly at the rated dv/dt. As soon as the switch voltage reaches the V_S value, the diode D_1 is turned on and the current through the switch and the inductor begins to decrease. The inductor energy ($1/2\ LI_0^2$) is dumped over to the switch. Hence, the switch voltage continues to rise beyond V_S. The falling rate of switch current is retarded. The voltage across the switch drops back to V_S as soon as the inductor deenergizes completely.

A finite resistance R takes the capacitor energy in the form of heat out of the system, thus retarding the voltage rise beyond V_S.

2.15.3 Overvoltage protection snubber

So far the role of parasitic inductance of the conductors in the switching process has been ignored. The inductor in the turn-on snubber of Figure 2.29(a) may as well be the parasitic inductance except that it must be placed in series with the input dc power supply, as shown in Figure 2.30(a). Following the reasoning given in the previous section, we know that a turn-off will produce a voltage spike on the switch. The circuit of Figure 2.30(b) protects the switch from over voltage. Prior to turnoff, the snubber capacitor is charged to V_S through R. Diode is reverse-biased. During the turnoff, Diode D_2 clamps the switch to V_S.

2.16 ZERO-CURRENT SWITCHING

Turning on and off a switch at zero current is the surest way of minimizing switching losses. An inductor in series with the switch will ensure zero current turnon because the current through the inductor cannot build instantaneously. However, the series inductor makes turnoff at zero current impossible. Figure 2.31 shows a simple modification that will enable both transitions at zero currents.

At turnon, the inductor permits slow buildup of the current through the switch. During turnoff, the inductor dumps its energy on the capacitor through the diode, thus enabling the switch to turn off at no current. The energy on the capacitor is either dissipated in a resistance (not shown) or diverted to other circuits (also not shown).

2.17 ZERO-VOLTAGE SWITCHING

Turning on and off a switch at zero voltage is also the surest way of minimizing switching losses. A capacitor in parallel with the switch will ensure zero voltage turnon because the voltage on the capacitor cannot build instantaneously. However, the parallel capacitor makes turnoff at zero voltage impossible. Figure 2.32 shows a simple modification that will enable both transitions at zero voltage.

At turnon, the diode conducts and the capacitor permits slow buildup of the voltage on the switch. During turnoff, the diode prevents the capacitor to discharge through the switch. The energy on the capacitor is either dissipated in a resistance (not shown) or diverted to other circuits (also not shown).

2.18 PSPICE SIMULATION

Simulate a step-down transformer 10:1. Apply a 10-Volt peak, a 100-Hz sine-wave voltage source

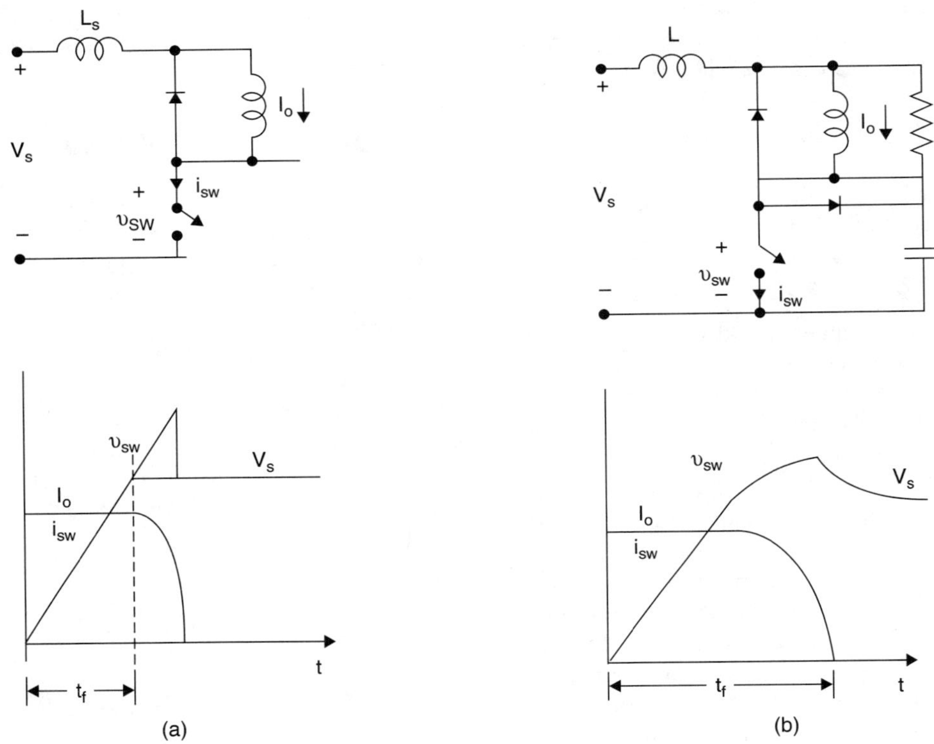

Figure 2.30
(a) Switching with parasitic inductance L_S without overvoltage snubber and (b) switching with overvoltage snubber

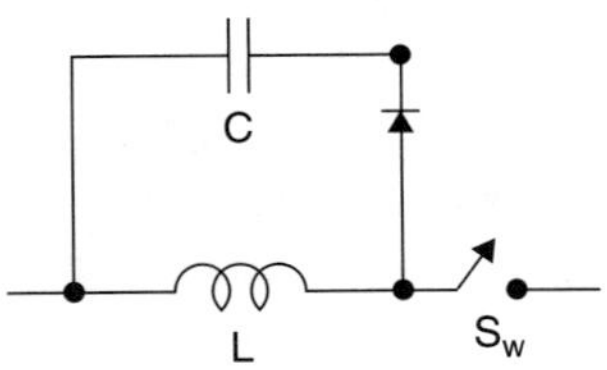

Figure 2.31
Zero-current switching

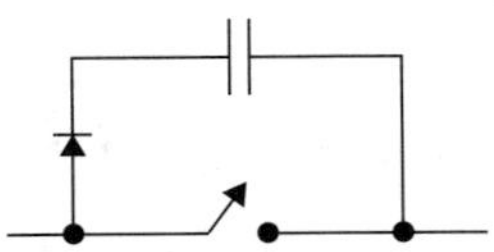

Figure 2.32
Zero-voltage switching

and a load resistance of 10 ohms on the secondary winding. Plot the voltage and current waveforms in the primary and secondary windings.

We have used the XT_nonlinear component as the transformer with the coupling coefficient of 1.0 (ideal coupling) and 1000 turns on the primary and 10 turns on the secondary winding. A resistance R_S is necessary for converging the solution. We have used a small value of 1 ohm as R_S.

Note the voltage on the secondary winding is identical to the primary voltage waveform but with a peak voltage of 1 volt. The current in the secondary winding has peak value of 100 mA whereas the primary current has a small peak value of 1 mA.

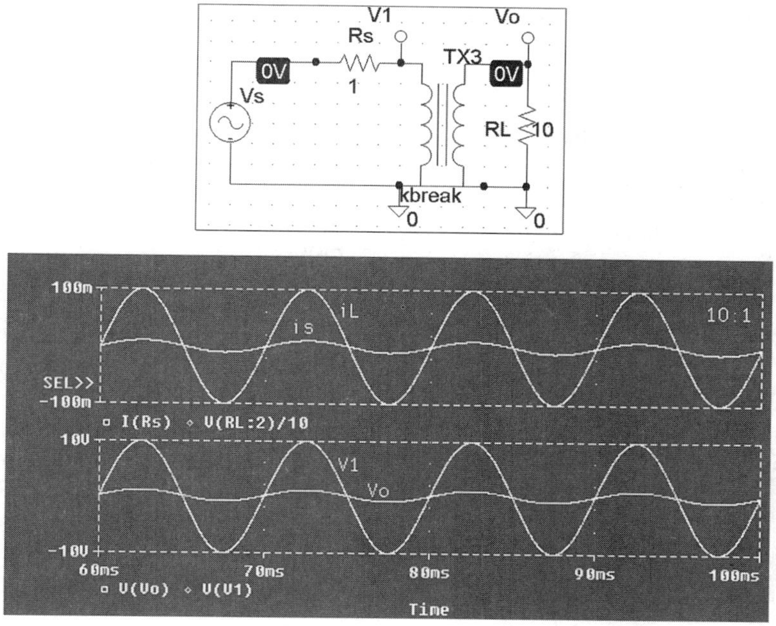

Figure 2.33
Voltage and current waveforms in a 10:1 step-down transformer

References

1. Ned Mohan, Tore M. Undeland, and William P. Robbins, *Power Electronics: Converters, Applications and Design,* 2nd Ed. , John Wiley & Sons, Inc., 1995.

2. John G. Kassakian, Martin F. Sclecht, and G. Vergassian, *Principles of Power Electronics,* Addison-Wesley, Boston, 1991.

3. Keith Billings, *Switch Mode Power Supply Handbook,* McGraw-Hill Inc., New York, 1989.

4. *Ferrite Cores,* Magnetics, Division of Spang & Company, Butler, PA, 1994.

5. *Power Conversion & Line Filter Applications,* Catalog 4, Issue G, Micrometals, Anaheim, CA.

6. *Powder Cores, MPP and High Flux Cores for Filter and Inductor Applications,* Magnetics, Division of Spang & Company, Butler, PA, 1993.

Problems

2-1 Find the inductance of an inductor that has 42 turns on a ferrite ring core of inner, outer diameters and height of 25.40 mm, 38.10 mm and 19.05 mm respectively. The relative permeability of the toroid is 3000.

2-2 Prove that the ratio of air gap to the magnetic path length in an air-gapped inductor is

$$\frac{g}{l_c} = \frac{\mu_0}{B}\left[\frac{Ni}{l_c} - H_c\right]$$

2-3 Find the inductance in Problem 2-1 when the inductor is wound on the middle limb of an E-E core. The E-E core has the relative permeability of 500, the core area at the middle limb of 181 mm^2, the magnetic path length of 100.4 mm, and (a) the air gap is zero and (b) the air gap in all limbs is 1 mm.

2-4 Design the air-gapped filter inductor of Problem 2-1 if the peak-to-peak triangular ripple current is 100 mA or less superimposed on a dc current of 40 A. The switching frequency is 20 kHz. Determine the core and the copper wire size.

2-5 Repeat problem 2-4 for a switching frequency of 100 kHz. Note the difference in size.

2-6 Find the diameter of a solid round copper conductor whose resistance at 20 KHz is equal to the resistance of a 7-strand litz copper conductor of the dimension given in Figure P2.6.

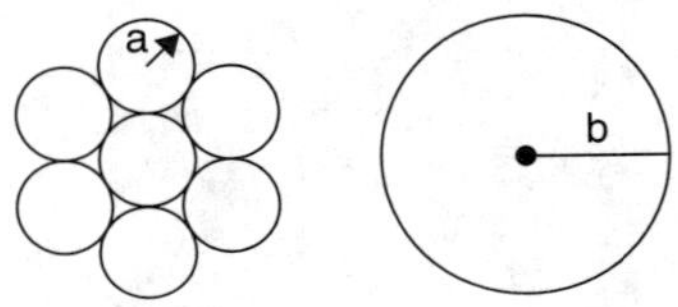

Figure P2.6
Litz wire and equivalent solid round conductor

2-7 Derive the equations (2-22) and (2-26).

2-8 Derive the equation (2-27).

2-9 A dc-dc down-converter (known as the Buck converter) is shown in Figure P2-9. Assume the diode is ideal. The switch voltage is constrained to rise or fall linearly at 50 V/mS. The switch current likewise is constrained to rise or fall linearly at 20 A/mS. The inductor is large enough at the switching frequency to maintain a ripple-free constant load current I_o=40 A. The switch is operated by a 50% duty cycle at 20 kHz.

a. Draw the voltage and current waveforms of the switch.
b. Find the switching losses.
c. Find the maximum frequency of switching.
d. Find the switching losses as a fraction of total power to the load at the maximum switching frequency.

2-10 Consider the switches in Problem 2-9 have a voltage drop of about 3 volts at the rated current and a reverse leakage current of 2 mA. Find the approximate additional losses in the converter operation at 20 kHz. Also find the total losses vis a vis the power delivered at the maximum switching frequency.

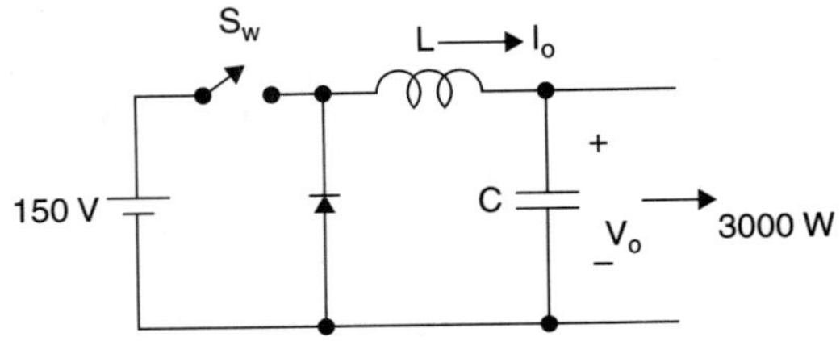

Figure P2.9
DC-DC down-converter

2-11 The circuit of Figure P2-9 is reorganized as a DC-DC up-converter (known as Boost converter) as shown in Figure 2P-11. The switch is operated by a 50% duty cycle at 20 KHz. DC current from the source is 20 A.

a. Draw the voltage and current waveforms of the switch.
b. Find the switching losses.
c. Find the switching losses as a fraction of total power to the load.

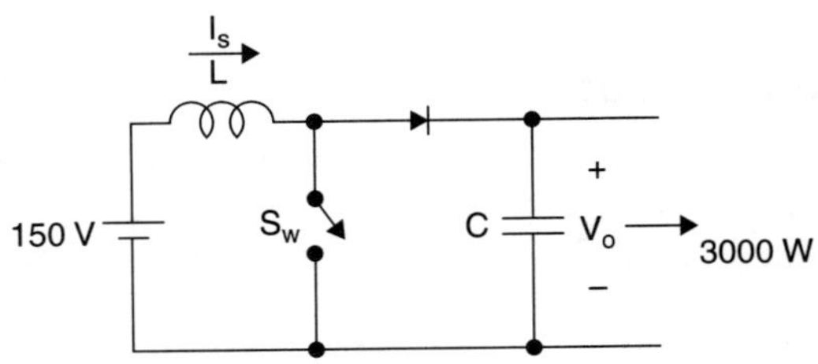

Figure P2.11
DC-DC up-converter

2-12 The circuit of Figure 2P.9 is reorganized as a dc-dc up/down-converter (known as the Buck-boost converter) as shown in Figure 2P.12. The switch is operated by a 40% duty cycle at 20 KHz. The dc output voltage is 100 V.

a. Draw the voltage and current waveforms of the switch.
b. Find the switching losses.
c. Find the switching losses as a fraction of total power to the load.

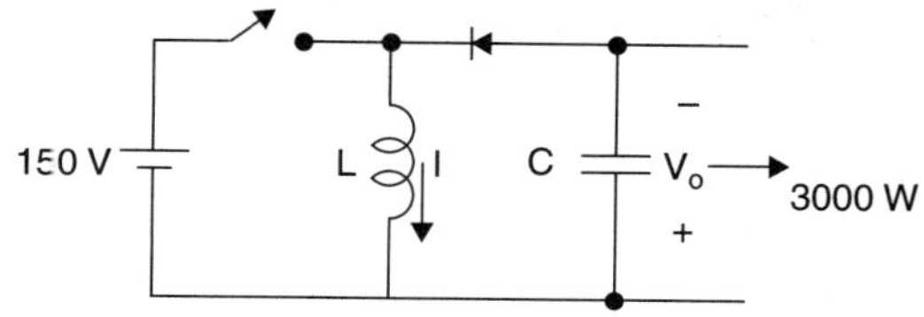

Figure 2P.12
Buck-boost converter

2-13 Consider the Buck converter of Problem 2-9. The ferrite core of the inductor has the B-H characteristics as given in Figure 2.12(b) with following parameters:
Initial permeability μ_i $= 3000\ \mu_0$
Saturation flux density B_0 $= 0.3$ Tesla
Remanent flux density B_r $= 0.15$ Tesla
Draw the voltage and current waveforms on the switch.

2-14 Analyze and plot the waveform of the voltage and the current through the switch in the turn-off snubber of the circuit in Figure 2.28(a) with a snubber capacitor smaller and more than the critical value given by equation (2-102).

2-15 A turn-on snubber is shown below. The load current is a dc 10 A. Select snubber inductor L_1 to limit the rate of rise of current through the switch to 10 A/μs.

2-16 The peak reverse recovery current in diode D_2 in Figure P2.15 occurs at 500 ns after the occurrence of the zero current through it. Determine the

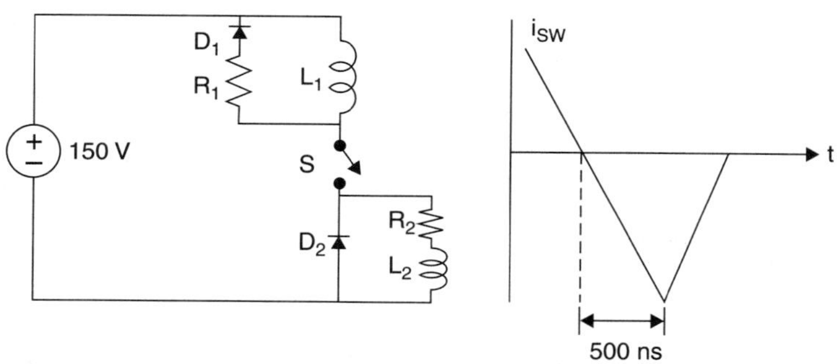

Figure P2.15

a. peak current pulse amplitude through the switch and

b. the peak energy stored in the turn-on snubber in a switching period.

2-17 A dc current of 40 amp is running through diode D and the inductor before the switch S is turned on. The peak reverse recovery current and the rate of decay of the reverse current are observed to be 25 A and 15 A/μs respectively. The on-state voltage drop is assumed to be 1.0 volts but neglect the off-state leakage. Assume inductor L is large.

a. Sketch the current waveform through the diode and determine t_{rr}.

b. Determine the maximum reverse voltage across the diode.

c. Estimate the excess charge recovered until the reverse current peak is reached.

d. Sketch the current through the switch and determine the maximum instantaneous current through the switch.

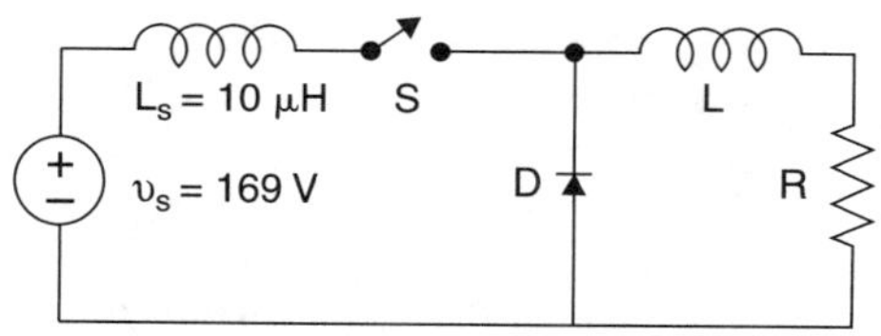

Figure P2.17

2-18 A 48-volts battery is switched for 20 μs across an ideal transformer but with the magnetizing inductance of $L_m = 100\ \mu H$, as shown in Figure P2.18. Plot the waveforms for the current from the battery, in the primary and secondary windings, and voltage on the secondary winding. Assume zero initial conditions.

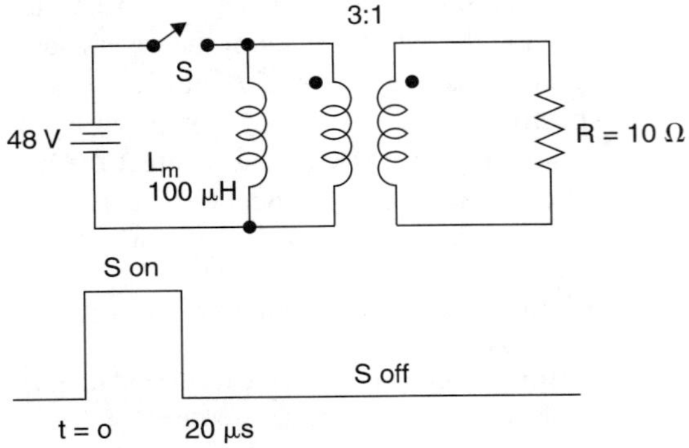

Figure P2.18

2-19 Consider a Ferrite E core pair with an air gap of 1 millimeter. Assume the core is not saturated.

a. Calculate the inductance with a 100 turns wound on the central limb.

b. Determine the maximum flux density if a triangular voltage waveform is applied at across the inductor.

2-20 Repeat Problem 2-19 if the core characteristics are linearized as below:

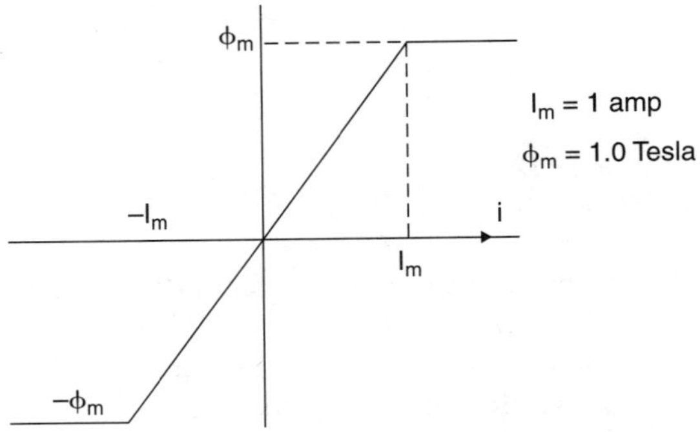

Figure P2.20

2-21 Design an LC filter that would suppress the third harmonic within 5% of the amplitude of the fundamental component. Assume ideal L and C. The output frequency is 400 Hz and the filter should be effective for a range of power from 1 KW to 9 KW.

2-22 Repeat Problem 2-21 for the inductor with the quality factor Q = 20.

2-23 Find the number of turns in a 0.1 mH toroidal inductor. The toroid specifications are: Outer diameter = 25 mm, Inner diameter = 20 mm, permeability μr = 1100.

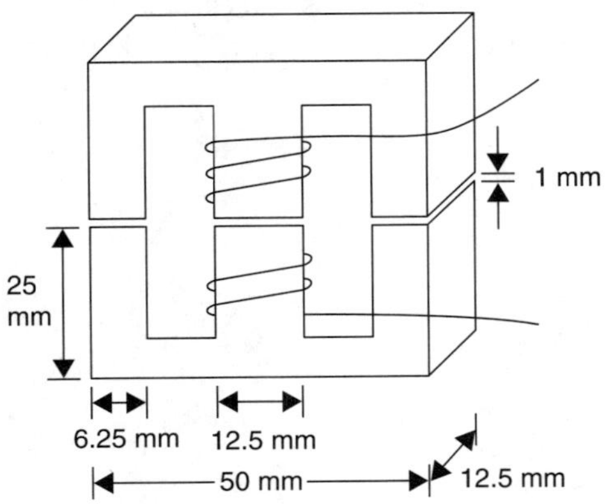

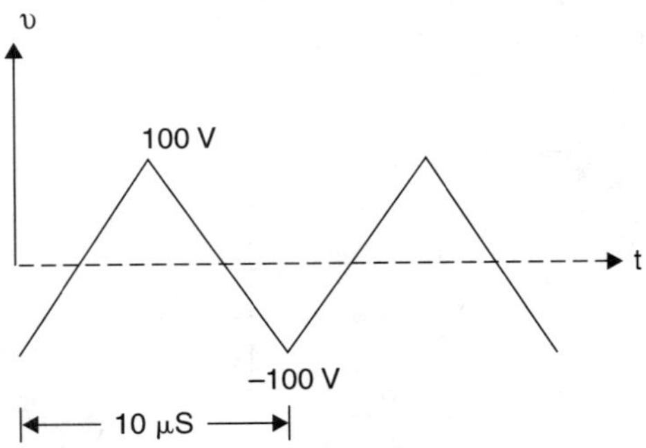

Figure P2.19

2-24 See the filter circuit of a dc to dc converter in Figure 2.15. The output voltage is 12 volts dc with zero ripple. Capacitor C is assumed to be infinitely large. DC current through the inductor is 5 A with a peak-peak current ripple of 0.2 A and the input voltage waveform from the dc-dc converter is a 24 volts square wave of 20 kHz. Design the filter inductor using Mn-Zn ferrite core with B_s= .3T and μ_r=150.

2-25 Design an inductor for ac application: L=100 μH, V_{rms}=120 V, I_{rms} = 4 A, f= 40 kHz, ΔT=50°C.

2-26 Design a two-winding single-phase transformer: rms primary voltage V_p= 110 V, rms secondary current I_p= 12 A, frequency = 100 kHz, turns ratio = 10:1.

CHAPTER 3
Power Diode

OBJECTIVES

1. The differences in the structure between the high-power diode and the low-power diode
2. Conductivity modulation in power diode
3. Equivalent circuit of power diode
4. The breakdown voltage and the leakage current of power diode
5. The switching characteristics of power diode
6. On-state losses power losses
7. Parallel and series connection of power diodes
8. Schottky power diodes

3.1 INTRODUCTION

Diodes in power electronic applications carry high currents, withstand high reverse voltages and should possess fast switching characteristics. These requirements render power diodes very different than the ordinary signal diodes. It is also difficult to achieve all the above three characteristics in a single device. Hence, several types of power diodes are available suited for particular applications.

The power diode is a high-current, high-voltage diode with medium fast switching characteristics. The Schottky Barrier diode uses a metal-semiconductor junction and has a lower on-state voltage than the p-i-n diodes. It has the high-current capability and fast switching characteristics but it is a low-voltage device and exhibits a much higher leakage current. These diodes are ideally suited for switch-mode power supplies. The on-state power losses, which are calculated by multiplying the on-state voltage drop by the on-state current, determine the package size of all diodes.

In order to understand the physics and the characteristics of the power diodes, and power semiconductor devices in general, we must learn the basics of current conduction in semiconductor materials and structures. Therefore, in this chapter, we begin with the basic principles of semiconductor physics and a generic p-n junction. Both the static characteristics and the switching characteristics of the p-n junction device will be covered. These principles will be applied to the transistors and thyristors in the two following chapters. The p-i-n diodes and Schottky Barrier diodes will be studied in that order.

This chapter will also discuss the parallel and series connection of diodes, which may be necessary to enhance the voltage and the on-state current capabilities of the diode respectively in some applications.

3.2 P-N JUNCTION

The junction of p and n type semiconductors is the basic structural component of semiconductor devices. A metallurgical junction of **p** and **n** semiconductors is formed by the deposition of layers or by controlled diffusion of p-type impurity into the n-semiconductor or vice versa. Electrons are the majority carrier on the n-side but the minority carrier on the p-side. Abrupt charge carrier densities exist across the junction before metallurgical joining.

After the metallurgical joining, majority carriers and minority carriers on both sides cross the junction powered by their respective thermal energies. However, the majority carrier transport dominates due to the high-density gradients. Holes, the majority carriers on the p-side diffuse to the n-side leaving behind an equal number of negatively charged *acceptor* atoms (called ions). Electrons, the majority carriers on the n-side, diffuse to the p-side leaving behind an equal number of positively charged ions. Thus a space charge is formed by the immobile charged ions on both sides of the metallurgical junction, see Figure 3.1. This region is called the space charge region or the depletion layer because of the desertion of free charge carriers. The space charge establishes the electric field and a barrier potential to discourage further diffusion of majority charge carriers across the junction. The electric field distribution inside the junction area is shown in Figure 3.1. The peak electric field at the metallurgical junction is given by

$$E_p = \frac{qN_D w_n}{\epsilon} = \frac{qN_A w_p}{\epsilon} \qquad (3.1)$$

where q is the charge of an electron. $\mathbf{N_A}$ and $\mathbf{N_D}$ are respectively the densities of the negatively and positively charged atoms on both sides of the junction. Epsilon ϵ is the dielectric constant of the semiconductor material. The w_n and w_p are the effective

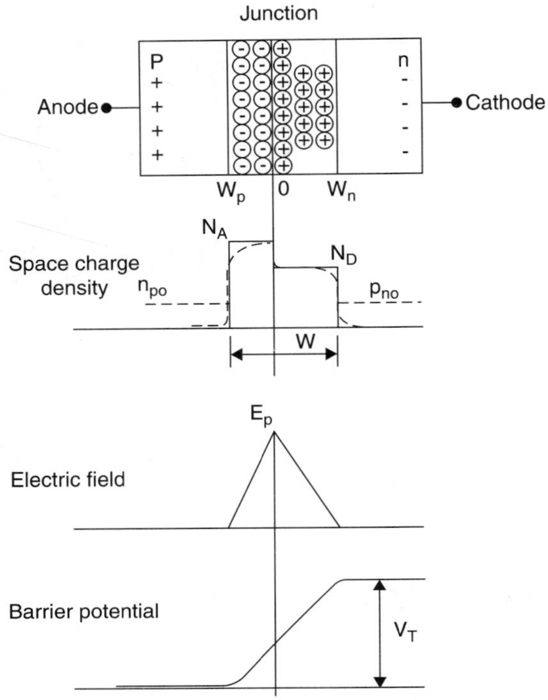

Figure 3.1
A p-n junction

widths of the space charge regions of the negatively charged ions and the positively charged ions respectively. In thermal equilibrium condition, the transportation of majority and minority carriers equalizes. The number of crossings across the junction increases with the rising temperature. The barrier potential is given by

$$V_T = \frac{kT}{q} \ln\left[\frac{N_A N_D}{n_i^2}\right] \qquad (3\text{-}2)$$

This potential is across the junction only and not accessible from outside the device. For $\mathbf{N_A = N_D = 10^{16}\,/cm^3}$ at room temperature 25°C, the barrier potential in silicon p-n junction is $V_T = 0.72$ volts.

The electrons that have crossed the junction into the p-side, accumulate beyond the immobile negatively charged ions and so do the holes on the n-side beyond the positively charged ions. The region of both types of ions is called the depletion or space charge region or layer. There are no charge carriers in this region. The diffused carriers constitute the excess minority carrier population on the respective sides of the junction. The spatial distribution of ions on both sides is shown by dotted lines and approximations by solid lines in Figure 3.1. The injected carriers on each side become the minority carriers. The injected minority carriers recombine with the majority carriers and establish the thermal equilibrium density levels.

Thus the metallurgical junction has given rise to two charge concentrations separated by the width of the depletion region. The depletion layer width or the junction width is the sum $W = w_p + w_n$. The p-side and n-side widths and the overall depletion layer width are

$$w_p = \sqrt{\frac{2\epsilon V_T}{qN_A\left(1 + \frac{N_A}{N_D}\right)}} \quad (3\text{-}3)$$

$$w_n = \sqrt{\frac{2\epsilon V_T}{qN_D\left(1 + \frac{N_D}{N_A}\right)}} \quad (3\text{-}4)$$

$$W = w_p + w_n = \sqrt{\frac{2\epsilon V_T}{q}\left[\frac{1}{N_A} + \frac{1}{N_D}\right]} \quad (3\text{-}5)$$

Width of the depletion layer on either side of the junction can be changed by varying the dopant density N_A or N_D. The smaller the dopant density, the depletion layer extends to a larger width in that region.

Peak electric field in terms of the barrier potential is obtained from eqs. (3-1) and (3-4):

$$E_p = \sqrt{\frac{2qV_T}{\epsilon}\left[\frac{N_A N_D}{N_A + N_D}\right]} \quad (3\text{-}6)$$

The depletion region consists of two charges of opposite polarities spacially separated from each other. This is a capacitorlike structure. The space charge of one polarity is approximated as

$$Q_j = qA\,N_D w_n = A\sqrt{2q\epsilon V_T\left[\frac{N_A N_D}{N_A + N_D}\right]} \quad (3\text{-}7)$$

The physically separated oppositely charged space charges across the **p-n** junction give rise to a built-in capacitance, called the junction capacitance C_j. This capacitance is calculated as the ratio of the incremental space charge to the incremental change in the barrier potential.

$$C_j = \frac{dQ_j}{dV_T} = A\sqrt{\frac{q\epsilon}{2V_T}\left[\frac{N_A N_D}{N_A + N_D}\right]} \quad (3\text{-}8)$$

Example 3-1 Depletion Widths at Different Doping Levels

For a silicon p-n junction with following parameters:

$\epsilon = 1.035 \times 10^{-12}$ *F/cm*, $k = 1.38 \times 10^{-23}$ *J/K*,
$q = 1.6 \times 10^{-19}$ *Coulomb*
$T = 273 + 25 = 298$ *K*, $n_i = 10^{10}$ cm^{-3}

Find the depletion layer widths on the p and n sides and the total junction widths for the following doping levels: (a) $N_A = N_D = 10^{16}$ cm^{-3}*, p-n junction, (b)* $N_A = 10^{17}$*,* $N_D = 10^{14}$ cm^{-3}*, p-n⁻ junction and (c)* $N_A = 10^{19}$*,* $N_D = 1 \times 10^{14}$ cm^{-3}*,* $p^+ - n^-$ *junction.*

The MATLAB solution is

```
>>epsilon=1.035e-12; k=1.6e-23; T=273+25;
  ni=1e10;
>>q=1.6e-19; NA=1e16; ND=1e16;
>>VT=(k*T/q)*log(NA*ND/ni^ 2)
  %Threshold voltage, eq. (3-11)
>>Wp=sqrt(2*epsilon*VT/(q*NA*(1+NA/ND)))
  %Depletion layer width on the p-side, eq.        (3-12)
>>Wn=sqrt(2*epsilon*VT/(q*ND*(1+ND/NA)))
  % Depletion layer width on the n-side, eq.       (3-13)
>>W=Wp+Wn % Total junction width
```

(a) >>NA=1e16 ; ND=1e16 ; **VT=0.8234 V**
Wp=2.3079 e-5 cm Wn=2.3079 e-5
W=4.6158 e-5

(b) >>NA=1e17 ; ND=1e14 ; **VT=0.7548 V**
Wp=3.1233 e-7 cm Wn=3.1233 e-4
W=3.1265 e-4

W_P is much smaller than W_n. Total junction width has increased.

(c) >>NA=1e19 ; ND=1e14 ; **VT=0.8920 V**
Wp=3.3971 e-9 cm Wn=3.3971 e-4 W=3.3972 e-4

W_P is thinner but W_n has not changed significantly. The depletion layer width on the n-side, hence, the total junction width is primarily determined by the lightly doped side.

3.2.1 Current through the p-n junction

Depletion layer in a p-n junction device has much higher resistance in comparison to the bulk regions of the n and p sides. Hence, when an external voltage v_d is applied to this device, positive polarity to the p-side (anode), as shown in Figure 3.3, it appears entirely across the depletion layer. It establishes an electric field, which opposes the built-in electric field in the space charge region. Consequently, the barrier potential decreases to $V_T - v_d$. This is called the *forward biasing* of the p-n junction. Width of the depletion layer on the p-side, w_p, decreases, see equation (3-3). Width of the depletion layer on the n-side, w_n, also decreases, see equation (3-4). The total junction width W decreases:

$$W(v_d) = W(0)\sqrt{1 - \frac{v_d}{V_T}},$$

$$\text{where } W(0) = \sqrt{\frac{2\epsilon V_T}{q}\left[\frac{1}{N_A} + \frac{1}{N_D}\right]} \tag{3-9a}$$

The lowering of the barrier potential upsets the equilibrium between the drift and diffusion in favor of diffusion. The externally imposed positive polarity on the p-side pushes the holes in the bulk region to drift towards the metallurgical junction. Some of these holes combine with the negative ions of the space charge region on the p-side of the junction reducing the effective width w_p. The excessive influx of the holes causes more holes to cross the junction. These holes enter the n-side as the excess minority carriers. The minority carriers eventually recombine with the majority carriers, the electrons, as they diffuse further into the bulk region of the n-side. The minority carrier density has the exponential distribution, see Figure 3.2. Similar phenomenon occurs to the electrons, which enter as the influx on the n-side due to the negative polarity. Densities of the injected minority carriers on the edges of the depletion region are given as the function of the forward bias voltage v_d:

$$p_n(0) = p_{n0}e^{qv_d/kT}, \quad n_p(0) = n_{p0}\, e^{qv_d/kT} \tag{3-9b}$$

where $p_{no} = n_i^2/N_D$, and $n_{po} = n_i^2/N_A$ are the thermal equilibrium densities of the minority carriers in the bulk regions on the n-side and p-side respectively.

Majority carrier transport overtakes the minority carrier transport across the junction, the later being stagnant at a given temperature. A net current flows from the anode (p-side) to the cathode (n-side).

In the steady-state forward bias condition, the excess carrier distributions, shown in Figure 3.2, neither grow nor decay in time. The recombined carriers are replaced by injected carriers. The forward current density in terms of the stored charges is given by

$$J_f = \frac{Q_n}{A\tau_n} + \frac{Q_p}{A\tau_p} \tag{3-10}$$

where A is the area of the p-n junction and Q_n and Q_p are the excess minority charges on the n-side and p-side respectively. These charges are obtained from the areas under the excess minority carrier density distribution:

$$Q_n = qA[n_p(0) - n_{p0}]L_n \ : \ n_{p0} = n_i^2/N_A \tag{3-11}$$

$$Q_p = qA[p_n(0) - p_{n0}]L_p \ : \ p_{n0} = n_i^2/N_D \tag{3-12}$$

L_n and L_p are the characteristic decay lengths of the excess electron density on the p-side and the excess holes on the n-side respectively.

When the external voltage of opposite polarity is applied to this p-n device, as shown in Figure 3.2, it enhances the barrier potential to V_T+v_d. This is referred as the *reverse biasing* of the p-n junction. Externally imposed polarities on the two sides pull the majority charge carriers in the bulk region away from the metallurgical junction. The outward drift causes the width of the depletion layer to increase.

A negligible number of majority charge carriers (holes on the p-side and electrons on the n-side)

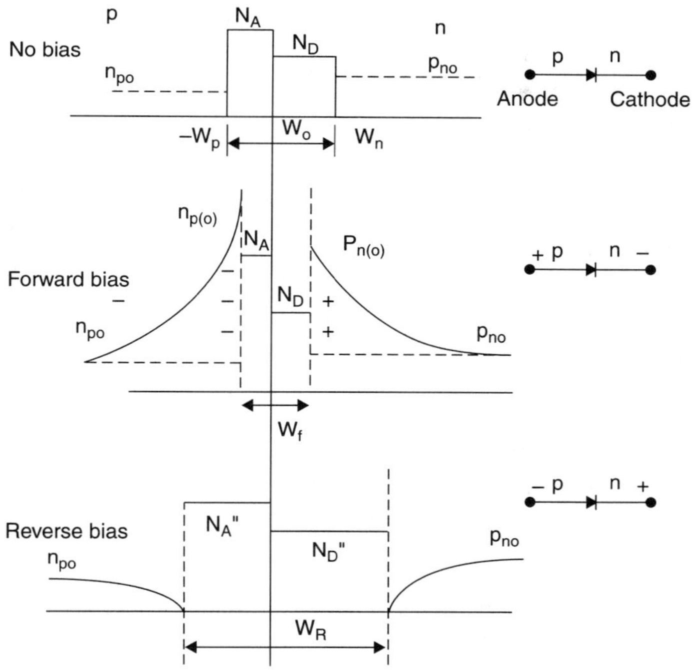

Figure 3.2
Biasing of p-n junction

cross the junction. Hence, the minority carrier densities on both edges of the p-n junction are nearly zero, see Figure 3.2. The total charge carrier transport is predominantly due to the minority carriers. The minority carrier transport is independent of the applied voltage and dependent only on the temperature. A constant reverse current flows in the device, called the reverse saturation current.

Substituting equation (3-9) and (3-11) in the equation (3-10), the forward current can be shown to be

$$I_f = I_s(e^{qv_d/kT} - 1)\ :$$

$$I_s = qn_i^2 A\left[\frac{L_n}{N_A\tau_n} + \frac{L_p}{N_D\tau_p}\right] \qquad (3\text{-}13)$$

The above equation gives the i-v characteristics of the p-n diode. I_s is the reverse saturation current. This is basically the current due to the transport of minority carriers across the p-n junction. It is called the saturation current because it is independent of the reverse voltage amplitude to fairly high levels. Since n_i has a strong dependence on the temperature, I_s is very sensitive to the temperature rise and thus it limits the high temperature operation of diodes. Figure 3.3 shows the i-v characteristics of a diode using MATLAB for a diode with $I_s = 10$ nA and $kT/q = 26$ mV.

The incremental capacitance C_j in terms of the forward bias voltage is obtained from equation (3-17)

$$C_j = \frac{dQ_j}{dv_d} = \frac{C_{j0}}{\sqrt{1 - v_d/V_T}}\ :$$

$$C_{j0} = \sqrt{\frac{q\epsilon}{2V_T}\left[\frac{N_A N_D}{N_A + N_D}\right]} \qquad (3\text{-}14)$$

The diode is thus also a voltage-controlled capacitor. The plot of the normalized capacitance C_j/C_{j0} versus the forward bias voltage v_d is shown in Figure 3.4.

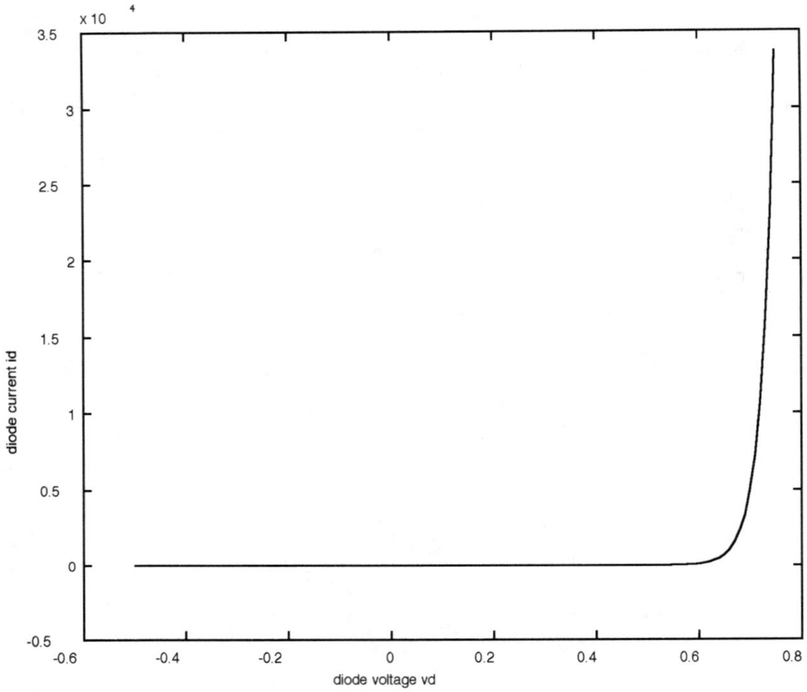

Figure 3.3
The i-v characteristics of a p-n junction diode

3.2.2 Reverse breakdown of p-n junction

There is a physical limit on the reverse voltage that can be applied on a p-n junction. At high electric fields a free electron acquires high energy, **eE** and upon impact with an atom, can liberate an extra electron. This process can snowball quickly and produce a large number of free electrons for current conduction in the reverse biased junction. This process is called *avalanche breakdown.* The threshold electric field $\mathbf{E_{BR}}$ is typically 200 kV/cm in power semiconductor devices. The voltage, at which the threshold electric field is achieved, is called the *breakdown voltage* $\mathbf{V_B}$. It can be obtained from equation (3-6) by substituting

$V_T = V_{BRAv}$ and $E_P = E_{BR}$:

$$V_{BRAv} \approx \frac{\epsilon E_{BR}^2}{2q}\left[\frac{1}{N_A} + \frac{1}{N_D}\right] \tag{3-15}$$

Under reverse bias, the depletion layer widens significantly. If the edge of depletion layer reaches the edge of the device, the reverse current begins to increase very rapidly. This condition is called the *punch through condition.* The breakdown voltage in this condition can be obtained from equation (3-4) by substituting $V_T = V_{BRPt}$, $w_n = l_n$ and $w_p = l_p$:

$$V_{BRPt} \approx \frac{qN_D l_n^2}{2\epsilon}\left[1 + \frac{N_D}{N_A}\right] \approx \frac{qN_A l_p^2}{2\epsilon}\left[1 + \frac{N_A}{N_D}\right] \tag{3-16}$$

where l_n and l_p are the widths (or thickness) of the n-side and p-side respectively. The lower of the above two types of breakdown voltages is designated as the limiting breakdown voltage. Most of the devices are designed to avoid the punch through condition.

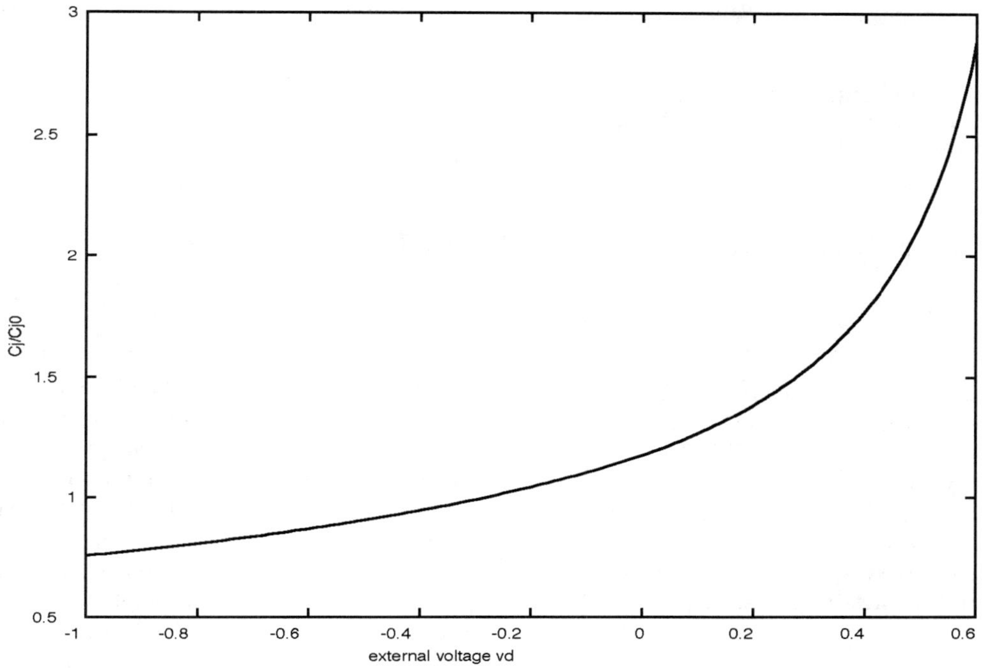

Figure 3.4
Capacitance C_j/C_{j0} versus the externally applied forward voltage on the diode

The above device closely depicts a low power general-purpose signal diode used for signal processing. The p-side is connected to the anode and the n-side to the cathode terminals. The overall i-v characteristics of a low power diode are shown in Figure 3.5.

Example 3-2 Breakdown Voltage at Different Doping Levels

For a silicon p-n junction of Example 3-1, find the avalanche breakdown voltages.

Add following lines to the previous MATLAB solution

```
>>EBR=2e5;
>>VBRAv=(epsilon*EBR^ 2/(2*q))*(1/NA+1/ND)
```

(a) >>NA=1e16; ND=1e16;

VBRav=25.875 V

(b) >>NA=1e17; ND=1e14;

VBRav=1295.0 V

Light doping of the n-side results in a very large increase in the breakdown voltage.

(c) >>NA=1e19; ND=1e14;

VBRav=1293.8 V

Heavy doping of the p-side does not materially alter the breakdown voltage.

Example 3-3 The p-n Junction Widths in Different Bias Conditions

For the silicon p-n junction of Example 3-1 with $N_A = 10^{17}$, $N_D = 10^{14}$ cm^{-3}, calculate the junction widths and the incremental capacitance at following external diode bias voltages: (a) $v_d = 0$ V, (b) $v_d = 0.6$ V and (c) $v_d = -100$ V.

Add following instructions to the MATLAB solution

```
NA=1e17; ND=1e14; vd=0;
W0=sqrt((2*epsilon*VT/q)*(1/NA+1/ND)); % junction
  width at zero bias, eq. (3-18a)
```

```
W=W0*sqrt(1-vd/VT) % junction width at the forward
  bias vd, eq. (3-18a)
Cj0=sqrt((q*epsilon/(2*VT))*NA*ND/(NA+ND));
  % incremental capacitance at zero bias, eq. (3-23)
Cj=Cj0/sqrt(1-vd/VT)% incremental capacitance at the
  forward bias vd, eq. (3-23)
```

(a) >>vd=0;

W=3.1265 e-4 cm Cj=3.3104 e-9

(b) >>vd= 0.6 ;

W=1.4158 e-4 cm Cj=7.3102 e-9

W decreases and the incremental capacitance increases.

(c) vd= −100.0 ;

W=0.0036 cm Cj=1.2938 e-10

W increases very significantly and the incremental capacitance decreases.

3.2.3 Equivalent circuit of a low-power diode

For drawing the equivalent circuit of the low-power diode, we would linearize the i-v characteristics in three segments: one to the right of the threshold voltage $\mathbf{V_T}$, the second between $\mathbf{V_T}$ and the reverse breakdown voltage $\mathbf{V_{BR}}$ and the third to the left of $\mathbf{V_{BR}}$. The linearized diode characteristics are shown by the dashed lines in Figure 3.5. The equivalent circuit in different bias conditions is given in Figure 3.6.

The reverse saturation current $\mathbf{I_s}$ is the reverse current due to the thermal equilibrium densities of minority carriers, the electrons on the p-side and the holes on the n-side. The forward current becomes appreciable after the threshold voltage $\mathbf{V_T}$. In the breakdown region, the current increases rapidly with an increase in the reverse bias voltage. The forward current increases exponentially with the increase in the forward voltage. The diode remains off for all voltages less than $\mathbf{V_T = 0.7}$ volts. The threshold voltage $\mathbf{V_T}$ decreases by approximately by 2.5 mV for each degree rise in temperature. The forward resistance is the reciprocal of the slope of the curve in the first quadrant, is approximately zero. The reverse saturation current in silicon diodes is tens of nano-amps, and doubles in every ten degrees rise of temperature. The reverse breakdown voltage $\mathbf{V_{BR}}$ is typically hundreds of volts.

3.3 POWER DIODE

A power diode is designed for high forward current and high reverse breakdown voltage. The area of p-n junction in power diodes is much larger than in a signal diode because it is designed for large current flow. The power diode consists of three semiconductor regions: the p^+, n^- and n^+ regions. The power diode fabrication begins with a heavily doped n^+ type substrate, see Figure 3.7. A thin layer (the epitaxial layer) of n^- (lightly doped) is grown by diffusing the n-type impurity on the top surface. The doping of this layer is very light, nearly intrinsic, hence it is also referred to by the **i-layer** and the device as the **p-i-n diode.** Another layer of a heavily doped p^+ material is grown on the top of the n^- layer. The p^+ layer forms the anode, the n^+ substrate forms the cathode. This device contains two junctions: p^+-n^- and n^--n^+ junctions. Figure 3.7 also shows the space charge density distribution in thermal equilibrium.

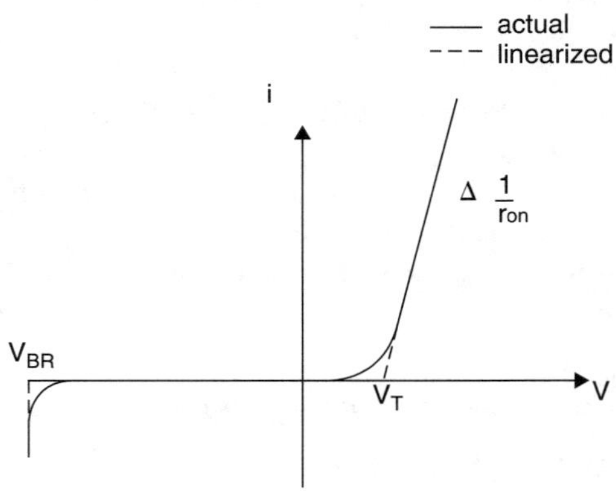

Figure 3.5
The i-v characteristics of a low-power diode

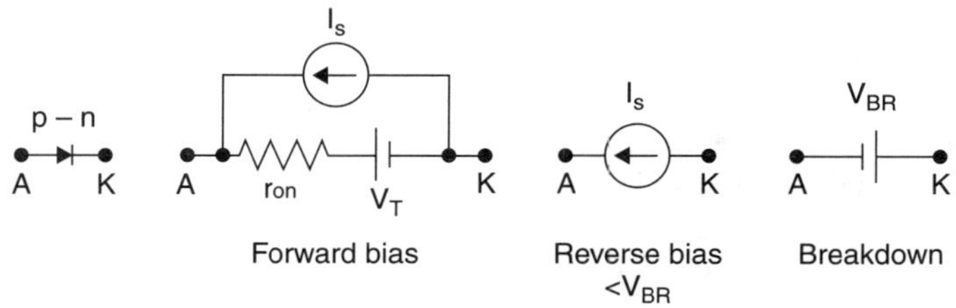

Figure 3.6
Equivalent circuit of a low-power diode

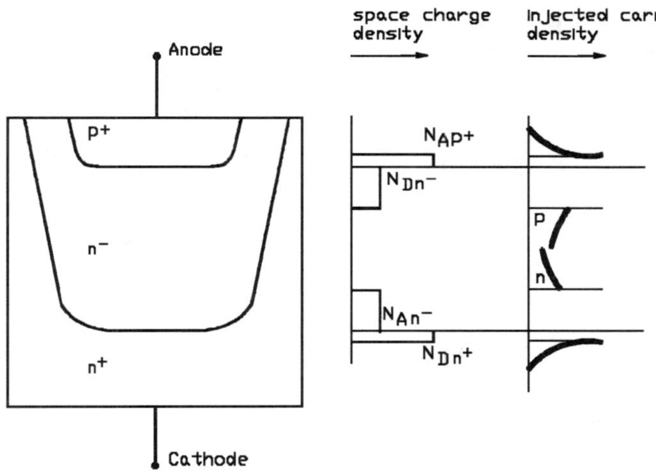

Figure 3.7
The power diode

3.3.1 Conductivity modulation in the p-i-n diode

Since both the p^+ and n^+ layers are heavily doped and the n^- layer is nearly intrinsic, the depletion layers at both junctions extend almost entirely in the latter. In the forward bias condition, a large number of holes from the p^+-side drift and cross to the n^--side (intrinsic region). A large number of electrons from the n^+-side also cross to the n^--side. Thus there is a big influx of charge carriers of both types in the nearly intrinsic middle layer, see Figure 3.8. The large influx is also called *high-level injection.* The influx is so large that the thermal equilibrium charge densities in the bulk region of the i-layer are also altered: $p_{n-} >> p_{n0}$ and $n_{n-} >> n_{n0}$. The excess charges from both junctions into the i-layer are also equal; $p_{n-} = n_{n-} = n$. Furthermore, at high-level injection, the edge densities are approximately equal to the bulk densities, consequently a nearly flat charge density distribution is established in the i-layer. The conductivity, given by equation (C-7), undergoes a big change. On the other hand, in the reverse bias condition, the carrier density decreases considerably when the excess carriers on the borders of the two junctions are swept away.

Resistance of the i-layer becomes very small (high conductivity) in forward bias and very large (low conductivity) in reverse bias. This phenomenon is referred as the *conductivity modulation.* High conductivity allows the diodes to carry a high current density in the forward bias. However, the heavy injection of minority carriers in the i-layer poses the problem of delayed charge removal during the reverse bias, leading to switching delays.

3.3.2 On-state losses

The on-state losses are calculated as the product of the on-state voltage and the on-state current in the forward bias condition. The on-state voltage drop in the p-i-n diode is

$$v_d = v_{j1} + v_i + v_{j2} \tag{3-17}$$

where v_{j1} and v_{j2} are respectively the forward voltage across the $p^+ - n^-$ and $n^- - n^+$ junctions. v_i is the voltage drop in the i-layer.

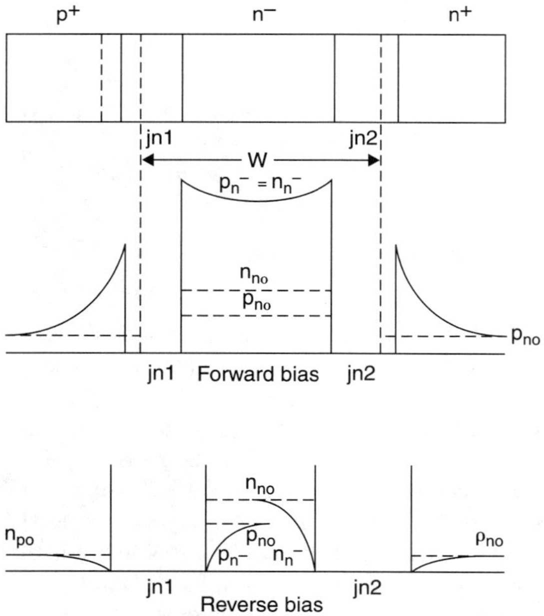

Figure 3.8
Conductivity modulation in the p-i-n diode in the forward and reverse bias conditions

In high-level injection where the minority carrier density injected in the i-layer is well above the majority carrier density levels, the voltage drop in the i-layer is (see Baliga)

$$\begin{aligned} v_i &\approx \frac{3kT}{q}\left(\frac{W}{2L_i}\right)^2 \quad \text{for} \quad \frac{W}{2} \le L_i \\ &\approx \frac{3\pi kT}{8q} e^{w/2L_i} \quad \text{for} \quad \frac{W}{2} \ge L_i \end{aligned} \tag{3-18}$$

where L_i is the diffusion length of the charge carriers in the i-layer in high-level injection. It must be noted that the voltage drop v_i is independent of the current density and, thus, the on-state current itself. The on-state losses are determined by the on-state voltage drop only.

Diodes with high reverse breakdown voltages, which are achieved by increasing the width of the i-layer, have significantly higher values of v_i. For medium reverse breakdown voltages (<1000 V), the v_i is less than 50 mV at room temperature (T=300 K), which is very small compared to the on-state voltage on the diode v_d. Ignoring v_i, the junction voltages on the two junctions are

$$v_{j1} = v_{j2} \approx v_d / 2 \tag{3-19}$$

Thus, the threshold voltage in the p-i-n diode is twice the threshold voltage of a single p-n junction. The threshold voltages in p-i-n diodes are in excess of 1.0 volt.

The forward current conduction is a function of the injection level of charge carriers in the i-region (see Baliga):

$$\begin{aligned} I &\propto e^{qv_d/2kT} \quad \textit{at extremely low current} \\ I &\propto e^{qv_d/kT} \quad \textit{at low level injection} \\ I &\propto e^{qv_d/2kT} \quad \textit{at high level injection} \end{aligned} \tag{3-20}$$

At very high current densities, the deviation from the exponential behavior occurs. The on-state voltage drop increases rapidly with the increase in the current density.

3.3.3 p-i-n diode in reverse bias

In the reverse bias, the excess carrier densities deplete on the junction boundaries. Widths of both junctions increase, extending almost entirely in the i-region. The reverse breakdown can occur in one of the two ways:

1. The electric field intensity crosses the breakdown limit (**E_{BR}=200 kV/cm**), the *avalanche breakdown,* or
2. the boundaries of the depletion layers of the two junctions meet, a phenomenon called the *punch through.*

In the first case, the breakdown voltage is inversely proportional to the donor impurity density, as obtained from assuming $N_A >> N_D$ in equation (3-15):

$$V_{BRAv} \approx \frac{\epsilon E^2_{BR}}{2qN_D} \qquad (3\text{-}21)$$

N_D is the donor density of the i-region. High breakdown voltage is achieved by low donor density.

In the second case, the breakdown voltage **V_{BR}** is obtained from equation (3-16) by assuming $N_A >> N_D$ and substituting l_n = W/2 where W is the width of the i-region :

$$V_{BRPt} \approx \frac{qN_D l_n^2}{2\epsilon} \approx \frac{qN_D W^2}{8\epsilon} \qquad (3\text{-}22)$$

The other half width W/2 contains the extension of the i-n^+ junction. The punch through voltage V_{BRPt} is proportional to the square of the half width of the i-region. Hence, a wide n^--region or i-region is the characteristic of the high voltage power diodes. However, a large width of the i-region increases the transition time, hence, high voltage characteristics are obtained at the cost of the speed. The on-state voltage drop v_d, on the contrary, can be reduced by decreasing the width of the i-layer, see equation (3-18). Thus, there is a trade-off between the high V_{BR} and the on-state voltage drop v_d .

3.3.4 Reverse leakage current

The leakage current of the p-i-n diode consists of three components, see Figure 3.9:

$$I_r = I_{rP^+} + I_{ri} + I_{rN^+} \qquad (3\text{-}23)$$

where $I_{rP}{}^+$: the diffusion leakage current in the p^+-region
$I_{rP}{}^+$: the space charge leakage current in the i-region
$I_{rP}{}^+$: the diffusion leakage current in the n^+-region

The first and the last components are due to the generation of electron-hole pairs by the thermal energy within the p^+-region and n^+-region respectively. The middle component is the result of the generation of electron-hole pairs by the thermal energy within the i-region, which is totally deplete with free charge carriers. This leakage current is given by

$$I_{ri} = \frac{qWn_iA}{\tau} \quad where \quad W = \sqrt{\frac{2\epsilon V_R}{qN_D}} \qquad (3\text{-}24)$$

W is the width and n_i is the intrinsic electron density in the i-region. The τ is the carrier lifetime in the i-region. V_R is the reverse bias, which is almost entirely across the i-region. The leakage current increases with increasing reverse bias voltage until the depletion layer no longer expands due to the punch through. Owing to the strong dependence of n_i on the temperature, the space charge leakage current is also a strong function of the temperature.

The diffusion leakage current $I_{rP}{}^+$ is the consequence of the generation of the minority carriers by the thermal energy within the regions close to the boundary of the depletion region (comprising of the

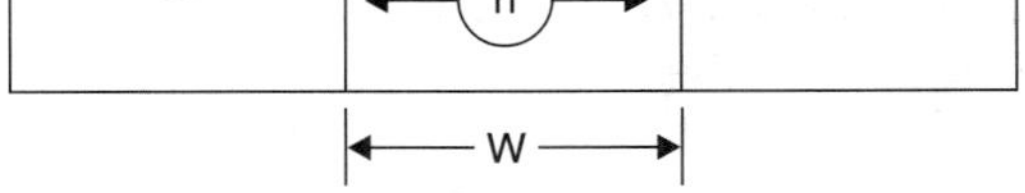

Figure 3.9
Leakage current flow in a reverse biased p-i-n diode

full i-region) on the p^+-side. These carriers diffuse to the depletion region and are then swept away by the electric field through it to the n^+-side. Similarly, the diffusion leakage current $I_{rN}{}^+$ is the consequence of the generation of the carriers by the thermal energy within the regions close to the boundary of the depletion region on the n^+-side. These carriers diffuse to the depletion region and are then swept away by the electric field through it to the p^+-side. These currents are given by

$$I_{rP^+} = \frac{qD_n n_i^2 A}{L_n N_A} \quad and \quad I_{rn^+} = \frac{qD_p n_i^2 A}{L_p N_D} \qquad (3\text{-}25)$$

Owing to the strong dependence of n_i on the temperature, the diffusion components of the leakage current are a stronger function of the temperature. The middle component is dominant below 100 degrees centigrade and the other two are dominant over it in the silicon devices.

3.3.5 Switching characteristics of p-i-n diode

Turning on a power diode involves bringing a large amount of charge in the i-layer. Turning off a diode involves removal of all the accumulated charge in the i-layer. The rates of accumulation and removal of charge are often determined by external factors such as the inductors in series or the capacitors in parallel. In addition the excessive charge has to flow through a crowded structure, hence the transition times are not negligible even in the face of an external step change in the voltage or current. As a simple approximation, assume the rate of rise or fall of the diode current to be linear.

Turn-On Transient

Assume that the diode is in the off-state for a long time. The minority carrier densities at the edges of the i-layer are well below the thermal equilibrium densities. A positive voltage $\mathbf{V_d}$ is applied on the diode terminals. The diode current begins to increase linearly (assumed). The turn-on time consists of two parts: $\mathbf{t_{on1}}$ and $\mathbf{t_{on2}}$, see Figure 3.10. During $\mathbf{t_{on1}}$, the forward current begins to build up linearly. The diode voltage increases linearly from the negative $\mathbf{V_R}$ level. During this process, the external voltage on the diode may cross zero and turns positive, but the voltage across each of the two junctions is still below the zero bias level. At time $\mathbf{t_{on1}}$, the junction voltage reaches the zero bias level and so do the charge carrier densities. The diode then crosses into the forward bias region. In the next phase, the charge begins to accumulate in the i-region, so the resistance drops bringing down the voltage drop even though the current is still building up. At time $\mathbf{t_{on2}}$, the current buildup is complete, the conductivity reaches the level determined by the amount of the forward current, and the voltage drop settles to a low value. The turn-on of the diode is complete.

Turn-Off Transient

Assume that the diode is in the on-state for a long time. The charge carrier densities of both types (electrons and holes) in the i-region are far above the thermal equilibrium densities. A negative voltage $\mathbf{V_R}$ is applied on the diode terminals. The diode current begins to decrease linearly (assumed). The turn-off time consists of three parts: $\mathbf{t_{off1}}$, $\mathbf{t_{off2}}$ and $\mathbf{t_{off3}}$. During $\mathbf{t_{off1}}$, as the diode current decreases linearly (an assumption for analysis purposes), the on-state voltage across the diodes decreases slightly. The current flows as long as the charge carriers on both sides of the junction are above the thermal equilibrium levels. Until the charge carriers are swept to below the thermal equilibrium levels, the two junction voltages remain above the threshold level. The current continues to fall and even becomes negative. The overall voltage on the diode terminals also becomes negative. The current crosses the zero level at $\mathbf{t_{off1}}$. From then on the current becomes negative and continues to fall further. At $\mathbf{t_{off1}}$, the junctions becomes reverse biased. The charge carriers at the junctions continue to dwindle, and the charge carrier densities are no longer able to sustain the high negative current amplitudes, hence the current turns around. It takes until $\mathbf{t_{off2}}$ to bring the charge carrier densities to reduce to the thermal equilibrium level and the current to settle to the

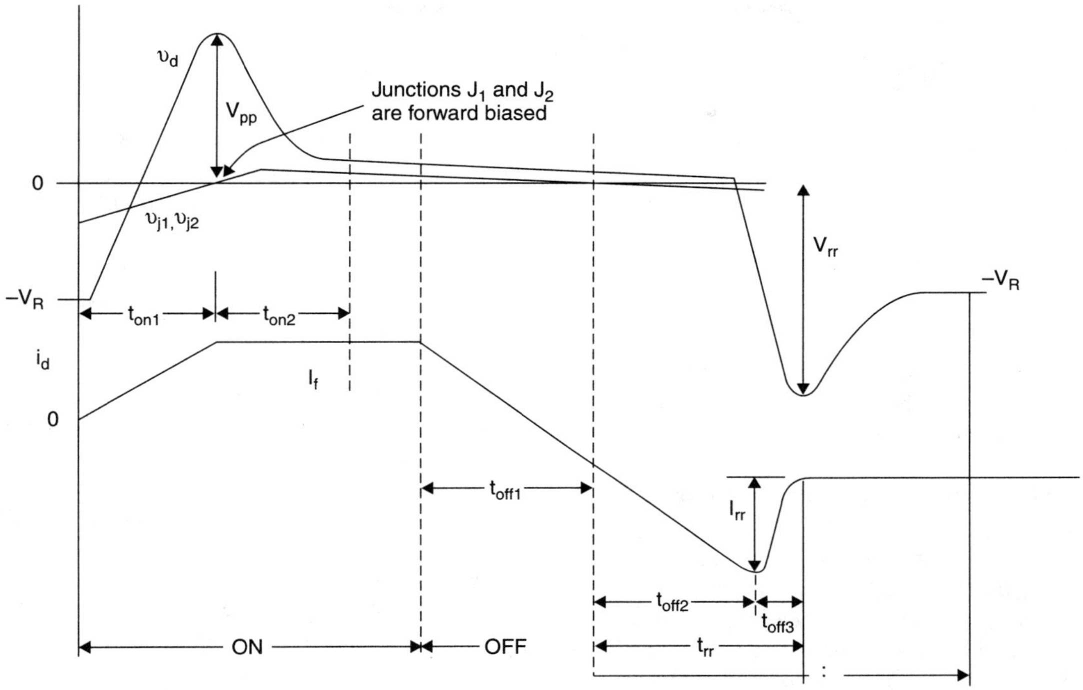

Figure 3.10
Turn-on and turn-off transients in the p-i-n diode

level determined by the reverse voltage applied. The sum $\mathbf{t_{off2}+t_{off3}}$ is called the reverse recovery time $\mathbf{t_{rr}}$. The time $\mathbf{t_{off3}}$ is generally smaller than $\mathbf{t_{off2}}$, hence $\mathbf{t_{rr}} \approx \mathbf{t_{off2}}$.

During the steady-state forward bias, the stored charge in the i-layer is, given by $Q_f = \tau I_f$, the product of the forward current I_f and the mean lifetime τ of the excess carriers in the i-layer. The stored charge $\mathbf{Q_f}$ does not change appreciably until the current direction reverses. Hence nearly the total stored charge is removed during $\mathbf{t_{rr}}$. Therefore,

$$Q_f = \int i_f \, dt \approx \frac{I_{rr} t_{rr}}{2} \tag{3-26}$$

$$t_{rr} = \frac{2Q_f}{I_{rr}} = \frac{2\tau I_f}{I_{rr}} \tag{3-27}$$

Furthermore, the peak reverse current $\mathbf{I_{rr}}$ is given by

$$I_{rr} \approx t_{rr} \frac{di_r}{dt} \tag{3-28}$$

Hence by combining eqs. (3-27) and (3-28),

$$I_{rr} = 2\frac{\tau I_f}{I_{rr}} \frac{di_r}{dt}$$

$$I_{rr}^2 = 2\tau I_f \frac{di_r}{dt}$$

$$I_{rr} \approx \sqrt{2\tau I_f (di_r/dt)} \tag{3-29}$$

where $$\tau = \frac{q}{kT} \frac{W^2}{\mu_p + \mu_n}; \qquad W = \frac{V_{BR}}{E_{BR}} \tag{3-30}$$

The peak reverse current is directly proportional to the forward current and inversely proportional to the rate of current removal. High breakdown voltage diodes have high values of τ, hence high values of $\mathbf{I_{rr}}$.

The reverse recovery time can be further derived as

$$t_{rr} = 2\tau \frac{I_f}{I_{rr}} = \sqrt{\frac{2\tau I_f}{di_r/dt}} \qquad (3\text{-}31)$$

Substituting τ from the eq. (3-30),

$$t_{rr} = AV_{BR}\sqrt{\frac{I_f}{di_r/dt}};$$

$$\text{where } A = \frac{1}{E_{BR}}\sqrt{\frac{q}{kT}\frac{2}{\mu_p + \mu_n}} \qquad (3\text{-}32)$$

The above equation further emphasizes the trade-off between the high breakdown voltage and the fast switching speed.

Example 3-4 Reverse Recovery in Power Diode

The reverse recovery time of a silicon power diode International Rectifier R23AF is $t_{rr} = 1\ \mu S$. The diode carries a forward current of 20 A for sufficient time. Find the peak reverse current if the breakdown voltage rating of the diode is 800 V.

The τ is obtained from equation (3-30)

$$\tau = \frac{q}{kT}\frac{(V_{BR}/E_{BR})^2}{\mu_n + \mu_p}$$

The peak reverse current is calculated from equation (3-31)

$$I_{rr} = \frac{2\tau}{t_{rr}} I_f$$

The MATLAB solution is

```
>>mun=1360; mup=490; k=1.6e−23; T=273+25;
  ni=1e10; q=1.6e−19; EBR=2e5;
>>VBR=800; Ifor=20; trr=1e−6;
>>tau=(q/(k*T))*(VBR/EBR)^ 2*(1/(mun+mup))
  % time constant τ
>>Irr=(2*tau/trr)*Ifor % Peak reverse current
```

tau=2.9022 e-7 sec. Irr=11.6089 A

3.4 PARALLEL AND SERIES CONNECTION OF DIODES

It is often required to add commercially available rectifier diodes in parallel to increase the forward current capacity of the combination. Similarly, the power rectifiers may be connected in series to bolster the reverse blocking voltage capacity.

3.4.1 Series connection of diodes

Consider the series connection of diodes in Figure 3.11(a). Due to the manufacturing tolerances, the i-v characteristics of the two diodes may differ slightly, as shown in Figure 3.11(b). For the same reverse current, the two diodes do not share the reverse bias voltage equally. The remedy is to connect an external resistance in parallel with each diode that carries a bleeding current, an order of magnitude higher than the diode current. The resistances in parallel should be of equal value if the sharing of the reverse voltage has to be equal.

Resistances alone may not equalize the sharing of the voltage transients. Hence, an additional branch is connected in parallel with each diode that contains a resistance in series with a capacitance, see the dashed connections in Figure 3.11(b).

Example 3-5 Series Connection of Power Diodes

Two diodes International Rectifier R23AF, 800 V are connected in series to share 1200 Volts dc. The reverse currents are 25 and 30 mA. Find the shunt resistance that would enable equal voltage sharing within (a) 5% and (b) 2%.

Given: $I_{r1} = 20$ mA, $I_{r2} = 25$ mA; $V_r = 1200$. From the circuit in Figure 3.11, the series current through the diode and resistor combine is,

$$I_r = I_{r1} + \frac{V_{r1}}{R} = I_{r2} + \frac{V_{r2}}{R} \quad \Rightarrow \quad R = \frac{V_{r1} - V_{r2}}{I_{r2} - I_{r1}}$$

where $V_r = V_{r1} + V_{r2}$.

Let the percentage difference in the two voltages be Vdiff = 0.05

Let us also assume for the design, V_{r2} = (1-Vdiff) V_{r1}.

Hence, $V_{r1} = V_r/(2\text{-Vdiff})$.

Implement above design using MATLAB .

```
>>Ir1=20e-3; Ir2=25e-3; Vr=1200; Vdiff=0.05;
>>Vr1=Vr/(2−Vdiff)
>>Vr2=Vr−Vr1
>>R=(Vr1−Vr2)/(Ir2−Ir1)
```

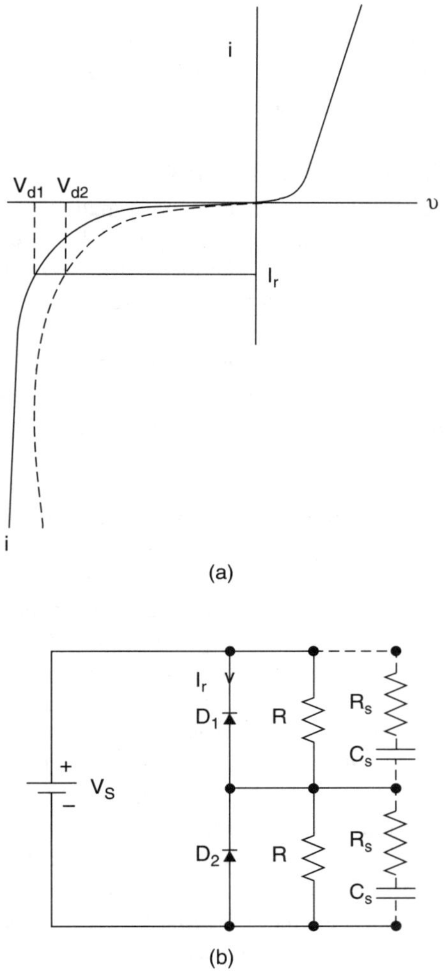

Figure 3.11
(a) Reverse i-v characteristics of diodes with manufacturing tolerances and (b) Voltage sharing by resistances R and transient sharing by R_S and C_S

(a) <u>5%</u> **Vr1=615.3846 V Vr2=584.6154 V**
R=6153.8 ohms
(b) <u>2%</u> **Vr1=606.0606 V Vr2=593.9394 V**
R=2424.2 ohms

Note that close matching requires lower shunt resistance and consequently more losses in the off-state. Select the resistance of the standard value that is slightly lower than the answer.

3.4.2 Parallel connection of diodes

Consider the parallel connection of diodes in the Figure 3.12(a). Due to manufacturing tolerances, the i-v characteristics of the two diodes in the forward bias region may differ slightly, as shown in Figure 3.12(b). For the same on-state voltage, the two diodes do not share the forward current equally. The remedy is connecting an external resistance in series with each diode that is an order of magnitude higher than the on-state resistance of the diode. The series resistances should be of equal value if the equal current sharing is desired. Inductors may have to be added in series with each resistance for equal sharing of current transients, see the dashed connections in Figure 3.12(b).

Example 3-6 Parallel Connection of Power Diodes

Two diodes International Rectifier R23AF, 800 V are connected in parallel to share 1000 Amps dc. The on-state resistances of the two diodes are 0.60 and 0.65 mohms. Find the series resistance (ignore inductances) that would enable equal current sharing within (a) 2% and (b) 1%.

Given: $r_{on1} = 0.60$ mohms, $r_{on2} = 0.65$ mohms, $I_F = 1000$ A.

From the circuit in Figure 3.12, the voltage drop on the diode and resistor combined is,

$$v_F = I_{F1}(r_{on1} + R) = (I_{F2}(r_{on2} + R) \Rightarrow$$
$$R = \frac{I_{F2}r_{on2} - I_{F1}r_{on1}}{I_{F1} - I_{F2}}$$

where $I_F = I_{F1} + I_{F2}$.

The percentage difference in the two currents is Idiff = 0.02

Let us also assume for the design, I_{F2} = (1-Idiff) I_{F1}.

Hence, $I_{F1} = I_F/(2\text{-Idiff})$.

Implement above design using MATLAB .

```
>>ron1=0.6e-3; ron2=0.65e-3; IFT=1000; Idiff=0.02;
>>IF1=IFT/(2-Idiff)
>>IF2=IFT-IF1
>>R=(IF2*ron2-IF1*ron1)/(IF1-IF2)
```

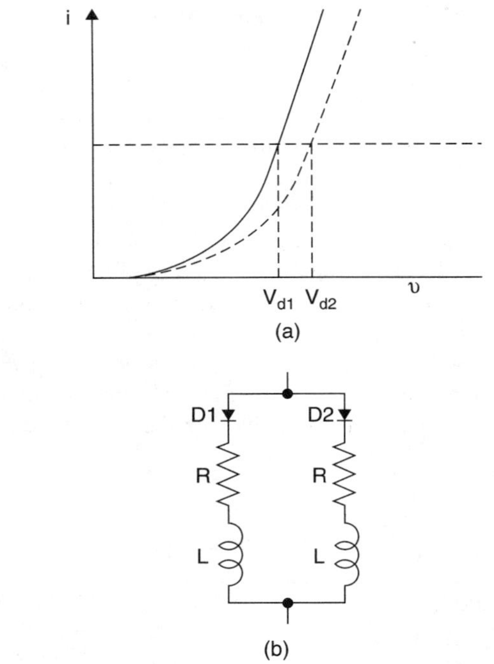

Figure 3.12
(a) Reverse i-v characteristics of diodes in the on-state with manufacturing tolerances and
(b) Current sharing by resistance R and transient sharing by R and L

(a) **2%** **IF1=505.0505 A** **IF2=494.9495 A**
R=0.0018 ohms
(b) **1%** **IF1=502.5126 A** **IF2=497.4874 A**
R=0.0044 ohms

Note that a higher series resistance increases the on-state losses. Select a resistance of the standard available value that is slightly higher than the answer.

3.5 SCHOTTKY BARRIER DIODE

The Schottky Barrier Diode (SBD) is a metal-semiconductor structure. A metal to n-semiconductor junction is preferred over a metal to p-semiconductor junction due to the higher mobility of electrons. The metal to semiconductor junction displays significantly different characteristics for different doping densities of the semiconductor. There are two types of metal to semiconductor junctions: (a) Schottky barrier junction and (b) ohmic junction. Before discussing a practical Schottky Barrier Diode, we will study both types of junctions.

3.5.1 Schottky barrier

A metal to n-semiconductor junction with a Schottky barrier is shown in Figure 3.13. The number of free electrons in the metal is far greater than in the n-semiconductor. The electrons in the metal have lower potential energy than their counterparts on the semiconductor region. Electrons in the n region are fewer but higher in potential energy. Electrons from the n region cross to the metal side leaving behind positively charged ions (space charge) on the n-side of the junction. The metal side acquires the negative charge and the n-side acquires the positive charge. Buildup of these charges on the two sides opposes the transfer of electrons across the metallurgical junction. The thermal equilibrium is established. The energy diagram of the metal-n junction in thermal equilibrium is shown in Figure 3.13.

A depletion layer is formed just like in a p-n junction, except the p-side is replaced by the metal. Most of the depletion layer is in the n-region. The space charge density is highest on the edge of the junction and exponentially decays away from the junction. Electrons in the metal face a steep energy barrier, called the Schottky barrier phi_b. Electrons in the bulk of the n-side face a barrier potential that is given by V_b where

$$qV_b = \phi_b - (E_n - E_m) \tag{3-33}$$

where E_n and E_m are the energies of the electrons in the n-region and the metal respectively.

The width of the zero bias depletion region is

$$W_0 = \sqrt{\frac{2\epsilon V_b}{qN_D}} \tag{3-34}$$

ϵ is the dielectric constant of the semiconductor material. N_D is the donor density in the n-region.

When the metal side is made positive with respect to the n-side, the potential barrier is lowered. The depletion layer shrinks. More electrons from

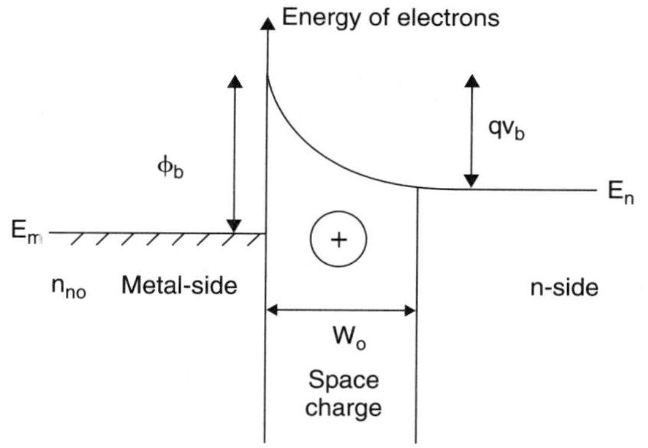

Figure 3.13
Energy diagram of a metal-n junction in thermal equilibrium

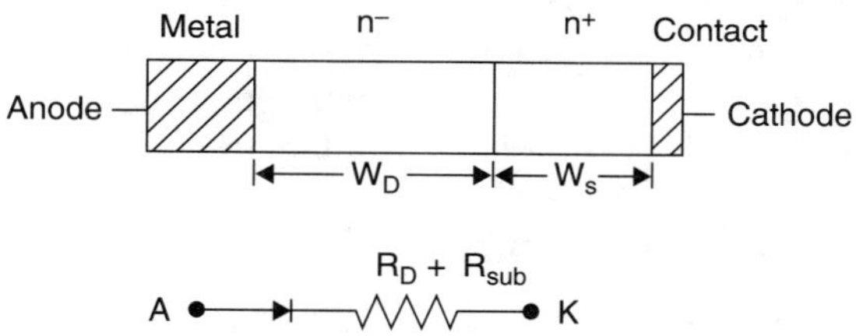

Figure 3.14
A Schottky Barrier Diode

the n-side cross over to the metal side. Note that electrons are the majority carrier on the n-side. The injected electrons on the metal side do not accumulate on the junction surface but instead immediately become the part of the free electrons in the metal. Hence the forward conduction of the Schottky Barrier Diode (SBD) involves majority carriers only. This is opposite of the p-i-n diode. The SBD is called the majority carrier device and the p-i-n diode is the minority carrier device.

In the reverse bias, width of the depletion region increases, almost entirely in the n-side. Like the p-n junction devices, the reverse blocking voltage of SBD are made higher by reducing the dopant density in the n-region. In practical SBD for power applications, the depletion region is made of a lightly doped n- material. Figure 3.14(a) shows such a **metal-n$^-$-n$^+$** SBD. The equivalent circuit of the SBD is shown in Figure 3.14(b). R_D and R_{sub} are the resistances of the n^- region and the n^+ substrate. The i-v characteristics are given by

$$I_f = I_S e^{qv_{fb}/kT} \quad : \quad v_{fb} = v_f - (R_D + R_{sub})I_f$$

$$: \quad I_S = R\,T^2 e^{-q\phi_b/kT} \qquad (3\text{-}35)$$

I_S is the reverse saturation current. R is the effective Richardson constant and T is the absolute temperature. The strong dependence of I_S on temperature and the barrier height is evident. The I_S in the silicon SBD is typically five orders of magnitude larger than that in the p-i-n diode.

On-state voltage is given by

$$v_f = \phi_b + \frac{kT}{q}\ln\left(\frac{I_f}{ART^2}\right) \qquad (3\text{-}36)$$

On-state voltage in the silicon SBD is also typically 0.3 volt less than that in the p-i-n diode. The on-state voltage decreases nearly linearly with increasing temperature.

The reverse blocking voltage can be increased by increasing the receptivity and the thickness of the drift region:

$$R_D = \frac{WA}{q\mu_n N_D} \propto (V_{BR})^{2.5} \qquad (3\text{-}37)$$

W is the width of the drift region in the breakdown situation. A is the area of the junction. The on-state voltage increases rapidly for high V_{BR} ratings of the SBD rectifiers. Thus, there is a trade-off between the on-state voltage (and hence losses) and the breakdown voltage.

Another trade-off is to be made between the leakage current I_S and the on-state voltage v_f. The on-state voltage can be decreased by selecting a metal that has a smaller Schottky barrier height ϕ_b but only at the cost of a sharp increase in the leakage current. The maximum operating temperature of the device also decreases. Furthermore, large leakage current in silicon SBD limits the maximum reverse blocking voltages to less than 100 volts.

Since the minority carriers are not involved in the forward bias condition, there is no stored charge in the n region. Consequently, the reverse recovery problems and the voltage over-shoot in the turn-on do not exist.

3.5.2 Ohmic junction

Unlike the previous type, the ohmic junction does not display rectifying properties. It shows purely resistive properties and, therefore, could be used as a port for an external electrical connection to semiconductor material in the semiconductor devices. The ohmic junction is formed by metallurgically joining a metal such as aluminum with a heavily doped n-semiconductor. Due to large dopant density N_D, as per equation (3-34), the depletion layer is very thin. Due to accumulation of large number of electrons in the thin depletion layer, the conductivity of the drift region enhances greatly. The large change in the conductivity undermines the Schottky barrier properties. The junction displays very low resistance, hence termed an ohmic contact.

3.5.3 High voltage Schottky barrier diode

The silicon SBD suffer from high leakage current that limits the breakdown voltages less than 100 volts. In addition, the high series resistance of the drift region at higher voltages limits the maximum value of the on-state current. The on-state resistance, also called the on-state specific resistance of the diode, is related to the breakdown voltage by the Baliga's figure of merit:

$$R_{on,\,sp} = \frac{4V_{BR}^2}{\epsilon\mu_n E_c^3} \tag{3-38}$$

where E_c is the critical electric field in the drift region in the breakdown situation. The specific on-resistance decreases with higher electron mobility and higher critical electric field strength. Another semiconductor material, GaAs, has the mobility 5.6 times higher and a higher electric field than silicon. These two factors result in the reduction of about 13 times in the on-specific resistance and a significant reduction in the on-state voltage. GaAs SBDs have been fabricated with V_{BR} up to 200 volts.

Silicon carbide, SiC, is a semiconductor material (6H-SiC in particular) of great promise for the fabrication of the SBDs because of much greater critical electrical field E_c. The on-specific resistance is about 200 times smaller in 6H-SiC than in silicon. The SiC SBDs can be fabricated up to V_{BR} of 5000 volts.

3.6 NEW TRENDS IN DIODES

The **junction barrier controlled Schottky diode (JBS)** is a Schottky device with a p-n junction integrated into the drift region. A p-n junction is used to reduce the leakage current by preventing Schottky barrier lowering during reverse bias. The leakage current is suppressed to a small value. Thus the Schottky barrier in JBS diodes can be made significantly smaller than for the conventional SBDs. This permits reduction in the on-state voltage even at high reverse blocking voltages. The on-state voltage is nearly reduced to half of that in conventional SBDs. JBS diodes with low on-state voltage are particularly very useful in VLSI technology where the power supply voltages are being reduced from 5 volts to 3 and lower volts in order to achieve high packaging density.

The **merged p-i-n/Schottky diode (MPS)** is another promising device in the high voltage SBD category. The drift region is lightly doped to increase the breakdown voltage. The p-n$^-$ junction in MPS, therefore, becomes forward biased in the on-state. Holes are injected into the drift region. Conductivity modulation takes place in a manner similar to the p-i-n diode. This allows large current flow via the Schottky region with a low resistance. However, the injection level that is needed to reduce the drift region resistance is not as large as in the p-i-n diodes. Hence the stored charge in MPS is much smaller than in p-i-n rectifiers. Thus MPS diodes have on-state characteristics similar to SBDs at low currents and like p-i-n diodes at high currents. See Figure 3.15 for a comparison of the on-state characteristics of MPS, conventional SBDs and p-i-n diodes.

3.7 DEVICE SPECIFICATION

In this section we give the important specifications that most of the manufacturers list in data sheets.

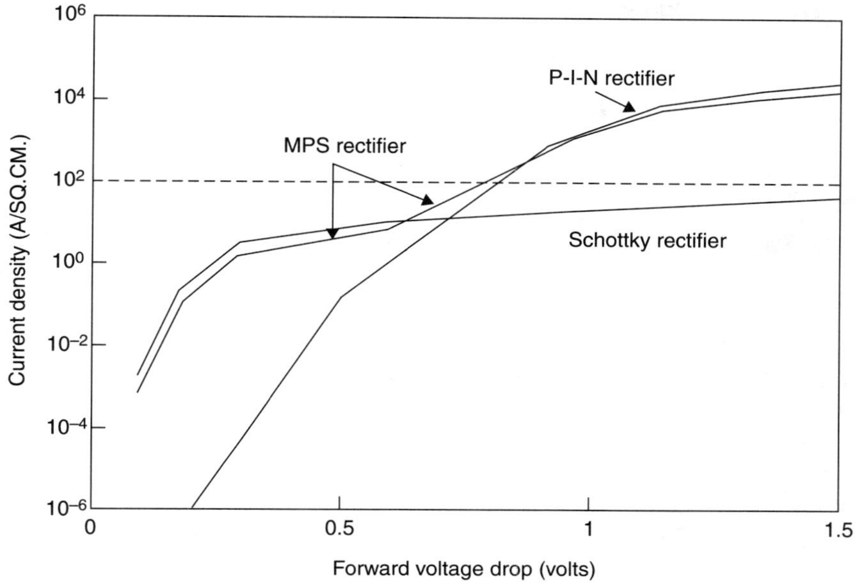

Figure 3.15
The on-state characteristics of MPS, conventional SBDs and p-i-n diodes

Forward characteristics:

Average Rectified Forward Current	$I_{F(AV)}$	This is the maximum average dc current that the diode can conduct in the forward bias.
Repetitive Peak Surge Current	I_{FSM}	This value is also used as the peak pulse current at a given pulse width.
Nonrepetitive Peak Surge Current	I_{FSM}	This is the one-time maximum surge current the diode can survive. It must be used as the worst-case transient value.
Forward Voltage	V_F:	The forward voltage drop that increases with the increase in the amount of the forward dc current. The maximum value may be used for the calculation of the on-state power loss.

Reverse Characteristics:

Peak Repetitive Reverse VoltageReverse Current	V_{RRM}	This is the maximum peak reverse voltage of a periodic waveform that the diode can withstand.
	I_R	The reverse current increases with the rise in the junction temperature T_J.

Switching Characteristics:

	t_A	Time to reach peak reverse current at a given di_F/dt.
	t_B	Time from peak reverse current I_{RM} to zero value.
Reverse Recovery Time	t_{rr}	Reverse recovery time at a given di_F/dt, sum of t_A+t_B.

Power and Thermal Characteristics:

Maximum Power Dissipation	P_D	This is the maximum value of the product of the voltage across the diode and the average dc current in the forward biased condition.
Operating Junction Temperature	T_J	The junction temperature of the thyristor as a result of the ambient and load conditions.
Case Temperature	T_C	The temperature of the thyristor case as a result of the ambient and load conditions.
Thermal Resistance, Case-to-Ambient	R_{CA}	The thermal resistance in steady-state from the case to the ambient.
Thermal Resistance, Junction-to-Ambient	R_{JA}	The thermal resistance in steady-state from the semiconductor junction to the ambient.
Thermal Resistance, Junction-to-Case	R_{JC}	The thermal resistance in steady-state from the semiconductor junction to the case.

3.8 PSPICE SIMULATION

In this section we will simulate the switching transients in a low-power signal diode and a medium-power diode, see Figure 3.16. Set the following:

Pulse generator V1 : 5 V V2: −5 V Td: 20ns Tr: 0.1ns Tf: 0.1ns PW: 2us PER: 4us

The simulation settings are: Print Step: 1ns Final time: 50us

Note the waveform v_d, the voltage across the medium-power diode 1N4002. The diode conducts for some time even after the pulse generator voltage has become −5 Volts. The low-power signal diode 1N4148 has negligible time of conduction in reverse bias. The reverse conduction time of 1N4148 can be observed by applying a low period V1 waveform.

3.9 SUMMARY

This chapter has explored the physics of p-n junction in semiconductor devices and the characteristics of the p-i-n power diodes. The power Schottky Barrier Diodes were also discussed. Design considerations were elaborated on for the series and parallel connection of diodes for the enhancement of the diodes capabilities. The main points are as follows:

1. Electrons and holes both take part in the current conduction (charge transport) in the semiconductor devices.
2. The current constitutes both the drift and the diffusion components. The drift is caused by the electric field and the diffusion is the consequence of the concentration gradients in the physical structure of the device.
3. Doping of the intrinsic (pure) semiconductor with p and n type impurities results in vastly different charge carrier densities.
4. A p-n junction is formed by metallurgically joining the p-doped semiconductor with an n-doped semiconductor. This junction has rectifying properties. A similar junction can also

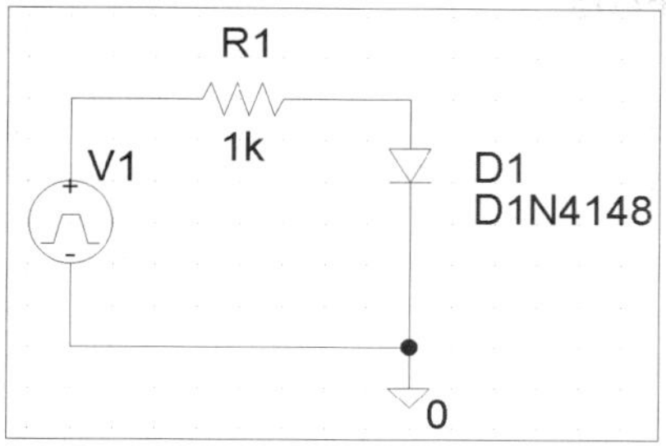

(a)

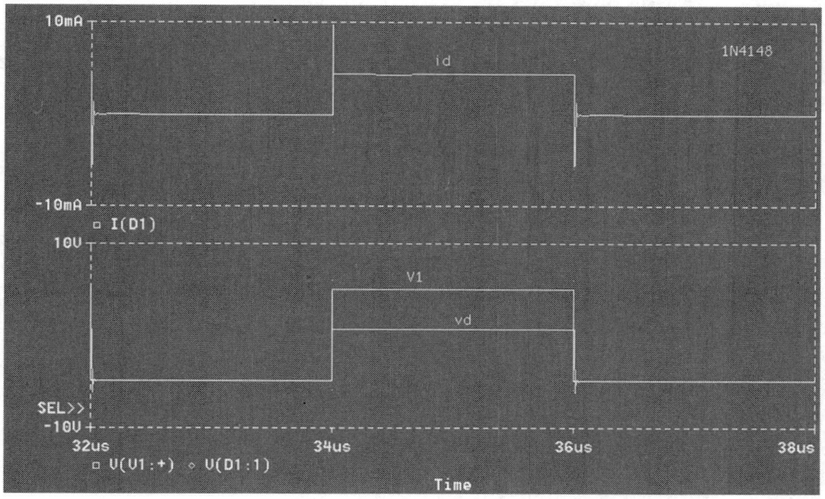

(b)

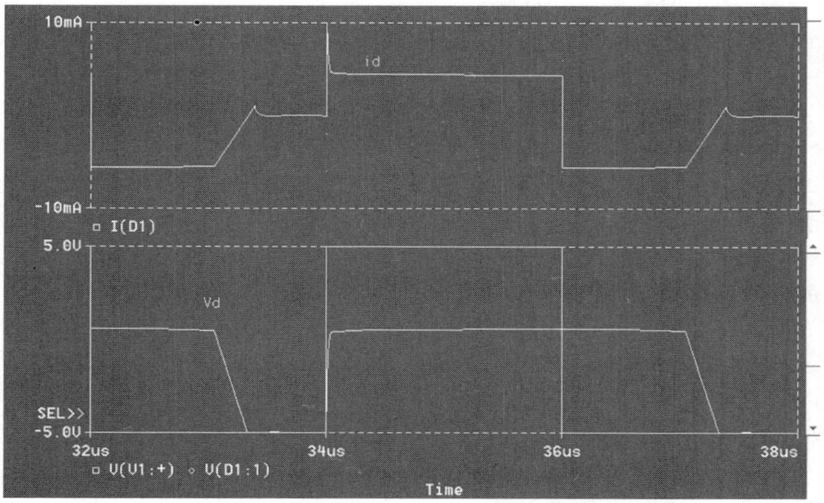

(c)

Figure 3.16
(a) Simulation circuit (b) switching waveforms in a low-power diode 1N4148 and (c) switching waveforms in a medium-power diode 1N4002

be formed with the metallurgical joining of one lightly doped semiconductor with a heavily doped semiconductor with only a single type of impurity. The metal to semiconductor junction, such as in a Schottky Barrier Diode, also has rectifying characteristics.

5. A potential barrier V_T is established across the p-n junction to balance the drift and diffusion components of the charge transport.

6. A depletion region forms on both sides of the p-n junction. The potential barrier V_T is across this region. It consists of immobile charged ions but no free charge carriers. The width of the depletion region on a side is inversely proportional to the dopant density on that side. The width on the metal side in a metal-semiconductor junction is negligibly small.

7. In the forward bias, the barrier potential is lowered, hence a large number of electrons and holes cross the junction resulting in a large current flow. The depletion width shrinks.

8. In the reverse bias, the potential barrier is enhanced, the depletion region widens and a small current flows. This current becomes saturated at small reverse voltages, called the reverse saturation current. The reverse saturation current nearly doubles for every 10-degree rise in temperature.

9. Avalanche breakdown occurs in the reverse bias condition when the maximum electric field at the junction exceeds a critical limit E_{BR}.

10. The punch through voltage breakdown occurs in the reverse bias condition when the depletion region spans the full physical width on any side of the junction.

11. High reverse voltage is supported by the i-region of the power diodes.

12. The breakdown voltage is inversely proportional to the dopant density and directly proportional to the width of the i-region in the p-i-n diodes.

13. The on-state voltage drop of p-i-n diodes is directly proportional to the width of the i-region. Thus, a high voltage diode would have high on-state voltage drop and consequently high on-state losses.

14. Conductivity modulation in the i-region increases the on-state forward current capability and minimizes the on-state losses.

15. The transit time is directly proportional to the width of the i-region, hence a large width diode (high breakdown voltage and high on-state losses) would have a large transit time and a slow switching speed.

16. Forward voltage of the diode may have a substantial overshoot in the turn-on transient.

17. Fast reverse recovery may cause large voltage spikes on the stray inductance.

18. The reverse recovery transient worsens at high reverse bias voltages.

19. The Schottky Barrier Diode has faster switching speeds but lower breakdown voltages than the p-i-n power diodes.

20. The Schottky Barrier Diodes have an insignificant reverse recovery problem.

21. The Schottky Barrier Diodes have lower on-state losses than p-i-n diodes.

References

1. Ned Mohan, Tore M. Undeland, and William P. Robbins, *Power Electronics: Converters, Applications and Design,* 2nd Ed., John Wiley & Sons, Inc., 1995.

2. B. Jayant Baliga, *Power Semiconductor Devices,* PWS Publishing Company, Boston, 1995.

3. John G. Kassakian, Martin F. Sclecht, and G. Vergassian, *Principles of Power Electronics,* Addison-Wesley, Boston, 1991.

Problems

3-1 Derive equations (3-14) and (3-18).

3-2 Two power diodes with following specifications are connected in parallel:
Diode 1: $v_d = 0.88 + 2.4 \times 10^{-4}\, i_d$
Diode 2: $v_d = 0.92 + 2.2 \times 10^{-4}\, i_d$
The current is equalized by placing a resistance R in series with each diode.
(a) Find the current in each diode without R for the total current of (i) 500 amps, (ii) 1000 amps and (iii) 1500 amps.
(b) Find R that would equalize current sharing to within 5% for a total current of 1000 amps.
(c) Find the sharing percentage at 500 and 1500 amps in part (b).

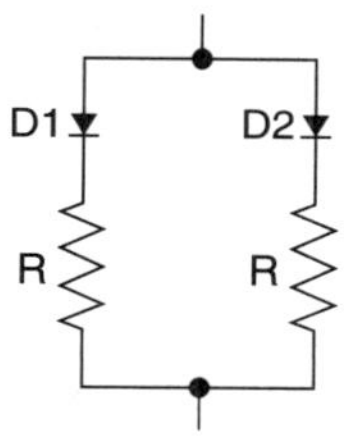

Figure P3.2

3-3 Calculate the leakage current of a Schottky barrier power rectifier at 125 °C, the reverse bias voltage of 1 V and 40 V. The rectifier has a breakdown voltage of 50 volts and the barrier height of 0.75 eV.

3-4 In the rectifier circuit of Figure P3-4, sketch the waveforms and determine the (a) repetitive peak current, (b) average current and (c) the rms current through the diodes. Assume identical power diodes.

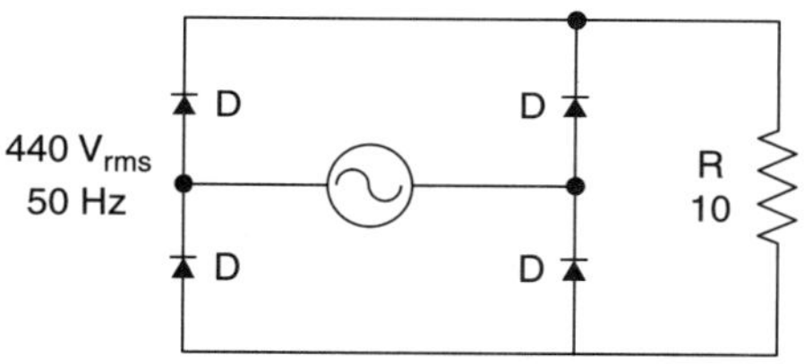

Figure P3.4

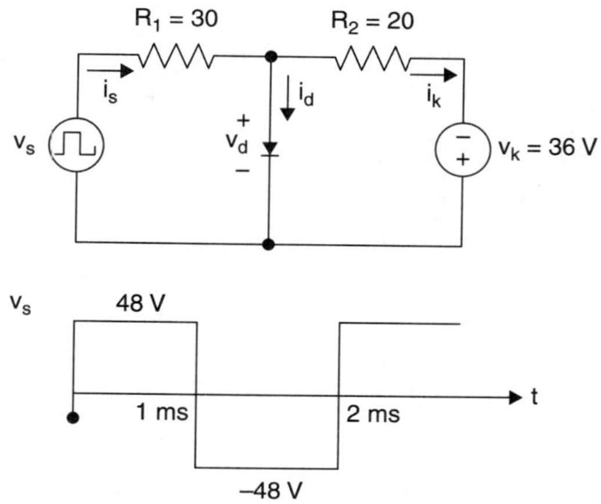

Figure P3.5

3-5 Plot the waveforms for the current in the power diode, the current from the pulse source and the current in the battery. The diode characteristics are given by
$v_d = 0.8 + 2.4 \times 10^{-4}\, i_d$

3-6 The two diodes in Problem 3-2 have the Peak reverse voltage rating of 1500 V and are connected in series to share 2000 Volts dc. The reverse currents are 35 and 45 mA. Find the shunt resistance that would enable equal voltage sharing within (a) 5% and (b) 2%.

3-7 The diodes in Figure P3-7 have reverse recovery time of $t_{rr} = 70\ \mu s$. Determine the effect of the reverse recovery time on the average dc output voltage for the supply frequencies
(a) 50 Hz
(b) 400 Hz and
(c) 1 kHz.

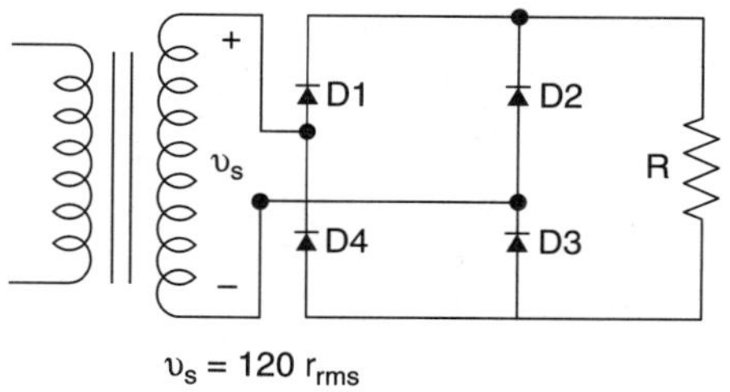

Figure P3.7

CHAPTER 4

Transistors: Level Triggered Switching Devices

OBJECTIVES:

1. The structure of a basic MOS cell in MOSFET
2. The physical structure, the equivalent circuit and the static characteristics of power MOSFET
3. The switching transients in power MOSFET
4. Parallel connection of MOSFETs
5. The structure of an IGBT cell
6. The physical structure, the equivalent circuit and the static characteristics of IGBT
7. The switching transients in IGBT
8. Complementary IGBT
9. Gate drive circuits for MOSFET and IGBT
10. Comparison of MOSFET and IGBT

4.1 INTRODUCTION

Bipolar power transistors have been known for a long time. You must have seen these devices in some other courses in the electrical engineering program of studies. Despite the attractive power ratings and reverse blocking voltage ratings, the transistors suffer from several problems: complex current driven circuits, second breakdown failure under the simultaneous application of high voltage and high current, thermal runaway problems and consequent problem in paralleling devices for current rating enhancement. Power BJTs are being fast replaced by metal-oxide semiconductor field-effect transistor MOSFET and insulated-gate bipolar transistor IGBT. Both types of devices have high input impedances, hence, require simpler voltage driven circuits. MOSFETs have completely taken over low-voltage, medium-power high-frequency switching applications. IGBTs are fast replacing BJTs in the high-voltage and high-power medium fast switching applications. MOSFETs do not have second breakdown problems as in BJTs and can be easily paralleled for current rating enhancement.

This chapter considers the basic physical mechanisms that govern operation, device characteristics, safe-operating area and other design considerations in MOSFETs and IGBTs. Power BJTs have not been discussed because the readers should have seen them in other courses on semiconductor devices.

4.2 BASIC MOS STRUCTURE

The Metal Oxide Semiconductor Field-Effect Transistor (MOSFET) is a normally off device in which the current is transported by majority carriers. A MOSFET, see the symbol in Figure 4.1, is faster than a BJT. However, in the on-state it behaves like a resistance, unlike a BJT, which exhibits a very low on-state voltage. A MOSFET has higher on-state losses than a BJT. Furthermore, the MOSFET has lower forward blocking voltage than a BJT (less than one hundred volts versus a few thousand volts).

Power MOSFETs are fabricated as a parallel combination of several identical cells as shown in Figure 4.2; each cell is a MOSFET device. Paralleling increases the current handling capability.

A basic cell may be of two types: n-channel and p-channel types. Figure 4.3 shows a basic n-channel cell.

The drain and source regions are made of heavily doped n^+ semiconductor that are diffused in a lightly doped p-substrate. The gate is a metal or polysilicon plate that is electrically isolated from the cell by an insulator SiO_2 layer.

There are two n-p junctions: one on the source side and the other on the drain side. This structure is like an npn bipolar transistor without external access to the base, both junctions reverse biased. Each junction has a depletion layer associated with it like the one seen in the previous chapter on diodes. In the unbiased condition, there is no voltage on the gate conductor. No current can flow from the drain to the source terminal. The MOSFET is in the *cut-off* region, the off-state of the MOSFET as a switch.

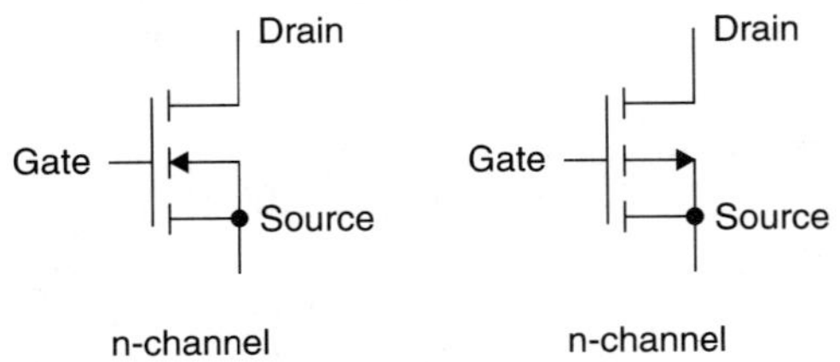

Figure 4.1
MOSFET symbols

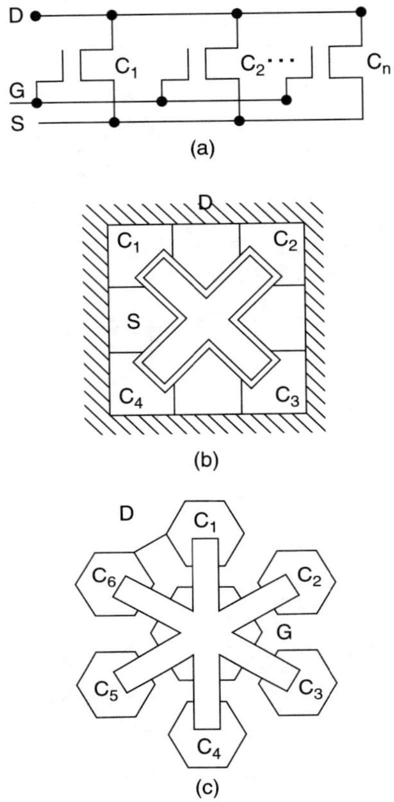

Figure 4.2
Power MOSFET as the parallel connection of several MOSFET cells

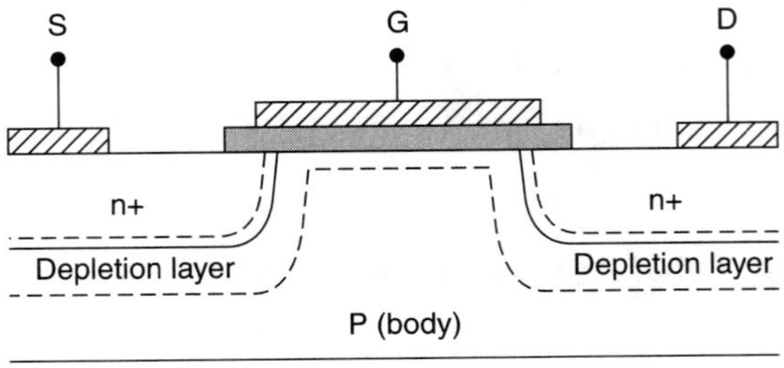

Figure 4.3
A basic enhancement mode n-channel MOS cell

When a small positive voltage is applied on the gate conductor with respect to the source or drain conductor, the positive charges accumulate on the conductor side of the SiO_2 layer (the insulator layer), see Figure 4.4(a). These positive charges create an electric field in the insulator layer and beyond in the p region. Holes in the p region drift away from the insulator-p-substrate interface leaving behind the negatively charged ions. Thus, a depletion layer is formed just beneath the insulator layer. The free electrons, however, drift towards the boundary of the insulator layer and the p-region. Being the p-region, the number of free electrons are much fewer. Current does not flow between the source and drain terminals. When the positive voltage on the gate is increased further, more and more free electrons from the bulk p-region drift to the region beneath the insulator layer. Beyond the threshold voltage $\mathbf{V_T}$, a significant number of electrons have accumulated. The region beneath the insulator layer now looks like a thin n-semiconductor layer. It is termed the inversion and the layer is called the inversion layer. This layer now acts as an n-channel between the two n^+ regions on the source and the drain sides. The thickness of the inversion layer increases with higher gate voltage.

Electrons can flow through the channel if a voltage is established across the n-channel between the source and drain regions externally. Electrons are the majority charge carriers in the source-channel-drain combination. The current transport is termed the *majority carrier transport,* which is an inherently faster transport mechanism than the minority carrier transport in bipolar devices.

Now consider that the drain is biased to be more positive than the source. A voltage gradient is established across the channel. The effective voltage difference between the channel and the gate varies as

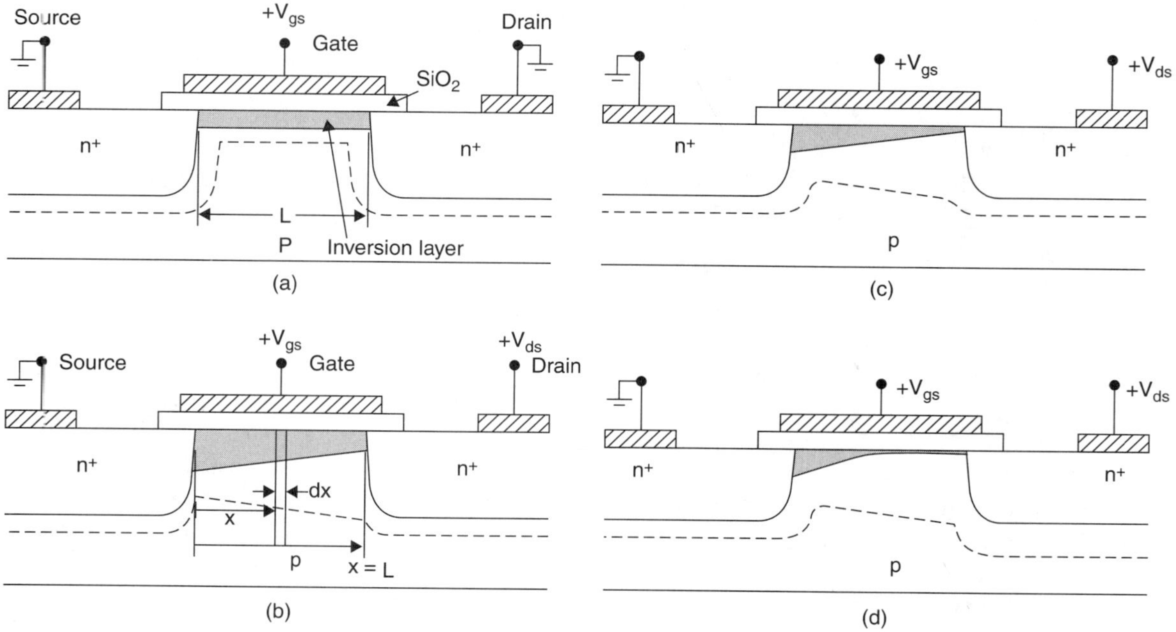

Figure 4.4
MOSFET channels in bias

$\mathbf{V_{gs}-V_{xs}}$, the $\mathbf{V_{gs}}$ and $\mathbf{V_{xs}}$ are the voltage from the gate to source and the voltage at the point x distance away from the source, all referenced to the source end, see Figure 4.4(b). Thickness of the inversion layer decreases as the point x moves away from the source end, being the minimum at the drain end.

This situation exists as long as the following condition is satisfied:

$$v_{gs} - v_{xs} \geq V_T \text{ where } v_{xs} = \frac{x}{L} v_{ds} \quad (4\text{-}1)$$

even at the drain end of the channel (x = L), hence,

$$v_{gs} \geq V_T + v_{ds} \text{ or } v_{ds} \leq v_{gs} - V_T \quad (4\text{-}2)$$

For a given v_{gs}, the channel size is fixed and so is the current from the drain to the source. The voltage drop in the thin dx segment of the channel at a distance x from the source end is

$$dv(x) = i_D \, dR_{ch}(x) \quad (4\text{-}3)$$

where the resistance $dR_{ch}(x)$ depends on the charge in the inversion layer by

$$dR_{ch} = \frac{dx}{Z \mu_{ns} Q_i(x)} \quad (4\text{-}4)$$

where Z is the width of the segment and the channel, μ_{ns} is the surface mobility of electrons and Q_i (x) is the inversion layer charge per unit area:

$$Q_i(x) = C_{ax}[v_{gs} - V_T - v(x)] \quad (4\text{-}5)$$

From these expressions,

$$\int_0^L i_D \, dx = -Z\mu_{ns} C_{ox} \int_0^{v_{ds}} (v_{gs} - V_T - v(x)) \, dv$$

$$i_D = \frac{\mu_{ns} C_{ox} Z}{L}[(v_{gs} - V_T)v_{ds} - v_{ds}^2] \quad (4\text{-}6)$$

where Z and L are the width and length of the channel, C_{ox} is the capacitance per unit area of the insulator layer and μ_{ns} is the surface mobility of electrons. For very low drain voltages,

$$i_D = \frac{\mu_{ns} C_{ox} Z}{L}(v_{gs} - V_T)v_{ds} \quad ; \quad v_{ds} << (v_{gs} - V_T) \quad (4\text{-}7)$$

The channel resistance in these conditions is

$$R_{ch} = \frac{v_{ds}}{i_D} = \frac{L}{Z \mu_{ns} C_{ox}(v_{gs} - V_T)} \quad (4\text{-}8)$$

In this situation, the n-channel behaves like a resistance. The MOSFET is said to operate in the *Ohmic region* or *linear region.* This is the on-state of the MOSFET when it is used as a switch. Power MOSFETs are designed with small length channels to achieve the low resistance during the on-state.

Transconductance of the MOS device in the linear region is obtained from differentiating equation (4-7):

$$g_m = \frac{di_D}{dv_{gs}} = \frac{\mu_{ns} C_{ox} Z}{L} v_{ds} \quad (4\text{-}9)$$

Since the inversion layer thins out towards the drain end, the current density increases and, therefore, the channel resistance increases near the drain end. This gives rise to an increasing voltage gradient **dv/dx,** and the electric field $\mathbf{E_x}$, in the channel towards the drain. Any further increase in the drain to source voltage is then given by

$$v_{ds} \geq v_{gs} - V_T \quad (4\text{-}10)$$

should make the inversion layer width and become zero near the drain end. However, high electric field sweeps away any electrons available in the vicinity, thus maintaining an orifice or a conduitlike structure even after the inversion layer thins out. The conduit is a minimum thickness inversion layer. The channel in this condition is said to have *pinched off,* see Figure 4.4(c).

As the drain voltage is increased further, the minimum thickness part of the channel advances towards the source end. The electric field at the drain end increases further, see Figure 4.4(d). The electron drift velocity linearly increases with the increase in the electric field, the electron mobility $\boldsymbol{\mu_e}$, up to a certain value only and saturates thereafter. Hence, the current becomes independent of any further increase in the drain voltage. MOSFET is then said to enter the *active region.* Current flow is then dependent only on the amount of the voltage on the gate with respect to the source. The drain current thus enters a saturation region with respect to the drain voltage.

$$i_D = \frac{\mu_{ns} C_{ox} Z}{2L}(v_{gs} - V_T)^2 \qquad (4\text{-}11)$$

Saturation current determines the maximum current that the channel can support at a given v_{gs}. Transconductance of the MOS device in this saturation region is obtained from differentiating eq. (4-12):

$$g_m = Z\,\mu_{ns} C_{ox}\frac{Z}{L}(v_{gs} - V_T) \qquad (4\text{-}12)$$

Figure 4.5 shows the complete i-v characteristics of the MOSFET cell.

A MOSFET could also be operated in the third quadrant, ($-V_{ds}$ and $-I_D$). When a negative voltage

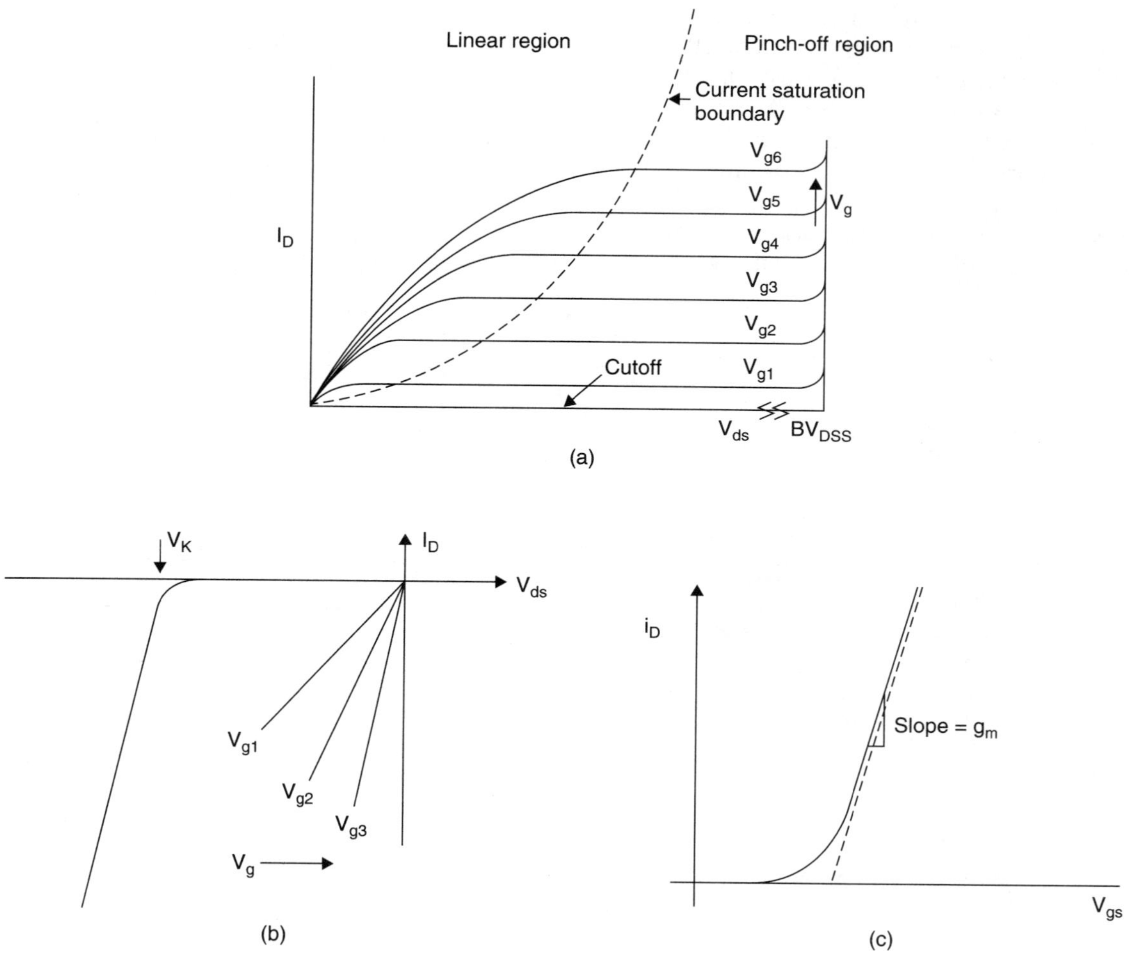

Figure 4.5
The i-v characteristics of the MOSFET cell in (a) in the first quadrant ($+V_{ds}$ and $+I_D$) and (b) i_D versus v_{gs} and (c) the third quadrant ($-V_{ds}$ and $-I_D$)

is applied with gate and source connected to the source, the p-n$^+$ junction is forward biased on the drain side. MOSFET behaves like a diode; that is, it conducts current from the source to the drain for $v_{SD} > 0.7$. This current flow is the result of injection of electrons (minority carriers) into the p-region from the n$^+$-drain. When a positive voltage is applied between the gate and the source, an n-channel is created between the drain and the source. This becomes an alternate path for the drain current, except that this current is as a result of the majority carrier transport through the channel. For this gate voltage, even for $v_{SD} << 0.7$, the device carries current. Minority carrier injection in the p-region at such low v_{DS} is very small. The entire current is due to the majority carrier transport through the channel. The device exhibits a very low resistance.

The device resistance decreases with the increase in the positive voltage at the gate, see Figure 4.5(c). MOSFET in this mode can be used as a replacement for diodes in low-voltage power supply applications. In these applications a positive gate signal must be provided in synchronism with the application of ac voltage at the drain to maintain it in the low forward drop on-state as well as in the blocking state. This device is called the *synchronous rectifier.* The synchronous rectifier retains the high speed switching characteristics of MOSFET.

4.3 POWER MOSFET

A power MOSFET is fabricated from several parallel connected cells organized in various geometrical designs. A single MOSFET cell is made up of four semiconductor layers: the n$^+$-n$^-$-p-n$^+$ structure as shown in Figure 4.6. The cell is grown by epitaxial diffusion process from an n$^+$ substrate.

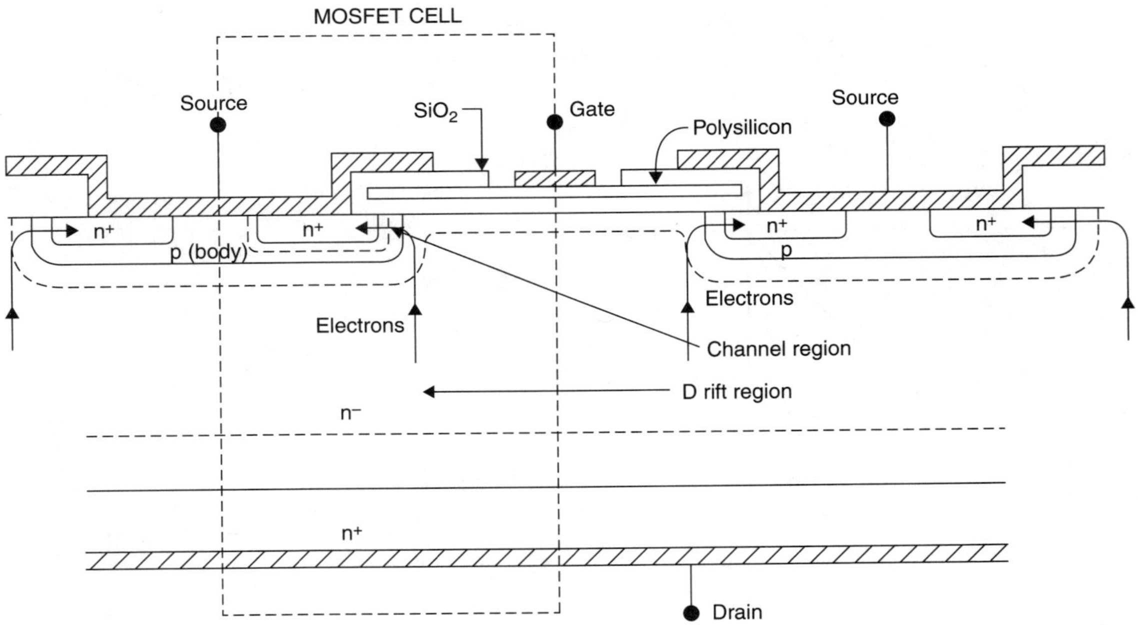

Figure 4.6
Power MOSFET

Typical dopant densities in various layers are given below:

n^+ layer:	10^{19} /cm^3
p layer:	10^{16} /cm^3
n^- layer:	$10^{14}-10^{15}$ /cm^3

The basic channel is in the p-region beneath the gate layer between the n^+ source and the n^- drain regions. The p-region is also called the *body region.* The bulk of the body region is electrically shorted to the source region, thus effectively shorting the n^+-p junction. The n^- layer, in between the n^+ region and p-region, serves as the *drift region* to accommodate the depletion layer width of the p-n^+ junction on the drain side. Like the power diodes, the breakdown voltage $\mathbf{V_{BR}}$ in MOSFET depends on the doping level and the width of n^- layer. MOSFET with high breakdown voltage has a low n^- dopant density and a wide-width n^- layer.

4.3.1 Body diode in power MOSFET

The MOSFET cell contains an integral pn^-n^+ (essentially p-i-n) diode from the p-base region to the drain terminal, see Figure 4.7(a). This is called the *body diode.* Figure 4.7(b) shows the body diode in the equivalent circuit between the drain and the source terminals. It allows reverse current to flow when the MOSFET is in the off-state. The p-base region is in contact with the source metal, hence, the p-i-n diode appears in anti-parallel with the MOSFET device.

Often the integral diodes are required to conduct current that matches with the power MOSFET rated current. The integral diode, however, exhibits *very slow reverse recovery* characteristics. The reverse recovery peak current I_p in the integral diode is generally high and increases with high di/dt at high switching speeds. Large I_P causes severe dissipation and stresses on other circuit elements involving a MOSFET.

4.3.2 Parasitic bipolar transistor

Besides the integral body diode, a parasitic BJT (n^-pn^+) also exists in the MOSFET structure, see Figure 4.7(a). The p-region serves as the base of this parasitic BJT. The BJT must be kept cut off at all times. This is done by shorting the base (p-body) to the emitter (n^+ source) by a common metal covering. The internal base of the transistor is connected to the source metal through a resistance R_B, the resistance of the lateral p-region. When the drain voltage is increased to near the avalanche breakdown voltage, current flows into the p-region of the BJT in addition to the normal MOSFET current through the n-channel. This avalanche current flows through the lateral p-region resistance R_B and produces a voltage drop that appears as the forward biasing voltage across the base to the emitter junction of the parasitic BJT. When this voltage increases beyond 0.7 volts, the parasitic BJT is no longer capable of supporting the p-base/n^- collector breakdown voltage BV_{DSS} $=BV_{CBO}$. This junction now can withstand only BV_{CEO}, which is about 60% of BV_{CBO}. This is referred to as the *second breakdown in MOSFET.*

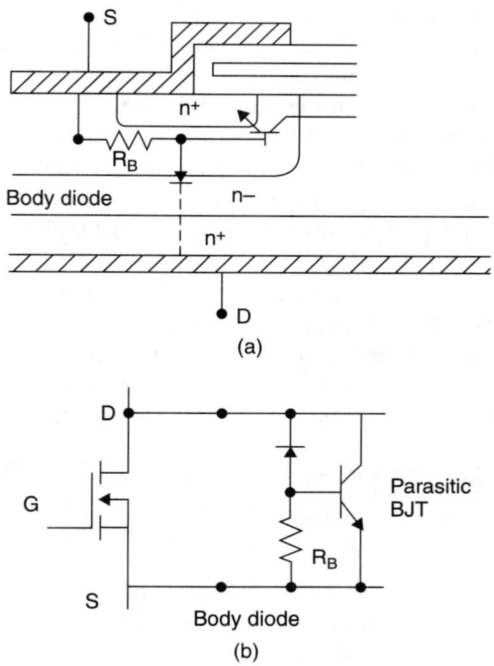

Figure 4.7
(a) Body diode and parasitic BJT and (b) The equivalent circuit of the power MOSFET

A parasitic capacitance C_{gd} exists between the internal base of the BJT and the drain. The high rate of rise of the drain voltage dv_{DS}/dt causes a voltage transient between the base and the emitter of the parasitic BJT. The BJT turns on and a latch-up condition develops. Thus, the parasitic BJT limits the maximum dv_{DS}/dt rating of the power MOSFET.

During the turn-off transient in the MOSFET, the reverse peak current through the body diode also flows through the resistance R_B. The voltage drop across R_B forward biases the base to the emitter junction of the parasitic BJT, resulting in further increase in the reverse peak current. In turn, the reverse recovery time of MOSFET also increases.

4.3.3 Equivalent circuit of MOSFET

The equivalent circuit of power MOSFET can be derived from the i-v characteristics.

Low Frequency Equivalent Circuit

The low frequency equivalent circuit model of the MOSFET in the on-state and active regions are given below in Figure 4.8.

Medium to High Frequency Equivalent Circuit

The parasitic capacitances in a MOSFET, shown in Figure 4.9(a), are:

- $\mathbf{C_{gd}}$: gate to the drain,
- $\mathbf{C_{gs}}$: gate to the source and
- $\mathbf{C_{ds}}$: drain to the source.

The gate to channel geometry is basically a capacitive structure. The gate to source capacitance is constituted of three components:

$$C_{gs} = C_{n+} + C_p + C_o \tag{4-13}$$

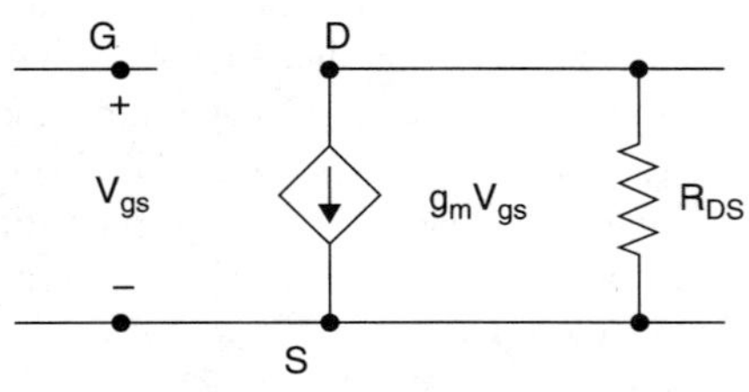

Figure 4.8
Low frequency equivalent circuit of MOSFET

Capacitances C_{n+}, C_p and C_o arise respectively from the overlap of the gate electrode with the n^+ source region, p-base region and the source metal. See Figure 4.9(b).

The gate to drain capacitance $\mathbf{C_{gd}}$ consists of two parts connected in series: the capacitance across the SiO_2 layer C_{ox} and across the depletion region C_S, as shown in Figure 4.9(b).

$$C_{gd} = \frac{C_{ox}C_S}{C_{ox} + C_S} \tag{4-14}$$

The oxide capacitance is given by

$$C_{ox} = \frac{\epsilon_{ox}}{t_{ox}} \tag{4-15}$$

ϵ_{ox} and t_{ox} are the dielectric constant and the thickness of the oxide layer respectively.

The depletion layer thickness and distribution varies very significantly with the gate to source voltage. Hence, the gate to drain capacitance C_{gd} varies with the gate to drain voltages as shown in Figure 4.9(c). It has a high value during the on-state and decreases with the increase in the drain to source voltage $\mathbf{v_{DS}}$. The dip in the curve corresponds to the maximum depletion in the width at the onset of the inversion layer below the oxide interface.

In the on-state, the drain to source capacitance $\mathbf{C_{ds}}$ is effectively shorted out by the low $\mathbf{r_{ds(on)}}$, but it appears in the off-state. $\mathbf{C_{gd}}$ and $\mathbf{C_{gs}}$ are included in all the three: the on-state, the active state and the off-state equivalent circuits.

Input capacitance at the gate of the MOSFET in an amplifier circuit includes the Miller's capacitance,

$$C_{input} = C_{gs} + C_m \text{ where}$$
$$C_m = (1 + g_m R_L)\, C_{gd} \tag{4-16}$$

where g_m is the transconductance and R_L is the load resistance to the drain terminal in an amplifier configuration.

The maximum operating frequency of the MOSFET can be defined as the frequency at which the input gate current becomes equal to the output drain current,

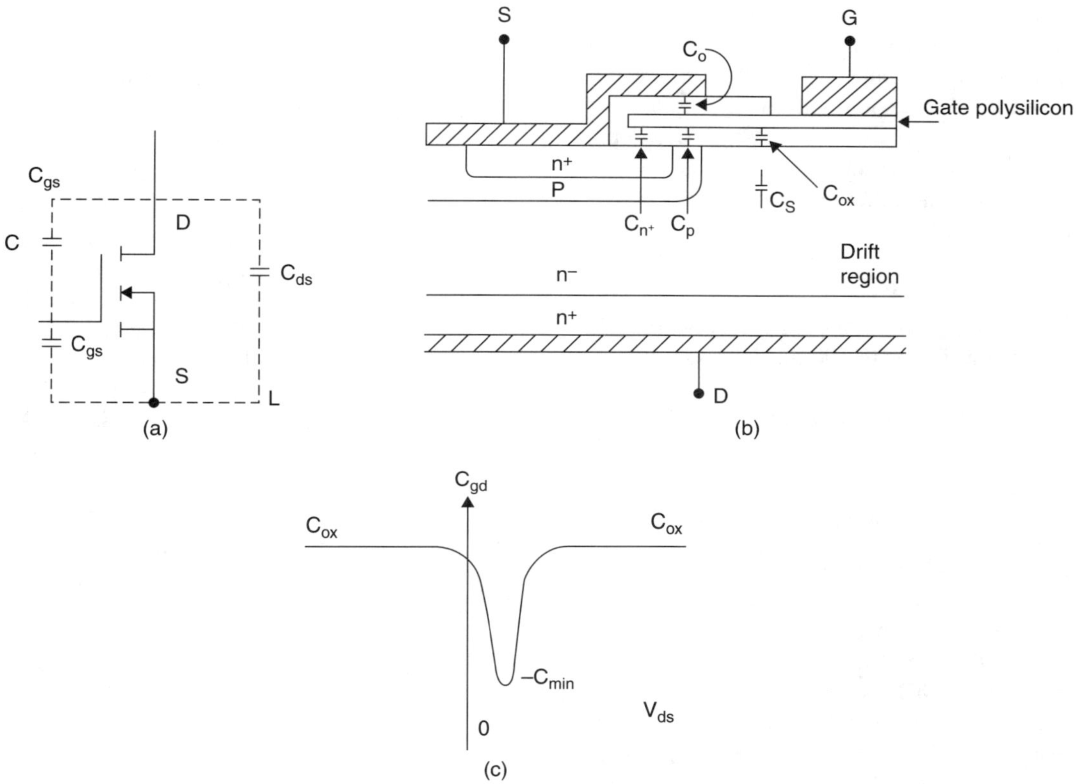

Figure 4.9
(a) Parasitic capacitances in MOSFET (b) inside the cell and (c) drift-region capacitance

$$i_{input} = 2\pi f\, C_{input}\, v_{gs},\ and\ i_{output} = g_m v_{gs} \qquad (4\text{-}17)$$

$$f_{\max} = \frac{1}{2\pi}\frac{g_m}{C_{input}} \qquad (4\text{-}18)$$

4.3.4 On-state characteristics

The threshold voltage for turning on the MOSFET as a switch and the on-state power losses are the main considerations in the designer's mind. The on-state losses in MOSFET are calculated as

$$P_{on} = I_D^2\, r_{DS(on)} \qquad (4\text{-}19)$$

In fact the on-state losses for a high voltage MOSFET are larger than in a similar BJT or a minority carrier device such as a BJT. Hence, the use of MOSFETs is restricted to low voltage applications (<100 volts).

Threshold Voltage

Threshold voltage V_T is the voltage on the gate electrode at which the strong inversion begins to occur in MOSFET. A high threshold voltage device requires a high voltage pulse to be applied at the gate to turn it on. At a too low V_T, the device can be turned on by noise signals. Typical values of V_T lie in the range between 2 and 3 volts.

On-State Resistance

In the on-state, the MOSFET operates in the *ohmic* mode. The on-state resistance $\mathbf{r_{DS(on)}}$ is the equivalent resistance between the source and drain terminals. It is comprised of resistances in the source, the inversion layer (the channel), the accumulation layer in the n^- -region, and the bulk region of the n^- region (also called the drift region) and the n^+ region. The constituent resistances are kept low by heavy doping levels in all regions except the n^- -region. Resistance of the drift region contributes most to the overall on-state resistance of the device []:

$$r_{DS(on)} \approx R_{n^-} \propto V_{BR}^{2.5} \qquad (4\text{-}20)$$

Typical $\mathbf{r_{ds(on)}}$ of a modern power MOSFET range from 9 milliohms to 100 milliohms.

4.3.5 Off-state characteristics

The off-state characteristics of importance are the breakdown voltages: the gate to source breakdown voltage and the drain to source breakdown voltage.

Breakdown Voltage
Gate-Source Breakdown $\mathbf{V_{gs(max)}}$
It is the maximum voltage the SiO_2 layer can withstand before breakdown. A typical breakdown electric field for SiO_2 is 5 to 10 Mega V/cm, translating into $\mathbf{V_{gs(max)}}$ of about 20 to 30 volts. The problem is that this much voltage can be easily acquired by static charges in an accidental careless handling of the device. Hence this device must be packed in a nonstatic package and must be handled only with a grounded band around the wrist. In the circuits a back to back zener diode protection at the gate is recommended.

Drain-Source Breakdown $\mathbf{BV_{DSS}}$
$\mathbf{BV_{DSS}}$ is the maximum permitted voltage between the drain and the source terminals with the gate shorted to the source. Voltages higher than this cause avalanche breakdown of the p-body to the n^- drain junction. Like the power diodes, the large breakdown voltage in MOSFET is achieved by low doping levels of the n^- region and longer path length of the charge carriers through the n^- region.

4.3.6 Safe operating area

Safe operating area defines the limit of operation of the device on the i_D-v_{DS} characteristics. See Figure 4.10. The following parameters determine the safe operating area:

$I_{D(max)}$: the maximum drain current
P_{max}: the maximum power dissipation and
V_{BDSS}: the forward blocking voltage

The maximum drain current $I_{D(max)}$ at low drain voltages is limited by the power dissipation rating. The maximum forward blocking voltage VB_{BDSS} at low drain currents is determined by the avalanche breakdown phenomenon. The maximum power dissipation limit is based on the maximum permissible temperature T_j on the semiconductor junctions in the device. The T_j depends on the amount of heat flowing across the junctions. A continuous power dissipation will cause a larger rise of junction temperature than an equal power for a short duration. Hence, the maximum power limit line, under the simultaneous application of high drain voltage and high drain current, is shifted upwards for shorter duration applications. The safe operating areas in MOSFETs are large square regions. The large square region of SOA means a minimal snubber circuit.

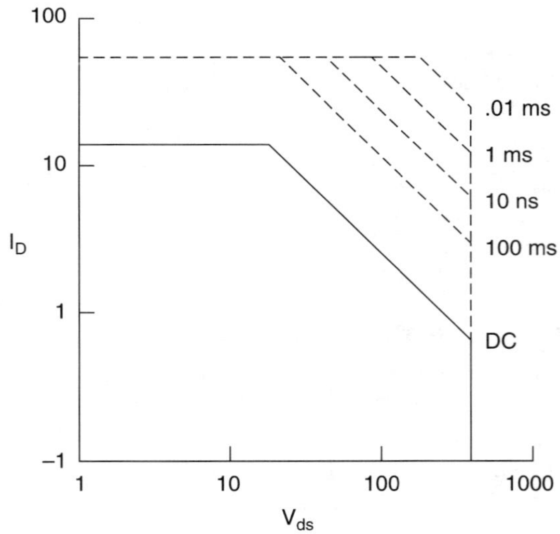

Figure 4.10
The Safe Operating Area

4.3.7 Effect of temperature

Power MOSFETs exhibits very good high temperature characteristics. Commercial devices are available with $T_{j,\ max}$ = 200 deg. C. The on-state resistance r_{DSon} increases with increase in temperature [1]:

$$r_{DSon}(T) = r_{DSon}(25°C)\left(\frac{T}{300}\right)^{2.3} \quad (4\text{-}20a)$$

where T is in Kelvin.

The electron and hole mobilities decrease with increasing temperature. Consequently the transconductance of MOSFET has a negative temperature coefficient:

$$g_m(T) = g_m(25°C)\left(\frac{T}{300}\right)^{-2.3} \quad (4\text{-}20b)$$

The threshold voltage V_T also has a negative temperature coefficient, typically −6 mV per deg. C.

Example 4-1 Plot r_{DSon} and g_m versus Temperature

Equations (4-20a) and (4-20b) are plotted using MATLAB. See Figure 4.11.

4.3.8 Switching characteristics

High speed switching may be accompanied by a very high rate of change dv_{ds}/dt, which can turn-on the parasitic BJT in the MOSFET leading to the problems of second breakdown and reduced safe operating area. This section will analyze the parameters that govern the transient behavior of the MOSFET in on and off states. For analysis, we consider a switching circuit as shown in Figure 4.12. The clamping diode is assumed to have no reverse recovery time. The load inductance is large enough to maintain a constant current I_O during both the turn-on and turn-off transients.

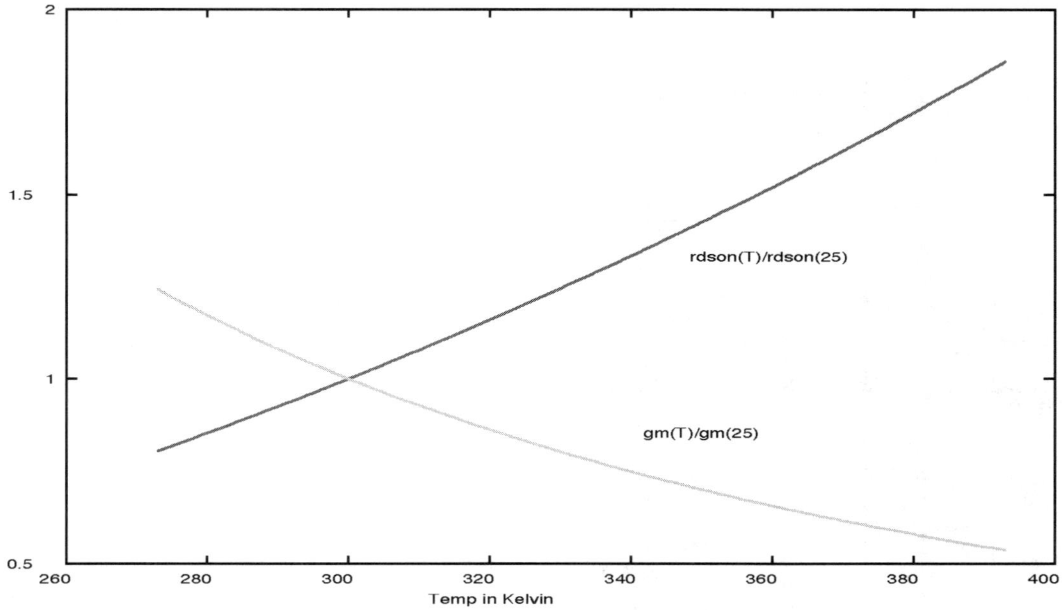

Figure 4.11
Effect of temperature on the r_{DSon} and g_m

Turn-On Transients

The turn-on transients of the MOSFET can be explained using four circuit modes. Figure 4.13(a) to (d) show the circuit modes and Figure 4.13(e) shows the current and voltage waveforms. Consider that the MOSFET is in the off-state for a sufficiently long time. The load current is freewheeling through the clamping diode. The initial conditions are: $v_{gs}=0$, $I_D=0$ and $v_{ds}=V_{DD}$.

Mode m1 begins when a step voltage $v_g=V_{DD}$ is applied to the gate resistance R_g. V_{DD} is much greater than the threshold voltage V_T. Capacitors C_{gs} and C_{gd} begin charging via resistor R_g. The gate to the source voltage remains below the threshold voltage V_T until time t_1. The voltage between the gate and the source of the MOSFET rises exponentially during this interval

$$v_{gs}(t) = V_{DD}\,[1 - e^{-t/\tau_1}]$$

where $$\tau_1 = R_g(C_{gs} + C_{gd}) \qquad (4\text{-}21)$$

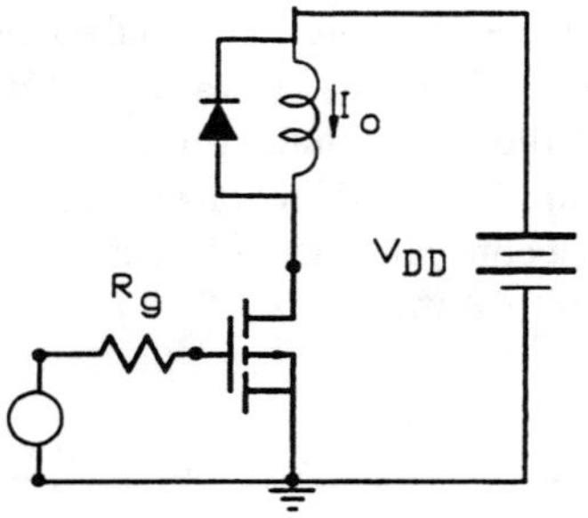

Figure 4.12
Circuit used for analyzing the switching transients.

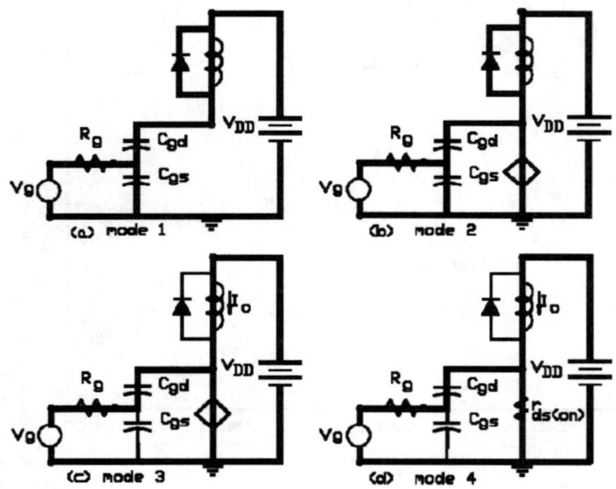

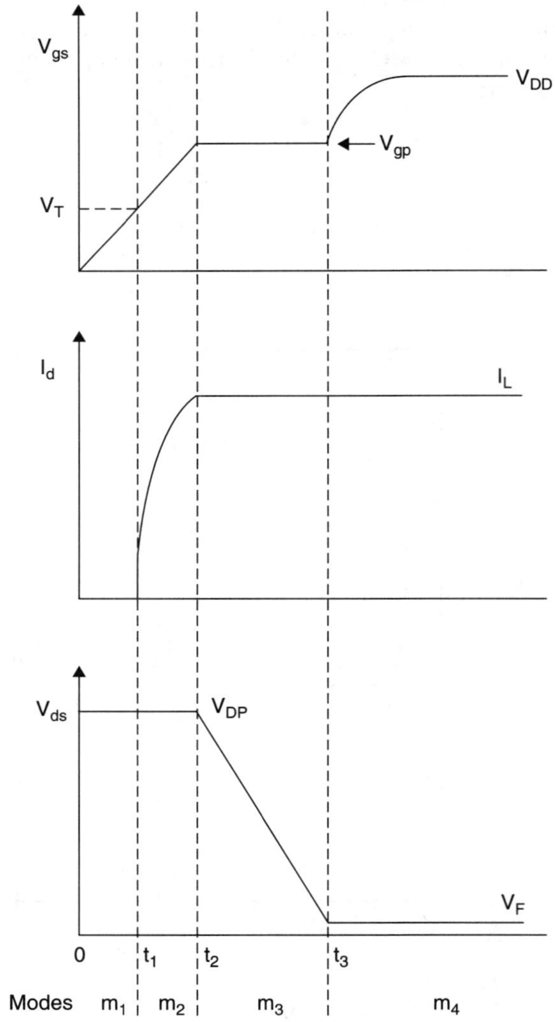

(e)

Figure 4.13
(a) to (d) Circuit modes in turn-on and (e) Waveforms

As long as the voltage $v_{gs} < V_T$, the MOSFET does not conduct current. The time to reach V_T is called the turn-on delay, t_1,

$$t_1 = R_g(C_{gs} + C_{gd}) \ln\left[\frac{1}{1 - V_T/V_{DD}}\right] \quad (4\text{-}22)$$

Mode m_2 begins after time t_1. The drain current begins to increase. The freewheeling current through the clamping diode begins to transfer to the drain circuit. Until the drain current equals the load current I_O, the diode D remains clamped. MOSFET supports full V_{DD}. MOSFET enters in the active region; however, since the load inductance is short circuited by the diode, there is no Miller gain and no Miller capacitance. The gate voltage will continue to rise as per equation (4-21). The drain current is assumed to be linear in the analysis, hence,

$$I_D(t) = g_m(v_{gs} - V_T) \quad (4\text{-}23)$$

The duration t_2-t_1 is given by

$$t_2 - t_1 = R_g(C_{gs} + C_{gd}) \ln\left[\frac{g_m V_{DD}}{g_m(V_{DD} - V_T) - I_O}\right] \quad (4\text{-}24)$$

At time t_2, I_O is completely transferred from the clamping diode to the MOSFET, the clamping diode is reverse biased and the MOSFET has a Miller gain.

The circuit mode m_3 begins after t_2. The drain voltage starts falling. The drain current being constant, the gate to source voltage is also constant at

$$V_{gp} = V_T + \frac{I_O}{g_m} \quad (4\text{-}25)$$

Current in the gate resistor R_g is

$$I_g = \frac{V_{DD} - V_{gp}}{R_g} \quad (4\text{-}26)$$

This current charges the capacitance C_{gd} linearly, hence, the drain voltage is given by

$$v_{DS} = V_{DD} - \frac{I_g}{C_{gd}}(t - t_2) \quad (4\text{-}27)$$

At time t_3, the v_{ds} decreases to the on-state voltage level V_F and the MOSFET enters the on-state. The duration of the fall of v_{ds} is calculated as

$$t_3 - t_2 = \frac{(V_{DD} - V_F)\, C_{gd}}{I_g} \quad (4\text{-}28)$$

After t_3 is in mode m_4, the MOSFET enters in the ohmic or linear region. The transconductance g_m is not constant and the gate voltage begins to increase, charging the capacitance C_{gs} and C_{gd} exponentially to V_{DD}.

A first order approximation of the exponential rise of the drain current by the linear rise may simplify calculations:

$$i_d \approx \frac{I_O}{t_2 - t_1}(t - t_1) \quad (4\text{-}28a)$$

Turn-Off Transients

The turn-off transient can also be explained using four circuit modes. Figure 4-14(a) shows the circuit modes and 4-14(b) shows the current and voltage waveforms. The MOSFET is assumed to be in the on-state for sufficiently long time. Initial conditions are: $v_{gs}=V_{DD}$, $I_D=I_O$ and $v_{ds}=V_F$.

Mode m1 begins as the voltage at the gate terminal is abruptly reduced to zero. The gate voltage v_{gs} begins to decrease

$$v_{gs}(t) = V_{DD}\, e^{-t/\tau_1} \textit{ where } \tau_1 = R_g(C_{gs} + C_{gd}) \quad (4\text{-}29)$$

No changes in the drain current or voltage occurs until the voltage $v_{gs} < V_{gp}$, which is necessary to maintain the load Current I_L. The time to reach V_{gp} is called the turn-off delay, t_1,

$$t_1 = R_g(C_{gs} + C_{gd}) \ln\left[\frac{V_{DD}}{V_T + I_O/g_m}\right] \quad (4\text{-}30)$$

V_{gp} is given by

$$V_{gp} = V_T + I_O/g_m \quad (4\text{-}31a)$$

Mode m_2 begins after time t_1. The drain current remains constant at I_O and so does the gate voltage $v_{gs}=V_{gp}$. The drain voltage begins to rise towards V_{DD}. Current in the gate resistor R_g is

$$I_g = \frac{V_{gp}}{R_g} = \frac{V_T + I_O/g_m}{R_g} \quad (4\text{-}31b)$$

This current discharges the capacitance C_{gd} linearly, hence, the drain voltage is given by

$$v_{DS} = V_F + \frac{I_g}{C_{gd}}(t - t_2) \tag{4-32}$$

At time t_2, v_{ds} reaches the supply voltage V_{DD}. The duration of the rise of v_{ds} is calculated as

$$t_2 - t_1 = \frac{(V_{DD} - V_F)C_{gd}}{I_g} \tag{4-33}$$

The stray inductance in series with the drain, which we have ignored so far, causes the drain voltage to exceed the V_{DD}.

At this point in time, the freewheeling diode will turn on. The gate voltage will now decrease exponentially,

$$v_{gs}(t) = V_{gp}\, e^{-t/\tau_1} \quad where\ \tau_1 = R_g(C_{gs} + C_{gd}) \tag{4-34}$$

The drain current also starts decreasing

$$I_D(t) = g_m(v_{gs} - V_T) \tag{4-35}$$

The drain current decreases to zero at time t_3. The duration t_3-t_2 is given by

$$t_3 - t_2 = R_g(C_{gs} + C_{gd}) \ln\left[\frac{I_O}{g_m V_T} + 1\right] \tag{4-36}$$

After time t_3, the gate voltage continues to decrease exponentially to zero.

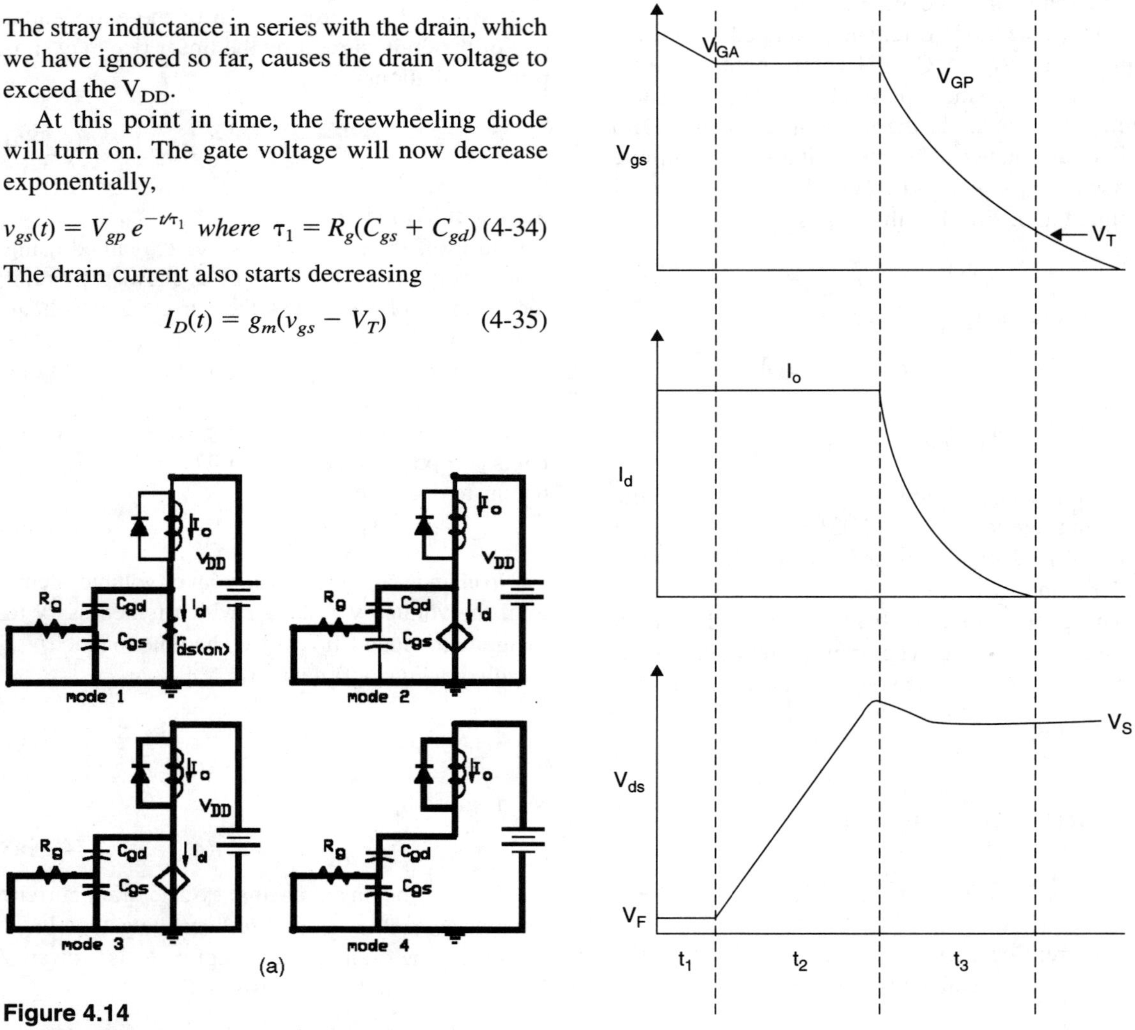

Figure 4.14
(a) Circuit modes in turn-off and (b) waveforms

A first order approximation of the exponential fall of the drain current by the linear fall may simplify calculations:

$$i_d \approx I_O - \frac{I_O}{t_3 - t_2}(t - t_2) \qquad \text{(4-36a)}$$

Example 4-2 Power Loss in a MOSFET Switching Circuit

The International Rectifier MOSFET IRF1010, 55 V is connected in the switching circuit of Figure 4.11 with the following components: $V_{DD} = 15$ V, $I_O = 40$ A. The clamping diode is assumed to be an ultrafast diode with zero reverse recovery time. Find the total losses in the MOSFET if switched at 50% duty cycle at 500 KHz.

The drain voltage and current in the example are drawn in Figure 4.15.

From the device specifications:

$g_m = 25$ Siemens, $r_{DSon} = 0.014$ ohms,
$V_T = 3$ V,

$C_{oss} = 1300$ pF (C_{ds}), $C_{iss} = 2500$ pF (C_{gs}) ,
$C_{rss} = 350$ pF (C_{gd})

turn-on: $t_{don} = 16$ nS (t_2 in Figure 4.12e) ,
$t_r = 120$ nS ($t_3 - t_2$ in eq. (4-24))

turn-off: $t_{doff} = 48$ nS (t_1 in Figure 4.13b) ,
$t_f = 80$ nS ($t_2 - t_1$ in eq. (4-35))

The rise time of the drain current is obtained from eq. (4-24): $t_2 - t_1 = 10.4$ nS.

The fall time of the drain current is obtained from eq. (4-38): $t_3 - t_2 = 12.2$ nS.

The switching period is T=1/500 KHz =2000 nS.

The on duration of the MOSFET is

$$t_{on} = T/2 - (t_{don} + (t_2 - t_1) + t_r + t_{doff}) = 1000 - (16 + 10.4 + 120 + 48) = 805.6 \text{ nS}$$

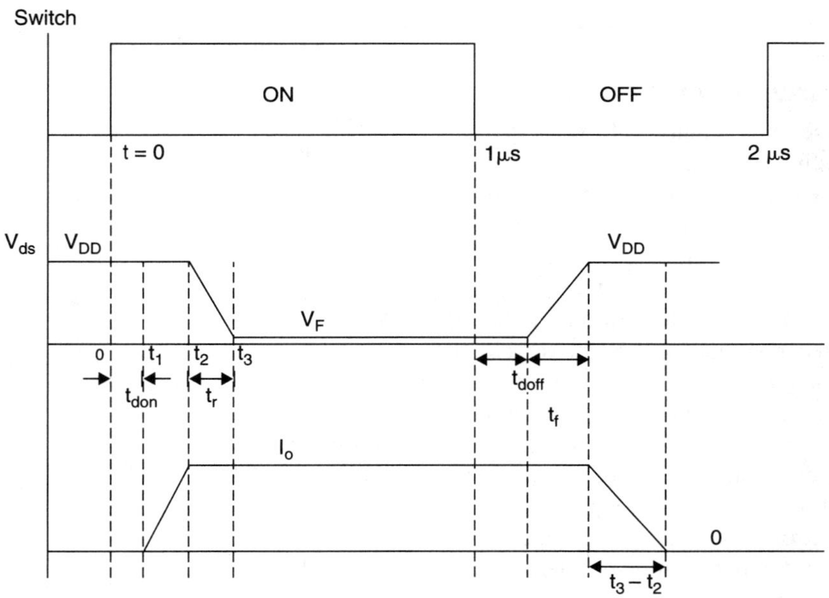

Figure 4.15
Switching waveform in Example 4-1

The on-state voltage from drain to source is

$$V_F = I_O\, r_{DSon} = 40 \times 0.014 = 0.56\ \text{V}$$

The power loss is calculated from

$$P_{Loss} = P_{sw\text{-}on} + P_{on} + P_{sw\text{-}off}$$

where
$$P_{sw-on} = \frac{1}{T}\left[V_{DD}\frac{I_O(t_2 - t_1)}{2} + I_O\left(\frac{(V_{DD} - V_F)t_r}{2} + V_F t_r\right)\right]$$

$$P_{on} = V_F I_O$$

$$P_{sw-off} = \frac{1}{T}\left[I_O\left(\frac{(V_{DD} - V_F)t_r}{2} + V_F t_r\right) + V_{DD}\frac{I_O(t_3 - t_2)}{2}\right]$$

The switching losses $= P_{sw\text{-}on} + P_{sw\text{-}off}$
$= 34.5730$ watts,
The on-state losses $= P_{on} = 9.0223$ watts
Total power Loss in the MOSFET
$= 43.5353$ watts

4.3.9 Gate drive circuits

A MOSFET does not require gate drive power; however, it requires the transfer of charge to and from the gate terminal for turning on and off respectively. For turning on, the gate voltage must be switched from a voltage below to a voltage above the threshold voltage V_T. The gate current required during the transitions is high and low during the on-state. The source resistance of the driver circuit should be very low during both transitions for fast transition times. Figure 4.16 shows the *Totem-pole driver* circuit.

The gate driver output pulse is referenced to the source terminal of the MOSFET. Input to the gate driver is the control pulses that are obtained from digital circuits or computers and are referenced to ground. In H-bridge (full bridge) converters, the source terminals of the upper two MOSFET elements, are elevated from electrical ground. Hence the gate driver output pulses to the upper two

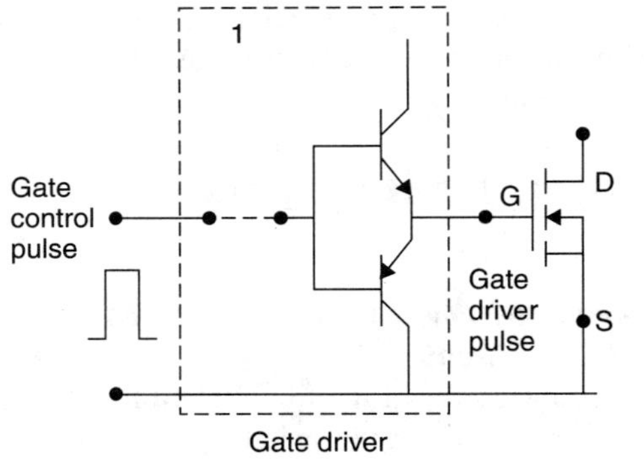

Figure 4.16
Totem-pole driver for MOSFET gate driver

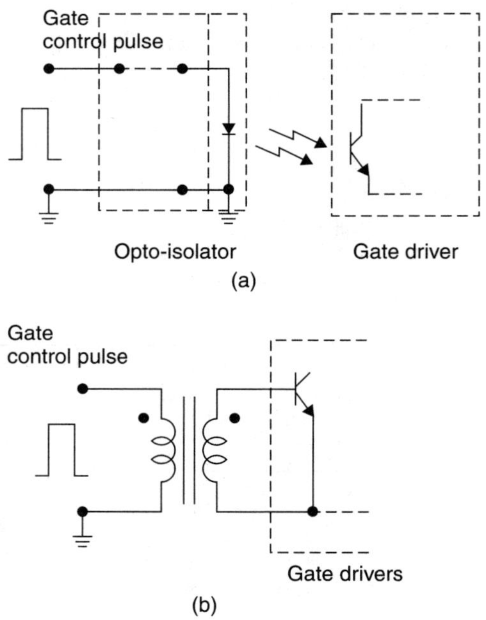

Figure 4.17
(a) Optical coupling and (b) transformer coupling of control pulses

MOSFETs are required to be isolated electrically from the gate control pulses. The electrical isolation is generally provided by optical or transformer coupling. Figure 4.17 (a) and (b) show two types of isolation.

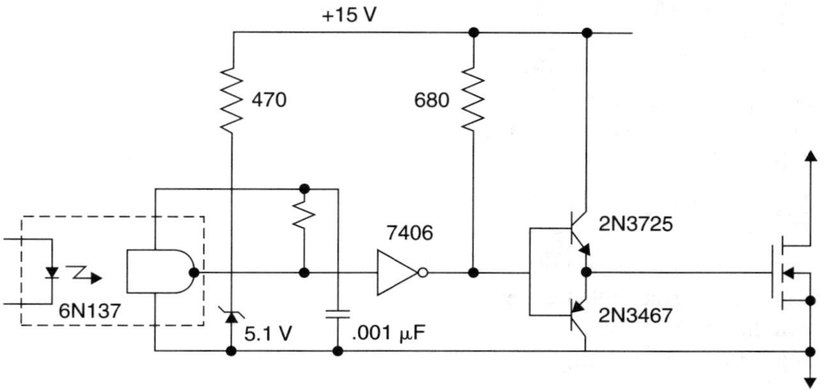

Figure 4.18
Optically isolated driver for MOSFET

Example 4-3 Logic level Gate Isolated Driver for Power MOSFET Switching Circuit

Design the 5-volt gate driver circuit for the switching circuit of Example 4-2.

The output stage must be a Totem-pole driver stage for driving the high input capacitance C_{gs}. The gate driver should apply a 15 Volt (equal to V_{DD}) for best results. The MOSFET is a level triggered device, hence, a dc connection is required from the logic level to the gate of the MOSFET. An optical isolator is the appropriate choice.

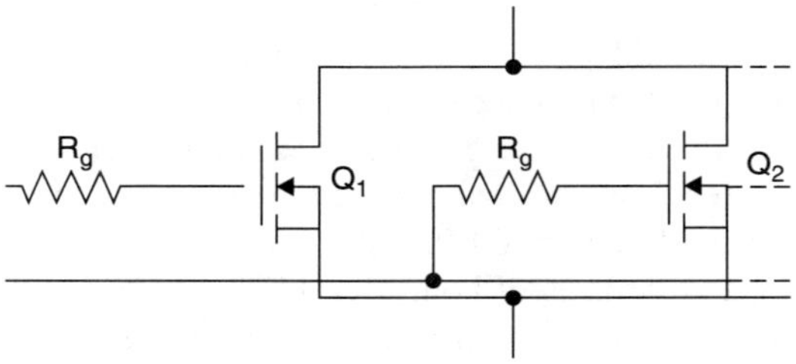

Figure 4.19
Parallel connection of MOSFETs

4.3.10 Parallel connection of MOSFETs

The power MOSFET is constructed like parallel connected MOSFET cells. Hence it lends itself naturally for parallel connection for increasing the forward current-carrying capability. Furthermore, MOSFET behaves like a resistance in the on-state and has a positive temperature coefficient. It means that in a high-power dissipation situation, the on-state resistance increases, which in turn reduces the amount of forward current. Hence, when a number of MOSFETs are connected in parallel for current sharing, drooping out of any device does not lead to more currents, the consequent thermal runaway and breakdown of the remaining devices. This phenomenon is the opposite of that encountered in power BJTs.

Stray inductances in the gate circuit of each MOSFET may differ due to fabrication tolerances. The stray inductances in conjunction with C_{gs} and C_{gd} can cause oscillations on the gate and possible unequal gate-to-source voltages. Equal sharing of the switching transients requires equalizing the gate to source voltage in each paralleled MOSFET. This is done by connecting equal damping resistances in series with the gate terminals, as shown in Figure 4.19. A symmetrical layout of paralleled MOSFETs helps greatly in this regard.

Example 4-4 Parallel Connection of Power MOSFETs

Two MOSFETs, International Rectifier IRF 1010, 55 V, are connected in parallel to share 100 amps dc. The device capacitances and the estimated stray inductances are as follows:

$C_{gs1} = 2500$ pF, $C_{gd1} = 350$ pF $\quad L_{g1} = 2$ nH
$C_{gs2} = 2450$ pF, $C_{gd2} = 320$ pF $\quad L_{g2} = 2.5$ nH

Find the series resistance in the gate that would enable equal voltage sharing at high frequencies. Also find the safe highest frequency.

The oscillating frequencies in the two MOSFETs are

$$f_1 = \frac{1}{2\pi\sqrt{L_{g1}(C_{gs1} + C_{gd1})}} = 6.6663 \times 10^7$$

$$f_2 = \frac{1}{2\pi\sqrt{L_{g2}(C_{gs2} + C_{gd2})}} = 6.0483 \times 10^7$$

We would place a low-pass filter such that it would be at least a decade lower in frequency to effectively equalize the gate voltages at high frequencies. Hence, the desired highest frequency of operation would be $f_h = 6$ Mhz. The series resistance in the gate terminals of both MOSFETs, R_g is calculated by using the highest gate capacitances as the following

$$R_g = \frac{1}{2\pi(C_{gs1} + C_{gd1})f_h}$$
$$= \frac{1}{2\pi \times (2500 + 350) \times 10^{-12} \times 6x10^6}$$
$$= 9.3 \text{ ohms}$$

Select a resistance of the standard available value that is slightly higher than the answer, say 10 ohms.

4.4 INSULATED GATE BIPOLAR TRANSISTOR (IGBT)

The insulated gate bipolar transistor, as the name suggests, is a bipolar transistor with an insulated base terminal. The insulated base terminal is called a gate. The insulated gate enables simpler gate drive circuits like those in MOSFET. An IGBT is a faster device than a BJT but slower than a MOSFET. The on-state losses are much smaller than a MOSFET but comparable to a BJT.

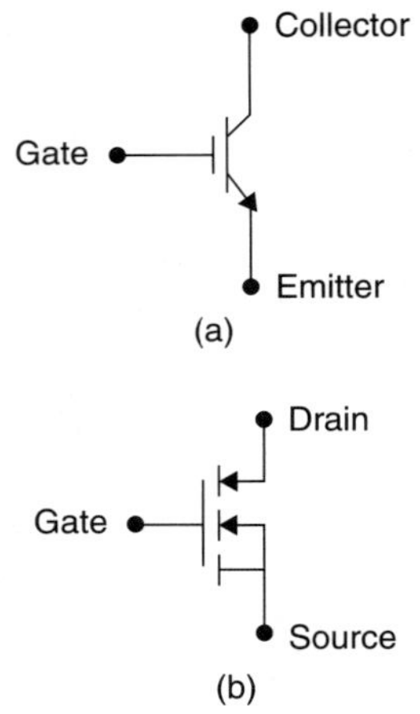

Figure 4.20
IGBT symbols

An IGBT is fabricated as a parallel combination of several identical cells with each cell being an IGBT device. Paralleling increases the current handling capability.

4.4.1 Basic IGBT cell

An IGBT is fabricated from several parallel connected cells organized in various geometrical designs. Figure 4.21 shows the cross-sectional view of a basic cell of n-channel IGBT.

An IGBT cell is very much similar to the MOSFET cell except the n^+ layer in the drain region is replaced by a p^+ layer and it is called the collector. The emitter region is made of heavily doped n^+ semiconductor. Like the MOSFET device, the IGBT cell also contains an enhancement MOS channel that exists in the p region beneath the gate layer between the n^+ emitter and the n^- region. Like in the MOSFETs, the bulk of the body region is electrically shorted to the emitter region, thus effectively shorting the n^+-p junction. In fact the IGBT struc-

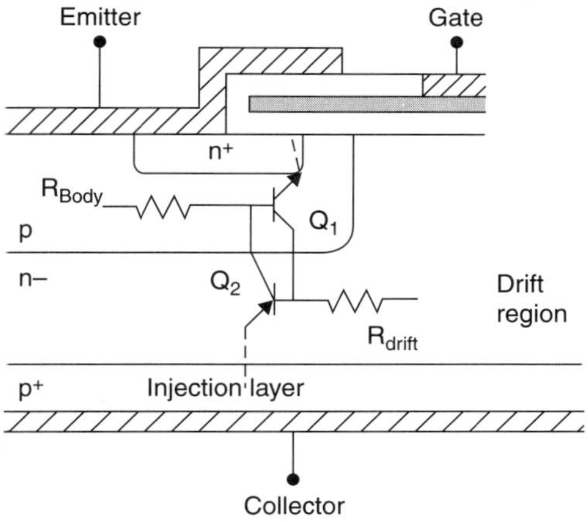

Figure 4.21
A basic n-channel IGBT cell

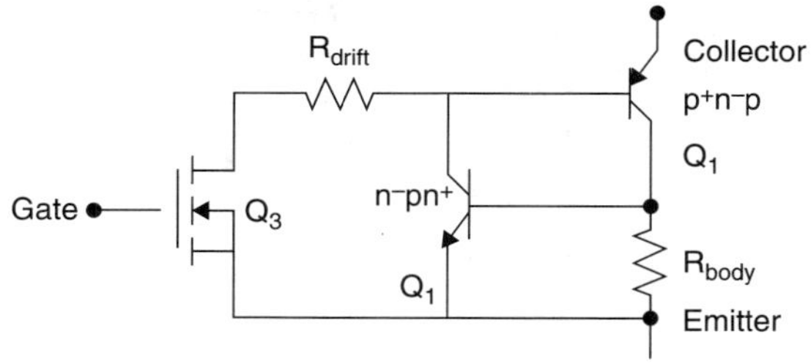

Figure 4.22
Equivalent device representation of an IGBT

ture is so similar to a MOSFET that a similar symbol is also used for the IGBT, given in Figure 4.20 (b). The n^- layer, in between the p^+ region and p region, serves as the drift region to accommodate the depletion layer width of the p-n^- junction.

IGBT can be thought of as the combination of a MOSFET and a p^an^-p transistor Q_2, see Figure 4.21 and 4.22. The base of the transistor Q_2 is connected to the drain (n^- region) of the MOSFET. The IGBT structure also has another inherent transistor structure n^-pn^+ Q_1 as shown in Figures 4.21 and 4.22. The intertwined connection of the two transistors Q_1 and Q_2 resembles a thyristorlike structure (to be discussed in a later chapter). This connection may give rise to latching up of the device to either a high-current mode or a zero-current mode. The latch up causes difficulty in turning off. Higher rates of $\mathbf{dv_{CE}/dt}$ during the turn-off (called the *reapplied rate*) can cause latch up. Special precautions are taken to avoid the latch up, either in the circuits or in the device geometry.

The emitter of transistor Q_2 is termed the collector of IGBT. The collector of Q_2, which is also the emitter of the transistor Q_1, is termed the emitter of IGBT.

There are two n-p junctions: one on the emitter side and the other on the collector side. Each junction has a depletion layer associated with it. In the unbiased conditions, there is no voltage on the gate conductor. The device structure is like an npn bipolar transistor without external access to the base, in which both junctions are reverse biased. No current can flow from the collector to the emitter. The MOSFET is in the **cut-off** region. This is the off-state of the IGBT as a switch.

Mode	***p^+n^-junction*** **E-B junction of Q_2**	***n^-p junction*** **E-B junction of Q_1**
Reverse Blocking (I_c=0)	reverse biased	forward biased
Forward Blocking (I_c=0)	forward biased	reverse biased
On-state and active state	forward biased	forward biased

An IGBT does not conduct current in the reverse blocking mode when a negative voltage is applied to the collector with respect to the emitter. The p^an^- junction is reverse biased. The depletion layer extends in the n^- drift region and it provides the reverse blocking voltage capability.

When a positive voltage is applied on the gate conductor with respect to the emitter conductor, the pn^- junction is reverse biased. The depletion layer extends in the n^- drift region and provides the forward blocking voltage capability. The forward and reverse blocking voltage capability of the IGBT cell

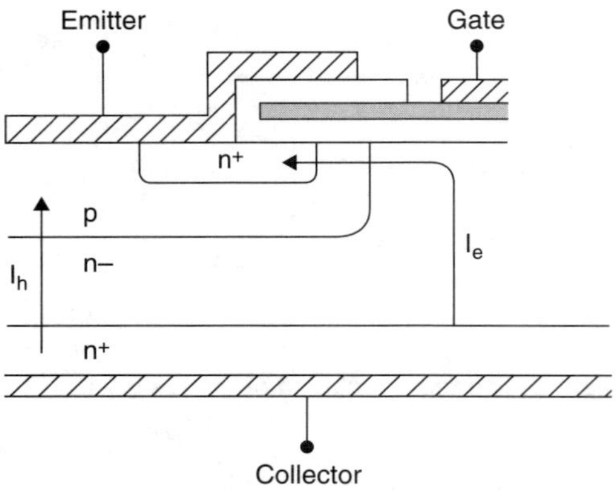

Figure 4.23
Currents in the on-state of the IGBT

is approximately equal because of the involvement of the common n^- drift region.

If the positive voltage at the gate is greater than the threshold voltage V_T, an n-channel is formed in the p-region just beneath the gate terminal in a manner similar to that observed in the MOSFET. Electrons flow from the n^+ emitter to the n^- drift region through the n-channel. These electrons constitute the base current of the $p^a n^- p$ transistor Q_2. The emitter of Q_2 (p^+ region) injects holes into the base (n^- region). Thus there is double injection into the n^- region: the electrons from the p-body region and the holes from the n^+ region. Density of the injected carriers increases with the increase in the positive bias on the collector of IGBT.

The density increases so much so that the injected carrier densities exceed the thermal equilibrium level doping densities. Conductivity of the n^- drift region increases very significantly. This situation is similar to the one we know in p-i-n diodes. As the consequence, the IGBT can be operated at high current density. In this regime of operation the IGBT is in the on-state. The IGBT current consists of two components, I_h: the current due to injected holes flowing across the transistor Q_2 and I_e and the current due to injected electrons flowing through the n-channel. See Figure 4.22 for the currents in the on-state of the IGBT.

$$I_h = \left[\frac{\alpha_2}{1 - \alpha_2}\right] I_e \tag{4-37}$$

α_2 is the current gain of transistor Q_2, typically 0.5. The electron current is given by eq. (4-11),

$$I_e = \frac{\mu_{ns} C_{ox} Z}{2L} (v_g - V_T)^2 \tag{4-38}$$

L is the channel length. The emitter current of IGBT, which is also the collector current I_C, is the sum,

$$I_C = I_E = I_h + I_e = \frac{1}{1 - \alpha_2} I_e \tag{4-39}$$

The emitter current is also the saturated collector current of the IGBT.

In the on-state, the n-channel in the MOSFET section of the IGBT behaves like a resistance. The on-state voltage drop of IGBT consists of three components:

$$V_{CE(on)} = I_{CE} R_{ch} + V_{drift} + V_{BE2} \tag{4-40}$$

the voltage drop in the n-channel, the voltage drop across the n^- drift region and the voltage drop across the forward biased base to emitter junction of the $p^a n^- p$ transistor Q_2. The V_{drift} in IGBT is much less than in MOSFET due to the significant increase in the conductivity due to double influx of charge carriers in the drift region. The conductivity increase is the main reason for the smaller on-state voltage drop of IGBT.

The transconductance of IGBT is obtained from differentiating eq. (4-39):

$$g_m = \frac{dI_C}{dv_{DS}} = \frac{1}{1 - \alpha_2} \frac{\mu_{ns} C_{ox} Z}{L} (v_g - V_T) \tag{4-41}$$

Comparison of eq. (4-41) with eq. (4-9) shows that the transconductance of IGBT is significantly larger than (typically 2 times) the g_m of the MOSFET. Figure 4.24 shows the complete i-v characteristics of the IGBT cell.

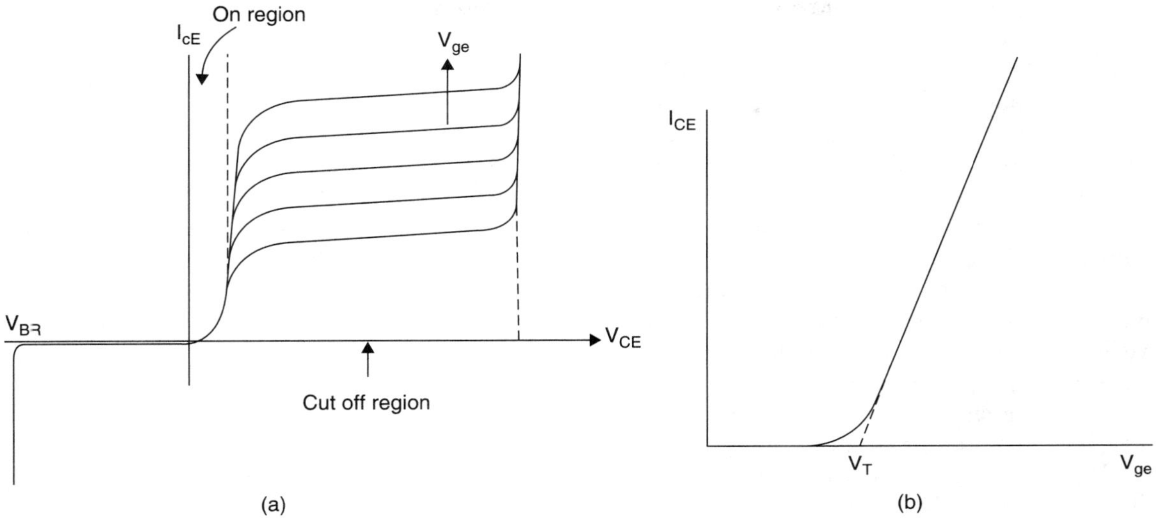

Figure 4.24
The i-v characteristics of the IGBT cell (a) I_C versus V_{CE} and (b) I_C versus V_g

4.4.2 Parasitic thyristor and the latch-up in IGBT

The Q_1 and Q_2 combination constitutes a parasitic thyristor in the IGBT structure. The maximum operating current of the IGBT is limited by the presence of the parasitic thyristor. In the on-state, the hole current I_h flows through the lateral p body region, passing through the resistance R_{body}. See Figure 4.23. The voltage drop across R_{body} may forward bias the base to emitter junction of transistor Q_1 at high current levels. At normal current level, the current gain of transistor Q_1, α_1, is very small, but it increases with the current level. If the sum of the current gains of both transistors $\alpha_1+\alpha_2$ exceeds unity, the inherent thyristor can enter the highly conducting state, called the latch up. The thyristor current, and hence the IGBT current, are no longer controllable from the gate terminal. The only way to turn off the IGBT is the forced commutation of current exactly as is done in conventional thyristors (to be discussed in the next chapter). The IGBT must be brought out of the latch up to prevent excessive power dissipation. The device manufacturer specifies the maximum collector current I_{DM} that can flow without latch up. Literature provides several ways of fabrication to avoid the latch up in IGBTs.

4.4.3 Device limits and safe operating area (SOA)

The following parameters determine the operational limits of the device:

$V_{ge(max)}$:	the maximum gate voltage
$I_{C(max)}$:	the maximum collector current
P_{max}:	the maximum power dissipation
$V_{CE(max)}$:	the forward blocking voltage in FBSOA and
V_{BR}:	the reverse breakdown voltage in RBSOA
$T_{j(max)}$:	the maximum permissible junction temperature

$V_{ge(max)}$ is the maximum gate voltage that is determined by the breakdown of the SiO_2 layer. $I_{CE(max)}$ is the maximum permissible collector current before the onset of latch up of the parasitic thyristor of the IGBT. IGBTs with $I_{CE(max)}$ of 200 to 400 amperes are commercially available.

The maximum collector to emitter voltage $V_{CE(max)}$ is determined by the breakdown voltage BV_{CEO} of the transistor Q_2. Devices with 1700 volts are commercially available.

The maximum junction temperature in commercially available IGBTs is 150° C.

The safe operating area (SOA) is defined as the area within the output characteristics of the IGBT where it can be operated without destructive failure due to power dissipation. In general there are three boundaries for the SOA: (a) $V_{CE(max)}$, (b) $I_{CE(max)}$ and (c) the boundary where both I_{CE} and V_{CE} become simultaneously large. The third situation is encountered during the turn-on and turn-off transitions. During the turn-on transient, a positive gate bias is applied that results in the simultaneous flow of the hole and electron currents in the presence of the high voltage across the device. This is called the *forward biased safe operating area FBSOA*. During the turn-off transient, the gate terminal is switched from positive to zero or a negative voltage that results in only the hole current in the presence of the high voltage across the device. This is called the *reverse biased safe operating area RBSOA*.

FBSOA : The forward bias safe operating area FBSOA of IGBTs is square for short switching times, identical to the FBSOA in MOSFETs. Thus it has a minimal need for snubbers.

RBSOA : The reverse bias safe operating area RBSOA, however, is different. The higher rate of $\mathbf{dv_{CE}/dt}$ during the turn-off (called the reapplied rate) flattens out the upper right-corner in RBSOA.

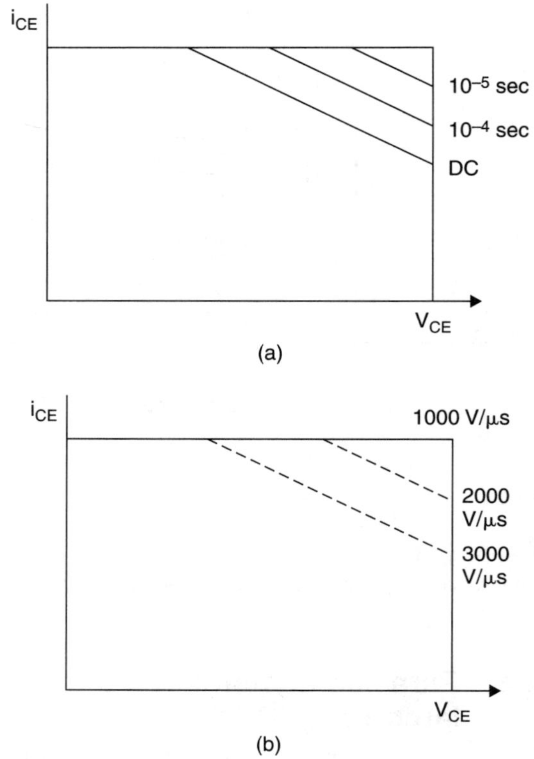

Figure 4.25
Safe Operating Area of IGBT (a) FBSOA and (b) RBSOA

4.4.4 Switching locus in I-V characteristics

To understand the locii of the IGBT during the transition to and from the on and off states, we would refer to the switching circuit shown in Figure 4.26(a). The current-voltage locus during the turn-on and turn-off transitions are shown in Figure 4.26(b). During the turn-on, while the collector withstands the supply voltage V_{CC}, the collector current overshoots due to the reverse recovery current of the freewheeling diode before settling down to the on-state value. In order to avoid the destructive failure of IGBT, the peak point A on the turn-on locus must be located within the FBSOA. The use of fast Schottky diodes as the freewheeling diode would avoid this problem.

During the turn-off, while the collector sustains a constant current, the collector voltage overshoots because of the stray inductance between the IGBT and the freewheeling diode. In order to avoid the destructive failure of IGBT, the peak point B on the turn-off locus must be located within the RBSOA.

4.4.5 Switching characteristics

In this section, we will analyze the parameters that govern the transient behavior of the IGBT in on and off states. High-speed switching may be accompanied by a very high rate of change dv_{CE}/dt, which can turn-on the parasitic thyristor in the IGBT. For analysis, consider the switching circuit of Figure 4.26. The freewheeling diode is assumed to have no reverse recovery time. The load inductance is large enough to maintain a constant current I_L during both turn-on and turn-off transitions.

Turn-On Transients

IGBT essentially acts like a MOSFET during most of the turn-on transition. Figure 4.27 shows the current and voltage waveforms.

Turn-Off Transients

The current and voltage waveforms in the turn-off transition are given in Figure 4.28. The transients up to time t_3 are identical to those observed in MOSFET. The transition of the collector current from I_L to zero takes two distinct time intervals instead of only one interval in the MOSFET. The first interval t_3-t_2 of rapid drop in the collector current corresponds to the MOSFET section of the IGBT. The

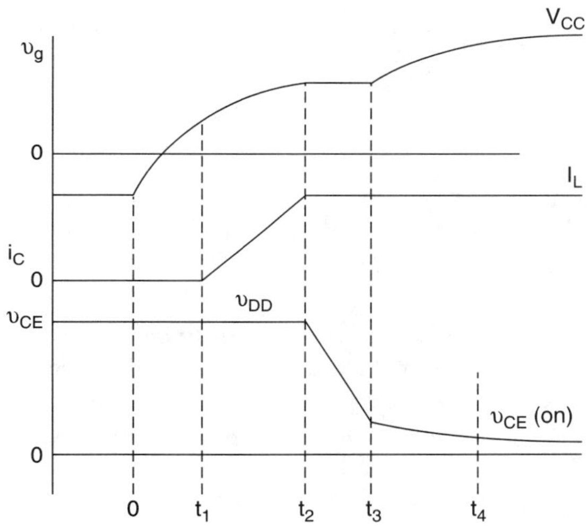

Figure 4.27
Turn-on transients in IGBT

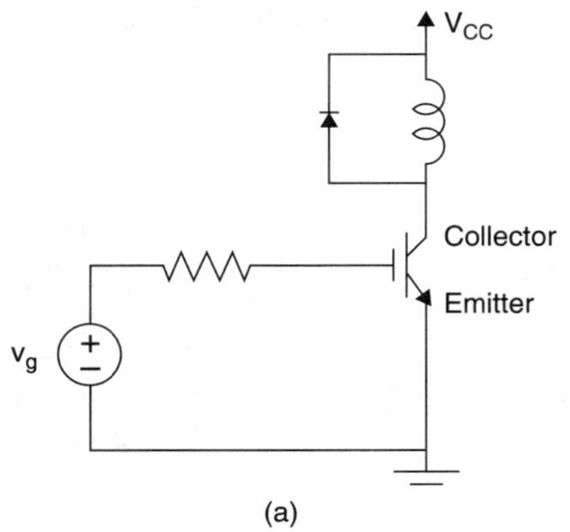

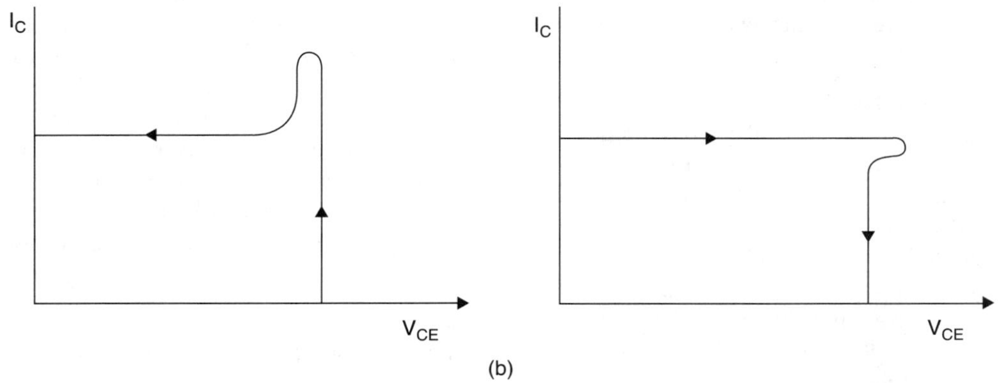

Figure 4.26
(a) Switching circuit for IGB and (b) switching low with an inductive load

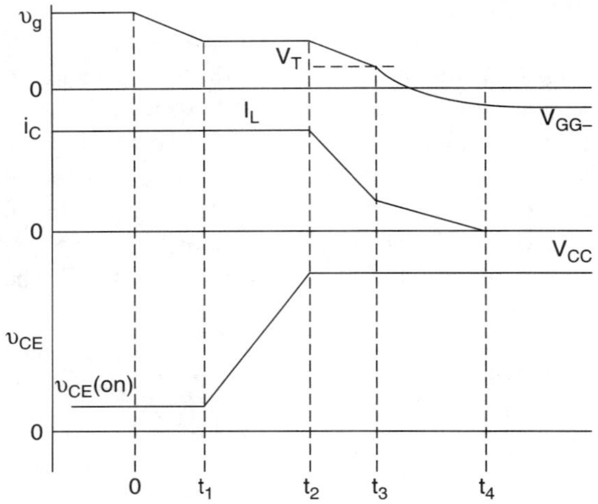

Figure 4.28
Turn-off transients in IGBT

second interval t_4-t_3 corresponds to the removal of injected holes and electrons in the n^- drift region. The excessive storage of the injected carriers is removed through recombination in the drift region. This is usually a slow removal, hence, it is referred to as a current tail.

4.4.6 Effect of temperature

The IGBT has excellent high temperature characteristics in forward conduction. IGBTs have been operated successfully with heatsink temperatures approaching 200°C.

On-State Characteristics

The on-state voltage drop in the IGBT consists of two components: the voltage drop in the n-channel and the voltage drop across the forward biased base to emitter junction of the p^+n^-p transistor Q_2. In the on-state, the n-channel in the MOSFET section of the IGBT behaves like a resistance. The channel resistance has a positive temperature coefficient. Hence, the voltage drop in the channel, $I_C R_{ch}$, has a positive temperature coefficient. The voltage drop across the emitter to base junction in the transistor Q_2 has a negative temperature coefficient. Consequently, the total on-state voltage of IGBT has low temperature dependence. In contrast, the on-state voltage in the power MOSFET consists of only the voltage drop in the channel, which exhibits very strong dependence on the temperature. Hence, the current handling capabilities in IGBT need not be derated at high ambient temperatures.

Furthermore, the temperature coefficient of the on-state voltage in IGBT is nearly zero at the current densities of 100 A/cm^2. A small positive temperature coefficient of the on-state voltage drop at higher current densities is beneficial in ensuring homogenous current distribution within chips and for achieving equal current sharing when paralleling IGBTs.

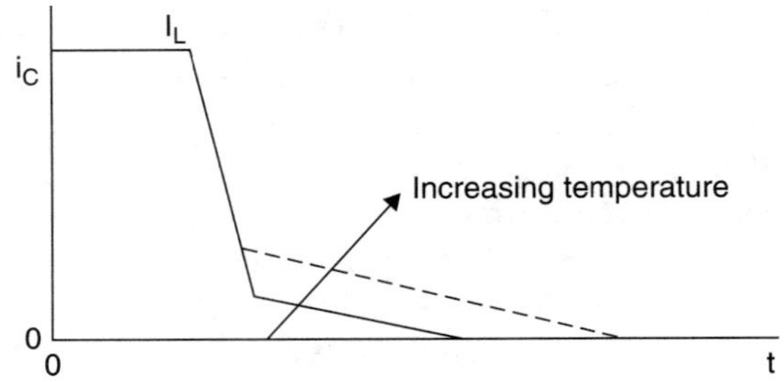

Figure 4.29
The effect of temperature on the collector current in IGBT during the turn-off transition

Switching Characteristics

The gate turn-off time in the IGBT is dominated by the tail in the collector current waveform. The tailing is due to slow removal of injected charge carriers by recombination in the n^- drift region. The temperature rise increases the minority carrier lifetime, thus, slowing down the recombination process, see Figure 4.29. The current tailing increases. For these reasons an n^+ buffer layer is inserted between the n^- layer and the p^+ layer on the collector side. The recombination process becomes faster in the buffer layer.

Latching Current Density

The latching current is dependent on the current gains of the two transistors in the IGBT cell, Q_1 and Q_2. Temperature rise increases both current gains leading to the decrease in the latching current.

4.4.7 Complementary IGBTs

Complementary p-channel IGBT is often needed in power circuits. One example is the bi-directional ac switch using an n-channel and a -channel IGBT as shown in Figure 4.30.

Because of lower mobility of holes, the p-channel IGBT exhibits slower switching speeds. However, the on-state voltage drop of both n-channel and p-channel IGBTs are nearly identical.

A comparison of the reverse blocking safe operating area RBSOA for the complementary types of IGBT are shown in Figure 4.31. The p-channel IGBT exhibits a superior current-latching current density but poor avalanche induced blocking voltage.

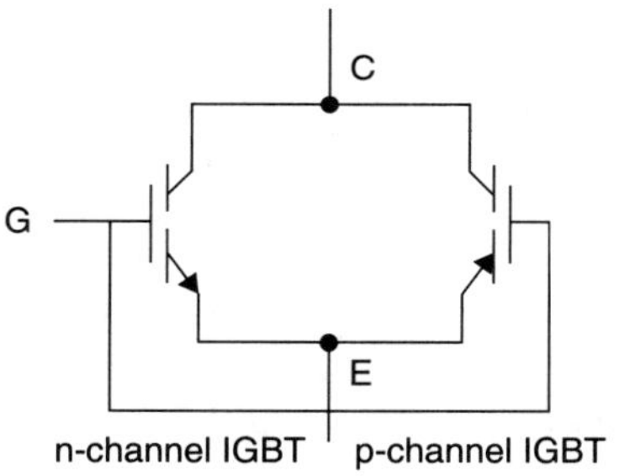

Figure 4.30
Bi-directional ac switch using an n-channel and a p-channel IGBT

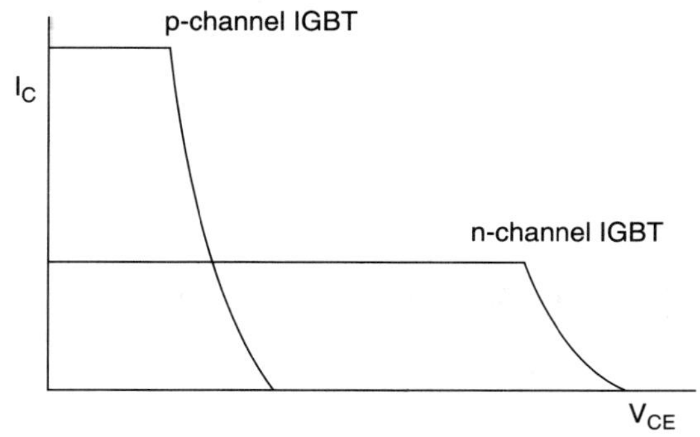

Figure 4.31
Comparison of RBSOA of complementary IGBTs

Example 4-5 Selection of IGBT and the Calculation of Power Loss

For the circuit shown in Figure 4.26 with the following specifications: V_{CC} = 350 V, I_O =4 A, select an appropriate IGBT. The clamping diode is assumed to be an ideal diode with zero reverse recovery time. The circuit is switched at 50% duty cycle at 40 Khz. Calculate the power losses.

The collector voltage and current in the example are drawn in Figure 4.32.

Harris Semiconductor IGBT, HGTP10N50C1, is selected, which has the following specifications:

Collector-Emitter breakdown voltage BV_{CES} : 500 V min

Maximum Continuous Collector current : 10 A

$V_{CE(on)} = 2.5$ V $V_{GE(TH)} = 3$ V, C_{oss} =50 pF (C_{CE}), C_{iss} =500 pF (C_{gE}) , C_{rss} = 30 pF (C_{gC})

g_m = 6.8 (estimated from graph in the data sheet)

turn-on: t_{don} = 50 nS (t_2 in Figure 4.13e), t_r =50 nS ($t_3 - t_2$ in equation (4-24)

turn-off: t_{doff} = 400 nS (t_1 in Figure 4.14b), t_f = 500 nS ($t_2 - t_1$ in equation (4-33)

The rise time of the collector current is obtained from an equation similar to equation (4-24):

$$t_2 - t_1 = 0.27 \text{ nS}.$$

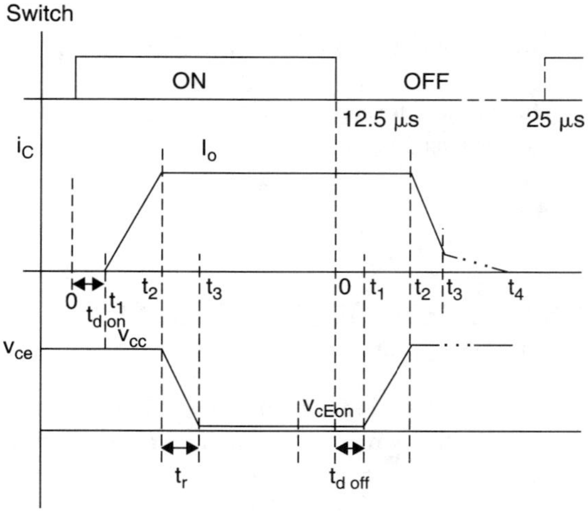

Figure 4.32
Switching waveform in Example 4-5

Let us assume that the current tail is approximately 10 times the fall time of the collector current. Hence, the total fall time of the collector current is obtained from equation (4-36):

$$t_4 - t_2 = 10 \times (t_3 - t_2) = 47.45 \text{ nS}.$$

The switching period is T=1/40 KHz =25 μS. The on duration of the IGBT is

$$\begin{aligned} t_{on} &= T/2 - (t_{don} + (t_2 - t_1) + t_r + t_{doff}) \\ &= 1000 - (50 + 0.27 + 50 + 400) \text{ nS} \\ &= 12.0\ \mu\text{S} \end{aligned}$$

The power loss is calculated from

$$P_{Loss} = P_{sw\text{-}on} + P_{on} + P_{sw\text{-}off}$$

where

$$P_{sw-on} = \frac{1}{T}\left[V_{CC}\frac{I_O(t_2 - t_1)}{2} + I_O\left(\frac{(V_{CC} - V_{CEon})t_r}{2} + V_{CEon}\, t_r\right)\right]$$

$$P_{on} = V_{CEon}\, I_O$$

$$P_{sw-off} = \frac{1}{T}\left[I_O\left(\frac{(V_{CC} - V_{CEon})t_r}{2} + V_{CEon}\, t_r\right) + V_{CC}\frac{I_O(t_4 - t_2)}{2}\right]$$

The switching losses $= P_{sw\text{-}on} + P_{sw\text{-}off} =$ 16.8462 watts,
The on-state losses $= P_{on} = 4.7999$ watts
Total power Loss in the MOSFET
= 21.6461 watts

4.5 COMPARISON OF MOSFET AND IGBT

In the power MOSFET, the decrease in the electron mobility with increasing temperature results in a rapid increase in the on-state resistance of the channel and hence the on-state voltage drop. In IGBT this increase is very small. The on-state voltage drop in power MOSFET increases by a factor of 3 between the room temperature and 200°C. Whereas in the identical conditions in the IGBT, the increment in the on-state voltage drop is very small. This factor is responsible for the maximum current rating of the power MOSFET to go down to 1/3 value at the highest temperature. This feature makes the IGBT extraordinarily well suited for applications, which require operation at high ambient temperatures. Furthermore, the current sharing in multiple paralleled IGBTs is far better than in power MOSFETs.

In power MOSFETs, the channel resistance also increases sharply with the increase in the breakdown voltage rating. The on-state voltage drop for high-voltage MOSFETs is significantly higher than for the low-voltage MOSFETs. In IGBTs, however, the on-state voltage drop of the MOSFET section is a small fraction of the total on-state voltage drop. Consequently the on-state voltage drop in IGBTs exhibit a relatively small increase with the increase in the high voltage rating. Figure 4.33 shows a comparison of forward characteristics in IGBTs and power MOSFETs.

The turn-on transients are identical in MOSFET and IGBT. In the turn-off transition, however, the IGBT has a large current tail due to slow removal of the charge carriers in the drift region through recombination.

The power MOSFET is suited for applications that require low blocking voltages and high operating frequencies. IGBT is the preferred device for applications that require high blocking voltages and lower operating frequencies.

4.6 DEVICE SPECIFICATIONS

This section gives the important specifications that most manufacturers list in their data sheets.

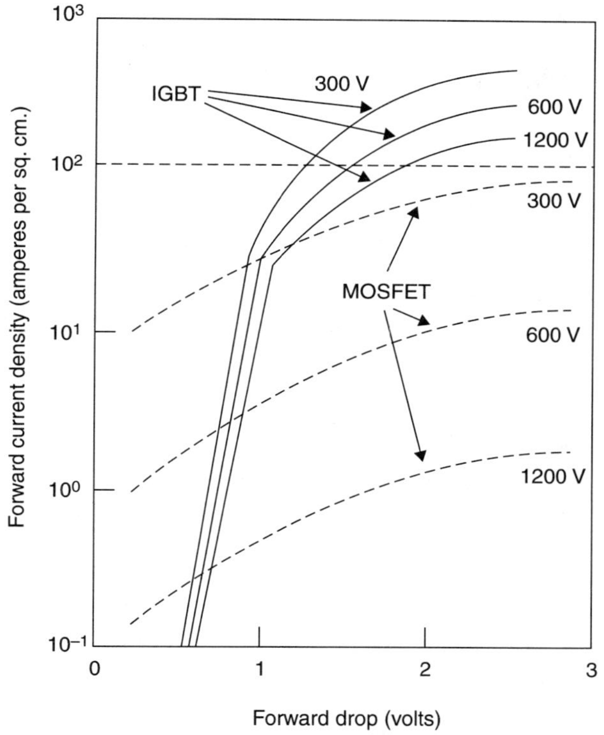

Figure 4.33
Forward characteristics in IGBTs and power MOSFETs.

MOSFET Specifications

Forward Characteristics:

Continuous Drain Current	i_D	The maximum value of the on-state dc current in the drain. It decreases with increase in temperature.
Peak Pulsed Drain Current	I_{DM}	The maximum allowable repetitive peak pulse current in the drain at a specific pulse width. It decreases with increasing temperature. It also decreases with increasing frequency of pulses. Check the curves.
On-State Resistance	$R_{DS(on)}$	The dc resistance between the drain and source in the on-state at a given gate-source voltage. It increases rapidly with increasing temperature.
On-State Drain Current	I_D	The maximum average on-state current through the thyristor under the stated conditions. This is also listed as the continuous current.
Drain-Source Breakdown Voltage	$V_{(BR)DSR}$	The breakdown voltage between the drain and the source when the gate is connected to the source by a given resistance.

Drain-Source Breakdown Voltage	$V_{(BR)DSS}$	The breakdown voltage between the drain and the source when the gate is shorted to the source.
Short-Circuit Output Capacitance	C_{OSS}	The capacitance between the drain and the source terminals with the gate short-circuited to the source for ac signals.

Reverse Blocking Characteristics:

Drain-Source Leakage Current	I_{DSS}	The maximum drain current when the gate is shorted to the source.
Gate-Source Leakage Current	I_{GSS}	The maximum leakage current from the gate to the source when the drain is shorted to the source.
	dv/dt	The maximum rate of rise of voltage between the drain and the source anode before the damage.

Gate Characteristics:

Gate Threshold Voltage	$V_{GS(th)}$	The threshold voltage between the gate and the source to turn on the MOSFET.
Short-Circuit Input Capacitance	C_{ISS}	The capacitance between the gate and the source terminals with drain shorted to source for ac signals.
Reverse Transfer Capacitance	C_{RSS}	The capacitance between the drain and the gate terminal with the source connected to the ground.
Reverse Gate Current	I_{GSS}	The direct current into the gate terminal when the drain is short-circuited to the source.
Gate-Source Breakdown Voltage	$V_{(BR)GSSF}$	The breakdown voltage between the gate and the source when the forward gate-source voltage is applied and the drain is short-circuited to the source.
Gate-Source Breakdown Voltage	$V_{(BR)GSSR}$	The breakdown voltage between the gate and the source when the reverse gate-source voltage is applied and the drain is short-circuited to the source.

IGBT Specifications

Forward Characteristics:

Continuous Collector Current	I_C	The maximum value of the on-state dc current in the collector. It decreases with increase in temperature.
Peak Pulsed Collector Current	I_{CM}	The maximum allowable repetitive peak pulse current in the collector at a specific pulse width. It decreases with increasing temperature. It also decreases with increasing frequency of pulses. Check the curves.
Collector-Emitter Saturation	$V_{CE(sat)}$	The voltage between the collector and the emitter in the on-state at a given gate-source voltage. It increases rapidly with increasing temperature.
On-State Drain Current	I_D	The maximum average on-state current the device may conduct under the stated conditions. This is also listed as the continuous current.
Collector-Emitter Breakdown Voltage	BV_{CES}	The breakdown voltage between the collector and the emitter when the gate is shorted to the emitter.
Collector-Gate Breakdown Voltage	BV_{CGR}	The breakdown voltage between the collector and the gate when the gate is connected to the emitter through a resistance.

Reverse Blocking Characteristics:

Parameter	Symbol	Description
Collector-Emitter Leakage Current	I_{CES}	The maximum collector current when the gate is shorted to the emitter.
Gate-Emitter Leakage Current	I_{GES}	The maximum leakage current from the gate to the emitter when the collector is shorted to the emitter.

Gate Characteristics:

Parameter	Symbol	Description
Gate Threshold Voltage	$V_{GE(th)}$	The threshold voltage between the gate and the emitter to turn on the IGBT.
Gate-Emitter Continuous Voltage	V_{GES}	The maximum dc voltage between the gate and the emitter of either polarity.
Gate-Emitter Pulsed Voltage	V_{GEM}	The maximum peak pulse voltage between the gate and the emitter of either polarity under the pulsed operation.
Gate-Source Breakdown Voltage	$V_{(BR)GSSR}$	The breakdown voltage between the gate and the source when the reverse gate-source voltage is applied and the drain is short-circuited to the source.

Switching Characteristics Common to MOSFET and IGBT:

Parameter	Symbol	Description
Turn-On Delay Time	$t_{d(on)}$	The delay between the application of the trigger and the start of the fall of the drain-source voltage.
Turn-Off Delay Time	$t_{d(off)}$	The delay between the application of the trigger and the start of the rise of the drain-source voltage.
Rise Time	t_r	The time interval between 10% to 90% on the rising edge of the drain-source voltage waveform.
Fall Time	t_f	The time interval between 10% to 90% on the falling edge of the drain-source voltage waveform.
Reverse Recovery Time	t_{rr}	The duration of removal of stored charge when the device is turned off.

Power and Thermal Characteristics Common to MOSFET and IGBT:

Parameter	Symbol	Description
Power Dissipation	P_D	The maximum allowed power dissipation in the MOSFET.
Power Derating Factor	W/° C	The P_d decreases with the increase in the temperature.
Operating Junction Temperature	T_J	The junction temperature of the device as a result of the ambient and load conditions.
Case Temperature	T_C	The temperature of the device case as a result of the ambient and load conditions.
Thermal resistance, Case-to-Ambient	$R_{\theta CA}$	The thermal resistance in steady-state from the case to the ambient.
Thermal resistance, Junction-to-Ambient	$R_{\theta JA}$	The thermal resistance in steady-state from the semiconductor junction to the ambient.
Thermal resistance, Junction-to-Case	$R_{\theta JC}$	The thermal resistance in steady-state from the semiconductor junction to the case.
Transient Thermal Impedance, Junction-to-Ambient	$Z_{\theta JA(t)}$	The transient thermal impedance from the junction to the ambient. It decreases with the decrease in the on-time in the pulsed operation.
Transient Thermal Impedance, Junction-to-Case	$Z_{\theta JC(t)}$	The transient thermal impedance from the junction to the case.

4.7 SUMMARY

Two very important transistor devices: MOSFET and IGBT are examined in this chapter. The important conclusions are as follows:

1. The MOSFET is a normally off device. It is turned on by the application of a positive gate to a source voltage.
2. The MOSFET is a majority carrier device. No charge is stored in the device structure, hence, no reverse recovery problem. Turns on and off faster than the minority carrier devices such as a BJT.
3. The MOSFET behaves like a low resistance in the on-state. The on-state losses are higher for high blocking voltage. Hence, low blocking voltage MOSFET are the preferred devices.
4. The MOSFET with high blocking voltage have long transition times, hence slower.
5. The on-state resistance has a positive temperature coefficient, which enables paralleling of devices for current rating enhancement.
6. The threshold voltage increases with temperature.
7. The MOSFET has a rectangular safe operating area. Hence, minimum snubbering is sufficient.
8. The MOSFET has a body diode connected in anti-parallel between the drain and source.
9. MOSFET also has a parasitic transistor that may latch up in the on-state, more likely in bridge configuration.
10. The IGBT is a BJT with an insulated base.
11. Unlike MOSFET, the IGBT has high reverse blocking voltage ratings.
12. IGBT contains a parasitic thyristor that may latch up in the on-state.
13. IGBT is faster than a BJT but slower than MOSFET.
14. The on-state losses of IGBT are much smaller than those of a MOSFET.
15. The IGBT has a rectangular SOA similar to the MOSFET.

References

1. Ned Mohan, Tore M. Undeland, and William P. Robbins, *Power Electronics: Converters, Applications and Design,* 2nd Ed. , John Wiley & Sons, Inc., 1995.

2. B. Jayant Baliga, *Power Semiconductor Devices,* PWS Publishing Company, Boston, 1995.

3. John G. Kassakian, Martin F. Sclecht, and G. Vergassian, *Principles of Power Electronics,* Addison-Wesley, Boston, 1991.

Problems

4-1 A voltage pulse is applied at the gate of the MOSFET as shown in Figure P4.1. The MOSFET specifications are:

$C_{gs} = 4500$ pF $C_{gd} = 220$ pF $C_{ds} = 750$ pF.

Determine the peak gate current during the turn-on.

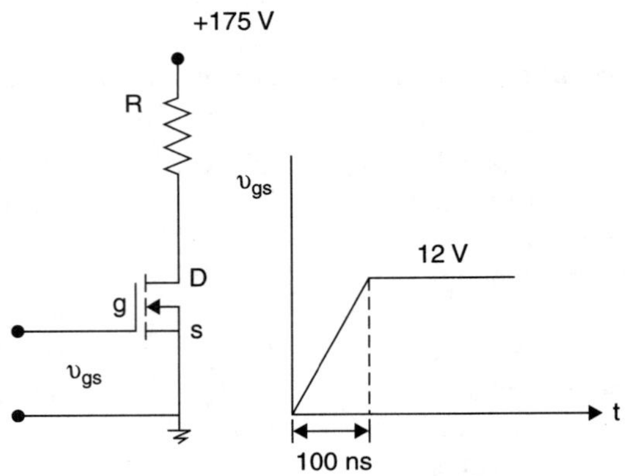

Figure P4.1

4-2 The input capacitance of the gate of the MOSFET of Figure P4.1 increases due to the Miller effect. Find it.

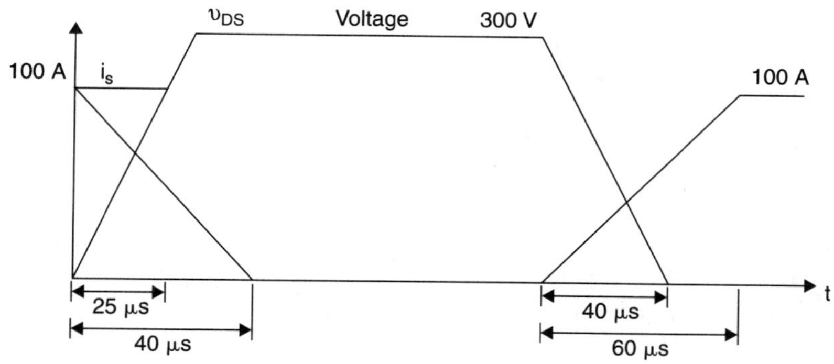

Figure P4.3

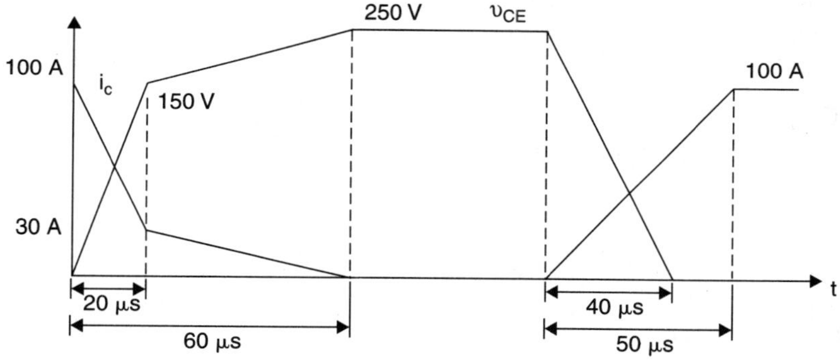

Figure P4.4

4-3 The turn-on and turn-off current and voltage waveform on a MOSFET switch are shown in Figure P4.3. Assume a switching frequency of 40 KHz. Determine the average power dissipation in the MOSFET.

4-4 The turn-on and turn-off current and voltage waveform on an IGBT switch with a resistive load are shown in Figure P4.4. Assume a switching frequency of 40 KHz. Determine the average power dissipation in the IGBT.

4-5 The turn-on and turn-off current and voltage waveform on an IGBT switch with an inductive load are shown in Figure P4.5. Assume a switching frequency of 40 KHz. Determine the average power dissipation in the IGBT.

4-6 The International Rectifier MOSFET IRF1010, 55 V is connected in the switching circuit of Figure 4.12 with the following components: $V_{DD} = 21$ V, $I_O = 25$ A. The clamping diode is assumed to be an ultrafast diode with zero reverse recovery time. Find the total losses in the MOSFET if switched at (a) 50% and (b) 30% duty cycle at 500 KHz.

4-7 Design the optically isolated 5-volt operated totem-pole gate driver circuit for the switching circuit of Example 4-2. The gate driver should apply a 15 Volt (equal to V_{DD}) for best results.

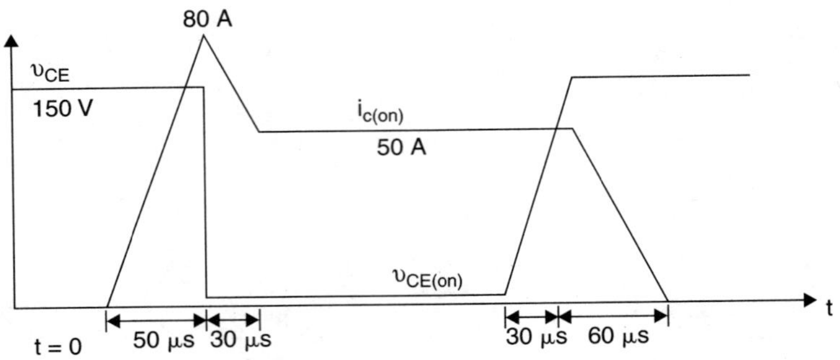

Figure P4.5

4-8 Two MOSFETs, International Rectifier IRF 1010, 55 V, are connected in parallel to share 80 Amps dc. The device capacitances and the estimated stray inductance are as follows:

$C_{gs1} = 3000$ pF, $C_{gd1} = 450$ pF $\quad L_{g1} = 2$ nH

$C_{gs2} = 3100$ pF, $C_{gd2} = 400$ pF $\quad L_{g2} = 2.5$ nH

4-9 For the circuit shown in Figure 4.26 with the following specifications: $V_{CC} = 300$ V, $I_O = 5$ A, select an appropriate IGBT. The clamping diode is assumed to be an ideal diode with zero reverse recovery time. The circuit is switched at 50% duty cycle at 40 KHz. Calculate the power losses.

CHAPTER 5

Thyristors: Pulse Triggered Devices

OBJECTIVES

1. The basic structure of a p-n-p-n device
2. The structure, the equivalent circuit and the characteristics of power thyristor
3. The switching transients in power thyristor
4. The maximum rate of rise of voltage and current in power thyristor
5. Series and parallel operation of thyristors
6. Triggering circuits for thyristor
7. Forced commutation (turning off) of thyristors
8. The structure, the equivalent circuit and the characteristics of gate turn-off (GTO) thyristor
9. The switching transients in gate turn-off thyristor
10. Gate drive circuits for GTO
11. Protection of thyristor and GTO
12. The characteristics of triac
13. The structure, the equivalent circuit and the characteristics of MOS controlled thyristor (MCT)
14. The switching transients in MCT

5.1 INTRODUCTION

Transistors require the input current or voltage trigger signal to be present continuously in order to keep them in the desired state. Thyristors of all types do not have this necessity. They require turning on and off pulses of the duration that is just sufficient to complete the transition from one state to another.

The conventional thyristor is also known by the name SCR, the *silicon controlled rectifier.* It is a three terminal diodelike device with a controlling gate terminal. The thyristor is identical to the power diode in reverse biasing conditions. In the forward bias conditions, the thyristor conduction can be blocked until a certain threshold current is injected into the gate terminal. The thyristor can be triggered into the on-state by the application of short duration current pulses at the gate terminal. Turning off a thyristor, however, is complicated. The thyristor has the best forward conduction characteristics and high-power handling capability. Because of these features, the thyristor is still the strong workhorse in power electronics.

The turning off problem of thyristors is taken care of in another variation, called the gate turn-off thyristor, GTO. A further improvement in the thyristor is obtained by adding the MOS controlled structure at the gate terminal. This device is called the MOS controlled thyristor, MCT. The insulated gate in MCT adds the voltage triggering capability to the low on-state voltage drop and high blocking voltage capabilities of thyristors.

At the end of this chapter, we will examine a popular variation of thyristors, called the triac, which is made of two thyristors integrated monolithically in antiparallel configuration. It is capable of controlling current in both polarities, thus, suited for ac applications.

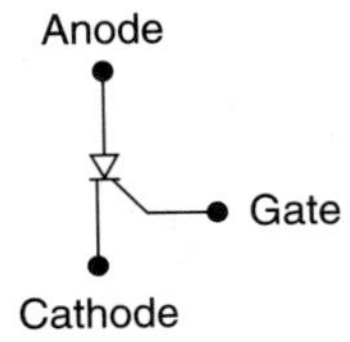

Figure 5.1
Thyristor symbol

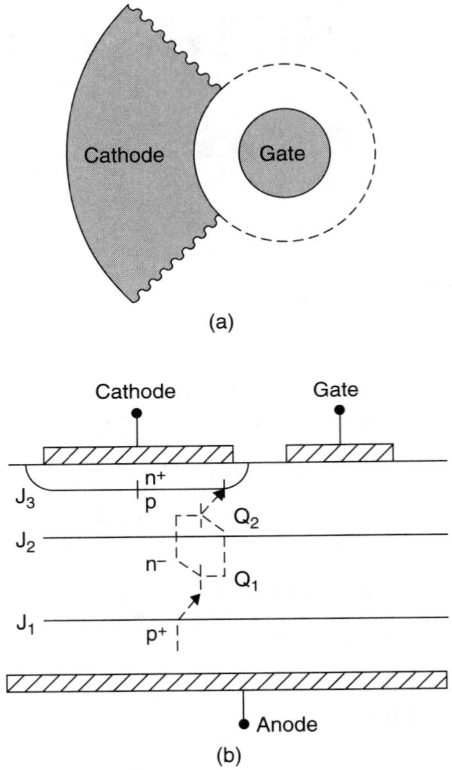

Figure 5.2
Thyristor (a) top view (b) side view

5.2 BASIC STRUCTURE

The thyristor is a normally off device in which the current is carried by majority carriers, see symbol in Figure 5.1. Unlike the transistors, thyristors require short pulse triggers. See Figure 5.1.

Figure 5.2 shows the top view and the side view of one of the many geometries.

The thyristor is made up of four semiconductor layers: the $p^+n^-pn^+$. The device is grown by gradual epitaxial diffusion process from a p^+ substrate that forms the anode part.

The typical dopant densities in various layers are given below:

p^+ layer:	10^{19} /cm^3
n^- layer:	$10^{13}-10^{14}$ /cm^3
p layer:	10^{17} /cm^3
n^+ layer:	10^{19} /cm^3

A thyristor structure has two inherent transistor elements $\mathbf{Q_1}$ and $\mathbf{Q_2}$ intertwined as shown in Figure 5.3. This connection is a positive feedback connection because the collector current of one is the base current of another. A small injection of external current at the gate terminal (the base of Q_2) produces a large collector current in Q_2. The collector current of Q_2 is also the base current of Q_1, hence it gives rise to an even larger current in the collector of Q_1. The collector current of Q_1 is also the base current of Q_2, hence it no longer needs the external current injection at the gate terminal.

A regenerative action sets in. The regenerative action stops only after both Q_1 and Q_2 reach the saturation states. The resulting state of the thyristor is the on-state.

A reverse phenomenon occurs when a small current is removed off of the gate terminal, leading the device to the off-state. However, the turning off of a thyristor is not as easy as it sounds, as we will see.

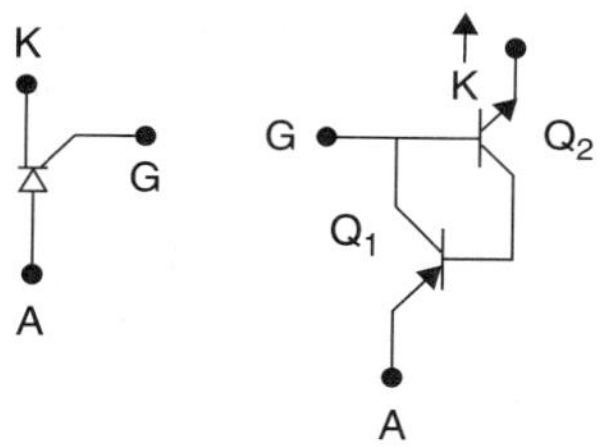

Figure 5.3
Two transistor equivalent representations of the thyristor

5.3 STATIC CHARACTERISTICS

The i-v characteristics in the static condition are given in Figure 5.4.

There are three states in the device operation:

State	Junction bias: J_1: $p^+ - n^-$	J_2: $n^- - p$	J_3: $p - n^+$
Reverse blocking	reverse	forward	reverse
Forward blocking	forward	reverse	forward
On-state	forward	forward	forward

Reverse Blocking State

The anode is negatively biased with respect to the cathode. The junctions J_1 and J_3 are reverse biased and J_2 is forward biased. The junction J_3 has a low breakdown voltage because of the high dopant densities, hence most of the reverse blocking voltage appears across the reverse biased junction J_1. The n^- region, being lightly doped, supports most of the

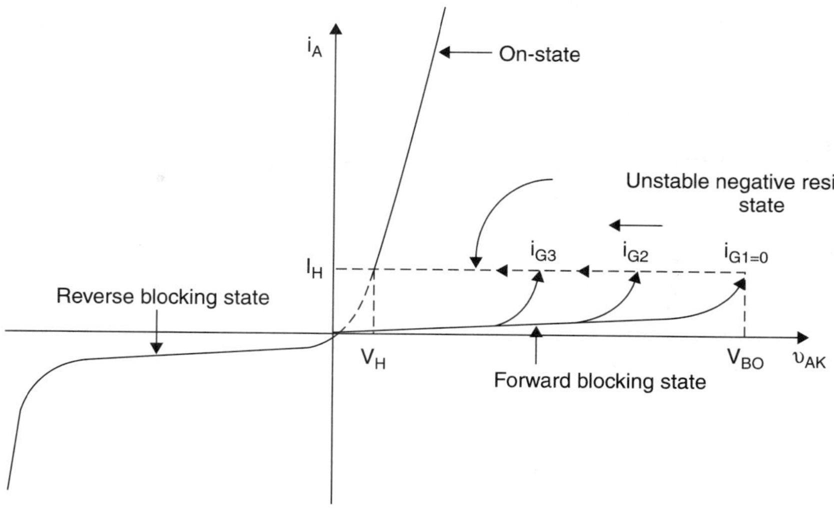

Figure 5.4
The i-v characteristics of the thyristor

depletion layer. Hence a large reverse breakdown voltage thyristor has a long (or wide) n^- region along with low doping levels.

Forward Blocking State

The anode is positively biased with respect to the cathode. The junctions J_1 and J_3 are forward biased and J_2 is reverse biased. Again the n^- region supports most of the depletion layer. Hence, the forward blocking voltage capability should be equal to the reverse blocking voltage capability. However, in practice, the forward blocking voltage is less than the reverse blocking voltage.

The condition of junctions in the forward blocking state suggests that the two transistors Q_1 and Q_2 are in active regions. Hence the collector currents can be expressed as the following:

$$I_{C1} = -\alpha_1 I_{E1} + I_{CO1} \tag{5-1}$$

$$I_{C2} = -\alpha_2 I_{E2} + I_{CO2} \tag{5-2}$$

The α's are the current gains, the I_E's the emitter currents and the I_{CO}'s are the reverse saturation currents of the collector-base junctions in the two transistors. Furthermore,

$$I_A = I_{E1}, I_K = -I_{E2} = I_A + I_G$$

hence,

$$I_A = \frac{\alpha_2 I_G + I_{CO1} + I_{CO2}}{1 - (\alpha_1 + \alpha_2)} \tag{5-3}$$

The above equation states that in order to have a small thyristor current in the forward blocking state, the sum $(\alpha_1 + \alpha_2)$ must be much smaller than unity.

On-State

Transition from the forward blocking state to the on-state takes place by a regenerative action, already described in the previous section. The regenerative action is possible only when the sum of the current gains $\alpha_1 + \alpha_2$ exceeds unity. This point is termed the *break-over point* and the corresponding anode to cathode voltage is called the *break-over voltage* V_{BO}.

The current gains α_1 and α_2 are of low value in the forward blocking state. The junction J_2 is reverse biased. Depletion layer of junction J_2 grows in both the n^--region and p-region as the anode to cathode voltage is increased. The depletion region grows more in the n^- layer. The width of both layers are effectively reduced, see Figure 5.5. The reduced n^--layer forms the base of p-n-p transistor

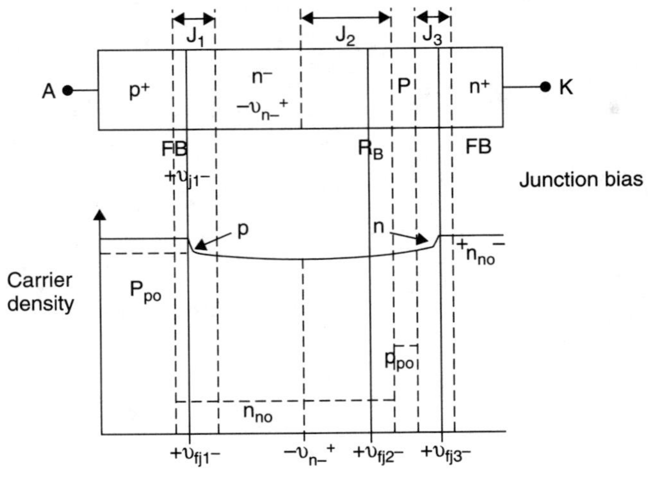

Figure 5.5
The on-state of a thyristor

Q_1, hence its current gain α_1 increases. The reduced p-layer forms the base of n-p-n transistor Q_2, hence its current gain α_2 also increases. The break-over point arrives when $\alpha_1 + \alpha_2$ is greater than one. The thyristor passes through a negative resistance state, which is an unstable state. Finally in the on-state, all three junctions are forward biased. The total on-state voltage drop of the thyristor is given by

$$V_f = V_{fJ1} + V_{fJ3} + V_{n-} - V_{fJ2} \qquad (5\text{-}4)$$

V_{fJ1}, V_{fJ2} and V_{fJ3} are the forward voltages across the junctions J_1, J_2 and J_3 respectively, typically 0.7-0.9 volt each. V_{n-} is the resistive voltage drop in the wide n^- drift region, typically a few tenths of a volt due to the large scale increase in the conductivity of the n^- region in the on-state.

In the stable on-state, there is a strong infusion of minority carriers (both electrons and holes) from the junctions J_1 and J_3 into the combined region now comprising of both n^- and p. The device now virtually behaves like a three layer structure, like a p-i-n power diode. There is a large stored charge in the middle layers and therefore gives rise to the conductivity modulation.

Like the power diodes, the stored charge in the middle layers must be removed during the turning off process, hence a reverse recovery phenomenon occurs in thyristors. The thyristor can be reversed only when the stored charge is reduced below the thermal equilibrium level. The thermal equilibrium level corresponds to the holding current $\mathbf{I_H}$ and the holding voltage $\mathbf{V_H}$. When this occurs, the thyristor will turn off by the regenerative action as mentioned before. The thyristor returns to the forward blocking state.

In a typical thyristor, the gate area is much smaller than the cathode area, refer to Figure 5.2. Hence, when a negative gate current is applied, it causes reverse bias only over a small area of the n^+-p junction J_3. Most of the current from the anode flows directly to the cathodes, see Figure 5.6. The gate region is quite far away from the center of the cathode regions, hence it is not able to affect the direct current flow from the anode to the cathode. The thyristor thus stays in the on-state irrespective of the presence of a negative current at the gate terminal.

Example 5-1 On-State Power Loss in Thyristors

The International Rectifier SE30EF8A, 800 V is connected in the switching circuit of Figure 5.7 where the ac source is $v_S = 169 \sin 2\pi 60t$. Find on-state losses in the thyristor if turned on at $\alpha = \pi/4$. Also find the power delivered to the resistive load.

From the device specifications:

$$V_{on} = 2.15\text{ V}, \qquad T = 1/60 = 16.67\text{ mS}$$

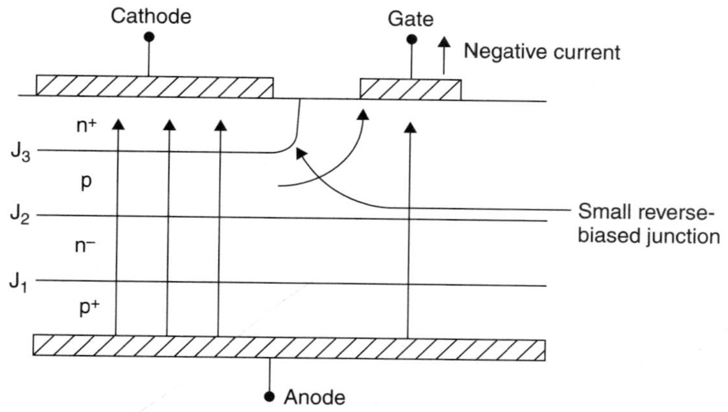

Figure 5.6
Thyristor when a negative current is applied to the gate

The source voltage, $v_S = 169 \sin 2\pi\, 60\, t$
The on-state voltage, $v_{ON} = 2.15 \sin 2\pi\, 60\, t$
$T/8 < t < T/2$

The on-state losses are

$$P_{ON} = \frac{1}{T}\int_{T/8}^{T/2} \frac{v_{ON}^2}{R}\, dt$$

The power delivered to the resistive load is

$$P_O = \frac{1}{T}\int_{T/8}^{T/2} \frac{v_s^2}{R}\, dt$$

$\mathbf{P_{ON}} = 1.05$ watts $\quad \mathbf{P_O} = 6.5338$ kW

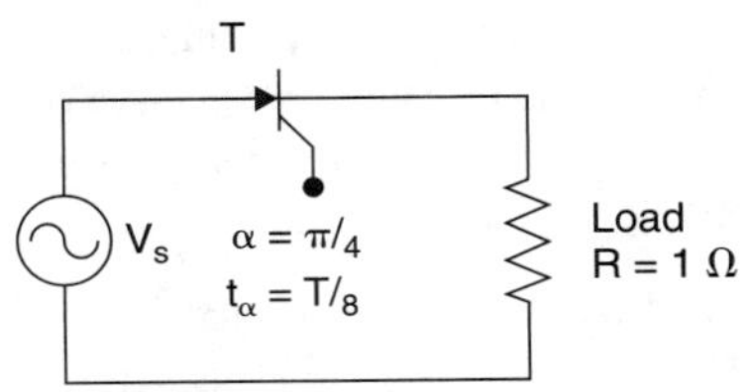

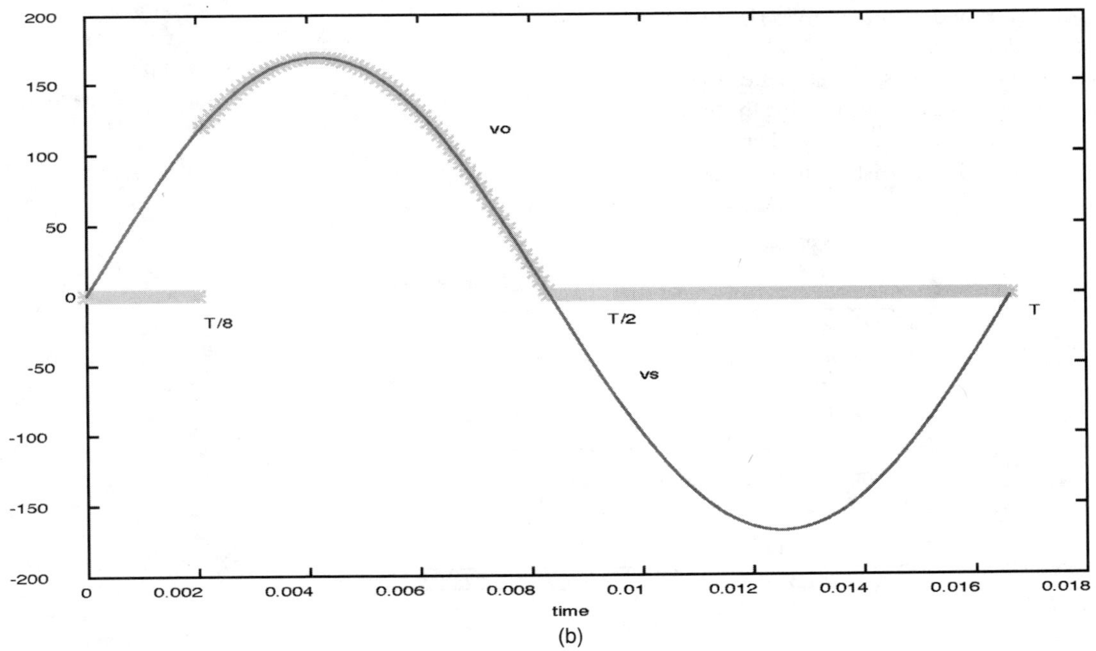

Figure 5.7
(a) A thyristor controlled power delivery (b) Voltage waveforms in Example 5-1

5.4 CATHODE-SHORTED STRUCTURE OF THYRISTORS

In the previous section, it was noted that the forward blocking capability of the practical thyristor is less than the reverse blocking capability. The forward blocking capability can be improved by reducing the current gain of the npn transistor, α_2 at the leakage currents during the forward blocking mode, while maintaining the high value at high currents in the on-state. This is done by short-circuiting the p-base region to the n^+ layer, as shown in Figure 5.8. This structure is referred to as the *cathode shorted structure.* The leakage current of the npn transistor, I_{CO2}, that flows in the p-base is diverted to the cathode at the metalized cathode shorts. This diversion of the leakage current around the n^+ emitter (of the npn transistor) suppresses the current gain α_2 at the low currents in the forward blocking mode. The resulting current gains of the transistors in the thyristor structure with cathode shorts are shown in Figure 5.9.

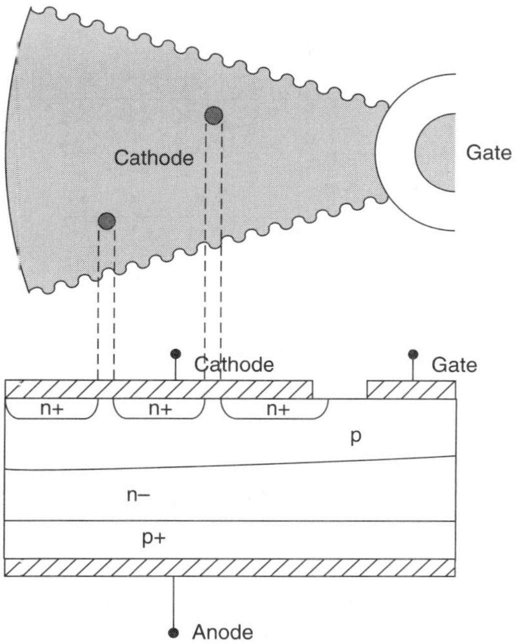

Figure 5.8
Cathode shorted structure of the thyristor

Reduction of the current gain brings the forward blocking voltage closer to the reverse blocking voltage.

5.5 EFFECT OF TEMPERATURE

The voltage blocking capability of thyristors degrades with increasing temperature, see Figure 5.10. The reverse blocking capability of a p-n junction diode is also shown in this figure for comparison. The reverse blocking capability of the p-n junction increases linearly with increasing temperature. In power thyristors, the reverse blocking capability increases initially but decreases rapidly at the elevated temperatures.

The forward blocking voltage capability, however, decreases linearly with increasing temperature.

5.6 SWITCHING CHARACTERISTICS

This section will discuss the mechanisms involved in the transitions of the thyristor from the forward blocking state to the on-state and from the on-state to the reverse blocking state. Along with these main issues, also to be discussed is the thyristor's behavior during high dV/dt and dI/dt conditions.

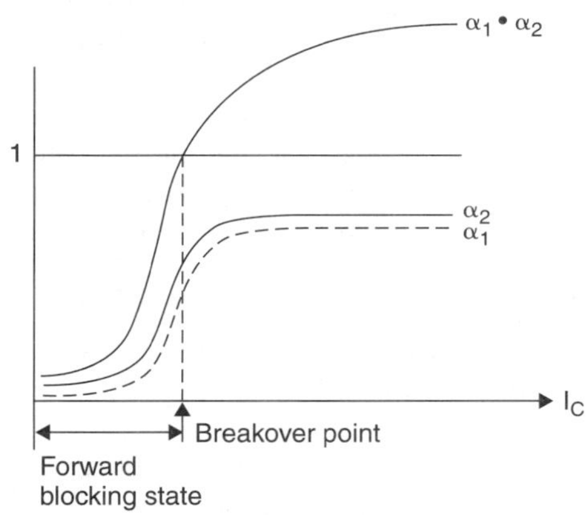

Figure 5.9
Variation of current gains as function of the collector current in the cathode shorted structure

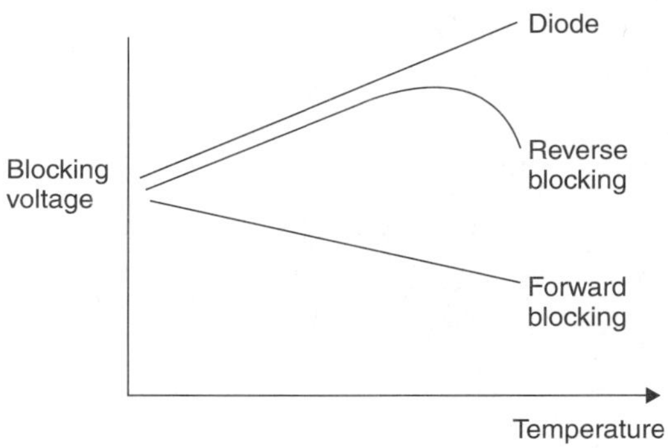

Figure 5.10
Blocking voltage capability of thyristors with respect to the temperature rise

5.6.1 dV/dt capability

In the blocking state of the thyristor, when a rapidly rising voltage is applied between the anode and the cathode, a current flows through the parasitic capacitances. When this current flows through the p-region, which is the base of the npn transistor Q_2, it can act as a gate drive current that can trigger the thyristor into the on-state. This current is given by

$$I_C = C_{J2} A_{J2} \left[\frac{dV_{J2}}{dt} \right] \tag{5-5}$$

C_{J2} and A_{J2} are the area and the capacitance per unit area of the junction J_2 respectively. Since J_2 is the only junction in reverse bias, most of the voltage from the anode to the cathode appears across the junction J_2, hence,

$$V_{J2} \approx V_{AK}$$

Furthermore, the junction capacitance C_{J2} is a strong function of the anode voltage V_{AK}. The vendor specified dV/dt capability of the thyristor is the safe value of dV_{AK}/dt for which the current I_C is smaller than the critical value for triggering the thyristor into the on-state. It can be further shown that

$$\left[\frac{dV}{dt} \right] \propto \frac{1}{\rho_{SB} L^2} \sqrt{\frac{V_{AK}}{N_D}} \tag{5-6}$$

where ρ_{SB} is the resistivity of the p-base region, L is the length of the n^+ -layer under the cathode metal and N_D is the doping concentration in the n^--base region. The dV/dt capability is increased by decreasing the

(a) Resistivity of the p-base region, that is, increasing the doping of the p-layer,

(b) Length of the n^+ layer or partitioning of the layer as is done in the cathode short structure, or

(c) Decreasing the doping concentration of the n^--base.

5.6.2 dI_F/dt capability

When the anode current in the on-state is made to rise rapidly, a localized hot plasma (spot) is formed just under the gate electrode, see Figure 5.11. The instantaneous power dissipation will be large and may damage the device structure in the plasma area.

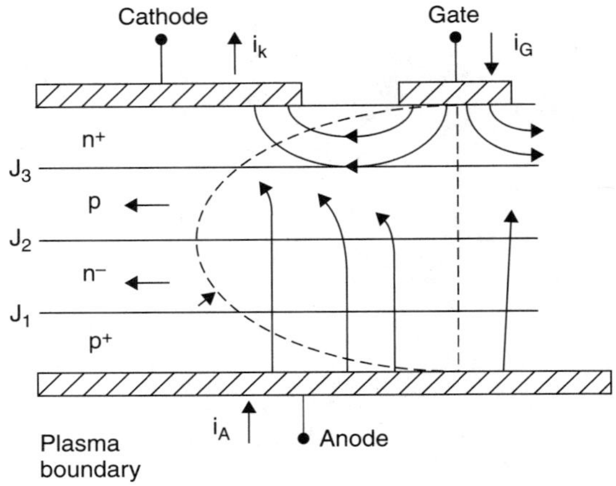

Figure 5.11
Hot plasma formation due to the application of high dI_F/dt in thyristors

5.6.3 Turn-on transients

The thyristor can be triggered from the blocking state to the on-state by the application of a positive gate current pulse. The gate current flows from the gate metal to the locations of the cathode shorts through the p-base region. The p-base region has a parasitic resistance R_S in the lateral direction, also called the pinch resistance. Flow of the adequate gate drive current through this resistance produces a voltage drop that forward biases the base-emitter junction (n^+-p junction) of the npn transistor Q_2. The conduction of current in Q_2 triggers the regenerative action, which was described earlier, until the thyristor settles firmly in the on-state. The turn-on transition consists of three segments, see Figure 5.12: (a) the delay time t_d, (b) rise time t_r and (c) spreading time t_S.

Delay Time This is the time elapsed between the application of the gate drive current pulse and the establishment of the regenerative action. It consists of the transit times t_p and t_n, of the electrons through the p-base region of the transistor Q_2 and

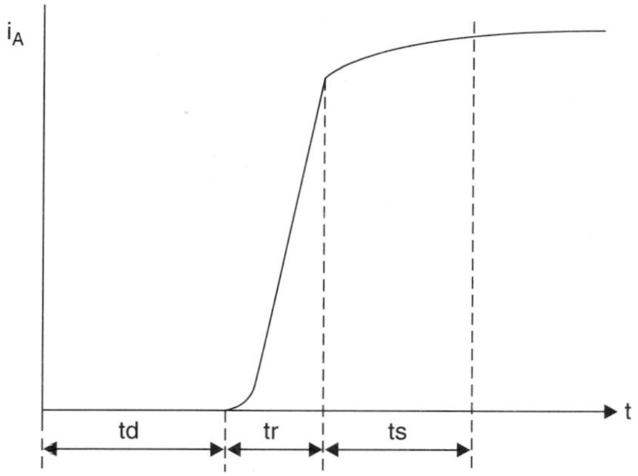

Figure 5.12
The turn-on transition in the thyristor

the holes through the n-base region of the transistor Q_1 respectively :

$$t_d = t_p + t_n \quad : \quad t_p = \frac{W_{BP}^2}{2D_n} \quad , \quad t_n = \frac{W_{BN}^2}{2D_p} \qquad (5\text{-}7)$$

where W_{BP} and W_{BN} are respectively the effective widths of the p-base and the n-base regions under the bias condition (the widths of the undepleted portions). D_n and D_p are the electron and hole diffusion constants respectively. Since the base width W_{BN} is much larger than W_{BP}, the delay time mostly consists of the transit time t_n of the holes through the n-base region. Furthermore, W_{BN} decreases with increasing forward blocking voltage. Hence, the transit time t_n is small for high voltage thyristors, typically 15 microseconds.

During the delay time, the anode current is zero, hence, the cathode current equals the gate current:

$$I_K = I_A + I_G, I_A = 0 \Rightarrow I_K = I_G \qquad (5\text{-}8)$$

Rise Time The rise time is related to the growth of the stored charge in both base regions. The resulting anode current increases exponentially to the steady-state value I_G/α_{1+}:

$$I_A = \frac{I_G}{\alpha_1}[e^{t/\sqrt{t_p t_n}} - 1]: \qquad (5\text{-}9)$$

The rise time is

$$t_r \approx 2.2\sqrt{t_p t_n} \qquad (5\text{-}10)$$

Spread Time In practical thyristors, the gate is located in the center of a wafer of a large diameter. The turn-on first occurs near the gate contacts, then spreads to the rest of the cathode area. The spreading velocity is known to be

$$v_s \propto \frac{I_A^{1/n}}{W_n}\sqrt{\tau_{HL}} \qquad (5\text{-}11)$$

The coefficient n is between 2 and 3. W_n is the actual width of the n-base region and the τ_{HL} is the high level life-time in the n-base region. Thus, the anode current will be distributed more rapidly for large I_A and high dI_A/dt, beneficial from the heat dissipation point of view. High voltage thyristors with large W_n have slow spread velocity, which in turn means that high blocking voltage capability and high dI_A/dt capability are two conflicting requirements. Similarly, the thyristors with fast turn-off time, which require low τ_{HL}, have slow spread velocity and large spread time.

5.6.4 Turn-off transients

The thyristor can be switched from the on-state to the forward blocking state by reducing the anode current below the holding current I_H. This is not easily possible and not preferred in power electronic circuits. Alternately, the thyristor can be relatively easily switched from the on-state to the reverse blocking state by reversing the polarity of the anode voltage. The second way occurs naturally in ac line operated power systems. Figure 5.13 shows the current and voltage waveforms for a thyristor in a resistive circuit when operated with an ac line source. The turn-on time is ignored in this illustration. The thyristor is triggered on at t_0 and remains in the on-state until T/2 when the anode voltage reverses polarity. The n-base region, however, has a large stored charge owing to the on-state bias. The thyristor will continue to conduct current irrespective of the direction of flow. All three junctions in the thyristor continue to remain forward biased. The anode current reverses direction to support the negative anode voltage. The stored charge removal is identical to what we have seen in the p-i-n diode except that here we have a larger number of junctions involved.

Figure 5.14 shows the exploded view of the reverse recovery waveform from T/2 onwards. All three junctions are forward biased until t_1. Electron and hole densities in the p-base and n-base regions are well above the thermal equilibrium levels. These densities at the junctions, however, decrease due to the charge removal by the negative current I_A,

$$i_A = \frac{v_s}{R_L} \quad for \quad \frac{T}{2} \leq t \leq t_1 \qquad (5\text{-}12)$$

At time t_1, the electron concentration at the junction J_3 goes below the thermal equilibrium density and becomes reverse biased. The junction J_3 can support the reverse breakdown voltage BV_{J3}, which is small (usually less than 50 volts) due to heavy doping in the n^+ region. This accounts for the small kink in the negative i_A in Figure 5.14. The anode current after t_1 is

$$i_A = \frac{v_s - BV_{J3}}{R_L} \quad for \quad t_1 \leq t \leq t_3 \qquad (5\text{-}13)$$

At time t_2, the hole concentration at the junction J_1 goes below the thermal equilibrium density and J_1

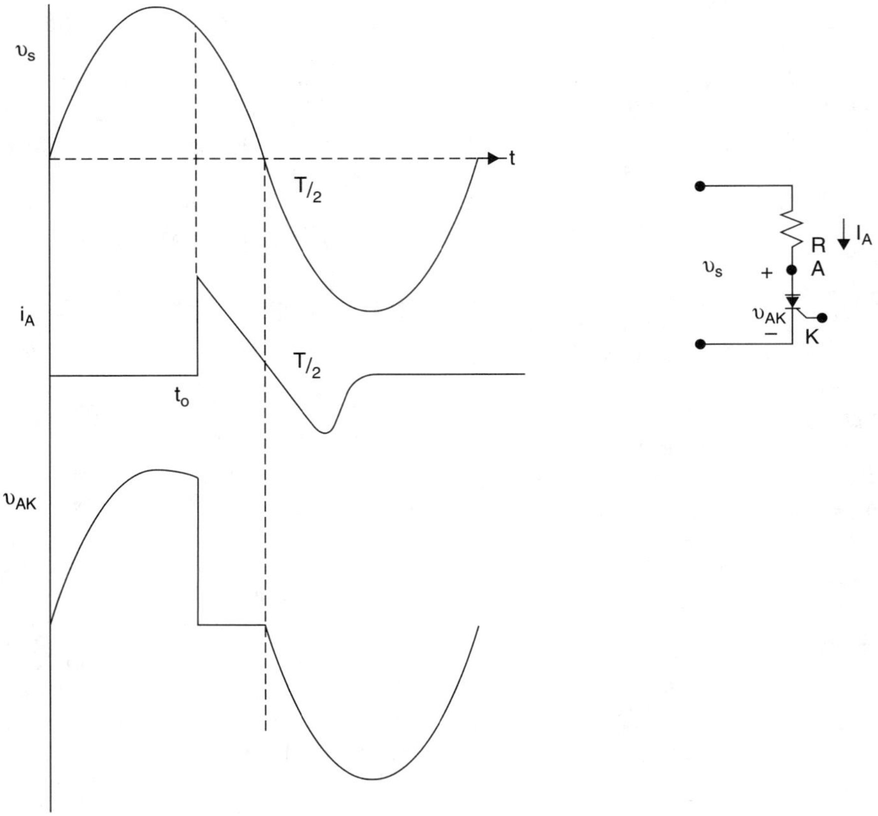

Figure 5.13
The current and voltage waveforms for the thyristor when operated with an ac line source

becomes reverse biased. The junction J_1 can support high reverse breakdown voltage, which is nearly equal to the total reverse blocking voltage capability of the thyristor due to light doping in the n^- region. However, since, the applied negative anode voltage in our circuit is below the reverse capability, the anode voltage increases to the applied voltage V_s after the time t_2. The anode current drops rapidly. The stored charge is being continuously drained. At time t_3, the junction J_2 is also reverse biased and the stored charge is confined to the n^- region only. After t_3, the stored charge is slowly removed through recombination. The reverse recovery is said to be complete at time t_4. The turn-off time is t_4-T/2.

If the line frequency is increased such that the negative half period is over before the recombination is complete, several things happen:

(a) The anode voltage becomes positive and the thyristor experiences the highest dV_A/dt, this is also referred to as the *reapplication of voltage* to the thyristor,

(b) The thyristor enters the forward blocking regime while the stored charge is still present in the n-base region,

(c) The stored charge in the p-base region of transistor Q_2 may trigger the thyristor into the on-state at time T, well before the actual trigger is applied at the gate at $T + t_0$.

The gate control is lost. Thus the highest operating frequency of thyristors is limited by the reverse recovery time, typically to 1 KHz.

5.7 SERIES OPERATION OF THYRISTORS

No two thyristors have identical characteristics. High voltage applications may require a series connection of thyristors. Equal sharing of reverse blocking voltages does not mean that the forward blocking voltage will also be shared equally. We would illustrate this with the example of two series connected thyristors, T_1 and T_2 of different characteristics, as shown in Figure 5.15. The parallel resistance R is selected such that the current through it is an order of magnitude higher than the reverse current of the thyristors.

Let I_{T1} and I_{T2} be the currents through the thyristors T_1 and T_2 respectively during the forward

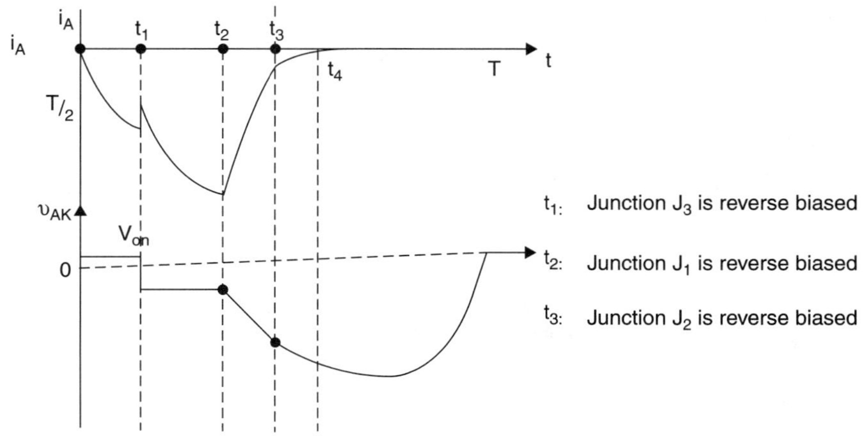

Figure 5.14
The turn-off transition in the thyristor

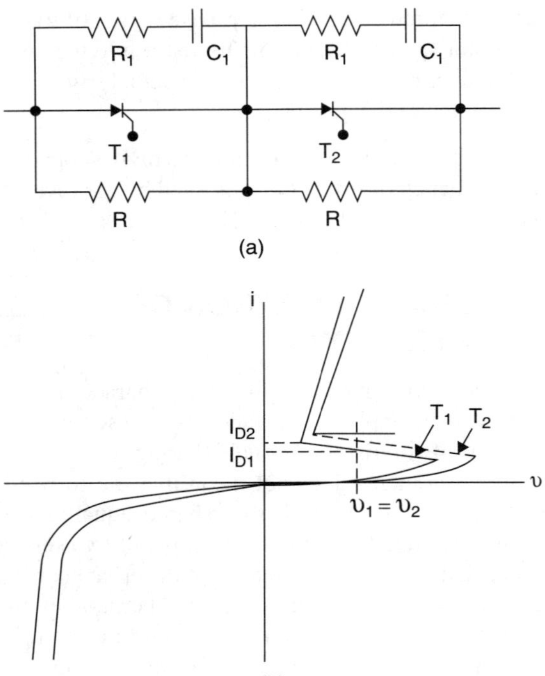

Figure 5.15
(a) Two series connected thyristors and (b) The i-v characteristics of two thyristors

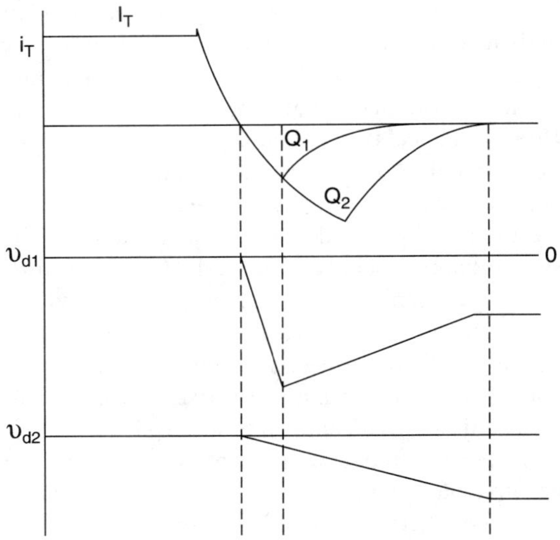

Figure 5.16
Reverse recovery time and voltage sharing

blocking condition. Then, the voltage across the thyristor T_1 is calculated as,

$$I_{T1} = I - I_1, I_{T2} = I - I_2, V_{T1} = I_1 R$$
$$\Delta I_T = I_{T1} - I_{T2} = I_2 - I_1$$
$$V_S = V_{T1} + I_2 R = V_{T1} + (I_1 + \Delta I_T)R \quad (5\text{-}14)$$
$$= 2V_{T1} + \Delta I_T R$$
$$\therefore V_{T1} = \frac{V_S}{2} - \frac{1}{2}\Delta I_T R$$

In the worst case where $I_{D1} = 0$, the forward blocking voltage across the thyristor T_1 is

$$V_{T1(\max)} = \frac{V_S}{2} - \frac{I_{T2}R}{2} \quad (5\text{-}15)$$

During the turn-off, the differences in the stored charges cause differences in the sharing of the reverse blocking voltage. The thyristor with the least recovered charge will face the highest voltage in the transient condition. C_1 across each thyristor ensures equal sharing of transient voltages. R_1 limits the discharge current through C_1, see Figure 5.16. The R_1C_1 combination also provides protection against high dv/dt conditions. The transient voltage across thyristor T_1 can be determined from the following:

$$\Delta V_{T1} = R\,\Delta I_T = \frac{Q_2 - Q_1}{C_1} = \frac{\Delta Q}{C_1} \quad (5\text{-}16)$$

where Q_1 and Q_2 are the stored charges of T_1 and T_2 respectively. Substituting equation (5-16) into (5-15),

$$V_{T1} = \frac{V_S}{2} - \frac{\Delta Q}{C_1} \quad (5\text{-}17)$$

The worst-case transient voltage sharing for $Q_1 = 0$,

$$V_{T1} = \frac{V_S}{2} - \frac{Q_2}{C_1} \quad (5\text{-}18)$$

Example 5-2 Series Connection of Thyristors
Two thyristors, International Rectifier S30EF8A, 800 V, 480 A and 520 A, are connected in series to share 1500 V. The thyristor specifications are:

maximum reverse leakage current difference $\Delta I_T = 40\ mA$

maximum stored charge difference $\Delta Q = 48\ \mu C$

Find the parallel resistance R and the capacitor C that would enable equal voltage sharing within 5%.

Differential voltage on the thyristors $= \Delta V_T = 0.05 \times V_S/2 = 37.5$ V

The series current through the diode and resistor combined is,

$$R = \Delta V_T / \Delta I_T = 937.5 \text{ ohms}$$

The shunt capacitor is determined from equation (5-18):

$$C_1 = \Delta Q / \Delta V_T = 48 \times 10^{-6} / 37.5 = 1.28\ \mu F.$$

The resistor R_1 in series with C_1 is for limiting the discharge current from the capacitor.

5.8 PARALLEL OPERATION OF THYRISTORS

Parallel connection of thyristors may be required in very high current applications. The thyristors may not share current equally due to the differences in the forward on-state characteristics. If one thyristor carries more current than others, its power dissipation increases, the junction temperature rises, the on-state voltage decreases. This in turn would increase its current share, thus leading to a thermal runaway condition and the consequent device failure. The thermal runaway may be avoided by common heat sink to equalize the temperatures of the paralleled thyristors.

Additionally a small series resistance in series with each thyristor forces equal current sharing, see Figure 5.17. The series inductance in each branch helps equalize transient current sharing during turn-on and protects against high di/dt conditions. Magnetic coupling of these inductors helps further. If the current through T_1 increases, a voltage of opposite polarity is induced in the inductor in series with T_2, thereby increasing the current in T_2.

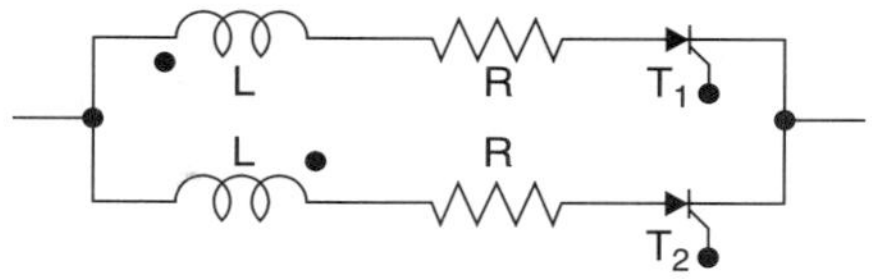

Figure 5.17
Current equalization in paralleled thyristors

5.9 THYRISTOR TRIGGERING CIRCUITS

The turn-on process requires that sufficient charge be supplied to the p-base region in order to increase α_2 to the point of regeneration ($\alpha_1 + \alpha_2 = 1$). Once the thyristor is triggered on, the regenerative action takes over and does not need any more triggering stimulus. The gate current pulse for turning on a thyristor needs to be a short pulse only. However, the gate current pulse must be of an area large enough to charge the base to the regenerative threshold. Typical pulse width is in the range of 5-50 μS.

The amplitude of current pulses must be such that it should be able to trigger the thyristor into the on-state at both the minimum and maximum operating temperatures. The manufacturers specify V_{GT} and I_{GT}, the minimum gate voltage and gate current respectively to trigger the thyristor on. Note that V_{GT} is not the voltage at I_{GT}. See Figure 5.18. The current pulse height should be slightly more than I_{G2}.

The following sections discuss both the nonisolated and isolated gate drive circuits.

5.9.1 Nonisolated gate drive circuits

Figure 5.19 shows a triggering generator circuit that has fast rise and fast fall times. Figure 5.19 also shows the generated current gate pulse. The pulse consists of a positive pulse with an initial overshoot, over a dc pedestal followed by a negative short pulse. The triggering waveform has the following desirable features:

1. The pedestal provides the gate an initial charge, which enables the thyristors to support maximum di/dt at turn on.

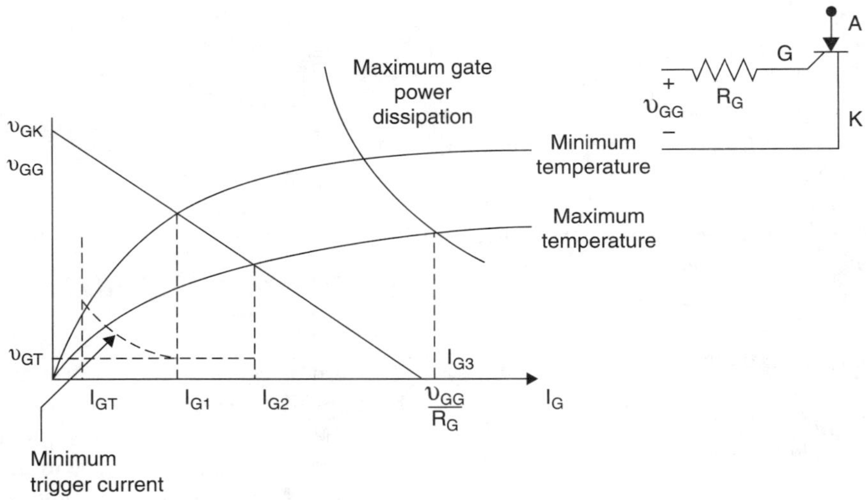

Figure 5.18
Determination of the gate current pulse amplitude in the on trigger for thyristors

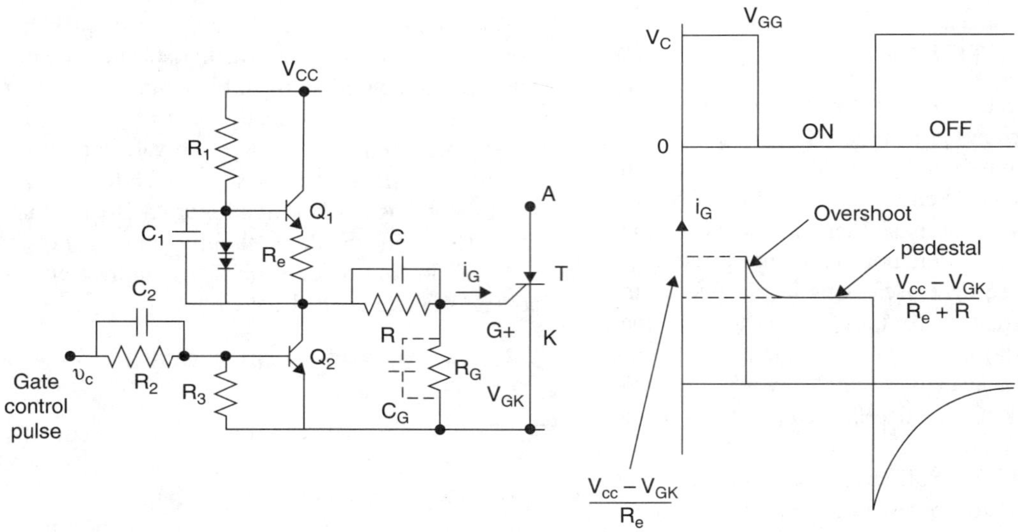

Figure 5.19
Thyristor gate drive circuit

2. The initial overshoot permits the anode current to fall below the holding current, which is helpful in low current situations.
3. The negative short pulse assists the thyristor in supporting high reapplied dv/dt in high frequency applications.
4. For an inductive load, the initial overshoot provides the gate current until the cathode current latches, thus, the thyristor can be avoided from premature turn off.

The resistor R_G subdues noise and high dv/dt at turn on. A capacitor across R_G would enhance this feature. The two diodes at the base of transistor Q_1 determine the overshoot in current i_G.

5.9.2 Optically isolated gate drive

Converters using thyristors, as shall be shown later, require isolation of triggering pulses, that is, the reference terminal cathode to not be grounded. Opto-coupling provides good isolation. The opto-coupler device may be a single transistor, a Darlington pair, a thyristor, or a triac. Figure 5.20 shows two opto-coupled gate trigger circuits; the transistor circuit needs an extra power supply and the opto-thyristor circuit does not. The opto-thyristor must have the same voltage ratings as the main thyristor. Both types of circuits have the danger of unexpected turn-on of the thyristor by the stray capacitance between the optically isolated components. Some manufacturers provide a *Faraday shield* between the input and output of the opto-coupler.

5.9.3 Transformer-coupled gate drive circuits

The pulse transformers are specially suited for thyristor drive circuits. Figure 5.21 shows a circuit that uses a transformer-coupled gate drive. The interwinding capacitance plays the same devastating role as the input-output capacitance in the opto-coupled drives. A *Faraday shield* between the primary and the secondary windings helps. The two

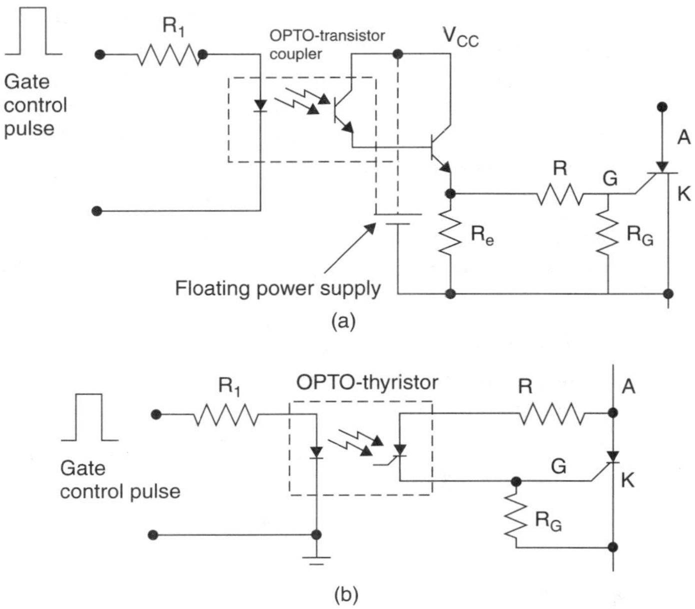

Figure 5.20
The gate trigger circuit using (a) opto-transistor and (b) opto-thyristor

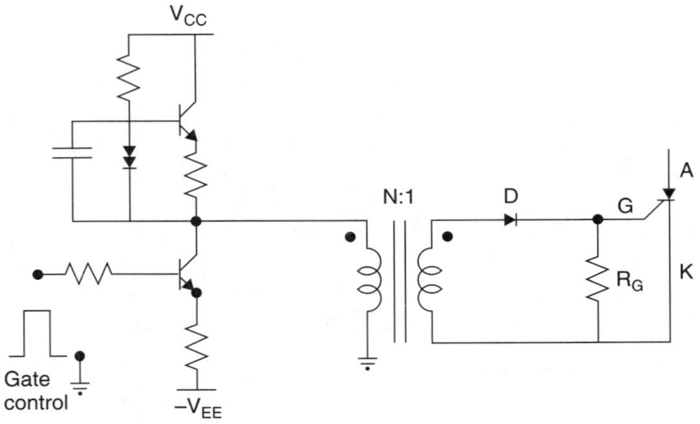

Figure 5.21
The transformer-coupled gate trigger circuit

diodes at the base of transistor Q_1 determine the overshoot in current i_G.

5.10 FORCED COMMUTATION OF THYRISTORS

Forced commutation is the name of the process of turning off a conducting thyristor. The commutation circuit performs two functions:

1. to force the current in a conducting thyristor to zero and
2. to ensure that the thyristor remains reverse biased for a time greater than the reverse recovery time of the device.

There are basically two types of commutation techniques: the impulse commutation and the resonant commutation. The impulse commutation is a hard commutation. The resonant commutation is a soft commutation. In both techniques, the energy is dissipated in the thyristor and the associated components.

5.10.1 Impulse commutation

The impulse commutation involves application of an abrupt reverse voltage to the thyristor. The basic principle of commutation is illustrated by using the schematic given in Figure 5.22(a), which is a step-down dc-dc converter using a thyristor T_1. The forced commutation is performed by the capacitor C and a switch: Before commutation, the thyristor T_2 is off and the capacitor C is prior charged to a negative voltage $-V_C$, the charging circuit is not shown to avoid complexity. Figure 5.22(b) shows the current flow in various stages of commutation.

Assume that the thyristor is on for a sufficient amount of time and the diode D is reverse biased. When turned on, the switch applies $-V_C$ across the thyristor T_1. The stored charge in the thyristor T_1 is removed instantly. The current immediately commutates from the thyristor to the capacitor. After that the load current charges the capacitor linearly from $-V_C$ to $+V_S$. At the end of charging, the diode D turns on and the capacitor is clamped. The thyristor is reverse biased for a time t_{off}, where

$$t_{off} = \frac{CV_C}{I_O} \tag{5-19}$$

The t_{off} is designed to be more than the reverse recovery time t_{rr}.

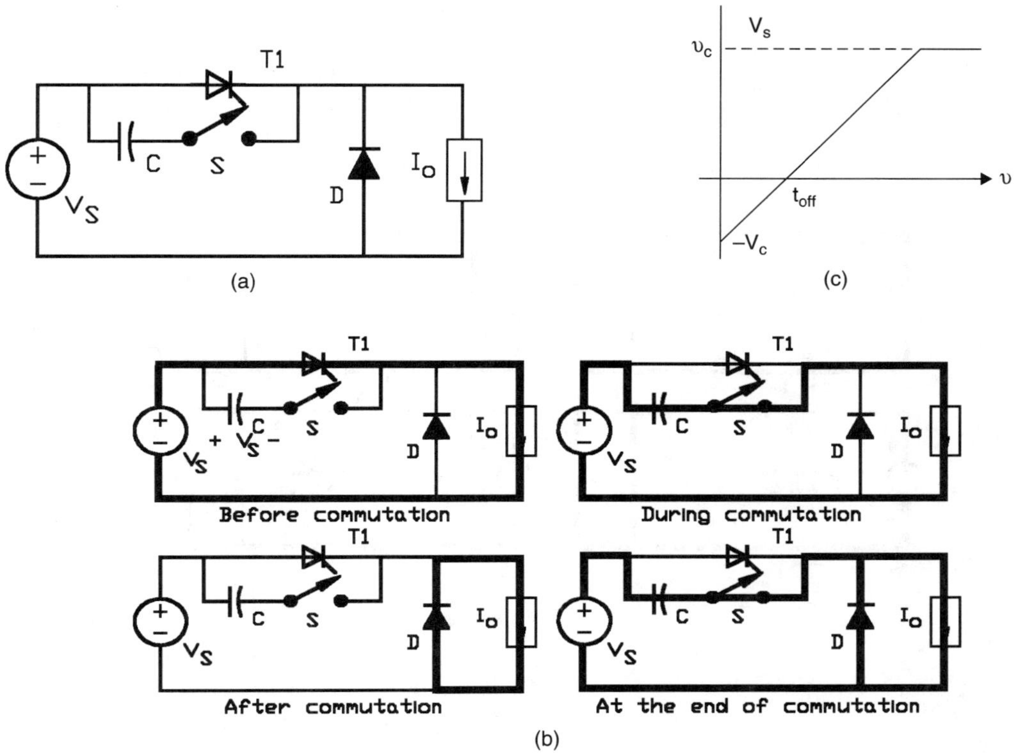

Figure 5.22
Voltage commutation of thyristor: (a) basic circuit (b) current flow during commutation and (c) waveforms

5.10.2 Resonant commutation

The resonant commutation circuit forces the thyristor's current to zero at a finite rate. The schematic is given in Figure 5.23(a). The capacitor C, L and the switch form the commutation circuit. The capacitor C is initially charged to V_S, again the charging circuit not shown. When the switch is turned on, the capacitor C resonates with L. The current through the inductor increases in negative direction sinusoidally, thus increasing the T_1 current over the I_O pedestal. The inductor current cruises up to I_O and the T_1 current cruises down to zero at t_{off}, at which point the T_1 current naturally ceases. The current through the capacitor then becomes constant at I_O value, which charges it linearly until it equals the source voltage V_S at time t_3. Any further increase in the capacitor voltage is to be done at the reduced current, which turns on the freewheeling diode D. The voltage across thyristor T_1 clamps to V_S once D turns on. The inductor current continues to decrease until it reaches zero. The commutation is complete.

The switches in both techniques are implemented by another thyristor and a diode in antiparallel.

Example 5-3 Comparison of Impulse and Resonant Commutation

The parameters in the impulse commutation circuit of Figure 5.22 and the resonant commutation circuit of Figure 5.23 are:

$$I_O = 50\,A, \quad v_S = 169 \sin 2\pi 60t, \text{ and } |V_C| = 169\,V$$

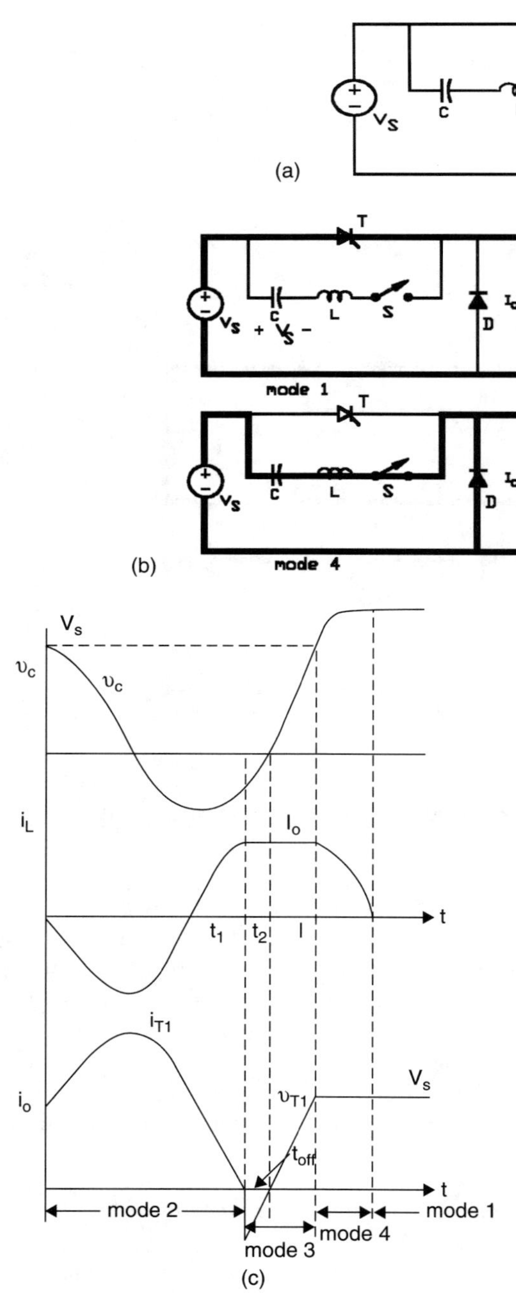

Figure 5.23
Resonant commutation of thyristor: (a) Basic circuit, (b) Circuit modes and (c) Waveforms

Find the C and L components in both circuits to achieve a turn-off time of 20 μS. Design the resonant frequency to be 30 kHz.

Impulse commutation :
C is obtained from equation (5-19)

$$C = \frac{t_{off} I_O}{V_C} = \frac{20 \times 10^{-6} \times 50}{169} = 5.92\ \mu F$$

Resonant commutation :
The capacitor voltage, the inductor current and the thyristor in the resonant circuit (to be derived later in Chapter 11) are given by

$$v_C = V_S \cos \omega_O t \quad : \quad \omega_O = \frac{1}{2\,\pi\, 25 \times 10^3}$$

$$i_L = -\frac{V_s}{\omega_o L} \sin \omega_o t$$

$$i_T = I_o + -\frac{V_s}{\omega_o L} \sin \omega_o t$$

The thyristor commutates at t = t_{off} where the thyristor current becomes zero, hence,

$$0 = I_O + -\frac{V_S}{\omega_O L} \sin \omega_O t_{off}$$

$$L = -\frac{V_S}{\omega_O I_O} \sin \omega_O t_{off} = 1.05\ \mu H$$

C is obtained from $C = \frac{1}{\omega_O^2 L} = 2.67\ \mu F$

5.11 THYRISTOR SNUBBER CIRCUITS

The thyristor requires strong snubbering to limit

1. the voltage and dV/dt on the thyristor during turn-off and during the reapplication of forward voltage
2. the current and di/dt during turn on

The peak reverse current in the reverse recovery transition of the thyristor poses a serious problem, especially in the presence of the line conductor inductance and any current and di/dt limiting inductor. The snubber circuit for the thyristor is shown in Figure 5.24. The series inductance L_S limits both the turn-on current magnitude and the rate of rise di/dt. The shunt capacitor C_S limits both the voltage magnitude and the rate dV/dt during turn-off.

The energy loss in each snubbering cycle is

$$W_{loss} = 0.5\ C_S\ V_R^2$$

V_R is the reverse voltage on the thyristor. Excess limiting using high value C_S increases the snubbering loss. Resistance R_S dissipates the lost energy during the snubbering process. This energy can also be recovered. Readers should refer to the literature on the topic.

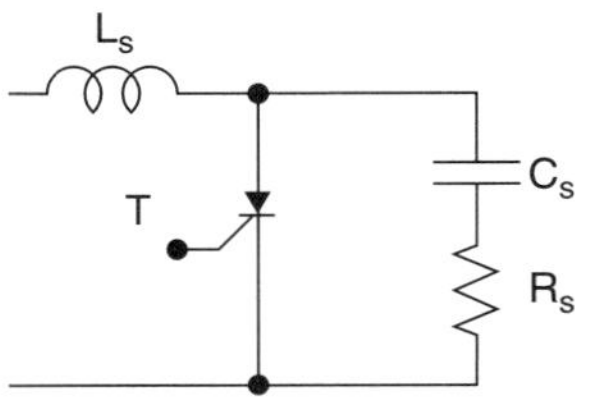

Figure 5.24
Turn-off snubbers for thyristors

Example 5-4 Snubber Design

Components in the snubber circuit of Figure 5.26 are:
$R = 5$ ohms, $L = 1\ \mu H$, $C = 0.1\ \mu F$,
V_S : a step of dc voltage 350 V,
Thyristor: International Rectifier S30EF80A.
The series resistance R_S is selected to limit the maximum current through the capacitor to 500 A. Determine the snubber inductance and capacitance if the maximum permitted values of di_T/dt and dV_T/dt are 350 A/µS and 350 V/µS respectively. Ignore the switching times.

R_S is selected from the capacitor current limitation: $R_S = 350$ V /500 A = 0.7 ohms.

Turn-on: The inductor current is zero prior to turn-on, hence, the maximum di_T/dt occurs at t = 0. The snubber inductance is then calculated as

$$L \approx \frac{V_S}{\left.\frac{di_T}{dt}\right|_{max}} = \frac{350}{350\ A/\mu S} = 1\ \mu H$$

Turn-off: The maximum dv_T/dt on the thyristor occurs at t = 0 when the inductor voltage is zero, the voltage on the capacitor is zero and the current through the capacitor is determined by the load resistance R. The snubber capacitance is calculated from

$$C = \approx \frac{i_C}{\left.\frac{dv_T}{dt}\right|_{max}} \approx \frac{350/5}{350\ V/\mu S} = 0.2\ \mu F$$

5.12 GATE TURN-OFF THYRISTORS (GTO)

Thyristors are suitable for ac line operated systems but not suitable for dc power systems owing to the turn-off problems. The operation in dc applications requires the capability of transition from the on-state to the forward blocking state. The gate turn-off thyristor (GTO) is the answer. The GTO requires a positive current pulse at the gate for turn-on and the negative current pulse for turn-off.

Some of the structural differences vis. a vis. the thyristor are as follows:

1. The p layer is thinner than in thyristor,
2. Cathode periphery is maximized by building cathode as small islands,
3. The gate to cathode distance is minimized, and
4. The n^- layer is shorted to the anode (called the anode-short).

Figure 5.25 shows the symbol and the structure of a GTO. The anode short is implemented by diffusing n^+ beads directly below the anode islands. This shorting is equivalent to shorting the base and emitter of the inherent transistor Q_1 in the thyristor structure, refer to Figure 5.2. The shorting speeds up the turn-off process. It effectively shorts the junction J_1 which, if you recall from the previous section on the thyristor, was responsible for withstanding almost total reverse blocking voltage.

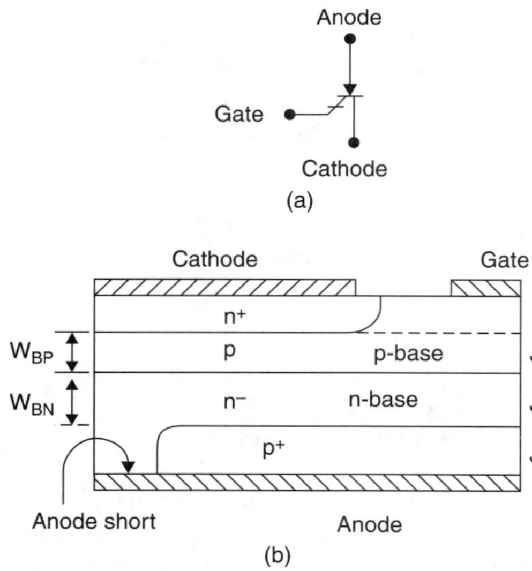

Figure 5.25
The symbol and structure of a GTO

Hence the price is paid in terms of the reduced reverse blocking voltage capability (approximately 20 to 30 volts). Since the GTO is primarily intended for dc applications, the reduced reverse blocking capability is not important.

The island structure of GTOs makes it necessary to apply higher positive gate current pulses than a thyristor for turn-on. In fact, a continuous gate current is required to ensure complete turn-on of all cathode islands.

The turn-off gate current diverts the holes from the p-base region to the gate terminal, see Figure 5.26. This reduces the current gain α_2 of the npn transistor Q_2 and breaks the regenerative cycle. The minimum base current for conducting the current in the npn transistor Q_2 is $(1-\alpha_2)I_K$. The condition for obtaining the turn-off is

$$\alpha_1 I_A - I_G < (1 - \alpha_2)\, I_K \tag{5-20}$$

Using equation (5-20) and $I_K = I_A - I_G$:

$$I_G > \frac{(\alpha_1 + \alpha_2 - 1)}{\alpha_2} I_A \tag{5-21}$$

The turn-off gain β_{max} is defined as the ratio

$$\beta_{max} = \frac{I_A}{I_G} = \frac{\alpha_2}{(\alpha_1 + \alpha_2 - 1)} \tag{5-22}$$

A large value of β_{max} is desirable. Typical value is in the range of 2 to 10. The turn-off current gain is small, hence large negative current pulses are needed for turning off. But a large amount of nega-

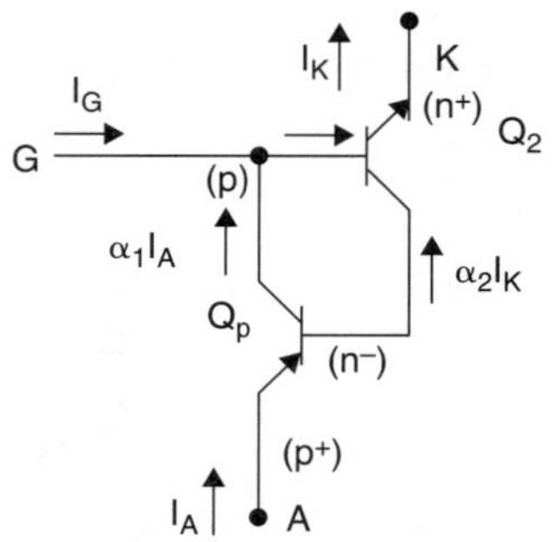

Figure 5.26
Turn-off gate current

tive gate current causes current crowding, thus the maximum current a GTO can turn off is limited.

The i-v characteristics of a GTO are similar to those of the thyristor.

5.12.1 Switching characteristics

The turn-on transition of a GTO is identical to the thyristor. We would therefore, concentrate only on the turn-off transition.

The turn-off time consists of three intervals:

1. storage time t_S,
2. the fall time t_f and
3. the current tailing time, see Figure 5.27.

Storage Time

The storage time is actually the delay time between the application of the reverse (negative) gate current pulse and the beginning of the decrement in the anode current. The stored charge in the p-base region is extracted until the junction J_2 becomes reverse biased across its entire area. During this interval, the cathode current that flows through a part of the n^+ emitter, is approximately constant.

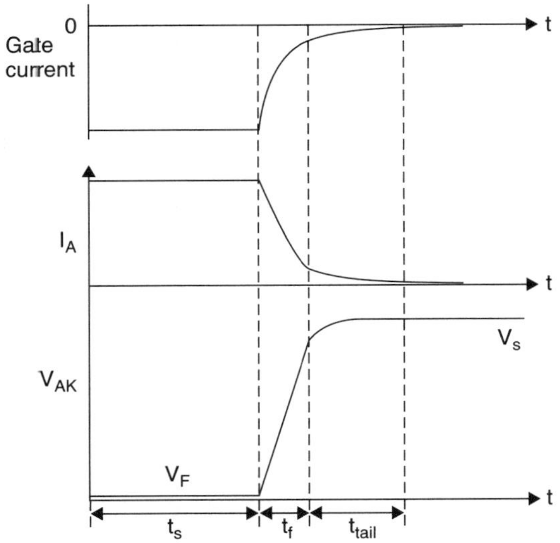

Figure 5.27
Current and voltage waveforms during the turn-off for a GTO

$$t_S \propto \beta \frac{W_{BP}}{W_{BP} + W_{BN}} \tag{5-23}$$

Where β is the turn-off gain, W_{BP} and W_{BN} are the thickness of the p-base and n-base regions respectively. The storage time can be reduced by performing the turn-off with a lower turn-off gain, which means by increasing the negative gate drive current.

It is customary to drive the gate by a linearly increasing current ramp. The storage time in this drive is

$$t_S \propto W_{BP}^2 \frac{I_K}{R} \tag{5-24}$$

where I_K is the cathode current and R is the gate current ramp rate. The storage time can be reduced by increasing the current ramp rate R.

Fall Time

At the end of the storage time interval, junction J_2 becomes reverse biased across the entire area under the n^+ emitter. The n-base region, however, still contains a high concentration of the carriers, the holes and electrons, at the end of the storage time interval t_S. The depletion layer of the reverse biased junction J_2 begins to form in the n-base region gradually as the carriers are being swept out. This process is complete in the time t_f, which is

$$t_f \propto \frac{1}{W_{BP} + W_{BN}} \sqrt{\frac{2\epsilon V_s}{qN_D}} \tag{5-25}$$

N_D is the dopant concentration in the n-base region and V_S is the steady-state voltage between the anode and the cathode. The fall time increases with the source voltage V_S.

Current Tail

At the end of time t_f, the depletion layer reaches its maximum width in the n-base region. This maximum value is dictated by the breakdown considerations. A small part of the stored charge is still left in the n-base region, which is removed by the slow recombination process. Anode-short in the GTO structure expedites the removal of this remnant charge.

The storage time t_S is usually the longest of the three times in the turn-off.

5.12.2 Maximum controllable current in GTOs

The maximum controllable current in GTO is defined as the highest anode current that can be turned off under the gate control. A large value allows the GTO to be operated over a wider range of the anode current level in the controlled turn-off. The GTO is designed for an allowable peak current that is less than the maximum controllable current by a safety factor.

5.12.3 GTO gate drives

GTO gate drive circuits differ from thyristor circuits in the following features:

1. A continuous gate current (the current pedestal) is necessary throughout the on-time of the GTO. This pedestal prevents the GTO from dropping out of conduction at low cathode currents.
2. A high amplitude (but low energy) negative current is required for turn off. The peak value is determined by the turn-off gain and the value of the cathode current switched. The rate of fall of the negative turn-off pulse must be less than the maximum specified di/dt. An inductor in series with the gate circuit usually serves this purpose. The drive circuits are abundantly available in the literature.

5.12.4 Protection of GTOs

It is strongly recommended that the GTO must not be turned on for a specified time after turn-off. Otherwise local hot spots will appear and destroy the device structure. Some excess carriers remain in the n-base region for a fairly long time and become the source of high conductivity areas under a few cathode islands. Hence, if an attempt is made to turn on the GTO before all the excess carriers are recombined, then the current would be confined to these high conductivity areas. Thus hot spots would develop in the GTO structure and the device may be destroyed.

Similarly, it is also strongly recommended that the GTO must be maintained in the on-state for a specified time before the turn-off is initiated. Again, the reason is to prevent the development of hot spots in the GTO structure.

The GTO must also be protected against over current. The over current is detected by current sensing. If the overcorrect is more than the maximum allowed peak current but less than the maximum controllable current, then the GTO can be protected by turning off with a negative current at the gate. If the overcorrect is more than even the maximum controllable current, then an attempt to protect the GTO by applying a negative current at the gate would result in the device failure. In this case the GTO is protected by a *crowbarring technique,* shown in the circuit of Figure 5.28. The thyristor in parallel with the GTO is turned on upon the detection of the over current to blow the fuse. In the case of bridge connected GTOs, such as in the full-bridge converter in single-phase or three-phase, all GTOs are turned on simultaneously to blow the fuse.

The GTO normally requires snubbers: the turn-on snubber to limit the **di/dt** and the turn-off snubber to limit the **dv/dt.** The turn-on snubber is in series and the turn-off snubber is in parallel, see Figure 5.29.

GTOs with high forward blocking voltages and high current handling capabilities are commercially available. The typical blocking voltage of 4500 volts, with a maximum controllable current of 1000 amperes, is achievable. GTOs are typically used in

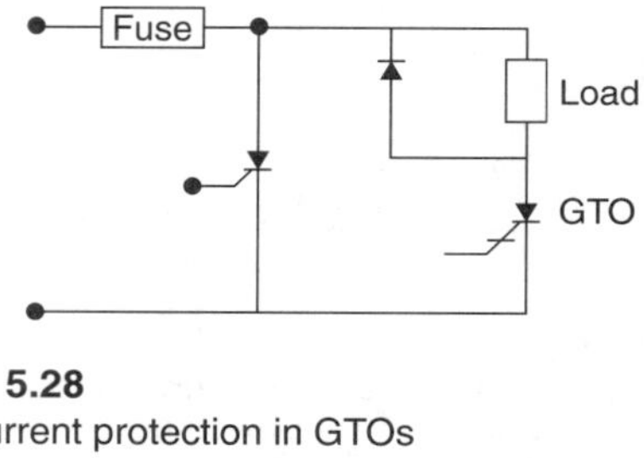

Figure 5.28
Over current protection in GTOs

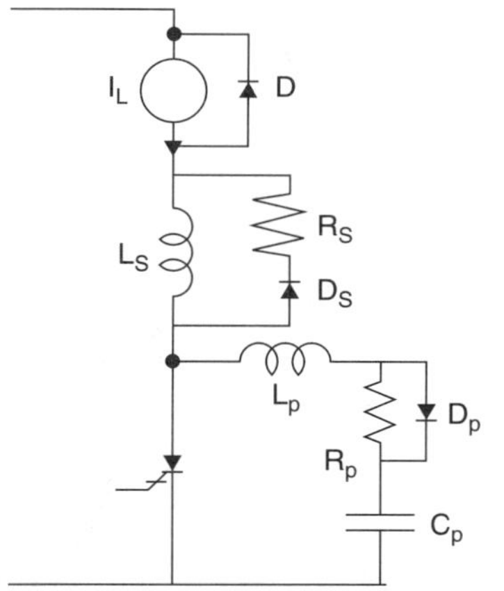

Figure 5.29
GTO with snubbers

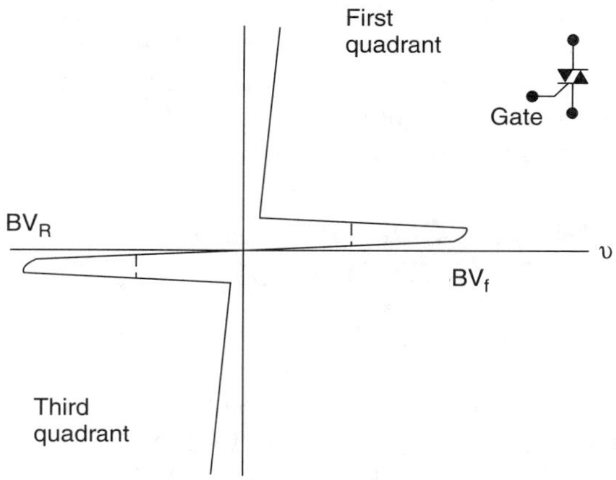

Figure 5.30
The i-v characteristics of a triac

the area of traction drives, such as those used in electric cars and electric locomotives.

5.13 TRIACS

The triac is a device that contains two thyristors connected back to back. It has bidirectional voltage blocking capability and bidirectional current conduction capability. The Triac is fully controlled from a single gate terminal. The i-v characteristics of a triac are shown in Figure 5.30. The triac has an approximately equal forward and reverse breakdown voltage and the gate triggered current conduction in both the first and the third quadrant with a low on-state voltage drop.

The triac structure is shown in Figure 5.31. The triac structure is basically an integration of two back-to-back thyristors ($n^+pn^-p^+$ and $p^+n^-pn^+$) into a monolithic structure. The integration is achieved by replacing the p^+ anode region of thyristor 1 with a second p-base region and by the diffusion of an additional n^+ emitter region on the back surface of the silicon wafer. This results in the formation of an additional junction J_5. The n^- region (n-base region) supports the blocking voltage in both directions. The on-state current in the forward and reverse directions flows through two separate areas within the structure.

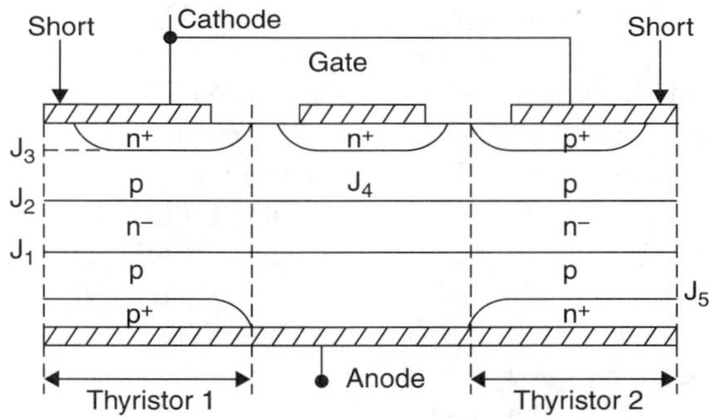

Figure 5.31
Triac structure

dV/dt Limitation

Consider that the triac is operating in the on-state in the first quadrant. When the source voltage reverses from positive to negative, thyristor 1 undergoes through the reverse recovery process and thyristor 2 enters the forward blocking mode. As a consequence, a high dV/dt is applied to thyristor 2. While this happens, a large concentration of carriers exist in the common n-base region. Hence it becomes a case of reapplied dV/dt stress on thyristor 2. The carriers in the n-base region cause the triggering of thyristor 2 in the on-state without the application of the gate control signal. As the operating frequency of the AC source voltage is increased, the reapplied dV/dt stress also increases. Thus, the maximum operating frequency of the triac is substantially lower than that for power thyristors.

5.14 MOS CONTROLLED THYRISTOR (MCT)

The MOS controlled thyristor is a thyristor with an insulated gate terminal. The insulated gate simplifies the drive circuits. MCT has low on-state losses and high current capabilities of thyristors and the simpler drive characteristics and faster switching speed of MOSFETs. An MCT is constructed as the parallel connection of many small MCT cells. An MCT cell is shown in Figure 5.32(b).

The MCT structure has four inherent transistors. The two transistors pnp $\mathbf{Q_1}$ and npn $\mathbf{Q_2}$ constitute the thyristor and the two MOSFETs, n-channel MOSFET $\mathbf{Q_4}$ and p-channel MOSFET $\mathbf{Q_3}$ connected to the bases of Q_1 and Q_2 respectively. The n-channel MOSFET is responsible for turn-on and p-channel MOSFET is responsible for turn-off. All MCT cells contain the p-channel MOSFET but only a few (1 in 20) n-channel MOSFETs. The n-channel MOSFET (the on-MOSFET) is on the outer periphery of the MCT cell. An n-channel MOSFET (the off-MOSFET) may help in turning more than one MCT cell in the vicinity because some of the excess carriers in the n-channel may diffuse to the nearby cells and provide the trigger for turn-on. Hence only a few seed n-channel MOSFETs are needed in the whole structure.

The p-channel off-MOSFET is integrated within the p-base region of the thyristor structure. An n-region is diffused in the p-base region alongside the n^+ island. This n-region forms the substrate for the p-channel MOSFET. A p^+ island is diffused under the cathode metal and the gate insulation, which serves as the source of the p-channel MOSFET. The source of the MOSFET is shorted to the n^+ cathode of the thyristor. The p-channel forms a short circuit between the n^+ emitter and the p-base of the npn transistor Q_2 in the thyristor structure.

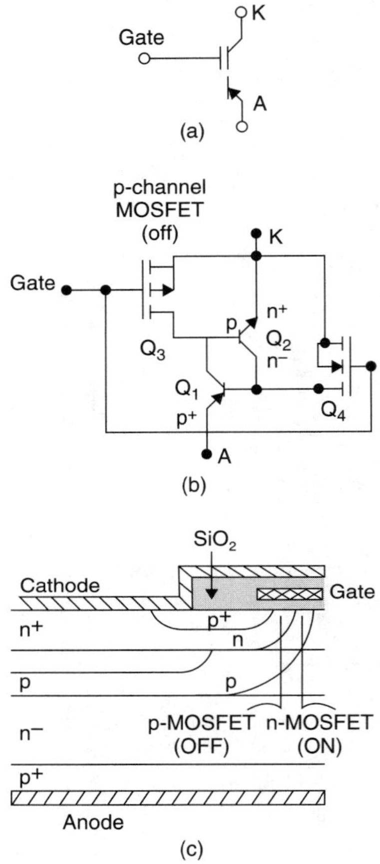

Figure 5.32
(a) The symbol (b) the equivalent circuit and (c) the structure of the MCT with p-channel MOSFET for turn-off and n-channel MOSFET for turn-on

The n-channel on-MOSFET is also integrated within the p-base region of the thyristor structure, see Figure 5.32(c). The p-base forms the substrate for the n-channel MOSFET. The n-channel is formed by inversion in the p-region below the gate electrode. The gate electrode is also common to the p-channel off-MOSFET. The n region through the n^+ region forms the source, and the n^- region, which is also the n-base, serves as the drain of the p-channel MOSFET. The p-channel when formed serves as the source of holes for the base of the pnp transistor Q_1.

5.14.1 Maximum controllable current density

The maximum controllable current density is the highest current density that can be turned off with the gate voltage control. Its maximum controllable current density is given by

$$J_{mCC} \propto \frac{(V_{GK} - V_T)}{\alpha_1} \tag{5-26}$$

where V_{GK} is the gate to cathode voltage, V_T is the threshold gate voltage for turn-on and α_1 is the current gain of the pnp transistor Q_1. At small anode voltages, the J_{mcc} of 1000 A/cm^2 are easily achievable. However, it has been found that the J_{mcc} decreases very rapidly with the increase in the anode voltage if no snubbers are used for the MCT. If a capacitive snubber is used across the MCT, a high value of J_{mcc} can be maintained besides reducing the dV/dt and achieving a safe turn-off, refer to [2].

5.14.2 MCT characteristics

The i-v characteristics in the static condition are similar to those of thyristors and GTOs, hence, not repeated here. Like the GTO, the MCT has negligible reverse voltage blocking capability (typically 25 volts). A pedestal voltage level must be maintained on the gate of MCT for proper triggering.

MCT behaves like a thyristor in the on-state because the off-MOSFET has a high impedance in this operating condition (negative voltage at the gate terminal). However, the on-state voltage drop of an actual MCT device has been observed to exceed those for the thyristor. An MCT with the blocking voltage of 2000 has a typical on-state voltage drop of 4 volts.

The on-state voltage of an MCT is lower than that in the IGBT. However, the MCT does not have a current saturation feature and the excellent forward-biased safe operating area (FBSOA) of the IGBT. The absence of any FBSOA for MCT is also a problem because this feature is useful for short-circuit protection in power electronic circuits. MCT is preferable to GTO due to better turn-off characteristics.

The dv/dt limitation of MCT is identical to those in thyristors and GTO. However, it is essential to ensure that the off-MOSFET is in the conducting state prior to turning off. It is accomplished by using a pedestal of positive voltage at the gate of MCT to maintain it in the conducting state. If the gate signal is allowed to go to zero during the blocking state, then even a moderate value of dv/dt can turn on the MCT. The typical dv/dt ratings of 1000 V/mS are commercially available.

The di/dt limitations of MCT are the result of current crowding due to the decrement in the on-state voltage with increasing temperature. The commercial MCTs are available up to 100 A/μS.

5.14.3 Switching characteristics

MCT is capable of switching from the on-state to the off-state and vice versa with typical switching times of 1 μSec. The switching waveforms are shown in Figure 5.33.

Turn-On for the MCT

The MCT is turned on by driving the n-channel MOSFET Q_4 into the conducting state and the p-channel MOSFET Q_3 into the blocking state.

Assume that MCT in Figure 5.33(a) is in the off-state for a long time. A positive voltage is applied at the gate terminal with respect to the cathode. The n-channel forms by inversion within the p-base region directly below the gate electrode. Electrons flow through the n-channel, causing a conventional current to flow from the drain (the base of Q_1) to the source (the cathode of MCT) of n-channel MOSFET Q_4. The pnp transistor Q_1 conducts current that in turn provides the base current to the npn transistor Q_2. The transistor Q_2 supplies the base current to the pnp transistor Q_1. Thus, a regenerative action sets in and the thyristor comprising of Q_1 and Q_2 is driven into the on-state. The positive voltage at the gate keeps the other p-channel MOSFET Q_4 in the blocking state.

An important advantage of the MOS gate structure for turning on the thyristor is that the current used to trigger the device is not provided by the gate circuit. This current is derived from

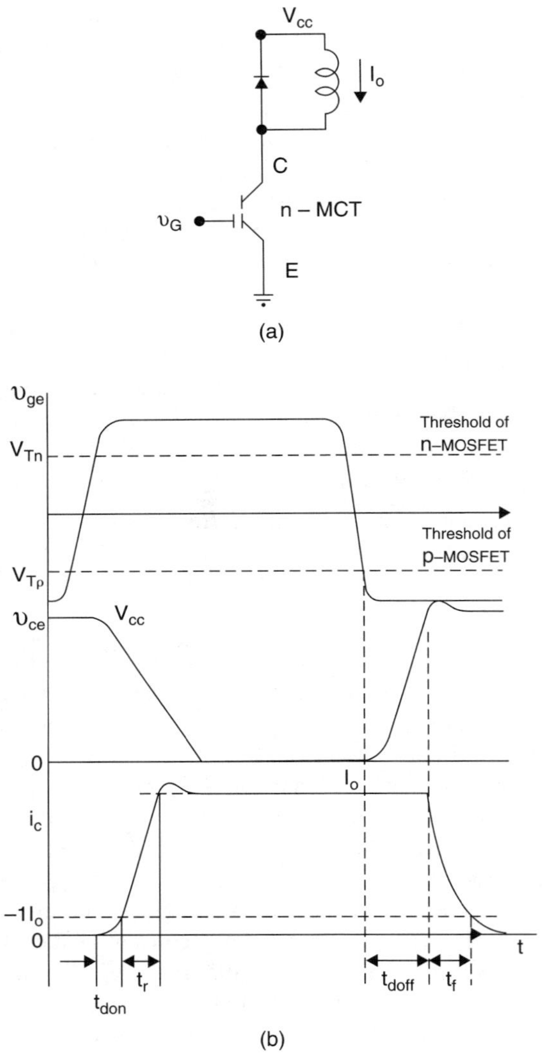

Figure 5.33
(a) Switching circuit and (b) waveforms in an MCT

the anode circuit. Consequently, a high dV/dt capability can be achieved without influencing the gate drive requirements.

The turn-on time consists of two parts: the delay time t_d and the current rise time t_r. The delay time consists of the time for the injection of excess carriers before the regenerative action begins. The rise time is the time of spreading the plasma of the excess carrier under the gate electrode. The MCT current rises very rapidly due to the regenerative action. The di/dt rates in the range of 1000 A/μS are easily achievable. Hence, stray connector inductance in the circuit can act as the turn-on snubber to limit the rate of rise of current.

Turn-Off for the MCT

In order to turn off the thyristor, the combination of Q_1 and Q_2, the sum of current gains is $\alpha_1 + \alpha_2 << 1$. Hence it is necessary to reduce either or at least one of the two current gains. The base of the npn transistor Q_2 has a higher dopant concentration and a thinner base width than the base of the pnp transistor Q_1, hence the current gain α_2 is larger than α_1. The current gain α_2 is the most appropriate candidate for this operation of current gain reduction. The current gain is reduced by shorting the base terminal of the transistor to the emitter. Hence, the base to emitter junction of the npn transistor Q_2 is shorted by the off-MOSFET, the p-channel MOSFET.

Assume that the MCT in Figure 5.33(a) is in the on-state for a long time. A negative voltage is applied at the gate terminal with respect to the cathode. The p-channel MOSFET Q_3 conducts current and effectively shorts the base-emitter junction of the npn transistor Q_2. The transistor Q_2 turns off and the transistor Q_1 turns off through a regenerative action. The negative voltage at the gate keeps the other n-channel MOSFET Q_4 in the blocking state.

The turn-off time also consists of two parts: the delay time t_d and the current fall time t_f. The delay time consists of the time for the anode to cathode voltage to reach its off-state value. The fall time constitutes the time of recombination in the base regions. The anode short, as explained in the GTO section, helps to reduce the turn-off time but only at the cost of increasing the on-state voltage drop.

5.14.4 Effect of temperature

Like in thyristors, the on-state voltage drop decreases with the rise in the temperature. The reduced on-state voltage promotes localization of the current flow in the multicell device structure or paralleled MCTs, resulting in the thermal runaway and consequent device failure.

Like in the IGBT, the turn-off time increases with increasing temperature.

The maximum controllable current density degrades rapidly, see Figure 5.34, with an increase in temperature. This factor is the most significant limitation on the high temperature operation of MCTs.

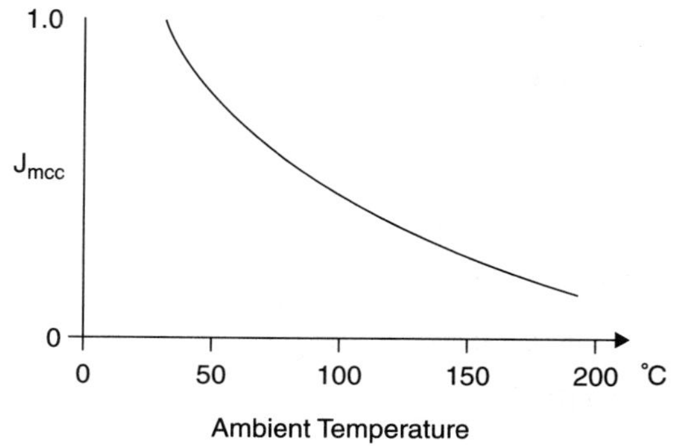

Figure 5.34
Degradation in the maximum controllable current density for the complementary MCT ($p^+np^-n^+$ with n-channel turn-off MOSFET)

5.15 DEVICE SPECIFICATION

This section gives the important specifications that most of the manufacturers list in the data sheet.

Thyristor and GTO Specifications

Forward characteristics:

Forward Current RMS	$i_{T(RMS)}$	The maximum value of the on-state rms current the thyristor may conduct.
Peak Forward Surge Current	I_{TSM}	The maximum allowable nonrepetitive surge current the thyristor may withstand at a specific pulse width.
Peak Forward Pulse Current	I_{TM}	The maximum allowable repetitive pulse current the thyristor may withstand. It decreases with increasing temperature. It also decreases with increasing frequency of pulses. Check the curves.
Average On-State Current	$I_{T(AV)}$	The maximum average on-state current the thyristor may conduct under the stated conditions. This is also listed as the continuous current.
Peak On-State Voltage	V_{TM}	The maximum voltage difference between the anode and the cathode at the stated conditions.
Holding Current	I_H	The value of the forward anode current that allows the thyristor to remain in conduction. Below this value the thyristor will return to a forward blocking state.
	I^2t	The maximum forward nonrepetitive over-current capability. Usually specified for $\frac{1}{2}$ cycle for 60 Hz operation.
	di/dt	The maximum rate of change of current through the thyristor.
Peak Repetitive Forward Blocking Voltage	V_{DRM}	The maximum value of repetitive forward voltage between the anode and the cathode that would not switch on the thyristor.

Reverse Characteristics:

Repetitive Reverse Blocking Voltage	V_{RRM}	The maximum value of repetitive reverse voltage between the anode and the cathode.
Peak Reverse Blocking Current	I_{RRM}	The maximum value of current that will flow at V_{RRM} and a specified temperature.
	dv/dt	The maximum rate of rise of voltage between the anode and the cathode before the damage.

Switching Characteristics:

Turn-On Time	t_{gt}	The time interval between a specified point on the gate trigger pulse to the transition from the off-state to the on-state.
Turn-Off Time	t_q	The time of recovery from the instant when the thyristor current has decreased to zero to the instant at which the thyristor is capable of supporting a new waveform without turning on again.

Gate Characteristics:

Gate Trigger Current	I_{GT}	The maximum value of gate current required to switch the thyristor from the off-state to the on-state under specified conditions. It decreases with increasing temperature and with increasing on-time in pulsed operation.
Gate Trigger Voltage	V_{GT}	The maximum value of voltage from the gate to the cathode required to switch the thyristor from the off-state to the on-state under the specified conditions. It decreases with increasing temperature.
Forward Peak Gate Current	I_{GFM}	The maximum gate current that may be applied to enable the thyristor to conduct. It decreases with increasing temperature.
Peak Gate Power	P_{GM}	The maximum allowable value of the power that may be dissipated between the gate and the cathode terminals.
Forward Average Gate Power	$P_{G(AV)}$	The maximum allowable value of the power, averaged over a period, that may be dissipated between the gate and the cathode terminals.
Peak Gate Power	P_{GM}	The maximum allowable value of the power that may be dissipated between the gate and the cathode terminals.
Forward Average Gate Power	$P_{G(AV)}$	The maximum allowable value of the power, averaged over a period, that may be dissipated between the gate and the cathode terminals.
Peak Gate Voltage Forward	V_{FGM}	The maximum peak value of voltage allowed between the gate and the cathode in forward bias.
Peak Gate Voltage Reverse	V_{RGM}	The maximum peak value of voltage allowed between the gate and the cathode in reverse bias.

Power and Thermal Characteristics:

Operating Junction Temperature	T_J	The junction temperature of the thyristor as a result of the ambient and load conditions.
Case Temperature	T_C	The temperature of the thyristor case as a result of the ambient and load conditions.
Thermal resistance, Case-to-Ambient	$R_{\theta CA}$	The thermal resistance in steady-state from the case to the ambient.
Thermal resistance, Junction-to-Ambient	$R_{\theta JA}$	The thermal resistance in steady-state from the semiconductor junction to the ambient.
Thermal resistance, Junction-to-Case	$R_{\theta JC}$	The thermal resistance in steady-state from the semiconductor junction to the case.
Transient Thermal Impedance, Junction-to-Ambient	$Z_{\theta JA(t)}$	The transient thermal impedance from the junction to the ambient. It decreases with the decrease in the on-time in the pulsed operation.
Transient Thermal Impedance, Junction-to-Case	$Z_{\theta JC(t)}$	The transient thermal impedance from the junction to the case.

MCT Specifications:

The gate characteristics of MCTs differ from those of the thyristor and GTO because of the MOS type construction of the gate.

Gate Characteristics:

Gate-Anode Leakage current	I_{GAS}	Leakage current from the gate to the anode under the stated conditions.
Input Capacitance	C_{iSS}	Input capacitance between the gate and the anode.

5.16 SUMMARY

This chapter has explored the device structure and characteristics of pulse triggered devices such as the thyristors. The important observations are as follows:

1. The thyristor is a four layer device.
2. The thyristor is a minority carrier device, has the highest reverse blocking voltage and forward current capabilities.
3. A current pulse applied to the gate will latch the thyristor into the on-state. The gate, however, cannot turn off the thyristor. The anode to cathode power supply polarity is reversed to turn it off.
4. The thyristor has lowest on-state voltage and, hence, the lowest losses in the on-state.
5. The thyristor can fail in high dV/dt and dI/dt conditions. Careful snubbering is required.
6. The forward voltage should not be reapplied to the thyristor soon after the turn-off. The rate of reapplied voltage should not be too

high, otherwise the thyristor may be triggered back into the on-state.

7. The thyristor has significant reverse recovery problems.
8. The GTO is a thyristor that can be triggered from the gate.
9. The GTO has very limited reverse blocking voltage ratings.
10. High negative gate current pulse is required for turn-off.
11. The maximum anode current that a GTO can turn off is limited.
12. Strong snubbering is required for GTOs.
13. The MCT is a GTO with MOS controlled gate.

References

1. Ned Mohan, Tore M. Undeland, and William P. Robbins, *Power Electronics: Converters, Applications and Design,* 2nd Ed., John Wiley & Sons, Inc., 1995.

2. B. Jayant Baliga, *Power Semiconductor Devices,* PWS Publishing Company, Boston, 1995.

3. John G. Kassakian, Martin F. Sclecht, and G. Vergassian, *Principles of Power Electronics,* Addison-Wesley, Boston, 1991.

Problems

5-1 The thyristor in the circuit shown below has a holding current of 20 mA. A 50 μs wide gate trigger pulse is applied at the delay angle of $\pi/2$. Will the thyristor turn on? Explain your answer. If it does not turn on, it is suggested to add a resistor in parallel with the R-L load. Find R that would just enable the thyristor to turn on.

5-2 Find the minimum pulse width of the gate pulse required to turn on the thyristor if the inductor is doubled in the circuit of Figure P5.1.

5-3 The thyristor in the circuit of Figure P5-1 has an i^2t rating of 20 amp^{2}s. An accidental short circuit occurs across (a) the resistor or (b) the inductor. Find the shortest fault clearance time needed in both cases to prevent damage to the thyristor.

5-4 The simplified forward characteristics of the thyristor are shown in Figure P5.4. Estimate

a. the average power loss and the average rms current in the thyristor in the following conditions:

b. A dc current of 50 amps for half period of the ac voltage source.

c. A dc current of 25 amps for 1/3 period of the ac voltage source.

d. Find the A dc current of 35 amps for 2/3 period of the ac voltage source.

5-5 The di/dt rating of the thyristor in Figure P5.5 is specified to be 50 A/μs. Find the minimum value of L to protect the thyristor from excessive di/dt.

5-6 The dv/dt rating of the thyristor in Figure P5.6 is specified to be 500 V/μs. Find the minimum value of C to protect the thyristor from excessive dv/dt.

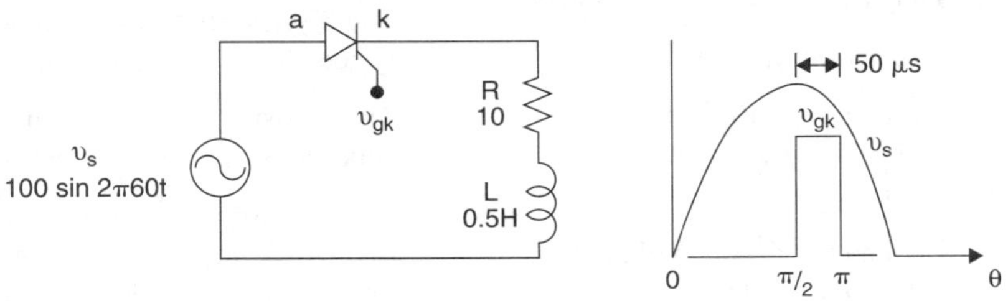

Figure P5.1

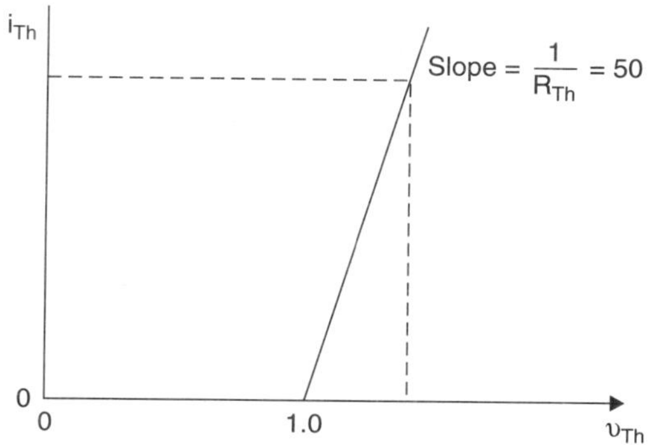

Figure P5.4

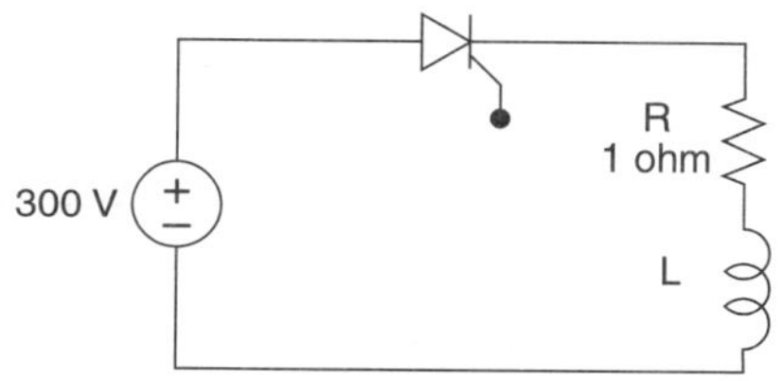

Figure P5.5

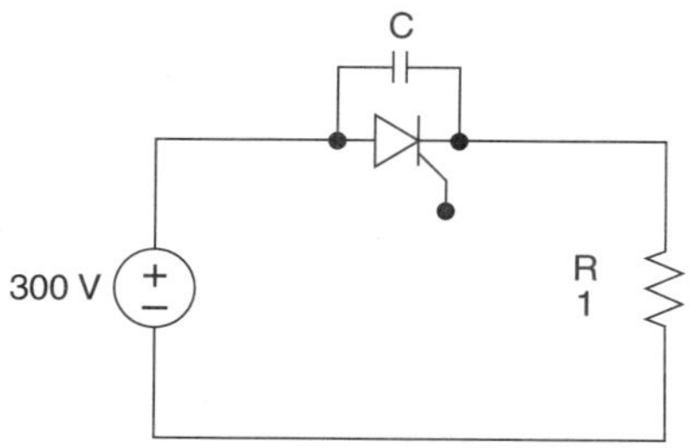

Figure P5.6

5-7 The International Rectifier SE30EF8A, 800 V is connected in the switching circuit of Figure P5.7 where the ac source is $v_S = 280 \sin 2\pi 50t$. Find on-state losses in the thyristor if turned on at $\alpha = \pi/3$. Also find the power delivered to the resistive load.

5-8 The parameters in the impulse commutation circuit of Figure 5.22 and the resonant commutation circuit of Figure 5.23 are: $I_O = 40$ A, $v_S = 350 \sin 2\pi 60t$, and $|V_C| = 350$ V. Find the C and L components in both circuits to achieve a turn-off time of 10 μS. Design the resonant frequency to be 20 kHz.

5-9 Design the snubber circuit of Figure 5.26 where R = 10 ohms, L = 2 μH, C = 0.1 μF, V_S : a step of dc voltage 350 V, Thyristor: International Rectifier S30EF80A.
The series resistance R_S is selected to limit the maximum current through the capacitor to 500 A. Determine the snubber inductance and capacitance if the maximum permitted values of di_T/dt and dV_T/dt are 250 A/μS and 350 V/μS respectively. Ignore the switching times.

5-10 The junction capacitance of a thyristor is $C_{j2} = 15$ pf and is assumed to be independent of the reverse blocking voltage. The thyristor can be turned on at 15 mA. Determine the critical value of dv/dt at which the thyristor would turn on.

5-11 The thyristor of Problem 5-10 and used in the circuit of Figure P5.11 is $C_{j2} = 15$ pf. Determine the value of the capacitance C_S so that the thyristor would not be turned due to dv/dt.

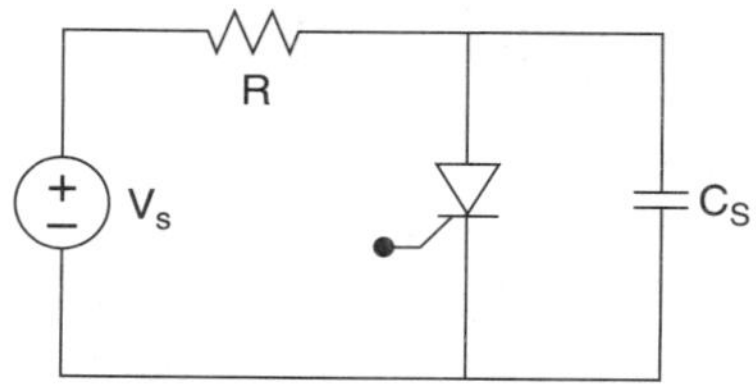

Figure P5.11

PART THREE

POWER ELECTRONIC CONVERTER CIRCUITS

CHAPTER 6
dc to Controlled dc

OBJECTIVES

1. The difference between the uncontrolled and the controlled conversion from dc to dc
2. The operational characteristics of Buck, Boost and Buck-Boost converter and their derived topologies.
3. Transformer-isolated dc to dc converters
4. The continuous and discontinuous conduction modes of operation of dc to dc converters

6.1 INTRODUCTION

The dc-dc converter converts a dc power source to another dc source with different terminal specifications. The dc-dc converters change dc to ac first and then change ac back to dc. The dc source is often the uncontrolled dc voltage with ripple from ac to dc rectifier.

The dc-dc converters are widely used in the switch-mode power supplies and dc motor drive applications. Some of these converters, especially in power supplies, have an isolation transformer. The dc-dc converters are also used as interfaces between the dc systems of different voltage levels.

The step-down and step-up voltage converters are the two basic converter topologies. These are referred as the Buck and Boost converters respectively. It must, however, be kept in mind that a step-down voltage converter is also a step-up current converter and vice versa because the input power must equal the output power. The Buck-Boost converter does both stepping up and down action. The ČUK converter is a dual of the Buck-Boost converter. All these converters have single, two and four quadrant variations in topologies.

The voltage source driven dc-dc converts are more popular than the current driven converters, hence, this chapter will concentrate on the voltage driven converters. All converts use effective filtering on both the input and the output to reduce the ac components from going outside the converters.

6.2 UNCONTROLLED AND CONTROLLED CONVERSION

DC to DC converters can be categorized in four types:

Voltage Source Converters

dc voltage to dc voltage

dc voltage to dc current

Current Source Converters

dc current to dc voltage

dc current to dc current

These two types employ dual topologies: voltage source/sink replaced by current source/sink, inductor replaced by capacitor and series connection replaced by parallel connection. This chapter will focus on the voltage source converters only.

It is worth mentioning that the current supplied by a dc voltage source may have an ac waveform, but it must contain a nonzero dc component that alone is responsible for the net power flow from the source to the sink. The same is true for the current waveform into a voltage sink. The dual is also true when the currents are interchanged by the voltages and the voltages by the currents.

Power conversion from a dc voltage source to a dc current sink is of the simplest form. See Figure 6.1(a). A voltage source provides the current demanded by the current sink. The ideal current sink can withstand the terminal voltage set by the source. Hence there is no need of any interfacing device. The power flow from the dc voltage source to the dc current sink need not be controlled.

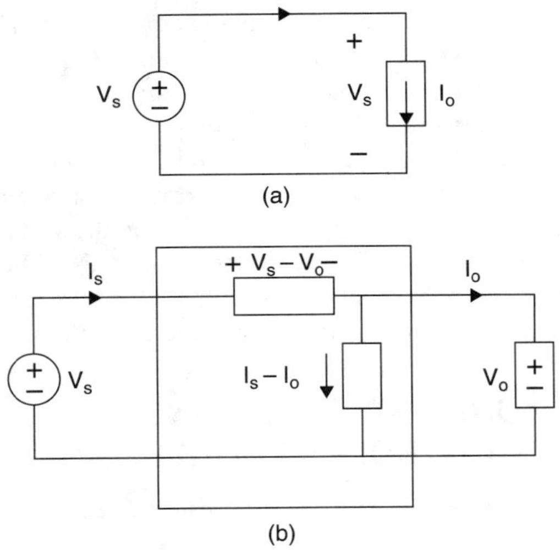

Figure 6.1
(a) Uncontrolled conversion: dc voltage source to dc current sink and (b) controlled conversion: dc voltage source to dc voltage sink

The controlled power transfer is required when following two situations or a combination of them arises:

1. either the dc voltage at the output terminals is different from the dc source voltage, or
2. the output dc current is different from the dc current from the source.

An example is shown in Figure 6.1(b) of a dc voltage to dc voltage converter that incorporates both situations. An intermediate element is employed in series to withstand the difference in voltage and a shunt element in parallel to store the difference in currents.

It must be noted that in Figure 6.1(b) the output power V_kI_k is different from the input power V_sI_s. The difference is stored in the elements added. In most of the applications in power electronics, the output power and the input power are equal, hence, either the series and shunt elements do not store energy or one element stores while the other delivers energy such that the net energy stored is zero.

The intermediate series controlling element must have two properties:

a. withstand a dc voltage equal to the difference between the source and sink voltages, and

b. be capable of carrying dc current as demanded by the sink.

Likewise, the intermediate shunt controlling element must have two properties:

a. sink a dc current equal to the difference between the source and sink currents, and

b. withstand a dc voltage (the output voltage).

An inductor can support the dc current but not the nonzero average voltage. A capacitor can support the nonzero average voltage but not the dc current. Only a switch can sustain both the nonzero voltage simultaneously with a nonzero current. Hence, the controlled dc to dc conversion requires a switch. The switch is turned on and off periodically, thus converting the input dc to an ac voltage waveform. Hence the switch should be followed by an averager (a low-pass filter) that can also support a nonzero average current.

An inductor fulfills the requirement. Figure 6.2 shows a generic dc voltage to dc voltage converter.

Several topologies, namely the Buck, the Boost and the derivatives, can be generated by placing the inductor in the three different legs of the switch. The two-position switch is synthesized from a one-position switch and a diode or two one-position switches, turned on alternately.

The converter topology for the energy transfer from a dc current source to a dc current sink is the dual of the topology of Figure 6.2 and is given in Figure 6.3.

A further classification of dc-dc converters is based on the number of quadrants of converter operation on the v_O-i_O plane. Above topologies are essentially the *half-bridge* and, hence, the *single quadrant* topologies. The two-position switch in the above converters may also be substituted by a network of switches for the *full-bridge* two and four quadrant converters. The full-bridge converter will be explored in the later sections of this chapter.

Before advancing further, let us examine the filtering considerations in the dc to dc converter. The output voltage just after the switch is a pulsating dc waveform. Hence, the output voltage consists of a dc component with a very significant amount of voltage ripple. When the load is resistive, the output current waveform has similar current ripple. The ripple is reduced by introducing an inductor in series with the switch, as is shown in Figure 6.2(b). Likewise, the switch renders the current from the source to be a pulsating dc waveform, a dc component with a very significant ripple. This ripple is reduced by connecting a capacitor at the source terminals. The dc to dc converter with input and output filters is shown in Figure 6.4. The input side filter smoothens the ac components of current drawn from the source.

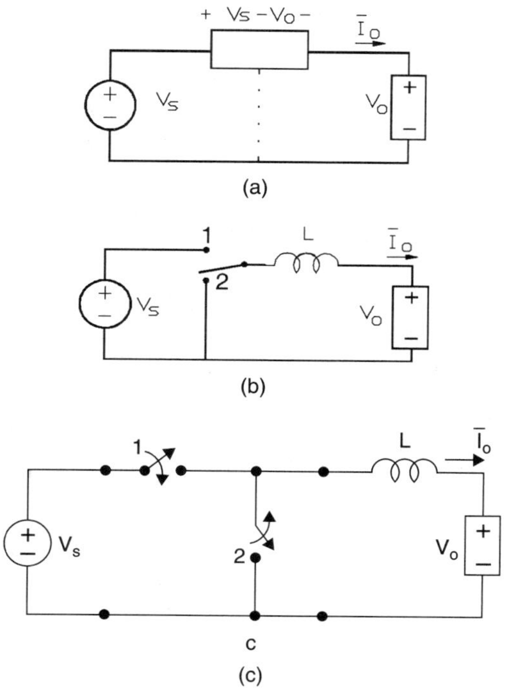

Figure 6.2
The dc voltage to dc voltage converter (a) System (b) generic circuit and (c) the implementation of the two-position switch.

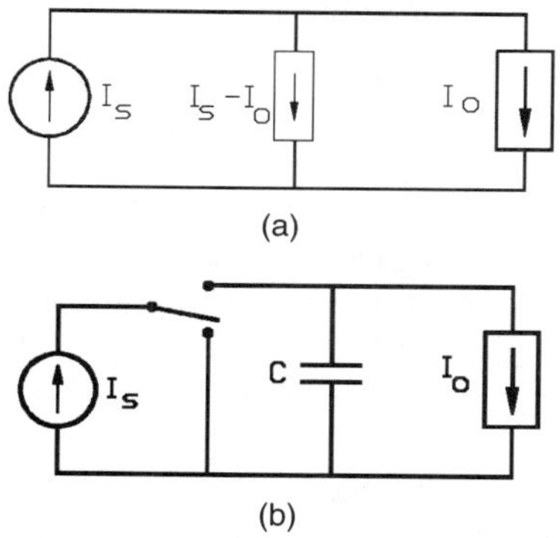

Figure 6.3
The dc current to dc current converter (a) system and (b) generic circuit.

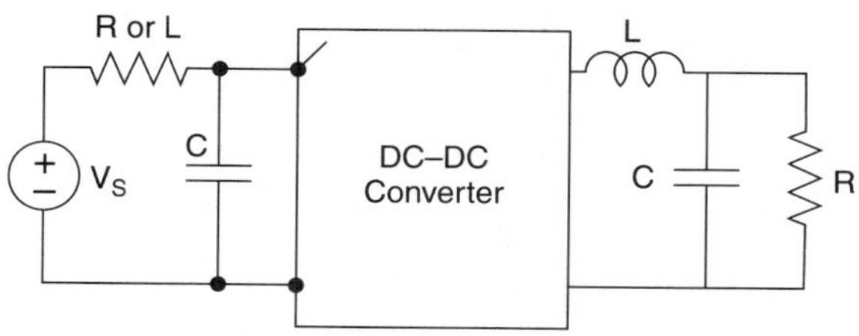

Figure 6.4
The dc voltage to dc voltage converter with filters.

6.2.1 Performance measures

This section will focus on the steady-state operation of both the Voltage Source Converters (VSC) and the Current Source Converters (CSC). The input source is assumed to be an ideal dc source V_S or I_S.

The following discussion about the dc-dc converters is meant for single quadrant converters, but it is also valid for two and four quadrant converters.

Output Characteristics

Voltage transfer ratio, defined as the ratio of dc output voltage to the dc input voltage, is an important measurement of the Voltage Source Converters:

$$T_{vv} = \frac{\bar{V}_O}{V_S} \tag{6-1}$$

Input Characteristics

Current reflection ratio, defined as the ratio of the input dc current to the average value of the output current, is calculated from

$$R_{ii} = \frac{\bar{I}_S}{\bar{I}_O} \tag{6-2}$$

However, if we ignore the switching losses and the losses in parasitic elements, then from the condition that the input power equals the output power we find

$$R_{ii} = T_{vv} \tag{6-3}$$

The $\mathbf{R_{ii}}$ being the measure of source utilization, this condition states that the source is utilized poorly at low voltage transfer ratios. This is true for all dc-dc converter topologies.

Control Characteristics

Control parameters are used to close the loop for regulation and control of one of several output variables V_O, I_O or the output power P_O. For an optimal performance of the overall system, the smallest range of variation of control parameter is desired for a large variation in the output variable. This amounts to selecting the region of highest slope on the plot of T_{vv} versus the control parameter in Voltage Source Converters.

6.3 BUCK CONVERTER

The Buck converter is a voltage step-down and current step-up converter. The topology is shown in Figure 6.5. This topology is the same as shown in Figure 6.2(b) except the inductor in the series with the voltage sink is made infinitely large and both elements are modeled by an ideal current sink. The two-position switch is synthesized from a switch and a diode. The switch is turned on for a time τ periodically at a rate 1/Ts.

A prominent application of the Buck converter is a dc regulated power supply in which the output voltage is regulated against the variations in the load resistance and the input voltages. These power supplies are used in computers and portable instruments in the medical and communication fields.

We will consider two modes of operations of the Buck converter in which the output current waveform is either continuous or discontinuous in time over a switching period.

6.3.1 Continuous conduction mode (CCM)

The ideal current sink in the output represents an infinitely large inductor in series with the load resistance. This is a hypothetical circuit but its discussion provides a good understanding of the Buck converter operation.

The steady-state operation consists of two circuit modes, shown in Figure 6.6.

As the switch is turned on in mode 1, the diode is reverse biased and the current flows through the inductor into the voltage sink. After a time τ, the switch is turned off. The inductor current then free-

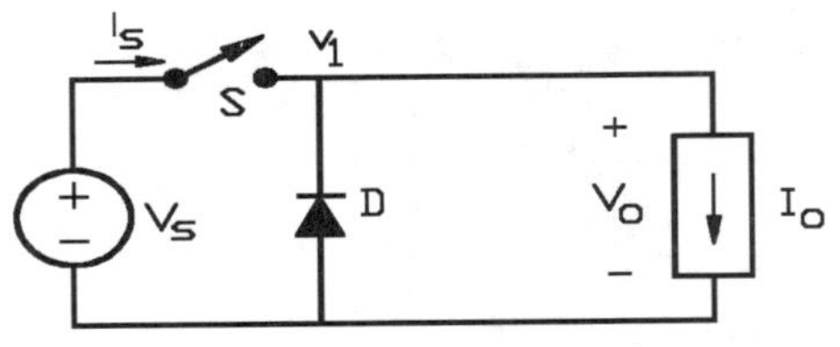

Figure 6.5
Buck converter topology

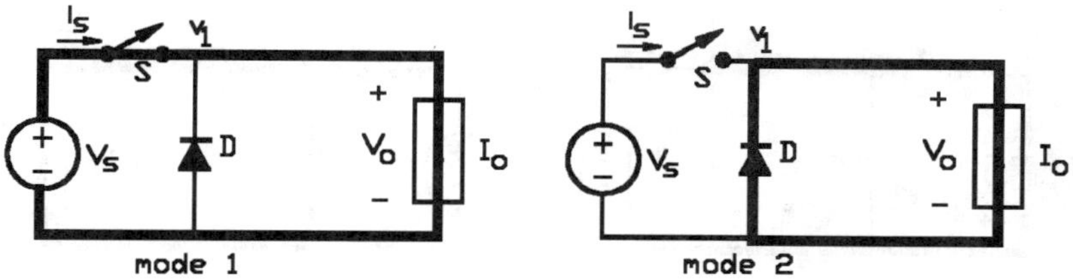

Figure 6.6
Circuit modes in steady-state operation

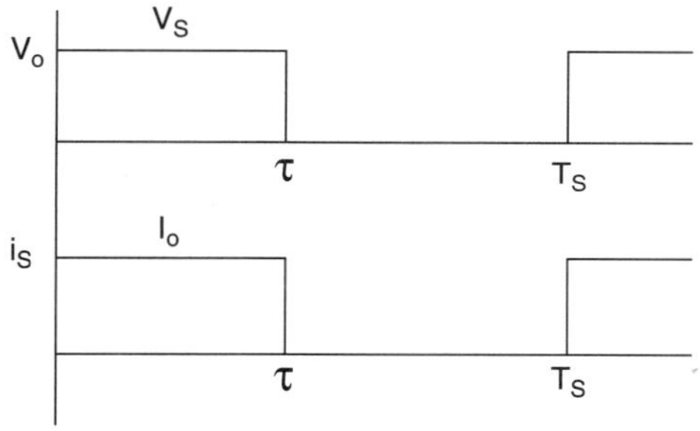

Figure 6.7
Waveforms

wheels through the diode as shown in the circuit of mode 2. The second mode is terminated at T_S when the switch is turned on again. The current and voltage waveforms are shown in Figure 6.7.

The average output voltage is

$$\overline{V}_O = V_S \frac{\tau}{T_S} = D\, V_S$$

$$T_{vv} = \frac{\overline{V}_O}{V_S} = D \qquad (6\text{-}4)$$

or, $$T_{vv} = \tau f_S$$

D is called the duty ratio. The duty ratio D is also the turn ratio of the *equivalent dc transformer.* The turn ratio of the equivalent dc transformer can be controlled electronically in a range of 0-1.

The dc output voltage is a linear function of the on-time of the switch, the pulse width, and the switching frequency (inverse function of the switching period T_s). The output voltage control using pulse width is known as the PWM control. The output voltage can also be controlled by switching frequency. The PWM control is more popular.

If the duty ratio D is made a linear function of v_c, a control voltage

$$D = k\, v_c$$

$$\overline{V}_O = (kV_S)v_c \qquad (6\text{-}5)$$

The output voltage is then a linear function of the control voltage. This is also the principle of switch-mode linear amplifier. The gain of this amplifier is determined by the input dc source voltage. There will be more about this in the chapter on dc to ac inverters.

Neglecting the power losses in the circuit elements, we could use the equation of the balance of power:

$$V_S \overline{I}_S = \overline{V}_O I_O \qquad (6\text{-}6)$$

where I_S is the average current from the dc source. Hence,

$$\frac{I_O}{\overline{I}_S} = \frac{V_S}{\overline{V}_O} = \frac{1}{D} \qquad (6\text{-}7)$$

Although the dc output voltage is smaller than the dc input voltage, the dc output current is larger than the input dc current (a step-up action in current).

Low Current Ripple

The ideal current sink of Figure 6.5 is replaced by a large but finite value inductor in series with a voltage sink V_O as shown in Figure 6.8. The inductor current, which is also the output current, waveform is an ac current superimposed on a dc level I_O.

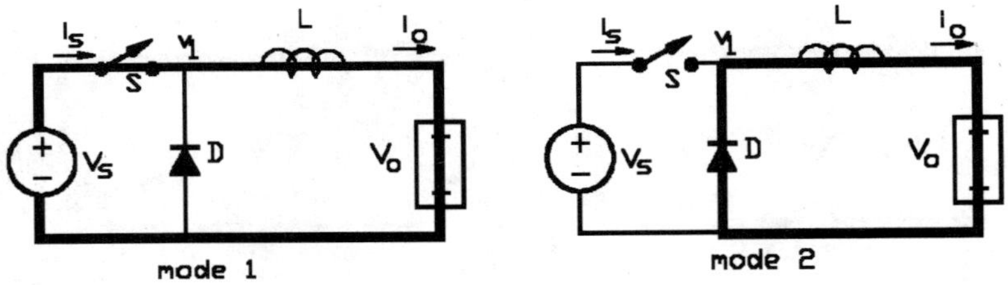

Figure 6.8
Buck converter with moderate inductor

The average output voltage is still given by equation (6-4) as the average dc voltage on the large inductor is zero in steady-state. The output dc voltage is independent of the current drawn by the load resistance.

$$\bar{I}_O = \bar{I}_L \tag{6-8}$$

The instantaneous output current is given by the following:

$$mode\ 1: L\frac{di_O}{dt} = V_S - \bar{V}_O;$$

$$mode\ 2: L\frac{di_O}{dt} = -\bar{V}_O \tag{6-9}$$

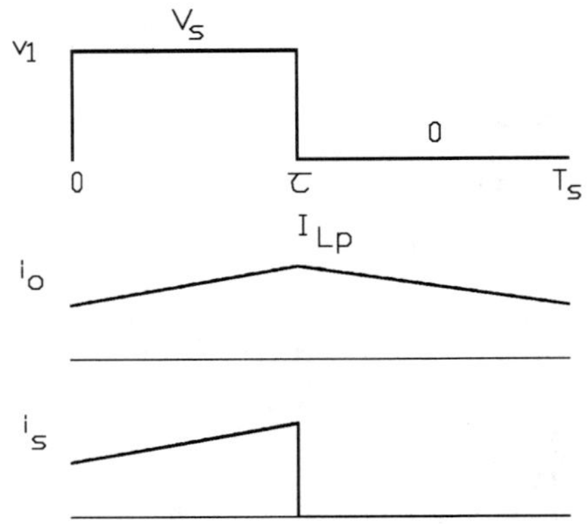

Figure 6.9
Waveforms

The current and voltage waveforms are shown in Figure 6.9. The instantaneous inductor current ramps up during the on-time of the switch and ramps down during the off-time. Average value of the inductor current is also the average output current:

$$\bar{I}_O = \frac{1}{T_S}\left[\int_0^{\tau} i_L\, dt + \int_{\tau}^{T_S} i_L\, dt\right] \tag{6-10}$$

The continuous mode of operation exists as long as the instantaneous inductor current at the beginning of the switching period is above zero. In the boundary case, the peak inductor current is calculated from the mode 2 circuit equation:

$$I_{LP} = \frac{\bar{V}_O}{L}(T_S - \tau)$$

I_{LP} is also the peak to peak ripple current.

$$\Delta I_L = I_{LP} = \frac{\bar{V}_O T_S}{L}(1 - D)$$

$$= \frac{V_S T_S}{L} D(1 - D) \tag{6-11}$$

The current ripple is independent of the magnitude of the load current.

The maximum value of the peak-to-peak current is

$$\Delta I_{L\max} = \frac{V_S T_S}{4L} \ at\ D = 0.5 \tag{6-11a}$$

The boundary value of the average inductor (output) current is calculated from the following:

$$\bar{I}_{OB} = \frac{1}{T_S}\left[\frac{1}{2} I_{LP} T_S\right] = \frac{V_S T_S}{2L} D(1 - D) \tag{6-12}$$

Maximum value of the boundary value output current is

$$I_{OB\max} = \frac{V_S T_S}{8L} \text{ at } D = 0.5 \qquad (6\text{-}12a)$$

For the control characteristics plots, the current is normalized as follows:

$$I_{nOB} = \frac{I_{OB}}{V_S T_S/2L} = D(1 - D) \qquad (6\text{-}13)$$

```
D=linspace(0,1,100); % array for duty ratio
Tvv=D; % Voltage transfer ratio
InoB=D .*( ones(size(D))-D); % normalized output
   current in the boundary case
plot (D,Tvv, D, InoB)
```

It must be noted that the minimum current for CCM mode is highest at 50% duty ratio. The voltage sink in the output is comprised of a load resistance R in parallel with an infinitely large capacitance. The corresponding value of the boundary value load resistance is:

$$R_B = \frac{\overline{V}_O}{I_{OB}} = \frac{2L}{T_S(1 - D)} \qquad (6\text{-}14)$$

For normal CCM operation, the load resistance must be lower than the boundary value R_B. In other words, a bleeder resistance of value R_B is required in parallel with a variable load resistance to ensure CCM operation.

Low Voltage Ripple

The assumption of ideal voltage sink will be dropped and a large value capacitor in parallel

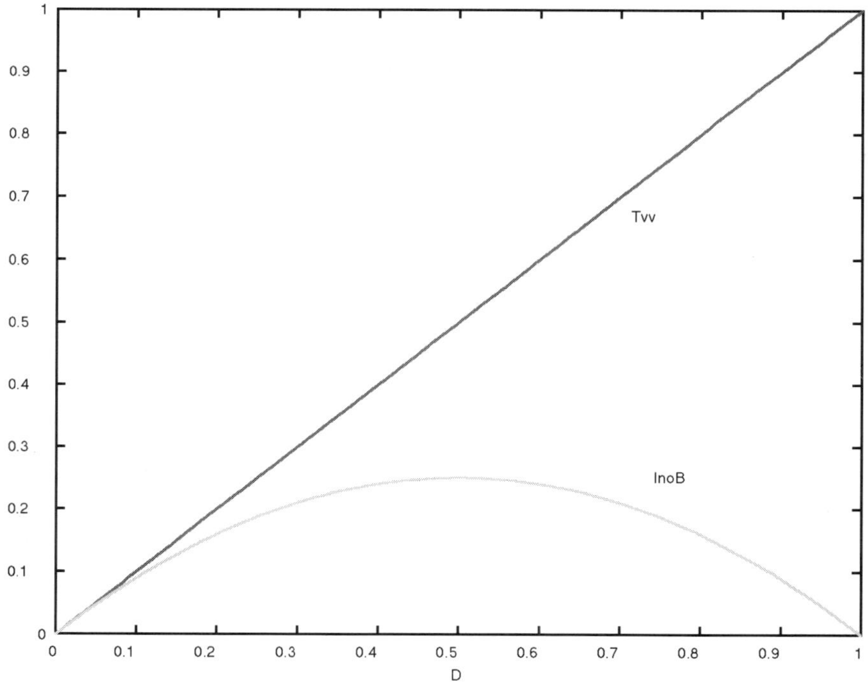

Figure 6.10
Control characteristics

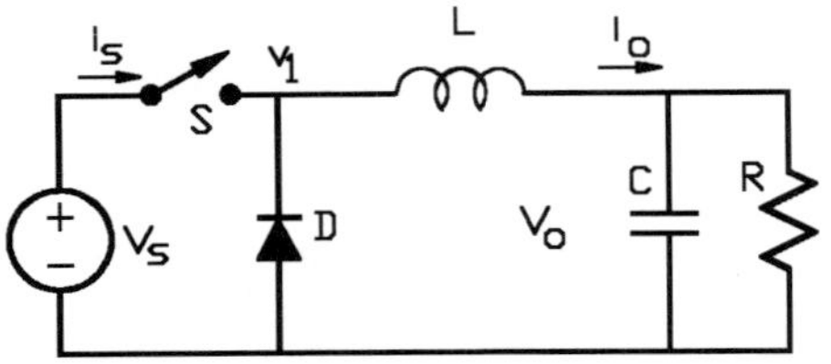

Figure 6.11
Buck converter with finite inductor and capacitor

with the load resistance will be used instead. See Figure 6.11. To simplify the analysis, make the following assumptions:

1. The peak-to-peak voltage ripple is a small percentage of the dc voltage V_O.
2. All the ripple current through the inductor flows into the filter capacitor and not through the load resistance.

The dc inductor current is equal to the average dc current in the load resistance because the average current in the capacitor must be zero.

The ripple current throughout the inductor, with dc component removed for the sake of simplified analysis, is depicted in Figure 6.12. The areas above and below the zero line are equal. The positive ripple current increases charge by ΔQ, and hence the voltage by ΔV_O across the filter capacitor and the negative ripple current decreases it. The generated voltage ripple is shown in Figure 6.12. The peak-to-peak ripple voltage on the capacitor is

$$\Delta V_O = \frac{\Delta Q}{C} = \frac{1}{C}\frac{1}{2}\frac{\Delta I_L}{2}\frac{T_S}{2} = \frac{\Delta I_L T_S}{8C} \quad (6\text{-}15)$$

Using equation (6-11),

$$\Delta V_O = \frac{T_S}{8C}\frac{\bar{V}_O T_S}{L}(1 - D)$$

$$\frac{\Delta V_O}{\bar{V}_O} = \frac{T_S^2}{8LC}(1 - D) = \frac{\pi^2}{2}\left[\frac{f_C}{f_S}\right]^2 (1 - D) \quad (6\text{-}16)$$

where $f_S = 1/T_S$, and $f_C = 1/2\pi\sqrt{LC}$

It should be noted again that the voltage ripple is independent of the load resistance or load current.

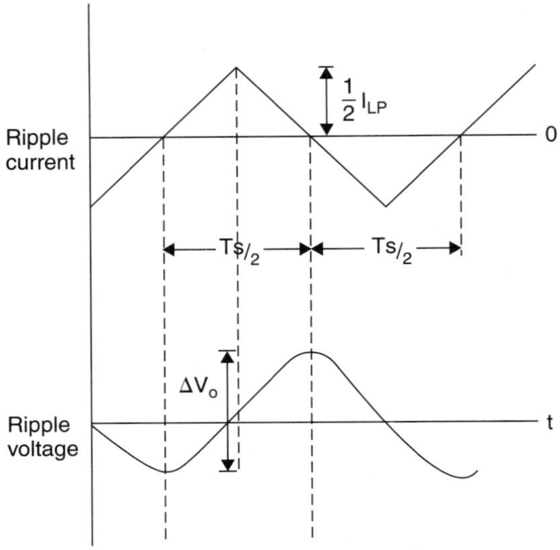

Figure 6.12
Ripple current and ripple voltage waveform

A low value of f_C/f_S (a factor of at least 10) will provide low voltage ripple.

Switch Stress

The knowledge of the switch stresses is very important for a robust design. Besides the switching losses, as seen in Chapter 2, there are two more stresses that should be taken into consideration: (a) peak voltage and (b) peak current through the switch.

PEAK VOLTAGE STRESS ON THE SWITCH

$$V_{sw\,max} = V_S$$

PEAK CURRENT THROUGH THE SWITCH

Peak current through the switch is the dc output current and half of the maximum value of the peak-to-peak ripple current

$$I_{sw\,max} = \bar{I}_O + \frac{1}{2}(\Delta I_{LP})_{max}$$

$$= \bar{I}_O + \frac{V_S T_S}{8L}$$

6.3.2 Discontinuous conduction mode (DCM)

The output current waveform becomes discontinuous if the converter is operated with the load current that is smaller than the boundary value current at the operating pulse width. One of the prominent applications of the Buck converter in DCM is the single quadrant control of a dc motor in forward motion only.

See Figure 6.13 for the current waveform. The steady-state operation in this mode of operation contains a third circuit mode in which after the current has ramped down to zero, no further current flows in the inductor, but it continues to flow in the load resistance, supplied by C. The C being large, the drop in the capacitor voltage is ignored.

Peak current in the inductor is calculated from

$$I_{Lp} = \frac{V_S - \bar{V}_O}{L}\tau_1$$

or, alternately calculated from

$$I_{Lp} = \frac{\bar{V}_O}{L}\tau_2 \qquad (6\text{-}17)$$

Equating the above two equations, gives

$$\frac{\bar{V}_O}{V_S} = \frac{\tau_1}{\tau_1 + \tau_2} = \frac{D}{1 - \Delta} \qquad (6\text{-}18)$$

where

$$D = \frac{\tau_1}{T_S}, \Delta = \frac{\tau_3}{T_S} \quad \text{and} \quad \tau_1 + \tau_2 + \tau_3 = T_S$$

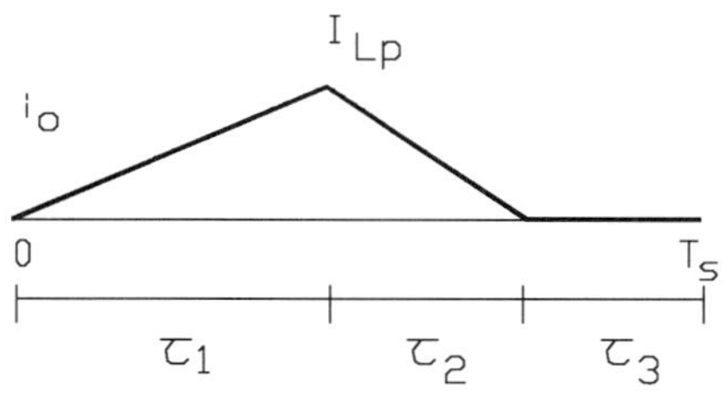

Figure 6.13
Current waveform in the discontinuous conduction mode

The peak inductor current is determined from

$$I_{LP} = \frac{\bar{V}_O}{L}\tau_2 = \frac{\bar{V}_O T_S}{L}(1 - \Delta - D)$$
$$= \frac{V_S T_S}{L}\frac{D}{(1 - \Delta)}(1 - \Delta - D) \qquad (6\text{-}19)$$

The average current in the inductor, which is also the average output current in the load resistor, is calculated from the following:

$$\bar{I}_O = \frac{1}{2}I_{Lp}\frac{\tau_1 + \tau_2}{T_S} = \frac{1}{2}I_{Lp}\frac{T_S - \tau_3}{T_S}$$
$$= \frac{1}{2}I_{Lp}(1 - \Delta) \qquad (6\text{-}20)$$
$$= \frac{V_S T_S}{2L}D(1 - \Delta - D)$$

Normalizing the output current by the boundary value current from equation (6-12),

$$I_{nO} = \frac{\bar{I}_O}{I_{OB}} = 1 - \frac{\Delta}{1 - D} \qquad (6\text{-}21)$$
$$\textit{and}\ \Delta = (1 - D)(1 - I_{nO})$$

For an open circuit condition (V_O=V_s), the $\Delta = 1-D$.

Substituting the above equation into equation (6-18) we can get the voltage transfer constant as a function of the normalized output current I_n and the duty ratio D, see Figure 6.14a.

```
D = 0.2:.2:1; % duty ratio
Tvv = [ ];In = [ ];
In1 = linspace(0,1,101);% output current less than
   the boundary value current
In2 = linspace(1,1.5,101); % output current more than
   the boundary value current
for n = 1:5
   delta = (1−D(n)).*(1−In1);
   Tvv1 = D(n)./(1−delta);% Tvv in DCM
   Tvv2 = ones(size(In2))*D(n);%Tvv in CCM
   In = [In; In1, In2]; % normalized output current in
   combined DCM and CCM
   Tvv = [Tvv ;Tvv1, Tvv2];
end
plot (In',Tvv')
```

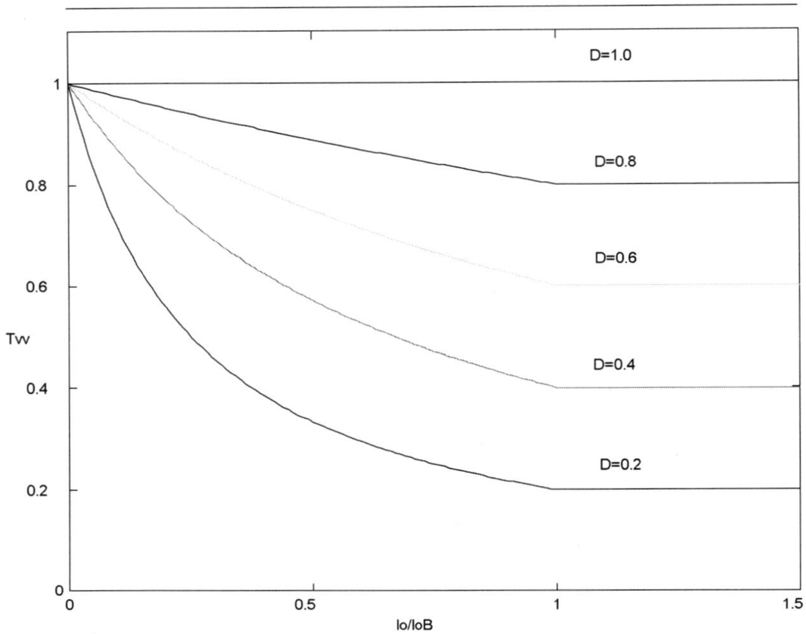

Figure 6.14a
Voltage transfer constant vs. output current normalized to the boundary value

In dc power supply applications, the output voltage is usually regulated against the variations in the input voltage and in the load current by varying the duty ratio. A plot of duty ratio D versus the output current normalized with respect to the boundary value current with Tvv as the parameter are shown in Figure 6.14b.

Example 6-1

A dc-dc converter converts a source of 48 V to 12 V. The converter components are R = 10 ohms, L = 5 mH, and f_s = 40 kHz. Plot the output voltage and current waveforms and determine

a. *the mode of operation,*

b. *the amount of current ripple,*

c. *the filter capacitor value if the voltage ripple is to be less than 1% of the output voltage, and*

d. *the inductor for which the converter would just enter in the discontinuous conduction mode.*

To determine the mode of operation, we find the duty ratio and the corresponding boundary value current from equations (6-4) and (6-12),

$$D = \frac{\bar{V}_O}{V_S} = 0.25 \quad , \quad I_{OB} = 0.0225\,A$$

The dc output current is $I_O = V_O/R = 1.2$ A $>> I_{OB}$, hence the converter is operating in the continuous conduction mode.

The peak-to-peak ripple current is calculated using equation (6-11)

$$\Delta I_L = 0.045A$$

To plot the output current waveform, which is also the inductor current, we need the lowest value of the waveform,

$$I_{LO} = \bar{I}_O - \frac{1}{2}\Delta I_L$$

$$I_{LO} = 1.2 - \frac{1}{2}0.045 = 1.1775A$$

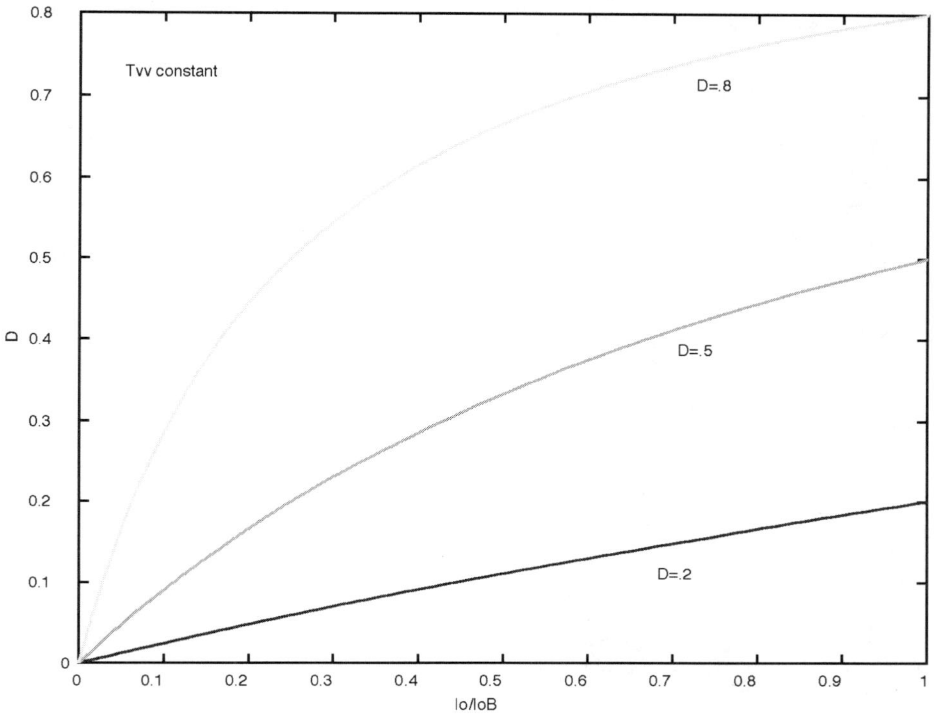

Figure 6.14b
Duty ratio vs. output current normalized to the boundary value

The waveforms are plotted using MATLAB:

```
Vs = 48;Vo = 12;R = 10; Ts = 1/40000;L = .001;
D = Vo/Vs; tau = D*Ts; Io = Vo/R;
t1 = 0:Ts/500:tau; t2 = tau:Ts/500:Ts;
   t = [t1 t2];
vo = [Vs*ones(size(t1)) zeros(size(t2))];
vl = vo−Vo;
I_0 = Io−Vo*Ts*(1−D)/(2*L);
io = (integ(vl/L,t)+IL0);
plot (t,vo/Vs, t,vl/Vs, t,io)
```

The filter capacitor value is calculated from equation (6-16)

$$\frac{\Delta V_O}{V_O} = 0.01 = \frac{\pi^2}{2}\left[\frac{f_c}{40{,}000}\right]^2(1 - .25)$$

The filter corner frequency f_c is 2.079 kHz or less. Selecting $f_c = 2$ kHz, the capacitor value is calculated as

$$C = \frac{1}{4\pi^2 L f_c^2} = 3.18\ \mu\text{F}$$

The inductor value for the boundary case arises when the output current I_O equals the boundary value I_{OB}. Hence, from equation (6-12), the inductor is L=0.5 mH.

6.4 BOOST CONVERTERS

The Boost converter is a voltage step-up and current step-down converter. The topology is shown in Figure 6.16(a). The current source is synthesized from a dc voltage source by adding a large

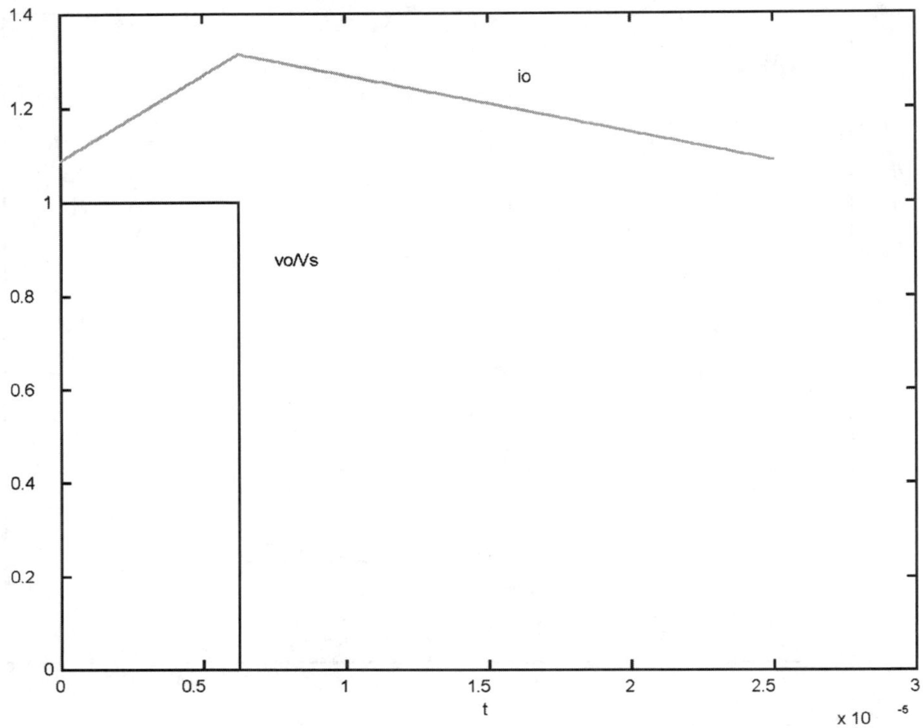

Figure 6.15
Inductor current and voltage waveforms in Example 6-1

value inductor in series. In dc regulated power supply applications, the voltage sink represents a large value capacitor in parallel with the load resistance. In the application of the dc motor control, the voltage sink represents the back emf of the dc motor. Figure 6.16(b) shows a practical realization of the Boost topology.

The switching action produces a pulsating current. The capacitive filter smoothens the pulsating current and provides a dc voltage to the load.

The capacitor is assumed to be large enough to maintain a minimum-ripple voltage $\mathbf{V_o}$ on the load resistance. The source current waveform over a switching period is continuous or discontinuous dependent on the value of the inductor and the load resistance.

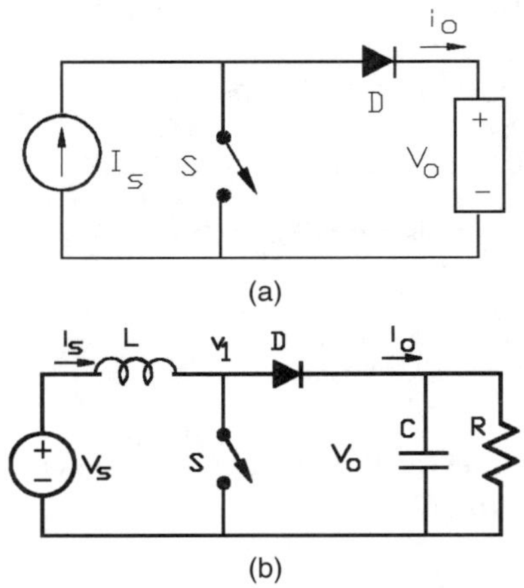

Figure 6.16
(a) Boost topology and (b) Circuit

6.4.1 Continuous conduction mode operation

The steady-state operation consists of two circuit modes as shown in Figure 6.17. As the switch is closed in mode 1, the diode is reverse biased and current flows through the inductor and the switch. After the time τ, the switch is turned off. The inductor current then flows through the diode into the voltage sink as shown in the circuit of mode 2. The second mode is terminated at T_S when the switch is turned on again. The current and voltage waveforms are shown in Figure 6.18.

The inductor current waveform is an ac current superimposed on a dc level. The dc inductor current is equal to the average dc current in the load resistance because the average current in the capacitor is zero. The average output voltage is calculated from the average output current:

$$\overline{V}_O = \overline{I}_O R$$

The average voltage across the inductor is zero, hence,

$$V_S\tau + (V_S - \overline{V}_O)(T_S - \tau) = 0 \qquad (6\text{-}22)$$

$$T_{vv} = \frac{\overline{V}_O}{V_S} = \frac{1}{1 - D}\,;\, D = \frac{\tau}{T_S}$$

Assuming lossless conversion, the equation of the balance of power holds true:

$$V_S \overline{I}_S = \overline{V}_O \overline{I}_O$$

where I_S is the average current from the dc source. Hence,

$$\frac{\overline{I}_O}{\overline{I}_S} = \frac{V_S}{\overline{V}_O} = 1 - D \qquad (6\text{-}23)$$

Although the dc output voltage is larger than the dc input voltage, the dc output current is smaller than the input dc current (a step-down action in current). Furthermore, the instantaneous source (also inductor) current is given by the following:

$$mode\ 1{:}\ L\frac{di_L}{dt} = V_S\,;$$

$$mode\ 2{:}\ L\frac{di_L}{dt} = -(V_S - \overline{V}_O) \qquad (6\text{-}24)$$

The instantaneous inductor current ramps up during the on time of the switch and ramps down during the off time. The output current is zero during τ and equal to the inductor current in the remaining period.

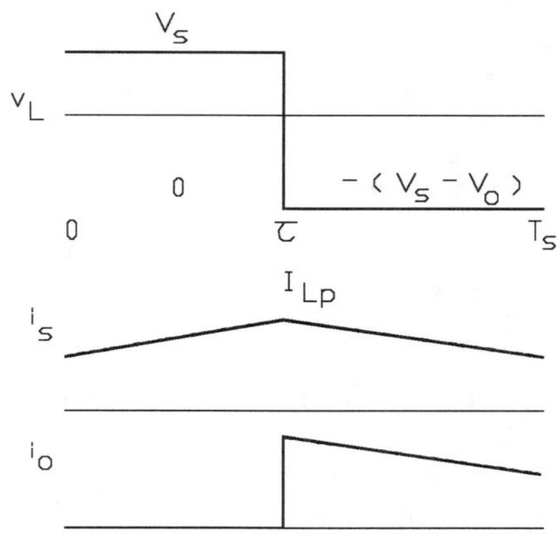

Figure 6.18
Waveforms in the Boost converter

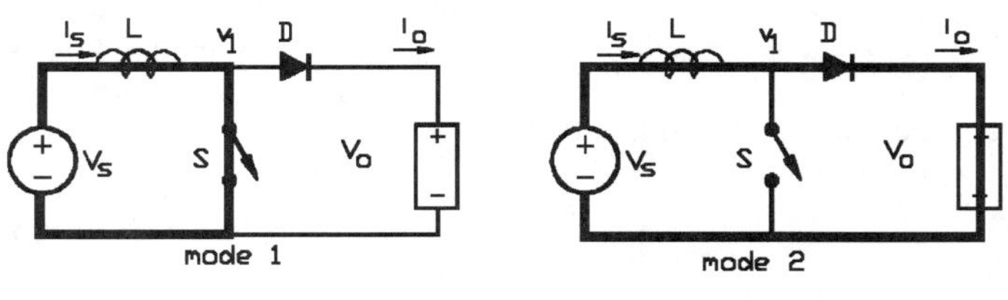

Figure 6.17
Boost converter in CCM operation

```
Ts = 1; VS = 100;D = 0.4*Ts;R = 1;
   L = 1;fs = 1/Ts;
t1 = linspace(0, D*Ts, 100);
   t2 = linspace(D*Ts: Ts:100);
t = [t1 t2];
vs1 = [VS*ones(size(t1)) zeros(size(t2)) ];
Vo = mean(vs1);
vo = [Vo*ones(size(t))];
io1 = [vs1-vo] .*t1/(L);
Ilm = [VS-Vo] *D*Ts/(L);
io2 = Ilm-vo.* (t1-D*Ts)/L;
io = [io1 io2];
Io = mean(io);
plot(t,vs1, t,io)
```

The continuous mode of operation exists as long as the inductor current at the beginning of the switching period is above zero. The peak inductor current is calculated from the circuit equation of mode 1 in equation (6-24):

$$I_{Lp} = \frac{V_S}{L}\tau$$

I_{LP} is also the peak-to-peak .

$$\Delta I_L = I_{Lp} = \frac{V_S T_S}{L} D \qquad (6\text{-}25)$$

The current ripple is independent of the load current.

Average output current for the boundary case is calculated from the following:

$$I_{OB} = \frac{1}{T_S}\left[\frac{1}{2} I_{LP}(T_S - \tau)\right]$$

$$= \frac{V_S T_S}{2L} D(1 - D) \qquad (6\text{-}26)$$

For control characteristics plots, the boundary value current is normalized as follows:

$$I_{nOB} = \frac{I_{OB}}{V_S T_S/2L} = D(1 - D) \qquad (6\text{-}27)$$

The converter characteristics in CCM are given in Figure 6.19.

```
D = linspace(0,1,100); %array for duty ratio
Tvv = 1./(1-D); %Voltage transfer ratio
Inob = D .*( ones(size(D))-D); %normalized output
   current in the boundary case
plot (D,Tvv, D,Inob)
```

Note that the plot of boundary value current I_{nOB} versus the pulse-width D is identical to that in the Buck converter; however, the minimum load resistance R_{nb} is very different. The converter forward transfer ratio T_{vv} can be ideally infinite for D = 1; however, the parasitic elements as capacitance across the open switch would bring it down. Furthermore, the high slope dT_{vv}/dD is obtainable at more than 50% duty ratios. High slope enables efficient control when D is used as the feedback control parameter.

Low Voltage Ripple

The assumption of ideal voltage sink will be dropped and instead a large value capacitor in parallel with the load resistance will be used. To simplify the analysis, the following assumptions will be made:

1. The peak-to-peak voltage ripple is a small percentage of the dc voltage V_O.
2. All the ripple current through the diode flows into the filter capacitor and not through the load resistance.

The dc current through the diode equals the average dc current in the load resistance because the average current in the capacitor must be zero.

The ripple current throughout the diode, with the dc component removed for the sake of simplified analysis, is depicted in Figure 6.20. The areas above and below the zero line are equal. The positive ripple current increases charge, and hence the voltage across the filter capacitor and the negative ripple current decreases it. Thus produced voltage ripple is shown in Figure 6.20. The peak-to-peak ripple voltage on the capacitor is

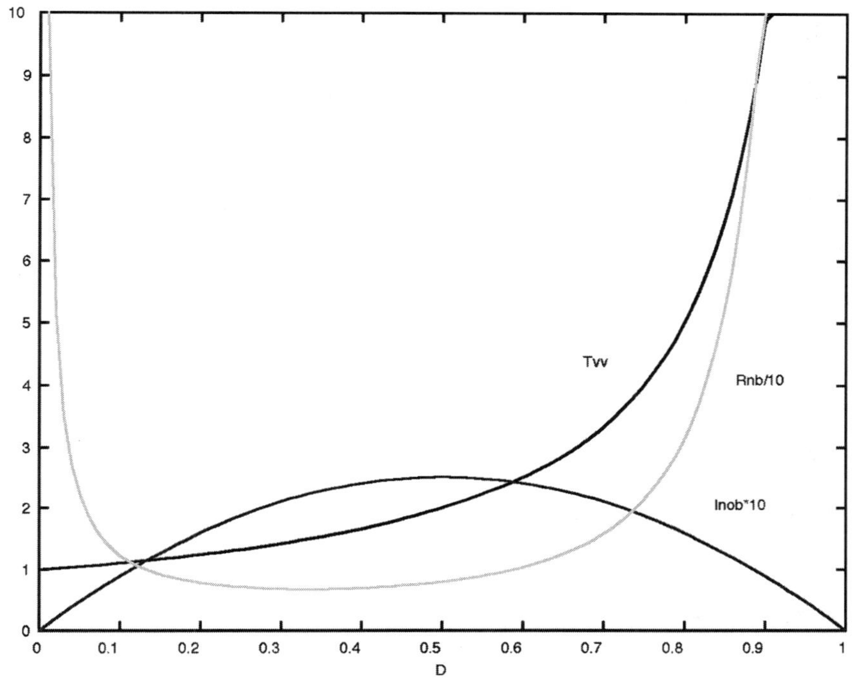

Figure 6.19
Characteristics of Boost converter in CCM

$$\Delta V_O = \frac{\Delta Q}{C} = \frac{\bar{I}_O \tau}{C} = \frac{\bar{V}_O}{R}\frac{DT_S}{C}$$

$$\frac{\Delta V_O}{\bar{V}_O} = \frac{DT_S}{RC} = \frac{D f_C}{f_S} \tag{6-28}$$

where $f_S = 1/T_S$, and $f_C = 1/RC$

A low value of f_C/f_S (a factor of at least 10) will provide low voltage ripple.

Switch Stress

PEAK VOLTAGE STRESS ON THE SWITCH:

$$V_{sw\ max} = V_O$$

PEAK CURRENT THROUGH THE SWITCH:

Peak current through the switch is the dc output current and half of the maximum value of the peak-to-peak ripple current

$$I_{sw\ max} = \bar{I}_O + \frac{1}{2}(\Delta I_L)_{max}$$

$$= \bar{I}_O + \frac{V_S T_S}{8L}$$

6.4.2 Discontinuous conduction mode operation (DCM)

Like the Buck converter, the Boost converter operates in the discontinuous conduction mode for load resistances higher than a boundary value $\mathbf{R_B}$. The steady-state operation in this mode of operation contains a third circuit mode in which no current flows in the inductor. See the current waveform in Figure 6.21. The average output voltage is obtained from the condition that the average on the inductor is zero:

$$V_S \tau_1 + (V_S - V_O)\tau_2 = 0$$

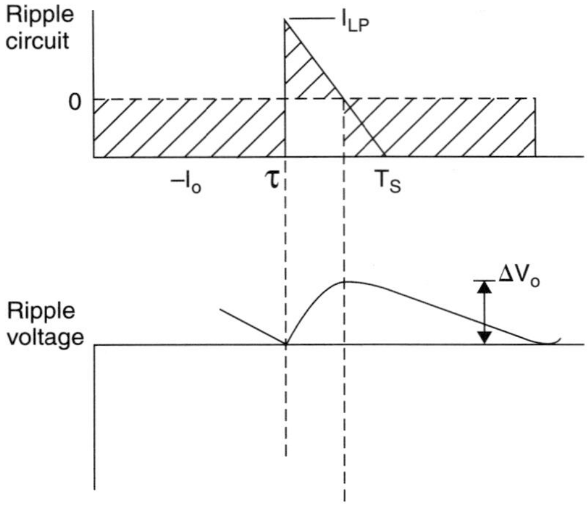

Figure 6.20
Consider the steady-state of the Buck regulator with Vs 5 125V, L 5 125mH, R 5 47 ohms, D 5 0.6 and fs 5 20 kHz. Assume a large C.
(a) Calculate the load voltage and
(b) calculate the power delivered by the dc voltage source.

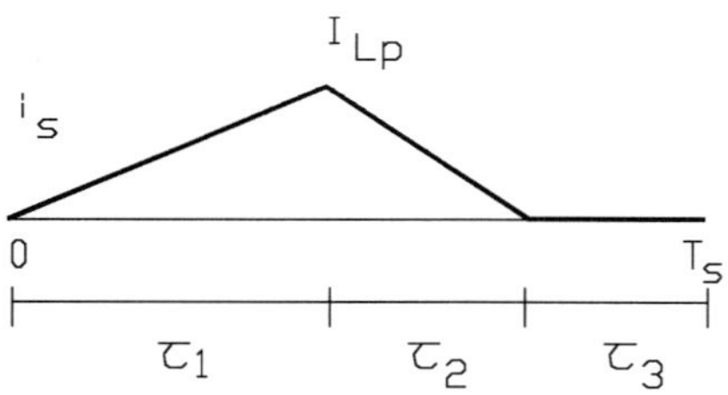

Figure 6.21
Source current waveform in DCM operation

$$T_{vv} = \frac{\overline{V}_O}{V_S} = \frac{\tau_1 + \tau_2}{\tau_2} = \frac{1 - \Delta}{1 - D - \Delta} \quad (6\text{-}29)$$

$$D = \frac{\tau_1}{T_S}, \Delta = \frac{\tau_3}{T_S}, \textit{ and } \tau_1 + \tau_2 + \tau_3 = T_S$$

The peak output current, which is also the peak inductor current, is given by equation (6-25).

The average output current is

$$\overline{I}_O = \frac{1}{2} I_{LP} \frac{\tau_2}{T_S} = \frac{V_S T_S}{2L} D(1 - D - \Delta) \quad (6\text{-}30)$$

The output current normalized by the boundary value current is

$$I_{nO} = \frac{\overline{I}_O}{I_{OB}} = 1 - \frac{\Delta}{1 - D} \quad (6\text{-}31)$$

The plot of the voltage transfer ratio T_{vv} versus the output dc current normalized to the boundary value, Io/IoB, with D as the parameter, are shown in Figure 6.22a.

In dc power supply applications, the output voltage is usually regulated against the variations in the input voltage and the load current by controlling the duty ratio. A plot of duty ratio D versus the output current normalized with respect to the boundary value current with Tvv as the parameter is shown in Figure 6.22b.

Example 6-2

In a Boost converter, the duty ratio is adjusted to regulate the output voltage to 48 V. The input dc voltage varies from 10 to 30 V. The converter is switched at 40 kHz. The maximum output power is rated at 100 watts. The converter is operated in CCM. Find

a. *the minimum value of inductor L, and*

b. *value of filter C for a voltage ripple of 1% or less at this value of L.*

$V_O = 48$ V, $T_S = 25$ S , $I_{omax} = 100$ W /48 V $= 2.083$ A, R $= 48 / 2.083 = 23.04$ ohms

The range of duty ratios D1-D2 for input voltages (10 to 30 V) is obtained from equation (6-25),

$$D_1 = 1 - \frac{V_{S1}}{V_O} = 1 - \frac{10}{48} = 0.7917$$

$$D_2 = 1 - \frac{V_{S2}}{V_O} = 1 - \frac{30}{48} = 0.375$$

The maximum output current I_{omax} must be equal to the maximum boundary value current for the range of duty ratios (D = .375 to .7917) in order to maintain the CCM operation. Maximum boundary value current $I_{OB\ max}$ occurs for D = .5 (see equation 6-26), which lies in the said range. $I_{OB\ max}$ is

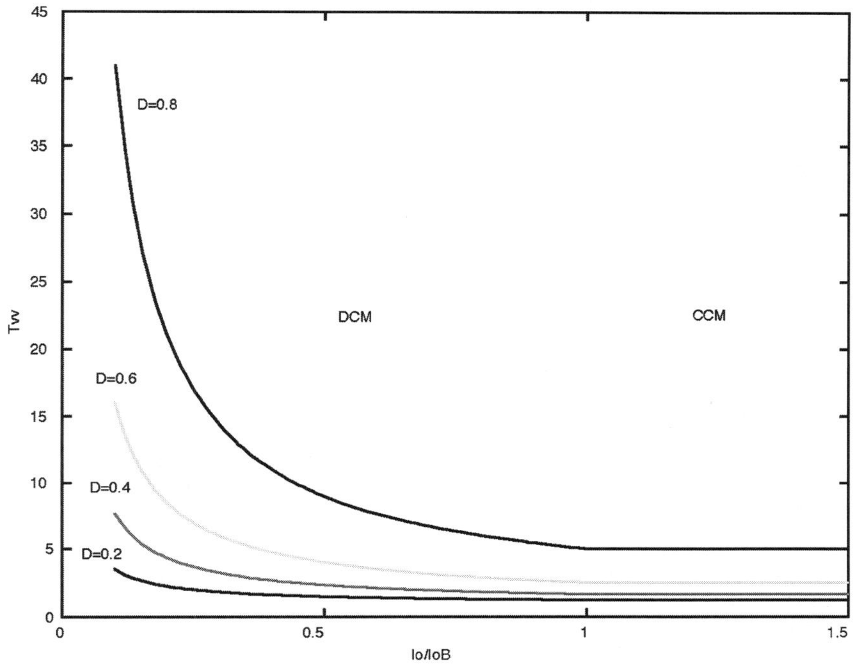

Figure 6.22a
Voltage transfer constant vs. output current normalized to the boundary value

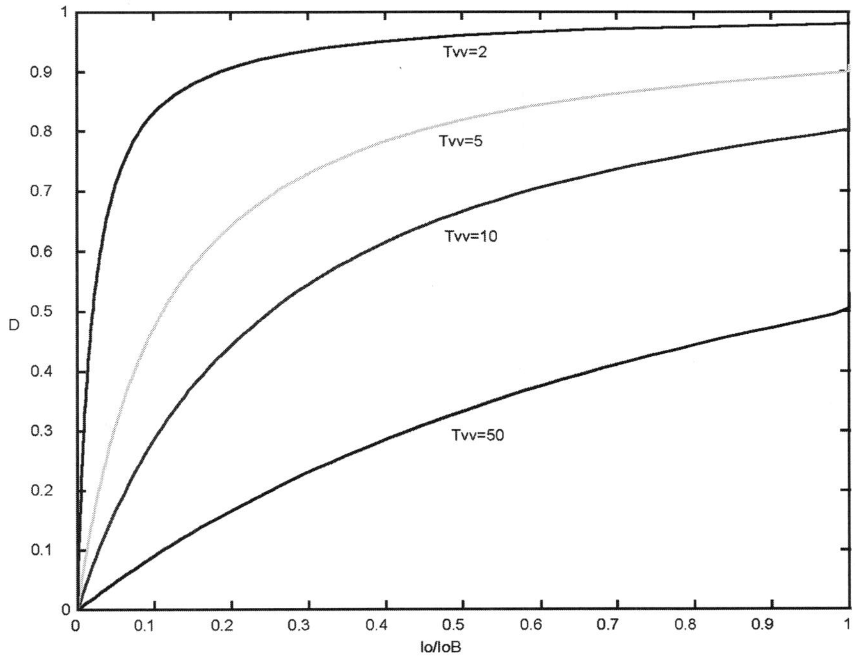

Figure 6.22b
Duty ratio vs. output current normalized to the boundary value

$$I_{OB\,\max} = \frac{V_S T_S}{8L} \textit{ for } D = 0.5$$

Hence, the minimum value of inductor to maintain $I_{OB\,\max}$ is

$$L = \frac{V_S T_S}{8 I_{OB\,\max}} = 36\ \mu H$$

The voltage ripple is required to be 1%, hence, from equation (6-28), the corner frequency of the RC filter and the necessary capacitance are calculated as

$$\frac{\Delta V_O}{V_O} = 0.01 = 2\pi \frac{f_C}{f_S}, f_C = 63.66\ Hz$$

$$C = \frac{1}{2\pi f_C R} = 108.5\ \mu F$$

6.5 BUCK-BOOST CONVERTERS

As the name suggests, this converter is made of a Buck topology in cascade with a Boost topology as shown in Figure 6.23a. It can provide both a step-down and a step-up function. The two switches can be combined into a single switch as shown in Figure 6.23b. This combination, however, requires the polarity reversal of the output voltage sink. The two-position switch is implemented by a unidirectional one-position switch and a diode, as shown in Figure 6.23c. The generalized topology is shown in Figure 6.23d where the inductor is replaced by a current sink and the filter capacitor in parallel with the load resistance by a voltage sink.

Retain the inductor in place of the current sink for further discussion. The source current waveform over a switching period is continuous or discontinuous dependent on the value of the inductor and the load resistance.

6.5.1 Continuous conduction mode operation

The steady-state operation consists of two circuit modes, shown in Figure 6.24.

As the switch is turned on in mode 1, the diode is reverse biased and the current flows through the inductor and the switch. After time τ, the switch is turned off. The inductor current then flows through the diode into the voltage sink as shown

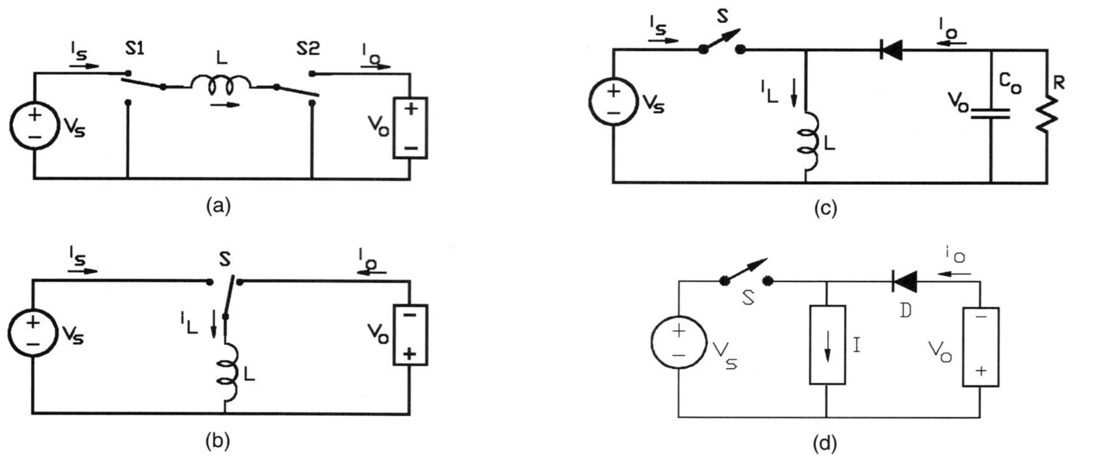

Figure 6.23
(a) Buck converter followed by the Boost converter (b) topology using a 3-position switch (c) the 3-position switch is synthesized from a 2-position switch and a diode and (d) the generalized topology.

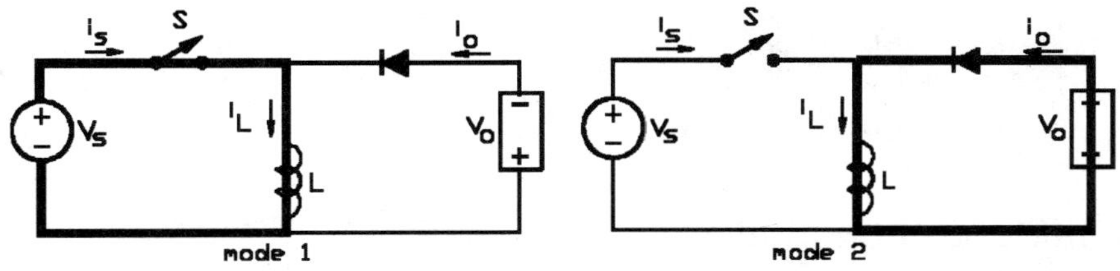

Figure 6.24
Circuit modes in Buck-Boost converter

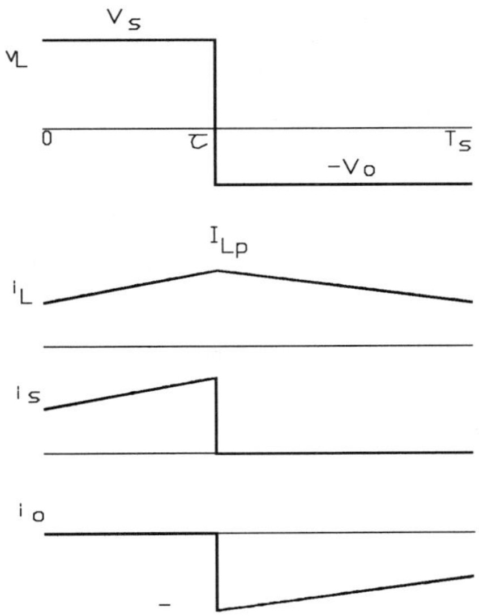

Figure 6.25
Waveforms in Buck-Boost converter

in the circuit of mode 2. The second mode is terminated at T_S when the switch is turned on again. The current and voltage waveforms are shown in Figure 6.25.

The average voltage on the inductor is zero, hence

$$V_S\tau + (-\bar{V}_O)(T_S - \tau) = 0$$

$$T_{vv} = \frac{\bar{V}_O}{V_S} = \frac{D}{(1 - D)} \tag{6-32}$$

Figure 6.26 shows the plot of T_{vv} versus D. Voltage step-up action is possible for $D < 0.5$ and step-up action for $D > 0.5$.

Assuming a lossless converter, the ratio of dc output current to the dc input current is

$$\frac{\bar{I}_O}{\bar{I}_S} = \frac{1 - D}{D} \tag{6-33a}$$

The CCM operation is possible as long as the lowest value of the inductor current is greater than zero. In the boundary case, the peak current is

$$I_{LP} = \frac{V_S}{L}\tau = \frac{V_S T_S}{L} D \tag{6-33b}$$

The boundary value of the output current for continuous mode of operation is calculated as

$$\bar{I}_{oB} = \frac{1}{T_S}\left[\frac{1}{2} I_{Lp} (T_S - \tau)\right]$$

$$= \frac{V_S T_S}{2L} D(1 - D) \tag{6-34a}$$

Note that the boundary value of the output current is identical in the Buck, Boost and Buck-Boost converters. The boundary value of the output current vs. D is shown in Figure 6.26.

Switch Stress

PEAK VOLTAGE STRESS ON THE SWITCH:

$$V_{sw\ max} = V_S + V_O \tag{6-34b}$$

PEAK CURRENT THROUGH THE SWITCH:

Peak current through the switch is the dc output current and half of the maximum value of the peak-to-peak ripple current

$$I_{sw\,max} = \bar{I}_O + \frac{1}{2}(\Delta I_L)_{max} = \bar{I}_O + \left(\frac{V_S T_S}{2L} D\right)_{max}$$

$$= \bar{I}_O + \frac{V_S T_S}{2L}$$

6.5.2 Discontinuous conduction mode operation

The waveforms are shown in Figure 6.27. The average output voltage is obtained from the condition that the average on the inductor is zero:

$$V_S^s \tau_1 + (-\bar{V}_O)\tau_2 = 0$$

$$T_{vv} = \frac{\bar{V}_O}{V_S} = \frac{\tau_1}{\tau_2} = \frac{D}{1 - D - \Delta} \qquad (6\text{-}35)$$

where D and Δ are as defined by equation (6-29). The average output current is calculated from the following:

$$\bar{I}_O = \frac{1}{2} I_{Lp} \frac{\tau_2}{T_S} = \frac{V_S T_S}{2L} D \frac{\tau_2}{T_S}$$

$$= \frac{V_S T_S}{2L} D(1 - D - \Delta) \qquad (6\text{-}36)$$

Normalizing the output current with respect to the boundary value current,

$$I_n = \frac{\bar{I}_O}{I_{OB}} = 1 - \frac{\Delta}{1 - D} \qquad (6\text{-}37)$$

$$and\ \Delta = (1 - D)(1 - I_n)$$

Substituting above equation into equation (6-35) we can get the voltage transfer constant as a function of the normalized output current In and the duty ratio D, see Figure 6.28a.

A plot of duty ratio D versus the output current normalized with respect to the boundary value current with Tvv as the parameter is shown in Figure 6.28b.

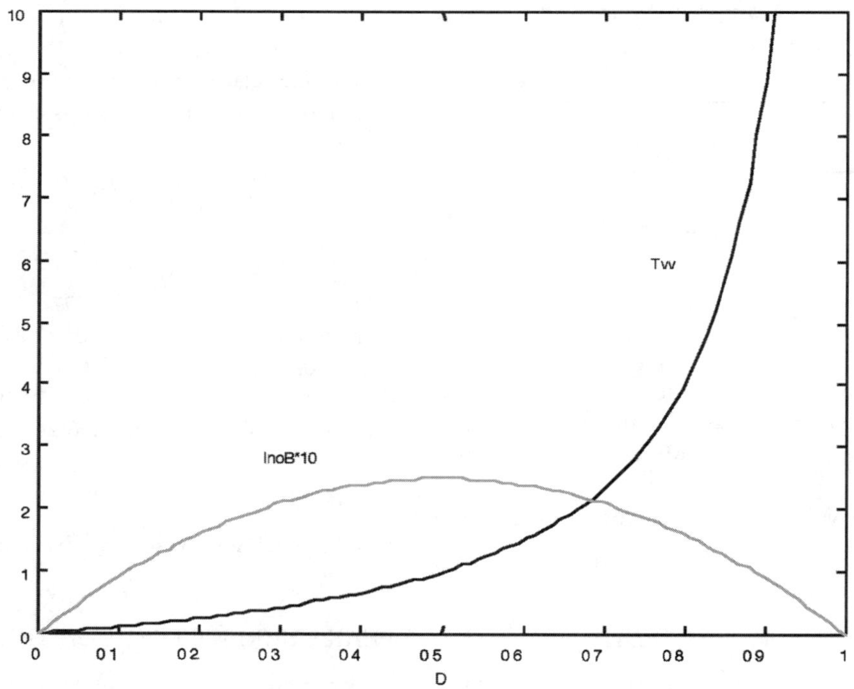

Figure 6.26
Voltage transfer ratio and the boundary value of the output current vs. the duty ratio in CCM operation of Buck-Boost operation.

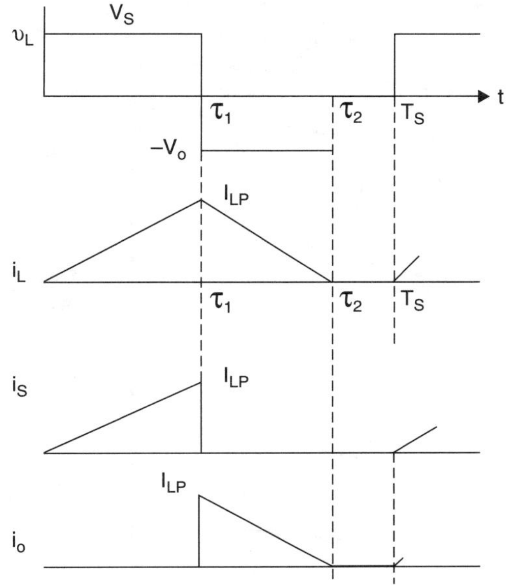

Figure 6.27
Waveforms in Buck-Boost converter in DCM operation

Example 6-3

Repeat the problem in Example 6-2 when the output voltage is specified to be 24 V.

Since the output voltage is both lower and higher than the input voltage, we use the Buck-Boost converter.

$$V_O = 24\text{ V},\ T_S = 25\ \mu\text{S},\ I_{omax} = 100\text{ W} / 24\text{ V} = 4.167\text{ A},\ R = 24 / 4.167 = 5.76\text{ ohms}$$

The range of duty ratios D1-D2 for input voltages (10 to 30 V) can be obtained by using equation (6-31),

$$D_1 = \frac{\overline{V}_O}{\overline{V}_O + V_{S1}} = \frac{24}{24 + 10} = 0.706$$

$$D_2 = \frac{\overline{V}_O}{\overline{V}_O + V_{S2}} = \frac{24}{24 + 30} = 0.444$$

The maximum output current I_{omax} must be equal to the maximum boundary value current for the range of duty ratios (D = .444 to .706) in order to maintain

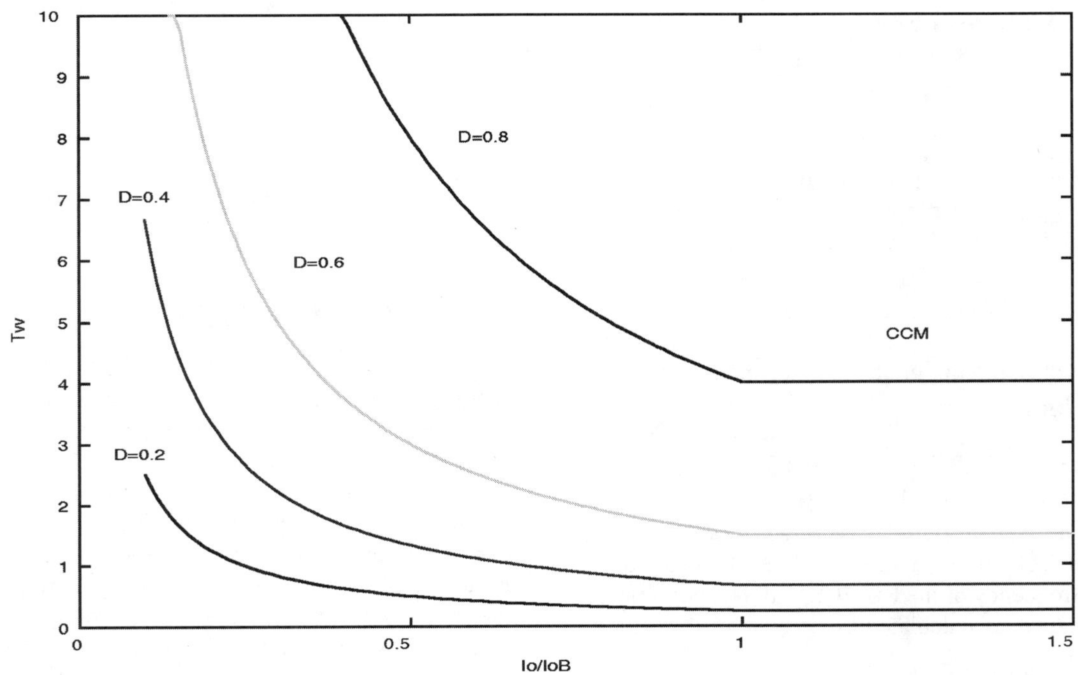

Figure 6.28a
(a) Voltage transfer constant vs. output current normalized to the boundary value

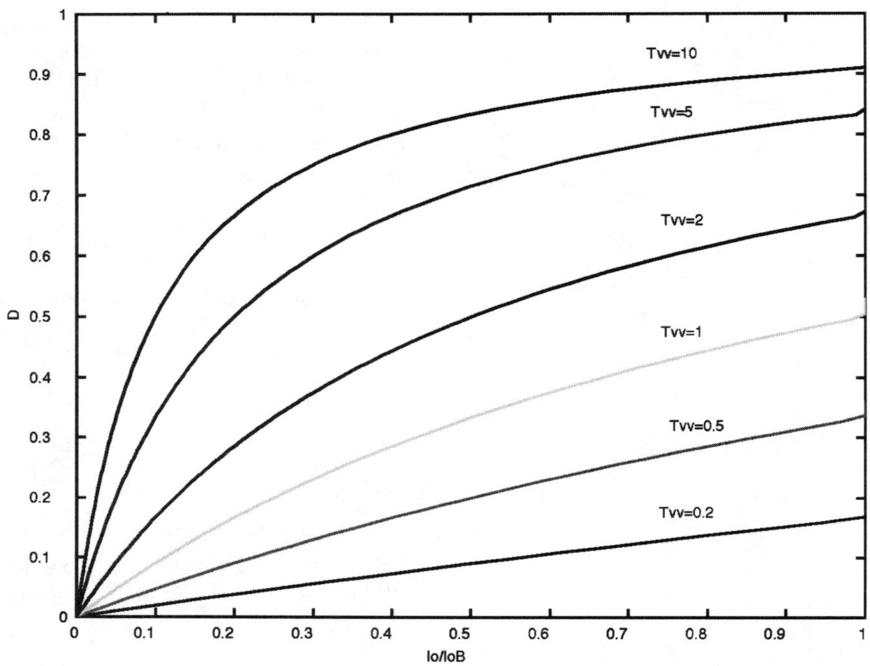

Figure 6.28 ***(continued)***
(b) Duty ratio vs. output current normalized to the boundary value

the CCM operation. Maximum boundary value current $I_{OB\ max}$ occurs for D = .5 (see eq. 6-33), which lies in the said range. $I_{OB\ max}$ is

$$I_{OB\max} = \frac{V_S T_S}{8L} for\ D = 0.5$$

Hence, the minimum value of inductor to maintain $I_{OB\ max}$ is

$$L = \frac{V_S T_S}{8 I_{OB\max}} = 18\ \mu H$$

The voltage ripple is required to be 1%, hence, from Problem 6-2 at the end of the chapter, the corner frequency of the RC filter and the necessary capacitance are calculated as

$$\frac{\Delta V_O}{V_O} = 0.01 = 2\pi \frac{f_C}{f_S}\quad ,\quad f_C = 63.66\ Hz$$

$$C = \frac{1}{2\pi f_C R} = 434\ \mu F$$

6.6 ČUK CONVERTER

The ČUK converter is the dual of the Buck-Boost topology. The voltage source and sink are replaced by current counterparts. The inductor is replaced by the capacitor, as shown in Figure 6.29a. Like its dual, the ČUK converter can provide both the step-down and the step-up voltages. The two switches can be combined into a single switch as shown in Figure 6.29b. This combination, however, requires the polarity reversal of the output voltage sink. The two-position switch is implemented by a unidirectional one-position switch and a diode, as shown in Figure 6.29c. The generalized topology

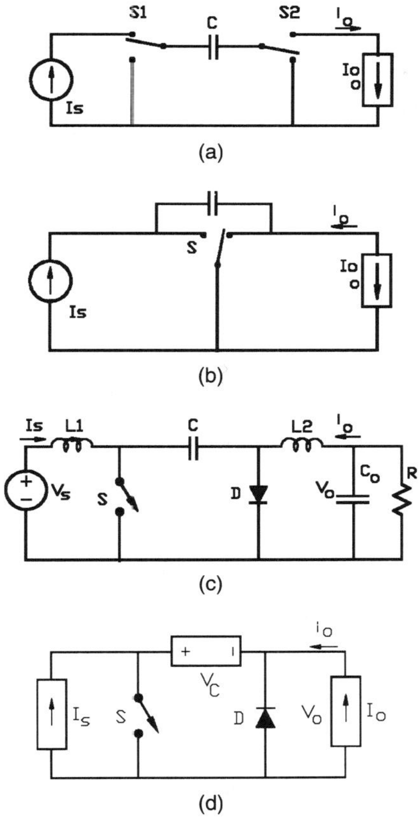

Figure 6.29
(a) The ČUK converter as the dual of the Buck-Boost converter (b) topology using a 3-position switch (c) the 3-position switch is synthesized from a 2-position switch and a diode and (d) the generalized topology

is shown in Figure 6.29d by substituting current sinks in place of the inductors L_1 and L_2 and the voltage sink in place of the capacitor. The output voltage is of reverse polarity the same as that in the Buck-Boost topology.

In the steady state, the inductors cannot support any dc voltages, hence, the average voltage on the capacitor is equal to the sum of the input dc voltage and the average output voltage.

$$\overline{V}_C = V_S + \overline{V}_O \tag{6-38}$$

If the capacitor C is chosen to be a large value, the voltage on it will be constant throughout the period.

Only the CCM operation will be investigated. The DCM operation and other details are well reported in literature.

The steady-state CCM operation can be described by two circuit modes, as shown in Figure 6.30.

As the switch is turned on in mode 1, the diode is reverse biased and the current in inductors L_1 and L_2 flow jointly through the switch. The switch is turned off after a time τ. The sum of inductor currents then flows through the diode, as shown in the circuit of mode 2. The second mode is terminated at T_S when the switch is turned on again. The current and voltage waveforms are shown in Figure 6.31.

The average output voltage is obtained from equating the average voltage on any one of the inductors to zero. For example, from the average voltage condition on L_1:

$$V_S \tau - \overline{V}_O (T_S - \tau) = 0$$
$$T_{vv} = \frac{\overline{V}_O}{V_S} = \frac{D}{1 - D} \tag{6-39}$$

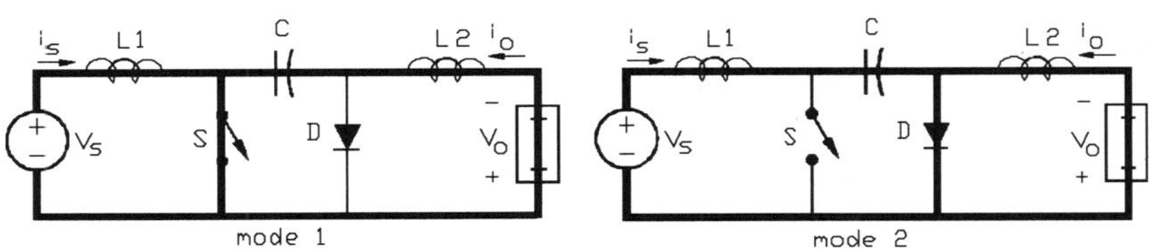

Figure 6.30

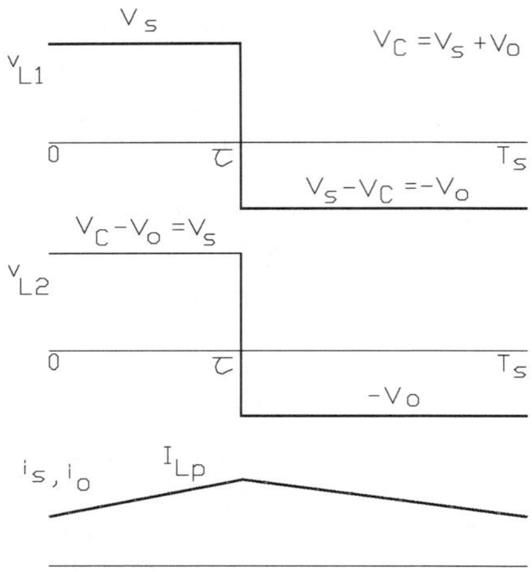

Figure 6.31
Waveforms in ČUK converter

The analysis of ČUK converter is similar to the analysis of the Buck-Boost converter, hence, left for the readers to perform. The current waveforms in both the source ($\mathbf{i_{L1}}$) and in the load ($\mathbf{i_{L2}}$) are ripple free in the continuous current mode of operation. This is a significant advantage of the ČUK topology over the Buck-Boost topology. This advantage translates in terms of low filtering requirements on both sides.

6.7 TWO-QUADRANT CONVERTERS

The two-quadrant dc-dc converter enables the return of energy back to the source if the load does not need it. A prominent application is the dynamic braking of dc motors.

Buck Topology

Buck topology of the two-quadrant dc-dc converter is shown in Figure 6.32a. The switch conducts current in the forward direction and the antiparallel diode permits current flow in the reverse direction. The circuit modes are shown in Figure 6.32b and the waveforms in Figure 6.32c. The average output current may thus have both polarities. The converter operates in CCM only.

Buck-Boost Topology

An example of the two-quadrant Buck-Boost converter is shown in Figure 6.33. It is suggested that readers draw the waveforms and analyze it.

6.8 FULL-BRIDGE CONVERTER

The full-bridge topology constitutes the four-quadrant dc-dc converter. The converter permits the output voltage of both polarities and each one with positive and negative polarity currents. Power flows in both directions. A prominent application is the forward and reverse motion with dynamic braking of dc motors, such as in the elevator drive.

The Buck topology of the four-quadrant dc-dc converter is shown in Figure 6.34. Switches conduct current in the forward direction and the antiparallel diodes permit the reverse current flow. This converter always operates in the Continuous Conduction Mode.

If the current in the switch-diode combination prior to turning on the switch flows in the reverse direction, the switch will not conduct. Hence, a turned on switch may or may not conduct current depending on the direction of the current flow.

The sequence of turning on switches in the bridge determines the waveform of the output voltage across the voltage sink. There are two sequences that are generally used:

(i) Bipolar switching: Switches are turned on exclusively in pairs. One of the pair constitutes switches S1 and S3 and the other S2 and S4. When the S1-S3 pair is turned on, the S2-S4 pair is turned off and vice versa. In practice, however, there is a blanking interval in which all four switches are off, thus avoiding the accidental shorting of source via the wrong combination. This switching sequence, also referred as PWM switching, produces a bipolar voltage waveform on the voltage sink as shall be seen shortly.

(ii) Unipolar switching: Switches are turned on exclusively in singles. This switching sequence produces a unipolar voltage waveform on the voltage sink. This sequence is also referred to as double PWM switching.

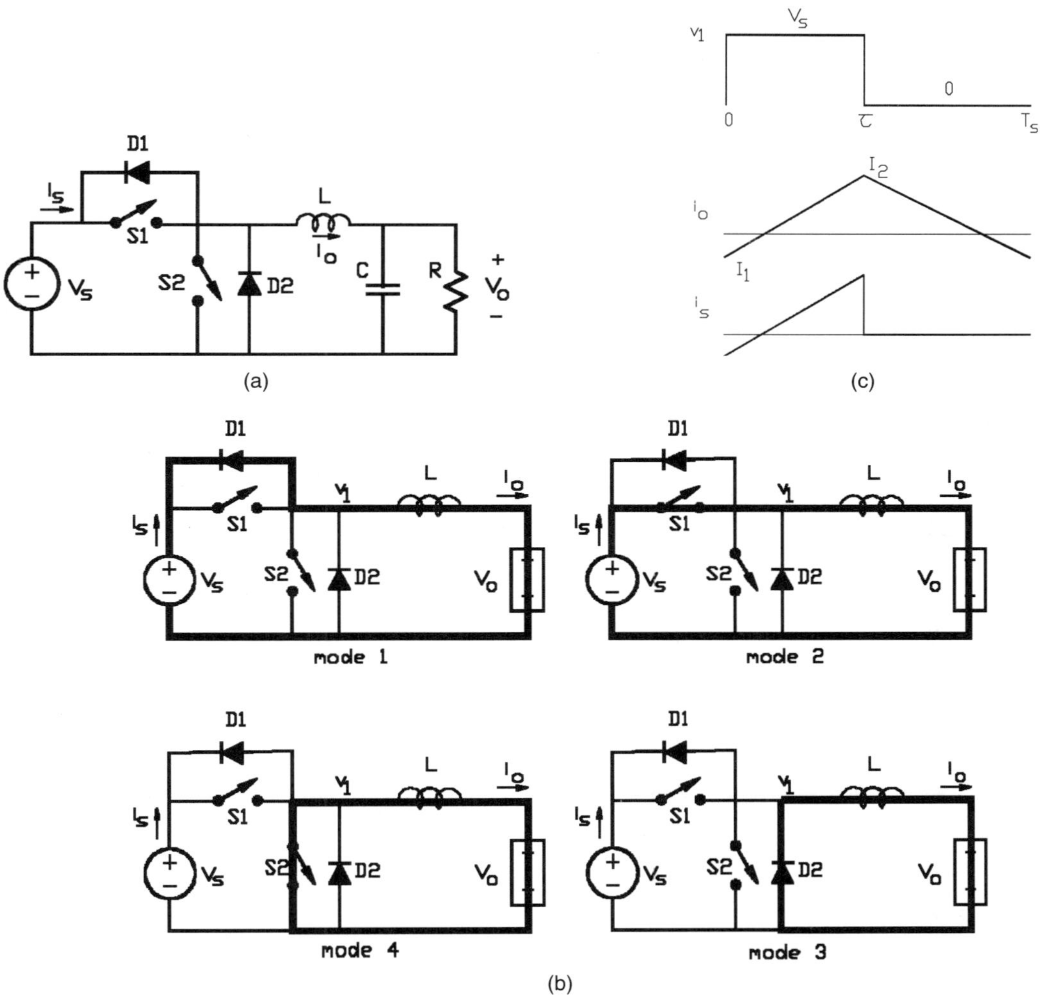

Figure 6.32
(a) Two-quadrant Buck topology (b) Circuit modes and (c) Waveforms

6.8.1 Full-bridge converter with bipolar switching

Circuit modes are shown in Figure 6.35a and the waveforms in Figure 6.35b. Prior to turning on the switch pair S1-S3, the inductor current is flowing in the reverse direction. Hence, when the switch pair is turned on, the switches do not conduct the inductor current; instead, it flows through diodes D1 and D3, as shown in mode 1. A voltage V_S-V_O is applied across the inductor, which forces the inductor current to increase linearly as a ramp. As the inductor current becomes zero, it is diverted through the switch pair S1-S3. This is the situation in circuit mode 2, which ends as the switch pair S1-S3 is turned off and S2-S4 is turned on after time τ. The voltage V_S-V_O is applied on the inductor throughout mode 1 and mode 2.

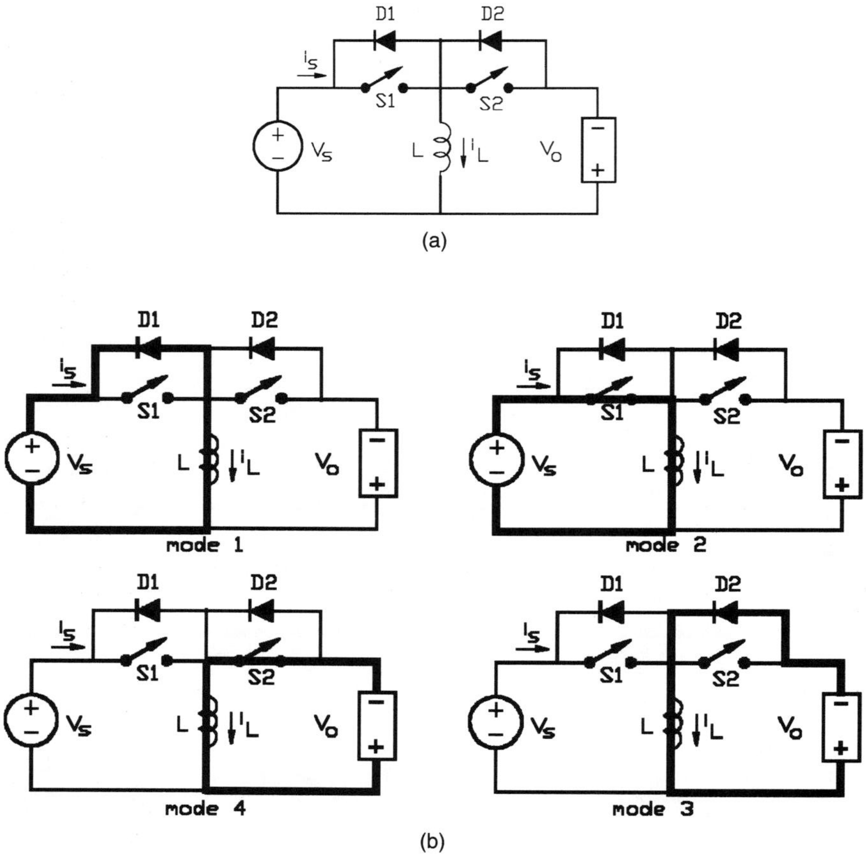

Figure 6.33
(a) Two-quadrant Buck-Boost converter (b) Circuit modes

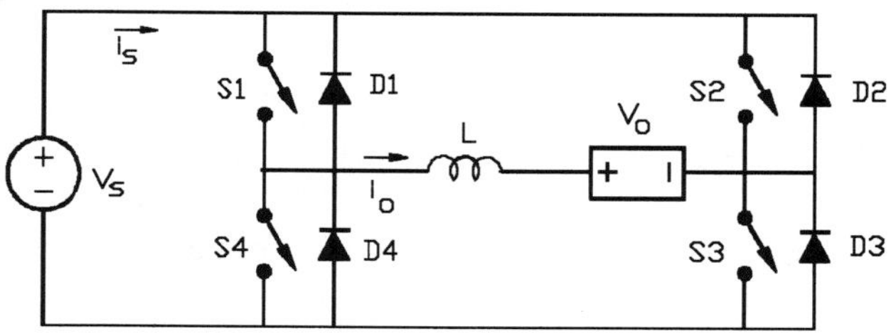

Figure 6.34
Full-bridge converter

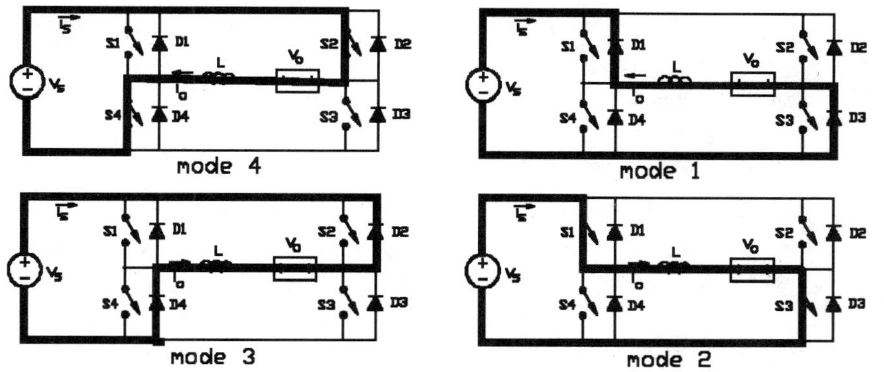

(a)

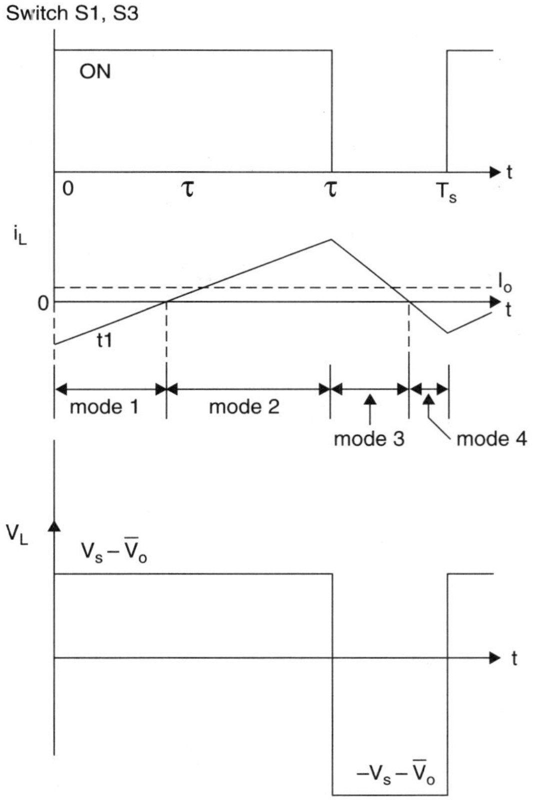

(b)

Figure 6.35
Circuit modes in bipolar switching

When the switch pair S2-S4 is turned on, the switches do not conduct the inductor current. Instead, it flows through diodes D2 and D4, as shown in mode 3. A voltage $-V_S$-V_O is applied across the inductor, whereupon the inductor current decreases linearly in a downward ramp. After the inductor current becomes zero, it is diverted through the switch pair S2-S4. This is the circuit mode 4, which ends when the switch pair S2-S4 is turned off and S1-S3 is turned on again after one switching period T_S. The voltage $-V_S$-V_O is applied on the inductor throughout mode 3 and mode 4. The waveforms are shown in Figure 6.35b.

The average output voltage on the inductor is zero, hence,

$$(V_S - \overline{V}_O)\tau + (-V_S - \overline{V}_O)(T_S - \tau) = 0$$

$$\overline{V}_O = V_S(2D - 1), \qquad T_{vv} = 2D - 1$$

$$\textit{where } D = \tau/T_S \tag{6-40}$$

The τ is the duration of the on-time of the switch S1. Plot of T_{vv} versus D is shown in Figure 6.36. The output voltage varies from $-V_S$ to $+V_S$ for D = 0 to 1. The voltage transfer ratio is also independent of the load current amplitude an d its polarity.

The current ripple is shown in Figure 6.35b. Peak-to-peak ripple is

$$\Delta I_O = \frac{V_S - \overline{V}_O}{L}\tau = \frac{2V_S T_S}{L} D(1 - D) \tag{6-41}$$

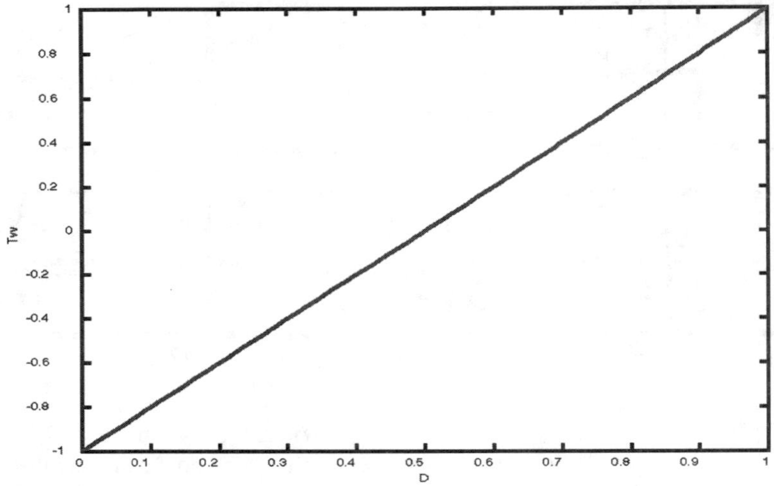

Figure 6.36
Voltage transfer ratio vs. duty ratio D in the full-bridge converter with bipolar switching

Ripple frequency is the switching frequency $1/T_S$. The ripple current amplitude is independent of the load resistance.

If the duty ratio is made a linear function of a control voltage v_c,

$$D = a\,v_c + b$$

where a and b are constants, the output voltage is given by

$$\overline{V}_O = V_S(2a\,v_c + 2b - 1)$$

If constant b is selected such that $2b-1 = 0$,

$$\overline{V}_O = 2aV_S\,v_c = k\,v_c \qquad (6\text{-}42)$$

The dc output voltage is a linear function of the control voltage v_c. This is the equation of the linear switch-mode amplifier, the voltage gain k is a function of the source voltage V_S. The switching frequency should be several times higher than the highest frequency of the control voltage v_c. D as a function of the control voltage is then

$$D = a\,v_c + .5 \qquad (6\text{-}43)$$

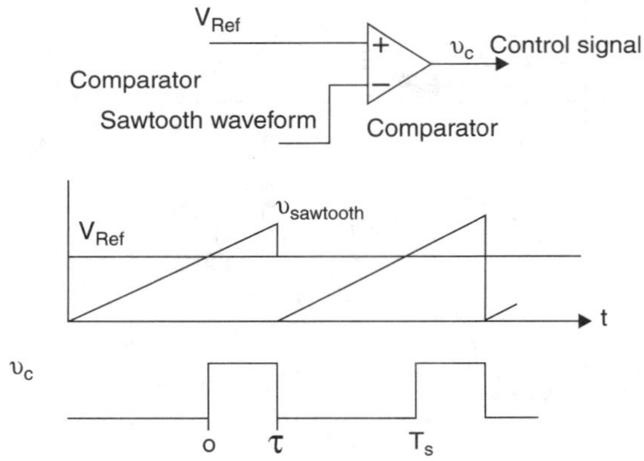

Figure 6.37
The generation of duty ratio controlling waveform

Above function can be realized by programming in computer controlled systems or a circuit shown in Figure 6.37. The reference signal is a triangular waveform of V_p peak amplitude and frequency equal to the switching frequency $1/T_S$. The control voltage v_c is a dc level varying from $-V_p$ to $+V_p$. Details are reported in literature abundantly.

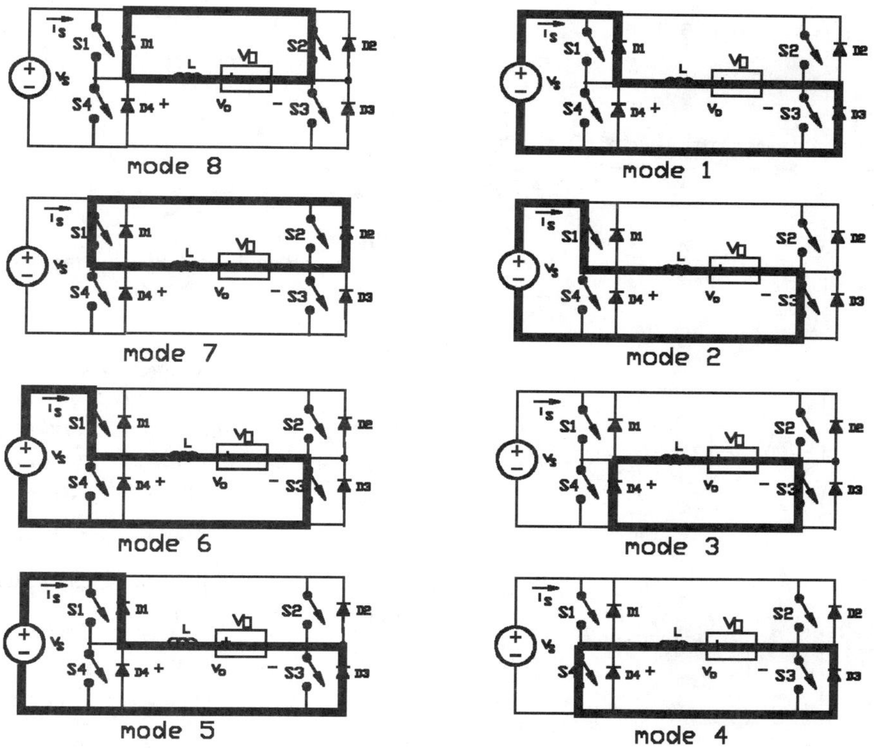

Figure 6.38
Circuit modes in full converter with unipolar switching

6.8.2 Full-bridge converter with unipolar switching

It is not necessary to use the switches in pairs. It may be noted that the switches, either S1-S4 or S2-S3 combinations should never be turned on simultaneously. However, the simultaneous turning on of switch combinations S1-S2 and S3-S4 does not cause any harm but only renders zero voltage on the inductor and voltage sink combination. Utilizing this information, a new switching sequence is adopted, as shown below:

	S1	**S2**	**S3**	**S4**	
mode 1,2	on	—	on	—	
mode 3,4	—	—	on	—	
mode 5,6	on	—	on	—	
mode 7,8	on	on	—	—	'—' : off-state

The circuit modes are shown in Figure 6.38 and the waveforms are shown in Figure 6.39. The output voltage waveform has two identical pulses in a switching period, hence,

$$\tau_1 = \tau_3 \quad \textit{and} \quad \tau_2 = \tau_4$$
$$\tau = \tau_1 + \tau_3 + \tau_4 = 2\tau_1 + \tau_4 \ \textit{and} \ \tau_1 + \tau_2 = T_S/2$$

τ is the duration of the on-time of the switch S1.

The average output voltage is

$$\overline{V}_O = V_S(\tau_1 + \tau_3) / T_S$$
$$\overline{V}_O = V_S(2D - 1), \qquad T_{vv} = 2D - 1 \quad \text{(6-44)}$$

where $D = \tau/T_S$

The voltage transfer ratio Tvv is same as that in the bipolar switching case.

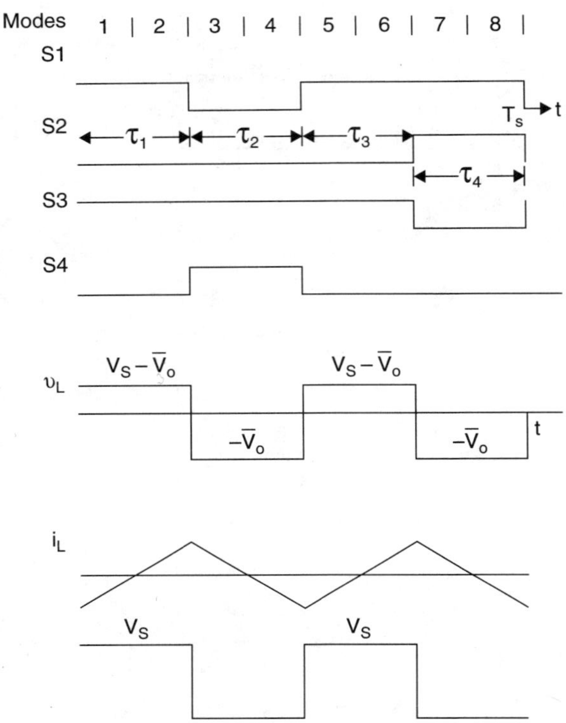

Figure 6.39
Waveforms in the full-bridge converter with unipolar switching

Current ripple is shown in Figure 6.39. The peak-to-peak ripple is

$$\Delta I_O = \frac{V_S - \bar{V}_O}{L} \tau_1$$

$$= \frac{V_S T_S}{L}(2D - 1)(1 - D) \qquad (6\text{-}45)$$

Ripple frequency is double that of the switching frequency $2/T_S$. The ripple current amplitude is independent of the load resistance and is less than that in the bipolar switching case. Hence, from ripple point of view, the unipolar switching is preferable.

Example 6-4

In a full-bridge converter, the duty ratio is adjusted to regulate the output voltage to 12 V. The input dc voltage is fixed 30 V. The converter is switched at 40 kHz. Find the pulse width of the switching waveform applied at the switch S1 and the peak-to-peak ripple in the output voltage for (a) bipolar switching, and (b) unipolar switching.

$$V_O = 12 \text{ V}, T_S = 25 \ \mu\text{S}$$

The duty ratio in both types of switching is calculated from equation (6-43),

$$D = \frac{\bar{V}_O + V_S}{2V_S} = \frac{12 + 30}{2 \times 30} = 0.7$$

The pulse width of the switching waveform of S1 = D T_S = 0.7×25 = 17.5 μS.

a. *Bipolar switching:* The voltage ripple in the output (see problem 6-11a) is

$$\frac{\Delta V_O}{\bar{V}_O} = \frac{DT_S^2}{4LC}\left(\frac{1 - D}{2D - 1}\right) = 0.082$$

b. *Unipolar switching:* The voltage ripple in the output (see problem 6-11b) is

$$\frac{\Delta V_O}{\bar{V}_O} = \frac{T_S^2}{16\,LC}(1 - D) = 0.012$$

The ripple content is about $\frac{1}{7}$ in the unipolar switching case.

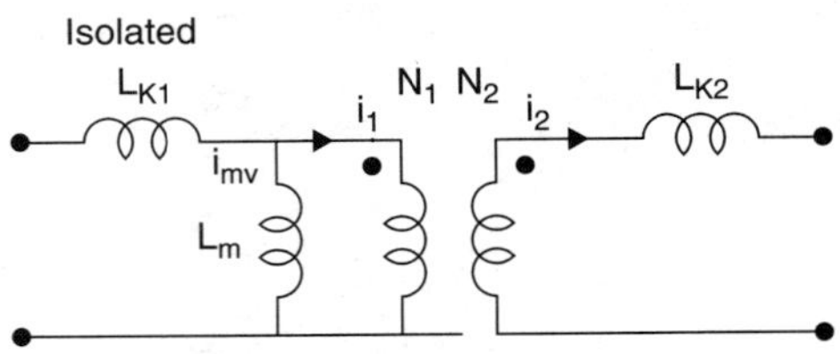

Figure 6.40
A simplified equivalent circuit of the isolation transformer

6.9 ISOLATED DC-DC CONVERTERS

A transformer can be inserted in dc-dc converter topologies: Buck, Boost and Buck-Boost to provide electrical isolation between the input and the output. Electrical isolation becomes necessary in many practical applications such as the switching power supplies. Furthermore, the ac waveforms can be stepped up or down by using the transformer with a view to reduce the component stresses that result when the voltage transfer ratio T_{vv} is far from unity. There would be absolutely no complications if the transformer is ideal, that is, it has zero leakage inductance and it needs zero magnetizing current (infinite magnetizing inductance). An ideal transformer does not store or consume energy. Inclusion of practical transformers needs special mechanism to overcome the nonidealizes. A simplified equivalent circuit of the transformer is discussed in Chapter 2 and is redrawn in Figure 6.4.

The following sections will discuss some of the important isolated topologies.

6.9.1 Isolated buck converter (forward converter)

This converter is derived from the Buck topology by inserting after the switch an isolation transformer and a diode D1.

With Zero Leakage Inductance ($L_K = 0$)

The equivalent circuit of the conceptual isolated Buck converter with the magnetizing inductance L_m is shown in Figure 6.41(a). Power flows directly from the source to the load when the switch is turned on, hence the converter is also called the forward converter.

With the ideal transformer (infinite L_m), the transfer ratio now includes the turns ratio:

Continuous Conduction mode:

$$T_{vv} = \frac{N_2}{N_1} D \qquad (6\text{-}46)$$

Discontinuous Conduction Mode:

$$T_{vv} = \frac{N_2}{N_1} \frac{D}{1 - \Delta} \qquad (6\text{-}47)$$

The transformer is not required to store energy. However, in a practical transformer, a part of the energy is stored in the magnetizing inductance L_m.

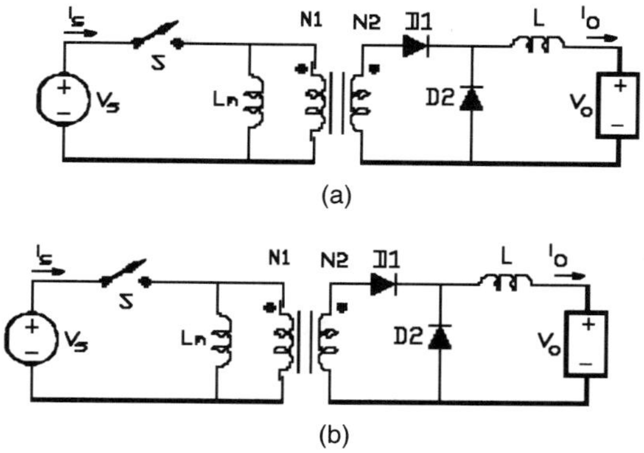

Figure 6.41
Equivalent circuit of (a) The conceptual isolated Buck converter (b) Isolated Buck converter with demagnetization

This energy must be either removed, burned in to a resistance, or a path is to be provided for the magnetizing current to freewheel during the off duration of the switch.

During the on period of the switch, the magnetizing current i_m increases linearly due to the application of dc voltage V_S. When the switch is turned off, a high voltage transient would develop on the primary winding of the transformer. Since, the transformer is still magnetized, a high voltage transient would be generated on the secondary winding and cause damage. Hence, during turn off the continuity of i_m must be maintained, and during the off period, a negative dc voltage source of suitable amplitude, see Figure 6.41(b), must be applied across L_m to force i_m to decrease linearly. The current i_m should be made to come down to zero during the off duration. Diode D1 prevents the negative voltage, induced on the secondary side during energy removal, from affecting the output voltage and the current. Incomplete demagnetization of the transformer would lead to dc magnetizing current and inefficient transformer operation. Figure 6.42 shows a circuit that implements this strategy. A third winding of N_3 turns is wound on the transformer for demagnetizing. The magnetizing energy is returned back to the source only. We would discuss the second circuit in

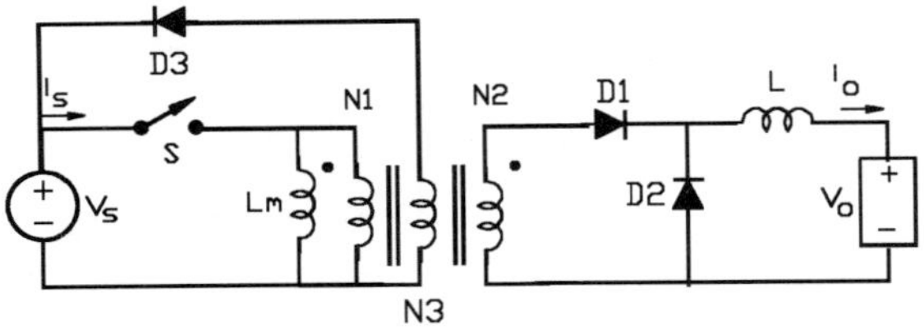

Figure 6.42
Isolated Buck converter with demagnetizing winding in the transformer

detail as it would enable a good understanding of multiwinding transformation.

Refer to the circuit modes in the steady-state operation and the waveforms in Figure 6.43(a) and (b). In the mode m1, dc voltage V_S is applied across L_m, causing the magnetizing current i_m to ramp upwards. A voltage $(N_2/N_1)V_S$ is induced on the secondary winding. Diode D1 is forward biased and D3 is turned off. Mode m2 begins when the switch is turned off. The following transformer equations apply in mode m2:

$$i_1 = -i_m \; and \; N_1 i_1 + N_2 i_2 + N_3 i_3 = 0 \; and \; i_2 = 0$$

$$hence \; i_3 = \frac{N_1}{N_3} i_m \qquad (6\text{-}48)$$

The third winding thus returns energy from the transformer back to the source during the de-energizing interval. Diode D3 is turned on and V_S is applied on the third winding. Consequently a $-(N_2/N_3)V_S$ is induced on the secondary winding. The output current free-wheels through the diode D2 and the diode D1 is reverse biased. V_S opposes current I_3. Hence, i_3 and consequently i_m ramps downwards until both become zero at time t_2. The transformer action stops as soon as i_m becomes zero. Diode D3 is also turned off and the voltage on all transformer windings become zero. The transformer remains de-energized during mode m3. The voltage across the inductor L in the output circuit during the combined duration of modes m2 and m3 is $-V_O$. The voltage inductor across the inductor L is same as in a nonisolated Buck converter, hence

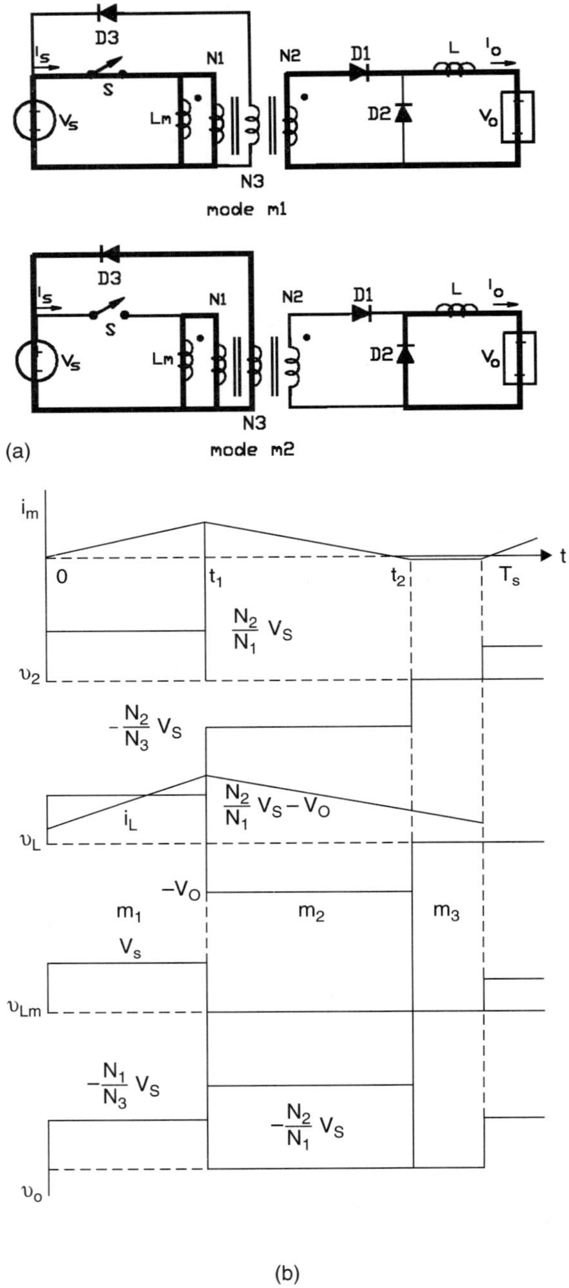

Figure 6.43
(a) Circuit modes in steady-state operation and (b) Waveforms

the inductor current waveform is also the same. The voltage transfer ratio T_{vv} also remains unchanged. However, the duty ratio becomes restricted because the off duration of the switch must be greater than the demagnetizing duration $t_m = t_2 - t_1$. The duration t_m can be determined from the condition that the average voltage across L_m over a switching period must equal zero:

$$V_S t_1 - \frac{N_1}{N_3} V_S t_m = 0 \tag{6-49}$$

$$t_m = \frac{N_3}{N_1} t_1 = \frac{N_3}{N_1} DT_S$$

and using equation (6-49) and $t_1 = D\,T_S$,

$$t_m \le T_S - t_1 \quad hence \quad D \le \frac{N_1}{N_1 + N_3} \tag{6-50}$$

A small value of N_3 (few turns) would increase demagnetizing voltage across I_m, reduce the demagnetizing interval t_m and allow the maximum value of D closer to unity. The third winding does not carry high current, hence it can be of a smaller gauge than the other two windings. In practice, N_3 is made equal to N_1. The duty ratio D is thus limited to 50%.

Component Stresses

Maximum voltage and the current through the switch are:

$$V_{sw,\max} = V_S + \frac{N_1}{N_3} V_S = \left(1 + \frac{N_1}{N_3}\right) V_S \tag{6-51}$$

$$I_{sw,\max} = i_{m,peak} + \frac{N_2}{N_1} i_{L,\max}$$

The inductor current i_L waveform is similar to that in the nonisolated buck converter. The low number of turns N_3 decreases the demagnetizing interval but increases the maximum voltage on the switch. Hence, $N_3 = N_1$ is a good compromise.

With Leakage Inductance

We would now include the leakage inductance of the transformer in the equivalent circuit of the converter of the previous section with the demagnetizing winding as shown in Figure 6.44(a). The leakage inductance on the primary and the secondary windings is modeled by L_K. Assume the filter inductor L to be infinitely large so that the output current is a ripple free dc of I_O value.

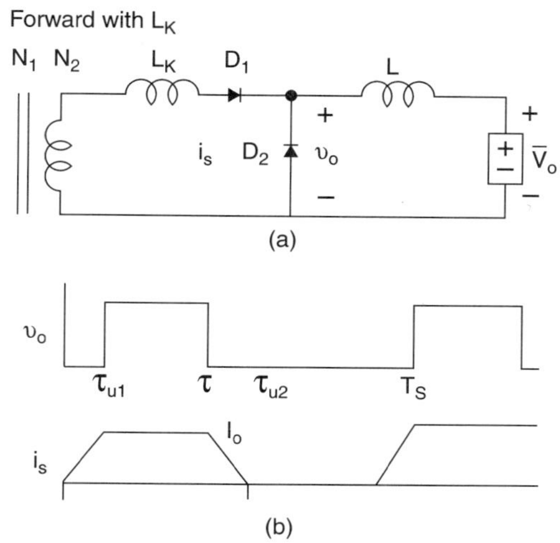

Figure 6.44
(a) Isolated Buck converter with leakage inductance and (b) Waveforms

L_k prevents abrupt increase in the current through diode D1 when the switch is turned on in the mode m1. Instead both diodes D1 and D2 (previously conducting) conduct simultaneously for an interval τ_{u1}. During this interval, the current commutates from D2 to D1. Thereafter, D1 alone conducts.

Similarly L_k prevents the abrupt termination of current in the diode D1 when the switch is turned off in the mode m2. Instead both diodes D1 and D2 conduct simultaneously for an interval τ_{u2}. During this interval the current commutates from D1 to D2. Thereafter D2 alone conducts. Also assume that the demagnetization takes more time than the commutation of current. The sum $\tau_{u1} + \tau_{u2}$ is called the commutation interval.

During the commutation interval τ_{u1}

$$L_k \frac{di_S}{dt} = \frac{N_2}{N_1} V_S \Rightarrow I_O = \frac{N_1}{N_2} \frac{V_S}{L_k} \tau_{u1} \tag{6-52}$$

During the commutation interval τ_{u2}

$$L_k \frac{di_S}{dt} = -\frac{N_2}{N_3} V_S \Rightarrow -I_O = -\frac{N_3}{N_2} \frac{V_S}{L_k} \tau_{u2} \quad (6\text{-}53)$$

The total commutation interval is

$$\tau_u = \tau_{u1} + \tau_{u2} = \frac{N_1 + N_3}{N_2} \frac{L_k I_O}{V_S} \quad (6\text{-}54)$$

The dc output voltage is obtained as the average of the output voltage waveform

$$\overline{V}_O = \frac{N_2(t_1 - \tau_{u1})}{N_1 \quad T_S} V_S = \frac{N_2}{N_1} DV_S - \frac{L_k}{T_S V_S} I_O \quad (6\text{-}55)$$

The dc output voltage drops as the output current increases. The leakage inductance behaves like the source impedance in the equivalent circuit as shown below in Figure 6.45:

Example 6-5

Find the commutation time, the duty ratio and the maximum limit on the duty ratio in the Isolated Buck converter with the following specifications:

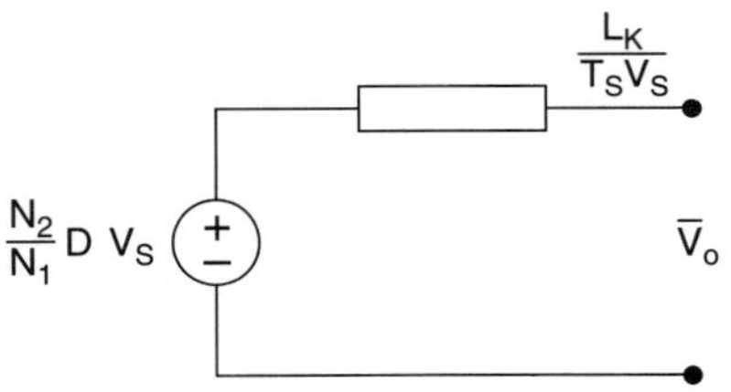

Figure 6.45
Equivalent circuit on the output terminals of the Isolated Buck converter with leakage inductance

$V_S = 50\ V,\ V_O = 5\ V,\ N1 = N_3 = 50,\ N_2 = 10,$
$L_k = 100\ nH,\ f_s = 100\ kHz,\ I_O = 100\ A$

$T_S = 10\ \mu S$. Using equation (6-54) the commutation interval is $\tau_u = 2\ \mu S$.

The duty ratio for 5 V output is obtained from equation (6-55) D = 0.502

For a given I_O, the maximum permissible on-time of the switch is

$$t_{1,\max} = T_S - \tau_{u2}$$

$$\text{Hence} \quad D_{\max} = 1 - \frac{\tau_{u2}}{T_S} = 1 - \frac{N_3}{N_2} \frac{L_k}{T_S} \frac{I_O}{V_S}$$

$$D_{\max} = 0.9$$

Two-Switch Isolated Buck Converter topology with similar performance is shown in Figure 6.46. Both switches are turned on and off simultaneously. When both switches are turned off, the magnetizing current i_m returns to the source through the diodes of the bridge. This circuit does not need a special arrangement for de-energizing. In the off-state, the two switches share the source voltage equally, hence, the voltage rating of the switches are half of that in a single switch topology.

Half-Bridge Isolated Buck Converter is shown in Figure 6.47(a). The dc source voltage is divided in two halves using two high value capacitors as shown in Figure 6.47(b). The two switches are turned on alternately for an interval τ during a switching period as shown in Figure 6.47(c). Thus there are two identical pulses in a switching period. The duty ratio is defined as

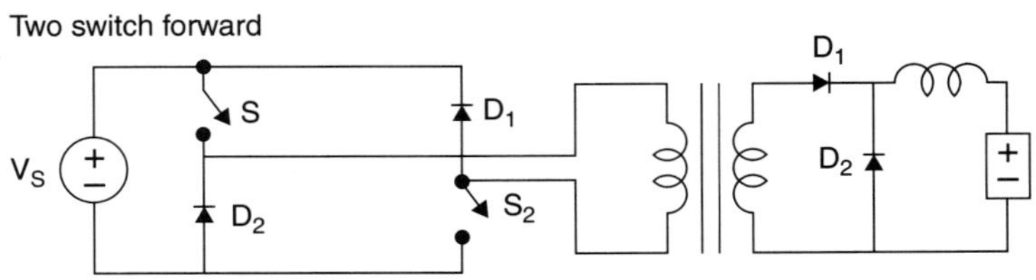

Figure 6.46
Two-switch isolated Buck converter

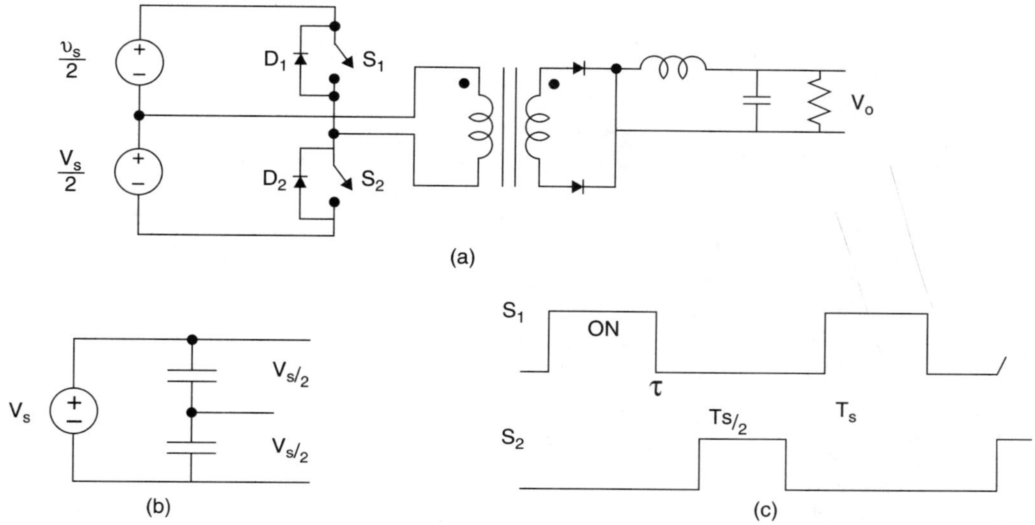

Figure 6.47
(a) Half-bridge isolated Buck converter topology, (b) Halving of a single dc voltage source, and (c) Waveforms

$$D = \frac{\tau}{T_S/2} \qquad (6\text{-}56)$$

The voltage transfer ratio has similar expression as in the isolated Buck converter with single switch. The antiparallel diodes in this circuit do not conduct current. They are used for protecting the switches against voltage transients only. The full-bridge (four-switch) converter topology is also in use. It is left for the readers as an exercise.

Push-Pull Converter is another variation of a two-switch isolated Buck converter, shown in Figure 6.48(a). The two switches are turned on alternately for an interval τ during a switching period generating two identical pulses in a switching period such as in the half-bridge converter.

When switch S1 is turned on for the first half of the switching period, a positive voltage is induced on the secondary winding at the anode of D1 and a negative voltage on the anode of D2. D1 conducts and D2 is reverse biased. The voltage across the inductor L is

$$v_L = \frac{N_2}{N_1} V_S - \overline{V}o \quad for \quad 0 \langle t \leq \tau \qquad (6\text{-}57)$$

After an interval τ, switch S1 is turned off. Since no voltage is induced in the secondary winding in this condition, both diodes D_A and D_B conduct and force the voltage on the inductor L to be $-V_O$. See the waveform in Figure 6.48(b). From the condition that the average voltage on the inductor L must be zero in each half switching period, we obtain

$$T_{vv} = \frac{\overline{V}_O}{V_S} = \frac{N_2}{N_1} D \qquad (6\text{-}58)$$

Readers are encouraged to see the literature for more details.

6.9.2 Isolated buck-boost (flyback) converter

The flyback converter is the Buck-Boost converter with the inductor replaced by a transformer. The equivalent circuit is shown in Figure 6.49 with the magnetizing L_m, ideal transformer, and the voltage sink as the model of the filter capacitor in parallel with the load resistance.

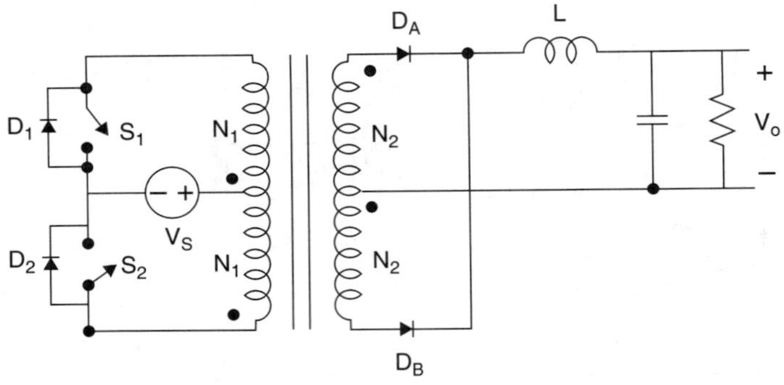

(a)

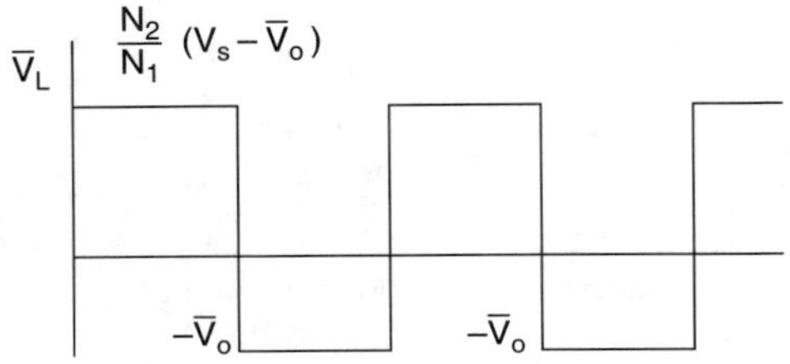

(b)

Figure 6.48
(a) Push-pull converter topology, and (b) Waveforms

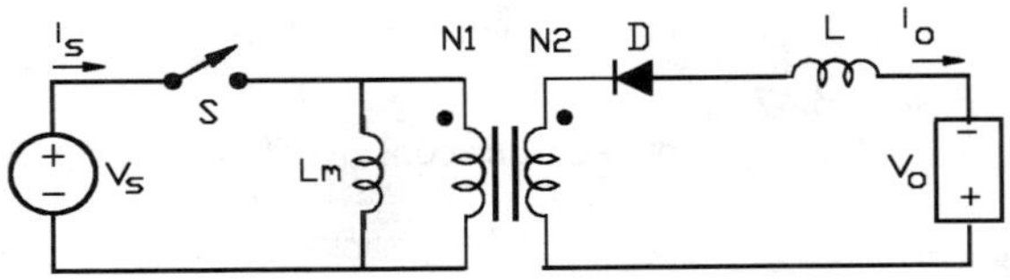

Figure 6.49
Flyback converter

The flyback converter may operate in both the Continuous Conduction Mode and the Discontinuous Conduction Mode. This section will concentrate on the CCM operation only.

In the CCM operation, the transformer remains magnetized throughout the switching period. The steady-state operation is explained with two circuit modes, as shown in Figure 6.50(a). In the circuit mode m1, turning on of the switch applies a positive voltage on Lm and the primary of the ideal transformer. A negative voltage is induced on the secondary winding, hence, the rectifier diode is reverse biased. No current flows in the secondary winding, hence, no current flows in the primary winding of the ideal transformer. The source supplies only the magnetizing current i_m. The magnetizing current increases linearly from an initial value to a peak value. See the waveform in Figure 6.50(b). The transformer stores energy $1/2\ L_m i^2_{m,peak}$.

In the mode m2, the switch is turned off. Current in the magnetizing inductance L_m, i_m may now be thought of as flowing into the primary winding of the ideal transformer. Thus the induced current in the secondary winding i_O flows through the diode to the voltage sink. The voltage across the secondary winding is V_O, which induces $(N_1/N_2)V_O$ across the primary winding and L_m. Current i_O decreases linearly. The stored energy of the transformer is thus delivered to the voltage sink. Correspondingly, the current i_m also decreases linearly. Mode m2 ends when the switch is turned back on. The converter is said to operate in the Continuous Conduction Mode (CCM) if the stored energy is not completely exhausted during the switch off condition.

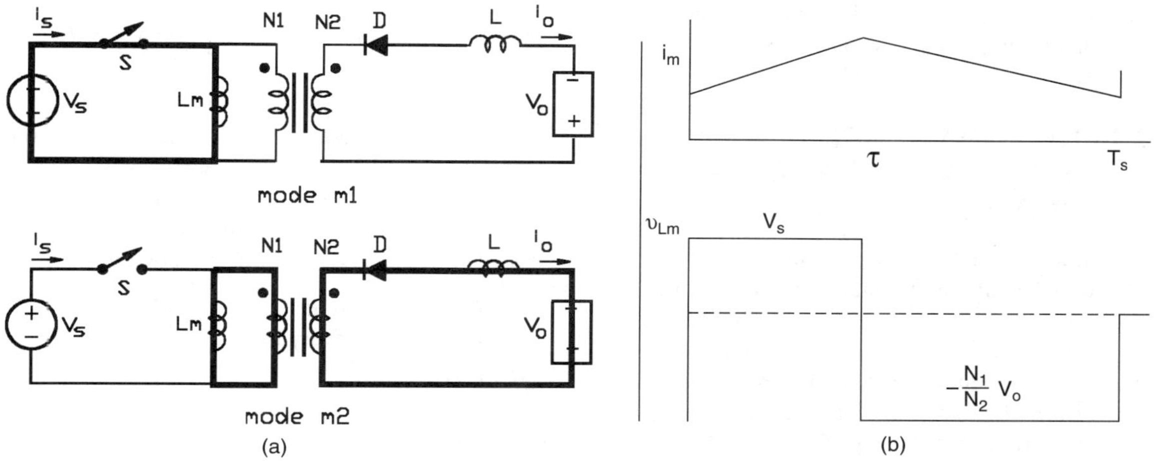

Figure 6.50
(a) Circuit modes and (b) Waveforms in steady-state

The flyback converter stores energy during the switch on duration and delivers it to the load during the switch off period. Unlike the forward converter, there is no direct energy transfer from the source to the load, hence, this converter is said to be the indirect converter. It is also called a flyback converter because the actual energy transaction to the load takes place during the flyback of the switch from the off to the on state. This converter is used for generating high voltages in the television and oscilloscope circuits. Unlike the forward converter, in which the transformer is designed for low magnetizing current (high L_m), the flyback transformers are designed for high i_m.

The voltage transfer ratio in the CCM operation can be obtained from the condition of zero average voltage on L_m:

$$V_S\tau - \frac{N_1}{N_2}\overline{V}o\,(T_S - \tau) = 0 \tag{6-59}$$

$$T_{vv} = \frac{\overline{V}_O}{V_S} = \frac{N_2}{N_1}\frac{D}{1-D}, \quad where \quad D = \frac{\tau}{T_S}$$

Readers are advised to do a similar analysis of the Discontinuous Conduction Mode, see problem 6-4.

Component Stresses

Maximum voltage and the current through the switch are:

$$V_{sw,\,\max} = V_S + \frac{N_1}{N_2}\overline{V}_O \tag{6-60}$$

$$I_{sw,\,\max} = I_{m,peak}$$

6.10 MULTI-OUTPUT DC-DC CONVERTER

The switch-mode dc power supplies in instrumentation often require more than one output voltage, each loaded with different current magnitudes. Rather than make a separate power supply for each, a transformer with multiple output windings is used. Each output winding has its own diode rectifier and filter components. The control, however, is provided by a single parameter, say the pulse width or the switching frequency. Only one output voltage may be regulated in this case. Hence, often an output with the most critical regulation requirements is feedback controlled. Load variation in the regulated output also causes the voltage variation in other output circuits even at constant load conditions. This problem is referred to as the *cross regulation problem.* This problem is explored in detail in the chapter on Resonant Converters.

6.11 DESIGN EXERCISE: SWITCHING POWER SUPPLY

Design a 48 V, 10 A power supply from a 5 V dc source. Allow an input voltage drooping of 10% in long time. The output voltage ripple should be less than 1% of the dc output voltage magnitude. The switching frequency is required to be 100 KHz or more. Limit the maximum load current to be 15 A.

Use the flyback topology of Figure 6.49 for reasons of economy as it contains a fewer number of components. The converter should be designed with a power MOSFET as the switch and a fast recovery power diode in the secondary side. Assume in the beginning that the on-state voltage drops in the MOSFET and the diode, $V_{on} = V_F = 1$ V. Refine your assumption in the second revision when the actual devices are selected.

The input voltage variation is 5 V to 4.5 V dc. A nominal value of 5 V would be used for the design at 50% duty ratio. The source voltage and the output voltages include the on-state voltages of the switching device and the diode, hence,

$$\overline{V_S} \Rightarrow V_S - V_{ON} = 5 - 1 = 4 \qquad \textit{and}$$
$$\overline{V_O} \Rightarrow V_O + V_F = 48 + 1 = 49$$

The load resistance is

$$R = \frac{\overline{V_O}}{I_O} = \frac{48}{10} = 4.8\ \textit{ohms}$$

The transformer turn ratio is obtained from equation (6-59):

$$(V_S - V_{ON})D - \frac{N_1}{N_2}(\overline{V_O} + V_F)(1 - D) = 0$$

$$(5 - 1)(0.5) - \frac{N_1}{N_2}(48 + 1)(1 - 0.5) = 0$$

$$\frac{N_2}{N_1} = 12.25 \approx 13$$

Diode Selection:

Peak reverse voltage $V_R = 2(V_O + V_F) = 2(48 + 1) = 98$ V

Peak maximum forward current $I_{F\,max} = 15$ A

Select a fast recovery diode with specifications better than 100 V and 15 A. A Schottky would be a good choice, but high voltage Schottky diodes are expensive. Choose the International Rectifier 25CPF20, 20 A 200 V Ultrafast recovery diode.

The forward voltage is $V_F = 0.98$ V
The reverse recovery time $t_{rr} = 50$ ns

MOSFET Selection:

The output voltage and currents referred to the primary side are

$$V_O' = (48 + 1)/13 = 3.77 \qquad \text{and}$$
$$I_O' = 10 \times 13 = 130 \text{ A.}$$
$$I_{O\,max}' = 15 \times 13 = 195 \text{ A}$$

The maximum voltage and maximum current through the switch are

$$V_{sw\,max} = V_S + V_O' = 5 + 3.77 = 8.77 \text{ V}$$
$$I_{sw\,max} = I_{O\,max}' = 195 \text{ A}$$

Select the MOSFET with specifications better than 8.77 V and 195 A. Choose a set of two MOSFETs in parallel, International Rectifier RFA100N05E, 100 A, 50 V, 10 mohm. Readers are encouraged to design the parallel connection on the lines of Example 4-4 in Chapter 4.

The on-state voltage of the selected MOSFET (at 100 A) is

$V_{ON} = 100 \times 0.010 = 1$ Volt
Turn-on time = 60 ns
Turn-off time = 100 ns

We can safely select a switching frequency of 200 KHz.

$$T_S = 1/\,200 \text{ KHz} = 5\ \mu\text{S}$$

The range of D using the revised values of the turn ratio, V_{ON} and V_F is calculated as

0.49 at $V_S = 5$ V to 0.52 at $V_S = 4.5$ V.

The peak-peak ripple In the output voltage is $V_r = 0.01\times48 = 0.48$ volts. The ripple output current (the peak-peak value) is

$$I_{LP} = \frac{V_r}{R_O} = \frac{0.48}{4.8} = 0.1A$$

Referred to the primary winding

$$I'_{LP} = .1 \times 13 = 1.3 \text{ A.}$$

The inductor L is determined from equation (6-33b):

$$L > \frac{V_S T_S D_{max}}{I'_{LP}} = \frac{5 \times 5 \times 10^{-6} 0.52}{1.3} = 10\ \mu H$$

The calculation of the switching losses in the MOSFET are left as an exercise for the readers, see problem at the end of the chapter.

6.12 PSPICE SIMULATION

PSPICE Example 6-1

Consider the dc-dc Buck converter of Example 6-1. Use MOSFET IRF150 for the switch and a 1N4148 signal diode as the freewheeling diode (a high-speed diode will be better). The converter specifications are

DC source = 48 V,	R = 10 ohms
L = 5 mH	Duty ratio D = 0.25
Switching frequency = 25 kHz	

Perform a transient analysis for at least 100 switching periods and plot the steady-state waveforms for the following in the last two periods:

a. the current io through the load resistor RL
b. the voltage vd at the cathode of the freewheeling diode
c. the dc output voltage V_O at the load resistor

The MOSFET S1 is switched by an external pulse source V_C as shown in Figure 6.51(a). The pulse waveform is transformer coupled or optically coupled in the practical implementation of the circuit.

Set the pulse source V_C as follows:

Lower level V1	= 0 V	Upper level V2	= 10 V
Pulse period	= 25 μs (40 kHz)	Pulse width (on-time)	= 6.25 μs (D = 0.25)
Delay time TD	= 0	Rise time	= 0.1 ns
Fall time	= 0.1 ns		

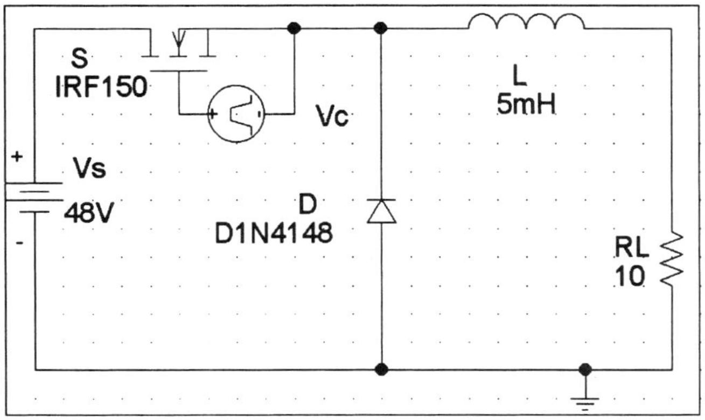

(a)

Figure 6.51
(a) Schematic

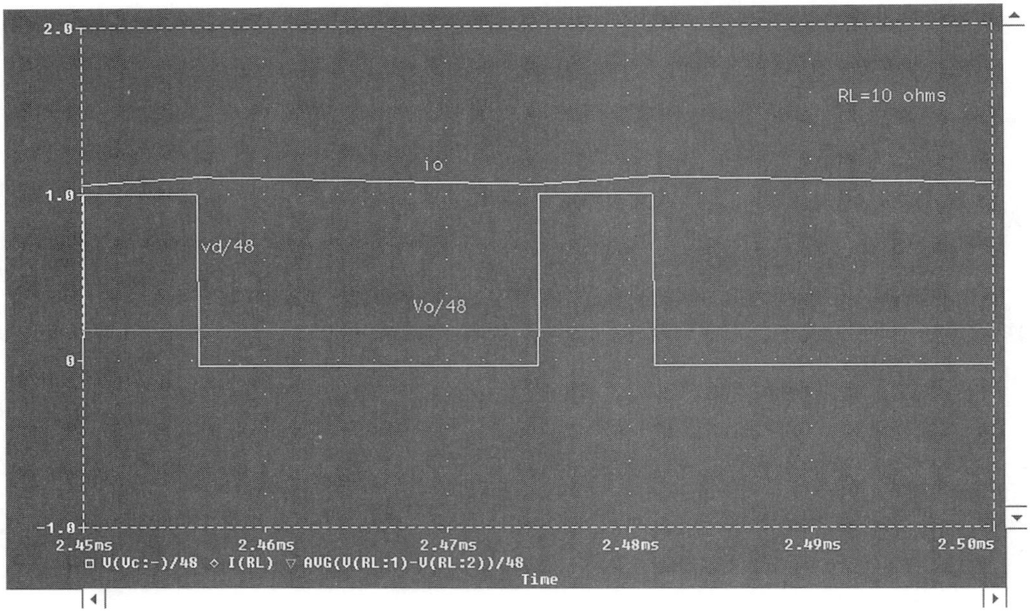

Figure 6.51 ***(continued)***
(b) Waveforms

Use the simulation step in the range: 1 ns to 100 ns. Organize your plots by displaying the ratios vd/24 and $V_O/24$ as the functions of time. Compare these waveforms with those obtained in Example 6-1.

6.13 SUMMARY

This chapter has focused on the voltage source driven dc-dc converters. Voltage step-down (Buck), voltage step-up (Boost) and the Buck-Boost combination converters have been analyzed in continuous and discontinuous current conduction modes. The single-switch, two-switch and four-switch topologies have been analyzed. The transformer isolated topologies were also analyzed in a separate section. The analysis consisted of deriving the control characteristics and the component stress calculations. Important conclusions are as follows:

1. The voltage step-down converter is also a current step-up converter and conversely, a voltage step-up converter is also the current step-down converter.
2. The transformer isolation permits low component stresses in the dc-dc converters versus the nonisolated versions.

References

1. Philip T. Krein, *Elements of Power Electronics,* Oxford University Press, New York, 1998.

2. Ned Mohan, Tore M. Undeland, and William P. Robbins, *Power Electronics: Converters, Applications and Design,* 2nd Ed., John Wiley & Sons, Inc., 1995.

3. Joseph Vithayathil, *Power Electronics: Principles and Applications,* McGraw-Hill Inc., New York, 1995.

4. John G. Kassakian, Martin F. Sclecht, and G. Vergassian, *Principles of Power Electronics,* Addison-Wesley, Boston, 1991.

5. Kjeld Thorborg, *Power Electronics,* Prentice-Hall Inc., 1988.

6. Daniel M. Mitchell, *Switching Regulator Analysis,* McGraw-Hill Book Company, New York, 1988.

7. Richard G. Hoft, *Semiconductor Power Electronics,* Van Nostrand ReinHold Company, New York, 1986.

8. G.C. Chryssis, *High Frequency Switching Power Supplies,* 2nd ed., McGraw-Hill, New York, 1989.

9. Motorola, *Switchmode Designer's Guide,* Motorola, manual SG79/D, 1993.

10. A.I. Pressman, *Switching Power Supply Design,* McGraw-Hill, New York, 1991.

11. R.P. Severns, E.J. Bloom, *Modern DC to DC Switchmode Power Converter Circuits,* Van Nostrand, New York, 1985.

12. G.W. Wester and R.D. Middlebrook, "Low-frequency characterization of dc-dc converters," *IEEE Trans. Aero. Elec. Sys.,* Vol. AES-9, No., 3, pp. 376–385, 1973.

13. S.B. Dewan, G.R. Slemon, and A. Straughen, *Power Semiconductor Drives.* Wiley, New York, 1984.

14. G.K. Dubey, S.R. Doradla, A. Joshi, and R.M.K. Sinha, *Thyristorized Power Controllers.* Wiley, New York, 1986.

15. A.W. Kelley and W.F. Yadusky, "Rectifier design for minimum line-current harmonics and maximum power factor," *IEEE Trans. Power Electronics,* Vol. 7. No. 2, pp. 332–341, April 1992.

16. C.K. Tse, S.C. Wong, and M.H.L. Chow, "On lossless switched-capacitor power converters," *IEEE Trans. Power Electronics,* Vol. 10, No. 3, pp. 286–291, May 1995.

17. S. Cuk and R.D. Middlebrook, "A general unified approach to modeling switching dc to dc converters in discontinuous conduction mode," *IEEE. Power Electronics Specialists Conference Record,* pp. 36–57, 1977.

18. V. Voperian, "Simplified analysis of PWM converters using model of PWM switch, Part I and Part II," *IEEE. Trans. Aerospace Electronics Systems,* Vol. 26, No. 3, pp. 497–505, May 1990.

Problems

6-1 A Buck converter operating in CCM has the following parameters:
$V_S = 48$ V $L = 250$ uH $R_L = 2$ ohm
$f_s = 20$ kHz and $D = 0.4$.

a. Determine V_O.
b. Sketch the inductor current i_L.
c. Determine at what output voltage the converter would move to DCM operation.

6-2 Calculate the peak-to-peak voltage ripple in the output of the converter of the problem 6-1 for the filter capacitor of 470 μF.

6-3 A Buck converter supplies 5-V to a load of 0.05 ohms from a 24-V source having an internal resistance of 0.1 ohm. Find the duty ratio D.

6-4 Derive the expressions of the peak-to-peak voltage ripple with respect to the dc output voltage in the Buck converter in DCM operation.

6-5 A Boost converter operating in CCM has following parameters:
$V_S = 5$ V $L = 250\ \mu$H $R_L = 20$ ohm
$C = 470\ \mu$F $f_s = 20$ kHz $D = 0.4$.

a. Determine V_O.
b. Sketch the inductor current i_L.
c. Determine the peak-peak voltage ripple in the output.
d. Determine at what output voltage the converter would move to DCM operation.

6-6 A 90-V dc source is used to charge a 150-V battery. Use the Boost converter in CCM operation at a switching frequency of 50 kHz. Find the inductor L that will give the peak-peak current ripple of less than 100 mA in a 50-A load current. Also find the current rating of the fuse that should be included on the source side.

6-7 Include the source resistance R_S in the Buck-Boost converter. Assume the inductor L to be infinitely large. Find

a. the voltage transfer ratio as a function of duty ratio and R_L/R_S.

b. the duty ratio for which the output voltage is maximum, and

c. the duty ratio for which the maximum power is transferred to the load.

6-8 A Buck-Boost converter operating in CCM has following parameters:

$V_S = 12$ V $L = 250\ \mu$H $R_L = 20$ ohms
$C = 470\ \mu$F
$f_s = 20$ kHz $D = 0.7$.

a. Determine V_O.

b. Sketch the inductor current i_L.

c. Determine the peak-peak voltage ripple in the output.

d. Find the minimum L for which the converter would move to DCM operation.

6-9 Draw the circuit of the full-bridge converter that uses Buck-Boost topology.

6-10 Using PSPICE, simulate the converter of Figure P6.10.

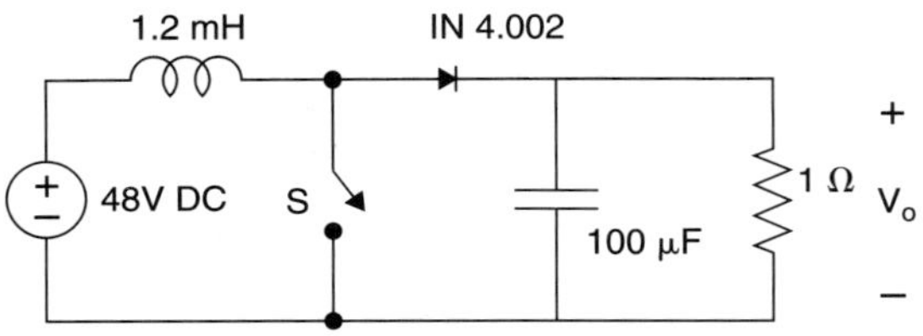

Figure P6.10

6-11 Derive the following expressions of the fractional peak-to-peak voltage ripple in the full-bridge dc-dc converter:

a. Bipolar switching:

$$\frac{\Delta V_O}{\overline{V}_O} = \frac{DT_S^2}{4LC}\left(\frac{1-D}{2D-1}\right)$$

b. Unipolar switching:

$$\frac{\Delta V_O}{\overline{V}_O} = \frac{T_S^2}{16\,LC}(1-D)$$

6-12 An isolated Buck converter with transformer turn ratio 2:1 supplies 12 V while the input voltage varies from 50 to 100 V. Find the minimum demagnetizing voltage V_c.

6-13 Find the dc voltage and the voltage ripple in the converter circuit of Figure P6.13. Determine and sketch the waveforms of i_s, i_m and i_O in the following converter:

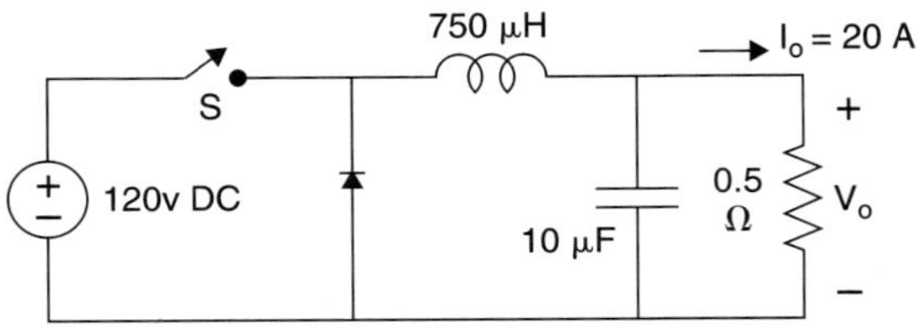

Figure P6.13

6-14 Draw the circuit of the forward converter with two switches.

6-15 Draw the circuit of the flyback converter with two switches.

6-16 Draw the circuit of the forward converter with two switches. Analyze the Discontinuous Conduction Mode operation of the flyback converter.

6-17 Analyze the flyback converter in CCM by including the transformer leakage inductance on the secondary winding.

6-18 Draw the flyback converter with two switches.

6-19 Assume a very large capacitor in the circuit shown below: (107, hof)

a. Plot the steady-state waveforms of I_S, i_O and I_c.

b. Calculate the power delivered by the dc source in the steady-state.

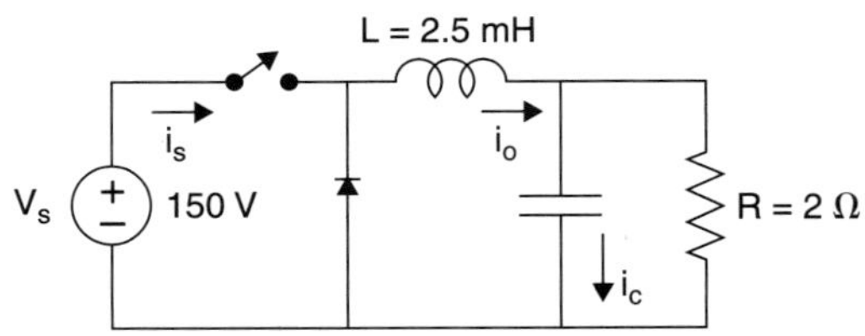

Figure P6.19

6-20 Consider the steady-state operation of the Buck regulator with V_s = 125 V, L = 125 μH, R = 47 ohms, D = 0.6 and f_S = 20 kHz. Assume a large C.

a. Calculate the load voltage.

b. Calculate the power delivered by the dc voltage source.

6-21 Repeat problem 6-20 with a capacitor of 10 μF.

6-22 An electric train is powered from a 1500 Vdc supply through a low-frequency dc-dc converter (also called the Chopper). The minimum on-time and off-time of the switch are 40 μs and 50 μs respectively.

a. Find the switching frequency required for the lowest output voltage of 15 Vdc.

b. Find the minimum output voltage if the maximum switching frequency is restricted to 2000 Hz.

c. Find the maximum duty cycle D at the switching frequency of 2000 Hz and the maximum output voltage possible.

6-23 An electric car uses a dc motor that is supplied power from a battery bank of 200 Vdc through a dc-dc converter. The converter operates at 2 KHz. The armature circuit of the motor has a resistance of 0.2 ohms and an inductance of 100 μH. The back emf at a speed of 40 miles per hour is 60 volts. The duty cycle D= 33.3%. Determine (a) the peak-peak ripple current and (b) the average dc current in the armature of the motor.

6-24 For the dc motor of problem 6-23, find the speed at which the back emf is 120 Vdc. What is the minimum duty cycle D below which the dc-dc converter will operate in the discontinuous conduction mode.

6-25 A two-switch isolated flyback converter is shown in Figure P6.25 with V_S = 150 V, N_p = 100 turns, N_S = 20 turns and D = 0.6. Determine the output voltage.

6-26 The ČUK converter has the following data: V_S = 150 V, f_S = 40 kHz, D = 0.2.

a. Determine the voltages V_O and V_C.

b. Plot the voltage waveforms on both inductors.

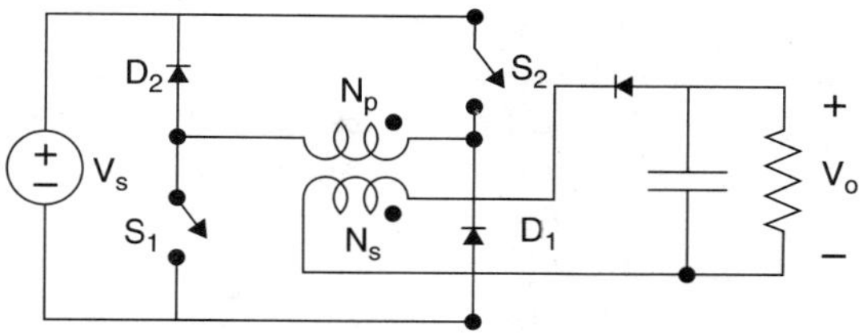

Figure P6.25

6-27 A boost converter has the following data: Switching frequency f_s = 100 kHz
V_s = 9 V, V_O = 48 V Duty cycle D = 0.6
L = 10 μH, C = 0.47 μF,
power delivered to load P_O = 50 W
Find the output voltage at P_o = 10 W.

6-28 A buck converter provides +9 V output from an input voltage in the range 12 to 28 V. The load may vary from 10 to 50 W. The converter is switched at 250 kHz.

a. Find the boundary value of L for a large value of C.

b. Find the largest value of L that would permit Discontinuous Conduction Mode operation.

6-29 In a full-bridge converter, the duty ratio is adjusted to regulate the output voltage to 12 volts. The input dc voltage varies in the range 15 V to 40 V. The converter is switched at 100 kHz. Find the range of pulse width of the switching waveform applied at the switch S1 and the peak-to-peak ripple in the output voltage with bipolar switching.

6-30 Repeat problem 6-29 for unipolar switching.

6-31 Find the commutation time, the duty ratio and the maximum limit on the duty ratio in the isolated Buck converter with the following specifications: V_S = 24 V, V_O = 5 V, N_1 = N_3 = 20, N_2 = 5, L_k = 1 μH, f_s = 40 kHz, I_O = 100 A .

6-32 Design a 48-V, 10-A power supply from a 12-V dc source. Allow an input voltage drooping of 10% in long time. The output voltage ripple should be less than 1% of the dc output voltage magnitude. The switching frequency is required to be 200 KHz or more. Limit the maximum load current to be 15 A.

6-33 A dc-dc converter converts a source of 175 volts to 12 volts. The converter components are I_O=2 A, L=100 μH, and f_s=100 kHz. Plot the output voltage and the current waveforms and determine (a) the mode of operation, (b) the amount of current ripple, (c) the filter capacitor value if the voltage ripple is to be less than 1% of the output voltage, and (d) the inductor for which the converter would just enter in the Discontinuous Conduction Mode.

6-34 In a Boost converter, the duty ratio is adjusted to regulate the output voltage to 90 volts. The input dc voltage varies from 10 to 28 V. The converter is switched at 40 kHz. The maximum output power is rated at 120 W. The converter is operated in CCM. Find

a. the minimum value of inductor L, and

b. value of filter C for a voltage ripple of 1% or less at this value of L.

6-35 Repeat the problem in Example 6-3 when the output voltage is specified to be 18 V.

6-36 Using PSPICE, simulate the following full-bridge converter of Figure 6.34a using (a) bipolar switching and (b) unipolar switching.

V_S = 48 V dc

Duty ratio of each switch = 0.25

T_S = 50 μS L=750 μH

CHAPTER 7
DC to Controlled AC

OBJECTIVES

1. Performance measures for Voltage Source and Current Source Inverters
2. Square-wave switching control of Voltage Source Inverter
3. Voltage Source Inverter with pulse-width (PWM) control in voltage and current mode
4. Current Source Inverter with PWM control
5. Adjustment of the harmonic profile of the inverter waveforms
6. Sine-PWM Inverter
7. Three-phase inverters
8. Rectifier mode operation of inverters

7.1 INTRODUCTION

In this chapter, we will learn about the dc-ac converters, conventionally called the inverters. Such converters are very popular in the battery-operated power systems such as the Uninterruptible Power Supplies (UPS) for hospitals, security complexes and ac motor drives. Low power level inverters are usually single-phase type and medium and high power inverters are three-phase type. We will look into the performance of dc-ac converters of several useful topologies. As in the previous chapter, the focus is on the study of the in-depth analysis of steady-state inverter performance. Furthermore, the focus will be on the inverter operation for low-frequency output voltages such as 50-60 Hz or 400 Hz in aircraft systems. Analyses of dc-ac converter circuits at low and high frequencies are very similar except that at high frequencies the parasitic capacitances and inductances, charge storage, and heat localizing problems in switching devices must be taken into account.

7.2 CONTROLLED INVERSION

The dc to ac inverter system is shown in Figure 7.1(a). Input is from a dc source and the output is desired to be a sinusoidal voltage or current with zero dc component, single-phase or three-phase signal. The load is a R-L-C load, an ac voltage sink or an ac current sink. Control parameter may be, as shall be seen later, an angle, a pulse width, a voltage or a current signal. The converter consists of switches, reactive components such as L, C, transformers and resistances.

Switches including the diodes are assumed to be ideal and unidirectional, that is, they have zero on-state losses, zero off-state losses, zero switching losses and unrestricted voltage or current carrying capabilities. Switching devices in a practical design example must be selected with these desired characteristics in mind. Inductors, capacitors and transformers are assumed to be ideal.

Electrical loads can be of three types: impedance load, ac voltage sink load or ac current sink load. Examples of impedance load are electromagnetic ac relays, electromagnetic pickup and lift equipments. The load comprises of a large value inductor. Examples of ac voltage sink are the constant speed drives for induction and synchronous motors. Examples of ac current sink are constant torque drives for the induction and synchronous motors.

Generally speaking, the input current is non-sinusoidal if the source is a dc voltage source and vice versa. The output voltage and the output current on the other hand, can be non-sinusoidal simultaneously. Hence, input and output low-pass filters are generally employed to reduce high frequency components of the non-sinusoidal signals on either side. A series inductor is a current smoother and a parallel capacitor is a voltage smoother.

Output current generally differs in phase with the output voltage. Hence, in a switching period, the polarities of current and voltage vary over all four quadrants, see Figure 7.2. Power flows from the converter to the load in the first and third quadrants and in the reverse direction in the second and fourth quadrants. A unidirectional switch in the inverters, therefore, has an inverse connected diode in parallel to allow bidirectional current flow.

Inverters derived from voltage source (VSI) are ideally suited for supplying power to a resistive and inductive load and the applications that require constant torque operation of ac motors. Current Source Inverters (CSI) are ideally suited for supplying power to largely capacitive loads and ac voltage buses. Inverters can be further classified on the ba-

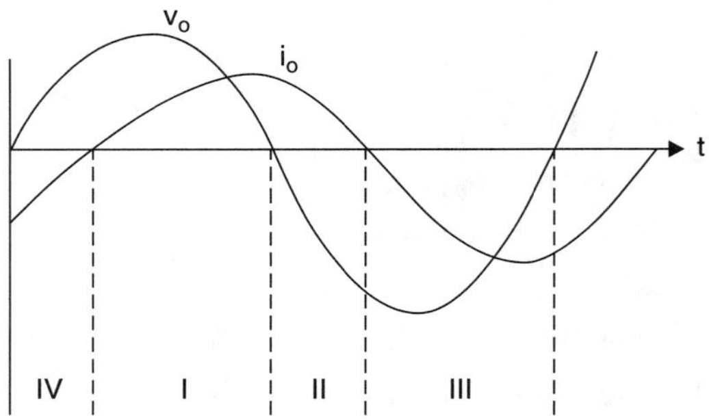

Figure 7.2
Output current and output voltage waveforms in an inverter

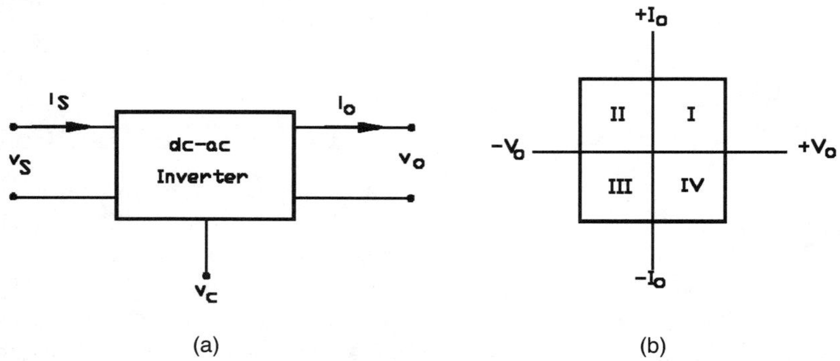

Figure 7.1
(a) A dc-ac inverter system and (b) operational quadrants on the v_o-i_o plane

sis of the mode of control of the output variable: voltage or current:

Inverter control in practice concerns three aspects:

fundamental frequency

amplitude

harmonic profile

Fundamental frequency is either equal to the switching frequency of the inverter (square-wave and PWM inverters) or its simple fraction ($\frac{1}{3}$, $\frac{1}{5}$ and so on) such as in Multiple PWM or Sine PWM inverters.

Power electronic applications require the amplitude control of output voltage or output current, called Voltage-Mode control and Current-Mode control respectively. An example of Voltage-Mode control is the adjustable speed drives of induction and synchronous motors. An example of Current-Mode control is a constant torque drive for induction motors. Output amplitude in both modes can be either controlled by varying the dc source voltage or by varying the pulse-width (pulse-width modulation, PWM). The inverters based on the control of dc source voltage are also termed *DC-Link Inverters.* These inverters use an ac to controlled dc converter on the input side. The inverter is square-wave controlled. This arrangement is expensive but exhibits a faster response time than PWM inverters. The control circuitry is also simpler.

Harmonic profiling is an extra feature added over the amplitude control. The objective of harmonic profiling is to bolster the amplitude of the fundamental component of the output voltage and the elimination or reduction of high order harmonics. One of the technique involves the notching or adding step pulses to a base pulse. The other technique involves multiple pulses per cycle of the fundamental, each pulse-width modulated to achieve desired wave-shaping of the output waveform (Sine-PWM). In the next sections, examples of several control techniques will be shown.

7.2.1 Performance measures

As in the previous chapter, the focus will be on the steady-state operation of both the Voltage Source Inverters (VSI) and Current Source Inverters (CSI). The input source is assumed to be an ideal dc source V_S or I_S. The output voltage or current is a single-phase periodic signal of frequency f_o ($\omega_o/2\pi$) and peak amplitude V_o or I_o.

Voltage Source Inverters (VSI)

Output Characteristics Consider some of the performance measures used in dc-ac converters. The Fourier expansion of output voltage waveform is given as follows:

$$v_O = \sum_{n=1,2,3,\ldots}^{\infty} V_{On} \sin n\omega_o t \tag{7-1}$$

where V_{on} is the peak amplitude of the n-th harmonic component. Harmonic content of the output voltage is the rms sum of all harmonics other than the fundamental component which is nothing but the difference of the output rms voltage V_O and the rms fundamental voltage $V_{O1}1$. Total Harmonic Distortion THD in the output voltage is defined as the ratio of the harmonic content to the rms fundamental voltage:

$$THD_O = \sqrt{\frac{\tilde{V}_O^2 - \tilde{V}_{O1}^2}{\tilde{V}_{O1}^2}} \tag{7-2}$$

Harmonic Factor V_n/V_1 is defined as the peak amplitude of n-th harmonic normalized to the peak amplitude of the fundamental. This factor when plotted versus the control parameter such as the pulse width in PWM type control gives the harmonic profile of the converter. Order of the lowest undesired harmonic is another important measure.

Voltage transfer ratio, defined as the ratio of the rms voltage of the fundamental component of the output voltage to the dc input voltage, is an important measurement of the VSI-VM inverters:

$$T_{vv} = \frac{\tilde{V}_{O1}}{V_S} \tag{7-3}$$

Output current $\mathbf{i_o}$ waveform and its rms value are determined from the load network. Voltage to current transfer ratio, defined as the ratio of rms value of the fundamental component of the output current to the dc input voltage, is an important measurement of the VSI-CM inverters:

$$T_{vi} = \frac{\tilde{I}_{O1}}{V_S} \tag{7-4}$$

High values of the slope of T_{vv} and T_{vi} versus the control parameter are desirable measures.

The output power consists of harmonics and is given by

$$P_O = \sum_{n=1,2,3,\ldots,\ldots} \tilde{V}_{On}\,\tilde{I}_{On}\cos\phi_n \tag{7-5}$$

where ϕ_n is the phase angle difference between the n-th harmonic voltage and the current waveforms.

Input Characteristics Input current $\mathbf{i_S}$ is a dc with ripple. The dc value of the input current is calculated from

$$\bar{I}_S = \frac{1}{2\pi}\int_{\theta}^{\theta+2\pi} i_S\, d\theta \tag{7-6}$$

Current reflection ratio, defined as the ratio of the input dc current to the rms value of the fundamental component of the output current, is calculated from

$$R_{ii} = \frac{\bar{I}_{S1}}{\tilde{I}_O} \tag{7-7}$$

R_{ii} parameter (always <1) is load dependent and it provides the measure of utilization of input source by the inverter. High values indicate better utilization.

Harmonic profile of the source current is given by

$$i_S = \sum_{n=1,2,3\ \ldots,\ldots} \tilde{I}_{Sn}\sin(n\theta + \phi_n) \tag{7-8}$$

where I_{Sn} and ϕ_n are respectively the peak amplitude of the n-th harmonic component and the phase angle of the n-th harmonic with respect to the fundamental component of the output sinusoidal voltage. Harmonic content of the input current is the rms sum of all harmonics, which is nothing but the difference of the rms source current I_S2 and the rms fundamental current $I_{S1}3$. Total Harmonic Distortion THD in the input current is defined as the ratio of the harmonic content to the rms fundamental current:

$$THDi = \sqrt{\frac{\tilde{I}_S^2 - \tilde{I}_{S1}^2}{\tilde{I}_{S1}^2}} \tag{7-9}$$

Input current ripple is defined as the difference of the instantaneous and the dc (average) input current:

$$i_{Sr} = i_S - \bar{I}_S \tag{7-10}$$

Peak-to-peak input current ripple is a measure of the inverter performance. Although no average power is drawn from the source at harmonic frequencies, the ripple current is responsible for resistive losses in the dc source and the conductor wires. It is especially a concern when the input dc source is an ac-dc rectifier.

The input power is given by

$$P_{in} = \bar{V}_S\,\bar{I}_S \tag{7-11}$$

The power efficiency of an inverter is defined as the ratio of the output power in the fundamental harmonic to the input power:

$$\eta = \frac{\tilde{V}_{O1}\tilde{I}_{O1}}{\bar{V}_S\bar{I}_S}\cos\phi_1 = \frac{T_{vv}}{R_{ii}}\cos\phi_1 \tag{7-12}$$

Control Characteristics Control parameters are used to close the loop for regulation and control of one of the output characteristics. For a good performance of the overall system, a smallest range of variation of control parameter is desired for a large variation in the output characteristics. This amounts to selecting the region of highest slope on the plot of T_{vv} versus the control parameter in VSI-VM inverters and R_{ii} versus the control parameter in VSI-CM inverters.

Current Source Inverters (CSI)

Fourier expansion of output current waveform is given as follows:

$$i_O = \sum I_{On}\sin(n\theta + \phi_n) \tag{7-13}$$

where I_{On} and ϕ_n are respectively the peak amplitude of the n-th harmonic component and the phase angle of the n-th harmonic with respect to the fundamental component of the output sinusoidal current.

Current to voltage transfer ratio T_{iv}, defined as the ratio of rms voltage of the fundamental harmonic of the output voltage waveform to the dc input current, is an important measure for CSI-VM inverters

$$T_{iv} = \frac{\tilde{V}_{O1}}{\bar{I}_S} \tag{7-14}$$

Output voltage $\mathbf{v_o}$ waveform and its rms value are determined from the load network. Current transfer ratio T_i, defined as the ratio of rms current of the fundamental harmonic of the output current waveform to the dc input current, is an important measure for CSI-CM inverters

$$T_{ii} = \frac{\tilde{I}_{O1}}{\bar{I}_S} \tag{7-15}$$

High values of the slope of T_{iv} and T_{ii} versus the control parameter are desirable measures.

Input voltage $\mathbf{v_S}$ is a dc with ripple. The dc value is calculated from

$$\bar{V}_S = \frac{1}{2\pi}\int_{\theta}^{\theta+2\pi} v_S \, d\theta \tag{7-16}$$

7.2.2 Inverter topologies

The dc-ac inverter of Figure 7.1 can be thought of as a three-position switch as shown in Figure 7.3(a). The load is connected to the pole of the switch and the first two positions are connected to dc voltage or dc current sources of opposite polarities. The third position is connected to the zero terminal of the source. The pole of the switch stays in the first and second positions for equal amounts of time in a switching period. The output waveform for zero amount of time on the third position is a square wave. The output waveform for nonzero time is pulse wave.

The three-position switch may be synthesized by two unidirectional electronic switches as shown in Figure 7.3(b). The third position of the switch is simulated by turning both switches off in VSI topologies and on in CSI topologies. This is the basic half-bridge inverter topology. The two diodes provide the path for current flow from the load to

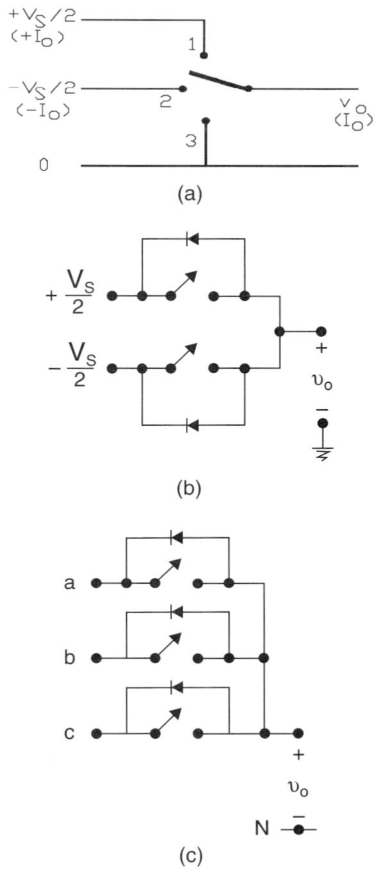

Figure 7.3
(a) The dc-ac inverter implemented by a 3-position switch (b) the 3-position switch synthesized from two 2-position switches and (c) three-phase inverter topology

the source as this may be required in an inverter operation involving more than one quadrant. Inverter supplying load to an R-L load operates in all four quadrants because the output voltage and current waveforms, for some segments in a switching period, have opposing polarities. A MOSFET has a built-in body diode that serves as the path for the current from the source to the drain. An IGBT, however, requires an external diode.

The 3-ϕ half-bridge inverter is made of three half-bridge inverters connected in parallel each

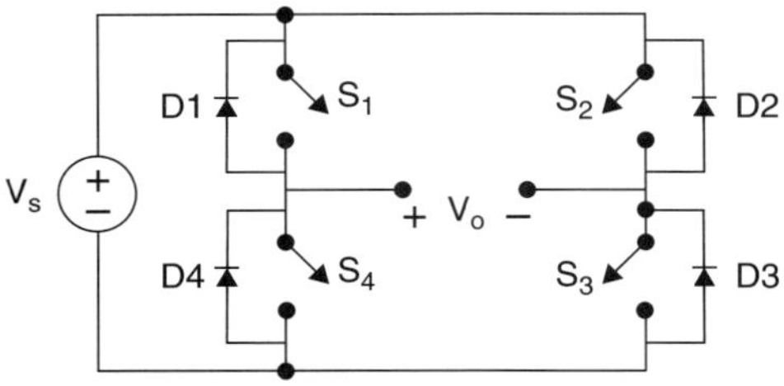

Figure 7.4
Full-bridge inverter topology

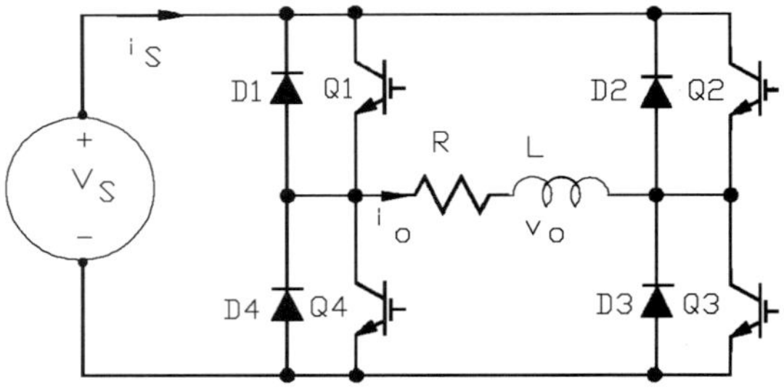

Figure 7.5
The 1-phase full-bridge VSI inverter

switched 120° apart. This topology is capable of handling high power with minimum component stresses. Furthermore, as shall be seen later, the triplen harmonics (3, 9, 15, and so on) are absent in this arrangement.

In this chapter, the focus will be on the voltage source driven topologies. The half-bridge topology of Figure 7.3(b) requires two separate voltage sources $V_S/2$. Figure 7.4 shows a full-bridge topology that requires a single voltage source. The full-bridge topology shall be studied in detail.

7.3 FULL-BRIDGE INVERTER (VSI) WITH SQUARE-WAVE SWITCHING

Popular dc-ac inverters in this category use full-bridge topologies. The topology is shown in Figure 7.5 using level-triggered switches such as the IGBTs and MOSFETs. This is the simplest topology. The switching scheme determines only the frequency of the output ac waveform. The output voltage amplitude is controlled by varying the amplitude of the source voltage.

This topology can also employ pulse-triggered switches such as the thyristors, GTOs and MCTs with appropriate gate control waveforms. Square-wave (or PWM) control needs forced commutation of switching devices, hence this topology is most suitable for GTOs, MOSFETs, IGBTs and MCTs.

7.3.1 Steady-state performance

The output voltage waveform is a square wave. The Fourier expansion of the output voltage waveform consists of only the sine terms and odd harmonics and the phase terms are absent.

$$v_O = \sum_{n=1,3,5,\ldots}^{\infty} \frac{4V_S}{n\pi} \sin n\theta : \quad \theta = \omega_S t \qquad (7\text{-}17)$$

The Fourier expansion of the output current i_o is

$$i_O = \sum_{n=1,3,5,\ldots}^{\infty} \frac{4V_S}{n\pi|Z_{\mathrm{Ln}}|} \sin(n\theta - \phi_n)$$

where

$$Z_{\mathrm{Ln}} = R + jn\omega_S L \qquad (7\text{-}18)$$

R is small in practical loads, hence it is ignored in our analysis. The amplitude of current harmonics then decreases as $1/n^2$. Power to the load decreases by $1/n^3$. The third harmonic current is one-tenth in amplitude in comparison to the fundamental component. Hence, the output current may be approximated by a pure sinusoid at the switching frequency as given below:

$$i_O = I_O \sin(\theta - \phi) \qquad (7\text{-}19)$$

where

$$I_O = \frac{4V_S}{\pi|Z_{L1}|}, \quad \theta = \omega_S t, \quad \phi = \tan^{-1}\frac{\omega_S L}{R}$$

Angle ϕ is the phase difference between the fundamental harmonic of the output voltage v_O waveform and the sinusoidal current.

The steady-state inverter operation can be explained by assuming two circuit modes as given in Figure 7.6. Mode 1 and mode 2 are split in two parts for the purpose of showing the current path.

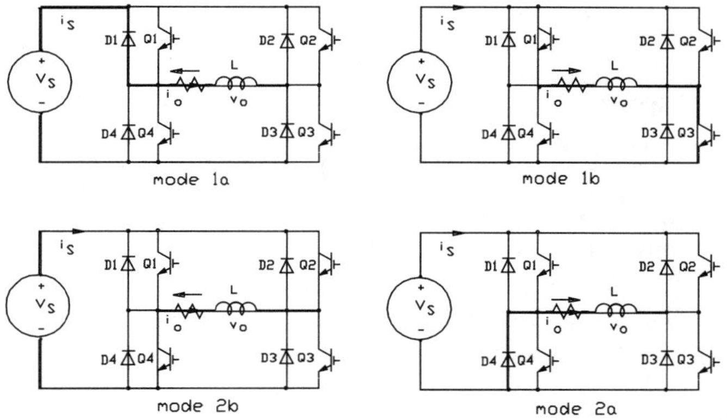

(a)

mode 1a $0 \leq \theta < \frac{\pi}{2}$ mode 1b $\frac{\pi}{2} \leq \theta < \pi$

mode 2a $\pi \leq \theta < \frac{3\pi}{2}$ mode 2b $\frac{3\pi}{2} \leq \theta° < 2\pi$

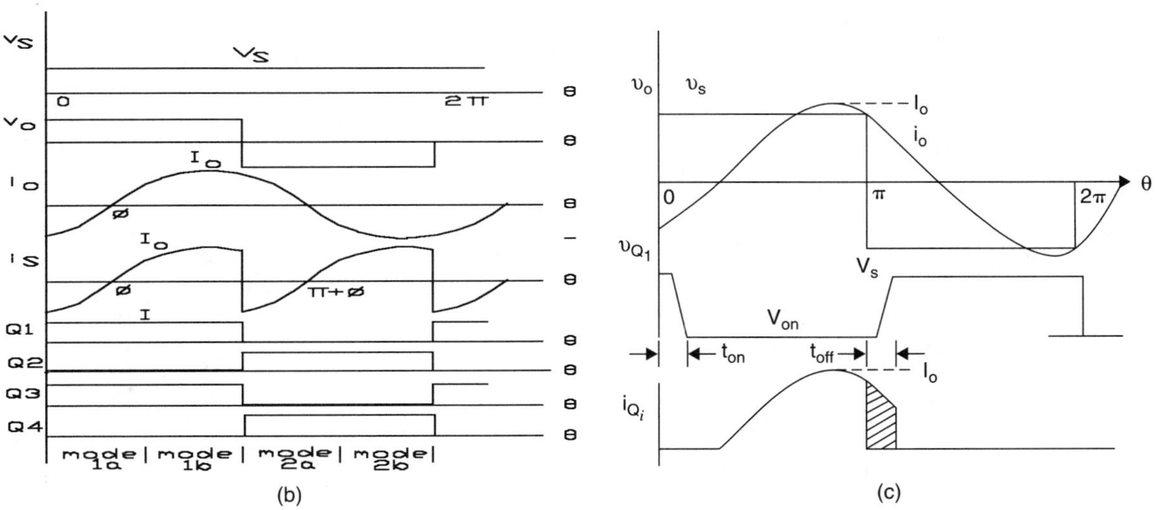

Figure 7.6
(a) Circuit modes (b) waveforms and (c) voltage and current through the switching transistor Q_1

In mode 1a, the IGBT Q_1 and Q_3 are turned on at zero conduction angle in a switching period. However, since the output current i_o is negative in mode 1a, it flows through diodes D3 and D1. The IGBTs Q_1 and Q_3 are therefore turned on at zero current (a very desirable feature because the switching losses are thus minimal). Mode 1b begins when the output current crosses the zero level at angle ϕ, at which point it is then steered through Q_1 and Q_3. A positive source voltage is applied to the load throughout mode 1a and mode 1b.

Mode 2a begins at the conduction angle π when Q_1 and Q_3 are force commutated (turned off forcibly) and Q_2 and Q_4 are turned on. The current in the inductive load maintains its direction and the amplitude in this transition. The current I_o flows through diode D2 and D4 back to the source. The IGBTs, even though turned on, do not conduct current. The negative source voltage is now applied to the load and remains so throughout mode 2a and mode 2b. Mode 2b begins at $\pi + \phi$ when the output current crosses the zero level, at which point it is then steered through Q_2 and Q_4. At the angle 2π, the IGBTs Q_2 and Q_4 are force commutated and Q_1 and Q_3 are turned on. The cycle repeats.

The voltage and current waveforms are shown in Figure 7.6(b). The waveforms have half-wave symmetry.

Total Harmonic Distortion THD in the output voltage is calculated using equation (7-2) :

$$THD = \sqrt{\frac{\pi^2}{8} - 1} = 0.4834 \qquad (7\text{-}20)$$

Third harmonic is the dominant lowest undesired harmonic.

The instantaneous power is

$$p(\theta) = v_O(\theta)\, i_O(\theta) \quad = V_S\, I_O \sin(\theta - \phi)$$

The negative power flow can be seen on the waveform, see Figure 7.7 in Example 7-1. The average power is calculated as

$$P_{av} = \frac{1}{\pi}\int_0^{\pi} V_S I_O\, sin\,(\theta - \phi)\, d\theta$$

$$= \frac{V_S I_O}{\sqrt{2}} \cos\phi = \frac{4V_S^2}{\pi|Z_{L1}|} \cos\phi$$

The voltage transfer ratio, defined as the ratio of rms voltage of the fundamental component of the output voltage waveform to the input dc voltage is

$$T_{vv} = \frac{\tilde{V}_{O1}}{V_S} = \frac{2\sqrt{2}}{\pi} = 0.9003 \qquad (7\text{-}21)$$

The voltage transfer ratio is constant because there is no amplitude control. The dc input current is calculated from

$$\bar{I}_S = \frac{1}{\pi}\int_0^{\pi} I_O \sin(\theta - \phi) d\theta$$

$$= \frac{2}{\pi} I_O \cos\phi \qquad (7\text{-}22)$$

The current reflection ratio is given by

$$T_{ii} = \frac{\bar{I}_S}{\tilde{I}_O} = \frac{2\sqrt{2}}{\pi} \cos\phi \qquad (7\text{-}23)$$

Switch Stress

The voltage and the current across the switching transistor Q_1 are drawn in Figure 7.6(c). The switch transition times are shown exaggerated with reference to the switching period. The switch is turned on at zero voltage as the diode D_1 is conducting at that instant. During the turn-off, however, Q_1 is turned off at nonzero current, the maximum value of which may be I_O. The turn-on and turn-off times include the delay and rise or fall times.

Maximum voltage across the switching transistor:	$2V_S$
Maximum current through the switching device:	I_O

The turn-off switching losses can be approximately calculated by assuming the switch current to be constant at $I_O \sin(\pi - \phi)$ and the switch voltage as a linearly rising voltage ramp:

$$P_{Q1} \approx \frac{1}{T_S} I_O \sin(\pi - \phi) \frac{1}{2} V_S t_{off}$$

$$\approx V_S I_O \sin\phi \frac{t_{off}}{2T_S} \qquad (7\text{-}23a)$$

$$P_{Q1,\max} = V_S I_O \frac{t_{off}}{2T_S}$$

Example 7-1 Square-Wave Switched Inverter

A square-wave switched dc-ac inverter is powered by a 169 V dc source and provides power to a load of 10 ohms in series with 10 mH. The inverter is switched at 400 Hz. Draw the output voltage and current waveforms, the instantaneous output power waveform and the harmonic profile of the output voltage up to five nonzero amplitude harmonics. Also determine the power delivered to the load.

```
dtheta = pi/500; VS = 169 ;
  phi = atan(2*pi*400*10e-3/10);
ZL1 = 10+j*2*pi*400*10e-3;
theta1 = [0:dtheta:pi]; %angles in mode 1
theta2 = [pi:dtheta:2*pi]; %angles in mode 2
theta = [theta1 theta2]; %angle array over a period
vo = [VS*ones(size(theta1)) -VS*ones(size(theta2))];
  %output voltage
Vorms = sqrt(mean(vo .^2)); %output rms voltage
[f,voh,ph] = harmonic(vo/VS, 10); % harmonic profile with
  10 harmonics
axis([0,10,0,1.4])
VO1 = voh(2);
vo1 = VO1*sin(theta);
IO = 4*VS/(pi*abs(ZL1));
io = IO*sin(theta-phi); %output current
p = vo.*io; % instantaneous power
P = mean(p) % average power
plot(theta, vo/VS, theta, vo1,'—', theta,io/IO,
  theta,p/max(p),'*')
```

(a)

(b)

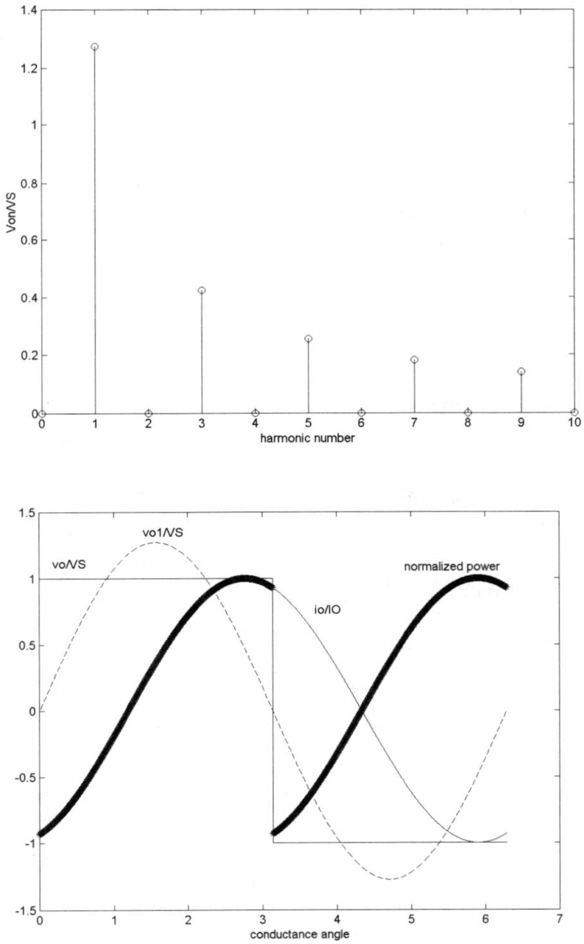

Figure 7.7
(a) Harmonic profile of the output voltage in square-wave switching (b) waveforms

7.3.2 Amplitude control in square-wave switching

The output voltage amplitude can be controlled by varying the amplitude of the source voltage. Usually a controlled ac-dc converter is used as the source to the dc-ac inverter as shown in Figure 7.8.

Overall voltage transfer ratio is calculated as the product of the voltage transfer ratios of the ac-dc converter and the dc-ac inverter:

$$T_{vv} = T_{vv1}\,T_{vv2} = 0.9003\,T_{vv1} \tag{7-24}$$

The overall current reflection ratio is calculated as the product of the current reflection ratios of the ac-dc converter and the dc-ac inverter:

$$R_{ii} = R_{ii1}R_{ii2} = 0.9003\,R_{ii1} \tag{7-25}$$

7.3.3 AC voltage sink load

Many of the industrial loads are inductance in series with an ac voltage sink such as the induction motor or the synchronous motor or a utility grid. The ac voltage sink is defined as

$$v_{ac} = V_{ac}\sin(\theta - \phi_1), \quad \theta = \omega_S t \tag{7-26}$$

Phase angle ϕ_1 is the phase difference between the ac voltage sink or source and the fundamental component of the output voltage of the bridge inverter. Also assume that the ac voltage sink is removing power out of the system at unity power factor, although this assumption is not necessary. Hence, the output current i_O is also in phase with v_{ac}.

Again, because of the phase difference in the inverter output voltage and the ac voltage sink waveforms, the inverter operates in more than one quadrant in a single switching period. Hence, the inverter topology should include the bidirectional switches. Figure 7.9(a) gives one such topology. The output voltage v_o and its waveform are determined solely by the switching network and, therefore, given by equation (7-17). The steady-state inverter operation can be explained by similar modes as found in the case of inductive load. The current and voltage waveforms are shown in Figure 7.9(b).

Output current through the inductor is obtained from

$$L\frac{di_O}{dt} = v_o - v_{ac} \tag{7-27}$$

Fundamental component of the output current phasor is

$$\tilde{I}_{O1}e^{-j\phi} = \frac{\tilde{V}_{O1} - \tilde{V}_{ac}}{j\omega_S L}$$

$$= \frac{1}{j\omega_S L}\left[\frac{2\sqrt{2}V_S}{\pi} - \tilde{V}_{ac}\,e^{-\phi}\right] \tag{7-28}$$

Comparing the real and imaginary parts,

$$\tilde{I}_{O1} = \frac{\tilde{V}_{ac}}{\omega_S L}\tan\phi\,, \quad \tilde{V}_{ac} = \frac{2\sqrt{2}}{\pi}V_S\cos\phi \tag{7-29}$$

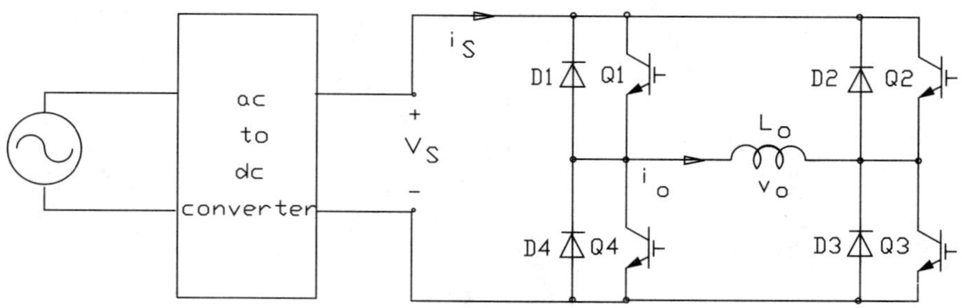

Figure 7.8
Amplitude control in square-wave switching

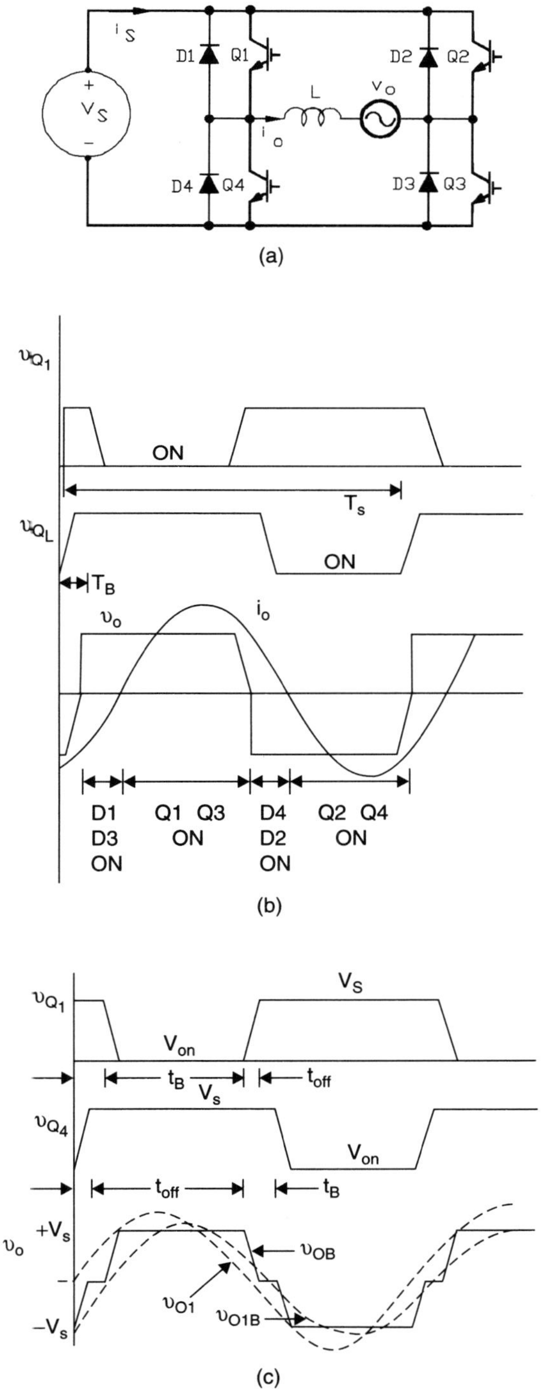

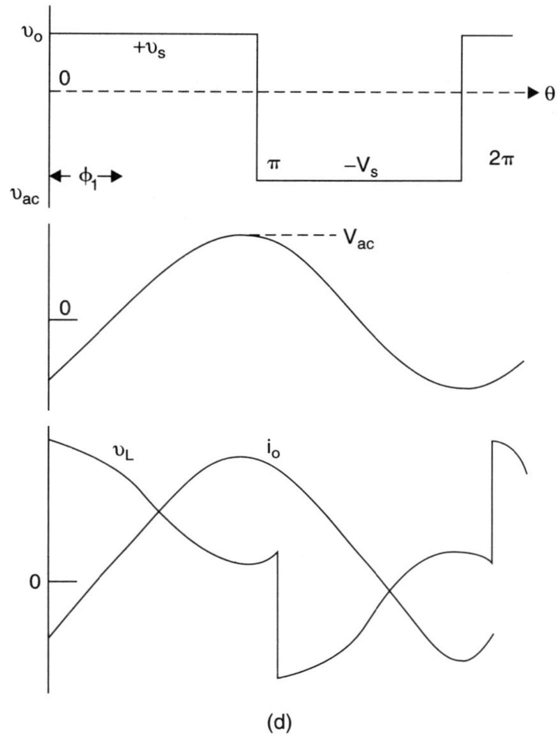

Figure 7.9
(a) PWM bridge-inverter with ac voltage sink or source
(b) current and voltage waveforms without blanking
(c) effect of blanking in R-load and (d) effect of blanking in R-L load

The fundamental component of the output phasor is then

$$\tilde{I}_{O1} = \frac{2\sqrt{2}}{\pi\omega_S L} V_S \sin\phi \tag{7-30}$$

The phase angle ϕ can be used as the control parameter.

The output power is

$$P_O = \tilde{V}_{ac}\,\tilde{I}_{O1} = \frac{4}{\pi^2\,\omega_S L} V_S^2 \sin 2\phi \tag{7-31}$$

Example 7-2 AC Voltage Sink Load

A 240 V, 400 Hz BUS is connected to the output of a 1-ϕ VSI through a 5 mH inductor. The inverter is powered by a 300 Vdc and delivers 1 kW power to the ac BUS. Find the phase angle ϕ_1 in the inverter operation.

$$v_{ac} = 240\sqrt{2}\sin(\theta - \phi_1)$$

From equation (7-31)

$$1000 = \tilde{V}_{ac}\tilde{I}_{O1} = 240 \times \tilde{I}_{O1} \Rightarrow \tilde{I}_{O1} = 4.167\,A$$

From equation (7-30)

$$4.167 = \frac{2\sqrt{2}}{\pi\, 2\,\pi\, 400 \times 5 \times 10^{-3}}\, 300 \sin\phi_1$$

$$\Rightarrow \phi_1 = 11.2°$$

7.3.4 Effect of blanking in square-wave switching

Simultaneous conduction of switch pair Q1-Q4 and Q2-Q3 in the inverter of Figure 7.9(a), must be avoided in all situations else the dc source would be "shorted." Switch Q1 should be turned on after a blanking time t_B and after the switch Q4 is turned off. The blanking time t_B must be greater than the turn-off time of Q4. Similar precaution must be observed in turning Q4 after turning off of Q1. Figure 7.9(c) and (d) show waveforms for exaggerated transition times for a purely resistive load and an R-L load respectively. Figure 7.9(c) shows the inverter output voltage v_{OB}, subscript B for blanking, its fundamental component v_{O1B}. It also shows the fundamental component in an ideal nonblanking situation, v_{O1}, for comparison. The magnitude of v_{O1B} is smaller than v_{O1} and has a phase difference.

Figure 7.9(d) shows the waveforms for an R-L load. Diodes D1-D3 turns on to carry the negative current as soon as the Q_2-Q_4 pair turns off completely. The Q_1-Q_3 pair, even though turned on after the blanking period, does not conduct current until the i_O becomes positive. The effect of blanking with inductive load is much less evident in comparison to that in the purely resistive load.

7.4 PULSE-WIDTH MODULATION (PWM) CONTROL OF VSI

The amplitude of the inverter output, see Figure 7.5, can be controlled by controlling the on-time of switch (Q-D combination) pairs S1-S4 and S2-S3 during half switching periods. This switching scheme introduces two intervals over the switching period in which all switches are off. The output voltage waveform contains two zero amplitude states. Inverters with PWM control do not require a preceding variable dc source or controlled ac-dc converter.

7.4.1 1-Phase full-bridge VSI with R-L load

Let the on-time of the switch pairs be denoted by the parameter delta, the pulse width. The output voltage waveform has quarter-wave symmetry and half-wave symmetry, the Fourier expansion of the output voltage waveform consists of only the sine terms and odd harmonics and the phase terms are missing.

$$v_O = \sum_{n=1,3,5,\ldots}^{\infty} \frac{4V_S}{n\pi}\sin\frac{n\delta}{2}\sin n\theta\ ,\ \theta = \omega_S t \qquad (7\text{-}32)$$

The Fourier expansion of the output current i_o is then

$$i_O = \sum_{n=1,3,5,\ldots}^{\infty} \frac{4V_S}{n\pi|Z_{Ln}|}\sin\frac{n\delta}{2}\sin(n\theta - \phi_n)$$

where

$$Z_{Ln} = R + j\,n\omega_S L \qquad (7\text{-}33)$$

Again, the power to the load at harmonic frequencies decreases in the order $1/n^3$. Third harmonic current is one-tenth in amplitude in comparison to the fundamental component. Hence, the output current may be approximated by a pure sinusoid at the switching frequency. Angle ϕ_1 is the phase difference between the fundamental harmonic of the output voltage waveform and the sinusoidal current.

$$i_O = I_O\sin(\theta - \phi_1),\ \textit{where}\ I_O = \frac{4V_S}{\pi|Z_{L1}|}\sin\frac{\delta}{2} \qquad (7\text{-}33a)$$

The steady-state inverter operation can be explained by assuming four circuit modes as given in Figure 7.10. Mode 1 and mode 3 are split in two parts for the purpose of showing the current path. In the mode 1a the IGBT Q_1 is already on and IGBT Q_3 is turned on at an angle $\pi/2$-$\delta/2$. The output current I_o is negative, hence it flows through diodes D3 and D1. Mode 1b begins when the output current crosses the zero level, at which point it is steered through Q_1 and Q_3. The current crosses at the angle ϕ_1. A positive voltage is applied to the load throughout mode 1a and mode 1b.

Mode 2 begins at angle $\pi/2 + \delta/2$ when Q_1 is force commutated (turned off) and Q_4 is turned on. However, due to its polarity, the current I_o flows through diode D4. The output voltage across the load is shorted out and the current I_o circulates freely.

Mode 3a begins at angle $3\pi/2$-$\delta/2$ when Q_3 is force commutated and Q_2 is turned on. Again, due to its polarity, the current I_o flows through diodes D4 and D2 until it reverses polarity. Then in mode 3b it is steered through Q_2 and Q_4. A negative voltage is applied across the load throughout mode 3a and mode 3b.

Mode 4 begins at angle $3\pi/2 + \delta/2$ when Q_4 is force commutated and Q_1 is turned on. However, due to its polarity, the current I_o flows through diode D1. The output voltage across the load is shorted out again and lasts up to $5\pi/2$-$\delta/2$. The voltage and current waveforms are shown in Figure 7.10b.

The output power is

$$
\begin{aligned}
P_O &= \tilde{V}_{O1}\tilde{I}_O \cos\phi_1 \\
&= \frac{4V_S}{\pi\sqrt{2}} \sin\frac{\delta}{2} \frac{I_O}{\sqrt{2}} \cos\phi_1 \qquad \text{(7-33b)} \\
&= \frac{8V_S^2}{\pi^2 |Z_{L1}|} \sin^2\frac{\delta}{2} \cos\phi_1
\end{aligned}
$$

Example 7-3 VSI with PWM Control

A PWM controlled inverter is loaded by 1 ohm in series with 5 mH inductor, powered by 169 Vdc and switched at 60 Hz. Draw the output voltage and current waveforms and the harmonic profile up to 30 harmonics.

```
dtheta = pi/200; VS = 169; R = 1; omegal = 2*pi*60*.005;
  %step angle, peak input voltage, peak
amplitude of fundamental output current, load resistance
  and load reactance
imped1 = R+i*omegal;
phi1 = atan(omegal/R); %lagging phase angle of the
  output current with respect to the fundamental
  component of the output voltage
delta = 2.0635; %pulse width
IO = (4*VS/(pi*abs(imped1)))*sin(delta/2);
  %peak amplitude of the output current waveform
theta1 = [pi/2-delta/2:dtheta:pi/2+delta/2];
  %angles in mode 1
theta2 = [pi/2+delta/2:dtheta:3*pi/2-delta/2];
  %angles in mode 2
theta3 = [3*pi/2-delta/2 : dtheta : 3*pi/2+delta/2];
  %angles in mode 3
theta4 = [3*pi/2+delta/2 : dtheta: 5*pi/2-delta/2];
  %angles in mode 4
theta = [theta1 theta2 theta3 theta4];
  %angle array over a period
vo = [VS*ones(size(theta1)), zeros(size (theta2)),
  -VS*ones(size(theta3)), zeros(size(theta4))];
%output voltage
% Building output current array
io1 = IO*sin(theta1-phi1);
io2 = IO*sin(theta2-phi1); io3 = IO*sin (theta3-phi1);
  io4 = IO*sin(theta4-phi1);
io = [io1 io2 io3 io4]; %output current
is = [io1 zeros(size(theta2)) -io3 zeros(size(theta4))];
  %input current
Vorms = sqrt(mean(vo .^2)); %output rms voltage
Iorms = IO/sqrt(2);
IS = mean(is);
[f,voh,ph] = harmonic(vo/VS);
  %amplitude and phase of fundamental component
axis([0,30,0,1])
vo1 = voh(2)*VS*sin(theta);
  %fundamental component of the output voltage
Vo1rms = voh(2)/sqrt(2);
Tvv = Vo1rms/VS; Tii = IS/Iorms;
td = (sqrt ( (Vorms*sqrt(2) /(voh(2)*VS) ) ^2 -1 ));
  % Total Harmonic Distortion
plot(theta,vo/VS,theta,io/IO,theta,is/IO,'*',
  theta,vo1/VS,'—')
```

Input Characteristics

The voltage transfer ratio, defined as the ratio of rms voltage of the fundamental component of the

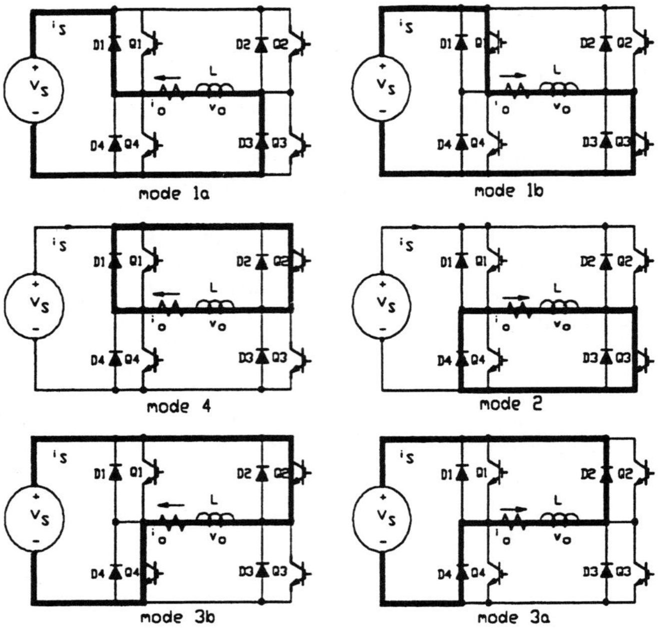

(a)

mode 1a $\frac{\pi}{2} - \frac{\delta}{2} \le \theta < \phi_1$ mode 2 $\frac{3\pi}{2} - \frac{\delta}{2} \le \theta < \frac{3\pi}{2} + \frac{\delta}{2}$ mode 3b $\pi + \phi_1 \le \theta < \frac{3\pi}{2} + \frac{\delta}{2}$

mode 1b $\phi_1 \le \theta < \frac{3\pi}{2} - \frac{\delta}{2}$ mode 3a $\frac{3\pi}{2} - \frac{\delta}{2} \le \theta < \pi + \phi_1$ mode 4 $\frac{3\pi}{2} + \frac{\delta}{2} \le \theta < \frac{5\pi}{2} - \frac{\delta}{2}$

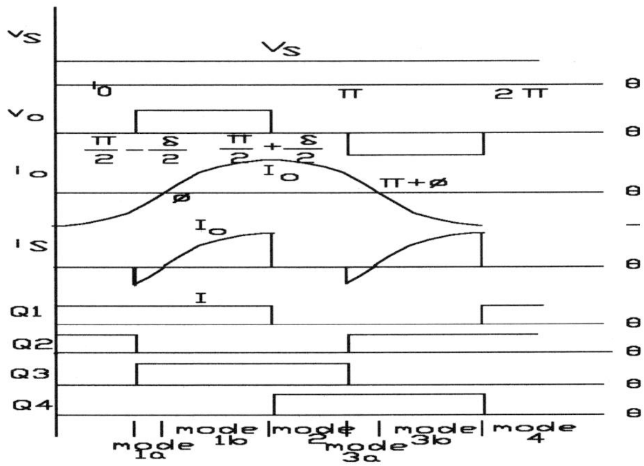

(b)

Figure 7.10
(a) Circuit modes and (b) waveforms

output voltage waveform to the input dc voltage, is given by

$$T_{vv} = \frac{\tilde{V}_{O1}}{V_S} = \frac{2\sqrt{2}}{\pi} \sin \frac{\delta}{2} \qquad (7\text{-}33c)$$

The input dc current and the current reflection ratio are

$$\bar{I}_S = \frac{1}{\pi} \int_{\pi/2-\delta/2}^{\pi/2+\delta/2} I_O \sin(\theta - \phi_1)\, d\theta$$

$$\bar{I}_S = \frac{2I_O}{\pi} \sin \frac{\delta}{2} \cos \phi_1$$

$$T_{ii} = \frac{\bar{I}_S}{\tilde{I}_O} = \frac{2\sqrt{2}}{\pi} \sin \frac{\delta}{2} \cos \phi_1 \qquad (7\text{-}33d)$$

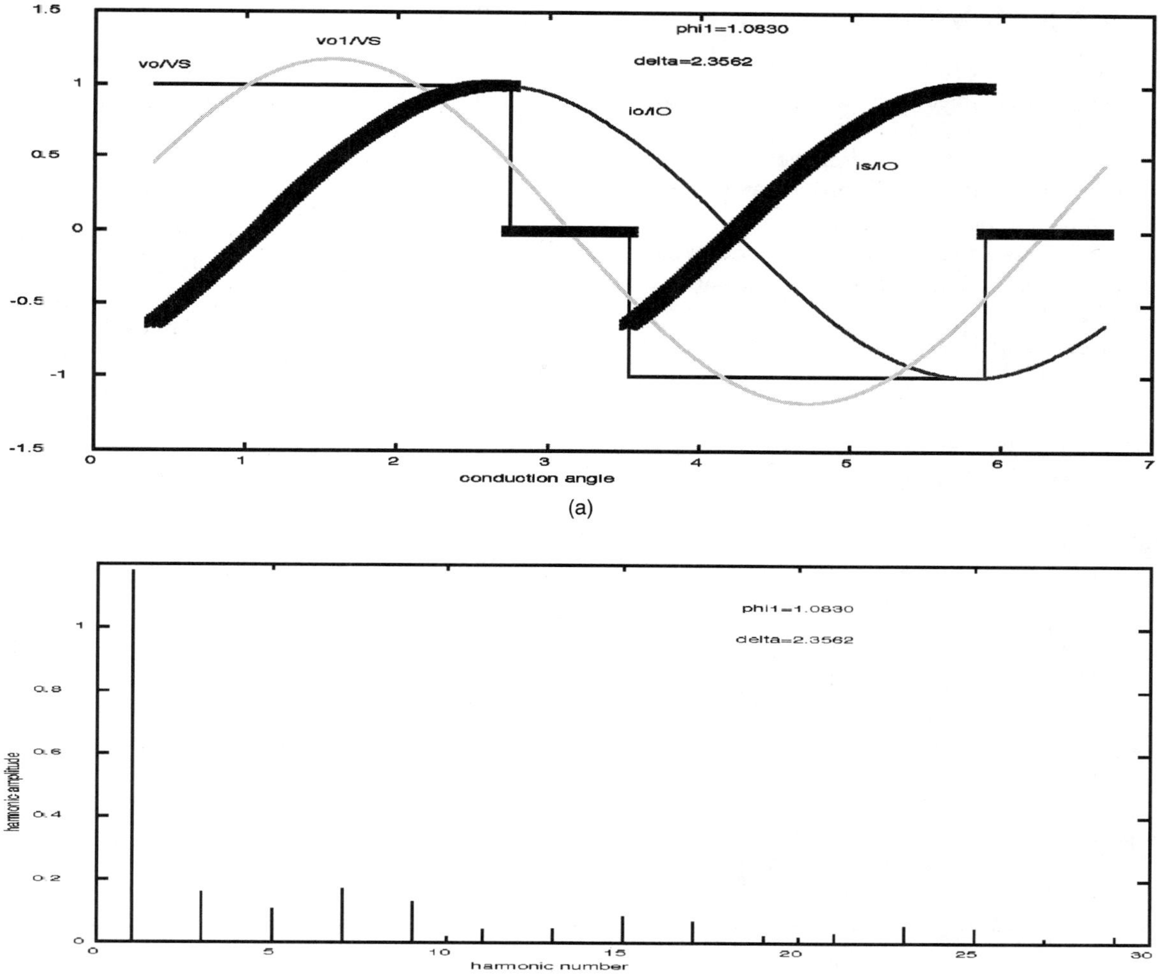

(a)

(b)

Figure 7.11
(a) Waveforms in PWM inverter and (b) harmonic profile of the output voltage

The control characteristics: the voltage transfer ratio, the reverse reflection ratio and the Total Harmonic Distortion in the output voltage versus the pulse width are given in Figure 7.12(a). Figure 7.12(b) shows the amplitudes of a few harmonics in the output voltage waveform versus the pulse width.

The third harmonic is the dominant lowest undesired harmonic.

7.5 CURRENT MODE CONTROL OF PWM VSI

In most of the VSI driven motor control applications, the actual objective is to control the current and thereby control the torque. In this mode of control, the output current is sensed, compared with a reference waveform. The feedback control is then used to obtain the desired current waveform. This topic is covered in some detail in a later chapter on converter control. Here, we would make only a few comments as follows:

Let the desired output current be

$$i_O = I_O \sin(\theta - \phi_1) \tag{7-34}$$

Peak amplitude of the fundamental component of the output voltage of the inverter is

$$V_{O1} = Z_L I_O = \frac{4V_S}{\pi} \sin\frac{\delta}{2} \tag{7-35}$$

Pulse width δ is determined from the above equation.

7.6 CURRENT SOURCE PWM FULL-BRIDGE INVERTER

The current source I_S is synthesized from a dc voltage source in series with a large inductor. The impedance of such an inductor at the desired fundamental frequency is generally at least ten times higher than the highest load impedance. A 1-phase full-bridge CSI is shown in Figure 7.13. The load is the ac voltage sink at the frequency ω_S and it is assumed that the fundamental component of the current to the sink is delivered at a phase angle ϕ_1. For a zero phase angle, the inverter operates in the first and the third quadrants and the current and voltage on the switches need only be unidirectional. For a

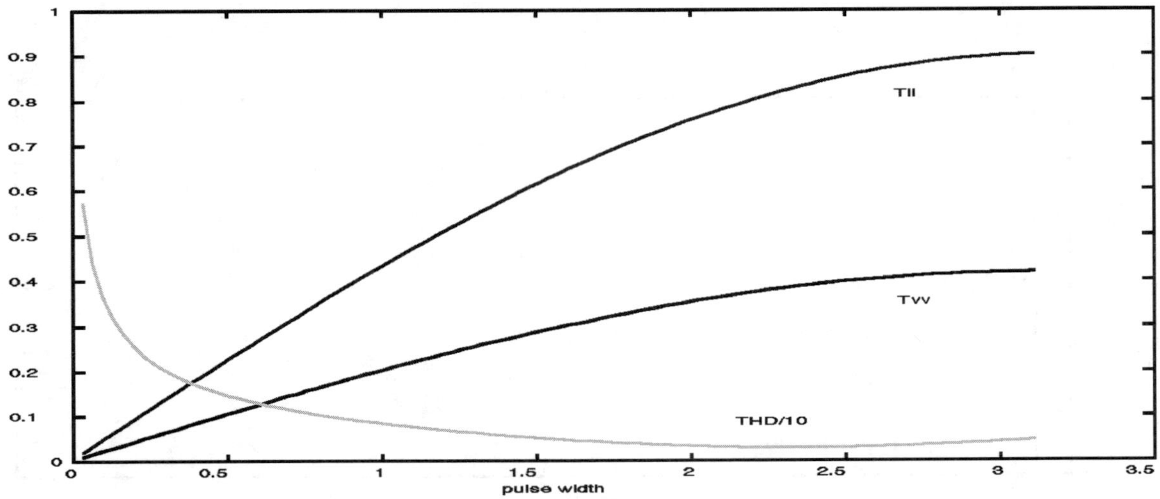

(a)

Figure 7.12
(a) Voltage transfer ratio and Total Harmonic Distortion as the functions of the pulse width

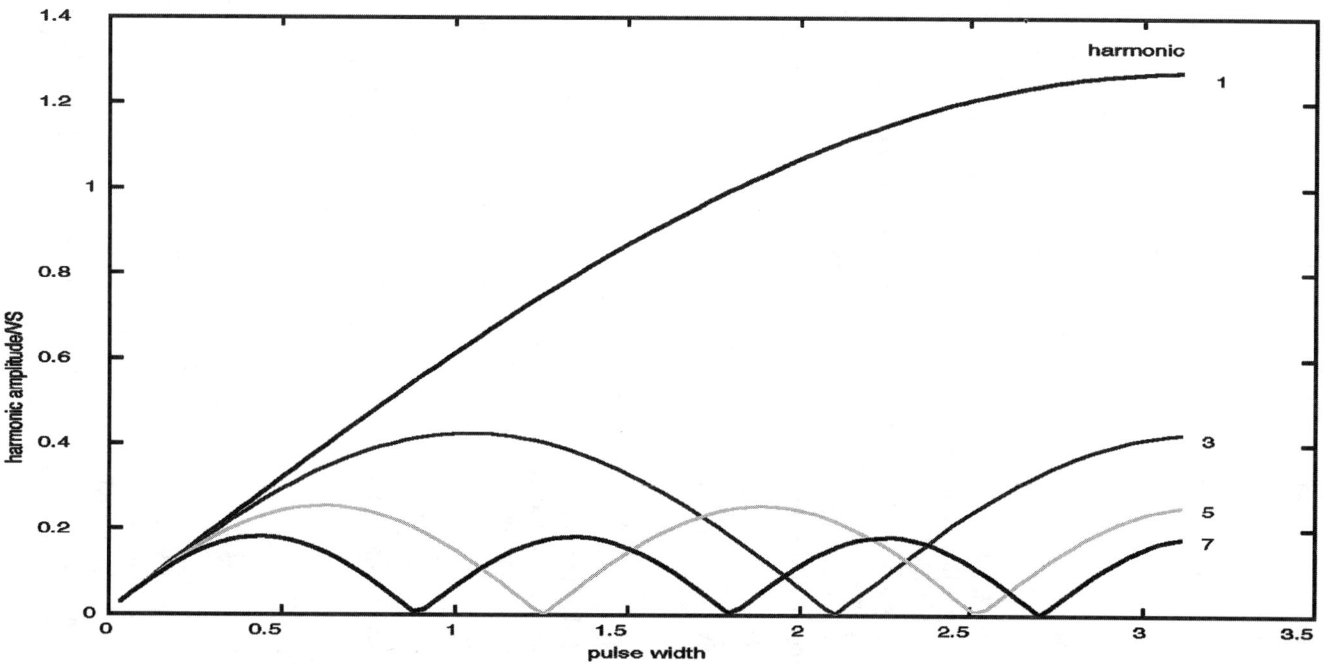

(b)

Figure 7.12 ***(continued)***
(b) Harmonic amplitudes of the output voltage waveform as the function of the pulse width

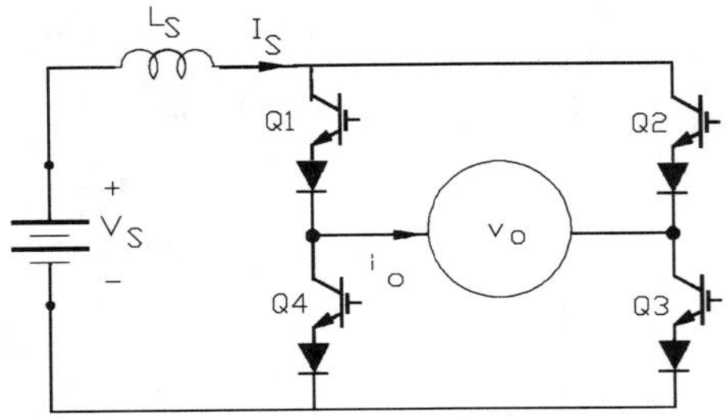

Figure 7.13
Current source full-bridge inverter

nonzero ϕ_1, the inverter operates in all four quadrants and the switches conduct unidirectional current but need bidirectional voltage blocking capability. The mode sequence and the waveforms for a single pulse width control are given respectively in Figure 7.14(a) and Figure 7.14(b). Output of CSI is a current waveform. The harmonic profile of the inverter output current is given by

$$i_O = \sum_{odd\,n} \frac{4\bar{I}_S}{n\pi} \sin\frac{n\delta}{2} \sin(n\theta) \quad :\theta = \omega_s t \qquad (7\text{-}36)$$

where ω_s is the switching frequency, which is also the fundamental frequency of the output waveform.

Current transfer ratio, defined as the ratio of rms value of the fundamental component of the output current to the input dc current, is calculated from

$$T_{ii} = \frac{\tilde{I}_{O1}}{I_S} = \frac{2\sqrt{2}}{\pi} \sin\frac{\delta}{2} \qquad (7\text{-}37)$$

The power delivered to the ac voltage sink is calculated as

$$P_O = V_S I_S = \tilde{V}_O \tilde{I}_{O1} \cos\phi_1$$
$$= \frac{2\sqrt{2}}{\pi} \tilde{V}_O I_S \cos\phi_1 \sin\delta/2 \qquad (7\text{-}38)$$

The output power can be controlled by controlling either pulse width δ or phase angle ϕ_1. The phase angle can be controlled through switch timing. The disadvantage of phase-angle control is that it reduces the power by reducing the output power factor. The switching losses in the inverter, however, remain constant; hence efficiency is also reduced simultaneously. A more desirable method is to maintain the angle ϕ_1 near zero and control power using pulse width δ.

The large inductor on the input side causes a slow start. However, the CSI has a built-in short-circuit protection because the current from the source is limited by the input inductor. It also provides shielding of output voltage against transients in the input dc voltage source, in the cases where the dc voltage is obtained from an ac source through a rectifier. CSI is an ideal choice for bus connected outputs and induction motor drives.

7.7 PRUNING OF HARMONIC PROFILE

Reducing the harmonic content of the output voltage or current is one of the important tasks of the designers. Harmonics appear as the undesirable noise. Harmonics reduce the power factor. They may excite mechanical resonance at detrimental frequencies or generate acoustical noise in ac motor drive systems.

As has been seen in the chapter on dc-dc converters, the ripple content (harmonics) is easily reduced by post-filtering after the converter. The post converter low-pass filter may be designed for cutoff frequencies that are several times lower than the switching frequency (usually in kilohertz). In the dc-ac inverters, however, the fundamental component of the output voltage or current is either the same as or a simple fraction of the switching frequency. Designing an effective low-pass filter with cutoff frequency between the fundamental frequency and the first undesired harmonic frequency is a difficult task. Hence, an active harmonic reduction strategy is often desirable. Three types of strategies are generally employed:

(i) Harmonic elimination by switch control,

(ii) Harmonic cancellation by vector adding the outputs of two or more inverters,

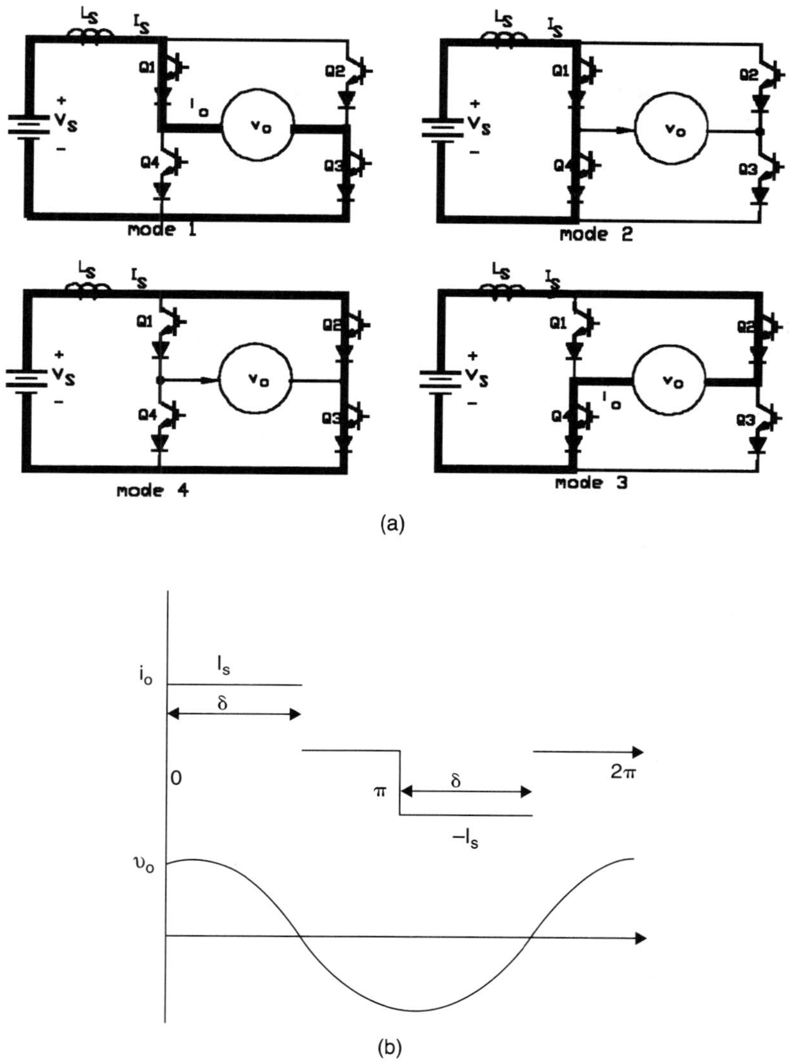

Figure 7.14
(a) Circuit modes in CSI and (b) Waveforms

(iii) To shift the first undesired harmonic frequency to high enough frequencies for easy filtering. This is done by wave-forming the output voltage or current by modulating the pulse width in a sinusoidal manner.

Harmonic Elimination

As it has been seen in the previous sections that in addition to the fundamental amplitude, the harmonic amplitudes are also the function of the pulse width in PWM inverters. For example, the third harmonic amplitude becomes zero at the pulse width $\delta=2\pi/3$. Selection of $\delta=2\pi/3$ renders the third harmonic amplitude to be zero. Hence the lowest undesired harmonic is moved up to five times the fundamental frequency. This makes it possible to design a better and smaller sized post-inverter filter. However, by fixing the pulse width we can no longer control the amplitude of the output voltage. The amplitude control can be introduced by adding or subtracting (notching) thin pulses in the output waveform. We first consider the notching of thin pulses.

7.7.1 Harmonic elimination by notching

Consider a thin pulse δ_k wide positioned centrally about the angle α_k from the zero conduction angle in a switching period, as shown in Figure 7.15. An identical negative pulse is positioned centrally about the angle $\pi + \alpha$.

The Fourier expansion of the thin pulse waveform is:

$$v_{kp} = \sum_{odd\ n} \frac{4V_S}{n\pi} \sin n\,\alpha_k \sin \frac{n\delta_k}{2} \sin n\theta \quad (7\text{-}39)$$

Notching is done by one of the two methods: (i) subtraction of the thin pulse train from the output voltage waveform or (ii) modification of the switching sequence. Amplitude of the n-th harmonic of the inverter output after notching is:

$$V_{On} = \frac{4V_S}{n\pi}\left[\sin \frac{n\delta}{2} - \sin n\alpha_k \sin \frac{n\delta_k}{2}\right] \quad (7\text{-}40)$$

The n-th harmonic can be eliminated by designing for $V_{on} = 0$ which gives the following condition:

$$\sin \frac{n\delta}{2} - \sin n\alpha_k \sin \frac{n\delta_k}{2} = 0 \quad (7\text{-}41)$$

The amplitude control parameter may be selected to be either the position angle α_k or the notch-pulse width δ_k. The above equation contains two variables, hence, one of them may be chosen arbitrarily and the other can be used as the control parameter.

Elimination of Third Harmonic

The third harmonic is the dominant and most undesirable harmonic. We will fix α_k and vary the notch-pulse width δ_k. Let us position the notch pulse at the center of the main pulse, that is, $\alpha_k = \pi/2$, see Figure 7.16. The above condition is then modified as

$$\sin \frac{n\delta}{2} - \sin \frac{n\pi}{2} \sin \frac{n\delta_k}{2} = 0 \ :\ n = odd \quad (7\text{-}42)$$

The above condition is satisfied when

$$0 < \delta_k < \delta \ \ and \ \ \delta_k = \delta \pm \frac{2\pi}{n} \ \ for \quad n = 3, 7, 11, \ldots$$

$$and \quad \delta_k = \delta \pm \frac{4\pi}{n} \quad for \quad n = 5, 9, 13, \ldots$$

The condition for elimination of the third harmonic is

$$\delta_k = \delta - \frac{2\pi}{3} \qquad for \quad \frac{2\pi}{3} < \delta \le \pi$$

Along with the third harmonic all harmonics that are the multiples of three (triplen, 3, 6, 9, 12, 15, and so on) are also eliminated. Hence, the ninth and fifteenth harmonics are deleted as well.

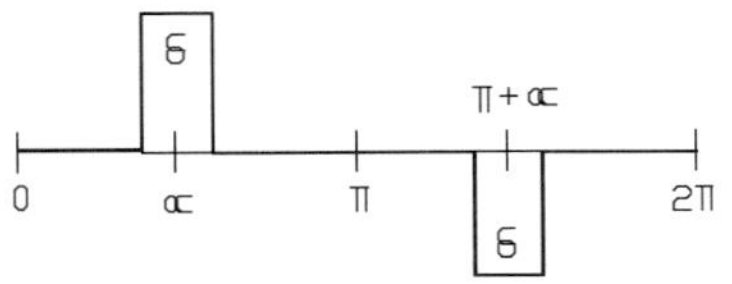

Figure 7.15
A thin pulse train

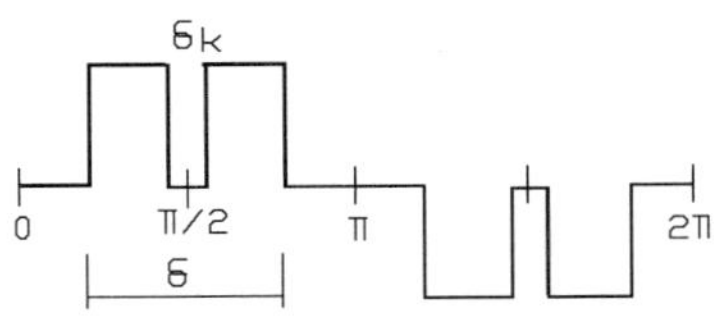

Figure 7.16
Notching of thin pulses from the PWM inverter

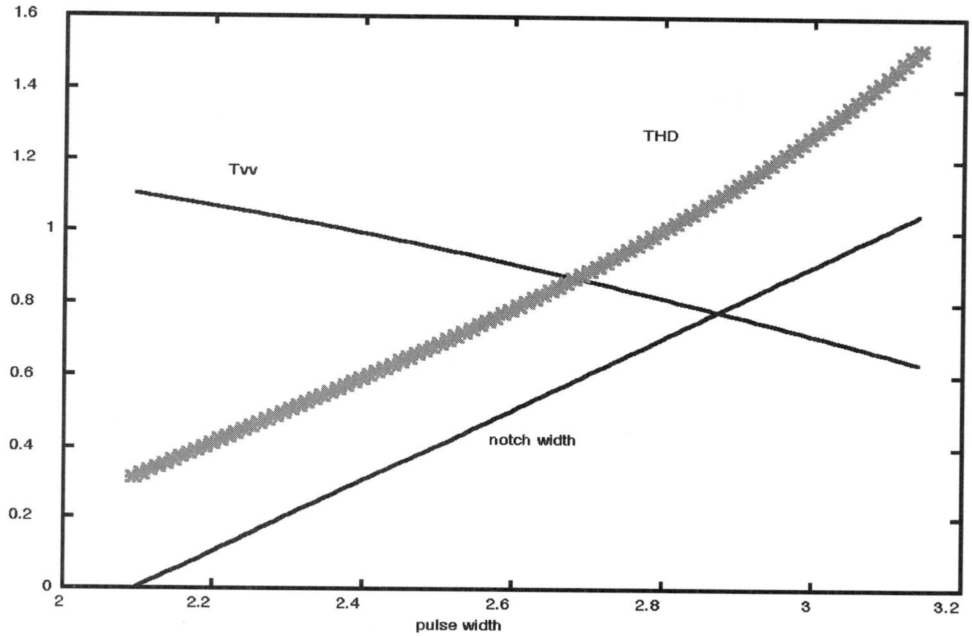

Figure 7.17
Voltage transfer ratio, THD and the width of the notch pulse as a function of the main pulse width with third harmonic elimination

Notching reduces the fundamental amplitude:

$$V_{O1} = \frac{4V_S}{\pi}\left[\sin\frac{\delta}{2} - \sin\frac{\delta_k}{2}\right] \qquad (7\text{-}43)$$

Hence, the Total Harmonic Distortion THD of the output voltage waveform increases (becomes poor). The number of switching increases, hence, the switching losses increase. The plots of δ_k, the voltage transfer ratio and the THD versus δ are shown in Figure 7.17.

Example 7-4

```
dtheta = pi/200; VS = 169; %step angle, dc source,
delta = [2*pi/3:pi/300:pi]; %array for main pulse width
[n1,r2] = size(delta);
Tvv = [ ]; deltak = [ ]; td = [ ];vor = [ ];
for n = 1:n2
   deltak1 = delta(n)−2*pi/3; % notch width
   deltak = [deltak deltak1];
   vo1r = 2*sqrt(2)*VS/pi *(sin(delta(n)/2) −sin(deltak1/2));
   %rms voltage of the fundamental harmonic after
   notching
   Tvv = [Tvv vo1r*sqrt(2)/VS]; % array for the voltage
   transfer ratio
   vor = VS*sqrt((delta(n)−deltak1)/pi); % rms voltage of the
   notched pulse waveform
   td = [ td (sqrt ( (vor^2−vo1r^2) /vo1r^2 ))]; %building
   array of THD
end
plot(delta,Tvv, delta,td,'*', delta, deltak)
```

Alternatively, we can fix the δ_k to be a certain angle and vary the position angle of the notch pulse. The output amplitude cannot be varied over a wide range. Hence, this method of harmonic elimination is used mainly in the constant voltage applications.

Elimination of Third and Fifth Harmonics Simultaneously

The conditions for eliminating the third and fifth harmonics together are:

$$\sin\frac{3\delta}{2} - \sin 3\,\alpha_k \sin\frac{3\delta_k}{2} = 0$$

$$\sin\frac{5\delta}{2} - \sin 5\,\alpha_k \sin\frac{5\delta_k}{2} = 0 \qquad (7\text{-}44)$$

Simultaneous control of the position angle and the width of the notch pulses eliminates both harmonics. An alternate method uses two identical notch pulses symmetrically placed around the center of the main pulse, as shown in Figure 7.18 below:

The position angle $\alpha_2 = \pi - \alpha_1$. The position angle α_1 and the pulse width δ_k can be determined graphically. Readers are encouraged to read the references for further details.

7.7.2 Harmonic elimination by stepping

Stepping involves adding a thin pulse to the main pulse via a transformer that has two primary windings and one secondary winding, as shown in Figure 7.19.

The expression for the amplitude of the n-th harmonic is:

$$v_{On} = \frac{4V_S}{n\pi}\left[\sin\frac{n\delta}{2} + \sin n\,\alpha_k \sin\frac{n\delta_k}{2}\right] \qquad (7\text{-}45)$$

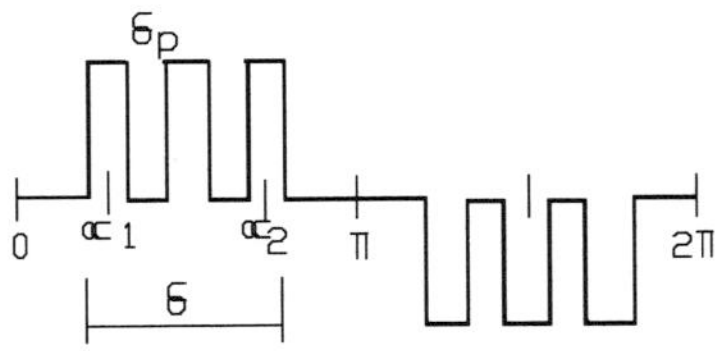

Figure 7.18
Simultaneous elimination of the third and fifth harmonics by notching

The step pulse is placed at the center of the main pulse, hence, select $\alpha_k = \pi/2$. Condition for the harmonic elimination is

$$\sin\frac{3\delta}{2} - \sin\frac{3\,\delta_k}{2} = 0 \qquad (7\text{-}46)$$

The above condition is satisfied when

$$0 < \delta_k < \delta \quad and \quad \frac{3\delta_k}{2} = \pi - \frac{3\delta}{2}$$

hence,

$$\delta_k = \frac{2\pi}{3} = \delta \qquad for \quad \frac{\pi}{3} < \delta \le \frac{2\pi}{3}$$

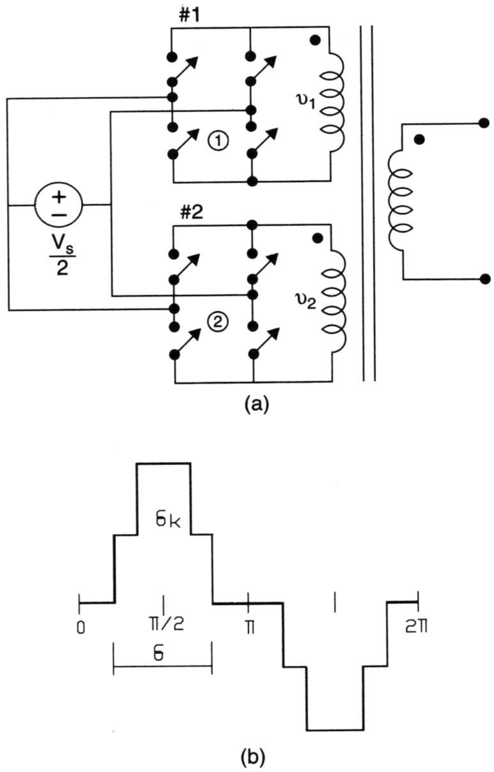

Figure 7.19
Harmonic elimination using step pulses (a) Circuit for adding a thin step pulse. H-bridge #2 generates thin pulse. (b) Stepped waveform

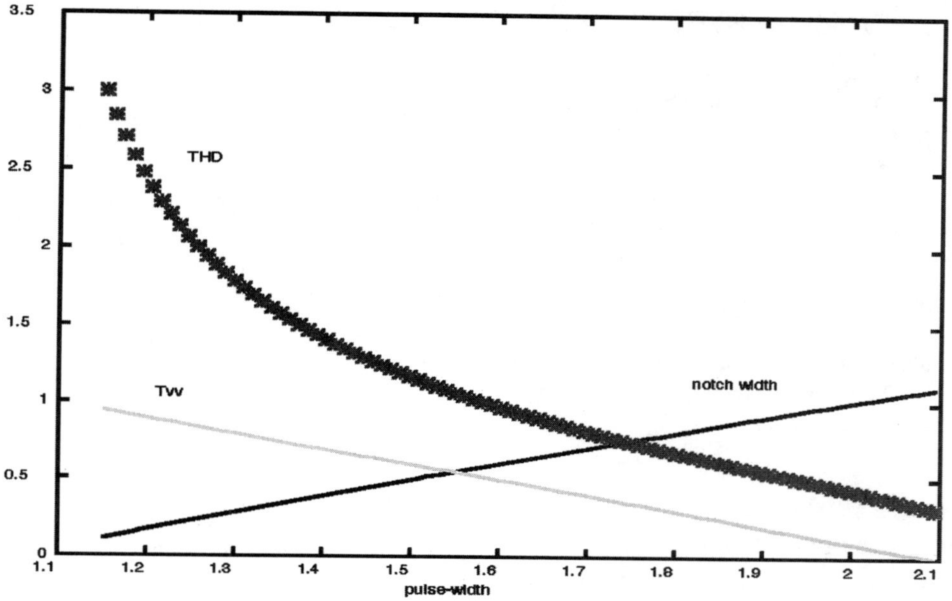

Figure 7.20
Elimination of third harmonic by stepping

Stepping increases the fundamental amplitude and therefore, the Total Harmonic Distortion THD of the output voltage decreases (improves).

The plots of δ_k, the voltage transfer ratio and the THD versus δ are shown in Figure 7.20.

Again, this method is used only in constant voltage applications.

Elimination of Third and Fifth Harmonics Simultaneously

The third and fifth harmonics together can be eliminated using a staircase as shown in Figure 7.21. The staircase waveform is synthesized by adding two pulses of differing widths but positioned at the same angle $\pi/2$. The n-th harmonic of the resulting step waveform is given by

$$v_{On} = \frac{4V_S}{n\pi}\left[\sin\frac{n\delta}{2} + \sin\frac{3\pi}{2}\sin\frac{3\delta_k}{2} + \sin\frac{5\pi}{2}\sin\frac{5\delta_k}{2}\right] \tag{7-47}$$

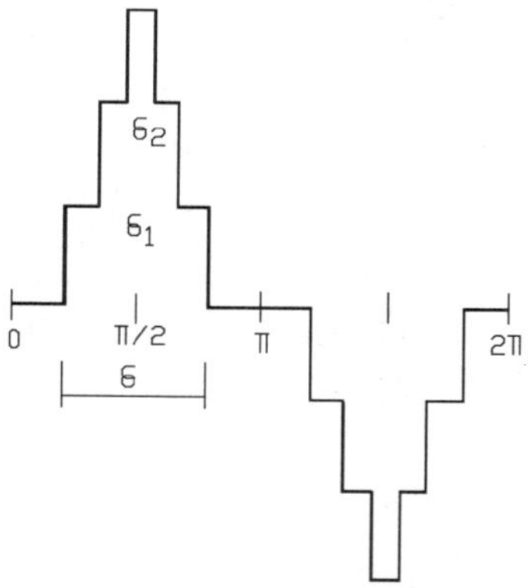

Figure 7.21
Simultaneous elimination of third and fifth harmonics using stepping

Set the amplitudes of the third and fifth harmonics equal to zero for eliminating them. Readers are encouraged to read the references for further details.

7.8 SINE-PWM INVERTER

This is the most popular PWM inverter. We will refer to the PWM inverter of Figure 7.5. It uses p pulses per half period ($T_O/2$) of the desired sinusoidal waveform of the output voltage. The pulse width of these pulses vary sinusoidal over the half period $T_O/2$. The switching frequency is 2p times the desired frequency f_O :

$$f_S = 2pf_O \tag{7-48}$$

These pulses may be unipolar (0 to V_S) or bipolar ($-V_S$ to $+V_S$). The following section will discuss the unipolar pulses.

7.8.1 Sine-PWM inverter using unipolar switching

The unipolar Sine-PWM inverter output may be considered as the summation of p pulse waveforms where each pulse has a positive and a negative pulse over the period T_O. Consider one such k-th pulse waveform with a pulse width δ_k and the position angle α_k with respect to the beginning of the period T_O and where

$$\alpha_k = (2k - 1)\frac{\pi}{2p} \text{ and } 0 < \delta_k < \frac{\pi}{p} \quad k = 1, 2, \ldots, p \tag{7-49}$$

The pulse widths vary as the sine of the position angle α_k (angle at the center of the pulse) in the half period $T_O/2$:

$$\delta_k = \delta_{kmax} \sin \alpha_k \tag{7-50}$$

The maximum pulse width occurs at the angles $\pi/2$ and $3\pi/2$ and is given by

$$\delta_{kmax} = a\frac{\pi}{p} \quad : \quad a \leq 1 \tag{7-51}$$

Parameter a is called the PWM constant.

Amplitude of n-th harmonic of the inverter output voltage waveform is:

$$V_{On} = \sum_{k=1}^{p} \frac{4V_S}{n\pi} \sin n\alpha_k \sin \frac{n\delta_k}{2} \tag{7-52}$$

The Sine-PWM pulses for the inverter with a = 0.8 and p = 7 are shown in Figure 7.22.

Example 7-5 Sine-PWM Inverter

```
dtheta = pi/500; VS = 169; p = 7;deltakm = 0.8*pi/p;
Tvv = [ ];theta = [ ]; vo = [ ]; dalpha = pi/p;
for k = 1:p
        alphak = (2*k-1)/(2*p) *pi;
          deltak = deltakm*sin(alphak);
        if(deltak > = pi/p-dtheta)
        deltak = pi/p; theta1 = [ ];theta3 = [ ];
     else
        theta1 = [alphak-dalpha/2:dtheta:
          alphak-deltak/2];
        theta3 = [alphak+deltak/2: dtheta:
          alphak+dalpha/2];
     end
     theta2 = [alphak-deltak/2:dtheta:
       alphak+deltak/2];
     theta = [theta theta1 theta2 theta3 ];
     vo1 = VS*[zeros(size(theta1)) ones(size(theta2))
       zeros(size(theta3))];
     vo = [vo vo1];
end
vo = [vo -vo]; theta = [theta theta+pi];
[f,voh,ph] = harmonic(vo/VS);
axis([0,100,0,1.2*max(voh)])
fun = voh(2)*(VS/sqrt(2))*sin(theta);
hold
plot(theta,vo,theta,fun,'—')
```

Amplitude of the n-th harmonic of the inverter output voltage waveform with Sine-PWM control is obtained from equation (7-50) and equation (7-51) and is given below:

$$v_{On} = \frac{4V_S}{n\pi} \sum_{k=1}^{p} \sin(n\,\alpha_k) \sin\left[\frac{n}{2}\delta_{max} \sin n\,\alpha_k\right] \tag{7-53}$$

The voltage transfer ratio T_{vv} is given by

$$T_{vv} = \frac{\tilde{V}_{O1}}{\overline{V}_S} = \frac{2\sqrt{2}}{\pi} \sum_{k=1}^{p} \sin \alpha_k \sin\left[\frac{\delta_{kmax}}{2} \sin \alpha_k\right] \tag{7-54}$$

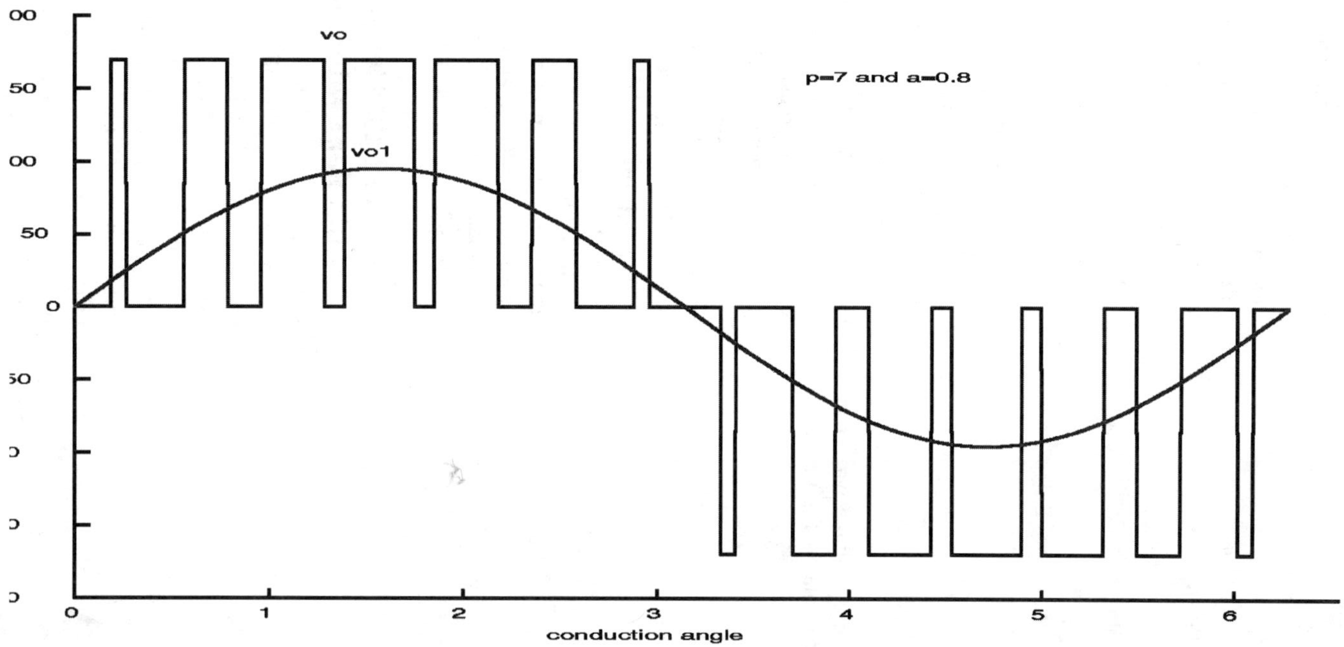

Figure 7.22
Sine-PWM pulses for the 1-phase full-inverter

For the large value of p, the δ_{kmax} is small and then the T_{VV} can be approximated as follows:

$$T_v \cong \frac{2\sqrt{2}}{\pi}\sum_{k=1}^{p} \sin\alpha_k \frac{\delta_{kmax}}{2}\sin\alpha_k$$

$$\cong \frac{\sqrt{2}}{\pi}\sum_{k=1}^{p}\delta_{kmax}\sin^2\alpha_k$$

$$\cong \frac{\sqrt{2}}{\pi}\delta_{kmax}\sum_{k=1}^{p}\sin^2\alpha_k$$

$$\cong \frac{1}{\sqrt{2}\pi}\delta_{kmax}\sum_{k=1}^{p}(1-\cos 2\alpha_k)$$

$$= \frac{p}{\sqrt{2}\pi}\delta_{kmax} = \frac{a}{\sqrt{2}} \qquad (7\text{-}55)$$

Summation of cosine of angles over a period equals zero.

The rms value of the output voltage waveform can also be similarly approximated for the large value of p:

$$\tilde{V}_O = \sqrt{V_S^2\frac{1}{\pi}\sum_{k=1}^{p}\delta_k}$$

$$= \sqrt{\frac{V_S^2}{\pi}\sum_{k=1}^{p}\delta_{kmax}\sin\alpha_k}$$

$$= V_S\sqrt{\frac{1}{\pi}\delta_{kmax}\sum_{k=1}^{p}\sin\alpha_k} \cong V_S\sqrt{\frac{1}{\pi}a\frac{\pi}{p}\sum_{k=1}^{p}\alpha_k}$$

$$\cong \sqrt{\frac{1}{p}a\sum_{k=1}^{p}(2k-1)\frac{\pi}{2p}}$$

$$= V_S\sqrt{\frac{\pi}{2p^2}a\,[p(p+1)-p]}$$

$$\cong V_S\sqrt{\frac{a\pi}{2}}$$

The output voltage THD is

$$THD = \sqrt{\frac{\tilde{V}_O^2-\tilde{V}_{O1}^2}{\tilde{V}_{O1}^2}} = \sqrt{\frac{V_S^2\dfrac{a\pi}{2} - V_S^2\dfrac{a^2}{2}}{V_S^2\dfrac{a^2}{2}}}$$

$$= \sqrt{\frac{\pi}{a}-1} \qquad (7\text{-}56)$$

Harmonic profile of Sine-PWM inverter is shown in Figure 7.23. The harmonics (side bands) are spaced at f_O interval and occur as band clusters. The band clusters are themselves spaced at $2pf_O$. This harmonic profile is far improved over the case with notching and stepping methods.

Theoretically, the harmonic numbers other than the fundamental component of the output voltage are given by

$$n = j\,(2\,p) \pm k \quad \textit{where} \quad j = 1, 2, 3, \ldots \quad \textit{and} \quad k = 1, 3, 5, \ldots \qquad (7\text{-}56a)$$

j is the cluster number and k is the side band number in the cluster.

The Sine-PWM with $a > 1$ is also possible and referred to as over-modulation. It amounts to wave-shaping of pulse widths by a non-sinusoidal waveform. In this type of control, the pulse at $\pi/2$ is wider than that in the case of Sine-PWM with $a \leq 1$. This will increase the amplitude of the fundamental component and hence will give a higher voltage transfer ratio T_{vv}. Lower order harmonic amplitudes, however, increase, hence THD increases. The output voltage waveform and the harmonic profile for $a = 1.2$ are shown in Figure 7.24(a) and (b).

More complex forms of Sine-PWM use a non-sinusoidal variation of pulse width aiming to increase the fundamental amplitude. The reader is recommended to refer to many such harmonic reduction schemes in the literature.

7.8.2 Sine-PWM inverter using bipolar switching

In bipolar switching, the turning off of switch pair S1-S3 in Figure 7.5 is followed by the turning on of the switch pair S2-S4, recalling that each switch is synthesized from a diode-transistor combination. As a result the output voltage switches between $+V_S$ and $-V_S$ levels. The output voltage waveform in bipolar switching is shown in Figure 7.25. The bipolar switching is also equivalent to notching out pulses of $2V_s$ amplitude with sinusoidally varying widths from a square-wave waveform.

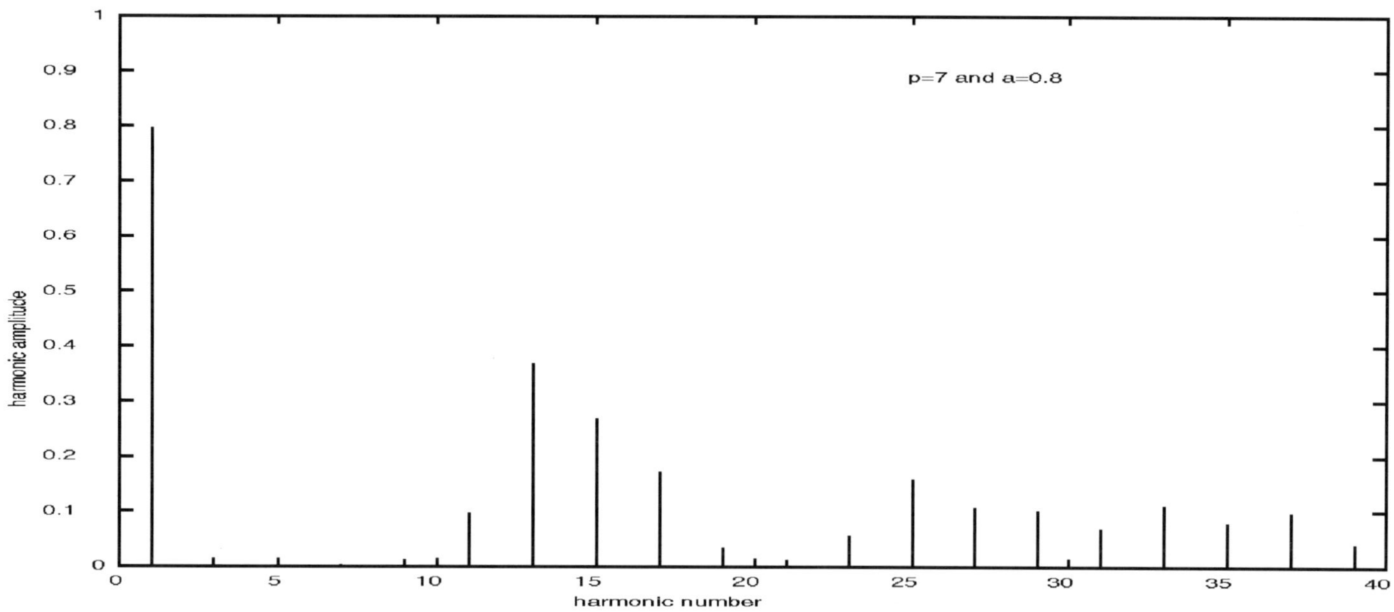

Figure 7.23
Harmonic profile of Sine-PWM inverter for $p = 7$ and $a = 0.8$

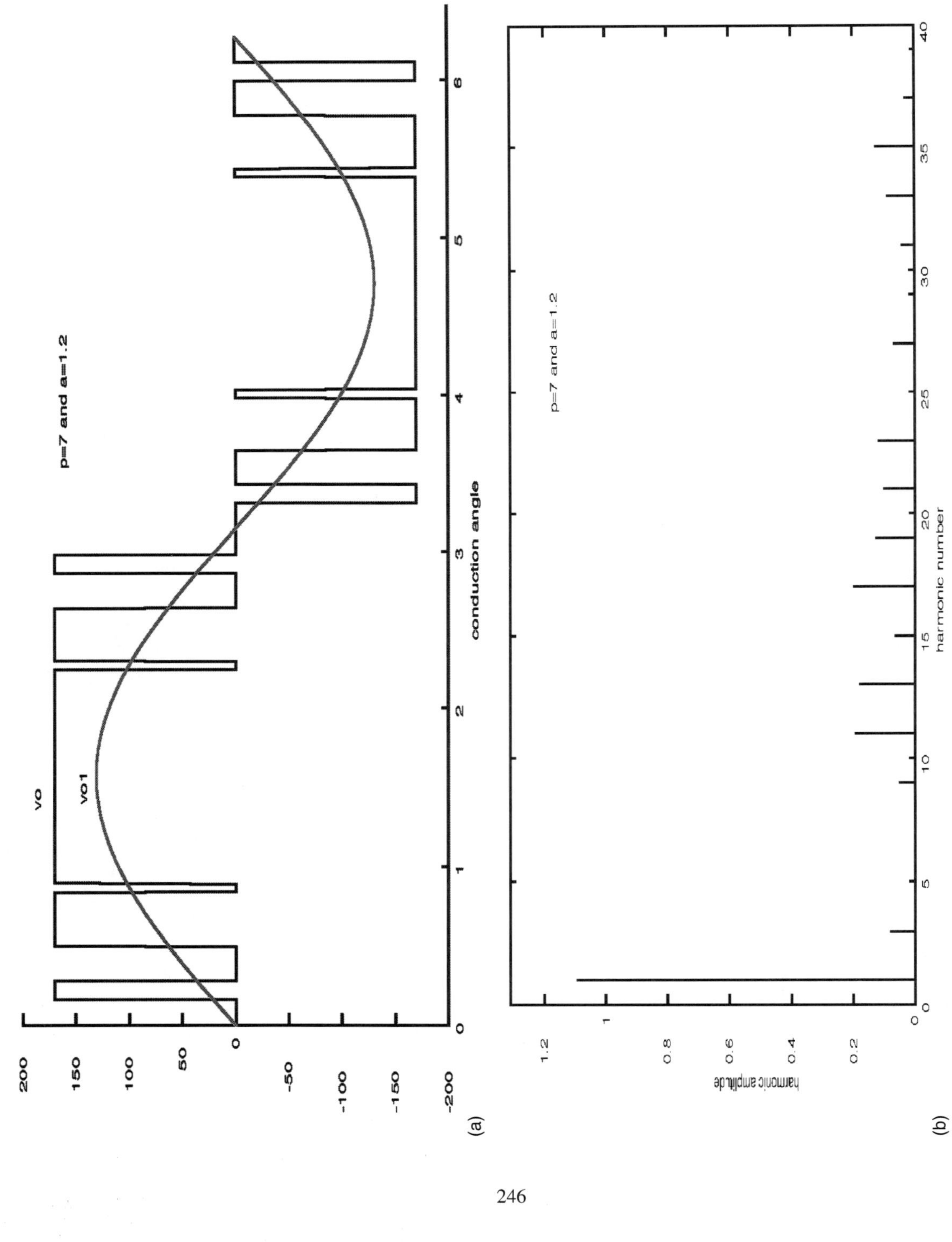

Figure 7.24
(a) Sine-PWM pulses for the 1-phase full-inverter p = 7 and a = 0.8 and (b) Harmonic profile of Sine-PWM inverter for p = 7 and a = 1.2

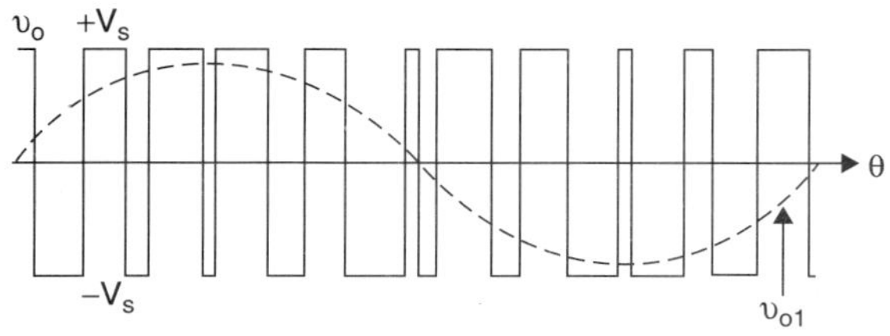

Figure 7.25
Sine-PWM with bipolar switching

The bipolar switching involves more switching than the unipolar switching, hence more switching losses, not to mention the addition of several blanking intervals to avoid shoot-through. The bipolar switching inverter is left as a problem at the end of the chapter.

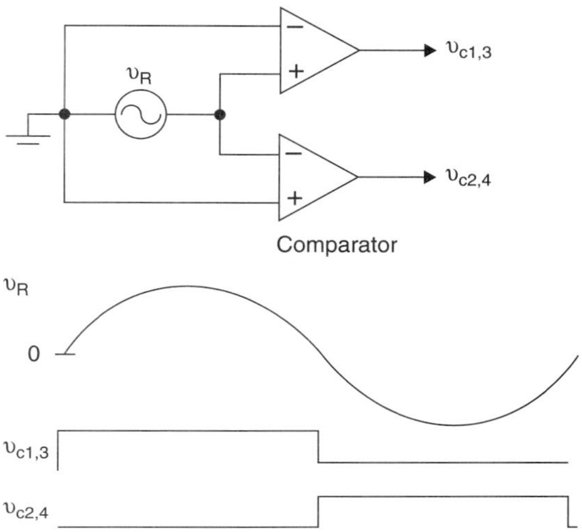

Figure 7.26
Control pulse generation in square-wave switching

7.9 CONTROL SIGNAL GENERATION

Before discussing more inverter topologies, we should pause and look at the generation of pulses that control the switches in the inverter. The control pulses could be generated using analog circuits or through software programming. The analog generation is suitable in simple inverter topologies such as the square-wave and single-pulse width mode control. The software generation is more suitable in Sine-PWM inverters. The software generation may use C++ or equivalent programming. The programming is beyond the scope of this book and will not be discussed.

The following discussion will focus on the generation of the control pulses in a one-phase full-bridge inverter using analog circuits.

(a) Pulses for Square-Wave Mode Control
Figure 7.26 shows the circuit for generating control pulses. The reference voltage v_R is a sine wave at a frequency that is the desired fundamental frequency of the inverter output voltage waveform. Outputs of the two comparators are applied to the gates of switch pairs S1-S3 and S2-S4 in the PWM inverter of Figure 7.5.

(b) Pulses for Single-PWM Control
Figure 7.27 shows the circuit for generating control pulses. The v_a is a triangular waveform of the same frequency as that of the reference voltage v_R. The control pulses for the switch pair S1-S3 are obtained as the output of the comparator, which compares the positive half of v_a with v_R. The control pulses for the switch pair S2-S4 are obtained from comparing v_R with the negative half of v_a waveform.

(c) Pulses for Sine-PWM Control
The instantaneous output voltage at the center of the k-th pulse is the local average of the switched waveform:

$$v_O(k) = d(k)V_S \quad : \quad d(k) = \frac{\delta_k}{\delta_{k\max}} \tag{7-57}$$

The function d(k) in the Sine-PWM inverter is a sinusoidal function of k.

Control pulses can also be generated in software. The hardware implementation has a fast response but the availability of modern high-speed Digital Signal Processing chips are fast bridging the gap.

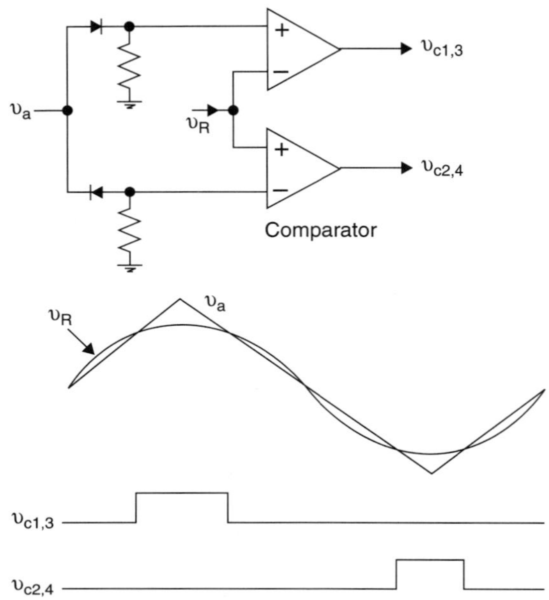

Figure 7.27
Control pulse generation in Single-PWM switching

7.10 3-PHASE FULL-BRIDGE INVERTER

The three-phase inverters are used for high power applications. Like in the single-phase inverters, the three-phase inverters are designed with square-wave control and single and multiple PWM control. In the square-wave control, the pulse width is fixed and the output voltage is controlled by the magnitude of the input voltage. In this type of control, the input dc is derived from an ac-dc converter. In the single and multiple PWM, the output voltage is varied by the pulse-width control and the harmonics are profiled by modulation of pulse width or switching frequency. Due to very improved harmonic profiling capability, the Sine-PWM technique is popular in modern inverters.

7.10.1 3-Phase full-bridge inverter with square-wave control

In this section the square-wave controlled three-phase inverter will be considered. Figure 7.28(a) shows a three-phase voltage source driven inverter. It is driving a Y-connected resistive load. Figure 7.28(b) shows the six circuit modes in which the inverter operates in a single switching period. The inverter output voltage waveforms are shown in Figure 7.28(c).

The switching control waveforms for the three legs of the three-phase bridge are shown in Figure 7.28(d).

The harmonic profile of the line-line voltage is

$$v_{l-1} = v_{ab} = \sum_{odd\ n} \frac{4\bar{V}_S}{n\pi}\sin\frac{n\pi}{3}\sin(n\theta) \quad (7\text{-}58)$$

$$v_{ph} = v_{aN} = \sum_{odd\ n} \frac{4\bar{V}_S/3}{n\pi}\sin n\theta + \sum_{odd\ n} \frac{4\bar{V}_S/3}{n\pi}\sin\frac{n\pi}{6}\sin n\theta_1 \quad (7\text{-}59)$$

The first term sin (np/3) becomes zero for n = 3, 9, 15, and so on, hence the triple harmonics do not exist. The fundamental component of the phase voltage leads the line-line voltage by $\pi/6$.

The voltage transfer ratios for line-line and phase voltages are

$$T_{vv,l-1} = \frac{\tilde{V}_{ab|n=1}}{V_S} = \frac{2\sqrt{2}}{\pi}\sin\frac{\pi}{3} = 0.7797 \quad (7\text{-}60)$$

$$T_{vv,ph} = \frac{\tilde{V}_{ph|n=1}}{V_S} = \frac{4}{3\sqrt{2}\pi}\left[\sin\frac{\pi}{3} + \sin\frac{\pi}{6}\right] = 0.4100 \quad (7\text{-}61)$$

The rms voltages are

$$\tilde{V}_{O,l} = \sqrt{\frac{1}{\pi}\int_0^{2\pi/3}\bar{V}_S^2\,d\theta} = 0.8165\bar{V}_S \quad (7\text{-}62)$$

$$\tilde{V}_{O,ph} = \sqrt{\frac{1}{\pi}\left[\int_0^{\pi/3}\left(\frac{\bar{V}_S}{3}\right)^2 d\theta + \int_{\pi/3}^{2\pi/3}\left(\frac{2\bar{V}_S}{3}\right)^2 d\theta + \int_{2\pi/3}^{\pi}\left(\frac{\bar{V}_S}{3}\right)^2 d\theta\right]} = 0.4714\bar{V}_S \quad (7\text{-}63)$$

THD of the output voltages are given by

$$THD_{l-1} = \sqrt{\frac{0.8165^2 - 0.7797^2}{0.7797^2}} = 0.3108$$

$$THD_{ph} = \sqrt{\frac{0.4714^2 - 0.4100^2}{0.4100^2}} = 0.5674 \quad (7\text{-}64)$$

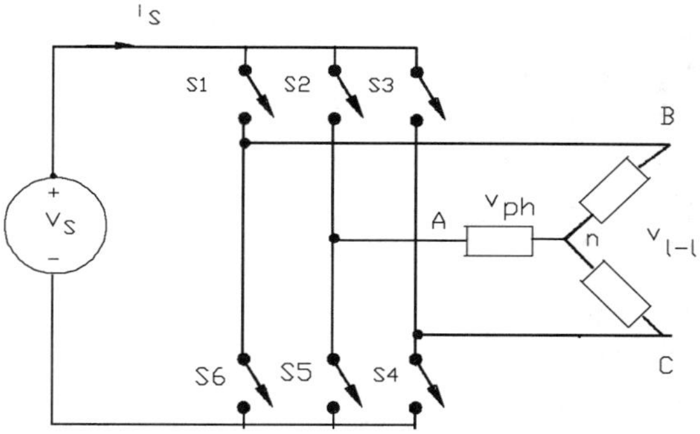

(a)

		S1	S2	S3	S4	S5	S6
Mode 1	$0 \leq \theta < \pi/3$	on	on	—	on	—	—
Mode 2	$\pi/3 \leq \theta < 2\pi/3$	on	—	—	on	on	—
Mode 3	$2\pi/3 \leq \theta < \pi$	on	—	on	—	on	—
Mode 4	$\pi \leq \theta < 4\pi/3$	—	—	on	—	on	on
Mode 5	$4\pi/3 \leq \theta < 5\pi/3$	—	on	on	—	—	on
Mode 6	$5\pi/3 \leq \theta < 2\pi$	—	on	—	on	—	on

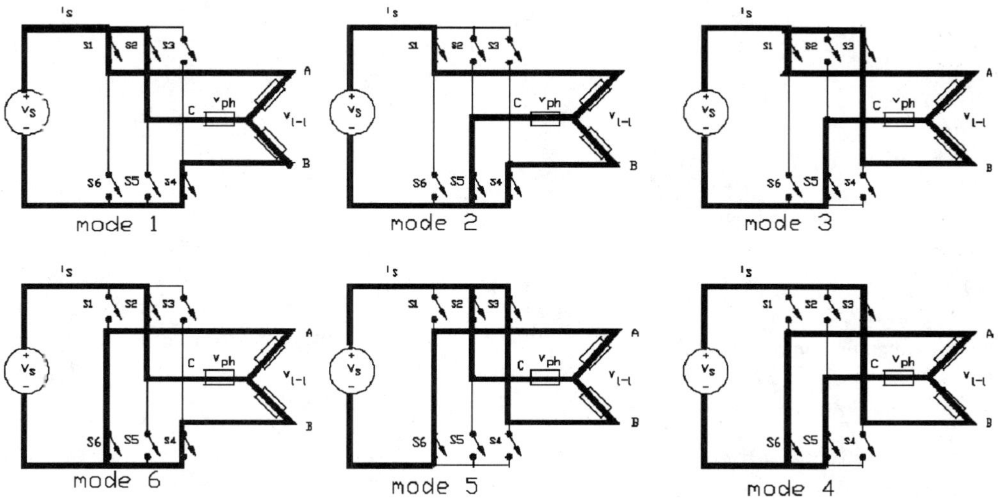

(b)

Figure 7.28
(a) 3-Phase square-wave switched Inverter and (b) circuit modes

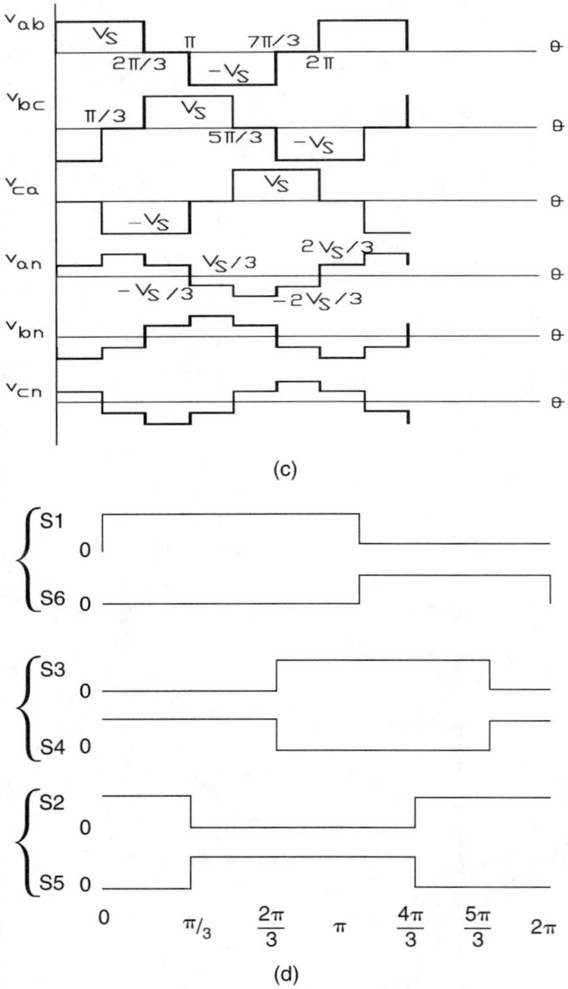

Figure 7.28
(c) waveforms and (d) switching control waveforms

Compare these THDs with 0.4834 for the single-phase full-bridge inverter with square-wave control (readers are suggested to verify). However, the absence of the third harmonic is the very significant advantage of three-phase inverters. The absence of the third harmonic reduces the peak-peak voltage ripple in the output voltage and pushes the ripple frequency higher. Hence smaller size inductors are sufficient for ripple reduction in the output waveform.

7.10.2 3-Phase bridge inverter with Sine-PWM control

The on time of switch pair in each of the three legs of the three-phase bridge inverter of Figure 7.28(a), is designed to vary sinusoidally. Consider the leg with the switch pair S1-S6. The on-time planning is as follows:

$$S1: \quad \delta_k = \delta_{k\max} \sin \alpha_k \qquad 0 \le \alpha_k < \pi$$
$$= 0 \qquad \textit{elsewhere}$$

$$S6: \quad \delta_k = \delta_{k\max} \sin (\alpha_k - \pi) = 0 \qquad \pi \le \alpha_k < 2\pi \quad \textit{elsewhere}$$

Similarly, the on-time plan for the other two legs are given below:

$$S3: \quad \delta_k = \delta_{k\max} \sin (\alpha_k - 2\pi/3) = 0 \qquad 2\pi/3 \le \alpha_k < 5\pi/3 \quad \textit{elsewhere}$$

$$S4: \quad \delta_k = \delta_{k\max} \sin (\alpha_k - 5\pi/3) = 0 \qquad 5\pi/3 \le \alpha_k < 2\pi + 2\pi/3 \quad \textit{elsewhere}$$

$$S2: \quad \delta_k = \delta_{k\max} \sin (\alpha_k - 4\pi/3) = 0 \qquad 4\pi/3 \le \alpha_k < 2\pi + \pi/3 \quad \textit{elsewhere}$$

$$S5: \quad \delta_k = \delta_{k\max} \sin (\alpha_k - \pi/3) = 0 \qquad \pi/3 \le \alpha_k < 4\pi/3 \quad \textit{elsewhere}$$

The line-to-line rms voltage can be shown to be

$$\tilde{V}_l = \frac{\sqrt{3}}{2\sqrt{2}} a V_S \qquad \textit{for} \quad a \le 1$$

For odd and triplen (multiples of three) values of 2p, the three-phase voltages are 120 degrees apart at all multiples of $2pf_S$ frequencies. Hence the line-line voltages would exhibit zero amplitudes at $2pf_S$, $4pf_S$, $6pf_s$, and so on. Thus, the most dominant harmonics are cancelled out.

7.10.3 DC source current in 3-phase full-bridge inverter

Consider a balanced Y-connected load to a balanced output of a three-phase inverter. Harmonics other than the fundamental components of the voltages and currents are ignored for the following discussion. All phase voltages are equal and the phase currents lag their respective phase voltages by equal

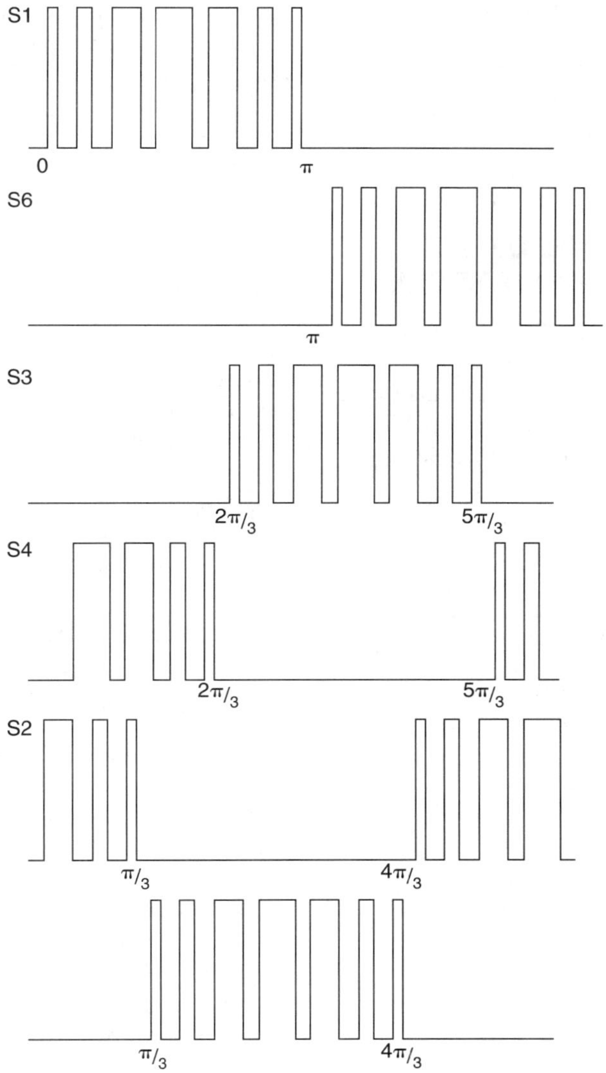

Figure 7.29
The on-time of switch pairs in the 3 legs of the 3-phase inverter

phase ϕ_1. The instantaneous power balance is established as follows:

$$\begin{aligned} V_S\, i_S(t) = {} & V_P \cos \omega_1 t\, I_O \cos(\omega_1 t + \phi_1) \\ & + V_P \cos(\omega_1 t - 2\pi/3) \\ & \quad I_O \cos(\omega_1 t - \pi/3 + \phi_1) \\ & + V_P \cos(\omega_1 t - 2\pi/3) \\ & \quad I_O \cos(\omega_1 t - 2\pi/3 + \phi_1) \end{aligned}$$

$$\begin{aligned} V_S\, i_S(t) = {} & \frac{3}{2} V_P I_O \cos(\phi_1) \\ & + V_P I_O [\cos(2\omega_1 t + \phi_1) \\ & + \cos(2\omega_1 t - \pi/3 + \phi_1) \\ & + \cos(2\omega_1 t - 2\pi/3 + \phi_1)] \end{aligned}$$

The sum of three terms involving the double frequency over a period is equal to zero, hence, the source current is a pure dc:

$$i_S(t) = \frac{3V_P}{2V_S} I_O \cos \phi_1$$

7.11 RECTIFIER MODE OF OPERATION OF INVERTERS

An inverter operates in the rectifier mode when the output voltage phasor V_{O1} and the output current phasor I_O have a phase difference greater than $\pi/2$. Figure 7.30 shows the schematic of a one-phase inverter driving an induction motor, modeled by a voltage sink E_O in series with an inductor L. In the motoring mode, the phasor V_{O1} is greater than the E_O phasor:

$$\bar{V}_{O1} = \bar{E}_O + j\omega L \bar{I}_O$$

In the motoring mode, the inverter output current I_O contains an in-phase component, $I_O \cos \phi_1$, with the inverter output voltage phasor V_{O1}. The inverter delivers a "positive" (forward) power to the induction motor. I_O lags V_{O1} by an angle less than $\pi/2$.

In the braking mode, the inverter output V_{O1} is forced to be smaller in magnitude than E_O. Consequently, the inverter output current I_O contains an out-of-phase component, $I_O \cos \phi_1$, with the voltage phasor V_{O1}. I_O differs in phase from V_{O1} by an angle that is more than $\pi/2$. The inverter delivers a "negative" (backward) power to the induction motor or equivalently saying that the motor delivers power back to the inverter.

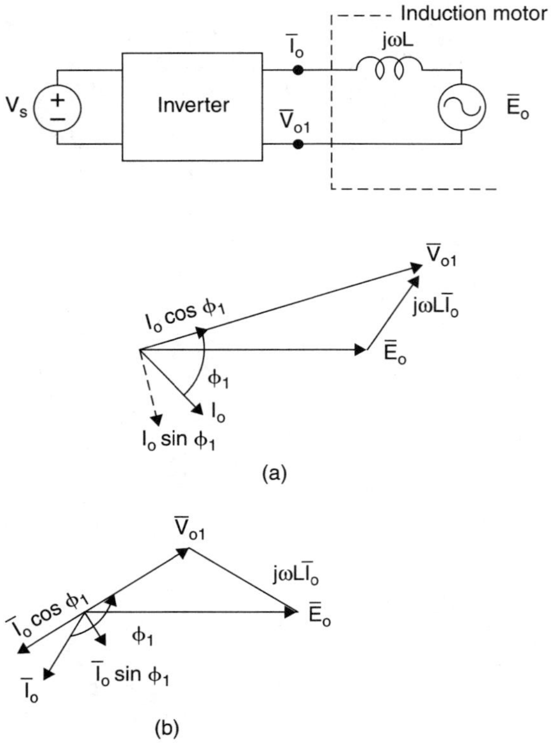

Figure 7.30
A 1-phase inverter driving an induction motor in (a) motoring mode and (b) in braking mode

7.12 DESIGN EXERCISE: 2 KW, 1-PHASE INVERTER FOR EMERGENCY APPLICATIONS

Design a one-phase 220 V, 50 Hz, 2 kW inverter from a 12-V dc battery. Allow an input voltage drooping of 10% in long time. THD should be less than 3% of the dc output voltage magnitude. The switching frequency is required to be 20 kHz or more, which is beyond the audible range of humans.

Use the voltage controlled Sine-PWM inverter topology with unipolar switching for reasons of minimum distortion and minimum filtering components. The converter will be designed with a MOSFET as the switch. Assume in the beginning the on-state voltage drops in the MOSFET, $V_{on} = V_F = 1$ V. We will refine our assumption in the second revision when we select the actual devices.

The number of pulses per half period of the output is obtained from

$$f_S = 2pf_O \quad \Rightarrow \quad p = \frac{20kHz}{2 \times 50} = 200$$

The input voltage variation is 12 V to 10.8 V dc. We would use the lowest value of V_S at the PWM factor a = 1.0 so that for nominal value of 12 V the PWM factor a is less than unity. The source voltage and the output voltages include the on-state voltages of the switching device, hence,

$$V_S \Rightarrow V_S - V_{ON} = 10.8 - 1 \times 2 = 8.8$$

The rms amplitude of the fundamental component of the output voltage is (equation (7-53) and equation (7-55))

$$\tilde{V}_{O1} \cong \frac{a}{\sqrt{2}} V_S \quad = \frac{1}{\sqrt{2}} 8.8 = 6.22$$

The transformer turn ratio is then 220/6.22 = 35.36. Select the turn ratio to be 36.

MOSFET Selection:

The output current referred to the primary side of the transformer is

$$\tilde{I}_O = \frac{2KW}{6.22} = 321.5\ A$$

The maximum voltage and maximum current through the switch are (equation (7-33a))

$$V_{sw\ max} = V_{Smax} = 12\ V$$
$$I_{sw\ max} = I_{Opeak} = 453.4\ A$$

Select the MOSFET with specifications better than 12 V and 453.4 A. The choice is a set of five MOSFETs in parallel, International Rectifier RFA100N05E, 100 A, 50 V, 10 mohm. Readers are encouraged to design the parallel connection on the lines of Example 4-4 in Chapter 4. The on-state voltage of the selected MOSFET at 100 A is 100×0.010 = 1 V. Hence we need not revise our initial estimate of V_{ON}.

7.13 PSPICE SIMULATION

Example PSPICE-7-1

Consider the PWM bridge inverter of Figure 7.8 with R-L load. Use MOSFET IRF150 with antiparallel connected 1N4148 diode for permitting reverse flow of current for the switches. A high-speed diode will be better. Use a single, centrally placed pulse of width = 8 μs for PWM control of the inverter. The inverter specifications are

DC source = 24 V, R = 10 ohms L = 5 mH
pulse width δ = 8 ms

Fundamental frequency of the inverter = 25 kHz

Perform a transient analysis for at least 100 switching periods and plot the steady-state waveforms for the following in the last two periods:

a. the current io through the load resistor RL
b. the voltage vo across the series combination of the load resistor RL and inductor L
c. the harmonic profile of the voltage waveform vo.

The MOSFETs S1-S4 are switched by four external pulse sources Vc1-Vc4 as shown in Figure 7.31(a). Vc1 controls the switching of S1 and S3.

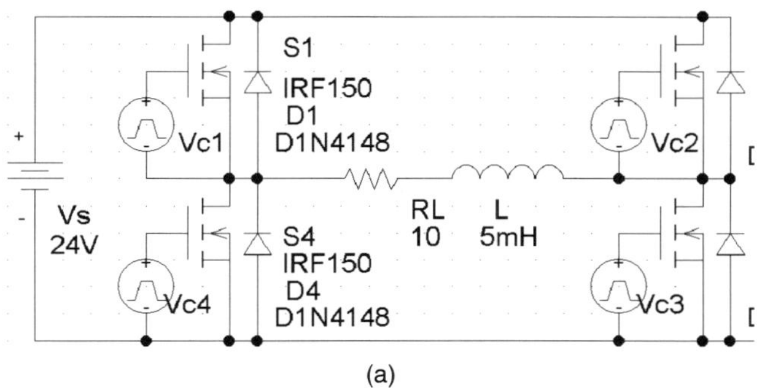

Figure 7.31
(a) Schematic

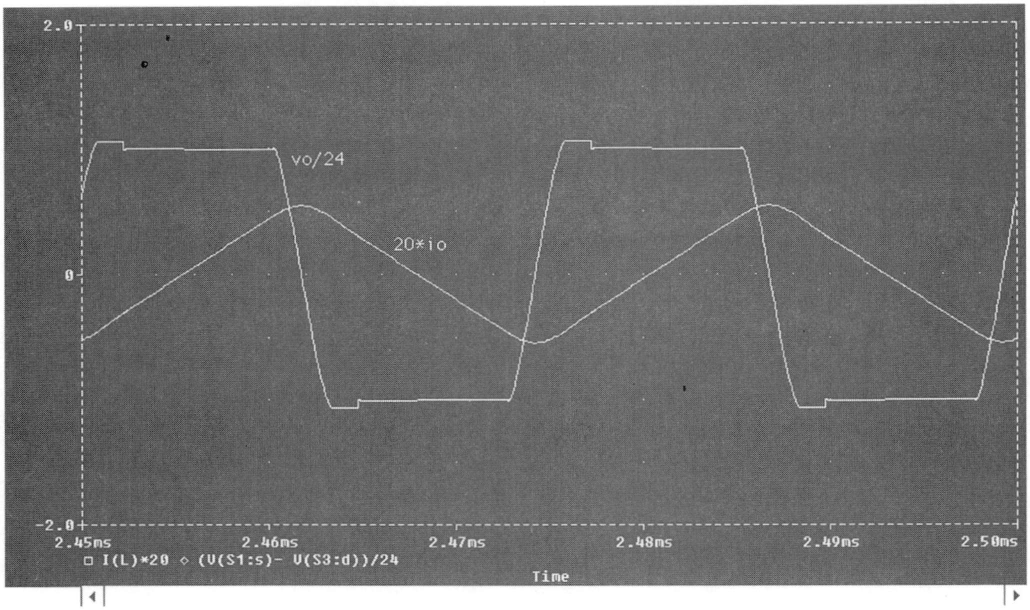

(b)

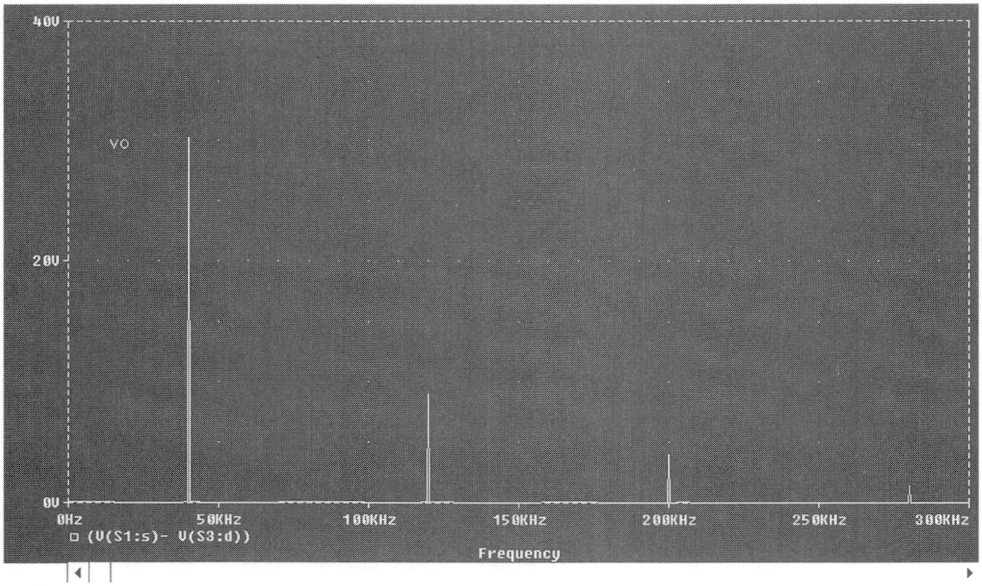

(c)

Figure 7.31
(b) waveforms and (c) the harmonic profile of the output voltage waveform

Vc2 controls the switching of S2 and S4. The pulse waveforms for the MOSFETs S1 and S2 are transformer coupled or optically coupled in the practical implementation of the circuit.

Set the pulse source Vc1 and Vc3 as follows:

Lower level V1	= 0 V
Pulse period	= 25 μs (40 kHz)
Delay time TD	= 2.25 μs
Fall time	= 0.1 ns

Upper level V2	= 10 V
Pulse width (on-time) δ	= 8 μs
Rise time	= 0.1 ns

Set the pulse source Vc2 and Vc4 for switching S2 and S4 with all settings of Vc1 and Vc3 except the delay time, TD = 14.75 μs.

Use the simulation step in the range: 1 ns to 100 ns. Organize your plots by displaying the ratios vo/24 and 20*io as the functions of time.

7.14 SUMMARY

1. The square-wave switching controls only the frequency of the output voltage. The source voltage controls the amplitude of the output voltage. Hence, usually a controlled ac-dc converter precedes such an inverter.
2. PWM control permits amplitude control. The ripple content is higher than in the square-wave switched inverter.
3. The Sine-PWM scheme permits the control of both the frequency and amplitude. The nonzero harmonic lie in the range of the switching frequency that can be made several times larger than the desired output frequency. Thus, ripple filtering is relatively easy.
4. The current-mode control permits the direct control of the output current. This scheme is used in servo drives.
5. The power flow in a dc-ac inverter can be reversed by operating it in the rectification mode. This method is utilized to save power during braking of inverter supplied induction motor drives.

References

1. Philip T. Krein, *Elements of Power Electronics,* Oxford University Press, New York, 1998.
2. Ned Mohan, Tore M. Undeland, and William P. Robbins, *Power Electronics: Converters, Applications and Design,* 2nd Ed., John Wiley & Sons, Inc., 1995.
3. Joseph Vithayathil, *Power Electronics: Principles and Applications,* McGraw-Hill Inc., New York, 1995.
4. Muhammad H. Rashid, *Power Electronics: Circuits, Devices and Applications,* 2nd Ed., Prentice-Hall Inc., New Jersey, 1993.
5. John G. Kassakian, Martin F. Sclecht, and G. Vergassian, *Principles of Power Electronics,* Addison-Wesley, Boston, 1991.
6. Kjeld Thorborg, *Power Electronics,* Prentice-Hall Inc., New Jersey, 1988.
7. Daniel M. Mitchell, *Switching Regulator Analysis,* McGraw-Hill Book Company, New York, 1988.
8. B. K. Bose, *Power Electronics and Ac Drives,* Prentice-Hall, Engelwood Cliffs, NJ, 1986.
9. D. M. Divan, T. A. Lipo, and T. G. Habetler, "PWM techniques for voltage source inverters," *IEEE Applied Power Electronics Conference,* 1990.
10. G. K. Dubey, S. R. Doradla, A. Joshi, and R. M. K. Sinha, *Thyristorized Power Controllers,* Wiley, New York, 1986.
11. M. Ehsani, "A tutorial on pulse width modulation techniques in inverters and choppers," *IEEE Applied Power Electronics Conference,* 1981.
12. D. C. Griffith, *Uninterruptible Power Supplies,* Marcel Dekker, New York, 1993.
13. P. Wood, *Switching Power Converters,* Van Nostrand Reinhold, New York, 1981.
14. P. Vas, *Vector Control of AC Machines,* Oxford, Clarendon Press, 1990.

Problems

7-1 Consider the inverter circuit of Figure P7.1. (111 hof) with V_S = 120 Vdc and L = 2 mH.

The switch S1 is turned on first and the initial load current is zero. The switches are turned on alternately by a square wave of frequency f_s = 1 kHz. Plot the steady-state waveforms for v_O, i_O, i_S, i_{SW1}, i_{D1}, v_{SW1}, i_{SW2}, i_{D2} and v_{SW2}.

7-2 Plot the harmonic spectrum of the output voltage and output current v_O and I_O in the above problem. Determine the THD in the output voltage.

7-3 Repeat problem 7-1 with the pulse waveform of duty ratio D = 0.25 at 1 kHz.

7-4 Consider the three-phase half-wave inverter circuit in Figure P7.4 with V_S = 150 Vdc and R = 10 ohms. (114 hof). Plot the voltage and current in all phases and all three line-line voltages, the source current i_S.

7-5 Repeat problem 7-4 if the resistor in each phase is replaced by a R-L load of 10 ohms-1 mH.

7-6 A half-bridge inverter is shown in Figure P7.6 with R = 10 ohms, L = 10 mH, and V_S=170 Vdc. Find the average dc current through the inductor. Plot the waveforms of v_O and i_O.

7-7 A PWM inverter is designed to operate at 75% duty cycle at the rated power. Find the duty cycle that would be required to obtain the output voltage to 10% more than the rated voltage.

7-8 A full-bridge inverter operates with D = 0.7, f_S = 60 Hz, V_S = 200 V, R = 10 ohms and L = 0.1 H.

a. Plot the waveforms for the voltage v_O and currents i_O and i_S.

b. Plot the harmonic profile of the current I_O.

c. Determine the voltage transfer ratio and dc current being supplied by the dc source.

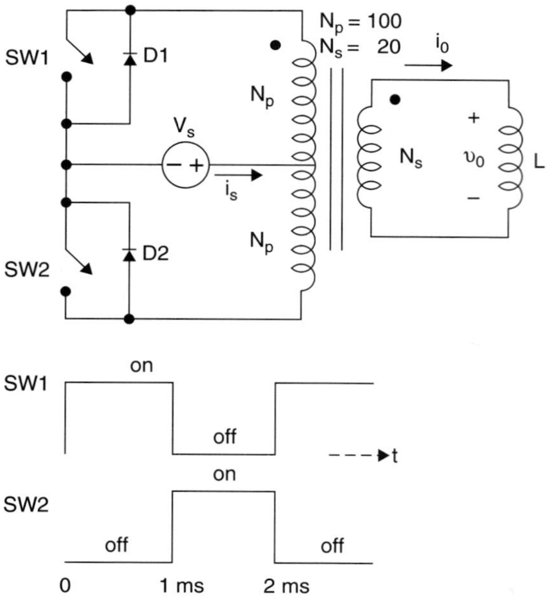

Figure P7.1

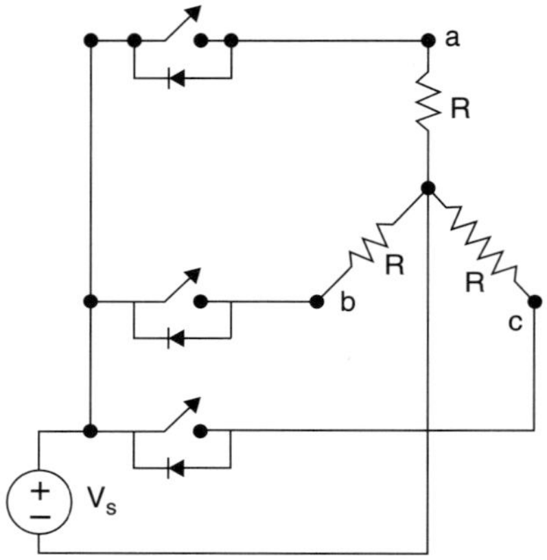

Figure P7.4

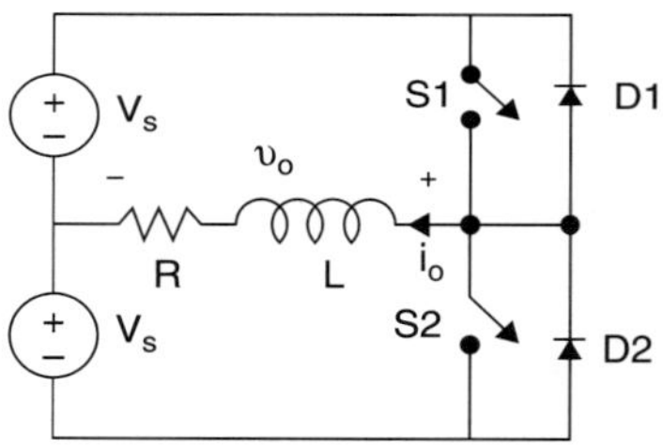

Figure P7.6

7-9 The three-phase square-wave inverter is supplied by a split dc voltage source with the midpoint connected to ground. The three-phase load is balanced Y-connected. $V_S/2 = 100$ Vdc and $f_O = 60$ Hz. Plot the voltage between the neutral of the load and the midpoint of the source.

7-10 A three-phase square-wave inverter is supplied by a 150 Vdc voltage source. The three-phase load is balanced Y-connected. $V_S/2 = 100$ Vdc, $f_O = 60$ Hz, R = 2 ohms and L = 10 mH.

a. Plot the current and voltage in any one phase.
b. Plot the harmonic profile of the current I_O.
c. Determine the voltage transfer ratio and dc current being supplied by the dc source.

7-11 The dc-link voltage of a three-phase Sine-PWM inverter is obtained from a three-phase diode bridge rectifier from a three-phase 208 V ac bus. The load is a Y connected motor. Determine

a. the rms phase voltage;
b. the rms voltage of the fundamental component of the motor phase voltage;
c. the THD of the phase voltage.

7-12 The dc-link voltage of a three-phase voltage source inverter is provided by a one-phase diode bridge rectifier from a 120 V 1-phase supply. The input side capacitor is 2000 μF. The ac motor is regeneratively braked, causing a reverse dc current of 20 A. Determine

a. The rate at which the dc-link voltage will rise due to braking;
b. The value of braking resistor needed on the dc side.

7-13 The number of pulses per half cycle of a Sine-PWM is 9.

a. Determine the rms value of the output voltage for a = 1.0
b. Plot the harmonic profile of the output voltage.
c. Calculate the THD in the output voltage.

7-14 Repeat problem 7-13 for a = 1.2 and 0.7.

7-15 A square-wave switched dc-ac inverter is powered by a 350-V dc source and provides the power to a load of 10 ohms in series with 10 mH. The inverter is switched at 400 Hz. Draw the output voltage and current waveforms, the instantaneous output power waveform and the harmonic profile of the output voltage up to five nonzero amplitude harmonics. Also determine the power delivered to the load.

7-16 A 150 V, 400 Hz bus is connected to the output of a one-phase VSI through a 5 mH inductor. The inverter is powered by a 300 V dc and delivers 1 kW power to the ac bus. Find the phase angle ϕ_1 in the inverter operation.

7-17 A PWM controlled inverter is loaded by 2 ohm in series with 2 mH inductor, powered by 170 Vdc and switched at 60 Hz. Draw the output voltage and current waveforms and the harmonic profile up to 30 harmonics.

7-18 Design a one-phase 220 V, 50 Hz, 5 kW inverter from a 12-V dc battery. Allow an input voltage drooping of 10% in long time. THD should be less than 3% of the dc output voltage magnitude. The switching frequency is required to be 20 kHz or more, which is beyond audible range of humans.

CHAPTER 8
AC to DC: No Control

OBJECTIVES

1. Half-wave and full-wave topologies of ac-to-dc rectifiers
2. Rectifier operation with R, L and C loads
3. Rectifier operation with dc and ac sinks in the load circuit
4. Rectifier operation when the load current is continuous or discontinuous in time
5. Rectifier operation with source side inductance
6. Three-phase rectifiers

8.1 INTRODUCTION

AC to DC converters without control are popularly known as rectifiers. The most commonly used ac source is 50 or 60 Hz voltage source which is available from the electric utility supply, also called the line source. The uncontrolled rectifiers are designed using diodes. The designs are inexpensive and popular in the industrial applications. In some of these rectifiers, the ac voltage from the electric utility is directly rectified without the use of an expensive and bulky transformer. In some applications, the dc voltage from the rectifier is connected to a dc bus for distribution to several different circuit systems, subsystems and other converters as loads. In other applications, the rectifiers supply power to inductive-resistive (motors) and capacitive-resistive (power supplies) loads.

The rectifiers are required to supply ripple-free dc voltage or dc current to the load. In practice, the peak-to-peak output ripple is designed to be as small as possible and the ripple frequency as large as possible. The rectifiers usually draw highly non-sinusoidal current from the electric utility supply, giving rise to poor power factors and thus poor efficiency. Improving power factor is a very important design objective. The power factor improvement techniques are discussed in detail later in chapter 12. Another design concern is the reduction of high frequency distortions in the line current, which are caused by switching loads or switch-mode converters as loads. The basic rectifier concepts, some typical rectifier topologies and distortion issues will be discussed in the chapter.

Applications

(a) *DC power supplies for computers and electronic equipments*

These applications generally involve low power levels (less than 500 W). The input is derived from a single-phase ac utility. The uncontrolled rectifier converts the ac into an inexpensive dc voltage. The uncontrolled dc voltage is then processed by a switch-mode dc-to-dc converter to provide multiple output voltages with minimum ripple content. These power supplies draw highly non-sinusoidal current, hence, the line quality issues are very critical.

(b) *Battery charging systems*

With more and more portable instrumentation, such as in wireless mobile communications and computers, the demand for inexpensive battery charging systems has exploded.

(c) *High voltage dc (HVDC) transmission converters*

Power transmission over high voltage dc lines is becoming more popular than on the ac lines. The ac voltages from power generating stations are rectified and connected to a HVDC bus line. This way there is no problem of differences in frequencies (such as 50 Hz or 60 Hz) or frequency transients. The British and French power networks are bussed together on a cross channel HVDC link. HVDC transmission is found to be more economical than the ac counterpart.

8.2 AC-DC CONVERTER TOPOLOGIES

The uncontrolled ac-dc converter is shown in Figure 8.1.

The line voltage is a three-phase ac supply at 50 or 60 Hz. Single-phase utilization uses one of the phase lines and one of the neutral lines. The single-phase applications may involve a step down or step up transformer. The two-phase ac supply may be obtained from a transformer with a center tap on the secondary winding. The three-phase applications involve a three-wire (Delta) connection or a four-wire (Y or Star) connection. Some applications involve an eight-phase ac supply by combining a Y-connected and the Delta connected transformer secondary windings.

The ac-dc converter topologies may be broadly classified in two categories:

Half-Wave:

1. 1-phase, 1-pulse ac-dc converter
2. 2-phase, 2-pulse ac-dc converter
3. 3-phase, 3-pulse ac-dc converter

Full-Wave:

4. 1-phase, 2-pulse ac-dc converter
5. 3-phase-Y, 6-pulse ac-dc converter
6. 3-phase-Δ, 6-pulse ac-dc converter

The number of pulses refers to the number of sinusoidal peaks in a period in the output voltage waveform. Topologies using generic switches with self-control are given in Figure 8.2.

Switches are ideal, unidirectional, unipolar and self-controlled, that is, no external control is needed to turn on or off. The switch turns on when the voltage across the switch tends to become more than zero (forward biased mode in diodes). Once turned on, the switch voltage is forced to zero. The switch turns off when either the voltage across it tends to become less than zero or when the current through the switch reduces to zero and tends to reverse polarity (reverse biased mode in diodes). The switching device is usually a diode, but it could also be a three-terminal device, wired as a two-terminal device. See Figure 8.3. Ideally the switch should not

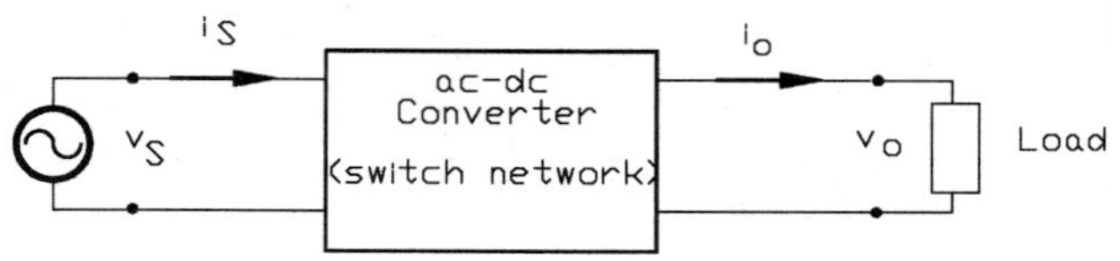

Figure 8.1
AC to DC converter system

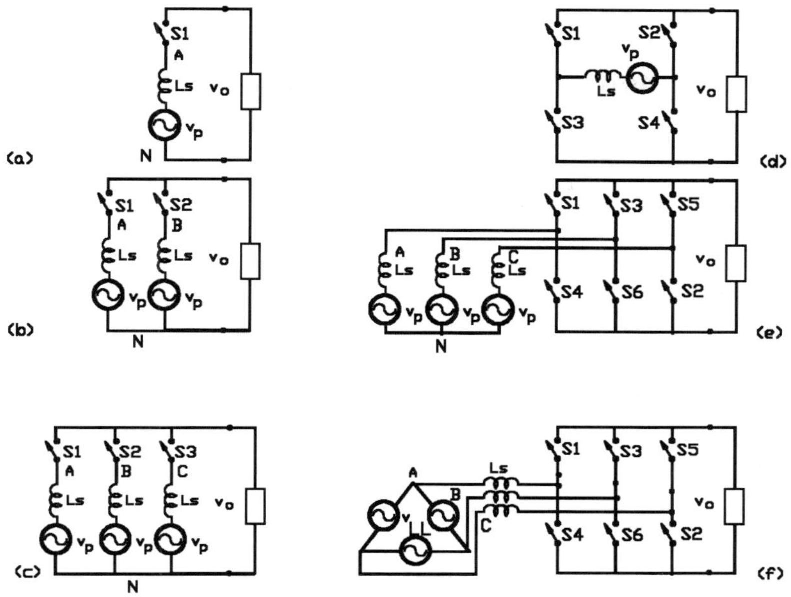

Figure 8.2
Rectifier topologies (a) 1-Phase half-wave (1-pulse), (b) 2-Phase half-wave (2-pulse), (c) 3-Phase half-wave (3-pulse), (d) 1-Phase full-wave (2-pulse) full-bridge, (e) 3-Phase -Y full-wave (8-pulse) full-bridge, and (f) 3-Phase-Δ full-wave (8-pulse) full-bridge

have any voltage across it in the on-state, but all devices in practice do have some voltage. The Schottky diodes and the diodes synthesized from the three-terminal devices have small on-state voltages. Ideally, the switching action should be independent of the frequency of the source. At line frequency, however, most of the diodes are close to ideal diodes in this regard.

In the half-wave topologies, the output voltage waveform contains one pulse per phase in a period. The switch in each phase is on in the positive half period and off during the negative half period of the input source voltage waveform of that phase. Thus, each phase supplies power during only half of the period.

In the full-wave topologies, the output voltage waveform contains two pulses per phase in a period. There are two switches for each phase. One of the switches is on in the positive half period and off during the negative half period of the input source voltage waveform of that phase. The other switch is in the opposite (complementary) state. Thus, each phase supplies power throughout the full period.

Half-wave topology has a lower count of semiconductor switches but require higher component stresses. The full-wave converter is capable of handling high power with minimum component stresses.

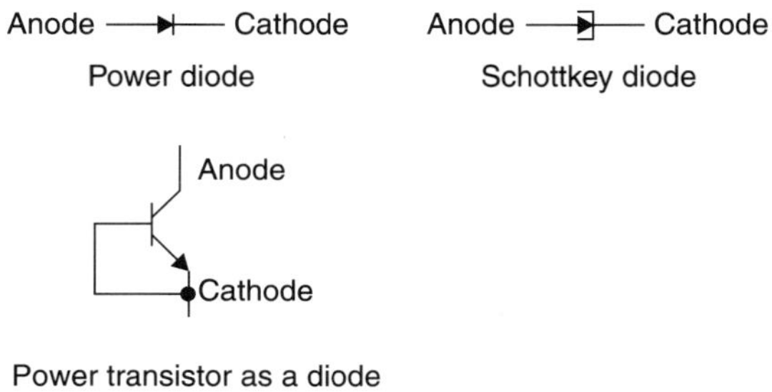

Figure 8.3
Power electronic switches in ac-dc converters with no control

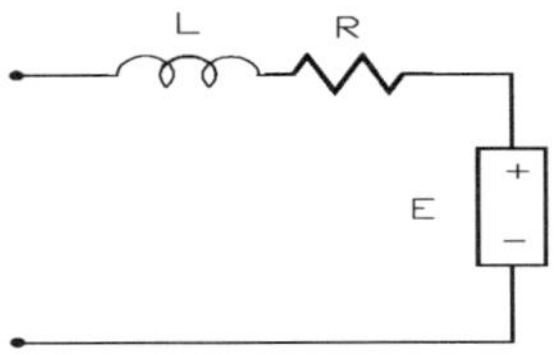

Figure 8.4
A general electrical load

A general electrical load is shown in Figure 8.4. Special cases are listed as follows:

1. **Resistive Load** (L=0, E=0)
2. **Inductive-Resistive** (medium L, low R and E=0): magnetic lift and relays
3. **Inductive-Voltage Sink** (medium L, low R and E): DC motors, HVDC bus and battery charging circuit
4. **Current Sink** (high L, low R and E): DC motors, heavy magnetic pick ups and relays
5. **Capacitive-Resistive** (L=0, R and E replaced by capacitor)
6. **Voltage Sink** (L=0, low R and high C): DC power supplies

8.2.1 Assumptions for analysis

1. A one-phase input voltage in the rest of this chapter is a sine wave of frequency f_s $(\omega_s/2\pi)$ and peak amplitude V_S as given below:

$$v_s = V_S \sin \theta \qquad : \theta \tag{8-1}$$

2. Switches including the diodes are assumed to be ideal and unidirectional, that is zero on-state losses, zero off-state losses, zero switching losses and unrestricted voltage or current carrying capacities. Switching devices in a practical design example must be selected with these desired characteristics in mind.
3. Inductors, capacitors and transformers are assumed to be ideal.
4. A large value inductor in series with a resistance may be assumed to be a constant current sink when

$$\frac{L}{R} \ll \frac{1}{f} \tag{8-2}$$

where f is the frequency of the output voltage source.

5. A large value capacitor in parallel with a resistance may be assumed to be a constant voltage sink when

$$\frac{1}{RC} \ll \frac{1}{f} \tag{8-3}$$

8.2.2 Performance characteristics

Converter performance consists of two parts: the transient and the steady-state part. The output voltage waveforms in both cases have high harmonic content, called ripple. In the steady state, all waveforms are stabilized and periodic, therefore easy to analyze. We will restrict our discussion to the steady-state performance in the rest of the chapter.

Forward Transfer Characteristics
The forward transfer characteristics describe the effect of output variables on the output variables. There are two performance parameters, which are used in the design of ac-dc converter systems:

Voltage Transfer Ratio

$$T_{vv} = \frac{\bar{V}_O}{\tilde{V}_S} \tag{8-4}$$

Voltage to Current Transfer Ratio

$$T_{vi} = \frac{\bar{I}_O}{\tilde{V}_S} \tag{8-5}$$

Reflective Characteristics
The reflective characteristics give the effect of output conditions on the source variable.

Current Reflection Ratio

$$R_{ii} = \frac{\tilde{I}_S}{\bar{I}_O} \tag{8-6}$$

It is desirable to have a small change in the output variable for a large variation in the output characteristics.

Harmonic Profile

The voltage and current waveforms at both the input and output ports are in general non-sinusoidal. The following measures are used for the undesirable Fourier components of a port variable:

Peak-to-peak ripple in the output voltage and the output current are denoted as $\mathbf{v_{or}}$ and $\mathbf{i_{or}}$ respectively.

Ripple frequency is the frequency of the voltage or current ripple waveform. It is a simple multiple of the output source frequency. A higher multiple is a design goal. A large ripple frequency can be filtered with relative ease and with small value and small size filter components.

Output current waveform in an ac-dc converter is non-sinusoidal. The Fourier expansion of output current is given as follows:

$$i_O = \bar{I}_O + \sum_{n=1,2,\ldots}^{\infty} I_{on} \sin(n\,\theta + \phi_n) \qquad (8\text{-}7)$$

where ϕ_n is the peak amplitude of the n-th harmonic component and the phase angle of the n-th harmonic with respect to the output sinusoidal voltage.

The Fourier expansion of input current is given as follows:

$$i_S = \sum_{n=1,2,\ldots}^{\infty} I_{Sn} \sin(n\,\theta + \phi_n) \qquad (8\text{-}7a)$$

where ϕ_n is the peak amplitude of the n-th harmonic component and the phase angle of the n-th harmonic with respect to the input sinusoidal voltage.

The phase angle of the fundamental component with respect to the output voltage waveform, ϕ_1, and the displacement factor DF, the cosine of ϕ_1, and the input power factor PF are three important performance measures.

$$DF = cos\,\phi_1 \qquad PF = \frac{\tilde{I}_{S1}}{\tilde{I}_S} \cos\phi_1 \qquad (8\text{-}8)$$

A low value of power factor indicates that the converter is using substantial reactive power. A unity power factor is desired.

Lowest undesired harmonic frequency is an important design criterion.

Harmonic Content of the source current is defined as the square root of the difference between the squares of the total rms source current and the rms fundamental current:

$$HC = \sqrt{\tilde{I}_S^2 - \tilde{I}_{S1}{}^2} \qquad (8\text{-}9)$$

Total Harmonic Distortion (THD) of the source current is defined as the ratio of the harmonic content to the rms fundamental current is calculated as following:

$$THD = \frac{HC}{\tilde{I}_{S1}} = \sqrt{\left(\frac{\tilde{I}_{S1}}{\tilde{I}_S}\right)^2 - 1} \qquad (8\text{-}9a)$$

The voltage and currents in the converter elements, especially, the switches may consists of high transients (other than the surges) which may be considerably higher than the normal ratings. These transients dictate the device selection and hence the cost.

Component stresses are measured as the ratio of the peak voltage and current values to the respective rms values on the elements.

8.3 1-PHASE HALF-WAVE RECTIFIER

A Half-Wave rectifier with different types of loads will be considered to illustrate basic fundamentals of rectification. Assume the source inductance $L_s=0$ for simplicity. Also assume that the switch is ideal, that is, the switch voltage is zero in on-state and zero current in off-state. The source voltage is assumed to be a sine wave of frequency ω and peak amplitude V_S as given in equation (8-1).

8.3.1 Resistive load

See Figure 8.5 for the schematic. The diode conducts in the positive half period only. The output current is in phase with the output voltage.

The diode voltage is

$$\begin{aligned} v_o &= V_S \sin\theta;\ 0 \qquad (8\text{-}10)\\ &= 0; \end{aligned}$$

The diode current is

$$i_d = \frac{v_o}{R} \tag{8-11}$$

The waveforms can be generated using MATLAB.

Example 8-1

Consider the ac-dc converter of Figure 8.5 with R = 2 ohms and 120 Vac, 60 Hz source. Draw the waveforms of the output voltage and the current delivered by the source.

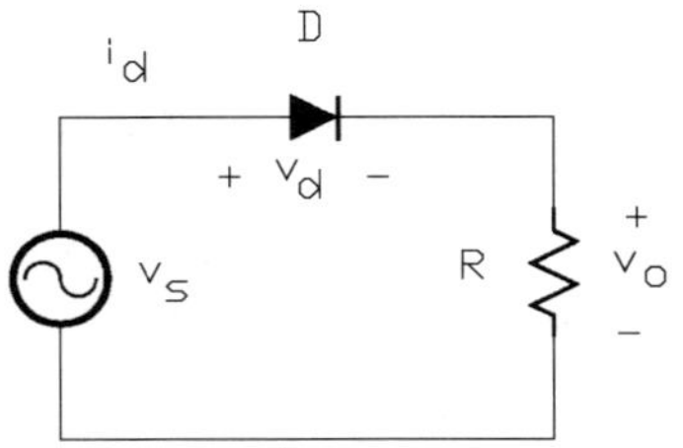

Figure 8.5
Half-wave rectifier with resistive load

MATLAB program for an example is given below:

```
VS=120*sqrt(2); R=2; % Peak source voltage and load
   resistance
dth=pi/512; % step increment in the conduction angle
th=0:dth:2*pi-dth; % conduction angles over a period
vs=VS *sin(th); % source voltage
vo=vs.*(vs>=0);% rectified output voltage: negative
   elements are made zero
id=vo/R; vd=vs-vo; % diode current and diode voltage
plot(th,vo,'*', th,id,th,vd)
```

8.3.2 Inductive load

We will first consider an ideal inductor load (with no resistance). See Figure 8.7 for the schematic. The diode current in the steady-state is obtained from

$$L\frac{di_d}{dt} = V_S \sin \omega t$$

$$\omega L\frac{di_d}{d\theta} = V_S \sin \theta, \quad \theta = \omega t \tag{8-12}$$

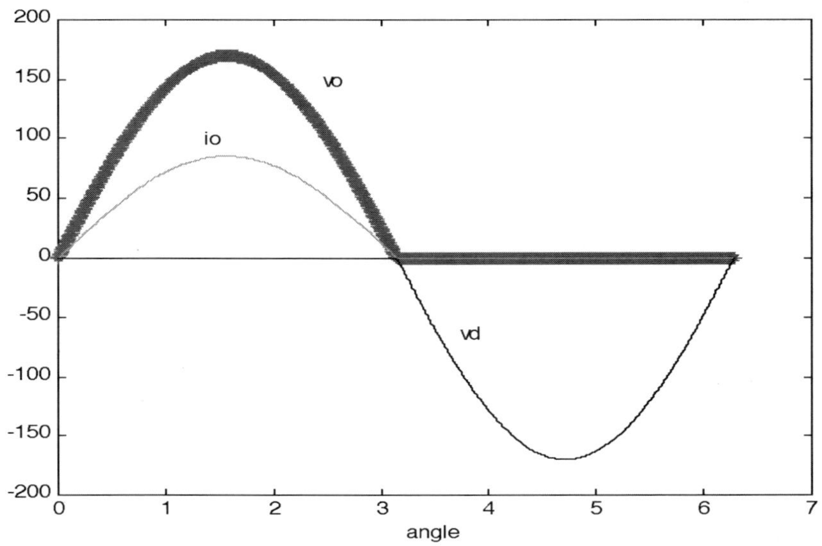

Figure 8.6
Waveforms

$$i_d = I_d(0) + \int_0^\theta \frac{V_S}{\omega L}\, d\theta \qquad (8\text{-}13)$$

The initial current $I_d(0)$ is zero. The source voltage becomes negative for angles greater than π; however, the diode must keep conducting to maintain the inductor current. The diode voltage must continue to be zero. As a consequence, the negative source voltage appears across the inductor. A negative voltage on the inductor causes the inductor current to decrease. Thus the diode conducts over the full period. The inductor current is maximum at the angle π. The output voltage is the replica of the source voltage and the average dc output voltage is zero.

The output current lags the output voltage by angle $\pi/2$. The peak current occurs at the zero output voltage. The output current is always positive and has a dc component. The peak current, hence, the peak-peak output current ripple is $V_S/\omega L$.

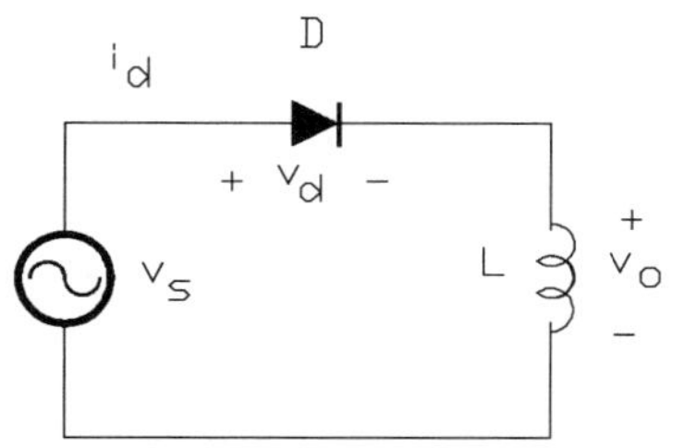

Figure 8.7
Half-wave rectifier with inductive load

Example 8-2

Replace the resistor in Example 8-1 by an inductor that has a reactance of 2 ohms at 60 Hz.

Add following lines to the program in the previous example.

```
omegaL=2; % 2πf times L
vo=vs; %
vd=vs−vo;
id=integ(vs/omegaL,th);
plot(th,vo,'*',th,id,th,vd)
```

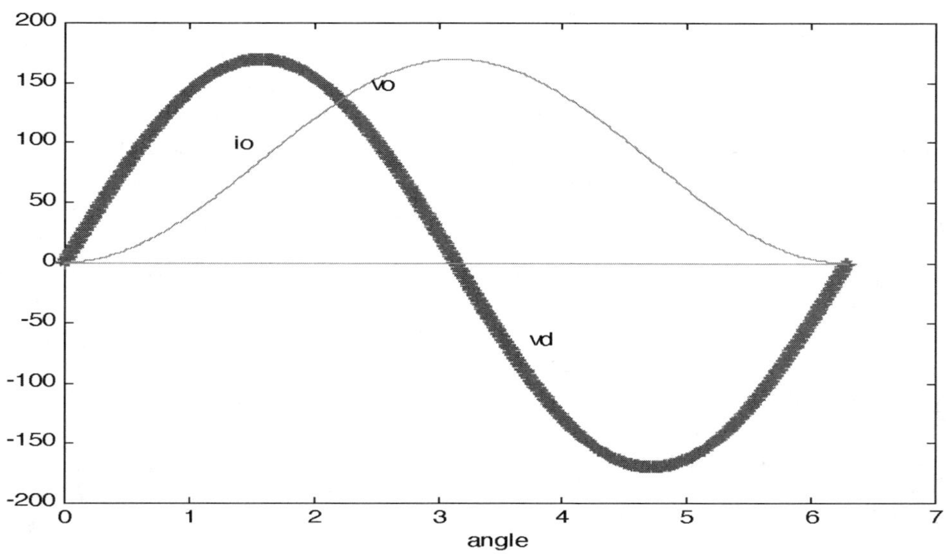

Figure 8.8
Waveforms

8.3.3 R-L load

Now consider a practical inductive load, inductance in series with a resistance. The resistance usually is the resistance of the inductor coil. See Figure 8.9 for the schematic. In the steady state, the diode conducts over more than the half period (until angle β). The output current is obtained from the solution of the differential equation

$$L\frac{di_d}{dt} + R\,i_d = V_S \sin \omega t$$

or

$$\omega L\frac{di_d}{d\theta} + R\,i_d = V_S \sin\theta, \quad 0 \le \theta < \beta \quad (8\text{-}14)$$

The diode current is, we assert

$$i_d = \frac{V_S}{\sqrt{R^2 + (\omega L)^2}}\left[\sin(\theta - \alpha) + e^{-R\theta/\omega L}\sin\alpha\right] + I_d(0)\, e^{-R\theta/\omega L}, \; 0 \le \theta < \beta$$

$$i_d = 0, \qquad \beta \le \theta < 2\pi \quad (8\text{-}15)$$

where

$$\alpha = \tan^{-1}\frac{\omega L}{R}$$

Output current lags the output voltage by an angle that is less than $\pi/2$. The peak current occurs at the instant where the difference of the source voltage and the voltage drop on the resistance is zero. The current continues to flow even after the source voltage has become zero and turned negative.

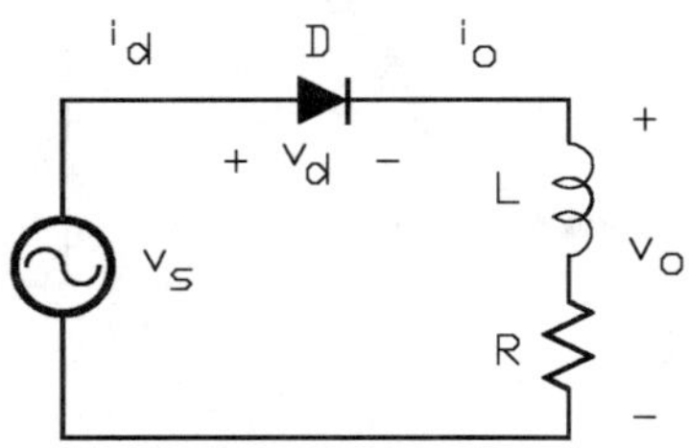

Figure 8.9
Half-Wave rectifier with R-L load

Example 8-3

Repeat Example 8-1 with both R = 2 ohms and the inductive reactance of 2 ohms.

Add following lines to the MATLAB program

```
alpha=atan(omegaL/R);
id=(VS/sqrt(R^2+omegaL^2))*(exp(−R*th/omegaL)*sin
  (alpha) +sin(th−alpha) );
id=id .*(id>=0);
vo=vs .*(id>=0);
vd=vs−vo;
plot(th,vo,'*',th,id,th,vd)
```

A large value of inductance reduces the peak-to-peak ripple current. For a very large inductance, it is fair to assume the output current to be dc with negligible ripple. Such a load is called the current sink. Furthermore, since the inductance cannot sustain an average dc voltage, the resistance R sustains the average output voltage of the converter circuit.

8.3.4 R-L load with freewheeling diode

Add a freewheeling diode D2 in the circuit of Figure 8.9 to allow the load current to be continuous and to prevent the output voltage from going negative. The modified circuit is shown in Figure 8.11. In steady state, there are two circuit modes based on the state of the two diodes. In the first mode m1, diode D1 conducts and D2 is reverse biased. The inductor and load resistors get power from the voltage source. The inductor current increases from the initial value $I_o(0)$ and peaks at angle π. Beyond that, mode m2 begins in which diode D1 is reverse biased and D2 conducts to permit the inductor current to freewheel. The inductor current will decrease due to loss in the load resistance. Assume that the load resistance is not too small to bring the inductor current to zero before the angle 2π. After 2π, mode m1 repeats and the inductor current increases again.

mode m1: The inductor current (or i_o) in mode m1 is determined from equation (8-14)

$$i_o = \frac{V_S}{\sqrt{R^2 + (\omega L)^2}}\left[\sin(\theta - \alpha) + e^{-R\theta/\omega L}\sin\alpha\right] + I_o(0)\, e^{-R\theta/\omega L}, \; 0 \le \theta < \pi \quad (8\text{-}16)$$

The inductor current at angle π is

$$I_o(\pi) = \frac{V_S}{\sqrt{R^2 + (\omega L)^2}} \sin \alpha[1 + e^{-R\pi/\omega L}] + I_o(0)\, e^{-R\pi/\omega L} \qquad (8\text{-}17)$$

mode m2: The inductor current in mode m2 is determined from

$$\omega L \frac{di_L}{d\theta} + R\, i_L = 0 \qquad (8\text{-}18)$$

$$i_o = I_o(\pi)\, e^{-\frac{R(\theta-\pi)}{\omega L}} \sin \alpha, \quad \pi \le \theta < 2\pi \qquad (8\text{-}19)$$

The inductor current at angle 2π equals $I_o(0)$, hence,

$$I_o(0) = \frac{V_S}{\sqrt{R^2 + (\omega L)^2}} \sin \alpha \frac{e^{-R\pi/\omega L}}{1 - e^{-R\pi/\omega L}} \qquad (8\text{-}20)$$

Thus the inductor (output) current has a dc component and ripple. The dc component is

$$\bar{I}_o = \frac{1}{2\pi} \int_0^{2\pi} i_o\, d\theta$$

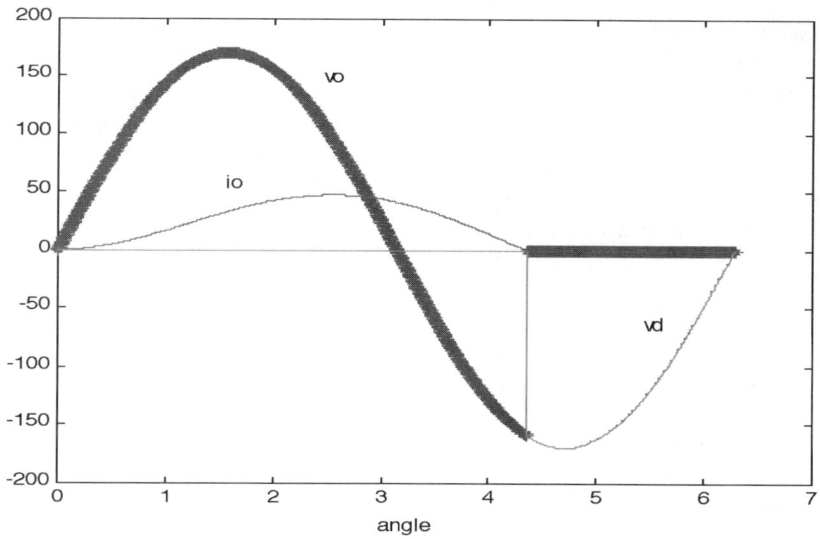

Figure 8.10
Waveforms

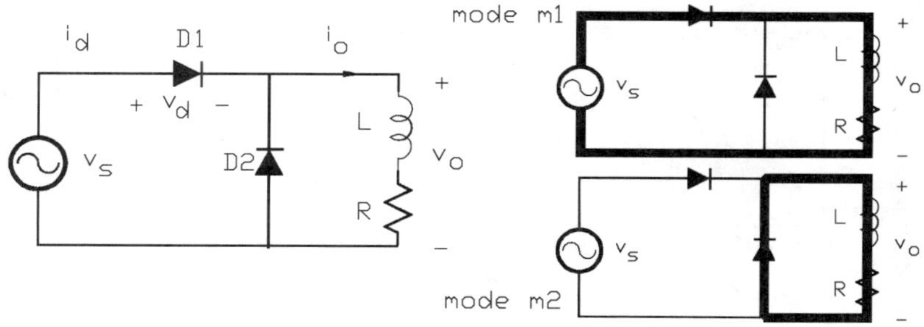

Figure 8.11
Half-wave rectifier with R-L load and freewheeling diode

and

$$peak - peak\ I_{or} = I_o(\pi) - I_o(0) = \frac{V_S\,\omega L}{R^2 + (\omega L)^2}$$

$$= \frac{V_S}{\omega L}; \qquad \omega L \ll R \qquad (8\text{-}21)$$

The ripple frequency is same as the source frequency. The current ripple is negligible for large value of L, the output current is constant dc as we know it in a current sink.

We would use a more methodical procedure for writing MATLAB for this case. Generate separate angle arrays for each circuit mode and write circuit equations mode-wise. Follow the steps carefully as we are going to use this mode-wise program extensively for all further MATLAB programming.

Example 8-4

Consider the converter of Figure 8.11 with R = 0.6, the inductive reactance of 2 ohms at 60 Hz and the ac source of 120 Vrms, 60 Hz. Draw the waveforms of the output voltage and the output current.

```
VS=120*sqrt(2); omegaL=2;R=.6; dth=pi/512;
  alpha=atan(omegaL/R);
Io0=(VS/sqrt(R^2+omegaL^2))*sin(alpha)*exp
  (-R*pi/omegaL)/(1-exp(-R*pi/omegaL));
% mode m1
th1=0:dth:pi-dth;
vs1=VS*sin(th1); vs2=VS *sin(th2); vs=[vs1 vs2];
V=VS/sqrt(R^2+omegaL^2);
io1=V*sin(th1-alpha)+(Io0+V*sin(alpha))*exp
  (-R*th1/omegaL); % output current in mode m1
vo1=io1*R;
Iopi=io1(size(th1,2)); % peak current in mode 1
is1=io1; % source current in mode m1
% mode m2
th2=pi:dth:2*pi-dth;
io2=Iopi*exp(-R*(th2-pi)/omegaL); vo2=io2*R;
is2=zeros(size(th2));
% one complete period
th=[th1 th2];
is=[is1 is2]; vo=[vo1 vo2]; io=[io1 io2];
```

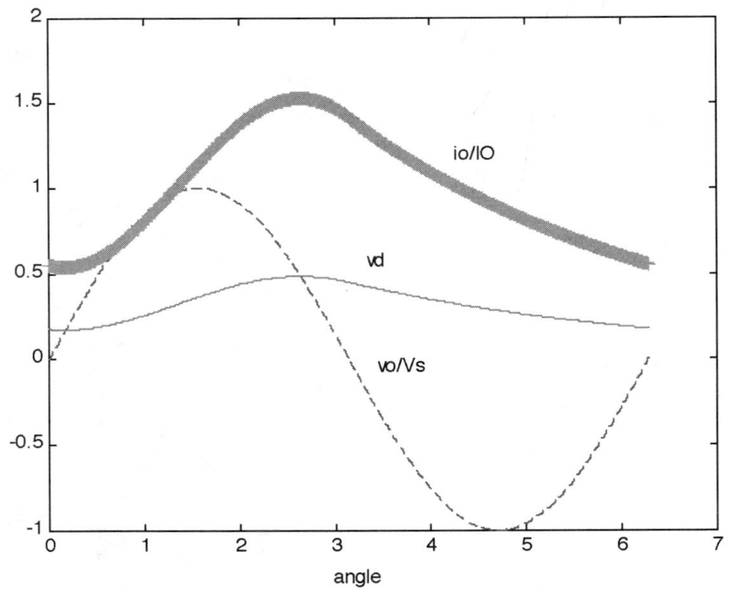

Figure 8.12
Waveforms

```
IO=mean(io); % dc average output
   current
plot(th,vs/VS,'--',th,vo/VS,th,io/IO,'+')
```

8.3.5 Capacitive load

First consider an ideal capacitor load (with no leakage resistance). See Figure 8.13 for the schematic and the waveforms. In the steady-state, the capacitor is charged to the peak input voltage and the diode does not conduct over the entire period. The peak voltage on the diode is twice that of the peak input voltage.

8.3.6 R-C load

Next consider a capacitance in parallel with a resistance. See Figure 8.14 for the schematic, two circuit modes in steady-state operation and the waveforms. In mode m1, the diode conducts for a short duration before the positive peak (θ_0 to $\pi/2$). The diode is reverse-biased and the capacitor is discharged over the rest of the period in circuit mode m2.

mode m1

$$v_o = V_S \sin\theta;\ \theta_0 \le \theta < \pi/2 \qquad (8\text{-}22)$$

$$i_d = C\frac{dv_o}{dt} + \frac{v_o}{R} \qquad (8\text{-}23)$$

The output current is, we assert

$$i_d = \frac{V_S}{\sqrt{R^2 + (1/\omega C)^2}} \sin(\theta - \theta_0 - \alpha);$$
$$\alpha = \tan^{-1}(1/\omega RC) \qquad (8\text{-}24)$$

mode m2

$$C\frac{dv_0}{dt} - \frac{v_o}{R} = 0: \quad \pi/2 \le \theta < 2\pi + \theta_0 \quad (8\text{-}25)$$

The output voltage is

$$v_o = V_S\left[1 - e^{-\frac{(\theta - \pi/2)}{\omega RC}}\right] \qquad (8\text{-}26)$$

The output current lasts until the input peak voltage is reached. Readers are encouraged to determine the exact solution using MATLAB.

The exact solution is complicated. We will approximate the voltage ripple waveform by a saw-tooth waveform with a period 2π.

The peak-to-peak voltage ripple is obtained from

$$C\frac{\Delta V_o}{T} = \frac{\bar{V}_o}{R} = \bar{I}_o$$

$$\Delta V_o = \frac{T}{C}\bar{I}_o = \frac{\bar{I}_o}{fC} \qquad (8\text{-}27)$$

The ripple frequency is same as the source frequency. The voltage ripple is negligible for the large value of C, and the output voltage is constant dc as we know it in a voltage sink.

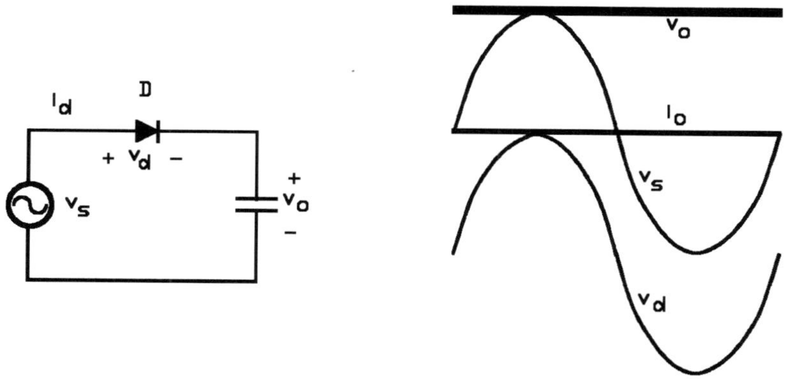

Figure 8.13
Half-wave rectifier with capacitor as the load

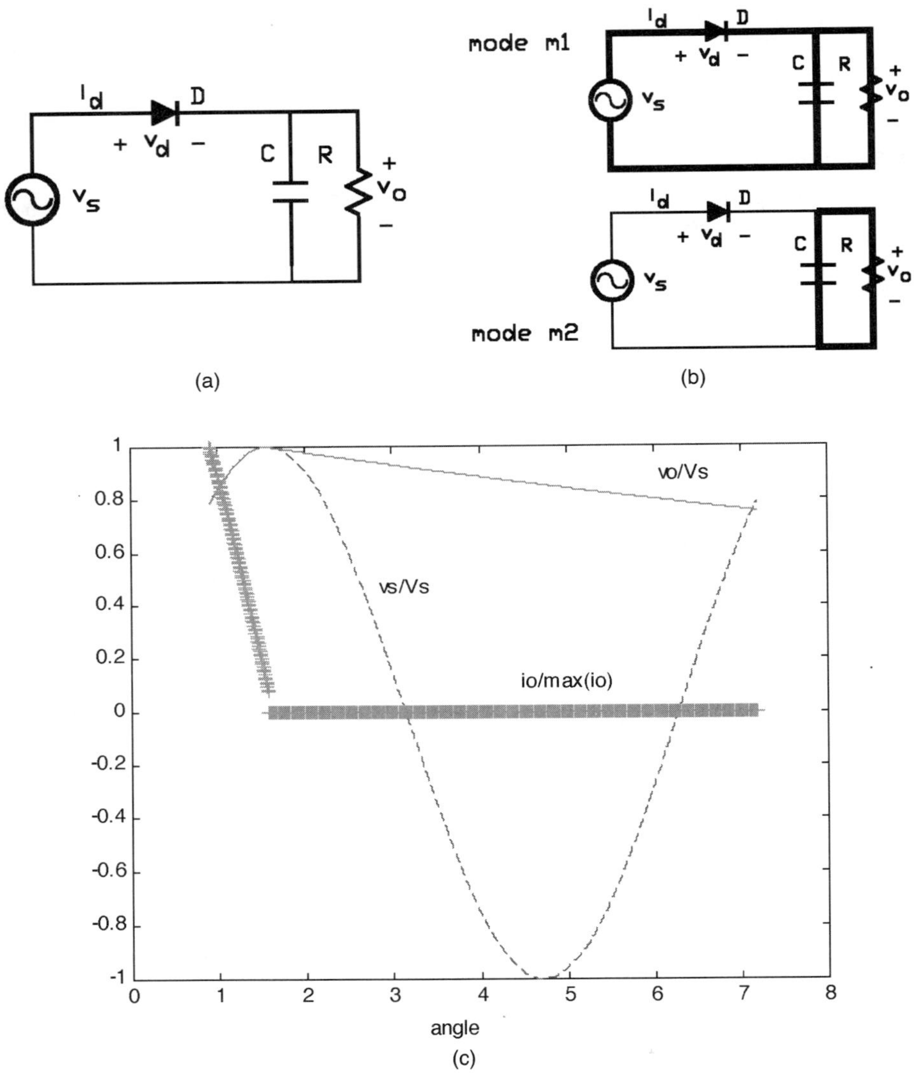

Figure 8.14
(a) Half-wave rectifier with R-C load (b) Circuit modes in steady-state operation (c) Waveforms

8.3.7 Voltage sink load

A voltage sink is connected to a rectifier through an inductance, which acts as a buffer to absorb the instantaneous difference in the voltage after the diode and the voltage sink. Refer to Figure 8.15 for the schematic. Assume that the diode conducts at angle α at which the source voltage exceeds the sink voltage V_o. This assumption is true only in certain conditions, which we will not worry about for this discussion. This assumption means that the initial output current is zero. Once the diode conducts, it continues to conduct even after the input voltage drops below $\mathbf{V_o}$.

$$\alpha = \sin^{-1}(V_o/V_S) \qquad (8\text{-}28)$$

$$\omega L \frac{di_d}{d\,\theta} = v_L = V_S \sin\theta - V_o \qquad (8\text{-}29)$$

The diode current begins with a zero slope and peaks at the angle at which the input voltage begins to drop below the $\mathbf{v_o}$. The diode current becomes zero at an angle β, which is less than 2π+α. The angle β is obtained by equating the average inductor voltage to zero.

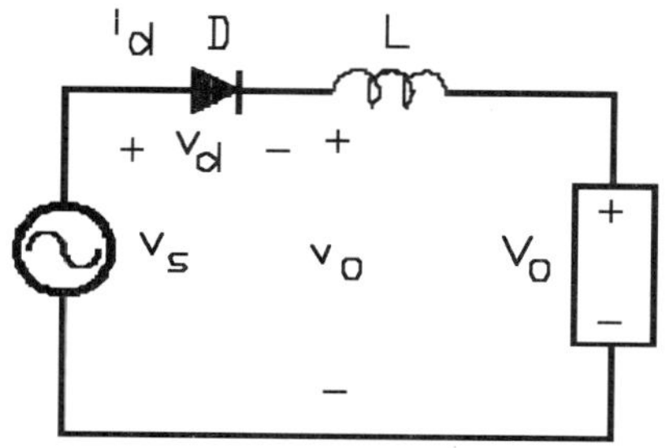

Figure 8.15
Half-wave rectifier with voltage sink load

Example 8-5

Consider the converter of Figure 8.15 with R = 0.6, the inductive reactance of 2 ohms at 60 Hz and the ac source of 120 Vrms, 60 Hz. The voltage sink is 90 Vdc. Draw the waveforms of the output voltage and the output current.

```
VS=120*sqrt(2); VO=90; omegaL=2;R=.6; dth=pi/512;
   alpha=asin(VO/VS);
th1=alpha:dth:2*pi-dth;
vs1=VS*sin(th1); id1=integ(((vs1-VO)/omegaL),th1);
th2=2*pi:dth:2*pi+alpha-dth;
vs2=VS *sin(th2); id2=zeros(size(th2));
th=[th1 th2];•=[vs1 vs2]; id=[id1.*(id1>=0) id2];
vo=[vs1.*(id1>0)+VO.*(id1<=0), VO*ones(size(th2))];
ID=mean(id); % Average output current
plot (th,vs/VS,'—',th,vo/VS, th, id/ID, '+')
```

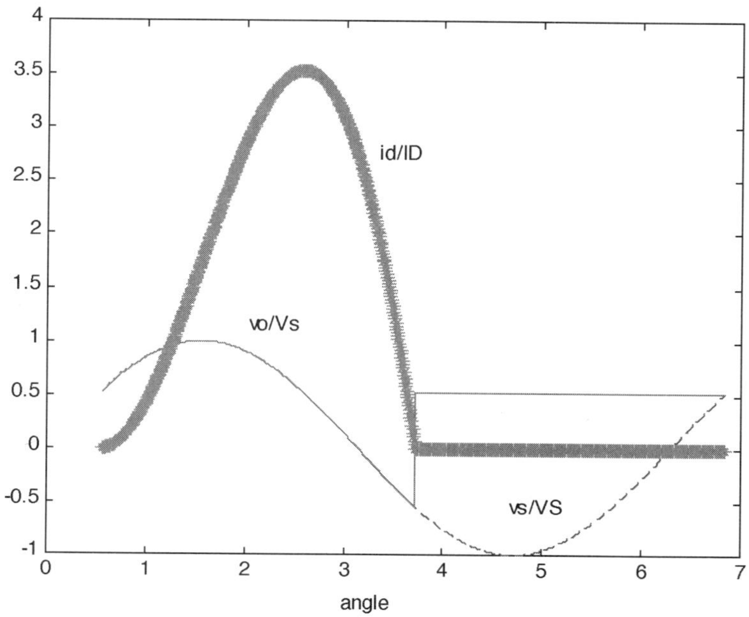

Figure 8.16
Waveforms

8.3.8 Voltage sink load with freewheeling diode

In steady state, there are two circuit modes based on the state of the two diodes. The first mode m1 begins at α, where the diode D1 conducts and D2 is reverse biased. The inductor and the sink get power from the voltage source.

The inductor current increases from the initial value $I_o(0)$ and peaks at angle π-α. Beyond that, the voltage across the inductor reverses polarity and the inductor current begins to decrease. At the angle π, the diode D2 reverses the diode D1. The inductor current now freewheels through D2. At this instant mode m1 ends and mode m2 begins.

In mode m2, the inductor current will decrease due to V_o. The inductor current drops to a positive value $I_o(0)$ at the angle 2π+α. After 2π+α, mode m1 repeats and inductor current increases again.

mode m1: The inductor current (or i_o) in mode m1 is, as determined from equation (8-25),

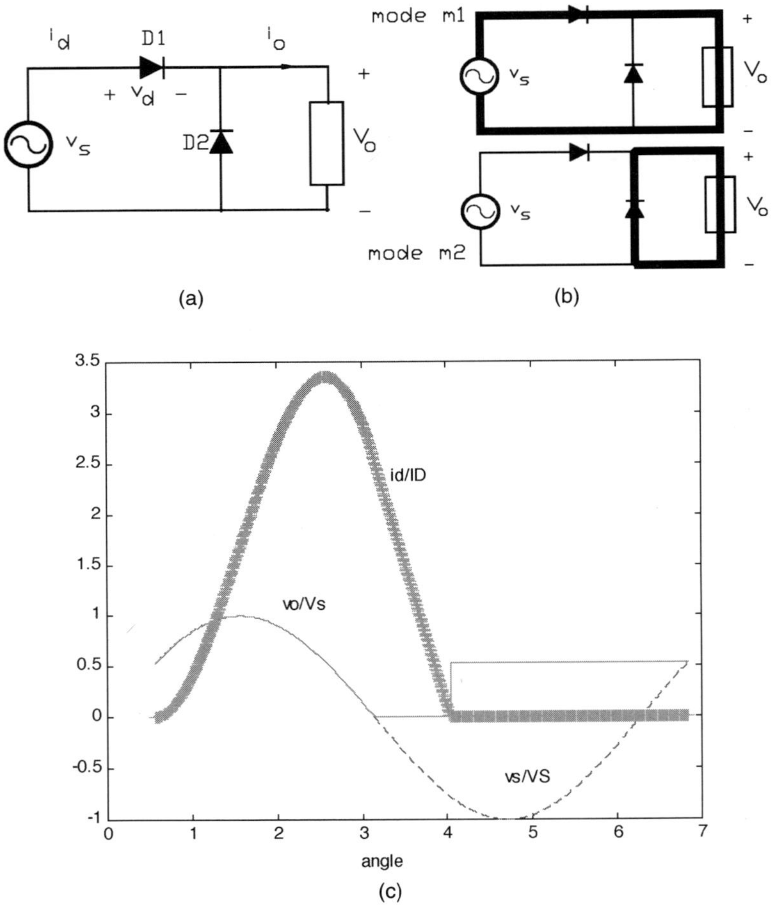

Figure 8.17
(a) Half-wave rectifier with voltage sink load and freewheeling diode (b) circuit modes in steady-state operation and (c) waveforms

$$i_o = I_o(\alpha) + \int_\alpha^\theta (V_S \sin\theta - V_o)\, d\theta$$

$$i_o = I_o(\alpha) + \frac{V_S}{\omega L}(\cos\alpha - \cos\theta) - \frac{V_o}{\omega L}(\theta - \alpha) \qquad (8\text{-}30)$$

The inductor current at angle π is

$$I_c(\pi) = I_o(\alpha) + \frac{V_S}{\omega L}(\cos\alpha + 1) - \frac{V_o}{\omega L}(\pi - \alpha) \qquad (8\text{-}31)$$

mode m2: The inductor current in mode m2 is determined from

$$\omega L \frac{di_L}{d\theta} = -V_o$$

$$i_o = I_o(\pi) - \frac{V_o}{\omega L}(\theta - \pi) \qquad (8\text{-}32)$$

At angle $2\pi+\alpha$, the final value of the inductor current is $I_o(\alpha)$, hence,

$$\cos\alpha = \frac{2\pi V_o}{V_S} - 1 \qquad (8\text{-}33)$$

Thus the inductor (output) current has a dc component and ripple. The dc component is

$$\bar{I}_o = \frac{1}{2\pi}\int_0^\pi i_o\, d\theta$$

$$peak - peak\ I_{or} = I_o(\pi) - I_o(0)$$

$$= \frac{V_S}{\omega L}(1 + \cos\alpha) - \frac{V_o}{\omega L}(\pi - \alpha) \qquad (8\text{-}34)$$

The ripple frequency is same as the source frequency. The current ripple is negligible for the large value of L and the output current is constant dc. Readers are encouraged to write the MATLAB program for this case.

8.4 1-PHASE FULL-BRIDGE AC-DC CONVERTER

A full-wave full-bridge converter is shown in Figure 8.18. The load may be purely resistive, inductive (a current sink), capacitive (a voltage sink)

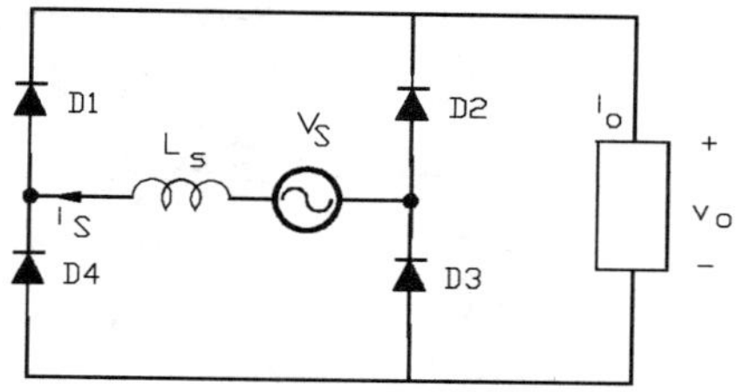

Figure 8.18
A 1-phase full-wave ac-dc rectifier

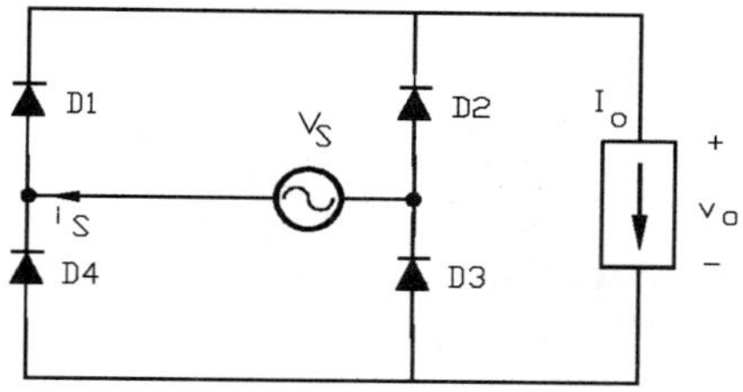

Figure 8.19
A 1-phase full-wave ac-dc rectifier with current sink load

or an inductor in series with a voltage sink or a combination. The source impedance consists of inductance in series with resistance. The inductance may be the self-inductance of long utility conductors or the leakage inductance of a utility transformer winding. Inductance is generally dominant over the resistance.

8.4.1 Current sink load

The general electrical load of Figure 8.4 with high value of inductor may be modeled by a current sink. This type of load is found in heavy magnetic pickup coils, relays and dc motors under constant torque application.

With Zero Source Inductance L_s

Consider the topology given in Figure 8.19 with a current sink as the load. In this section, however, ignore the source inductance.

The steady-state operation consists of two circuit modes over a period of the source waveform depending upon the state of the diodes **D1-D4,** shown

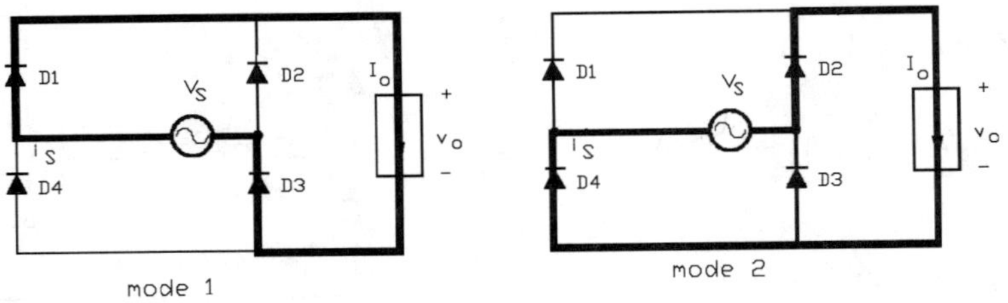

mode 1 $0 \le \theta < \pi$ **mode 2** $\pi \le \theta < 2\pi$

Figure 8.20
Circuit modes

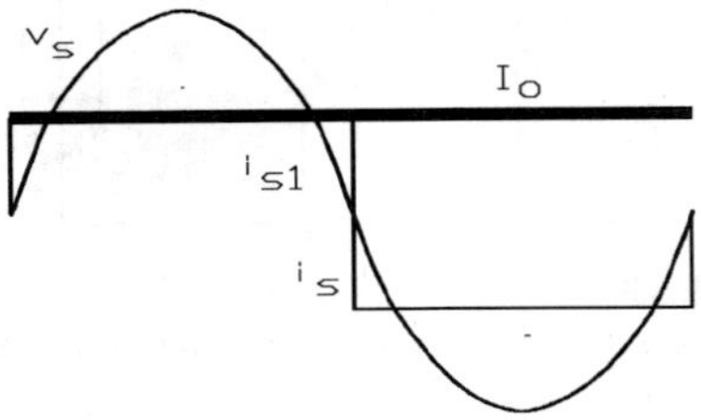

Figure 8.21
Waveforms

in Figure 8.20. Mode 1 begins when diode D_1 and D_3 turn on at zero angle. Current I_o flows through the path shown in Figure 8.20. After angle p, the input source voltage becomes negative, the mode 2 begins, the current is now steered through D_2 and D_4. Output voltage $\mathbf{v_o}$ and the input current $\mathbf{i_s}$ are shown in Figure 8.21.

Performance characteristics, mentioned in section 8.2.2, will be determined from the circuit modes, plotted and discussed. One way to analyze converter performance is to use equations, derive closed form solutions and then plot the graphs of the characteristics. We will use an alternative way, the numerical computer technique of determining characteristics from the knowledge of circuit modes, thus avoiding the derivation of complicated mathematics. As a first step in the numerical method, make arrays for the voltages and current waveforms of interest over a repetitive period of time or conduction angle θ. Smaller sub-arrays are constructed for each circuit mode separately and then concatenated to span a repetitive period. The resulting arrays are then processed to find average, rms values and Fourier coefficients and so on. The characteristics are then plotted and discussed. The numerical method provides a reasonably good engineering solution and design parameters.

Example 8-6

Following the MATLAB program generates the arrays theta, vs, vo, io and is which contain respectively the angles θ, input voltage v_S, output voltage v_o, output current i_o and the input current i_S for a $V_S = 120$ V rms and $I_o = 1.8$ amps. As a first step, small arrays are constructed for each circuit mode and then combined together to form the total array for each variable. An array spans a 2π angle. Run the program and confirm the waveforms.

```
VS=120*sqrt(2); IO=1.8; dth=pi/512;
th1=0:dth:pi−dth;th2=pi:dth:2*pi−dth; th=[th1 th2];
vs1=VS*sin(th1);vs2=VS *sin(th2); vs=[vs1 vs2];
vo1=vs1; vo2=-vs2; vo=[vo1 vo2];
io1=IO*ones(size(th1));io2=IO*ones(size(th2));
is1=io1; is2=−io2;
io=[io1 io2];
is=[is1 is2];
plot(th,vs/VS,'- -',th,vo/VS,'*',th,io/IO,'^', th,is/IO,'- -')
```

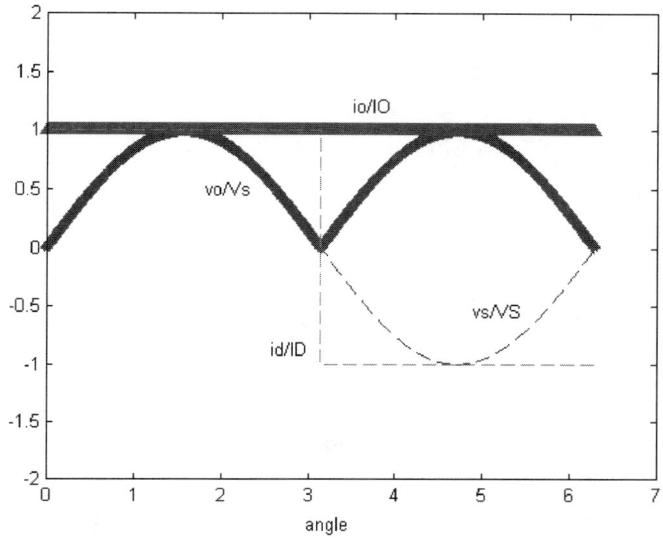

Figure 8.22
Waveforms in full-wave rectifier

Characteristics

Output dc voltage is calculated as the average of the output voltage waveform over a half period:

$$\bar{V}_o = \frac{1}{\pi}\int_0^{\pi} V_S \sin\theta \, d\theta = \frac{2V_S}{\pi} \qquad (8\text{-}35)$$

The voltage transfer ratio is given by:

$$T_{vv} = \frac{\bar{V}_o}{\tilde{V}_S} = \frac{2\sqrt{2}}{\pi} \qquad (8\text{-}36)$$

```
VO=mean(vo) % dc output voltage
Tvv=VO/(VS/sqrt(2)) % voltage transfer ratio
```

VO= 108.0376 Tvv= 0.9003

Input current is a square wave with I_o as the peak value. RMS input current is equal to the dc output current I_o. Current reflection ratio R_{ii} is unity. Following MATLAB instructions will calculate the rms input current:

```
isrms= sqrt (mean(is .^ 2)) % rms input current
```

isrms= 1.8000

The i_s waveform has half-wave symmetry because the waveform in the upper half period is the mirror image of the waveform in the lower half period. Hence the dc component is zero and even harmonics are absent in the Fourier expansion, given below:

$$i_s = \sum_{n=1,3,5,\ldots}^{\infty} \frac{4\bar{I}_o}{n\pi}\sin(n\theta) \qquad (8\text{-}37)$$

Figure 8.23 shows the amplitudes of several harmonics of the input current.

Example 8-7

Plot the frequency spectrum of the source current in Example 8-6. Calculate the rms fundamental current, displacement factor, power factor and THD of the source current waveform.

Add the following two MATLAB lines to the program in Example 8.2 for finding ten harmonics:

```
[f,ish,ph] = harmonic(is); %FFT analysis and harmonic
  profile
```

The fundamental component of the source current i_{s1} is in phase with the source voltage. Displacement

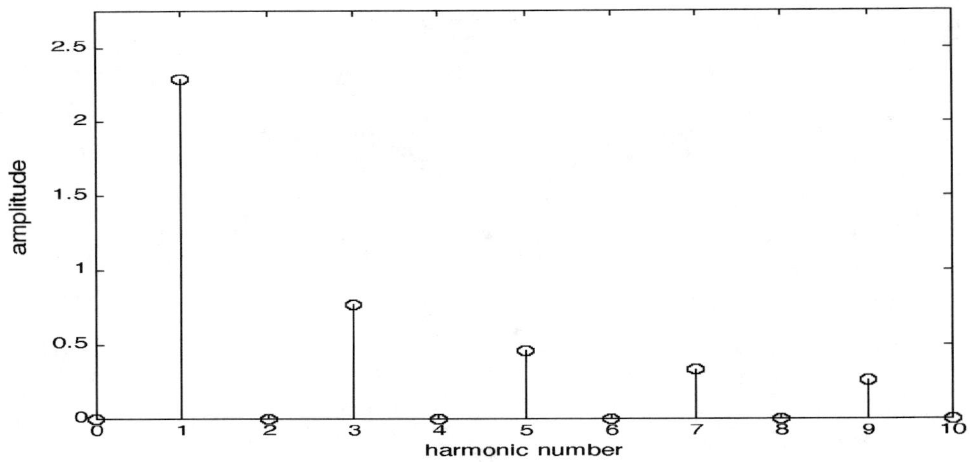

Figure 8.23
Harmonic profile of input current

factor DF is unity and the input power factor PF is calculated as

$$PF = \frac{2\sqrt{2}}{\pi}$$

Total harmonic distortion is found by using equation (8-9):

$$THD = \sqrt{\frac{\pi^2}{8} - 1} = 0.4834$$

```
is1rms=ish(2)/sqrt(2) %rms fundamental is1 current
df=cos(ph) %input displacement factor
pf=df*is1rms/isrms %iput power factor
td=sqrt((isrms/is1rms)^2-1) %total harmonic distortion
```

is1rms = 1.6206 df = 1.0000
pf = 0.9003 td = 0.4834

Ripple Current

This section will include the winding resistance R of the inductor that constitutes the current sink in Figure 8.19. The ripple current is assumed to be low and ωL/R>>1/f_s. Ignore the source inductance. The load current is assumed to be continuous, a dc component with ripple. The diode pair D1-D3 conducts the positive source current. The diode pair D2-D4 conducts the negative source current. The dc output voltage is as given by equation (8-35). In the steady state, the inductor cannot sustain any dc voltage, hence, the dc voltage in equation (8-35) appears across the resistance R. The dc load current then is

$$\bar{I}_o = \frac{\bar{V}_o}{R} = \frac{2V_S}{\pi R} = .6366\frac{V_S}{R} \tag{8-38}$$

The load current waveform, however, contains a ripple over the dc value. The dc voltage and the peak source voltages are related as

$$\bar{V}_o = V_S \sin\alpha \tag{8-39}$$

Hence, from equations (8-35) and (8-39) α=0.6901 radians. The instantaneous inductor current is obtained by integrating the voltage across the inductor,

$$i_o = \int_{\alpha}^{\theta} [|V_S \sin\theta| - \bar{V}_o]\, d\theta$$

$$i_o = I_{o1} + \frac{V_S}{\omega L}[\cos\alpha - \cos\theta] + \frac{\bar{V}_o}{\omega L}(\theta - \alpha) \tag{8-40}$$

The inductor voltage is zero at the angles α and π-α; consequently, the inductor current is minimum and maximum respectively at these angles. The ripple current is the difference between the maximum I_{om} and the minimum I_{o1}

$$I_{or} = I_{om} - I_{o1} = \frac{2V_S}{\omega L}[\cos\alpha - 1 + 2\alpha/\pi]$$

$$= 0.4212\frac{V_S}{\omega L} \quad (8\text{-}41)$$

The ripple current normalized to the short circuit current I_{osc} is

$$I_{nor} = \frac{I_{or}}{I_{osc}} = 0.4212 \; where \; I_{osc} = \frac{V_S}{\omega L} \quad (8\text{-}42)$$

The ripple frequency is twice the ac source frequency. The average dc current can also be approximated as

$$\bar{I}_o \approx \frac{I_{o1} + I_{om}}{2} = I_{o1} + .2106\, I_{osc} \quad (8\text{-}43)$$

Example 8-8

Calculate and display the ripple current in the converter in the previous example. R is 60 ohms. Ignore the source inductance L_s.

We will first make sure that equation (8-2) is satisfied:

$$\frac{\omega L}{R} = \frac{2 \times \pi \times 60 \times 0.1}{60} = 0.628 \ll \frac{1}{2f_s} = \frac{1}{30}$$

$$I_{or} = .4208 * \frac{120 \times \sqrt{2}}{2\pi \times 60 \times 0.1} = 1.8943\,A$$

```
VS=120*sqrt(2);omegaL=2*pi*60*.1;R=60;
Vodc=2*VS/pi; Iodc=Vodc/R;Iosc=VS/omegaL;
alpha=asin(Vodc/VS); Io1=Iodc-.2108*Iosc;
dth=pi/256;th=alpha:dth:2*pi+alpha-dth;
  vs=VS*sin(th);
vl=abs(vs)-Vodc; io=Io1+integ(vl/omegaL, th);
is=io.*(vs>0)-io.*(vs<0);
plot(th; vs/VS,'- -',th,io/Iodc,th, is/Iodc,'*')
```

With Source Inductance L_s

In previous sections, we have ignored the source resistance and source inductance in the ac input voltage source. In this section consider the effect of source side inductance in single-phase full converter with constant current output. Consider the converter shown in Figure 8.18 with source inductance

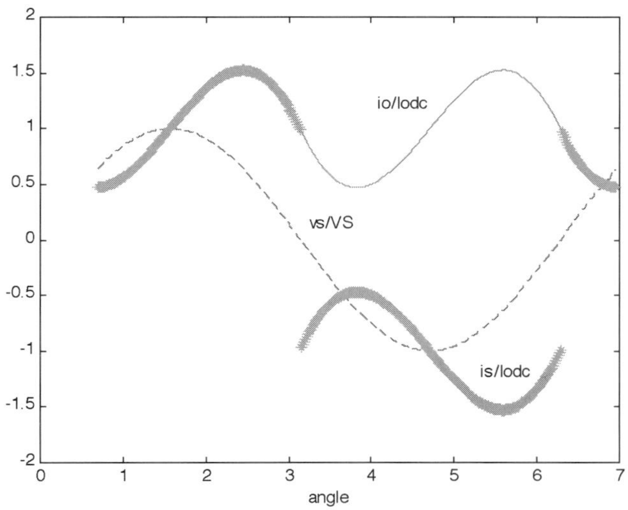

Figure 8.24
Waveforms

L_S. Steady-state converter operation is described by six circuit modes over a period beginning from u to $2\pi + u$, see Figure 8.25 (a).

Mode 1 begins when the input voltage is positive, and the diodes D_1 and D_3 begin to conduct current at angle u. Current I_o flows through the path shown in Figure 8.25 (b). The constant current does not develop any voltage across L_s, hence, the output voltage v_o follows the input voltage waveform. The input source voltage becomes negative at π and mode 2a begins. The current through the source inductance, however, cannot change instantaneously but now it begins to decrease. The output current sink now does not get its full amount from the source through D_1, hence the output current freewheels through diodes D_2 and D_3. The diodes D_2 and D_3 effectively short-circuit the output voltage. The source inductance L_s takes on all the negative voltage and maintains the diodes D_1 in conduction until its current becomes zero. The current in diode D_1 commutates, mode 2 ends and mode 2b begins. The source voltage forces diode D_4 to conduct and the current through the inductor L_s in mode 2b changes polarity and begins to increase. The output current still freewheels through diodes D_3 and D_2. This circuit mode exists until the inductor current builds up to $-$Io value at an angle $\pi + u$. Diode D3 commutates and the current flow is as shown in mode 3.

The mode 3 ends at 2π when the input source voltage becomes positive. In the following mode 4, the current through the source inductance, however, cannot change instantaneously but now it begins to decrease in amplitude. The output current sink now does not get its full amount from the source through D_2, hence the output current freewheels through diodes D_4 and D_1. The diodes D_4 and D_1 effectively short-circuit the output voltage again. The source inductance L_s takes on all the positive voltage and maintains the diode D_2 in conduction until its current becomes zero. The current in diode D_2 commutates, mode 4a ends and mode 4b begins. The source voltage forces diode D_3 to conduct, and the current through the inductor L_s in mode 4b changes polarity and begins to increase. The output current still freewheels through diodes D_4 and D_1. This circuit mode exists until the inductor current builds up to $+ I_o$ value at an angle $2\pi + u$. The diode D_4 commutates and the circuit comes into mode 1.

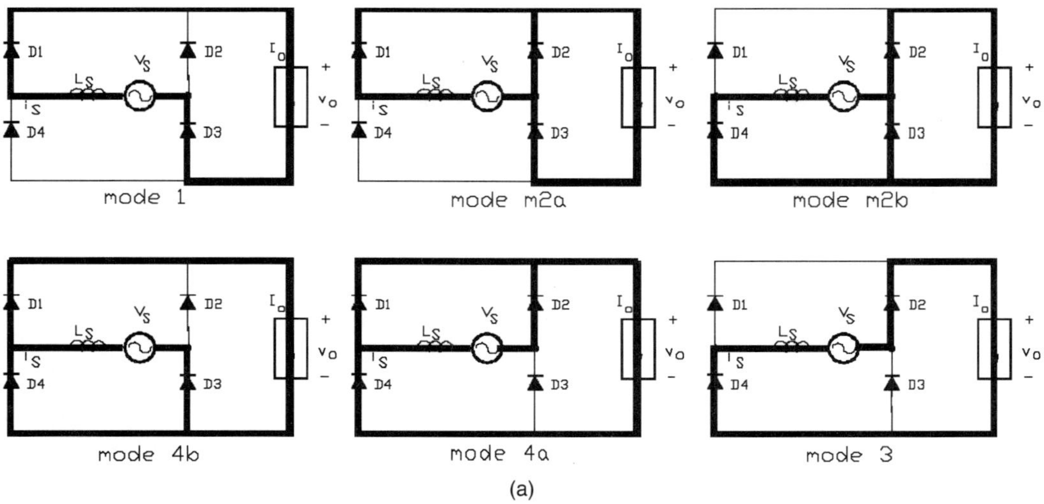

Figure 8.25
A 1-phase full-wave rectifier with source inductance (a) Circuit modes

The current in inductor L_S, which is also the input current i_s, in mode 2a and 2b is found from:

$$L_s \frac{di_L}{dt} = V_S \sin\theta, \quad \pi \le \theta < \pi + u \quad (8\text{-}44)$$

$$i_L = I_o + \frac{V_S}{\omega L_s}(1 + \cos\theta) \quad (8\text{-}45)$$

Inductor current varies from $+I_o$ at π angle to $-I_o$ at $\pi+u$, which gives

$$\cos u = 1 - 2I_{no}; \quad I_{no} = \frac{I_o}{I_{osc}}; \quad I_{osc} = \frac{V_S}{\omega L_s} \quad (8\text{-}46)$$

where I_{no} is the normalized output current. I_{osc} is the maximum current in the short circuit condition.

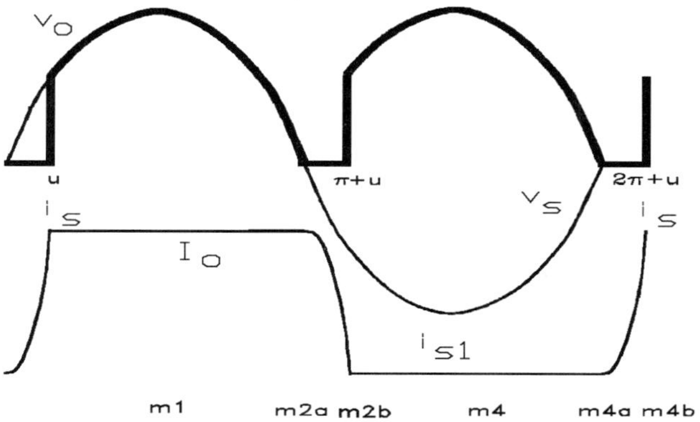

(b)

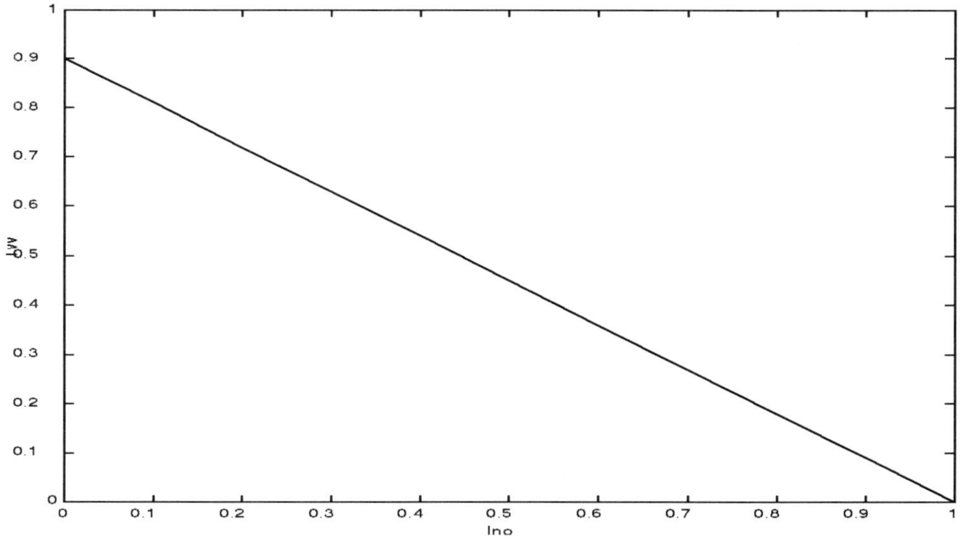

(c)

Figure 8.25 ***(continued)***
(b) waveforms (c) T_{vv} vs. normalized output current

DC output voltage is obtained from

$$\overline{V}_o = \frac{1}{\pi}\int_0^{\pi} V_S \sin\theta \, d\theta$$

$$= \frac{V_S}{\pi}(1 + \cos u) = \frac{2V_S}{\pi}(1 - I_{no}) \quad (8\text{-}47)$$

Using the above two equations, we can get the converter voltage transfer ratio as a function of normalized output current

$$T_{vv} = \frac{\overline{V}_o}{\tilde{V}_S} = \frac{2\sqrt{2}}{\pi}(1 - I_{no}) \qquad (8\text{-}48)$$

and

$$\overline{V}_o = \frac{2\sqrt{2}}{\pi}\tilde{V}_S - \frac{2\omega L_s}{\pi} I_o \qquad (8\text{-}49)$$

The T_{VV} versus I_O is given in Figure 8.25 (c). Unlike the converter with L_s=0, the dc output voltage is a function of the output current being delivered to the load. The effect of source inductance has resulted in introducing an equivalent source resistance of $2\omega L_s/p$ in the output circuit of the converter.

Example 8-9

Consider the circuit of Figure 8.18 with L_s = 100 mH and a 120 V rms, 60 Hz source. Find the maximum output current. Select the dc current lesser than the maximum and calculate the commutation angle and the performance measures. Draw the plot of output dc voltage versus the load current.

The no-load output voltage is obtained from equation (8-42) by putting I_o = 0.

The maximum limit on the output current, which is obtained from equation (8-40).

$$I_{omax} = \frac{\sqrt{2}\tilde{V}_S}{\omega L_s} = 4.50\ A$$

Select I_o=1.8 A for illustration. The rest of the problem is done using MATLAB. Some important MATLAB lines are given below. The rest of the program is left for the readers to fill.

```
IO=1.8; u=acos(1-2*INO) %commutation angle
dth=pi/512; %angle step
th1=u:dth:pi-dth; %mode m1
is1=IO*ones(size(th1);
th2=pi:dth:pi+u-dth; %mode m2
is2=IO+integ(vs2,th2)/omegaLs;
```

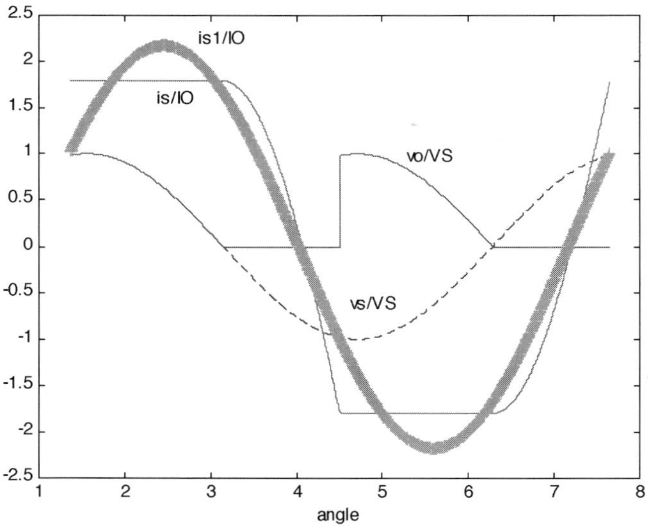

Figure 8.26
Waveforms in full-wave rectifier with source inductance

```
th3=pi+u:dth:2*pi-dth; %mode m3
th4=2*pi:dth:2*pi+u-dth; %mode m4
th=[th1 th2 th3 th4]; %one period
vs=[vs1 vs2 vs3 vs4]; %input source voltage
is=[is1 is2 is3 is4]; %input current
vo=[vs1, zeros(size(th2)), -vs3, zeros(size(th4))];
  %output voltage
[f,ish,ph]=harmonic(is); %find the dc and fundamental
  component of the input current
Rii=isrms/IO % current reflection ratio
phase1=ph-u %phase angle of is1
is1=ish(2)*sin(th+phase1); %array for is1
```

Iomax= 4.5016 u = 1.3692 VO = 65.1257
Tvv = 0.5427 isrms = 1.5718

Rii = .8732 is1rms = 1.5353
phase1 = −.8818 df = 0.6358
pf = 0.6210 td = 0.2192

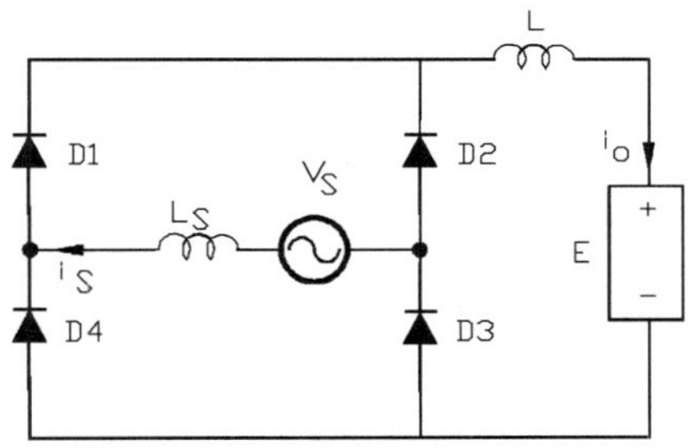

Figure 8.27
A full-wave rectifier with voltage sink load

By comparing the results with those in the previous section, notice that the voltage transfer ratio T_{vv}, the current reflection ratio R_{ii}, the displacement factor DF and the power factor PF have become worse. The total harmonic distortion THD has, however, improved.

8.4.2 1-Phase full-bridge converter with voltage sink load

Consider the topology given in Figure 8.27. The voltage sink E on the dc side may be a DC bus supply. It may be the back emf in the armature of a dc motor in forward motion. The converter is required to provide a dc current. Voltage sink E should be less than the peak input voltage V_S. In this section, assume that the winding resistance of inductor L is small and $\omega L/R$ is much greater than $1/f_s$. Also ignore the source inductance L_S.

Continuous Conduction Mode (CCM)

Consider the converter with source inductor $L_S=0$ as given in Figure 8.28. The inductance L on the load side is assumed to have a winding resistance R. The output current is continuous and has a dc component.

Diodes D1 and D3 conduct from 0 to π, and D2 and D4 conduct from π to 2π. The dc output voltage is given by equation (8-35). The inductor L

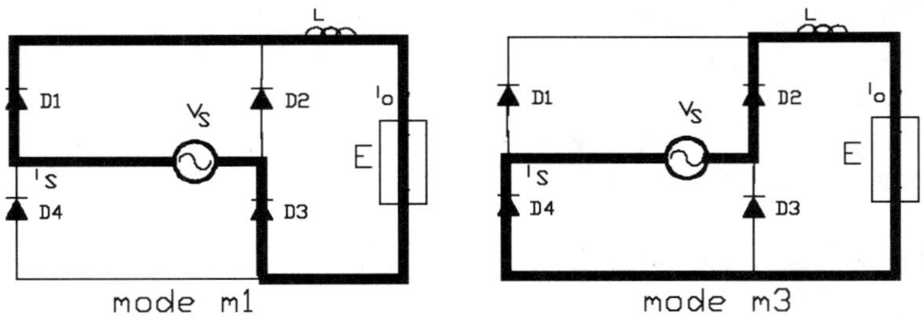

mode 1 $\alpha \leq \theta < \pi + \alpha$ **mode 2** $\pi + \alpha \leq \theta < 2\pi + \alpha$

Figure 8.28
Circuit modes in Continuous Conduction Mode

cannot sustain dc voltage in steady state, hence, voltage V_o-E appears across the winding resistance of the inductor L. The dc output current is

$$\bar{I}_o = \frac{\bar{V}_o - E}{R} \tag{8-50}$$

Large value of inductance L gives rise to low ripple current. The rectifier then operates into a current sink load (see section 8.4.1).

For medium and small values of L, the output current contains considerable ripple. The output current waveform is determined from the integration of the voltage across the inductor v_L

$$v_L = |V_S \sin\theta| - \bar{V}_o \tag{8-51}$$

The rest of the analysis is similar to the section "Ripple Current" on page 276. Minimum dc output current to ensure the CCM operation is obtained from equation (8-43) by substituting $I_{o1}=0$, hence,

$$\bar{I}_o \geq \bar{I}_{omin} \approx 0.2104\, I_{osc} \tag{8-51a}$$

From equations (8-50) and (8-51a), we assert,

$$\bar{V}_o - E \geq 0.2104 \frac{V_S}{\omega L / R}$$

$$E \leq V_S\left[\frac{2}{\pi} - \frac{0.2104}{\omega L / R}\right] \tag{8-51b}$$

Example 8-10

Consider the circuit of Figure 8.27 with L = 100 mH which is used as a battery charger. The source voltage 120 V rms, 60 Hz is charging a bank of batteries in series totaling 90 V dc. The output voltage may be assumed to constant during the charging period. The inductor has a winding resistance of 10 ohms. Find the dc output current, the peak-peak ripple and all performance measures.

The peak source voltage $V_S = 120\sqrt{2} = 169.7056$ V.

We should ensure if the CCM operation does indeed exist by using equation (8-51b)

$$E120\sqrt{2}\left[\frac{2}{\pi} - \frac{0.2104}{2\pi 60 \times 0.1/10}\right] = 98.5666\ V$$

The battery bank voltage is less than 98.5666 V, hence the converter indeed operates in CCM.

The peak-peak current ripple is found using equation (8-41),

$$I_{or} = .4208 \frac{120\sqrt{2}}{2\pi \times 60 \times 0.1} = 1.8943\ A$$

The rest of the problem is solved using MATLAB. Some important MATLAB lines are given below. The remaining program is left for the readers to write.

```
Vodc=2*VS/pi; Iosc=VS/omegaL;
alpha=asin((Vodc/VS));
Iodc=(Vodc−E)/R; Io1=Iodc−.2108*Iosc;
dth=pi/512; th=alpha:dth:2*pi+alpha−dth;
vl=abs(vs)−Vodc;
io=Io1+integ(vl/omegaL,th);
is=io.*(vs>0) − io.*(vs<0);
Ior=max(io)−min(io) % ripple current
```

Discontinuous Conduction Mode (DCM)

For voltage sink E more than the given in equation (8-51b), the output current is discontinuous. In this mode of operation, the output current is zero at the time of commutation from one diode pair to the other pair. Hence, the source inductance L_S may be merged with the dc side inductance L or vice versa for analysis purposes. We would retain L. The inductor L has a winding resistance R.

The steady state operation can be explained by four circuit modes.

mode 1 $\alpha \leq \theta < \beta$

mode 2 $\beta \leq \theta < \pi + \alpha$

mode 3 $\pi + \alpha \leq \theta < \pi + \beta$

mode 4 $\pi + \beta \leq \theta < 2\pi + \alpha$

Mode m1 begins at the angle α

$$\alpha = \sin^{-1}\frac{E}{V_S}$$

The output current i_o begins to build up from zero and flows from the source through the diodes D1 and D3, L_S and the load. Current i_o rises and then comes down to zero at an angle β. In the following mode m2, all diodes are reverse biased and i_o is zero. The polarity of the source voltage reverses at angle π. The next mode m3 begins when the

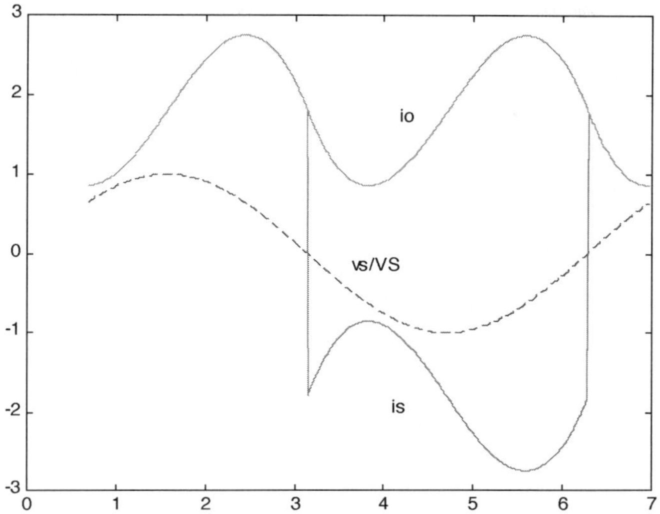

alpha=.6901 IO=1.8025 isrms=1.9262 is1rms=1.7325
phase1=−.3559 df=0.9373 pf=0.8431 td=.4860 Ior= 1.8953

Figure 8.29
Waveforms

diode pair D2 and D4 conduct at the angle $\pi+\alpha$. Mode m3 ends at $\pi+\beta$. In the following mode m4, all diodes are again reverse-biased.

In the steady-state, the average voltage across inductor L is zero, hence,

$$\int_{\alpha}^{\beta} v_L \, d\theta = \int (|V_S \sin\theta| - E)\, d\theta = 0 \qquad (8\text{-}52)$$

$$\cos\alpha - \cos\beta - \frac{E}{V_S}(\beta - \alpha) = 0 \qquad (8\text{-}53)$$

hence output current i_o in mode m1 is zero until an angle α and extinguishes at angle β, which is less than $\pi+\alpha$.

$$i_o = \int_{\alpha}^{\theta} \frac{v_L}{\omega L}\, d\theta \qquad (8\text{-}54)$$

$$i_o = \frac{V_S}{\omega L}(\cos\alpha - \cos\theta) - \frac{E}{\omega L}(\theta - \alpha) \qquad (8\text{-}55)$$

The output current i_o in mode m3 is the exact replica of that in mode m1.

Average output current is calculated from

$$\bar{I}_o = \frac{1}{\pi}\int_{\alpha}^{\beta} i_o \, d\theta \qquad (8\text{-}56)$$

Peak output current occurs at angle π-α and is given by

$$I_{opeak} = \frac{V_S}{\omega L} 2\cos\alpha + \frac{E}{\omega L}(2\alpha - \pi) \qquad (8\text{-}57)$$

This is also the peak-peak current ripple amplitude. A large inductor value ensures low current ripple.

The short circuit current is limited and is obtained by substituting E=0 (α=0 and β=π)

$$I_{osc} = \frac{V_S}{\omega L} \qquad (8\text{-}58)$$

Crest factor is defined as the ratio of peak output current to the average output current. It is given by

$$CF = \frac{I_{opeak}}{\bar{I}_o} \qquad (8\text{-}59)$$

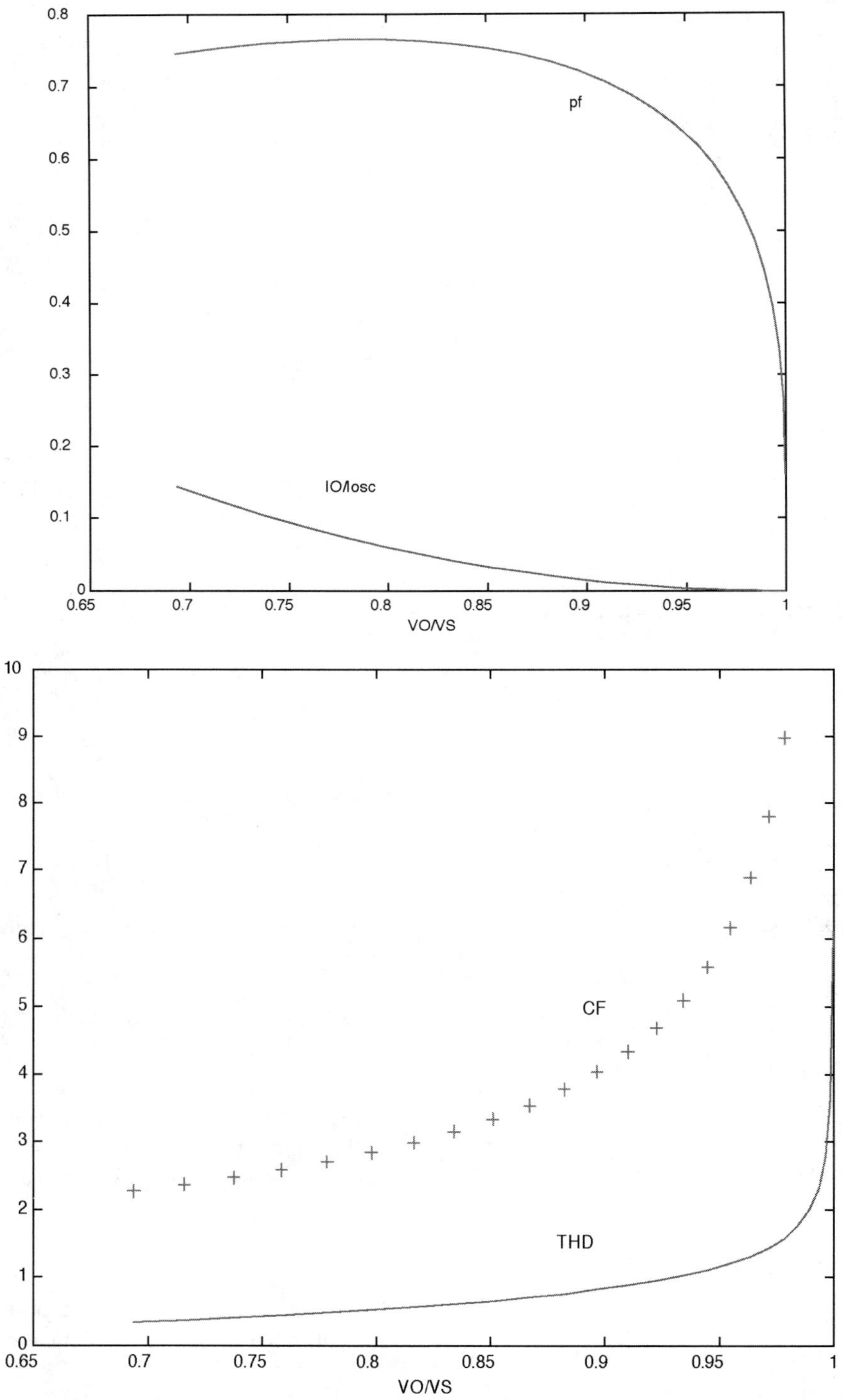

(a)

(b)

Figure 8.30
(a) Normalized output current and input power factor vs. the normalized output voltage in DCM and (b) crest factor and Total Harmonic Distortion vs. the normalized output voltage in DCM

The output current normalized to the short circuit current I_o/I_{osc} and the input power factor pf versus V_o/V_S is given in Figure 8.30(a). The THD and the crest factor CF versus V_o/V_S is given in Figure 8.30(b). The MATLAB program for obtaining the plots is given in the Appendix D.

The plots show that the average output current decreases as the output voltage is increased. The power factor remains nearly constant except near very high output voltages. The THD and the crest factor worsen at high output voltages. The output current becomes continuous when the extinction angle β is $\pi+\alpha$.

Example 8-11

The battery bank in the example is increased to 120 V. The inductor has a winding resistance of 10 ohms. Find the current ripple and the performance measures.

The peak source voltage $V_S = 120\sqrt{2} = 169.7056$ V. Angle α is

$$\alpha = \sin^{-1}\left[\frac{120}{169.7056}\right] = 0.7854\ rad$$

Some important MATLAB lines are given below. The remainder program is left for the readers to write.

```
E=120; alpha=asin(E/VS) dth=pi/512;
th1=alpha:dth:pi+alpha-dth;
th2=pi+alpha:dth:2*pi+alpha-dth;
vl1=abs(vs1)-E;
io1=integ(vl1/omegaL,th);
io1=io1.*(io1>0);
[f,ish,ph]=harmonic(is);
phase1=ph-alpha
is1=ish(2)*sin(th+phase1);
```

8.4.3 A 1-Phase full-bridge rectifier with L_S and voltage sink load

Consider the topology given in Figure 8.32(a). Voltage sink V_o is the model of a high value capacitor in parallel with a resistive load as is the case in dc power supply. It may be the back emf in the armature of a dc motor in forward motion.

For voltage sinks higher than specified by equation (8-51b) the converter operates in discontinuous

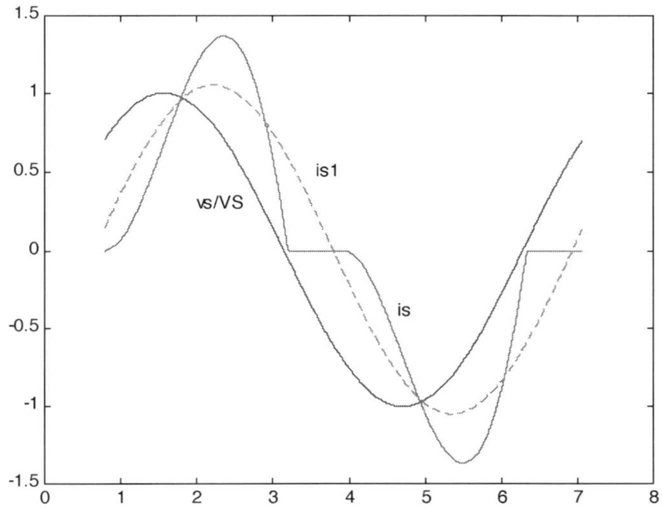

alpha=0.7854 IO=0.5897 isrms=0.7882 is1rms=0.7420 df=0.7966
pf=0.7499 phase1=–.6491 td=0.3585

Figure 8.31
Waveforms

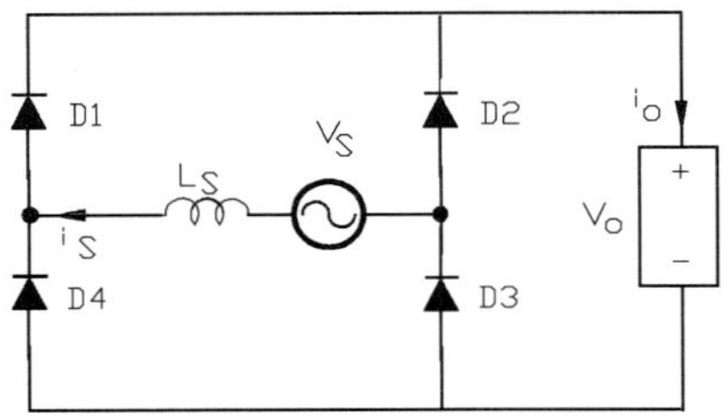

(a)

Figure 8.32
(a) Full-wave rectifier with voltage sink load

current mode. The converter response is exactly the same as in the DCM operation in the previous section.

For lower voltage sinks, the converter operates in a continuous current mode. There are two circuit modes in steady state similar to those in the previous section. The source current and, hence, the current in the load is shown in Figure 8.33.

Mode m1 begins at an angle α. The inductor current, which is also the output current i_o, begins to build up from zero and flows from the source through the diodes D1 and D3, L_S and the load. Current i_o rises and then comes down but does not become zero at the angle π. The source voltage reverses polarity at π, but since the current through L_S is positive, the mode m1 continues beyond π until the inductor current becomes zero at $\pi+\alpha$. See Figure 8.33. In the following mode m2, inductor current direction reverses and flows through the diode pair D2 and D4 and the load.

The voltage sink V_o is the average of the output voltage waveform after the diode network,

$$V_o = \frac{1}{\pi}\int_{\alpha}^{\pi+\alpha} V_s \sin\theta \, d\theta = \frac{2V_S}{\pi}\cos\alpha \qquad (8\text{-}60)$$

The instantaneous inductor current is obtained by integrating the voltage across the inductor (ignore the winding resistance of the inductor L_S),

$$i_L = \int_{\alpha}^{\theta} \frac{v_L}{\omega L}\, d\theta; \quad v_L = V_S \sin\theta - V_o \qquad (8\text{-}61)$$

The inductor current is maximum at the angle π-α_0,

$$I_{Lm} = \frac{1}{\omega L}\int_{\alpha}^{\pi-\alpha_0} (V_S \sin\theta - V_o)\, d\theta \qquad (8\text{-}62)$$

where

$$\alpha_0 = \sin^{-1}\frac{V_o}{V_S} \qquad (8\text{-}63)$$

The average dc current is

$$\bar{I}_o = \frac{1}{\pi}\ \pi + \alpha \int_{\alpha} i_L\, d\theta$$

$$\bar{I}_o = \frac{2V_S}{\pi\omega L}\sin\alpha \qquad (8\text{-}64)$$

Example 8-12

A single-phase 1 hp dc motor is driven by a 120 V 60 Hz supply. The armature inductance of the motor is 100 mH and the back emf at the desired speed is 60 V dc. Use the full converter of Figure 8.32(a). Draw the voltage and current waveforms using MATLAB. Also find the actual power being delivered (assume the losses to be zero), rms input current, input power factor and Total Harmonic Distortion THD.

Some important MATLAB lines are given below. The remainder program is left for the readers to write.

```
VO=60; alpha=acos((pi*VO/(2*VS)))
dth=pi/512; th1=alpha:dth:pi+alpha;
  th2=pi+alpha:dth:2*pi+alpha;
vl1=vs1−VO; io1=integ(vl1/omegaL,th);
```

8.5 LINE QUALITY ISSUES

The electric ac power supply serves many customers with varied types of loads. The electric distribution system consists of generating stations, utility lines and transformers. An ideal electric supply with zero source impedance and zero impedance of the utility lines and transformers would supply undistorted sine wave voltage to all customers. In practice, however, the electric generators, utility lines and the transformers have nonzero impedances and are the cause of distortions in the electric

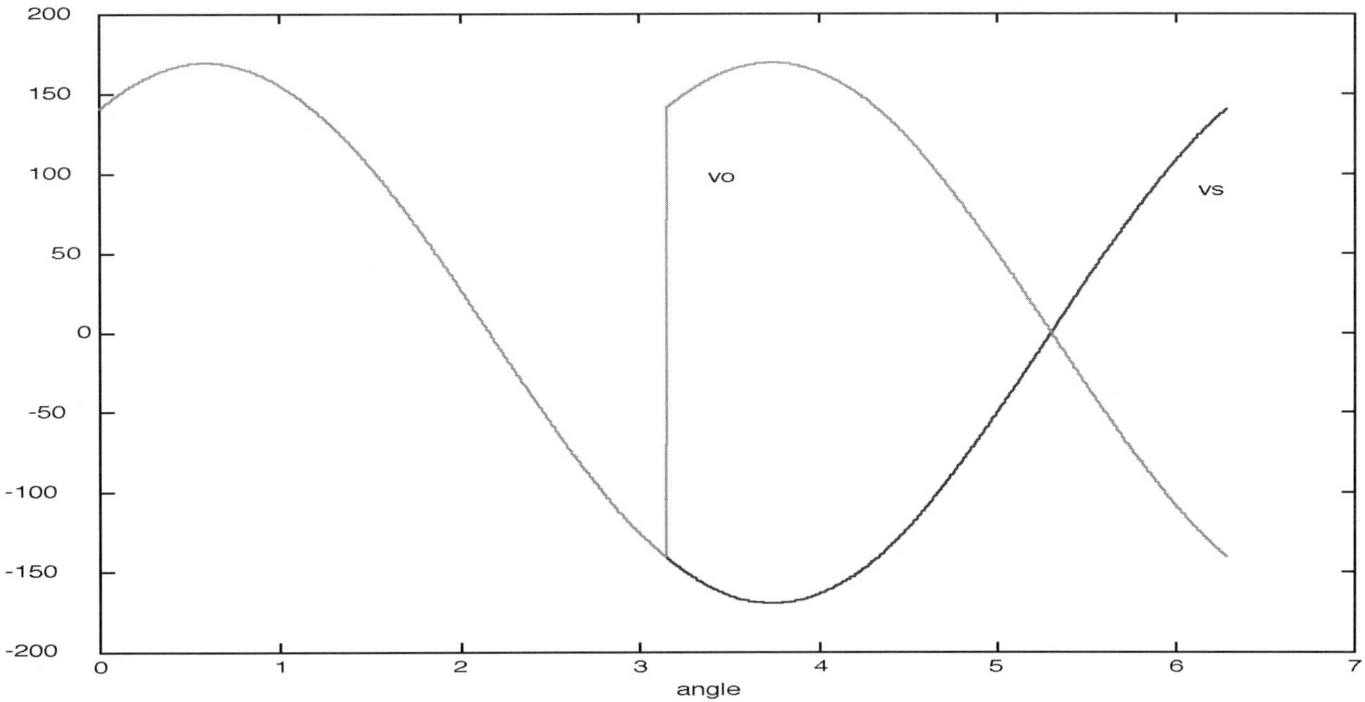

(b)

Figure 8.32 ***(continued)***
(b) Voltage waveforms

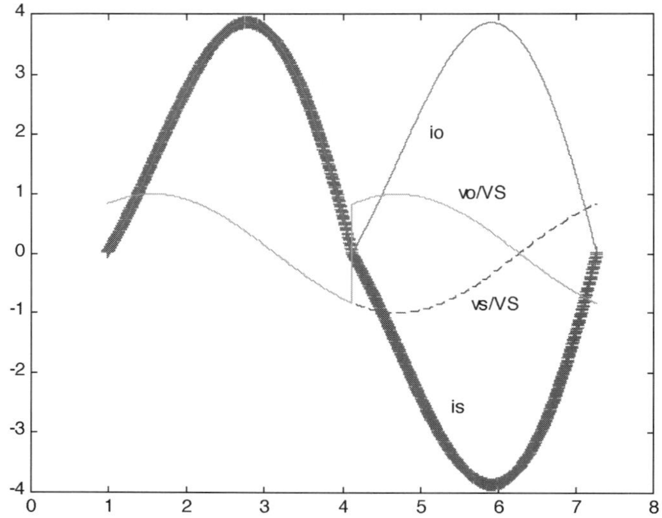

alpha=.9820 IO= 2.3898 isrms=2.6797 is1rms=2.6741
phase1=−1.1045 df=.4496 pf=.4486 td=.0649

Figure 8.33
Waveforms

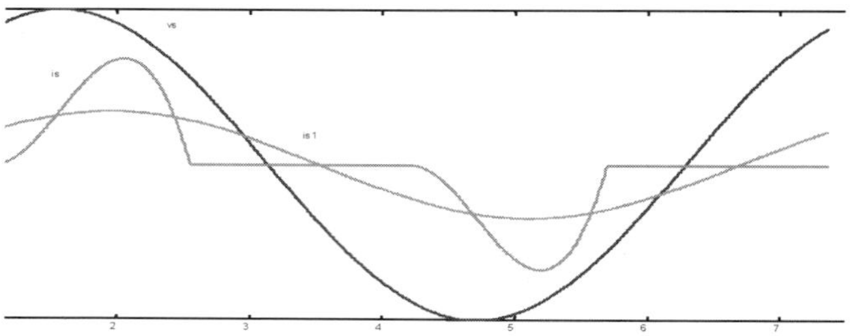

Figure 8.34
Line-current distortions in a 1-phase ac supply

supply. The line quality is becoming a very critical issue in the recent times. The source and line inductances play an important role in the line quality issues. We will illustrate a few effects in a single-phase ac supply system.

8.5.1 Line-current distortion

In the previous sections, we have seen that the source current in most of the rectifier circuits is very rich in harmonics. This is especially true in capacitive and voltage sink loads in the discontinuous conduction mode operation. Figure 8.34 shows a highly distorted source current in a rectifier with the voltage sink load. Even though the real power delivered by the ac source is contributed by the fundamental component of the source current, the rectifier diodes must withstand several times higher peak currents. The addition of series inductors to boost the source inductance minimizes the height of the current peaks. The height of the current peak is not the only problem. The source inductance also severely distorts the line voltage available to all customers.

8.5.2 Line-voltage distortion

The utility line supplies power to several loads simultaneously. This section will analyze a situation in which two loads share a single one-phase line ac source. The load 1 is assumed to be a current sink. The line inductance in this situation must be divided in two parts in each line: L_{S1} and L_{S2}, as shown in Figure 8.35(a). Usually L_{s1} is the source inductance and L_{s2} is the transformer leakage inductance feeding the converter rectifier. Assume that load 1 is drawing a constant current I_o. The load 2 is not drawing any current. We are interested in finding out the voltage available to load 2.

Consider the line voltage v_{AB} at the point of common coupling (PCC). There is one notch per half cycle in the waveform of v_{AB} as shown in Figure 8.35(a). The two notches occur at 0 and π angles each of duration equal to the commutation angle u.

The notch width u is obtained from equation (8-46)

$$u = \cos^{-1}\left[1 - \frac{2\omega(L_{S1} + L_{S2})I_o}{V_S}\right] \quad (8\text{-}65)$$

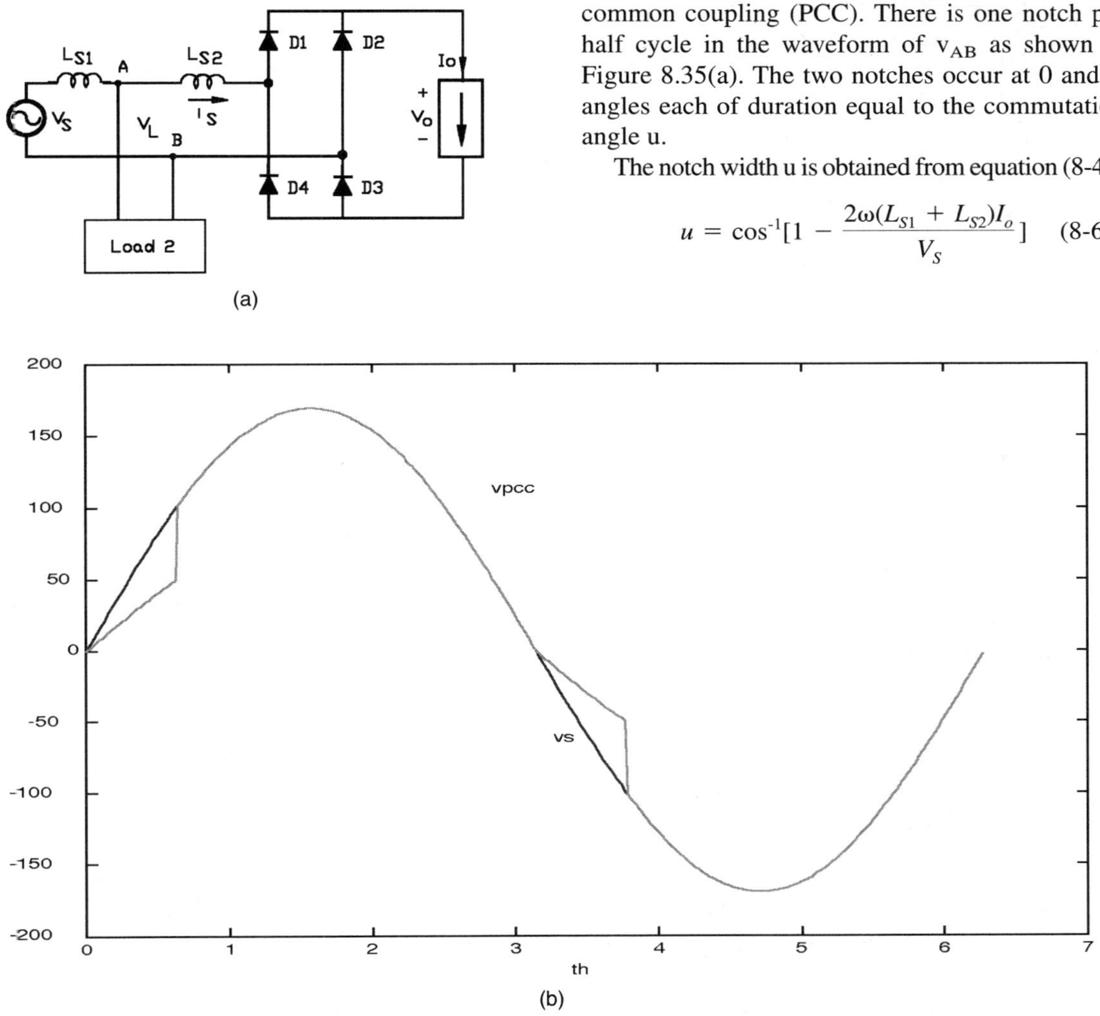

Figure 8.35
(a) Line-voltage distortion due to a current sink load and (b) VPCC waveform with current sink load

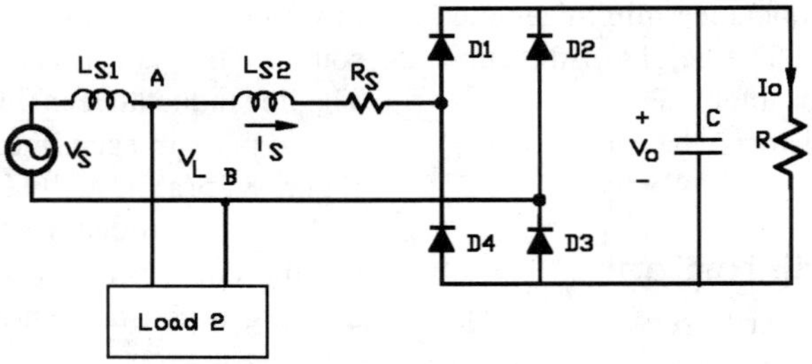

(c)

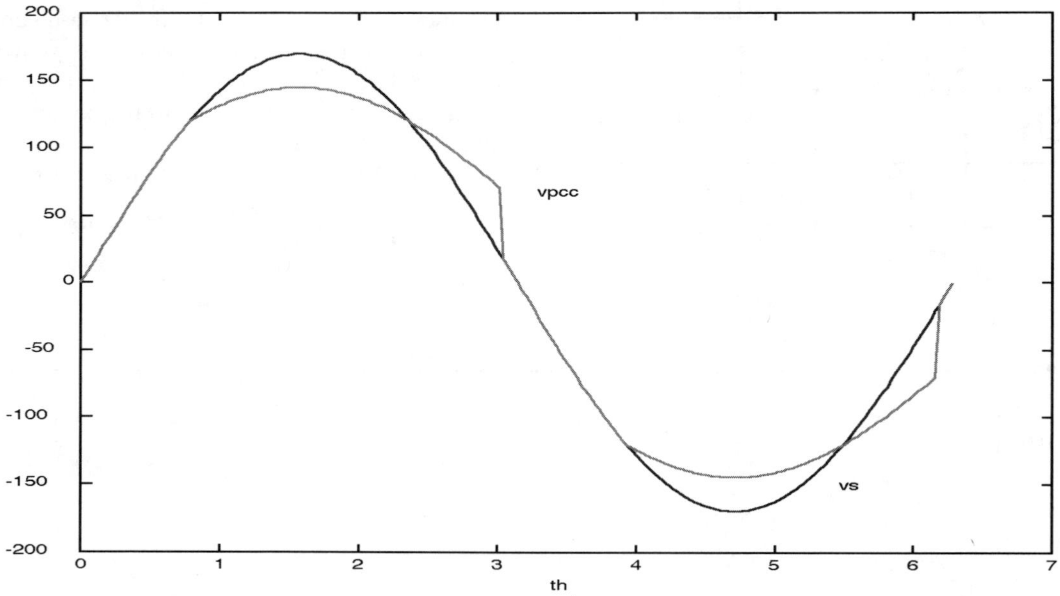

(d)

Figure 8.35 *(continued)*
(c) Line-voltage distortion due to a current sink load (d) VPCC waveform with voltage sink load

During the commutation angle, the line voltage is applied across the series combination of L_{S1} and L_{S2},

$$V_S \sin\theta = \omega(L_{S1} + L_{S2})\frac{di_o}{d\theta} \qquad (8\text{-}66)$$

The voltage at the PCC in a half period is given by

$$V_{AB} = V_S \sin\theta - \omega L_{S1}\frac{di_o}{d\theta}; \quad 0 \le \theta < u$$

$$V_{AB} = L_S \frac{2}{L_{S1} + L_{S2}} V_S \sin\theta; \quad 0 \le \theta < u$$

$$V_{AB} = V_S \sin\theta; \quad u \le \theta < \pi \qquad (8\text{-}67)$$

The voltage waveform at the PCC is shown in Figure 8.35(b). Note that the width u depends on the dc current in load 1. The notch depth in the voltage waveform is independent of the current in the load1 and depends only on the line inductance.

Consider the other situation as shown in Figure 8.35(c). Assume DCM operation. The output current i_o is obtained from equation (8-53)

$$i_o = \frac{V_S}{\omega(L_{S1} + L_{S2})}(\cos\alpha - \cos\theta) - \frac{V_o}{\omega(L_{S1} + L_{S2})}(\theta - \alpha) \qquad (8\text{-}68)$$

Notch width = the conduction angle β-α. Hence, during the conduction angle α to β, the voltage at the PCC in a half period is given by equation (8-67) and is

$$V_{AB} = V_S \sin\theta - \frac{L_{S1}}{L_{S1} + L_{S2}}[V_S \sin\theta - V_o]; \quad \alpha \le \theta < \beta$$

$$V_{AB} = \frac{L_{S2}}{L_{S1} + L_{S2}}\left[V_S \sin\theta - V_o \frac{L_{S1}}{L_{S2}}\right]; \quad \alpha \le \theta < \beta$$

$$V_{AB} = V_S \sin\theta;\ 0 \le \theta < \alpha \ \textit{and}\ \beta \le \theta < \pi \qquad (8\text{-}69)$$

From the previous section we know that the load current is a function of the dc voltage V_o. Hence, both the notch width, β-α, and the amplitude distortion are dependent on the current in load 1.

The line notching due to the load affects the quality of power available to other loads. The stray or parasitic capacitances further complicate by adding ringing along with notches. The inclusion of stray capacitance is too complicated for analysis and therefore better handled by simulation.

8.5.3 Neutral Current in a Three-Phase Four-Wire Utility

As seen in previous sections, the line current in single-phase rectifiers is highly non-sinusoidal in wave shape. This section investigates the effect of non-sinusoidal line currents in a three-phase four-wire system as given in Figure 8.36. For understanding the concepts, consider a balanced system in which all three phases have identical full-wave full-bridge rectifiers and all rectifiers have identical current sink loads. The line current in each phase is

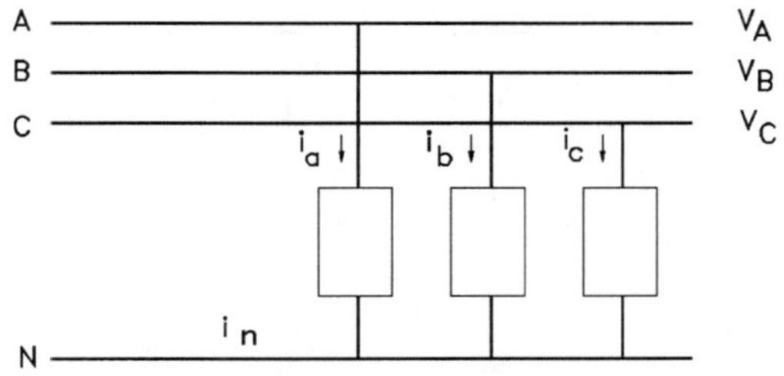

Figure 8.36
Current in the neutral line

$$i_A = \sum_{1,3,}^{\infty} \frac{4I_o}{n\pi} \sin(n\omega t + \phi_n)$$

$$i_B = \sum_{1,3,}^{\infty} \frac{4I_o}{n\pi} \sin\left(n\omega t + \phi_n - n\frac{2\pi}{3}\right)$$

$$i_C = \sum_{1,3,}^{\infty} \frac{4I_o}{n\pi} \sin\left(n\omega t + \phi_n - n\frac{4\pi}{3}\right) \qquad (8\text{-}70)$$

Note that the phase of the third harmonic current in all phases is ϕ_3. In fact, this is true for all triplen harmonics, 3rd, 9th, 15th, and so on.

The current in the fourth wire, the neutral line, is the vector sum of the three line currents

$$i_N = i_A + i_B + i_C \qquad (8\text{-}71)$$

Figure 8.37 shows the plot of the neutral current, a square wave of three times the frequency of the phase voltage. Neutral current harmonics are at the triplen of the source frequency. The rms neutral current is

$$\tilde{I}_N = I_o,\ \tilde{I}_{N3} = \frac{4I_o}{3} \qquad (8\text{-}72)$$

The third harmonic dominates over all triplens. Hence, the neutral wire conducts three times the third harmonic current of each individual line. The selection of the wire gauge of the neutral wire must take this factor into account.

Example 8-13

Consider that a three-phase four-wire utility is loaded with three identical 1 hp dc motors of Example 8-5 in each phase. Plot the third harmonic of neutral current and its rms value.

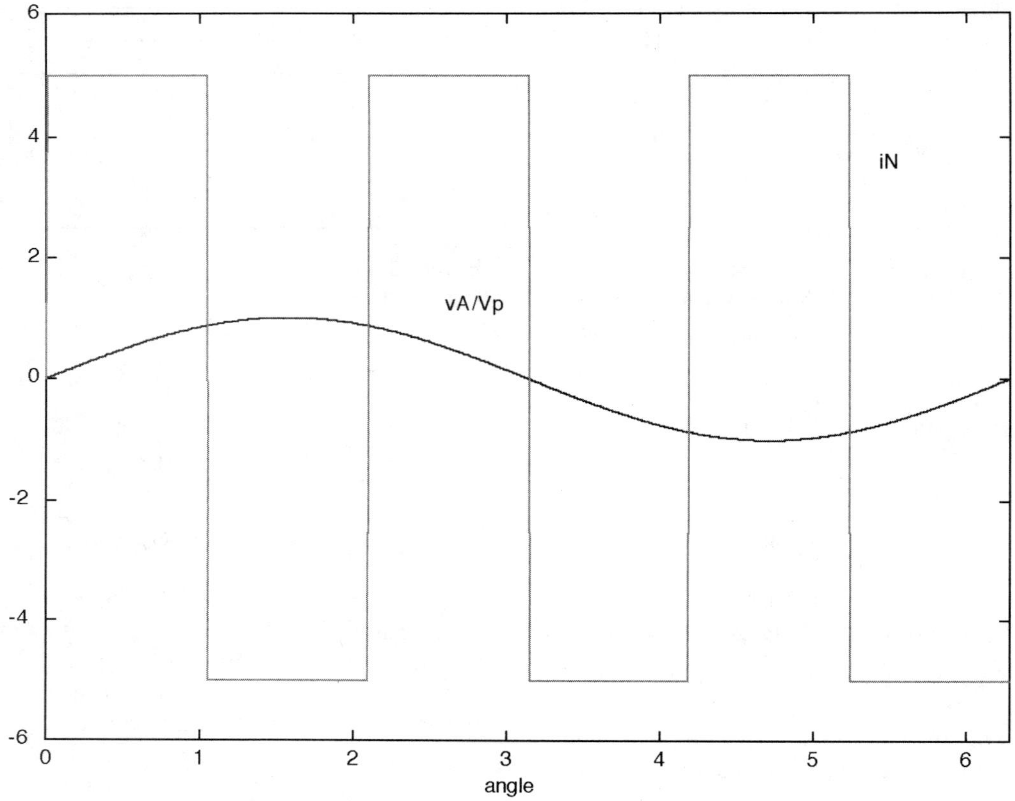

Figure 8.37
Neutral current

```
Vp=120; IO=5; dth=pi/512; th=0:dth:2*pi-dth;
vAN=Vp*sin(th); vBN=Vp*sin(th-2*pi/3);
   vCN=Vp*sin(th-4*pi/3);
iA=IO*((vAN>0)-(vAN<0));
   iB=IO*((vBN>0)-(vBN<0));
   iC=IO*((vCN>0)-(vCN<0));
iN=iA+iB+iC;
```

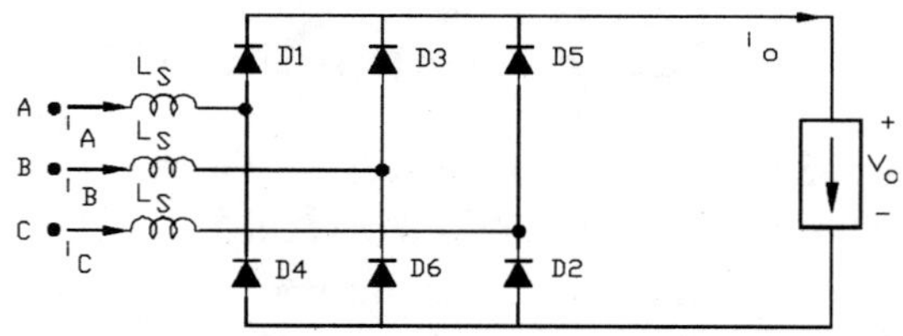

Figure 8.38
A 3-phase 6-pulse full-bridge rectifier

8.6 3-PHASE 6-PULSE FULL-WAVE RECTIFIER

The output voltages in single-phase converters have high ripple amplitude and low ripple frequency ($2f_s$). These limitations can be overcome or minimized by using multiphase input sources. Figure 8.38 shows a three-phase ac-dc rectifier with a highly inductive load. The line voltage is a sine wave of frequency ω and rms amplitude V_L.

The line voltages and line-line voltages are given below:

$$v_{AN} = \sqrt{2}\,\tilde{V}_P \sin\theta,\ v_{BN} = \sqrt{2}\,\tilde{V}_P \sin(\theta - 120°),$$
$$v_{CN} = \sqrt{2}\,\tilde{V}_P \sin(\theta - 240°)$$
$$v_{AB} = \sqrt{2}\,V_{LL} \sin(\theta + 30°),$$
$$v_{BC} = \sqrt{2}\,V_{LL} \sin(\theta - 90°),$$
$$v_{CA} = \sqrt{2}\,V_{LL} \sin(\theta - 210°) \qquad (8\text{-}73)$$

V_P is the line to neutral rms voltage and V_{LL} is the line-to-line rms voltage ($=\sqrt{3}V_P$). The load is assumed to be highly inductive, hence the output current I_o is constant and ripple free. The frequency of the voltage ripple is $6f_s$; therefore, the filtering components are small.

8.6.1 3-phase 6-pulse rectifier: constant output current

With Zero Source Inductance

Converter Operation The output voltage waveform consists of six pulses, each $\pi/3$ wide. Each pulse shows the response of a circuit mode in the steady state. The state of diodes and the transfer of currents are shown in the following table:

pulse number (mode)	conducting diode	current transition
1	D6	D5 to D1
2	D1	D6 to D2
3	D2	D1 to D3
4	D3	D2 to D4
5	D4	D3 to D5
6	D5	D4 to D6

The steady state operation consists of six circuit modes during a single period of the source waveform, depending upon the state of the thyristors as shown in Figure 8.39.

In the following discussion, we reference all angles with respect to the voltage waveform of phase A, v_{AN}. The line-to-line voltage v_{AB} leads v_{AN} by 30°. In the first mode, diode D6 was

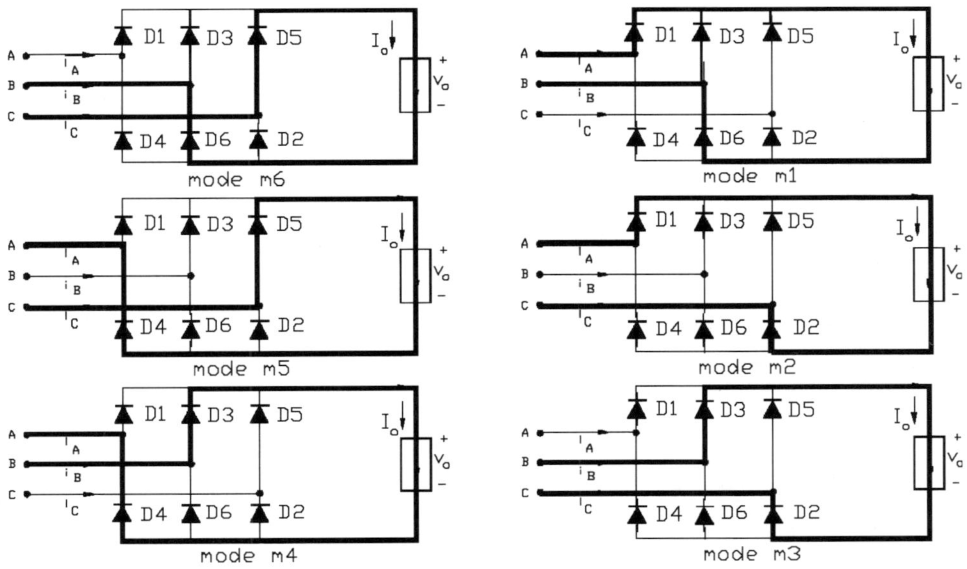

Figure 8.39
Circuit modes

already conducting and diode D1 begins to conduct at $\pi/6$. D5 is reversed biased and the output current is steered via D1. Current flows from phase A, through D1, load and D6 to phase B. At angle $\pi/2$ in mode 2, the diode D2 begins to conduct, which reverse biases D6 and steers the output current from phase A, through D1, load and D2 to phase C. At angle $5\pi/6$ in mode 3, diode D3 conducts, which reverse biases D1 and steers the output current from phase B, through D3, load and D2 to phase C. At angle $7\pi/6$ in mode 4, diode D4 conducts, which reverse biases D2 and steers the output current from phase B, through D3, load and D4 to phase A. At angle $9\pi/6$ in mode 5, diode D5 is turned on, which reverse biases D3 and steers the output current from phase C, through D5, load and D4 to phase A. Finally at angle $11\pi/6$ in mode 6, diode D6 conducts, which reverse biases D4 and steers the output current from phase C, through D5, load and D6 to phase B. Mode 6 ends at $2\pi+\pi/6$ when D1 is turned back on again. Output voltage $\mathbf{v_o}$ and the current through the load $\mathbf{i_o}$ are shown in Figure 8.40.

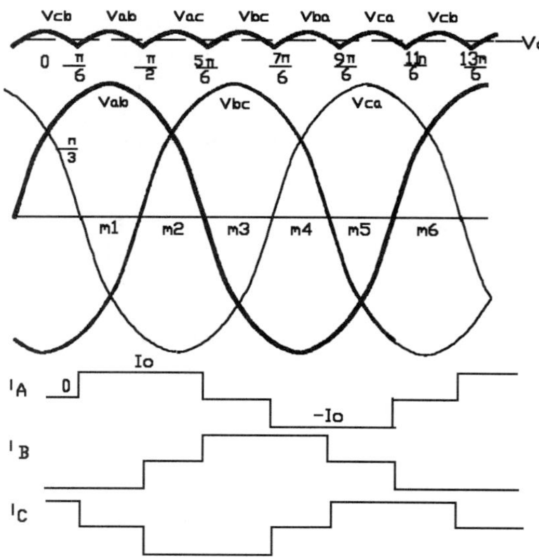

Figure 8.40
Waveforms

Characteristics Output dc voltage is calculated as the average of the output voltage waveform over a subperiod from $\pi/6$ to $\pi/2$:

$$v_o = v_{AB} = \sqrt{2}\, V_{LL} \sin(\theta + \pi/6); \quad \pi/6 \le \theta < \pi/2 \tag{8-74}$$

$$\overline{V}_o = \frac{1}{\pi/3}\int_{\pi/6}^{\pi/2} v_o\, d\theta = \frac{3\sqrt{2}}{\pi}\tilde{V}_{LL} \tag{8-75}$$

Voltage transfer ratio is found as

$$T_{vv} = \frac{\overline{V}_o}{\tilde{V}_{LL}} = \frac{3\sqrt{2}}{\pi} = 1.35 \tag{8-76}$$

Output voltage ripple frequency is $6f_s$.

RMS input line current and current transfer ratio are

$$\tilde{I}_L = \sqrt{\frac{1}{\pi}\int_{\pi/3}^{\pi} I_o^2\, d\theta} = \sqrt{2/3}\, I_o = 0.816\, I_o$$

$$= I_o\sqrt{\frac{2}{3}} \tag{8-77}$$

$$R_{ii} = \frac{\tilde{I}_L}{I_o} = \sqrt{\frac{2}{3}} \tag{8-78}$$

Example 8-14

Draw the phase current waveform of phase A in a four-wire three-phase, 120 V rms line-line voltage, 60 Hz, supply power to a current sink load of 5 A. Determine the rms current, the dc output current Io, phase angle, input power factor, THD and the rms fundamental current in phase A. Draw the frequency spectrum for the line current to at least 20 harmonics.

Some important MATLAB lines are given here. The rest of the program lines are left for the reader.

```
dth=pi/512; %step angle in angle array
th1=-pi/6:dth:pi/6-dth; th2=pi/6:dth:pi/3-pi/6-dth;
  th3=pi/3+pi/6:dth:2*pi/3+pi/6-dth;
th4=th1+pi; th5=th2+pi; th6=th3+pi; th=[th1 th2 th3
  th4 th5 th6];
vAB=sqrt(2)*VLL*sin(th+pi/6);
vBC=sqrt(2)*VLL*sin(th+pi/6-2*pi/3);
vCA=sqrt(2)*VLL*sin(th+pi/6-4*pi/3);
```

```
vAN= sqrt(2/3)*sin(theta);
vo1=sqrt(2)*VLL*sin(th2+pi/6);
  % output voltage in mode m1
isA=[zeros(th1) IO*ones(th2) IO*ones(th3) zeros(th4)
  -IO*ones(th5) -IO*ones(th6)];
[f,isAh,ph]=harmonic(isA); phase1= ph+pi/6;
  isA1=isAh(2)*sin(th+phase1);
```

VO = 162.0266 isArms = 4.0825 df = 1.0000
pf = .9558 isA1rms = 3.9020

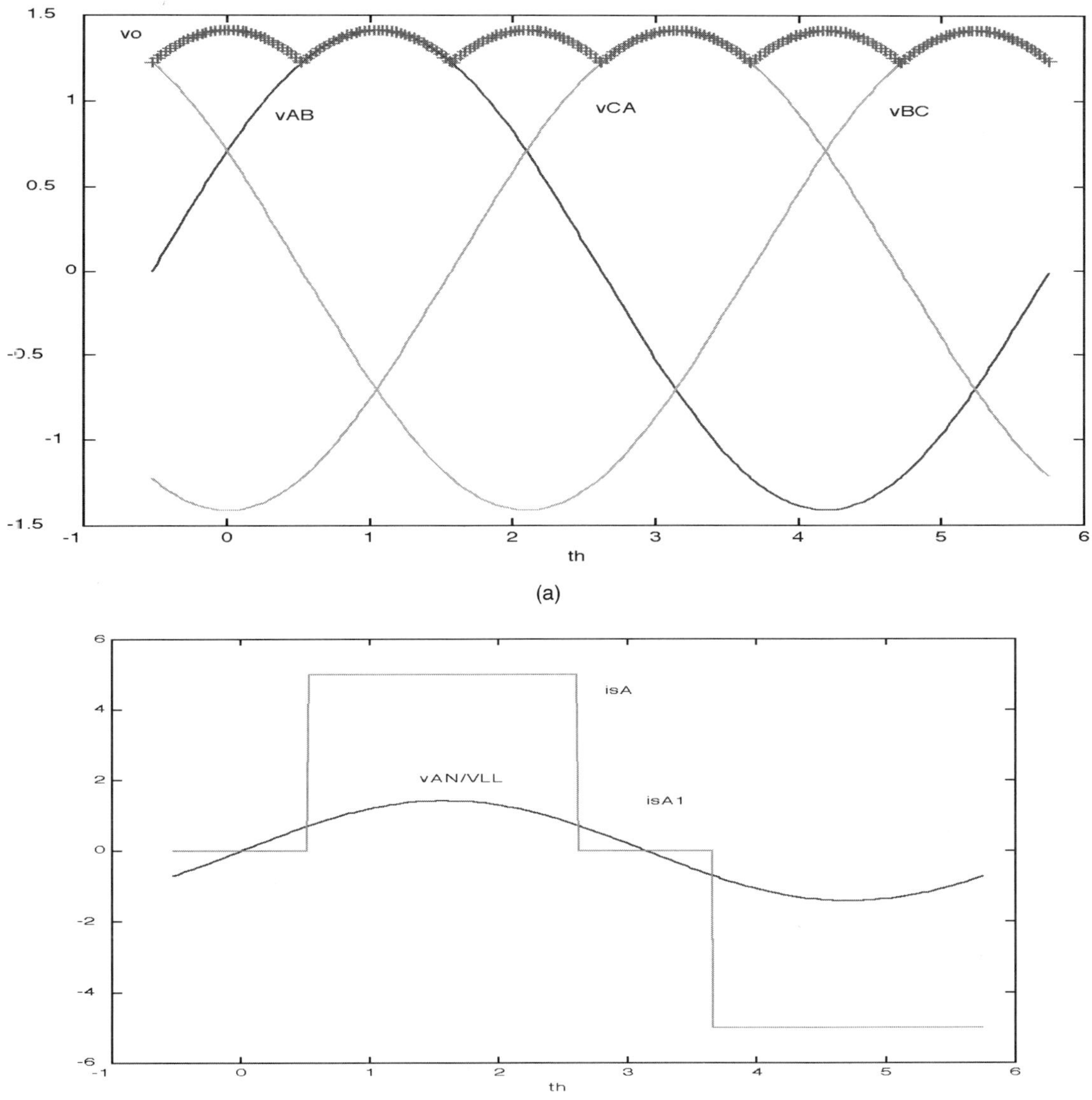

(a)

(b)

Figure 8.41
(a) Voltage waveforms and (b) current in phase A

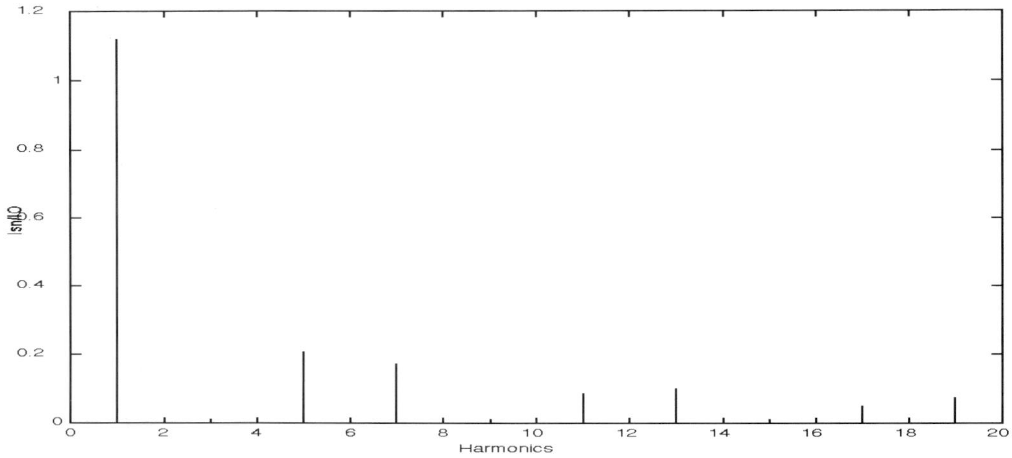

(c)

Figure 8.41 ***(continued)***
(c) Current in phase A

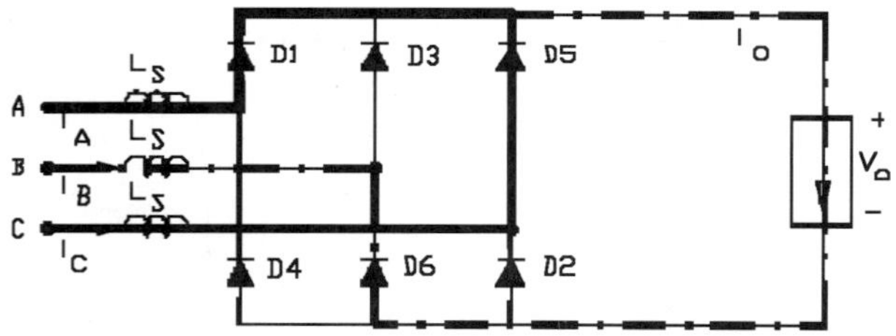

(a)

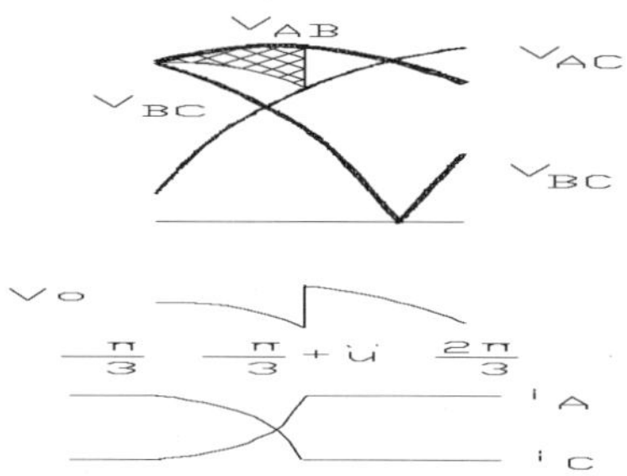

(b)

Figure 8.42
(a) Mode 1x (b) Partial output voltage waveform with L_S

With Source Inductance L_S

This section will consider the effect of source side inductance in a three-phase full-bridge rectifier with constant current output. Like the one-phase full converter, the source inductance affects only the output voltage waveform. It will suffice to investigate the waveforms in one pulse period. A new mode is created whenever current is steered between the diodes. For example, a new mode 1x is created when the current is steered away from diode D5 to D1, as given below and shown in Figure 8.42.

Prior to mode 1x, diodes D5 and D6 are conducting. D1 conducts at $\pi/3$. The source inductance L_S keeps D5 conducting. The current in D5 decreases and increases in D1. D5 will cease to conduct when the current through it builds up to I_o at $\pi/3+u$ where u is called the commutation angle. During the angle u, L_S in line B conducts constant current I_o and is effectively short circuited. Once D5 is reverse biased, D1 begins to conduct constant current I_o and is also effectively short circuited. It should be noted that the line voltage v_{AC} is shorted through two source inductances during this mode. This shorting causes what is normally called line notching in power lines and degrades the quality of the power from the utility.

Output voltage during mode 1x is given by

$$v_o = v_{AB} - L_S\frac{di_A}{dt} \quad (8\text{-}79)$$

The source inductance in line B carries constant current and, therefore, does not have any voltage across it.

The sum of line currents i_A and i_C equals the constant current I_o, hence

$$\frac{di_A}{dt} = -\frac{di_C}{dt}, \text{ and } v_{AC} = L_S\frac{di_A}{dt} - L_S\frac{di_C}{dt} \quad (8\text{-}80)$$

Therefore,

$$v_o = v_{AB} - \frac{v_{AC}}{2} = v_{AB} + \frac{v_{CA}}{2}$$

$$v_o = \sqrt{2}\tilde{V}_{LL}\sin(\theta + 30^\circ) + \frac{\tilde{V}_{LL}}{\sqrt{2}}\sin(\theta + 150^\circ) \quad (8\text{-}81)$$

Output voltage waveform is shown in Figure 8.43(b).

The output voltage over a pulse period is

$$v_o = \sqrt{2}\tilde{V}_{LL}\sin(\theta + 30^\circ) + \frac{\tilde{V}_{LL}}{\sqrt{2}}\sin(\theta + 150^\circ);$$

$$\pi/6 \le \theta < \pi/6 + u = \sqrt{2}\tilde{V}_{LL}\sin(\theta + 30^\circ);$$

$$\pi/6 + u \le \theta < \pi/2 \quad (8\text{-}82)$$

Output dc voltage is calculated as the average of the output voltage waveform over a subperiod from $\pi/6$ to $\pi/2$:

$$\bar{V}_o = \frac{3}{\pi}\left[\int_{\pi/6}^{\pi/6+u} v_o\, d\vartheta + \int_{\pi/6+u}^{\pi/2} v_o\, d\theta\right]$$

$$= \frac{3}{\pi\sqrt{2}}\tilde{V}_{LL}(1 + \cos u) \quad (8\text{-}83)$$

Line current i_C in the inductor L_S on line C in mode 1x is found from:

$$-L_S\frac{di_C}{dt} = \frac{v_{AC}}{2}$$

$$i_C = I_o - \int_{\pi/6} \frac{\sqrt{2}\tilde{V}_{LL}}{\omega L_S}\sin(\theta + 150^\circ)d\theta \quad (8\text{-}84)$$

Angle u is obtained by equating i_C to zero at $\pi/6+u$

$$\cos u = 1 - \frac{\sqrt{2}\omega L_S I_o}{\tilde{V}_{LL}} \quad (8\text{-}85)$$

Hence, the dc output voltage is

$$\bar{V}_o = \frac{3\sqrt{2}}{\pi}\tilde{V}_{LL} - \frac{3}{\pi}\omega L_S I_o$$

$$= 1.35V_{LL} - 0.95\,\omega L_S I_o \quad (8\text{-}86)$$

The dc voltage is now a function of the output current.

8.6.2 3-Phase 6-Pulse converter: constant output voltage

We will analyze an example of voltage sink load in Discontinuous Conduction Mode of operation where the voltage sink

$$V_o > \frac{2\sqrt{2}}{\pi}V_{LL} \quad (8\text{-}87)$$

See Figure 8.43(a). The diode sequencing is same as in Section 8.6.1.1. Each phase conducts in two consecutive mode periods in both polarities. For example, phase A conducts positive current in mode m1 and mode m2. Current in each of these modes is discontinuous. The source current in mode m1 is

$$i_{sA} = \frac{1}{L}\int_{\alpha}^{\beta}\frac{v_L}{\omega L}d\theta;\ \pi/6 \le \alpha < \beta \le \pi/2 \quad (8\text{-}88)$$

where $v_L = v_{AB} - V_o = \sqrt{2}V_{LL}\sin(\theta + \pi/6) - V_o$

Phase A current in mode m2 is also given by

$$i_{sA} = \frac{1}{L}\int_{\alpha'}^{\beta'}\frac{v_L}{\omega L}d\theta;\ \pi/2 \le \alpha' < \beta' \le 5\pi/6 \quad (8\text{-}89)$$

where $v_L = v_{AC} - V_o = \sqrt{2}V_{LL}\sin(\theta - \pi/6) - V_o$

Negative phase A current is the mirror image of the positive current.

Example 8-15

Draw the phase current waveform of phase A in a four-wire three-phase, 120 V rms line-line voltage, 60 Hz, supply power to 100 mH inductor and a

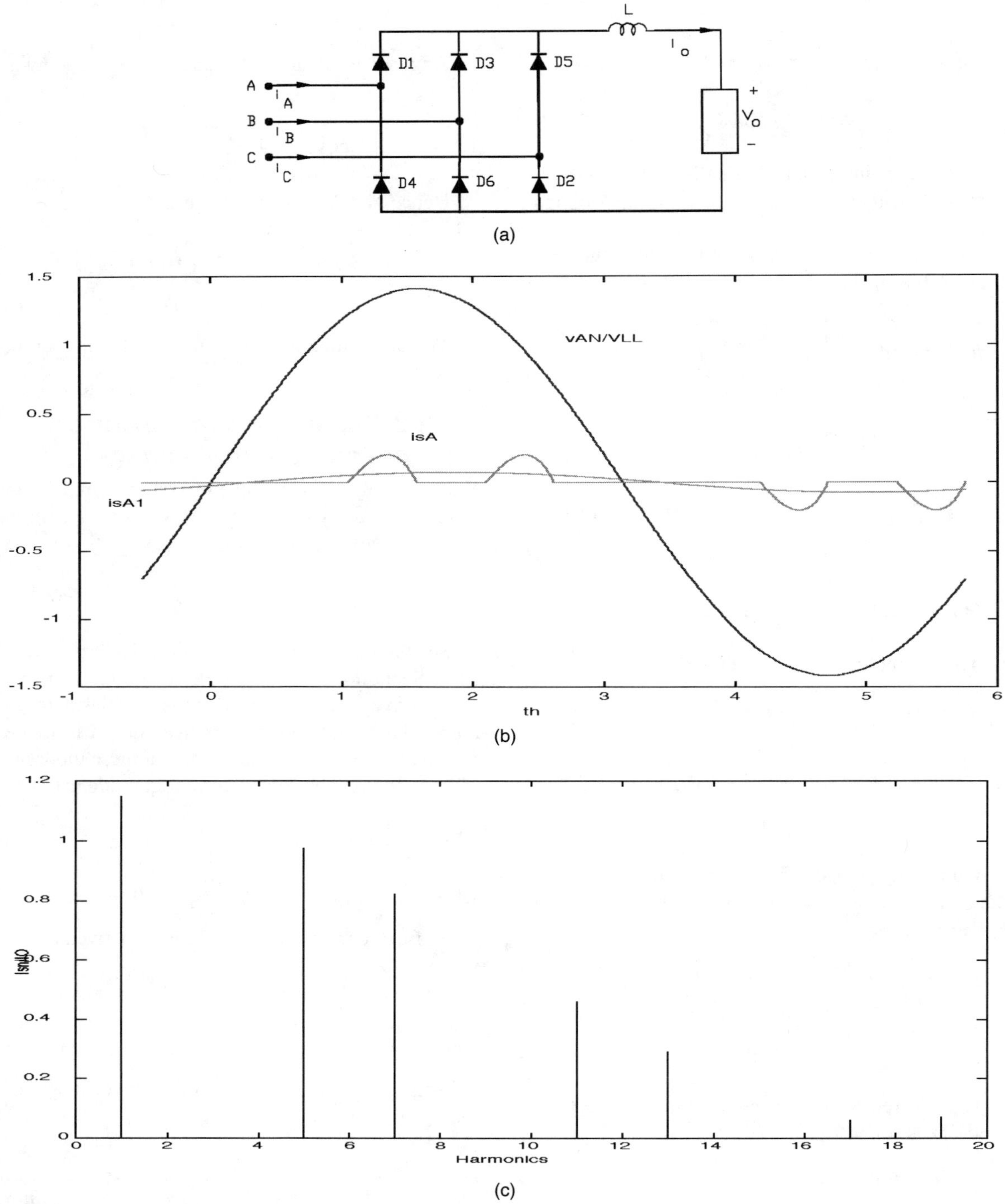

Figure 8.43
(a) 3-Phase rectifier with voltage sink load in DCM (b) current in phase A and (c) harmonic profile of current in phase A

voltage sink of 162 Vdc. Determine the rms current, the dc output current Io, the phase angle, the input power factor, the THD and the rms fundamental current in phase A. Draw the frequency spectrum of the line current up to at least 20 harmonics.

Some important MATLAB lines are given here. The rest of the program lines are left for the reader.

```
VO=162; vlA=sqrt(2)*VLL*sin (th2+pi/6)-VO;
isA1=integ(vlA/omegaL,th2);
isA1=isA1.*(isA1>0);
isA=[zeros(size(th1)) isA1 isA1 zeros(size(th4))
   -isA1 -isA1];
[f,isAh,ph]=harmonic(isA);
phase1=ph+pi/6
isA11=isAh(2)*sin(th+phase1);
```

isArms = .0841	**df=.9619**
IO = .0658	**pf=.6117**
isA1rms = .0535	**td=1.2135**

8.6.3 3-Phase 12-Pulse full-bridge rectifier

The ripple voltage can be further reduced and ripple frequency can be increased by using a six-phase input voltage source. The six-phase source can be synthesized from a three-phase source, a transformer with Y-connected secondary and another transformer with Δ-connected secondary, as shown in Figure 8.44. The secondary voltages of the two transformers differ by 30°, hence by adding them in series, a six-phase source becomes available.

8.7 TRANSIENT CONSIDERATIONS

Transient behavior at the time of turning on and turning off power to the rectifier circuits should be taken into account in the design process. At the turn on time, the capacitors are in the discharged state and the inductors are in the de-energized state. Application of power causes heavy inrush currents and high voltage transients on the devices connected to the capacitors and inductors respectively. Disconnecting the power causes similar transients.

Of particular interest is the inrush current in the rectifier with capacitive loads. Consider the worst case inrush current when the source voltage is at maximum V_S in a single-phase application. If the switching time is Δt, the inrush current is

$$i_{s0} = C\frac{V_S}{\Delta t}$$

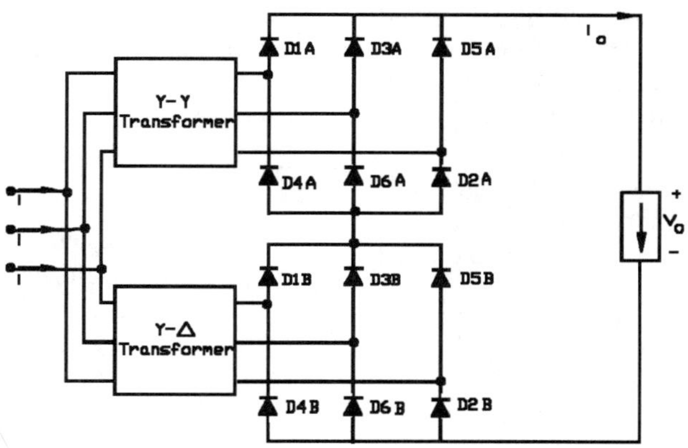

Figure 8.44
A 12-pulse converter

Alternatively, when the input voltage is zero but rising at the highest dv/dt, the inrush current is

$$i_{s0} = C\frac{V_S}{\omega}$$

Diode selection must take into account the inrush current capability. In the three-phase applications, the voltage V_S is replaced by $\sqrt{2}V_{LL}$, the line-to-line voltage.

8.8 DESIGN EXAMPLE

Design the charger for a 12-V automotive battery of 450 AH capacity in 8 hours.

The 450 ampere-hour capacity of the battery is the product of the constant discharge current of 45 A multiplied by the 10 hours in which the current remains constant. After 10 hours, the battery voltage drops to a voltage level that is detrimental to the life of battery. A completely discharged battery needs a constant current charge until its full voltage is built up. After the battery voltage is built up to full rated voltage, the charging circuit must either be disconnected or put in a constant voltage-trickle charge (low current) mode. The discharging and charging current of a battery is dependent on the maximum permitted inside temperature, hence, the charging current should not exceed too much over the discharging rate. Thus an eight-hour charge requires a constant current I_o= 56.25 A. We will use the single-phase rectifier of Figure 8.18 with source inductance. Select topology with source inductance for the reasons of low THD in the line current.

Use equation 8-47 such that the dc voltage available on the battery side is more than the voltage of the discharged battery (typically 90% of the rated value), hence,

$$\overline{V}_o = \frac{2V_S}{\pi}\left(1 - \frac{I_o}{I_{osc}}\right) \geq 0.9 \times 12$$

$$\frac{I_o}{I_{osc}} \leq .9$$

$$\frac{I_o \times 2\pi \times L_S}{V_S} \leq .9$$

Thus the source inductance should be less than or equal to 7.5 mH. The rectifier diodes and the source inductor should have the continuous current capacity more than I_{osc}=56.25/.9= 62.5 (usually 1.2-1.4 times). The maximum reverse voltage rating of the diode should be more than V_S=169.7056. A general purpose rectifier diode IR is a reasonable choice. Using the knowledge gained in Chapter 2, the wire gauge, the core and the number of turns are determined for the source inductance.

8.9 EFFECT OF DEVICE CHARACTERISTICS

In this chapter ideal diodes have been assumed, thus we avoided the discussion of the device characteristics such as the threshold voltage, forward resistance, diode capacitance and reverse recovery transition. The first two parameters do not significantly affect the performance. However, the rectifier design at high frequencies should take into account the reverse recovery transition and the capacitance of the reverse biased diodes. Inclusion of the capacitors turn simple diode rectifiers into switched L-C circuits. The scope of the text does not permit more discussion on these topics.

8.10 PSPICE SIMULATION

PSPICE Example 8-1 1-Phase AC-DC Converter with and without the Source Inductance

Consider the bridge rectifier of Figure 8.45(a) with source side inductance and the R-C load. Use 1N4002 as the rectifier. The uncontrolled rectifier specifications are

AC source = 24 V peak, 100 Hz
RL = 20 ohms C3 = 1000 μF

Perform a transient analysis for at least 5 periods and plot the waveforms of the output voltage and the source current for

(a) L_S = 0 and (b) L_S = 100 mH

Use the simulation step in the range: 1 ns to 100 ns.

Note that the source current contains thin and higher amplitude pulses for L_S=0. The source inductance L_S helps in smoothing the current pulses. Note the high inrush current in both cases.

PSPICE Example 8-2 3-Phase Voltage Rectifier with Voltage Sink

Consider the three-phase rectifier of Figure 8.46(a) with a dc voltage sink of 6 V and the R-L load. Use 1N4002 as the rectifiers. The uncontrolled rectifier specifications are

Phase AC source = 24 V peak, 100 Hz

R = 1 ohms L = 10 mH Voltage sink V_O: 6 V dc

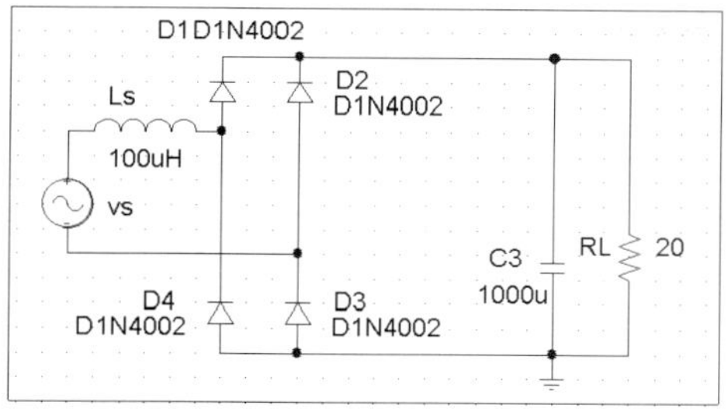

(a)

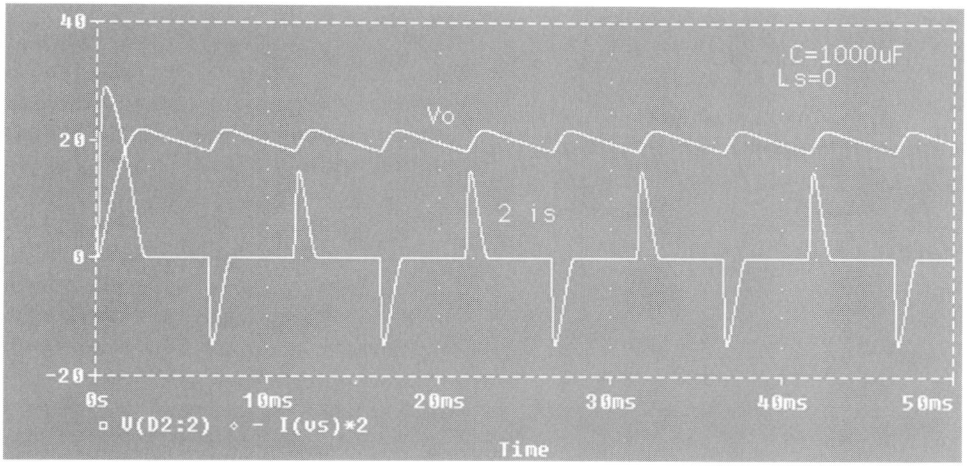

(b)

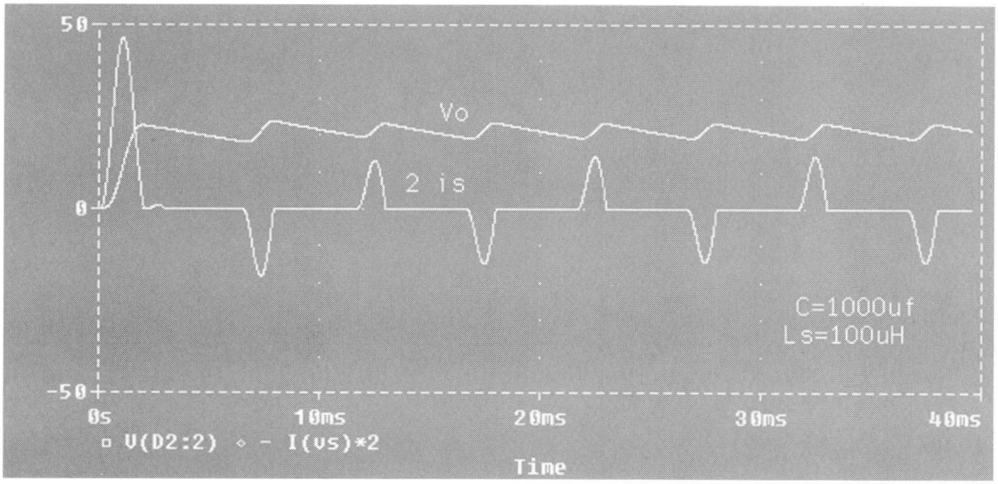

(c)

Figure 8.45
(a) Schematic, (b) waveforms for $L_S = 0$ and (c) waveforms for $L_S = 100\ \mu H$

Perform a transient analysis for at least 10 periods and plot the steady-state waveforms in the last two periods of the

(a) voltage and current in phase A,
(b) currents in phase A and phase B
(c) voltage at the output of the bridge rectifier, before the load resistor

Use the simulation step in the range: 1 ns to 100 ns

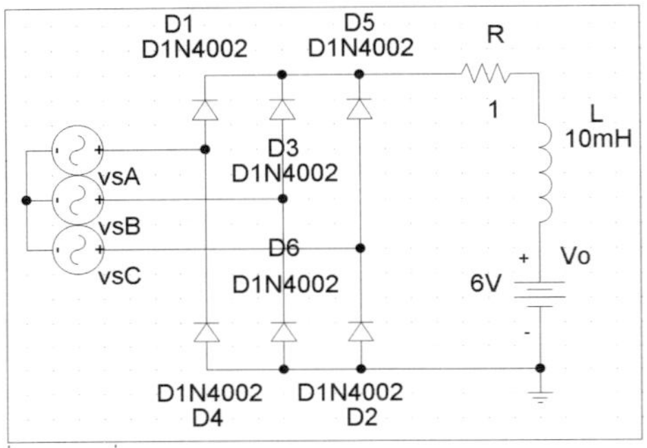

(a)

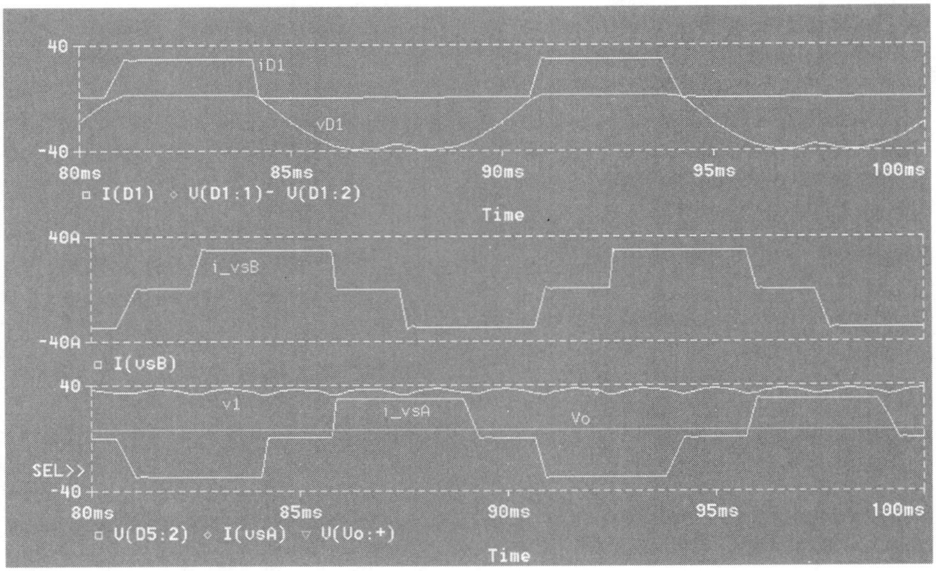

(b)

Figure 8.46
(a) Schematic and (b) waveforms of the 3-phase rectifier

8.11 SUMMARY

1. Line ac-to-dc converters without control provides an economical dc voltage source.
2. In diode rectifiers with capacitive and dc source loads, the line current may be discontinuous and cause poor (low) power factor operation. The line current has a high harmonic content.
3. The source inductances L_S gives rise to an effective source resistance, which is responsible for the drop in the output voltage with increasing current.
4. The source resistance is responsible for notching and other distortions in the ac line distribution system.
5. The neutral wire in a three-phase four-wire power system carries a large third harmonic current.
6. Ripple voltage and current are twice the line frequency in the case of a 1-phase full-wave

2-pulse rectifier and 6 times in a 3-phase full-wave 6-pulse rectifier. The ripple frequency may be further increased to 12 times by using the 12-pulse rectifier. The higher the ripple frequency, the better the filtering is possible on the output side with small size filter components.

References

1. Philip T. Krein, *Elements of Power Electronics,* Oxford University Press, New York, 1998.
2. Ned Mohan, Tore M. Undeland, and William P. Robbins, *Power Electronics: Converters, Applications and Design,* 2nd Ed. , John Wiley & Sons, Inc., 1995.
3. Joseph Vithayathil, *Power Electronics: Principles and Applications,* McGraw-Hill Inc., New York, 1995.
4. Muhammad H. Rashid, *Power Electronics: Circuits, Devices and Applications,* 2nd Ed., Prentice-Hall Inc., New Jersey, 1993.
5. John G. Kassakian, Martin F. Sclecht, and G. Vergassian, *Principles of Power Electronics,* Addison-Wesley, Boston, 1991.
6. S.B. Dewan, G.R. Slemon, and A. Straughen, *Power Semiconductor Drives.* Wiley, New York, 1984.
7. G.K. Dubey, S.R. Doradla, A. Joshi, and R.M.K. Sinha, *Thyristorized Power Controllers.* Wiley, New York, 1986.
8. A.W. Kelley and W.F. Yadusky, "Rectifier design for minimum line-current harmonics and maximum power factor," *IEEE Trans. Power Electronics,* Vol. 7. No. 2, pp. 332-341, April 1992.
9. C.K. Tse, S.C. Wong, and M.H.L. Chow, "On lossless switched-capacitor power converters," *IEEE Trans. Power Electronics,* Vol. 10, No. 3, pp. 286-291, May 1995.
10. S. Cuk and R.D. Middlebrook, "A general unified approach to modeling switching dc to dc converters in discontinuous conduction mode," *IEEE Power Electronics Specialists Conference Record,* pp. 36-57, 1977.
11. V. Voperian, "Simplified analysis of PWM converters using model of PWM switch, Part I and Part II," *IEEE. Trans. Aerospace Electronic Systems,* Vol. 26, No. 3, pp. 497-505, May 1990.

Problems

8-1 In the half-wave rectifier of Figure 8.9, 120 V rms at 60 Hz, L=30 mH and R=10 ohms. Calculate the rms source current and the power delivered. Plot the current i and the source voltage v_s waveforms.

8-2 Add a freewheeling diode and repeat problem 8-1 for two values of R=1 and 10 ohms.

8-3 In the half-wave rectifier of Figure 8.14, 220 V rms at 50 Hz and C=10 μF. Calculate the voltage ripple for load resistance R=100 and 1000 ohms. Determine the average power delivered and plot the output voltage, current i_d through the diode and the source voltage v_s waveforms.

8-4 In the half-wave rectifier of Figure 8.15, 120 V rms at 60 Hz, L=30 mH and V_o=120 V. Calculate the rms source current and the power delivered. Plot the current i and the source voltage v_s waveforms.

8-5 Add a freewheeling diode and repeat problem 8-4.

8-6 Verify using MATLAB waveforms of a half-wave rectifier with voltage sink load and freewheeling diode as shown in Figure 8.17b.

8-7 Analyze the full-wave rectifier of Figure 8.19, for a square wave of 120 V peak and a current sink load I_o. Determine the dc output voltage, the voltage transfer ratio, the input power factor and the Total Harmonic Distortion. Plot the waveforms of the source current, the source voltage and the output voltage.

8-8 Repeat problem 8-7 with a bipolar pulse waveform with a pulse width τ and period T_s.

8-9 Repeat problem 8-7 with a source impedance L_S=50 mH.

8-10 In the single-phase full-wave rectifier of Figure 8.19, 120 V rms, 60 Hz, and the winding resistance of the inductor in the current sink R=10 ohms. The ratio $\omega L/R >> 1/2f_s$ is assumed to be valid. Find the value of the inductance L to limit the current ripple to within 10% of the dc value.

8-11 Calculate the power dissipated in the rectifiers if the rectifier diodes have the on-state voltage drop of 1 V each.

8-12 Verify the results of problem 8-10 using PSPICE.

8-13 For problem 8-10, plot the percent peak-to-peak ripple current versus the ratio L/R. Limit the percent ripple in a range from 0 to 10%.

8-14 For the single phase rectifier of Figure 8.19 with a source inductance L_S, plot the commutation angle u versus the normalized output current I_{no}. Determine the maximum limit on the output current.

8-15 Plot the input power factor and the THD versus the commutation angle in problem 8-14.

8-16 Include source resistance R_S in place of L_S in the one-phase full-wave rectifier of Figure 8.19. Analyze and find the output waveforms. Compare them vis. a vis. the observations of Section 8.4.1.3.

8-17 Include both the source inductance L_S and dc side inductance L in Section 8.4.2.2 and compare the results.

8-18 The circuit shown below is a transformerless dc unregulated power supply. Ignore R_S and determine analytically the dc output voltage, the current ripple, the input power factor and the THD using MATLAB. Verify by simulation using PSPICE. Compare and comment.
vs=120 Vac, 60 Hz
L_S=200 mH, R_S=10 ohms
C=100 μF, R=50 ohms

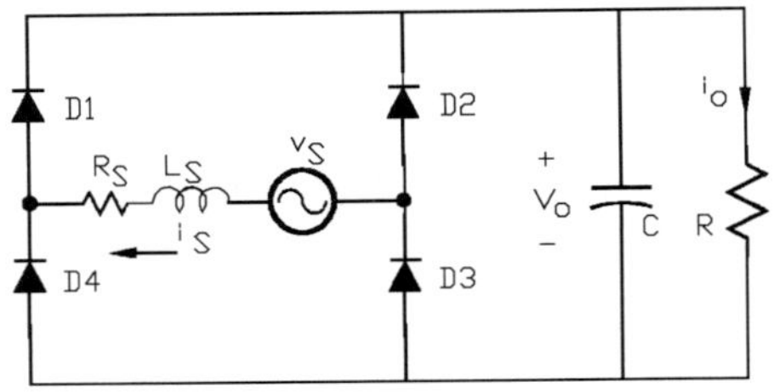

Figure P8.18

8-19 Verify the results of Example 8-6 using PSPICE by selecting a proper rectifier diode in place of ideal diodes.

8-20 Analyze the line-voltage distortion in a one-line supply due to a voltage sink load in CCM operation.

8-21 Draw the current waveform in the neutral wire of a three-phase four-wire system in a voltage source/source load with DCM operation.

8-22 Draw the current waveform in the neutral wire in an unbalanced single-phase loading with dc currents: phase A, 10 A, phase B, 20 A and phase C, zero current. Find the rms current in the neutral wire.

8-23 Draw the current waveform in the neutral wire in an unbalanced single-phase supply with equal dc current loads of 10 A in each phase but each phase voltage differing: phase A, 110 V rms; phase B, 120 V rms; and phase C, 120 V rms.

8-24 Find the expression of the peak-peak current ripple in the three-phase rectifier of Figure 8.38.

8-25 The three-phase rectifier circuit of Figure 8.38 is supplying approximately 2 kW load with 220 V rms at 50 Hz, L_S=7.5 μH, C=1000 μF. The load is a pure resistance of 50 ohms. Obtain the waveforms and simulate using PSPICE.

8-26 Find the maximum inrush current in a three-phase six-pulse rectifier connected to 208 V line and a 1100 μF on the load side. Assume that switch time is 1 ms.

8-27 Design the dc power supply for a dc motor from a 208 V rms three-phase ac supply that is drawing a dc current of 24 A and a back emf of 60 volts. The ripple current must be within 5% of the rated 24 A.

8-28 Consider the full-wave rectifier of Figure 8.19, with the ac source v_S = 200 sin t. The R-L load is connected in series with a dc voltage sink of 100 V. The R-L load is such that the dc current from the source is discontinuous.

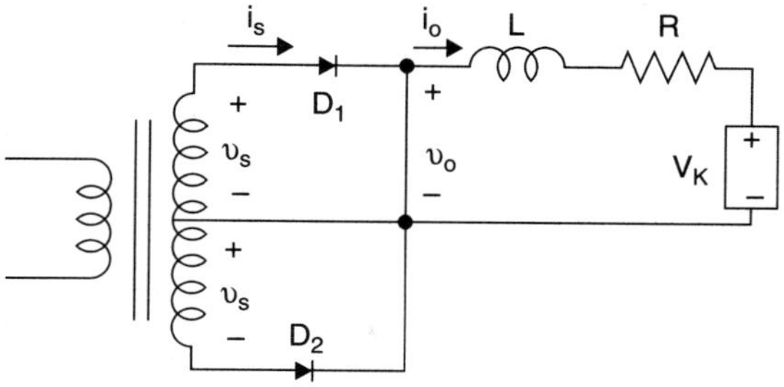

Figure P8.28

a. Plot the waveform of i_O and i_S.
b. Plot the harmonic profile of I_O and I_S.
c. Determine the average dc current I_O and rms current $I\tilde{}_S$.

8-29 A three-phase bridge rectifier has to supply 50 A at 450 V dc. What should be the ac input line voltage? Determine the current and voltage ratings of the diodes.

8-30 A three-phase bridge rectifier is supplied from a 460 V, 60 Hz 3-phase bus. The rectifier is loaded by 10 ohms in series with 2 mH inductor. Determine the frequency and amplitude of the fundamental component of the ripple in the output current.

8-31 A three-phase, half-wave rectifier is supplied from the secondary winding of a Y-connected transformer. The primary of the transformer is connected to a 660 Vrms (line) supply. The rectifer supplies a dc current of 20 A at 240 Vdc to a load. Determine the ratings of the primary and the secondary windings of the transformer.

CHAPTER 9
AC to DC: With Control

OBJECTIVES

1. Techniques of phase angle control, pulse width control, and Sine-PWM control
2. Generation of control signal
3. Converter operation with R, L and C loads
4. Converter operation with voltage and current sink loads
5. Converter operation when the output current is continuous and discontinuous in time
6. Converter operation with source-side inductance
7. Converter operation in the inversion mode
8. Series and parallel connection of converters
9. Three-phase controlled ac-dc converters

9.1 INTRODUCTION

We have seen in the previous chapter that the output voltage of the ac-dc converters using diodes is not controllable. That is because the diode is a self-controlled switch. Turning on a diode depends on the polarity of the cathode voltage. Turning off depends on the polarity of the device current. Output voltage can be controlled if the diode can be replaced by a switch whose turning on and off can be controlled by a variable control voltage that is different from the anode to cathode voltage. Three-terminal devices, such as the thyristors, bipolar transistors and MOSFETs, can be used as the controllable switches. This chapter is about the ac-dc converters that use these controllable switches. These converters are very popular in battery charger systems, electromagnet power supplies and variable speed dc drives ranging from the fractional horsepower to the several thousand horsepower level. Low power level converters are usually the single-phase type and medium and high power converters are the three-phase type.

This chapter will look into the performance of ac-dc converters of several useful topologies. The study of converter performance should normally include both the transient and the steady-state responses. The transient-state performance is very complicated and beyond the scope of this book. This chapter concentrates on the in-depth analysis of steady-state converter performance. Furthermore, the chapter will focus on the converter operation for low frequency input voltage sources such as the 50 to 60 Hz line or the 400 Hz line in aircraft systems. Analyses of ac-dc converter circuits at low and high frequencies are very similar except that at high frequencies the parasitic capacitances and inductances, charge storage and heat localizing problems in switching devices must be taken into account. The high-frequency ac-dc conversion considerations are discussed as the part of dc-dc converters in a later chapter. The last section gives the design examples of dc motor drives.

9.2 CONTROLLED AC-DC CONVERSION

A general ac to dc converter system is shown in Figure 9.1(a). Input may be a sinusoidal voltage source or a current source. Output is required to be either a dc voltage or a dc current. The control signal may be, as shall be seen later, an angle, a pulse width, a voltage or a current signal. Most of the practical sources in power electronics are voltage sources, and the converters are required to supply controllable dc currents or dc voltages or a combination of both. The converter consists of switches, reactive components like L, C, transformers and resistances.

The basic topologies are the same as in the ac-dc converters without control, see Section 6.2 in Chapter 6, except the switches are externally controlled devices. The electrical loads are similar. Assumptions for the performance analysis are also identical.

Ideally both output voltage and current must be dc and ripple free. In many applications it is sufficient to have at least one of them a constant dc. When both output voltage and output current are of similar polarity, the power flows from source to load (positive power delivery). The converter is then said to operate in either the first or third quadrant in the V_o-I_o space. When they are dissimilar, the power flows from the load back to the source (negative power delivery). The converter is then said to operate in either the second or fourth quadrant. The former is called the rectifier operation and the other the inverter operation. The regions of both operations are shown on the output plane in Figure 9.1(b). A dc motor using an ac-dc converter, moving in a forward direction, is the example of a first-quadrant operation. Movement in the reverse direction is the example of the third-quadrant operation. The dynamic breaking of the dc motor in forward motion requires the second-quadrant operation. The dynamic breaking in the reverse motion requires the fourth-quadrant operation. Operation of the second or fourth quadrant is referred to as the inverter operation. Some applications involve all four quadrants such as the electric elevator moving up or down and using dynamic breaking in both directions.

A full-wave converter is capable of handling high power with minimum component stresses. The half-bridge and full-bridge topologies have the identical power handling capability. A half-bridge topology has a lower count of semiconductor com-

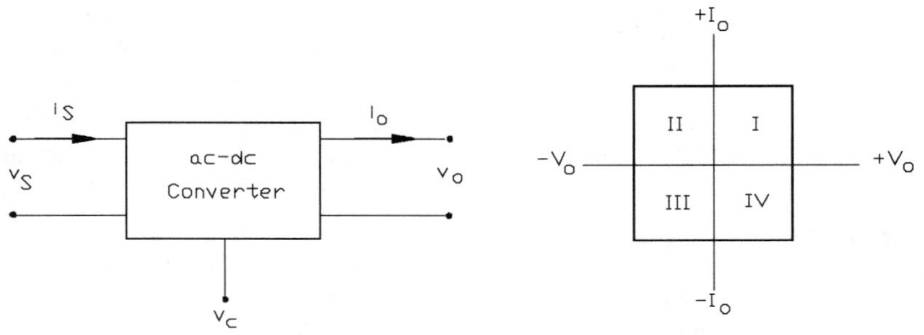

Figure 9.1
(a) An ac-dc converter system and (b) rectifier and inverter modes of operation

ponents but requires higher component stresses and a bulkier and underutilized transformer.

The switches unidirectionally conduct current in the direction of the arrow when turned on. The switch may be a three-terminal power electronic switching device such as the power bipolar transistor BJT, the Silicon Controlled Rectifier SCR, the Gate-Turn-Off thyristors GTO, the MOSFET, the IGBT, the MCT, etc. All switches have nonzero but small voltage across them in the on-state. The BJT, MOSFET and IGBTs are level-triggered devices. Thyristors (SCR), GTO and MCTs are pulse triggered devices. Bidirectional switches are synthesized by connecting two three-terminal devices in antiparallel. The other three-terminal device may also be substituted by a diode. The antiparallel diode may also be the body diode as found in the MOSFET devices.

Performance characteristics of controlled ac-dc converters are the same as those listed in Section 8.2.2 in the previous chapter. Control characteristics are added to determine the performance of the feedback control systems.

Control Characteristics

Control parameters are used to close the loop for regulation and control of one of the output characteristics. For a good performance of the overall system, a small range of variation of control parameter is desired for a large variation in the output characteristics. This amounts to selecting the region of highest slope on the plot of one of the forward characteristics versus the control parameter. The control parameter is usually the switching frequency, the on-time of switches or the delay in switching action or the overlapping period of simultaneous turn on or off of switches. More about the control is covered in Chapter 12.

9.2.1 Control techniques

Techniques of controlled conversion from ac to dc are illustrated by a simple half-wave circuit using a generic unidirectional switch shown in Figure 9.2. Switch S when turned on conducts current in the direction of the arrow. The output voltage waveform depends on the switch control waveform and the pulse-

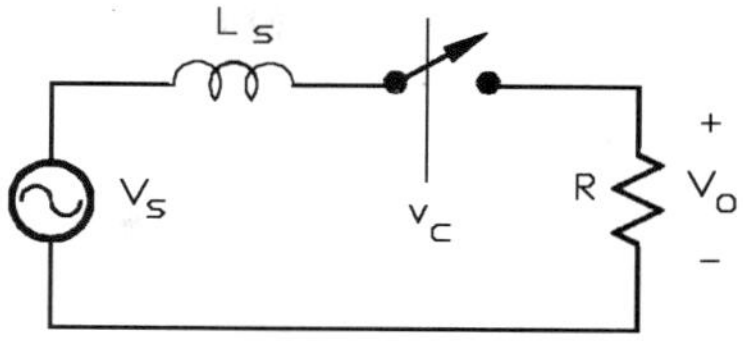

Figure 9.2
A half-wave ac-dc converter type of the unidirectional switch

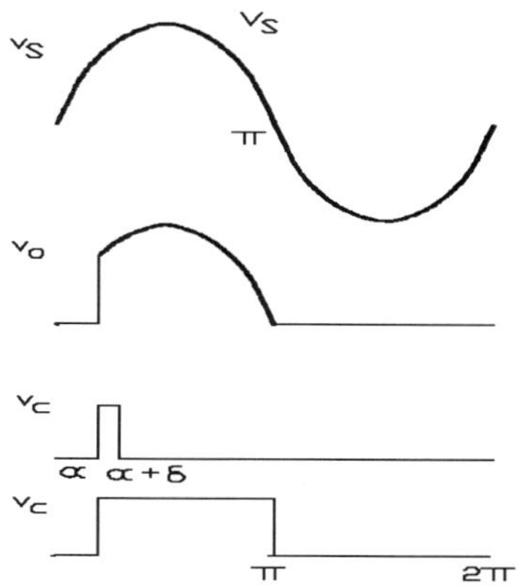

Figure 9.3
Firing angle control

triggered switch, such as a thyristor, or the level-triggered switch, such as a transistor. Usually the source has an inductance, such as the ac line inductance or the leakage inductance of the transformer. For analysis purposes, we will neglect this inductance.

Phase Angle Delay Control (Firing Angle Control)

Switch S is turned on by a single positive pulse per half period applied to its control terminal. The rising edge of the control pulse is delayed from the beginning of the period of the input voltage waveform by a controllable angle α, called the firing angle. In a pulse-triggered switch, such as a thyristor, the falling edge of the control pulse lies at an angle $\alpha+\delta$, as shown in Figure 9.3(c), where δ is a short angle. Once turned on, the switch remains on even if the triggering pulse has subsided. It can be turned off only when the current through it is reduced below the holding current (assumed zero here for simplicity). It should be kept in mind that current pulses are required for triggering in thyristors and GTOs. MCTs, MOSFETs and IGBTs require voltage pulses. Output voltage v_o is shown in Figure 9.3(b). In a level-triggered switch, such as a transistor, the falling edge lies at angle p, as shown in Figure 9.3(d).

Extinction Angle Control

The rising edge of the control pulse coincides with the beginning of the input voltage waveform. The falling edge lies at a controllable angle π-β. Angle β is called the extinction angle. In a pulse-triggered switch, the control pulse consists of two short pulses: one for turning on and the other for forced turn-off. The control pulses for level-triggered and pulse-triggered switches are shown in Figure 9.4(c) and (d). Output voltage v_o on the resistive load is shown in Figure 9.4(b).

Pulse Width Modulation (PWM) Control

The control pulse is symmetrically positioned with respect to the positive and negative peaks of the input voltage waveform. Pulse width δ is the control parameter. In the case of the pulse-triggered switch, like the extinction angle control, the control pulse consists of two short pulses. Control pulses for the level-triggered and pulse-triggered switches are shown in Figures 9.5(c) and (d). Output voltage v_o is shown in Figure 9.5(b).

Multiple Pulse Width Modulation (MPWM) Control

Control signals consist of p pulses per half period of the input voltage waveform. Both the pulse width δ and the number of pulses p can be used as the control parameters. There are two popular methods: Uniform multiple PWM, p pulses of uniform pulse-widths; and Sine-PWM, p pulses with sinusoidal variation of pulse width. This type of control is suited for level-triggered switches. Output voltage v_o in a three-pulse uniform width control is shown in Figure 9.6(b).

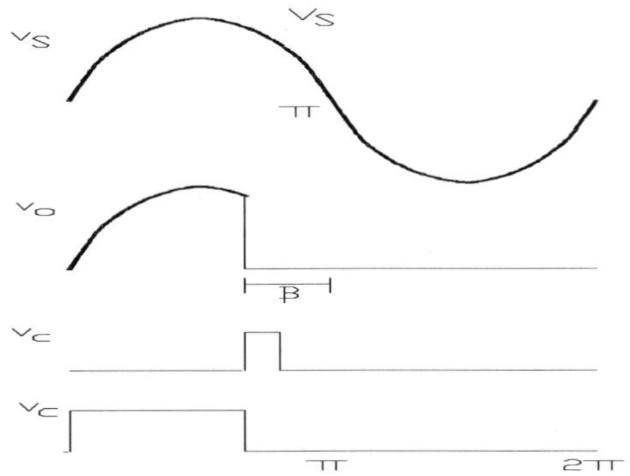

Figure 9.4
Extinction angle control

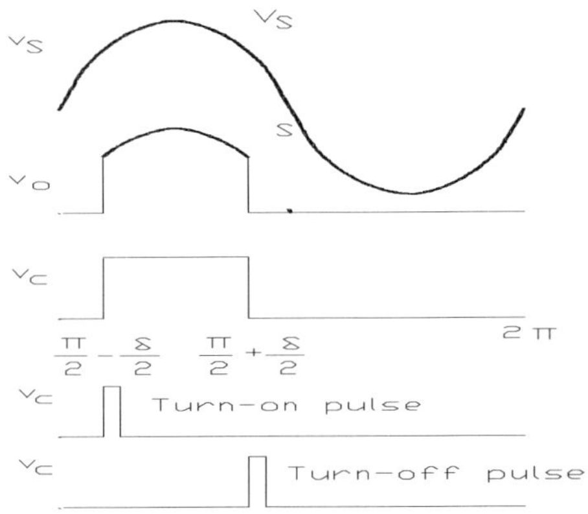

Figure 9.5
PWM control

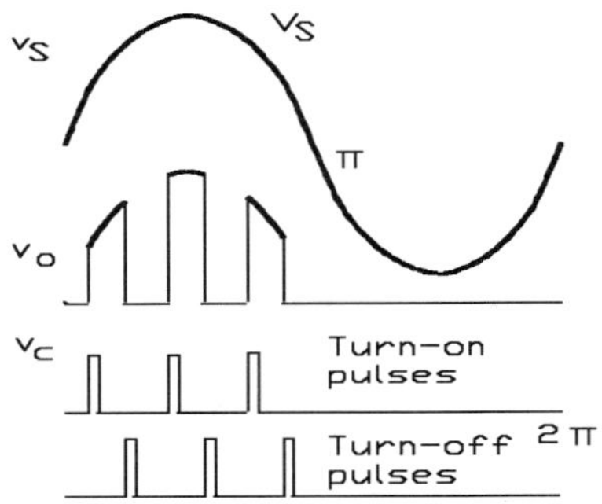

Figure 9.6
Multiple PWM control

9.2.2 Control signal generation

Control signals can be generated using either an analog circuit or by software programming on a dedicated hardware.

Analog Implementation of Control

A reference signal v_R and a control signal v_c are applied as inputs to an analog comparator, see Figure 9.7(a). The comparator output signal, v_{gate}, is used to control the gate of switching devices. Waveforms for various control techniques in full-wave converters are shown in Figures 9.7(b) to (f).

Phase delay angle:

$$\alpha = \frac{V_c}{V_R}\pi \tag{9-1}$$

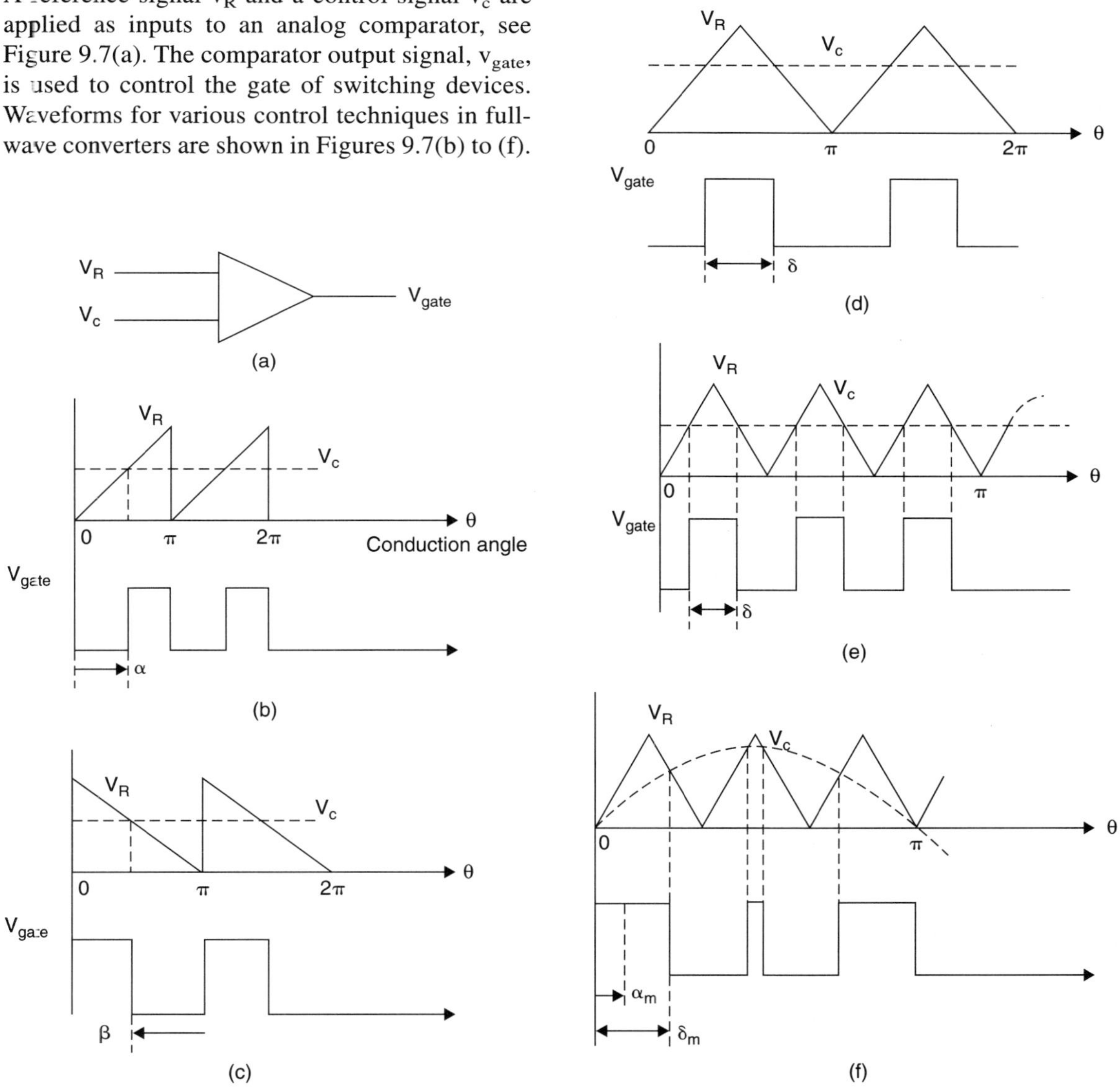

Figure 9.7
Control signal generation

Extinction angle:

$$\beta = \frac{V_c}{V_R}\pi \tag{9-2}$$

Pulse width:

$$\delta = \frac{V_c}{V_R}\frac{\pi}{2} \tag{9-3}$$

In the Uniform Multiple PWM Control Signal (Figure 9.7(e)), the reference signal is a triangular wave of frequency pf_s, where p is the number of pulses in a half period of the input ac voltage. The comparator output signal consists of p pulses of uniform pulse width δ in a half period of the input ac voltage. Pulse width δ and delay angle α_m for the m-th pulse are given by

$$\delta = \frac{V_c}{V_R}\frac{\pi}{2p} \tag{9-4}$$

$$\alpha_m = (2m-1)\frac{\pi}{2p} - \frac{\delta}{2} \tag{9-5}$$

In the Sine-PWM control signal (Figure 9.7(f)), the reference signal is a triangular wave of frequency pf_s, where p is the number of pulses in a half period of the input ac voltage. The control signal is a sinusoidal voltage with peak voltage V_c and frequency f_s. The comparator output signal consists of p pulses in a half period of the input ac voltage. Delay angle α_m and pulse width δ_m for the m-th pulse are given by

$$\frac{V_R}{V_c}\sin(\alpha_m) + \frac{2p}{\pi}\alpha_m - 1 = 0 \tag{9-6}$$

$$\frac{V_R}{V_c}\sin(\alpha_m + \delta_m) + \frac{2p}{\pi}\delta_m - 2 = 0 \tag{9-7}$$

9.3 1-PHASE AC-DC CONVERTERS

Popular ac-dc converters use full-bridge topologies. Full-bridge converters are designed for delivering constant but controllable dc current or dc voltage to the load. Electrical load may be a battery charger that needs unidirectional power flow from the ac source to the battery. In a dc motor load, power flows in both directions: from the source to the dc motor in forward or backward motoring and from the dc motor to the ac source in generator mode or braking mode. The former is called the single-quadrant operation and the later is called two-quadrant operation. Full-bridge converters are capable of operating in a single quadrant or two quadrants. The next sections will look into a full-converter with both types of loads: the constant current load and the constant voltage loads.

9.3.1 1-phase 2-pulse full-bridge converter with constant output current

Full-converter topology is shown in Figure 9.8 using pulse-triggered switches such as thyristors, GTOs and MCTs. Level-triggered switches such as the IGBTs and MOSFETs may also be used with proper gate control waveforms. The source inductance L_s represents the source inductance, utility conductor inductance and/or leakage inductance of the utility transformer. First consider a hypothetical situation $L_S = 0$ and then later include its effect.

Phase Angle Delay Control

Converter Operation Converter operation in steady-state is best described over a period that begins from the phase angle α to $2\pi+\alpha$. The steady-state operation involves two circuit modes during a single period of the source waveform depending upon the state of the switches, as shown in Figure 9.9.

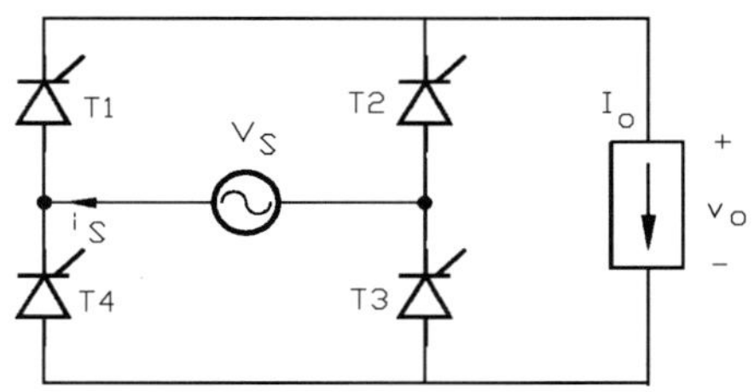

Figure 9.8
A 1-phase 2-pulse full-bridge converter topology

Mode 1 begins when thyristors T_1 and T_3 are turned on at an angle α by the control pulses applied at their gate terminals. During mode 1, thyristors T_1 and T_3 are in forward conduction mode and thyristors T_2 and T_4 are in reverse blocking mode. Current I_o flows through the path shown in Figure 9.9. After angle π, the input source voltage becomes negative but the thyristors T_1 and T_3 keep conducting. Note that the current sink is the model of a high value inductor, voltage across it can change instantaneously but the current cannot. Hence the output voltage v_o becomes negative and follows the input voltage v_S waveform. Note here that the current I_o is emanating from the positive end of the current sink, therefore, the current sink is acting as a generator, the second-quadrant operation. This is the case where the current sink (load inductor) is returning power back to the input source. The input source is supplying power to the load during α to π, referred to also as the rectifier operation. The load is returning power back to the input source during π to $\pi+\alpha$, referred to as the inverter operation.

Mode 2 begins when thyristors T_2 and T_4 are turned on at an angle $\pi+\alpha$ by the control pulses applied at their gate terminals. The current is steered away from thyristors T_1 and T_3 to T_2 and T_4 effecting a natural commutation. Now thyristors T_1 and T_3 are in reverse blocking mode. The converter operation in this mode is identical to that of mode 1 during the angle from $\pi+\alpha$ to $2\pi+\alpha$. The output voltage v_o and the input current i_s are shown in Figure 9.10(a).

The angle arrays for modes for a given $\alpha=\pi/4$ in MATLAB are given below:

```
dtheta=pi/512; alpha=pi/4;
theta1=alpha:dtheta:pi+alpha−dtheta;% mode m1
theta2=pi+alpha:dtheta:2*pi+alpha−dtheta; %mode m2
theta=[theta1 theta2]; % angle array
```

The input and output voltage arrays for a 120 Vac, 60 Hz source:

```
VS=120*sqrt(2); vs=VS*sin(theta); %input voltage array
vo=[vs(1:size(theta1,2)) −vs(size(theta1,2))];
  %output voltage
```

The input current array for a dc output current $I_o=50$ A:

```
IO=50; is=[IO*ones(size(theta1)) −IO*(size(theta2))];
  %source current
```

Analysis Performance characteristics, mentioned in Section 9.2, will be determined from the circuit modes, plotted and discussed. One way to analyze converter performance is to use equations (9-1) to (9-7) and derive closed form solutions and then plot the graphs of the characteristics. An alternative way to analyze converter performance is to use the numerical computer technique of determining

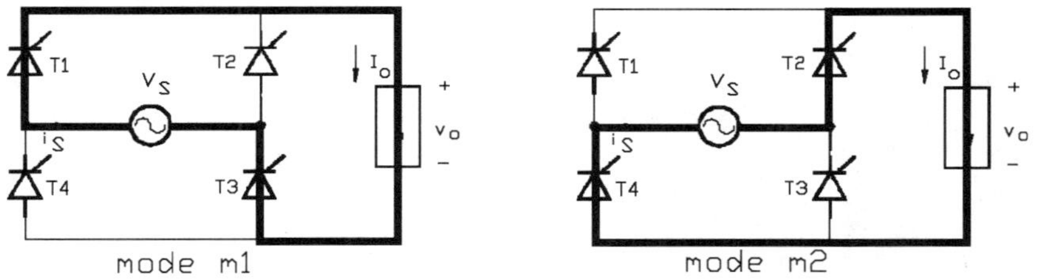

mode 1 $\alpha \le \theta < \pi + \alpha$ **mode 2** $\pi + \alpha \le \theta < 2\pi + \alpha$

Figure 9.9
Circuit modes

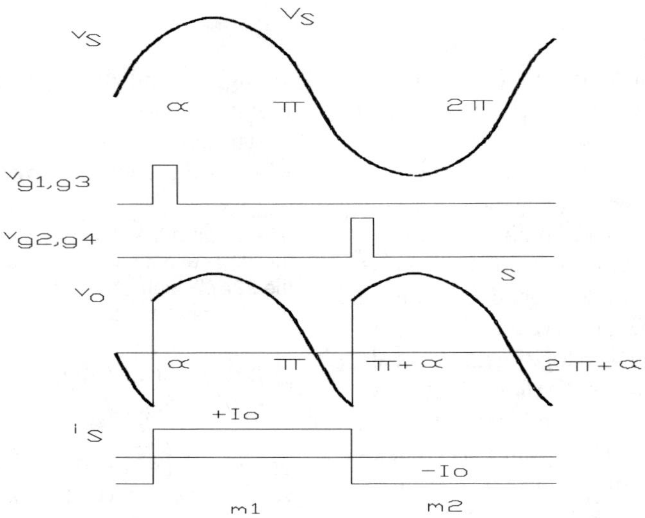

(a)

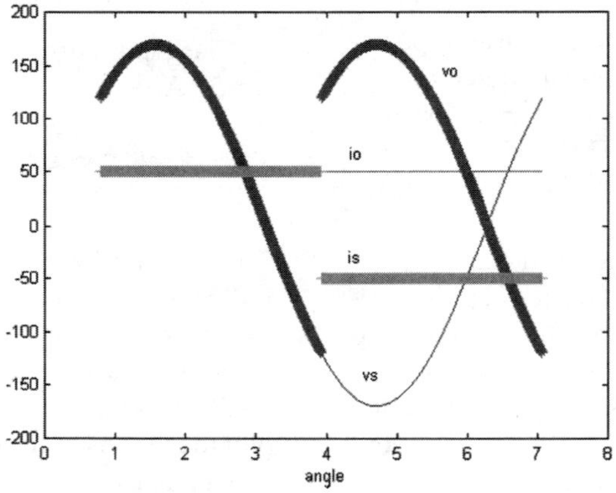

(b)

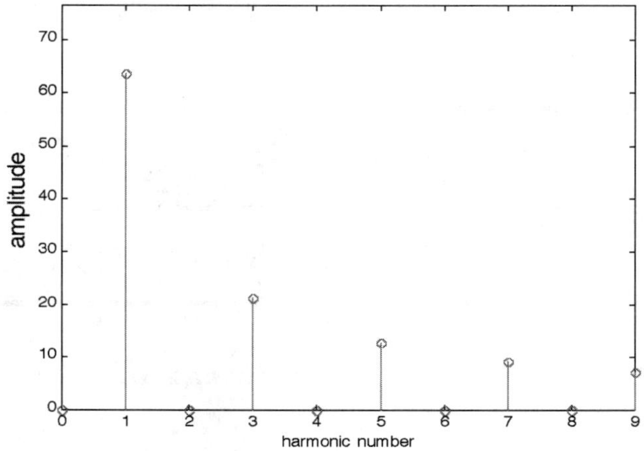

(c)

Figure 9.10
(a) Waveforms (b) waveforms and (c) frequency spectrum of input current

characteristics from the knowledge of circuit modes. As a first step in the numerical method, we make arrays for the voltages and current waveforms of interest over a repetitive period of time or a conduction angle θ. Smaller sub-arrays are constructed for each circuit mode separately and then concatenated to span a repetitive period. The resulting arrays are then processed to find average, rms values and Fourier coefficients, etc. The characteristics are then plotted and discussed. The numerical method provides a reasonably good engineering solution and design parameters.

Output Characteristics Output dc voltage is calculated as the average of the output voltage waveform over a half period:

$$\bar{V}_o = \frac{1}{\pi}\int_{\alpha}^{\pi+\alpha} V_S \sin\theta \, d\theta = \frac{2V_S}{\pi}\cos\alpha \quad (9\text{-}8)$$

The voltage transfer ratio, defined as the ratio of output dc voltage to the rms input voltage, is given by:

$$T_{vv} = \frac{\bar{V}_o}{\tilde{V}_S} = \frac{2\sqrt{2}}{\pi}\cos\alpha \quad (9\text{-}9)$$

```
VO =mean(vo) % dc output voltage
Tvv =VO/(VS/sqrt(2))
  %voltage transfer ratio
```

VO= 76.6285 Tvv= .6386

Input Characteristics Input current is a square wave with I_o as the peak value. The RMS input current is equal to the dc output current I_o:

$$\tilde{I}_S = \sqrt{\frac{1}{\pi}\int_{\alpha}^{\pi+\alpha} I_o^2 \, d\theta} = I_o$$

```
isrms= sqrt(mean(is .^ 2))
isrms= 50
```

Current reflection ratio R_{ii}, defined as the ratio of input rms current to the dc output current, is unity.

The i_s waveform has half-wave symmetry because the waveform in the upper half period is the mirror image of the waveform in the lower half period.

Hence, the dc component is zero and even harmonics are absent in the Fourier expansion, given below:

$$i_S = \sum_{n=1,3,5,\ldots}^{\infty} \frac{4I_o}{n\pi}\sin(n\theta - n\alpha) \quad (9\text{-}10)$$

RMS current and the phase angle of the fundamental harmonic with respect to the input voltage are

$$\frac{\tilde{I}_{S1}}{I_o} = \frac{4}{\pi}, \quad \phi_1 = -\alpha \, (lagging) \quad (9\text{-}11)$$

```
[f,ish,ph] = harmonic(is,10);
  %FFT analysis and harmonic profile
```

Displacement factor DF and input power factor PF are

$$DF = \cos\alpha$$

$$PF = \frac{2\sqrt{2}}{\pi}\cos\alpha \quad (9\text{-}12)$$

Total harmonic distortion is

$$THD = \sqrt{\frac{\pi^2}{8} - 1} = 0.4834 \quad (9\text{-}13)$$

```
phase1=ph−alpha %phase angle of the fundamental
  with respect to the input voltage
df=cos(phase1) %displacement power factor
pf=((ish(2)/sqrt(2))/isrms) * df %input power factor
td=sqrt ((isrms/(ish(2)/sqrt(2))) ^2 −1 )
  %total harmonic distortion
```

df=1.0000 pf=0.9003 td=0.4834

The power factor is poor for low output voltages where the angle α is in the vicinity of π/2. Since the rms input current and its rms fundamental component do not depend on the phase delay angle α, the THD is also independent of α.

Performance Discussion See Figure 9.11 for the plots of the performance parameters. T_{vv} versus the firing angle a. Note that the output voltage varies from 2/p at $\alpha = 0$, zero at $\alpha = \pi/2$, to $-$ 2/p at $\alpha = \pi$. The converter operates in the first quadrant for $\alpha = 0$ to $\pi/2$ where both the output voltage and the current are positive. This is referred to as the rectifier mode of operation. The converter operates in the second quadrant for $\alpha = \pi/2$ to π where the output voltage is negative and the output current is positive. This is referred to as the inverter mode of operation.

The slope of the curve dT_{vv}/da indicates the behavior of a feedback loop involving the firing angle α. A high slope is preferable because a small change in firing angle α causes a large variation in output voltage. The slope is high for the angles in the vicinity of $\pi/2$, where the output voltage are low. Hence, feedback loop operation is preferable only at low output voltages.

The phase angle ϕ_1 is negative; therefore, the input power factor is the lagging type. The power factor is high at small firing angles α (or near π) where the output voltages are high. The slope $dT_{vv}/d\alpha$ is, however, small at small α. This characteristic is at variance with the optimum feedback loop operation, where a high $dT_{vv}/d\alpha$ is preferred. An important characteristic of this converter is that the total harmonic distortion is independent of the firing angle and hence the output voltage.

Rectifier Versus Inverter The inverter operation needs some special attention. We note from the waveforms of v_o that the steady-state operation of a converter over a single half period alternates between the rectifier mode ($\alpha \leq \theta < \pi$) and the inverter mode ($\pi \leq \theta < \pi + \alpha$). In the first quadrant, where the overall converter operates as a rectifier, the rectifier operation is for a longer duration than the inverter operation. In the fourth quadrant, where the overall converter operates as an inverter, the rectifier operation is for a shorter duration than the inverter operation. A sustained operation in the fourth quadrant is possible only when the load contains an energy-giving component such as a battery, a dc generator or a photovoltaic cell. In the absence of such a device, the converter can operate as an inverter only transiently. Examples of transient operation are the deenergizing of the coil of an electromagnet and regenerative braking in dc motors.

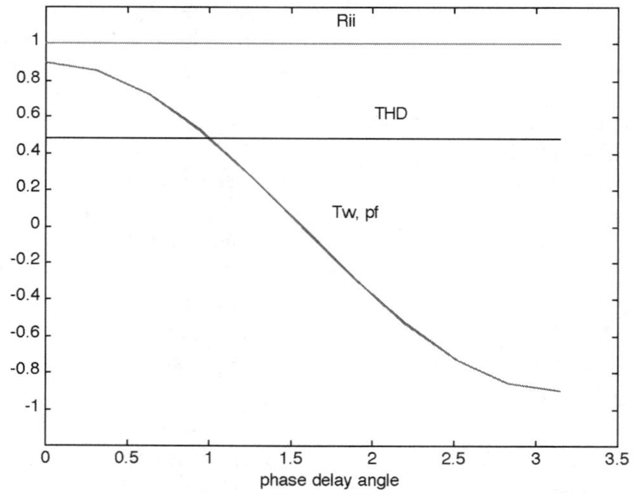

Figure 9.11
Performance characteristics

Example 9-1

The full-bridge converter of Figure 9.12, with an input voltage 120 V at 60 Hz, is required to provide a dc current of 20 A to a highly inductive load such as the coil of an electromagnet. The coil has an inductance of 0.5 H and a resistance of 3 ohms. Find the phase delay angle a, the rms input current, the input power factor and the Total Harmonic Distortion THD. Use the formulae and MATLAB.

In the steady-state, dc current in the load coil is determined solely by its resistance. Hence, the dc voltage needed to supply 20 A to the coil is obtained as

$$\text{dc output voltage } V_o\text{: } 20 \times 3 = 60\ V$$

Load current ripple is determined by the time constant of the load coil, L/R=0.5/3 which is >> 1/2×60 (half period). Hence, the load current can be approximated to be ripple free and we can use the equations developed in this section.

$$\text{Phase delay angle } \alpha\text{: } \alpha = \cos^{-1}\left[\frac{\pi}{2\sqrt{2}}\frac{\bar{V}_o}{\tilde{V}_S}\right]$$

All other parameters are calculated by a MATLAB program given below. Only important lines are given here. The reader is asked to fill in the rest.

```
dtheta=pi/512; VO=60; IO=20; VS=120*sqrt(2);
  alpha=acos(pi*VO/(2*VS))
theta1=[alpha:dtheta:pi+alpha];
  theta2=[pi+alpha:dtheta:2*pi+alpha];
theta =[theta1 theta2];
vo=[(VS * sin(theta1)) (-VS * sin(theta2))];
is=[(IO * ones(size(theta1))) (-IO * ones(size(theta2)))];
[f,ish,ph] = harmonic(is);
```

```
alpha = 0.9820                          df = 0.5579
isrms = 20                              pf = 0.5023
Rii = 1 Phase1=-.9879                   td = 0.4834
```

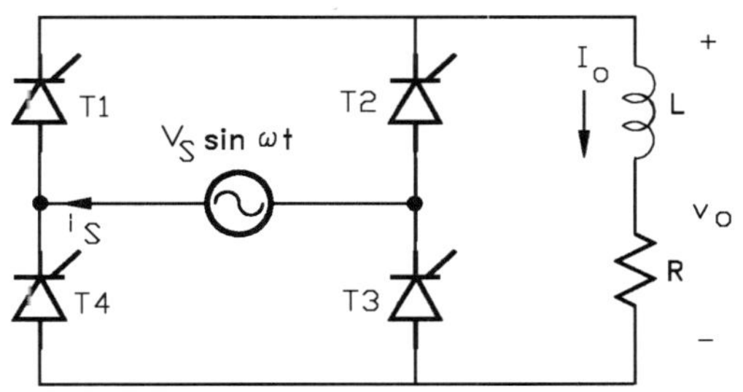

Figure 9.12
An electromagnet supply

Example 9-2

The coil in Example 9-1 must be de-energized after the intended work is completed. One way to do that is to connect the coil across a resistor and burn out the power. An alternative is to return the energy stored in the coil back to the ac source thus avoiding the losses and unwanted heating. The full-converter of Example 9-1 can also be used to demagnetize the coil. It can be done by operating the converter in the inverter mode at maximum negative output voltage for as long as it is necessary to bring down the coil current to zero. Find the firing angle and the time of the inverter operation.

The maximum negative output voltage in the inverter mode is obtained by substituting $\alpha=\pi$ in equation (9-14)

$$\bar{V}_o = \frac{2}{\pi}(\sqrt{2} \times 120)\cos\pi = 108.0380\ V$$

So angle α is changed to π instantly. However, the output dc voltage does not change instantaneously to the new value. It may take a full period to change. However, for our calculations we will ignore this delay and assume that the output voltage changes instantaneously from +60 V to −108.0380 V. Hence a step of −168.0380 V is imposed on the coil. The following equation will give the decay profile of the current in the coil, see Figure 9.13:

$$L\frac{di_o}{dt} + R_{i_o} = -168.0380$$

The coil current decays exponentially

$$i_o = \bar{I}_o - \frac{168.0380}{L}e^{-t/\tau}$$

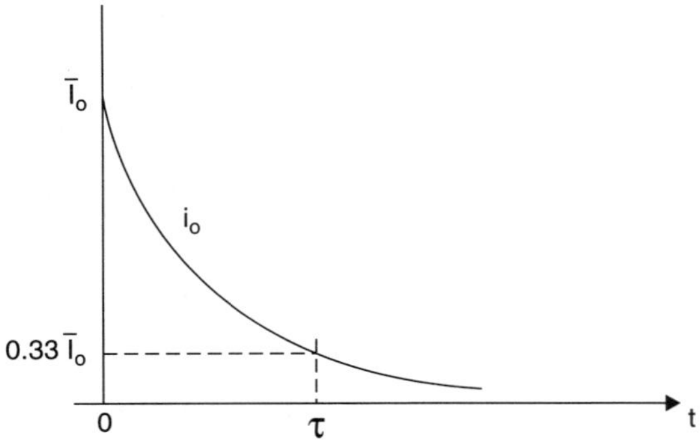

Figure 9.13
Load current decay in inverter mode

where the time constant τ is L/R, the coil inductance divided by the coil resistance.

$$\tau = \frac{L}{R} = \frac{0.5}{3} = 0.167$$

The coil current decays to zero at time t_0

$$t_o = \tau \ln \frac{168.0380}{\bar{I}_o} = 0.3547 \textit{ seconds}$$

This is equivalent to a little more than two cycles.

Extinction Angle Control

The extinction angle control requires forced commutation of switching devices. Hence it is more suitable for GTOs, MCTs, IGBTs and MOSFETs. The converter with control is capable of a two-quadrant operation. The output voltage $\mathbf{v_o}$ and the input current $\mathbf{i_s}$ are shown in Figure 9.14.

All characteristics of this topology are identical to those of the topology with firing angle control except that the phase angle ϕ_1 and hence the input power factor are the leading type. This type of control may be used for power factor improvement.

$$PF = \frac{2\sqrt{2}}{\pi} \cos \beta \; (leading) \qquad \text{(9-14)}$$

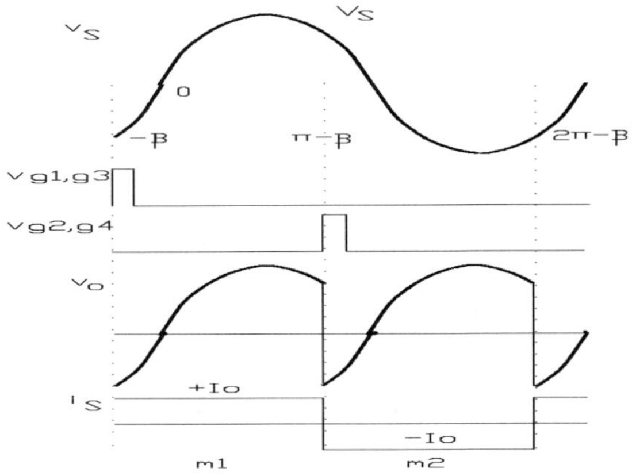

mode 1 $-\beta \le \theta < \pi - \beta$

mode 2 $\pi - \beta \le \theta < 2\pi - \beta$

Figure 9.14
Waveforms

PWM Control

PWM control requires forced commutation of switching devices, hence it is most suitable for GTOs, MOSFETs, IGBTs and MCTs. We will replace thyristors by IGBTs for illustration. A converter with this type of control is capable of single-quadrant operation. AC to dc converters with PWM control assumes four circuit modes in steady state as given in Figure 9.15(a). The voltage and current waveforms are shown in Figure 9.15(b).

Characteristics

Output dc voltage, current-to-current transfer ratio, rms current and its phase angle of the n-th harmonic with respect to the input voltage, displacement factor **DF,** input power factor, and Total Harmonic Distortion in the input current are given below:

The dc output voltage is

$$\bar{V}_o = \frac{1}{\pi} \int_{\pi/2 - \delta/2}^{\pi/2 + \delta/2} V_S \sin \theta \, d\theta$$

$$= \frac{2V_S}{\pi} \sin \frac{\delta}{2} \qquad \text{(9-15)}$$

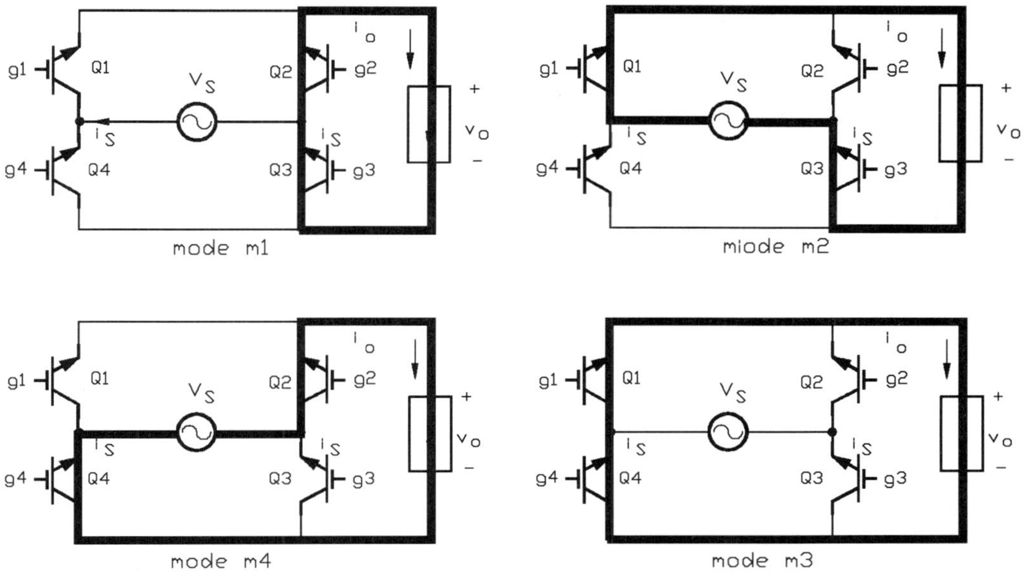

mode 1 $-\frac{\pi}{2}+\frac{\delta}{2}\le\theta<\frac{\pi}{2}-\frac{\delta}{2}$ **mode 2** $\frac{\pi}{2}-\frac{\delta}{2}\le\theta<\frac{\pi}{2}+\frac{\delta}{2}$

mode 4 $\frac{\pi}{2}+\frac{\delta}{2}\le\theta<\frac{3\pi}{2}-\frac{\delta}{2}$ **mode 3** $\frac{3\pi}{2}-\frac{\delta}{2}\le\theta<\frac{3\pi}{2}+\frac{\delta}{2}$

(a)

Figure 9.15
(a) Circuit modes

The voltage transfer ratio is

$$T_{vv} = \frac{\bar{V}_o}{\tilde{V}_S} = \frac{2\sqrt{2}}{\pi}\sin\frac{\delta}{2}, \quad 0 \le \delta < \pi/2 \tag{9-16}$$

The source current is bipolar pulse-wave with I_o peak value and pulse width δ, hence, the rms source current is

$$\tilde{I}_s = \sqrt{\frac{1}{\pi}\int_{\pi/2-\delta/2}^{\pi/2+\delta/2} I_O^2\, d\theta} = I_O\sqrt{\frac{\delta}{\pi}} \tag{9-17}$$

The current reflection ratio is

$$R_{ii} = \frac{\tilde{I}_S}{I_o} = \sqrt{\frac{\delta}{\pi}} \tag{9-18}$$

The harmonic profile of the source current is

$$i_s = \sum_{odd\ n} I_{sn}\sin(n\theta + \phi_n) \tag{9-19}$$

From Fourier analysis,

$$I_{sn} = \frac{4\,I_o}{n\pi}\sin\frac{n\delta}{2}, \quad \phi_n = 0,$$
$$n = 1, 3, 5, \ldots \tag{9-20}$$

$$DF = \cos\phi_1 = 1, \quad PF = \sqrt{\frac{8}{\pi\,\delta}}\sin\frac{\delta}{2} \tag{9-21}$$

$$THD = \sqrt{\frac{\pi\delta}{4(1-\cos\delta)} - 1} \tag{9-22}$$

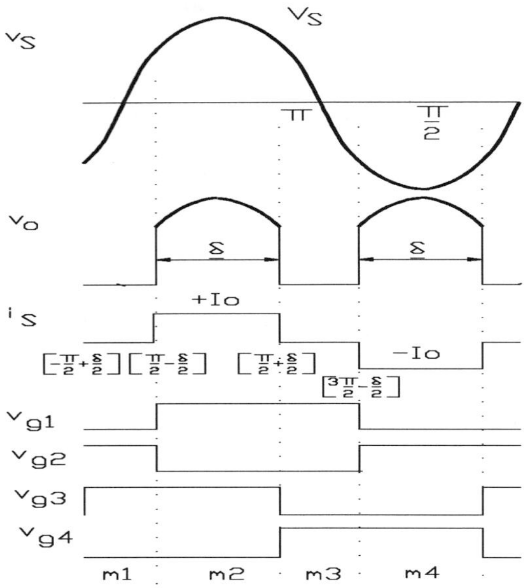

(b)

Figure 9.15 ***(continued)***
(b) Waveforms

Performance Discussion

See Figure 9.16 for the plots of the performance parameters versus the pulse width δ. Note that the output voltage varies from zero at $\delta = 0$ to $2\sqrt{2}/\pi$ at $\delta = \pi/2$. The converter operates in the first quadrant only because both the output voltage and the current are positive.

The slope of the curve $dT_{vv}/d\delta$ indicates the behavior of a feedback loop involving the pulse width angle δ. Like the phase delay angle and extinction angle controls, the slope is high for small pulse widths, where the output voltages are low. Hence, the feedback loop operation is preferred only at low output voltages.

Phase angle ϕ_1 is zero. Hence the input power factor in the PWM control is higher than those in firing angle and extinction angle control. The power factor is low at low output voltages. This characteristic is at variance with the optimum feedback loop operation. The Total Harmonic Distortion is very high at low output voltages. At high output voltages, however, THD is lower than in the converters with phase delay angle and extinction angle control.

Example 9-3

Repeat Example 9-1 with PWM control.

Peak source voltage is

$$V_S = 120\sqrt{2} = 169.7056$$

Pulse width (equation (9-23))

$$\delta = 2\sin^{-1}\left[\frac{\pi}{2}\frac{\overline{V}_o}{V_S}\right] = 1.1776 \textit{ radians}$$

All other parameters are obtained from equation (9-23). A partial MATLAB program is given below:

```
dtheta=pi/512; VO=60; IO=20; VS=120*sqrt(2);
  delta=2*asin(pi*VO/(2*VS))
theta1=[-pi/2+delta/2:dtheta:pi/2-delta/2];
  theta2=[pi/2-delta/2:dtheta:pi/2+delta/2];
theta3=[pi/2+delta/2 : dtheta : 3*pi/2-delta/2];
  theta4=[3*pi/2-delta/2 : dtheta: 3*pi/2+delta/2];
theta =[theta1 theta2 theta3 theta4];
vo=[zeros(size(theta1)) VS*sin(theta2)
  zeros(size(theta3)) -VS*sin(theta4)];
is=[zeros(size(theta1)) IO*ones (size(theta2))
  zeros(size(theta3)) -IO*ones(size(theta4))];
[f,ish,ph]=harmonic(is); %FFT analysis
```

delta = 1.1776	**df = 1.0000**
Tvv=.4989	**pf= 0.8158**
Rii = 0.6114	**td = 0.7089**
Phase1=.0014	

The power factor in PWM control is much higher than in Example 9-1 with firing angle control. Total Harmonic Distortion is, however, higher. A low R_{ii} in PWM indicates that the input source is not being utilized at its maximum.

Multiple PWM Control

Both the pulse width and the number of pulses per half cycle can be used as the control parameters. Lower order harmonics other than the fundamental

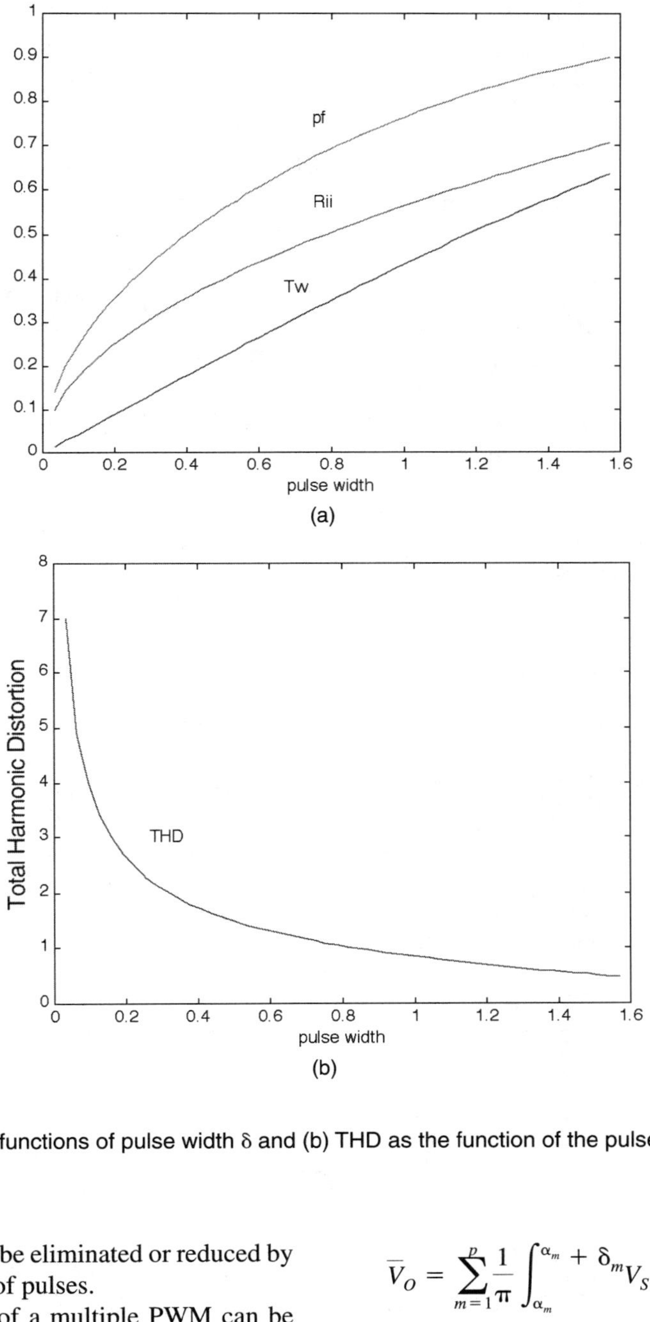

Figure 9.16
(a) Tvv, Rii and pf as the functions of pulse width δ and (b) THD as the function of the pulse width δ

of the input current can be eliminated or reduced by increasing the number of pulses.

The output voltage of a multiple PWM can be calculated by finding the average of a m-th pulse and then the sum over all such pulses:

$$\overline{V}_O = \sum_{m=1}^{p} \frac{1}{\pi} \int_{\alpha_m}^{\alpha_m + \delta_m} V_S \sin\theta \, d\theta$$

$$\overline{V}_o = \frac{V_S}{\pi} \sum_{m=1}^{p} [\cos\alpha_m - \cos(\alpha_m + \delta_m)] \quad (9\text{-}23)$$

where δ_m and α_m are respectively the pulse width and the delay angle of the m-th pulse.

Input current i_s waveform has both quarter-wave and half-wave symmetry; therefore, the Fourier expansion consists of odd harmonics and sine terms:

$$I_{sn} = \frac{2I_o}{n\pi}\sum_{m=1}^{p}[\cos n\alpha_m - \cos(\alpha_m + \delta_m)]$$

$$n = 1, 3, \ldots \tag{9-24}$$

In Uniform Multiple PWM, the pulse width and delay angles of the control pulses are calculated from equations (9-4) and (9-5). In Sine-PWM, the delay angles and pulse widths of the control pulses are obtained from the solution of the equations (9-6) and (9-7). The phase angle of the fundamental harmonic is zero in both types of control; therefore, the converters with these controls have high power factors. The advantage of an improved power factor must be weighed against the increase in harmonic distortion and switching losses. More details of the multiple PWM are beyond the scope of this book and hence not covered.

9.3.2 1-phase full-bridge converter with source inductance

The previous section ignored the source resistance and source inductance in the ac input voltage source. The source resistance, in general, does not affect the converter performance as significantly as does the source inductance. The source inductance may represent the inductance of long conductors or the leakage inductance of a transformer secondary. This section will consider the effect of source side inductance in a single-phase two-pulse full-bridge converter with constant current output.

Consider the converter shown in Figure 9.17 with source inductance L_S and thyristors. Assume for the following discussion that the load current I_o is well filtered and is constant. The steady-state converter operation is described by the four circuit modes, given below:

Mode 1 begins when thyristor T_1 and T_3 are conducting current at angle γ. Current I_o flows through the path shown in Figure 9.18(a). After angle π, the input source voltage becomes negative but the thyristors T_1 and T_3 keep conducting. The output voltage v_o becomes negative and follows the input voltage v_S waveform. Mode 2 begins when thyristors T_2 and T_4 are fired at an angle $\pi+\alpha$ by the control pulses applied at their gate terminals. The current through the inductor L_S cannot be steered away instantaneously, hence, all four thyristors are on simultaneously. This causes a short circuit across the output current sink. In the meanwhile, the currents through thyristors T_1 and T_3 decreases and the current through T_2 and T_4 increases. This mode lasts until the angle $\pi+\alpha$ when the current in thyristors T_1 and T_3 becomes zero, the current in L_S changes to $-I_o$ and the thyristors T_2 and T_4 conduct full load current.

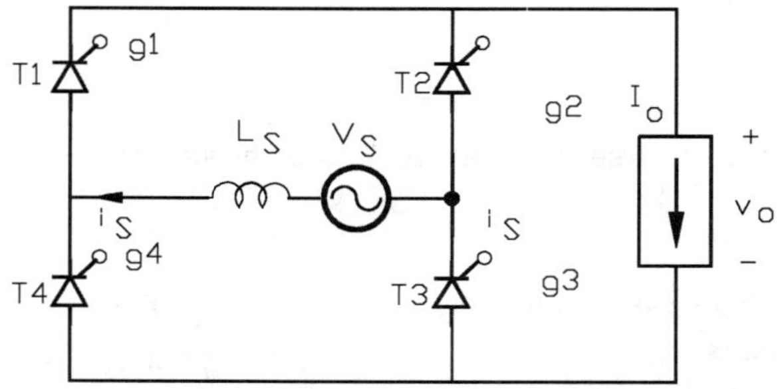

Figure 9.17
A 1-phase full-bridge converter with source inductance

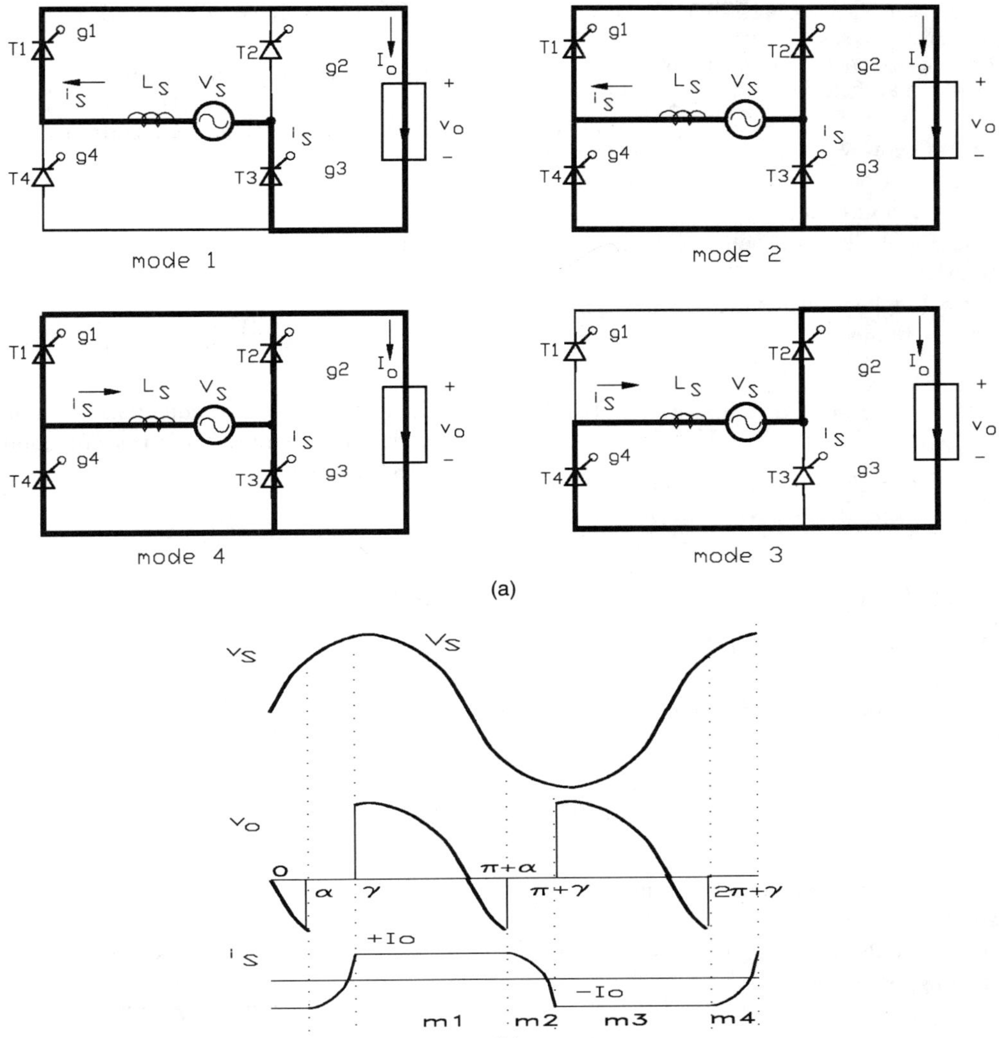

Figure 9.18
(a) Circuit modes (b) Waveforms

In the following mode m3, thyristors T_2 and T_4 conduct the output current until the angle $2\pi+\alpha$. Thyristors T_1 and T_3 are fired at $2\pi+\alpha$ in mode 4. However, the current in L_S takes time to commutate the current until angle $2\pi+\gamma$. The angle $\gamma-\alpha$ is called the commutation angle u. Output voltage $\mathbf{v_o}$ and the input current $\mathbf{i_s}$ are shown in Figure 9.18b. Note that the current transitions are smoother than without L_S.

For a given phase delay angle α the angle array for the four modes in MATLAB is

```
dtheta=pi/512;
theta1=gamma:dtheta:pi+alpha-dtheta;
```

```
theta2=pi+alpha:dtheta:pi+gamma−dtheta;
theta3=pi+gamma:dtheta:2*pi+alpha−dtheta;
theta4=2*pi+alpha:dtheta:2*pi+gamma+dtheta;
theta =[theta1 theta2 theta3 theta4];
```

We need to know γ (gamma) before entering the MATLAB instructions, hence, wait until we find an expression for it a little later.

Note that for angle $\gamma \geq \pi$, output voltage is zero for all values of α.

The current in the inductor L_S, which is also the input current i_s in mode 2, is

$$L_S = \frac{di_L}{dt} = V_S \sin\theta, \quad \pi + \alpha \leq \theta < \pi + \gamma \quad (9\text{-}25)$$

$$i_L = I_o - \frac{V_S}{\omega L_S}(\cos\alpha + \cos\theta) \quad (9\text{-}26)$$

```
is1=IO*ones(size(theta1)); % source current in mode 1
is2=IO−integ((vs(size(theta2))/omegaLs, theta2);
  % source current in mode 2
is=[is1 is2 −is1 −is2]; % source current in a period
```

Inductor current becomes $-I_o$ at the angle $\pi+\gamma$, which gives

$$\cos\gamma = \cos\alpha - 2I_{no}, \quad I_{no} = \frac{\omega L_S I_o}{V_S} \quad (9\text{-}27)$$

where I_{no} is the normalized output current.

The commutation angle or overlap angle u, defined as the angle during which the source current is commutating, is found from

$$u = \gamma - \alpha \quad (9\text{-}28)$$

```
gamma=acos(cos (alpha) − 2*Ino); u=gamma−alpha;
```

Angle γ must be less than or equal to π, which gives the maximum value of the normalized output current as (something which was not a constraint in the converter without L_s=0),

$$I_{nomax} = \frac{1 + \cos\alpha}{2} \quad (9\text{-}29)$$

The output voltage in MATLAB is

```
vo=[vs(size(theta1)), zeros(size(theta2)),
  vs(size(theta3)), zeros(size(theta4))];
```

DC output voltage is obtained from

$$\begin{aligned}\overline{V}_O &= \frac{1}{\pi}\int_{\alpha}^{\pi+\alpha} V_S \sin\theta \, d\theta \\ &= \frac{V_S}{\pi}[\cos\gamma - \cos(\pi + \alpha)] \\ &= \frac{2V_S}{\pi}(\cos\alpha - I_{no}) \end{aligned} \quad (9\text{-}30)$$

Using the above two equations, we can get the converter voltage transfer ratio as a function of normalized output current

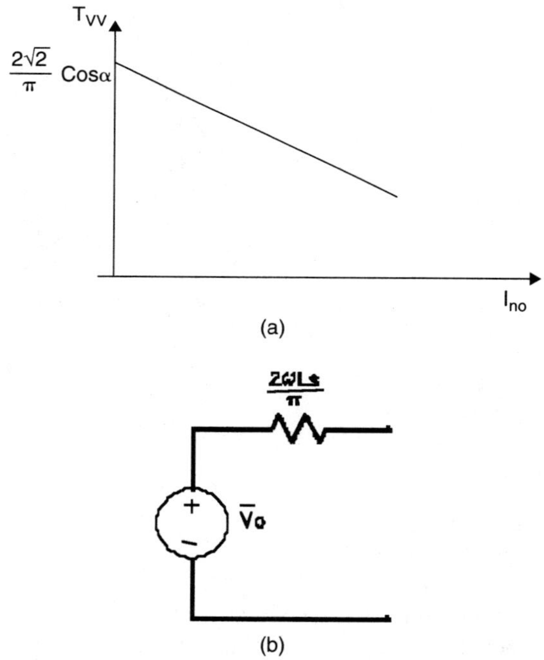

Figure 9.19
(a) Voltage transfer ratio versus the normalized output current and (b) Thevenin equivalent circuit on the output side

$$T_{vv} = \frac{\overline{V}_o}{\tilde{V}_S} = \frac{2\sqrt{2}}{\pi}(\cos\alpha - I_{no}) \qquad (9\text{-}31)$$

The harmonic profile of the source current is

```
[f,ish,ph] = harmonic(is,10);
  %FFT analysis and harmonic profile
phase1=ph−gamma; %lagging phase angle of the
  fundamental source current
df=cos(phase1); %input displacement factor
pf=((ish(2)/sqrt(2))/isrms) * df; %input power factor
td = (sqrt ( (isrms/(ish(2)/sqrt(2))) ^2 −1 ); %Total
  Harmonic Distortion
```

Unlike the converter with L_s=0, the dc output voltage is a function of the phase angle delay and the output current being delivered to the load. The effect of source inductance has resulted in introducing an equivalent source resistance of $2\omega L_s/\pi$ in the output circuit of the converter. The power factor is now the function of both α and the commutation angle u (hence, the output dc current I_o). Similarly, the rms fundamental component I_{s1} and the THD depend on both α and the commutation angle u.

Note that the I_{nomax} is a function of parameter α.

Example 9-4

Consider a full-converter of Figure 9.17 with L_S = 100 mH. Find the maximum output current if the no-load output voltage is desired to be 60 Vdc. Select a current lesser than maximum and calculate the performance measures. Draw the waveforms of the output voltage and the source current.

The no-load output voltage is obtained from equation (9-34) by putting I_o=0. The phase delay angle is

$$\alpha = \cos^{-1}\left[\frac{\pi}{2\sqrt{2}}\frac{\overline{V}_o}{\tilde{V}_S}\right]$$

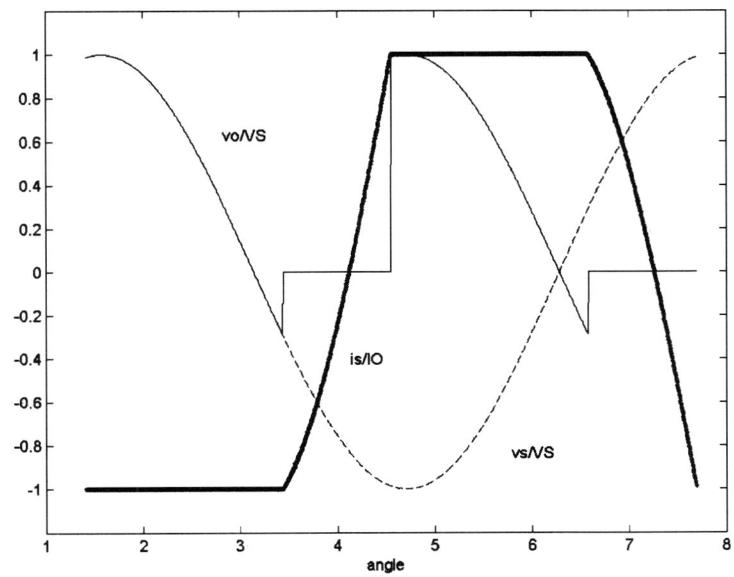

Figure 9.20
Waveforms

The maximum limit on the output current is obtained from equation (9-31)

$$\bar{I}_{omax} = \frac{(1 + \cos\alpha)}{2}\frac{\sqrt{2}\,\tilde{V}_S}{\omega\, L_s} = 3.50079\ A$$

We will select I_o=1.8 A for illustration. The rest of the problem is done using MATLAB. A partial program is given below:

```
dtheta=pi/512; VS=120*sqrt(2); VO=60; IO=1.8;
   omegaLs=2*pi*60*.1;
INO=omegaLs*IO/VS; alpha=acos(pi/2*VO/VS+INO );
   % phase delay angle
vs=VS*sin(theta);
   isrms=sqrt(mean(is .^2)); Rii= (isrms/IO);
```

isrms=1.5892	**df=.3047**
Rii=.8829	**pf=.2968**
phase1=–.50220	**td=.2329**

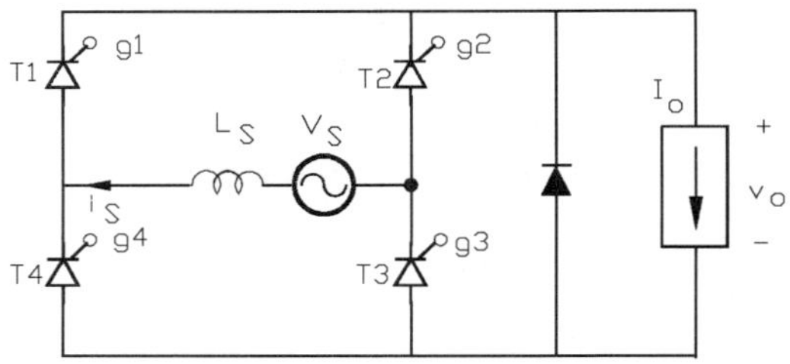

(a)

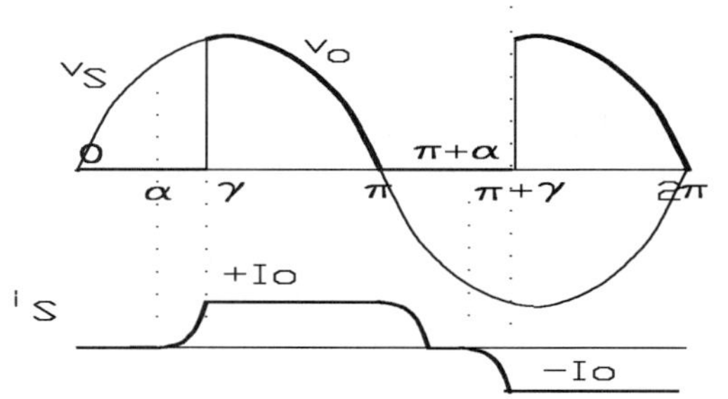

(b)

Figure 9.21
(a) A 1-phase full-bridge converter with source inductance and freewheeling diode (b) The source current

9.3.3 1-phase full-bridge converter with freewheeling diode

The freewheeling diode prevents the output voltage from reversing polarity in both half periods. Consider the converter shown with a freewheeling diode in Figure 9.21(a). Recall from the previous section that the current commutation takes place from one thyristor pair T1-T3 to another pair T2-T4 during the $\pi+\alpha$ to $\pi+\alpha+u$ interval. In this converter, however, current first commutates from thyristor pair T1-T3 to the freewheeling diode during the π to $\pi+u$ interval, and then from the freewheeling diode to thyristor pair T2-T4 during the $\pi+\alpha$ to $\pi+\alpha+u$ interval. Unlike the converter without the freewheeling diode, the output voltage does not change polarity in both half periods. However, the source current becomes discontinuous, see the waveform in Figure 9.21(b). Readers are encouraged to use MATLAB to determine the performance parameters.

These waveforms would have also been observed in a purely resistive load. The advantage of the freewheeling diode is that the output dc current has an independent path in the event of interruption in the line voltage supply. Without the freewheeling diode, the highly inductive output current would have to crash in any such interruption causing high transient voltages on the thyristors. Another advantage of the freewheeling diode is that the converter can be turned off by simply blocking the triggering pulses to the gates of the thyristors. If, for instance, T1-T3 are conducting when the triggering pulses to all thyristors are blocked, then they will continue to conduct until angle π. Then the current will commutate to the freewheeling diode. Without the freewheeling diode, the T1-T3 pair would continue to conduct because there is no other conduction path. Thyristors T1-T3 would have to be turned off by a special commutation circuit by interrupting the input ac source. The disadvantage is that the input power factor and the THD become dependent on the phase delay angle α.

9.3.4 1-phase full-bridge converter with voltage sink or source as the load

This section will analyze an ideal converter with zero source inductance L_S and lossless inductor L on the load side, as shown in Figure 9.22. The practical considerations will be added later on in the section. The converter uses phase angle delay control and is required to provide a dc current to the load. The voltage sink or source E may be a dc Bus supply, or the back emf in the armature of a dc motor in forward motion.

The converter is not controllable for phase delay angles less than α_0

$$\alpha_0 = \sin^{-1}\frac{V_o}{V_S} \tag{9-32}$$

For angles $\alpha \geq \alpha_0$, the converter may operate in any one of two modes: Continuous Current Mode (CCM) or Discontinuous Current Mode (DCM) based on the waveform of the output current.

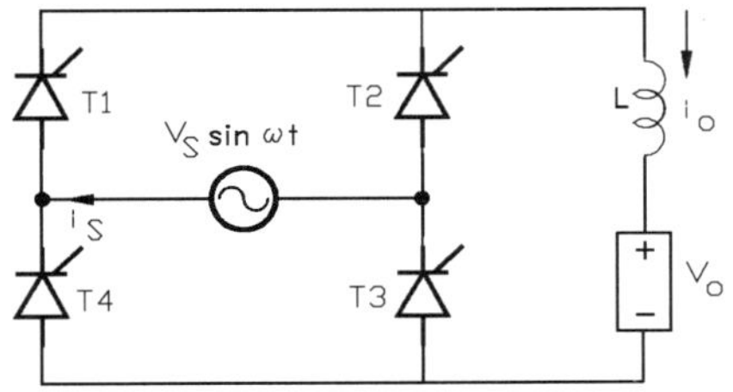

Figure 9.22
A 1-phase 2-pulse full-bridge converter topology

Continuous Conduction Mode (CCM)
The output current is continuous and has a dc component. The phase delay angle α is greater than or equal to α_0. Thyristors T1 and T3 conduct from α to $\pi+\alpha$ and T2 and T4 from $\pi+\alpha$ to $2\pi+\alpha$. In the steady state, the average voltage across the inductor is zero, hence the average dc value of the output voltage waveform, given by equation (9-8), must equal the voltage sink V_o,

$$V_o = \frac{2V_S}{\pi}\cos\alpha \tag{9-33}$$

For large values of L, the output current is constant with very low ripple content. The converter can be treated as having a current sink load for which the analysis is already given in Section 9.3.1.

For medium and small values of L, the output current contains considerable ripple. Converter operation is described by two circuit modes as shown in Figure 9.23(a). The voltage and current waveforms are shown in Figure 9.23(b). The output current waveform is determined from the integration of the voltage across the inductor v_L

$$i_o = I_0(\alpha) + \int_\alpha^\theta \frac{v_L}{\omega L}\, d\theta$$

$$i_o = I_o(\alpha) + \frac{V_S}{\omega L}(\cos\alpha - \cos\theta) - \frac{V_o}{\omega L}(\theta - \alpha) \tag{9-34}$$

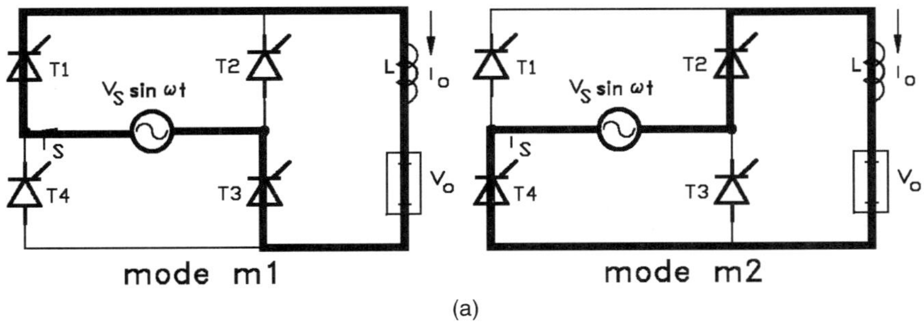

Figure 9.23
(a) Circuit modes

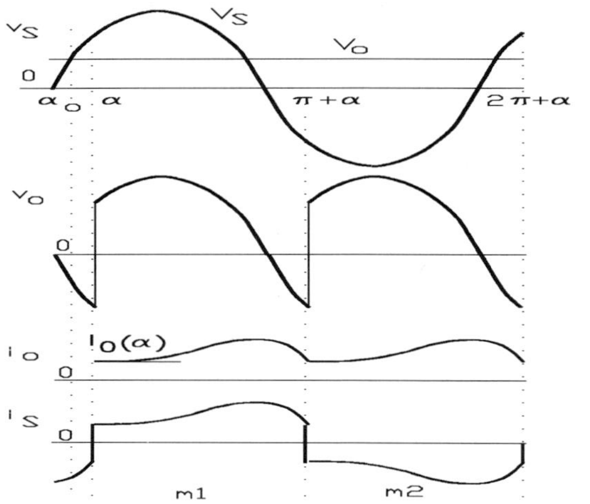

Figure 9.23 ***(continued)***
(b) Waveforms

where $I_o(\alpha)$ is the inductor current at angle α.

Average output current is calculated from:

$$\bar{I}_o = \frac{1}{\pi}\int_{\alpha}^{\pi+\alpha} i_o\, d\theta = I_o(\alpha) + \frac{2V_S}{\pi\omega L}\sin\alpha \quad (9\text{-}35)$$

The minimum dc current for CCM operation is maximum at $\alpha=\pi/2$ at which the output voltage is zero.

Peak output current occurs at angle $\pi-\alpha_0$ and is given by

$$I_{o\,peak} = I_o(\alpha) + \frac{V_S}{\omega L}(\cos\alpha + \cos\alpha_0) - \frac{V_o}{\omega L}(\pi - \alpha - \alpha_0) \quad (9\text{-}36)$$

Crest factor is defined as the ratio of peak output current to the average output current

$$CF = \frac{I_{opeak}}{\bar{I}_o} \quad (9\text{-}37)$$

Peak-to-peak current ripple is found from equations (9-33) and (9-38)

$$\Delta I_o = I_{opeak} - I_o(\alpha) = \frac{V_S}{\omega L}\left[\cos\alpha_0 + \frac{2}{\pi}(\alpha + \alpha_0)\cos\alpha\right] \quad (9\text{-}38)$$

A large inductor ensures a low boundary current and a low current ripple.

Since the initial current I_{o1} is positive at all instants of time, the converter is operating in continuous current mode. Output current becomes discontinuous when the desired output current is less than the boundary value current.

Practical Considerations

Inclusion of Winding Resistance R: The voltage sink V_o then includes the voltage drop across the winding resistance R of the inductor L. Let E be the actual voltage source. The dc output current then is

$$\bar{I}_o = \frac{V_o - E}{R} \quad (9\text{-}39)$$

Minimum dc current necessary for CCM operation is obtained by putting $I_o(\alpha)=0$.

$$\bar{I}_{omin} = \frac{V_o - E}{R} \geq \frac{2V_S}{\pi\,\omega\,L}\sin\alpha \quad (9\text{-}40a)$$

Hence the ratio E/V_S versus α is (see Figure 9.24 for $\omega L/R=1$),

$$\frac{E}{V_S} \leq \frac{2}{\pi}\left[\cos\alpha - \frac{\sin\alpha}{\omega\,L/R}\right] \quad (9\text{-}40b)$$

For large $\omega L/R$,

$$E \leq \frac{2V_S}{\pi}\cos\alpha$$

Inclusion of Source Inductance: Like in the case of Section 9.3.2, the nonzero source inductance gives rise to transition edges to the source current waveform. The readers are encouraged to do the detailed mode analysis. However, we will leave this topic here.

Example 9-5

A single-phase 1 hp dc motor is driven by a 120-V 60 Hz supply. The armature inductance of the motor is 100 mH and the back emf at the desired

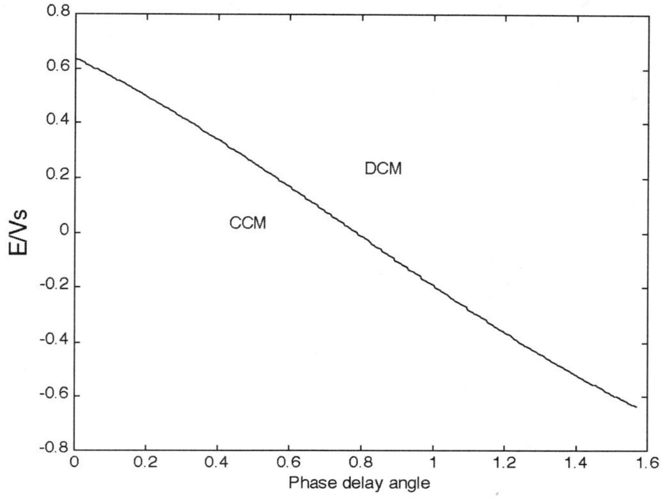

Figure 9.24
Voltage sink and phase delay angle

speed is 60 volts dc. Use the full converter of Figure 9.22. Draw the voltage and current waveforms using MATLAB. Also find the phase delay angle α, average output current, crest factor, rms input current, input power factor and Total Harmonic Distortion THD.

A partial MATLAB program is given below:

```
dtheta=pi/512; VS=120*sqrt(2);R=1;
PO=746; E=60; Iodc=PO/E; VO=E+Iodc*R;
alpha=acos(pi*VO/(2*VS)); omegal=2*pi*60*.1;
Iomin=2*VS*sin(alpha)/(pi*omegal); Io1=Iodc−Iomin;
    theta1=[alpha:dtheta:pi+alpha−dtheta];
    theta2=[pi+alpha:dtheta:2*pi+alpha−dtheta];
    theta =[theta1 theta2];
    vs=VS *sin(theta); vo=[VS*sin(theta1)
      −VS*sin(theta2)];
    Tvv=[Tvv VO/(VS/sqrt(2))];
    io1=Io1+integ((VS*sin(theta1) −VO)/omegal,theta);
      io1=io1 .* (io1>0);
%negative values in io1 array are converted to zero
  as negative current is not permitted through the
  switch S1.
  io=[io1 io1]; is=[io1 −io2];
  isrms=sqrt(mean(is .^2));
  Rii=[Rii (isrms/Iodc) ];
plot(theta, vs/VS,'—', theta,vo/VS,'*', theta,io/IO)
```

iodc=12.4333 alpha=.8360
isrms=12.4884
Rii=1.0041
phase1=−.8676 rad (lagging)
df=.6466
pf=.6033 td=.3856 CF=1.1179

Example 9-6

Find the dc motor current in Example 9-5 at which the motor current waveform becomes discontinuous at the specified speed (back emf E = 60 V). Verify the waveform using MATLAB.

The peak input voltage is

$$V_S = 120\sqrt{2}$$

The dc output voltage then is

$$\bar{V}_o = \frac{2V_S}{\pi}\cos\alpha = E + \bar{I}_{omin}R$$

$$\frac{2V_S}{\pi}\cos\alpha = E + \frac{2V_S}{\pi\,\omega L}\sin\alpha \times R$$

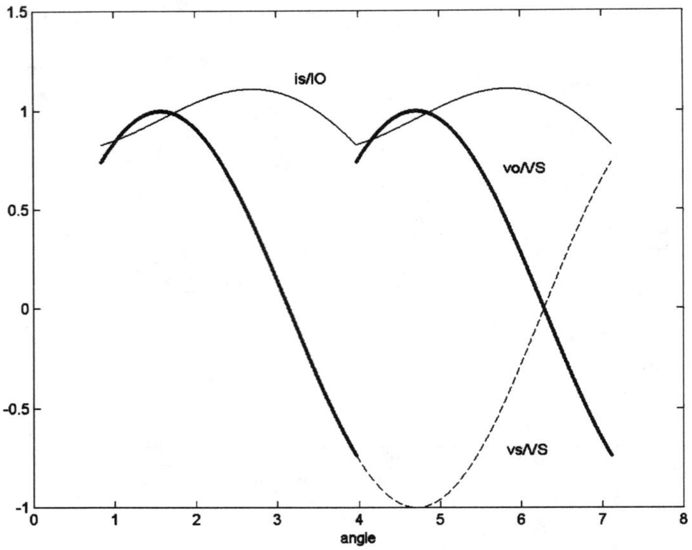

Figure 9.25
The waveforms of the source current and the output voltage

Substitute the values and solve for angle $\alpha=0.9572$. The dc motor current at this phase delay angle α is

$$\bar{I}_{omin} = \frac{2V_S}{\pi\omega L}\sin(.9572)$$

The partial MATLAB program is shown below:

```
%determination of alpha
a=0:dtheta:pi;
f=(2*VS/pi)*cos(a)−(2*VS/(pi*omegal/R))*sin(a)−E;
io1=Io1+integ((VS*sin(theta1) −VO)/omegal,theta);
  io1=io1 .*(io1>0);
[b,n]=min(f.^2);
alpha=a(n);
% DC motor current
Iomin=2*VS/(pi*omegal)*sin(alpha); VO=E+Iodc*R;
vs=VS * sin(theta) ;
vo=[VS*sin(theta1) −VS*sin(theta2)];
io=[io1 io1]; IO=mean(io);
plot(theta, vs/VS,'—', theta,vo/VS,'.', theta,io/IO)
```

Discontinuous Conduction Mode (DCM)

The source current i_s and the output current i_o are discontinuous for dc current I_o lower than the minimum current as given in equation (9-37). In this mode the currents i_o and i_s are zero at the time of commutation from one thyristor pair to the other pair. Hence, the source inductance L_S may be merged with the dc side inductance L or vice versa for analysis purpose. We would retain L. Once again, hold L to be an ideal inductor.

The steady-state operation can be explained by four circuit modes; mode 1 and mode 3 are identical to the modes as shown in Figure 9.23(a). No current flows in the other two modes.

Mode m1 begins at the angle α. The angle α is different from α_0.

The output current i_o begins to build up from zero and flows from the source through the thyristor T1 and T3, L and the load. Current i_o rises and then comes down to zero at an angle β. In the following mode m2, all thyristors are reverse blocked and i_o is zero. The polarity of the source voltage reverses at angle π. The next mode m3 begins when the thyristor pair T2 and T4 conduct at the angle $\pi+\alpha$. Mode m3 ends at $\pi+\beta$. In the following mode m4, all thyristors are again reverse-blocked.

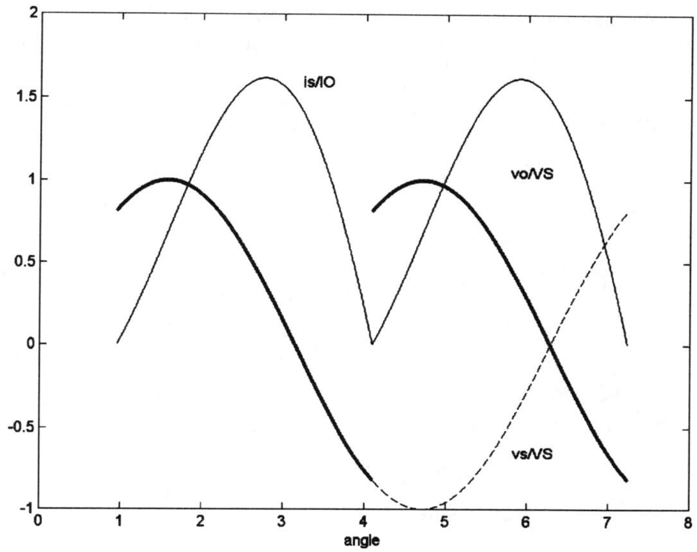

Figure 9.26
Waveforms in the boundary case

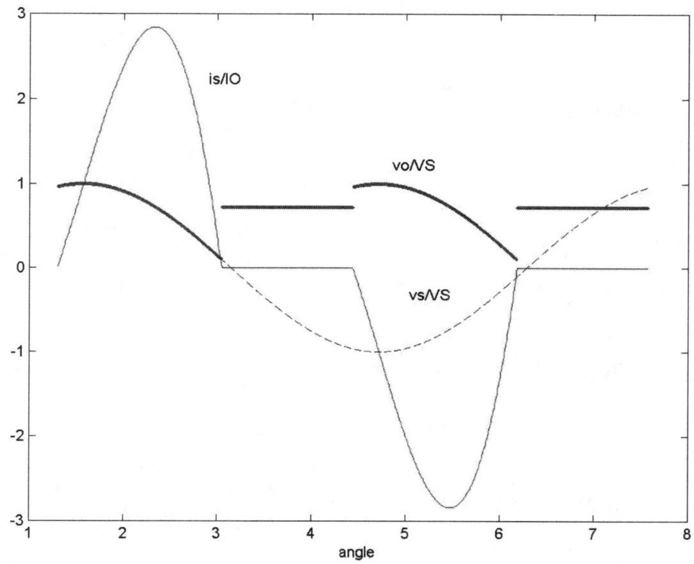

mode 1 $\alpha \le \theta < \beta$ **mode 2** $\beta \le \theta < \pi + \alpha$

mode 3 $\pi + \alpha \le \theta < \pi + \beta$ **mode 4** $\pi + \beta \le \theta < 2\pi + \alpha$

Figure 9.27
Waveforms in DCM operation

In the steady state, the average voltage across inductor L is zero, hence,

$$\int_{\alpha}^{\beta} v_L \, d\theta = \int_{\alpha}^{\beta} (V_S \sin\theta - V_O)\, d\theta = 0 \quad (9\text{-}41)$$

hence,

$$\cos\alpha - \cos\beta - \frac{V_o}{V_S}(\beta - \alpha) = 0 \quad (9\text{-}42)$$

The output voltage v_o is

$$v_o = V_S \sin\theta, \quad \textit{mode 1 and mode 3}$$
$$= V_o, \quad \textit{mode 2 and mode 4}$$

Since the inductor L cannot sustain a dc voltage, the average dc of the v_o waveform should equal V_o

$$V_O = \frac{1}{\pi}\left[\int_{\alpha}^{\beta} V_S \sin\theta \, d\theta + \int_{\beta}^{\pi+\alpha} V_O \, d\theta\right]$$

$$V_o = V_S \frac{\cos\alpha - \cos\beta}{\beta - \alpha} \quad (9\text{-}43)$$

Output current i_o in mode m1 is zero until an angle α and extinguishes at angle β, which is less than $\pi+\alpha$.

$$i_o = \int_{\alpha}^{\theta} \frac{v_L}{\omega L} d\theta \quad (9\text{-}44)$$

$$i_o = \frac{V_S}{\omega L}(\cos\alpha - \cos\theta) - \frac{V_o}{\omega L}(\theta - \alpha) \quad (9\text{-}45)$$

The output current i_o in mode m3 is the exact replica of that in mode m1.

The average output current is calculated from

$$\bar{I}_o = \frac{1}{\pi}\int_{\alpha}^{\beta} i_o \, d\theta \quad (9\text{-}46)$$

Peak output current occurs at angle $\pi - \alpha_0$

$$I_{o\,peak} = \frac{V_S}{\omega L}(\cos\alpha + \cos\alpha_0) - \frac{V_o}{\omega L}(\pi - \alpha_0 - \alpha) \quad (9\text{-}47)$$

This is also the peak-peak current ripple amplitude. A large inductor value ensures low current ripple.

The short-circuit current is limited and is obtained by substituting $V_o=0$ ($\alpha=0$ and $\beta=\pi$)

$$I_{osc} = \frac{V_S}{\omega L} \quad (9\text{-}48)$$

The crest factor is defined as the ratio of peak output current to the average output current. It is given by equation (9-39).

The winding resistance of the inductor L can be incorporated in the analysis in a similar way as in the previous section.

The output current becomes continuous when the extinction angle β is $\pi+\alpha$.

Example 9-7

Replace the motor of Example 9-5 to a 1/4 hp with all other parameters remaining unchanged. Draw the output current and the source current waveforms for an emf of 90 V.

```
dtheta=pi/512; VS=90*sqrt(2);R=1; omegal=2*pi*60*.1;
PO=746/4; E=90; Iodc=PO/E; VO=E+Iodc*R
alpha1=asin(Iodc*pi*omegal/(2*VS))
alpha0=asin(VO/VS)
alpha=max(alpha1, alpha0)
[beta]=betam(alpha,VO/VS);
    theta1=[alpha:dtheta:beta-dtheta];
    theta2=[beta:dtheta:pi+alpha-dtheta];
    theta3=[pi+alpha:dtheta:pi+beta-dtheta];
    theta4=[pi+beta:dtheta:2*pi+alpha-dtheta];
    theta =[theta1 theta2 theta3 theta4];
vs=VS * sin(theta) ;
vo=[VS*sin(theta1) VO*ones(size(theta2))
  -VS*sin(theta3) VO*ones(size(theta4)) ];
vl=[VS*sin(theta1)-VO zeros(size(theta2))
  -VS*sin(theta3)-VO zeros(size(theta4)) ];
io1=integ((VS*sin([theta1 theta2]) -VO)/omegal,[theta1
  theta2]);
io1=io1 .*(io1>0); io=[io1 io1 ]; IO=mean(io);
is=[io1 -io1]; isrms=sqrt(mean(is .^2));
plot(theta, vs/VS,'—', theta,vo/VS,':', theta, is/IO)
```

Plots in Figure 9.27 correspond to this example.

9.3.5 Ac-dc converter in inversion mode

Discussion about the inversion mode operation was initiated in an earlier section. Output voltage in the inverter is negative, hence, the power flows backwards, from the dc side (load side) to the ac side (source side). In this section we will analyze the inversion mode of operation in both constant current and constant voltage output situations with the source inductance included. We will consider inverter operation with the phase delay angle control only. Phase delay angle lies in the range π to 2π. The inverter, thus realized is commutated by the input ac voltage at the line frequency.

Inversion Mode in Constant Current Output Refer to the converter of Figure 9.17. The waveforms in the inversion mode are given in Figure 9.28.

We should ensure that the commutation must be completed before the angle π after which the polarity of the input line ac voltage is reversed.

$$\alpha + u = \gamma \leq \pi, \quad \pi/2 \leq \alpha \leq \pi \qquad (9\text{-}49)$$

If this precaution is not taken, the current steering from T1-T3 pair to T2-T4 pair will not take place at all. Instead, the current in T1-T3 pair will begin to roll back up. The T1-T3 pair will thus be always on.

Commutation Failure A failure situation is very common in practical line commutated inverter operation in which there may be spurious or inadvertent change in the phase delay angle from one half period to another. In order to understand this problem, let us assume the commutation failure as shown in Figure 9.29. The phase delay angle is changed from α_1 (for thyristors T1 and T3) in the first half period to α_2 (for thyristors T2 and T4) in the second half period. The phase delay angle reverts back to α_1 again in the third half period and beyond. Thyristors T1 and T3 are turned at angle α_1 but conduct full current only after $\alpha_1 + u$ in the first half period. In the second half period, thyristor pair T2 and T4 are turned on at $\pi+\alpha_2$. Current begins to increase in the T2-T4 pair and simultaneously decrease in the T1-T3 pair. The current transition will continue up to $\pi+\alpha_2+u$, which overshoots beyond 2π. After 2π, however, the source voltage becomes positive and would maintain T1 and T3 in conduction, thus preventing the current from diverting via the T2-T4 pair. Consequently, the current backs up in the T1-T3 pair and tails off in the T2-T4 pair. Finally, the T1-T3 pair conducts full current and the T2-T4 pair commutates. The T1-T3 pair continues to conduct beyond 3π into the fourth half period because the turn-on gate pulses are applied at $2\pi+\alpha_1$. The T2-T4 pair is turned on at angle $3\pi+\alpha_1$. At that instant, normal current transition takes place from

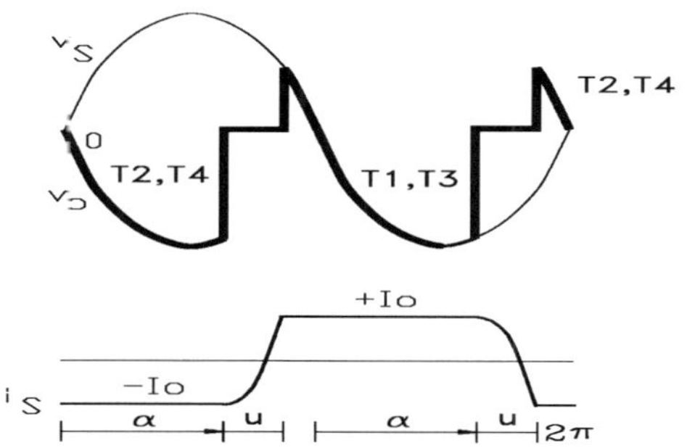

Figure 9.28
Inversion mode operation in constant current output situation

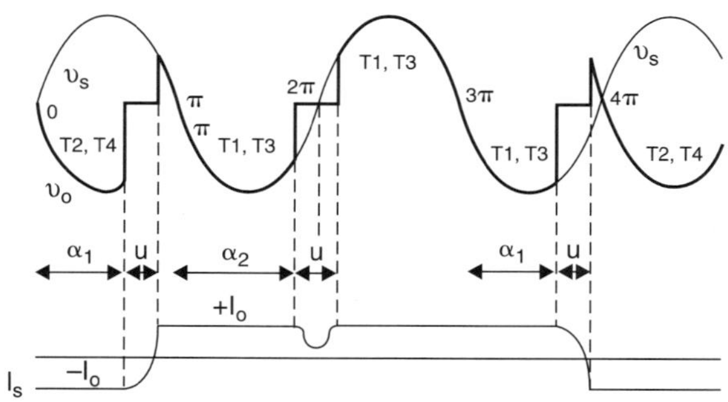

Figure 9.29
Commutation failure

the T1-T3 pair to the T2-T4 pair and continues over the commutation angle u.

A prolonged commutation failure causes differential heating problems on the thyristors in the bridge. The best we can do is to keep a margin over and above the commutation angle u. Hence,

$$\alpha_1 \geq \pi/2, \; and \quad \alpha_1 + u = \pi - \alpha_m \qquad (9\text{-}50)$$

α_m is the margin.

Inversion Mode in Constant Voltage Output

The inversion mode in the constant voltage output represents most of the practical situations. See Figure 9.22 for the inverter circuit. DC current flows from the dc side to the ac side. A large value of inductor L would render the output current to be a ripple free dc current. Functioning and waveforms would be similar to those in the constant output current case.

For very small values of L, the output current is discontinuous. There is no problem of commutation due to the source inductance L_s. For medium values of L, the output current is continuous, contains ripple and has a nonzero commutation angle. The source current waveform in this situation is given in Figure 9.30. The boundary values of currents and the corresponding values of L are determined by similar analysis as in Section 9.3.4.

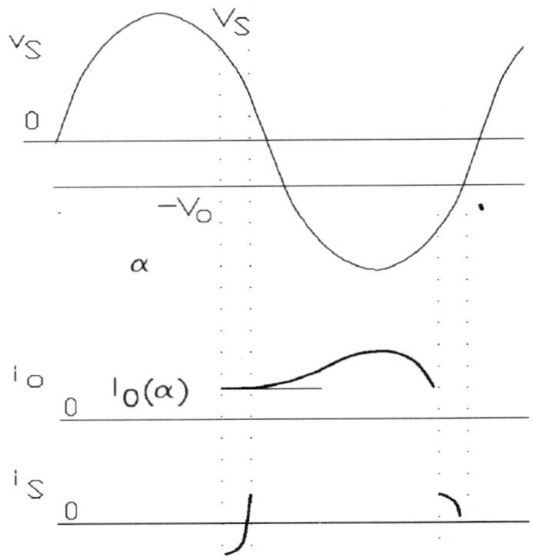

Figure 9.30
Inversion mode operation in constant voltage output situation

9.3.6 Series and parallel connections of 1-Phase full-bridge converters

A series connection of two or more full-bridge converters are used for obtaining higher dc voltages. Parallel connections are used for increasing current capability. This section will discuss a special parallel connection of full-bridge converters, also called dual connection, which will enable a four-quadrant operation. These converters are used in high-power variable dc motor drives.

See Figure 9.31 for the dual connection of two thyristor H-bridges A and B.

Inductance L_r is introduced to limit the circulating current due to any instantaneous outputs of two full-bridge converters (to be discussed later). For our present discussion we will ignore L_r by making it zero. Note carefully the polarity of the output voltage with respect to the polarity of thyristors in the thyristor bridge B. It can be easily inferred that if the bridge A is operating as a rectifier then bridge B must operate as an inverter and vice versa. Let the firing angles of the two full converters be α_A and α_B. The dc output voltages of the A and B converters are

$$\overline{V}_{A0} = \frac{2V_S}{\pi}\cos\alpha_A \; and \; \overline{V}_{B0} = -\frac{2V_S}{\pi}\cos\alpha_B \quad (9\text{-}51)$$

$$and \; \overline{V}_{A0} = \overline{V}_{B0} \; therefore \; \alpha_B = \pi - \alpha_A \qquad (9\text{-}52)$$

The instantaneous output voltage waveforms are different even though the average is the same. The voltage waveforms are shown in Figure 9.33. Note especially the output current vis.a.vis. the output voltage in bridge B.

$$\begin{aligned} v_{A0} &= V_S \sin\theta, \quad \alpha_A \leq \theta < \pi + \alpha_A \\ &= -V_S \sin\theta, \quad \pi + \alpha_A \leq \theta < 2\pi + \alpha_A \\ &= V_S \sin\theta, \quad \alpha_A \leq \theta < \pi - \alpha_A \end{aligned} \qquad (9\text{-}53)$$

$$\begin{aligned} v_{B0} &= -V_S \sin\theta, \quad \pi - \alpha_A \leq \theta < 2\pi - \alpha_A \\ &= V_S \sin\theta, \quad 2\pi - \alpha_A \leq \theta < 2\pi + \alpha_A \end{aligned} \quad (9\text{-}54)$$

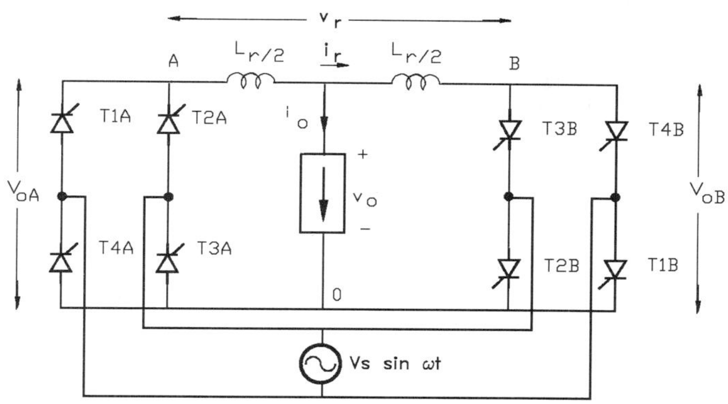

Figure 9.31
A dual converter

For positive load current I_o, the dual converter operates in the first and second quadrants on the V_o-I_o plane. For negative load current I_o, the dual converter operates in the third and fourth quadrants on the V_o-I_o plane.

Positive I_o:
For $\alpha=0$ to $\pi/2$, bridge A operates as a rectifier and supplies the total positive dc current $+I_o$ required by the load. Bridge B operates in the inverter mode but is idle and at zero current. For angles $\alpha=\pi/2$ to π, bridge A operates as the inverter and accepts the total positive dc current $+I_o$ returned by the load. Bridge B operates in the rectifier mode but again is idle. See Figure 9.32. The rectifier current paths are shown by solid thick lines and the zero-current paths by dotted thick lines.

Negative I_o:
For $\alpha=0$ to $\pi/2$, bridge A operates as the rectifier but is idle and supplies zero current. Bridge B operates in the inverter and accepts the total negative dc current $-I_o$ returned by the load. For angles $\alpha=\pi/2$ to π, bridge A operates as the invertor but is idle. Bridge B operates in the rectifier mode and supplies the total negative dc load current $-I_o$.

There are two problems in the above converter operation. The first problem is an infinitely high circulating current from one bridge to another in the case of a slight practical mismatch in firing angles. The other problem is that for any polarity of load current, only one bridge operates at any time. Any fast turnaround in the polarity of load current will be difficult due to the transient time involved in conducting the idle bridge.

Both these problems can be tackled by introducing a finite value of inductance L_r. The instantaneous differential voltage $v_{oA}-v_{oB}$ is nonzero only during $\pi-\alpha_A$ to $\pi+\alpha_A$, and $2\pi-\alpha_A$ to $2\pi+\alpha_A$. The circulating current i_r, which flows through inductor L_r only and not through the load, is also shown in the same figure

$$v_{A0} - v_{B0} = L_r \frac{di_r}{dt} \tag{9-55}$$

Circulating current i_r in the period $\theta=\alpha$ to $\pi-\alpha_A$ is

$$\begin{aligned} i_r &= \frac{1}{\omega L_r} \int_\alpha^\theta (v_{A0} - v_{B0})\, d\theta \\ &= -\frac{2V_S}{\omega L_r}(\cos\alpha - \cos\theta) \end{aligned} \tag{9-56}$$

and the peak value occurs at π

$$i_{rpeak} = \frac{2V_S}{\omega L_r}(1 - \cos\alpha) \tag{9-57}$$

The peak load current is important in the sense that it determines the peak current rating of thyristors

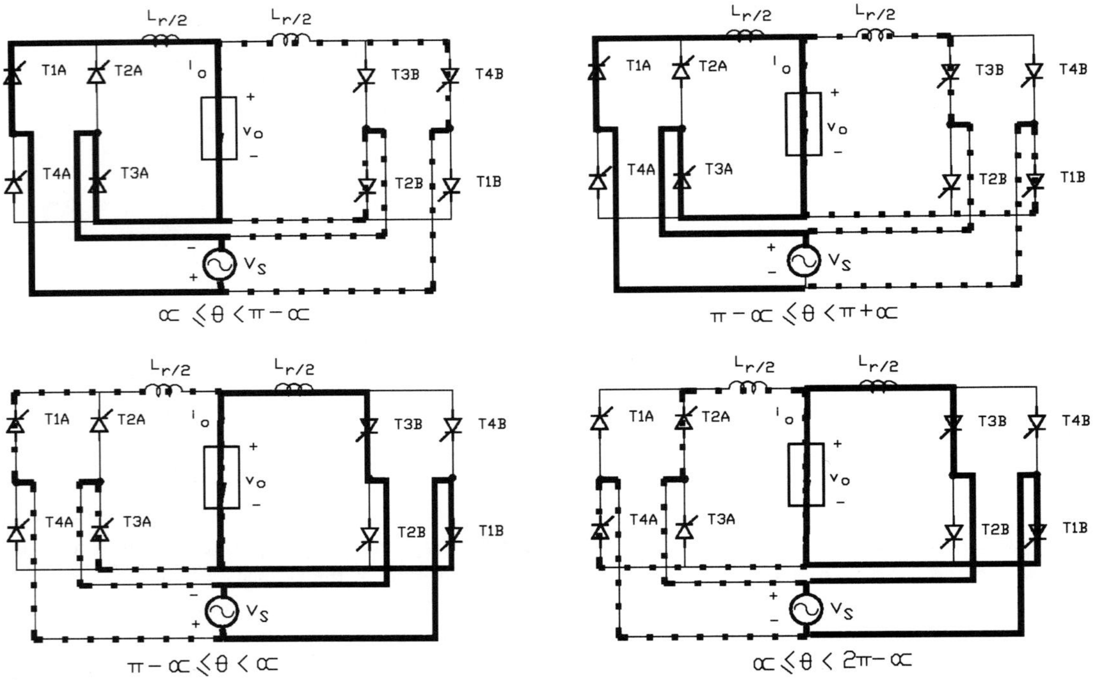

Figure 9.32
Circuit modes in a dual converter for positive output current

in each bridge. The circulating current maintains current in both bridges at all times.

We will now see how this dual converter can be used to control the speed and the braking of a dc motor in both directions. Assume that the field winding of the motor is separately excited by a constant dc voltage. For forward motoring, the converter is operated with $\alpha=0$ to $\pi/2$. The bulk of the dc motor current is supplied by bridge A and bridge B conducts circulating current. The bridge in forward braking, the converter in bridge B is operated with $\alpha=\pi/2$ to π. The dc motor returns the energy to the ac input source via bridge B while bridge A conducts the circulating current. For reverse motoring, the dc voltage supply to the field winding is reversed (make sure that you avoid opening the field winding). The converter is operated with $\alpha=0$ to $\pi/2$ for reverse motoring operation and with $\alpha=\pi/2$ to π for reverse braking. Bridges A and B perform reverse roles.

9.4 3-PHASE 6-PULSE FULL-BRIDGE CONVERTER

Output voltages in single-phase converters have high ripple amplitude and low ripple frequency ($2f_s$). These limitations can be overcome or minimized by using multiphase input sources. Figure 9.34 shows a three-phase ac-dc converter with a highly inductive load. The line voltage is a sine wave of frequency ω and rms amplitude V_L. The line voltages are given by equation (6-77) in a previous chapter.

The load is assumed to be highly inductive, hence the output current I_o is constant and ripple

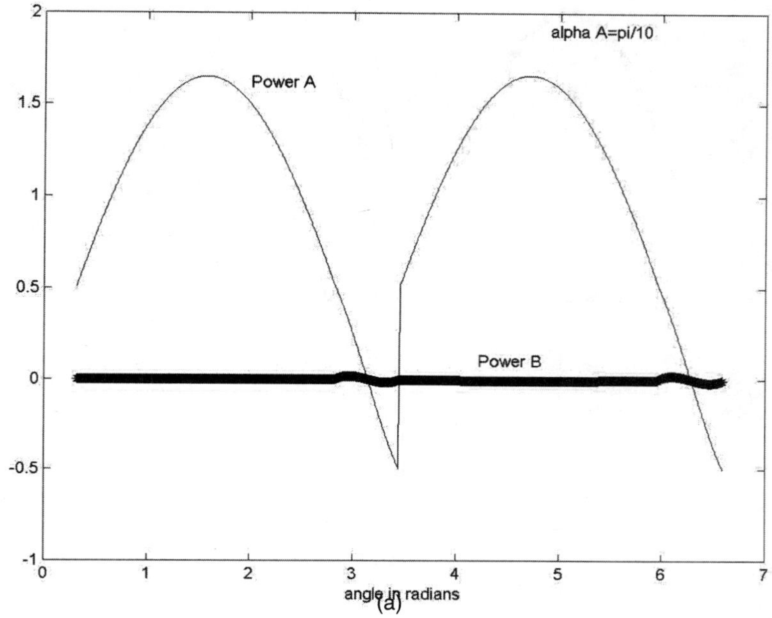

(a)

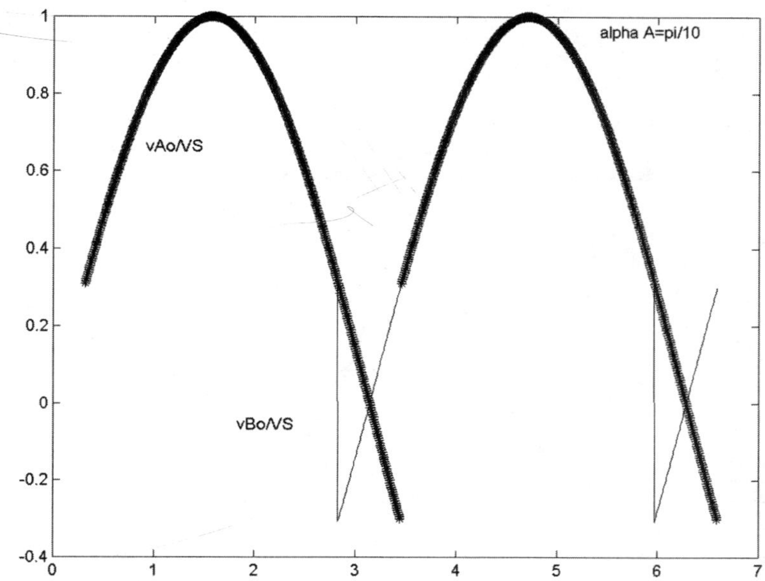

(b)

Figure 9.33
Waveforms in a dual converter

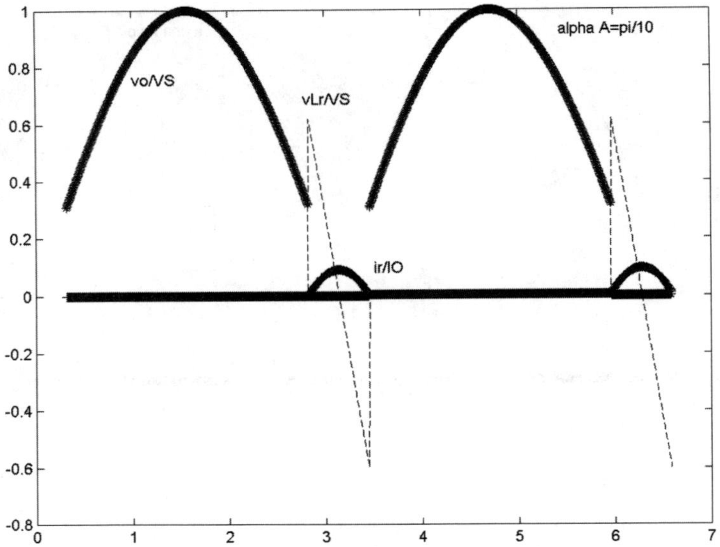

(c)

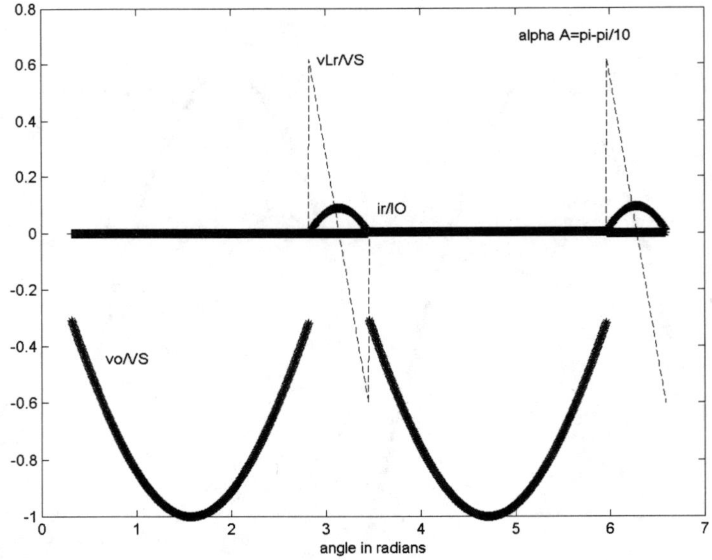

(d)

Figure 9.33 ***(continued)***

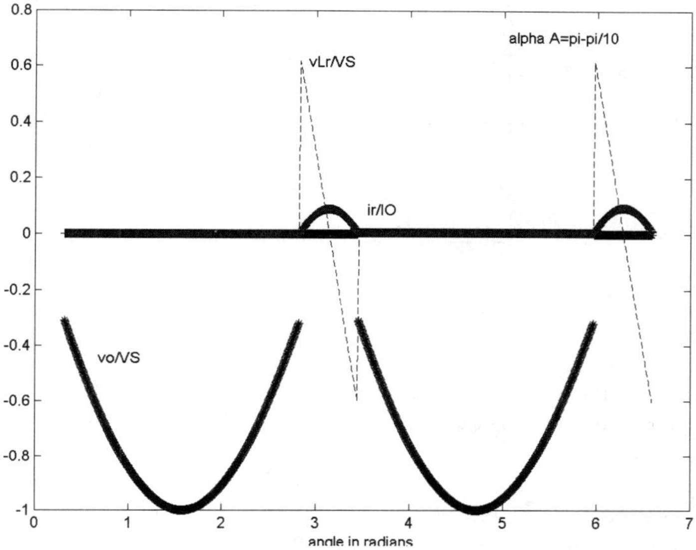

(e)

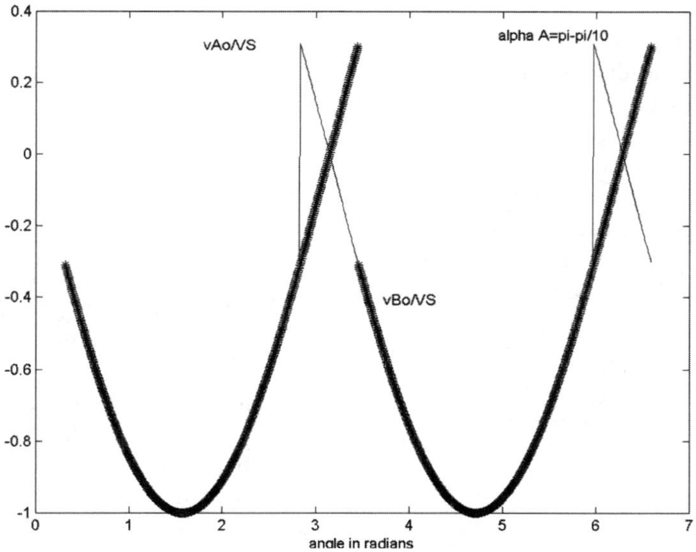

(f)

(a), (b), (c) $\alpha_A = \pi/10$, (d), (e), (f), $\alpha_A = \pi - \pi/10$

Figure 9.33 ***(continued)***

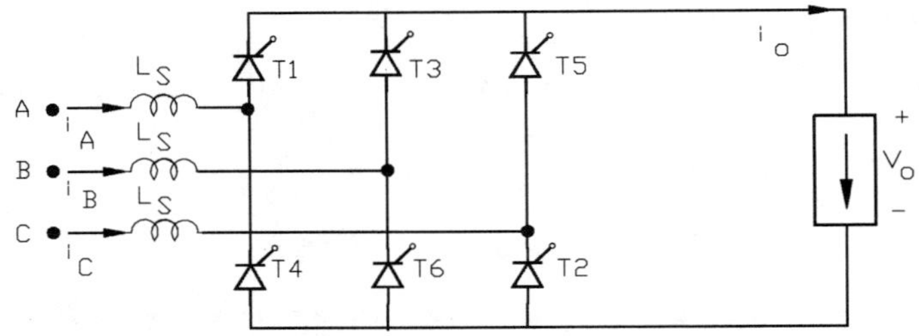

Figure 9.34
A 3-phase 6-pulse full-bridge converter topology

free. Thyristors are fired at the intervals of $\pi/3$. Frequency of the voltage ripple is $6f_s$, therefore, the filtering components are small.

9.4.1 3-phase 6-pulse full-bridge converter with zero source inductance

Converter Operation

If the gate currents were continuously applied, the thyristors would behave as diodes and their operation would be similar to that described in the previous chapter. Under this condition ($\alpha = 0$), the waveforms of the output voltage and the current in phase A are shown in Figure 9.35(a). The output waveform consists of six pulses, each $\pi/3$ wide. Each pulse shows the response of a circuit mode in the steady-state. The state of thyristors and transfer of currents are shown in the following table:

pulse number (mode)	conducting thyristor	current transition
1	T6	T5 to T1
2	T1	T6 to T2
3	T2	T1 to T3
4	T3	T2 to T4
5	T4	T3 to T5
6	T5	T4 to T6

The average output voltage V_o is given by

$$\overline{V}_o = \frac{3}{\pi}\sqrt{2}\,V_{LL} \;:\; V_{LL} = V_{ab} = V_{bc} = V_{ca} \quad (9\text{-}58)$$

V_{LL} is the line to line rms voltage. The output voltage can be controlled by delaying the current transition from one thyristor to another by an angle α. There are two gate control pulses per phase, six in total. The control angle α in each phase is referenced to $\pi/3$ and $4\pi/3$ angles.

gate turn-on pulse	current transition	angle
1	T5 to T1	$\pi/3 + \alpha$
2	T6 to T2	$2\pi/3 + \alpha$
3	T1 to T3	$\pi + \alpha$
4	T2 to T4	$4\pi/3 + \alpha$
5	T3 to T5	$5\pi/3 + \alpha$
6	T4 to T6	$2\pi + \alpha$

The steady-state operation consists of six circuit modes during a single period of the source waveform depending upon the state of the thyristors, as shown in Figure 9.35.

In the first mode, thyristor T6 was already on and thyristor T1 is turned on at $\pi/3 + \alpha$. T5 is reverse biased and the output current is steered via T1. Current flows from phase A, through T1, load and T6 to phase B. At angle $2\pi/3 + \alpha$ in mode 2, thyristor T2 is turned on, which reverse biases T6 and steers the output current from phase A, through T1, load and T2 to phase C. At angle $\pi + \alpha$ in mode 3, thyristor T3 is turned on, which reverse biases T1 and steers the output current from phase B, through T3, load and T2 to phase C. At angle $4\pi/3 + \alpha$ in mode 4, thyristor T4 is turned on, which re-

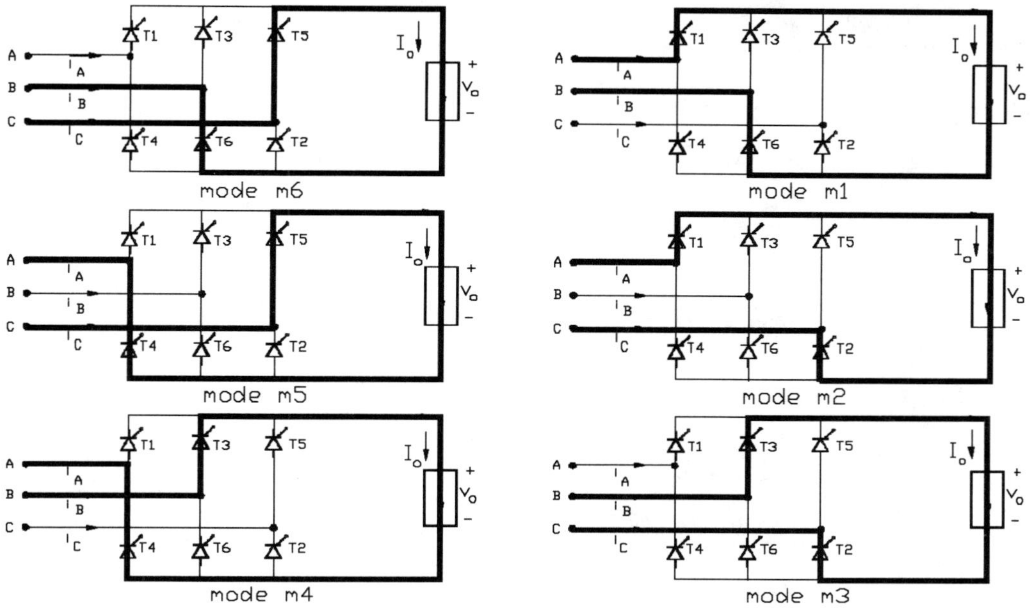

Figure 9.35
Circuit modes

verse biases T2 and steers the output current from phase B, through T3, load and T4 to phase A. At angle $5\pi/3 + \alpha$ in mode 5, thyristor T5 is turned on, which reverse biases T3 and steers the output current from phase C, through T5, load and T4 to phase A. Finally, at angle $2\pi + \alpha$ in mode 6, thyristor T6 is turned on, which reverse biases T4 and steers the output current from phase C, through T5, load and T6 to phase B. Mode 6 ends at $7\pi/3 + \alpha$ when T1 is turned back on again. Output voltage $\mathbf{v_o}$ and the current through the load $\mathbf{i_o}$ are shown in Figure 9.36.

Characteristics

Output dc voltage is calculated as the average of the output voltage waveform over a subperiod from $\pi/3 + \alpha$ to $2\pi/3 + \alpha$:

$$\overline{V}_o = \frac{3}{\pi}\int_{\pi/3+\alpha}^{2\pi/3+\alpha} \sqrt{2}\ \tilde{V}_{LL} \sin\theta d\theta$$
$$= \frac{3\sqrt{2}}{\pi}\tilde{V}_{LL}\cos\alpha \qquad (9\text{-}59)$$

Output dc voltage is positive and varies from $3V_L/\pi$ at angle $\alpha = 0$ to 0 at $\alpha = \pi/2$. The output dc voltage is negative for angles $\pi/2$ to π and the converter operates as an inverter. The converter can, however, operate steadily in inverter mode only when it consists of a device that can deliver energy, such as a dc motor in reverse motion or in regenerative braking mode. Readers are urged to sketch the output voltage waveforms for $\alpha = \pi/2$ and $\alpha = 3\pi/4$.

Voltage transfer ratio is found as

$$T_{vv} = \frac{\overline{V}_o}{\tilde{V}_L} = \frac{3\sqrt{2}}{\pi}\cos\alpha, \quad 0 \le \alpha < \pi/3 \qquad (9\text{-}60)$$

The output voltage ripple frequency is $6f_s$. Maximum ripple voltage occurs for $\alpha = \pi/2$.

RMS input line current and current transfer ratio are found as

$$\tilde{I}_L = \sqrt{\frac{1}{\pi}\int_{\pi/3+\alpha}^{\pi+\alpha} I_o^2\, d\theta}$$
$$= I_o\sqrt{\frac{2}{3}}, \qquad R_{ii} = \frac{\tilde{I}_L}{I_o} = \sqrt{\frac{2}{3}} \qquad (9\text{-}61)$$

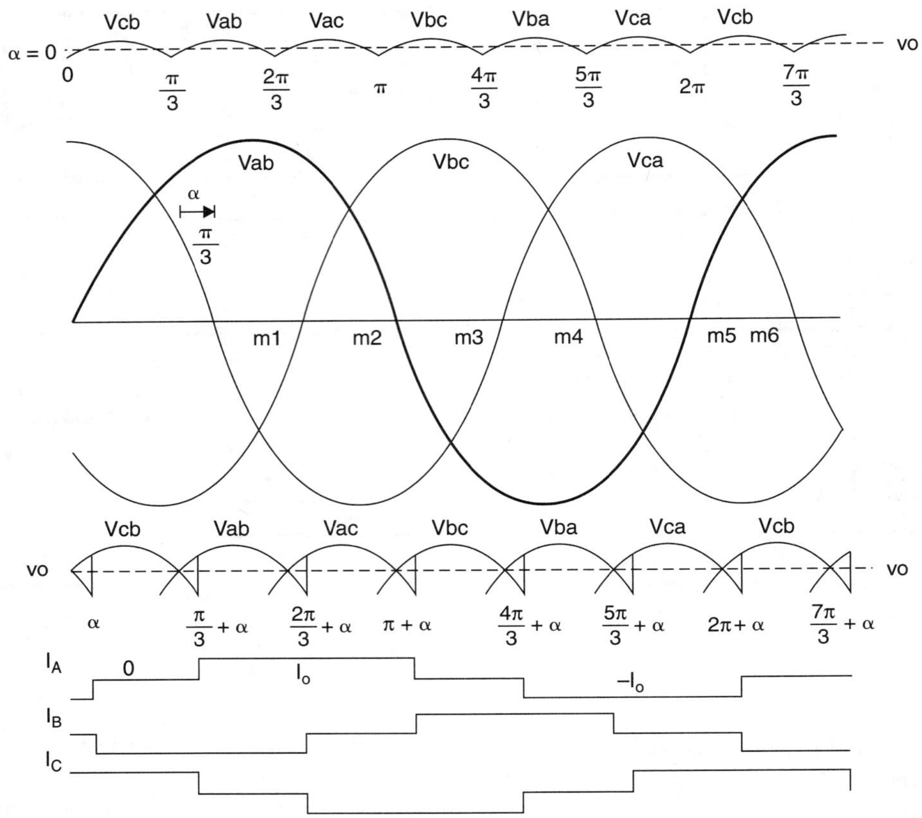

Figure 9.36
Waveforms

The plot of these parameters is given in Figure 9.37.

Comparison of these characteristics with those of a 1-phase full converter with current sink, Figure 9.8, shows that the voltage transfer ratio T_{vV} is higher, but the current reflection ratio m_i and the power factor are smaller (poorer). Total Harmonic Distortion is also higher.

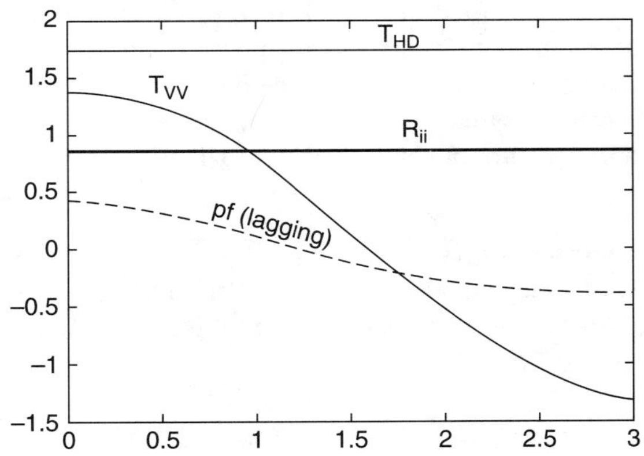

Figure 9.37
Performance characteristics of a 3-phase full converter

9.4.2 3-phase full-bridge converter with source inductance

This section will consider the effect of source side inductance in a three-phase full converter with constant current output. Like the one-phase full converter, the

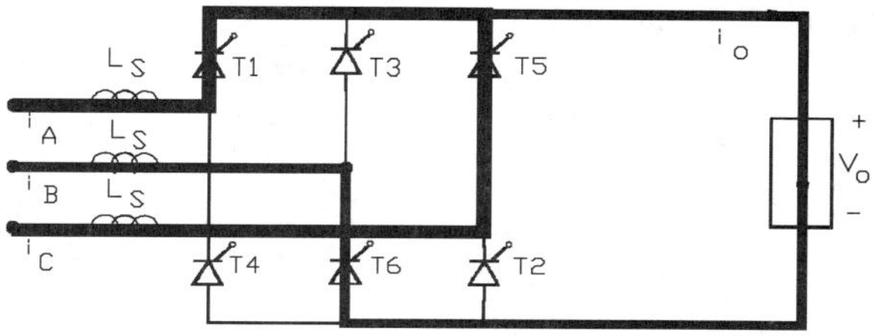

Figure 9.38
Mode 1x

source inductance affects only the output voltage waveform. It will suffice to investigate the waveforms in the vicinity of firing angle α. A new mode is created whenever the current is steered between thyristors. For example, a new mode 1x is created when the current is steered away from thyristor T5 to T1, as given below and shown in Figure 9.38.

Prior to mode 1x, thyristors T5 and T6 are on. T1 is fired on at $\alpha = \pi/3 + \alpha$. The source inductance L_S keeps T5 conducting. Current in T5 decreases and in T1 increases. T5 will cease to conduct when the current through builds up to I_o at $\pi/3 + \gamma$. During the angle $\gamma - \alpha$, L_S in line B conducts constant current I_o and is effectively short-circuited. Once T5 extinguishes, T1 begins to conduct constant current I_o and is also effectively short-circuited. It should be noted that the line voltage v_{AC} is shorted through two source inductances during this mode. This shorting causes what is normally called line notching in power lines and degrades the quality of the power from the utility.

Output voltage during in mode 1x is given by

$$v_o = v_{ab} - L_S \frac{di_A}{dt} \qquad (9\text{-}62)$$

The sum of line currents i_A and i_C equals the constant current I_o, hence

$$\frac{di_A}{dt} = -\frac{di_C}{dt}, \text{ and } v_{ac} = L_S \frac{di_A}{dt} - L_S \frac{di_C}{dt} \qquad (9\text{-}63)$$

Therefore,

$$v_o = v_{ab} - \frac{1}{2} v_{ac}$$

$$= \sqrt{2}\tilde{V}_L \sin\theta - \frac{1}{\sqrt{2}}\tilde{V}_L \sin(\theta - \frac{\pi}{3}) \qquad (9\text{-}64)$$

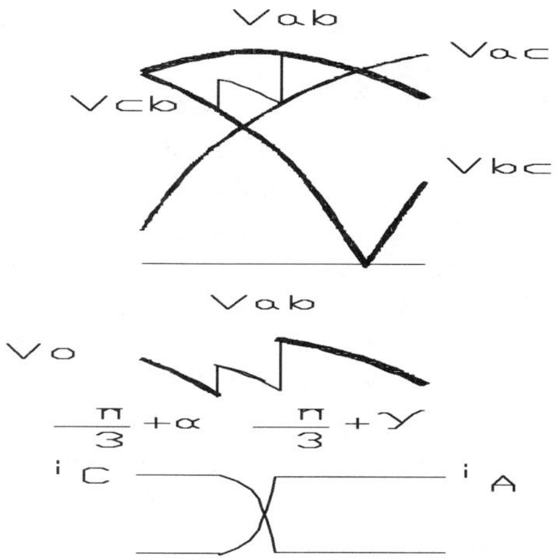

Figure 9.39
Partial output voltage waveform with L_S

Output voltage waveform is shown in Figure 9.39.

Output dc voltage is calculated as the average of the output voltage waveform over a subperiod from $\pi/3 + \alpha$ to $2\pi/3 + \alpha$:

$$\overline{V}_O = \frac{3}{\pi}\left[\int_{\pi/3+\alpha}^{\pi/3+\gamma} v_o \, d\theta + \int_{\pi/3+\gamma}^{2\pi/3+\alpha} v_o \, d\theta\right]$$

$$= \frac{3\tilde{V}_L}{\sqrt{2}\pi}(\cos\alpha + \cos\gamma) \qquad (9\text{-}65)$$

Line current i_C in the inductor L_S C in mode 1x is found from:

$$-L_S\frac{di_C}{dt} = \frac{1}{2}v_{ac} \tag{9-66}$$

$$i_C = I_o - \frac{\tilde{V}_L}{\sqrt{2}\omega L_S}(\cos\alpha - \cos(\theta - \pi/3)),$$
$$\pi/3 + \alpha \le \theta < \pi/3 + \gamma \tag{9-67}$$

Angle γ is obtained by equating i_C to zero at $\pi + \gamma$

$$\gamma = \cos^{-1}\left[\cos\alpha - \frac{\sqrt{2}\omega L_S I_o}{\tilde{V}_L}\right] \tag{9-68}$$

9.4.3 Cross coupling due to 3-phase full-bridge converters

The three-phase utility supplies power to several loads simultaneously. The line notching due to a load, explained in a previous section, affects the quality of power available to other loads. The stray or parasitic capacitances further complicate by adding ringing along with notches. The inclusion of stray capacitances is too complicated for analysis and therefore better handled by simulation.

Consider that two loads are sharing a single utility. The line inductance in this situation must be divided in two parts in each line: L_{S1} and L_{S2}, as shown in Figure 9.40. Usually L_{s1} is the source inductance and L_{s2} is the transformer leakage inductance feeding the converter-rectifier. Assume that load 1 has a 3-ϕ controlled rectifier and is drawing a constant current I_o. The load 2 is not drawing any current. We are interested in finding out the size of voltage notches in the voltage available to load 2.

Consider the line-to-line voltage v_{AB} at the point of common coupling (PCC). There are three notches per half cycle in the waveform of v_{AB} as shown in Figure 9.41. See Figure 9.42 for the following explanations. The first notch occurs at α where lines A and B are shorted in load 1. The current decreases in line A from $-I_o$ to 0 and the current in line B increases from 0 to $-I_o$. The second notch occurs when lines A and C are shorted at $\pi/3 + \alpha$. The current increases in line A from 0 to $+I_o$, and the current in line C decreases from I_o to 0. The third notch occurs when lines B and C are shorted at $2\pi/3 + \alpha$. The current decreases in line B from $-I_o$ to 0, and the current in line C increases from 0 to $-I_o$.

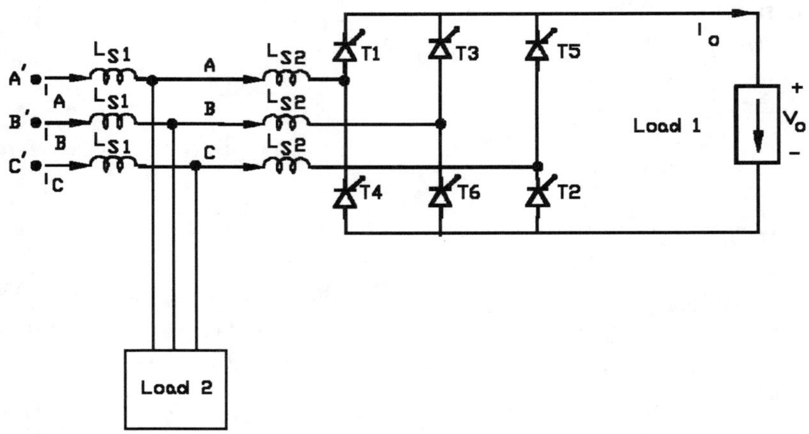

Figure 9.40
Two loads sharing a single utility

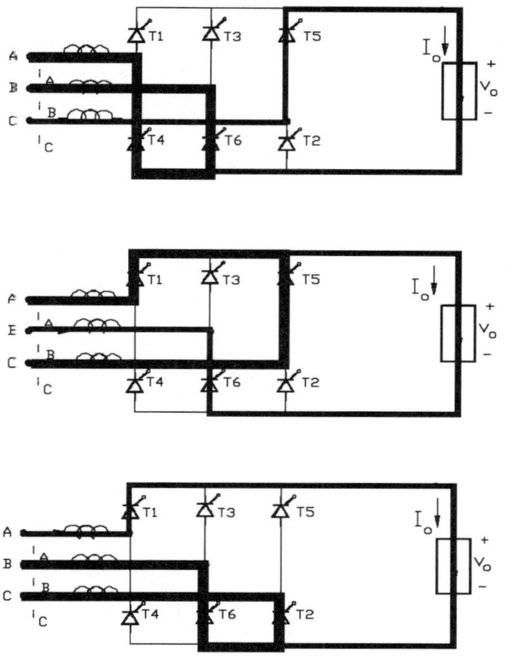

Figure 9.41
Current flow in circuit modes

Lines A and B shorted: $i_A = -I_o \_ 0,\ i_B = 0 \_ -I_o$
Lines A and C shorted: $i_A = -0 \_ I_o,\ i_C = I_o \_ 0$
Lines B and C shorted: $i_B = -I_o \_ 0,\ i_C = 0 \_ -I_o$
All notch widths equal $\gamma - \alpha$ given below:
Notch width:

$$\cos^{-1}\left[\frac{2\omega(L_{S1}+L_{S2})I_o}{\sqrt{2}\,\tilde{V}_L}+\cos\alpha\right]-2\alpha \tag{9-69}$$

Notch depths are given by
First notch at α:

$$-\frac{L_{S1}}{L_{S1}+L_{S2}}\sqrt{2}\,\tilde{V}_L\sin\alpha \tag{9-70}$$

Second notch at $\pi/3 + \alpha$:

$$-\frac{\sqrt{2}}{2}\tilde{V}_L\sin\alpha \tag{9-71}$$

Third notch at $2\pi/3 + \alpha$:

$$+\frac{\sqrt{2}}{2}\tilde{V}_L\sin\alpha \tag{9-72}$$

Note that the notch width depends on the dc current in load 1 I_o, but the notch depths are independent of I_o. IEEE standard 519 has prescribed limits for notch depths and notch widths in the utilities.

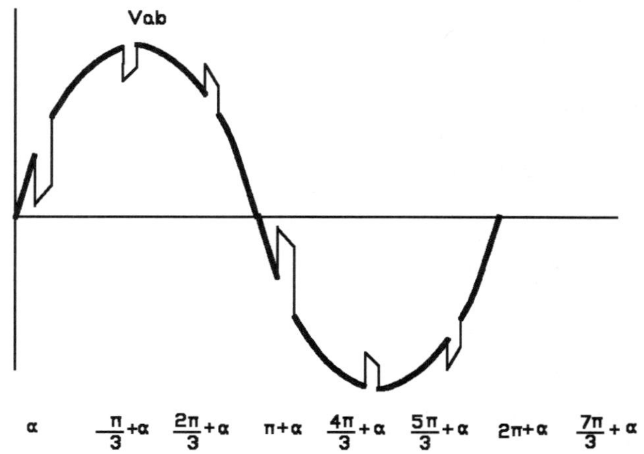

Figure 9.42
Line notching

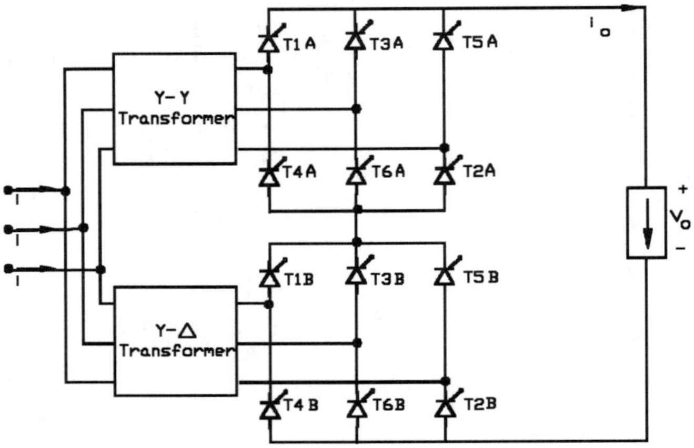

Figure 9.43
A 12-pulse converter

9.4.4 A 12-pulse 3-phase full-bridge converter

A three-phase bridge gives a six-pulse output voltage. The ripple voltage can be further reduced, and the ripple frequency can be increased by using a six-phase input voltage source. The six-phase source can be synthesized from a three-phase source, a transformer with Y-connected secondary and another transformer with Δ-connected secondary, as shown in Figure 9.43. The secondary voltages of the two transformers differ by 30°, hence by adding them in series, a six-phase source becomes available.

The line current waveforms are improved. Output voltage ripple is reduced. The ripple frequency is doubled (12 times the line frequency).

9.5 DESIGN EXERCISE: PWM DRIVE FOR DC MOTOR

9.5.1 Equivalent circuit of a dc motor

Consider a separately excited dc motor. The equivalent circuit is shown below:

The back emf, given by the following equation, is represented as a dc voltage sink.

$$e_b = k_e\, I_f\, \omega_m \qquad (9\text{-}73)$$

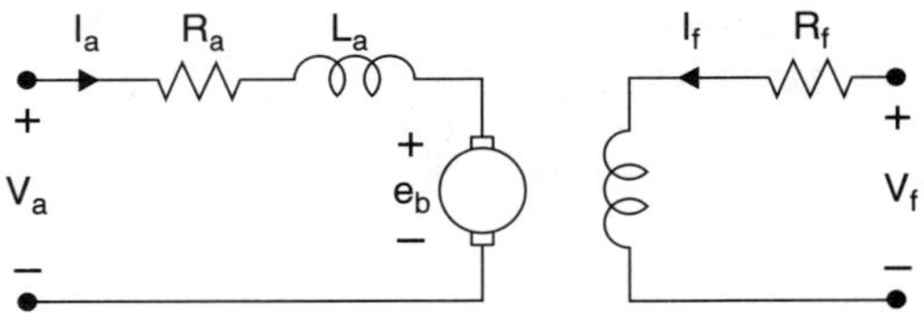

Figure 9.44
Equivalent circuit of a separately excited dc motor

where k_e is the motor electrical constant, I_f is the field current, and ω_m is the angular velocity of the motor. The torque developed in the motor T_m is given by

$$T_m = k_m\, I_f\, i_a \qquad (9\text{-}74)$$

where k_m is the motor torque constant and i_a the armature current. The electrical power consumed by the dc voltage sink is equal to the mechanical power developed.

9.5.2 1-phase adjustable speed drive (ASD)

Practical adjustable speed motor drives are used in dc motors rated for 15 hp or less. An ASD operates in the first and fourth quadrants. The first-quadrant operation is for the forward motoring and the fourth quadrant is for regenerative braking. Hence, a full-converter of Figure 9.22 with phase delay angle control is the most appropriate converter for this application.

Example 9-8

A 10 hp, 1700 rpm self-excited dc motor, driven from a 120 V 60 Hz source, has the following parameters:

$R_a = 0.02$ *ohms*	$L_a = 1$ *mH*
$R_f = 150$ *ohms*	$k_m = 0.5$ *N-m/A*^{2}
$k_e = 0.608$ *V/A.rad/sec.*	

The field is separately energized by a full-bridge rectified dc derived from the ac 120 V, 60 Hz source. The motor speed is to be varied from 10% to the rated speed at a constant torque. The maximum peak-to-peak torque ripple should not exceed 5% of the rated torque. Design an ac-dc converter for driving the motor.

Peak input voltage $V_S = \sqrt{2} \times 120 = 169.7056$ V
Load power $P_L = 10\times746 = 7460$ W
Rated speed $\omega_m = 2\pi\times1700/60 = 178.0236$ rad/s
Rated torque $T_L = 7460/178.0236 = 41.9046$ W/rad/s

The back emf of the dc motor is also the dc output voltage of the converter, given by equation (9-33). The maximum back emf, which corresponds to the rated speed, occurs at the minimum value of the phase delay angle, $\alpha_{min} = 0.576$ radians (33°). Allowing a 10% margin for parameter variations, we set the minimum phase delay angle $\alpha_{min} = 0.63$ radians. The maximum back emf then is

$$e_{bmax} = 2\times169.7056\times\cos(0.63)/\pi = 87.2976 \text{ V}$$

The field current for the motor is calculated from the e_{bmax} and equation (9-73):

$$e_{bmax} = k_e\, I_f\, \omega_m = 0.608\times I_f\times178.0236$$
$$I_f = 0.8065 \text{ A}$$

The field winding is energized by a full-bridge diode rectifier. The voltage applied to the field winding is calculated as

$$2\,V_S/\pi = 2\sqrt{2}\times120/\pi = 108.038 \text{ V}$$

It is reasonable to assume the voltage drop due to commutation in the bridge converter is negligible. The required resistance in the field circuit is obtained as

$$R_f = 169.7056/0.8065 = 210.4141 \text{ ohms}$$

Since the field winding resistance is only 150 ohms, an extra resistor of 60.4141 ohms must be added in series with the field winding. An adjustable 100 ohms, 100 W potentiometer will be very appropriate.

Assuming no losses in the brushes and no rotational losses, the output power is equal to the product of armature current and the motor speed. The dc armature current at rated speed is then found as

$$I_a = 7460/87.2976 = 85.4528 \text{ A}$$

The back emf at 10% rated speed is $e_{bmin} = 8.7298$ V. The corresponding phase delay angle α_{max} is calculated using equation (9-33),

$$\alpha_{max} = \cos^{-1} \pi\times8.7298/(2\times169.7056) = 1.4899 \text{ radians}$$

The minimum phase delay angle α_0, for which the converter is not controllable in the selected topology, is given by equation (9-32) and it must be less than the α_{min}:

$$\alpha_0 = \sin^{-1} 8.7298/169.7056 = 0.0515 \text{ radians}$$

This condition is satisfied.

The motor must be operated at a constant armature current in CCM to ensure operation at constant load torque. We should further ensure that this current is about 10% above the boundary value, given by equation (9-40a). Hence, the boundary value current is 76.9 A. The armature inductance for ensuring this condition is calculated as

$$L \geq \frac{2 \times 169.7056}{\pi \times 2\pi \times 60 \times 76.9} \sin(1.4899) = 3.71\ mH$$

The armature inductance also determines the peak-to-peak current ripple, which in turn determines the torque ripple. The torque ripple depends on the current ripple, which is maximum at the phase delay angles in the vicinity of $\pi/2$ where the motor runs at low speed. The lower limit on the armature inductance from this consideration is calculated from equation (9-38):

$$\Delta I_o = 0.05 \times I_o = 4.2726$$
$$L \geq 8.88 \text{ mH}$$

The inductance in the armature circuit is selected to be the highest of the above two values, 8.88 mH. Since the armature inductance is only 1 mH, an extra inductance of 7.88 mH is added in series with the armature of the dc motor.

The peak value of the armature current in the dc motor is also the peak current in the IGBT switches. The specifications of the IGBTs and diodes are calculated by allowing a safety margin and are given below:

Major Specifications of IGBTs:

Average and Repetitive Peak current ≥ 100 A

Repetitive maximum Drain to Source voltage ≥ 200 volts

Two modules of International Rectifier IRGT120 F06, a half-bridge module, meet the requirements.

Major Specifications of Diodes in the Field Circuit:

Average and Repetitive Peak current ≥ 1 A

Repetitive Peak Inverse Voltage ≥ 200 volts

International Rectifier 20D2 diodes meet the above requirements.

The feedback circuit block diagram is given in Figure 9.45. The full-bridge (also called H-bridge) of the IGBTs is driven by an H-bridge Driver IC. International Rectifier IR2110, an H-bridge Driver is an appropriate choice. The motor speed is sensed by a tachometer, compared to an adjustable voltage reference corresponding to the desired speed, and given as the input to a voltage-to-pulse width converter such as a Unitrode 3526A IC. Readers are urged to do a detailed design of the feedback circuit and simulate on PSPICE.

9.6 PSPICE SIMULATION

PSPICE Example 9-1 1-Phase AC to Controlled DC Converter with and Without the Source Inductance

Consider the bridge semi-converter with a freewheeling diode, as shown in Figure 9.46(a) with the source side inductance and the R-C load. Use 2N1595 as the thyristor and 1N4002 as the uncontrolled diode. The converter specifications are:

AC source = 24 V peak, 100 Hz RL = 1 ohm

L5 = 10 mH

Perform the transient analyses for at least five periods and plot the waveforms in the last period of the output voltage and the source current for the following values of the source inductance, (a) L6 = 0 and (b) L6 = 1 mH.

The thyristors X1 and X2 are switched by two external pulse sources V7 and V9 respectively as shown in Figure 9.46(a). The pulse sources V7 and V9 are transformer coupled or optically coupled in the practical implementation of the circuit.

Set the pulse source V7 as follows:

Lower level V1 = 0 V
Upper level V2 = 10 V
Pulse period = 10 ms

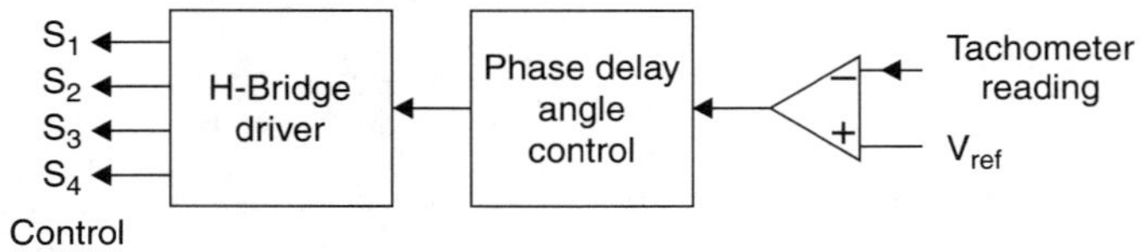

Figure 9.45
Feedback control circuit for dc motor drive

Pulse width (on-time) $=100\ \mu s$
Delay time TD $=2$ ms
Rise time $=0.1$ ns
Fall time $=0.1$ ns

Set the pulse source V9 with all settings of V7 except the delay time, TD=7 ms.

Use the simulation step in the range: 1 ns to 100 ns.

Note the half-period transitions in the source current become very smooth for L6 = 1 mH.

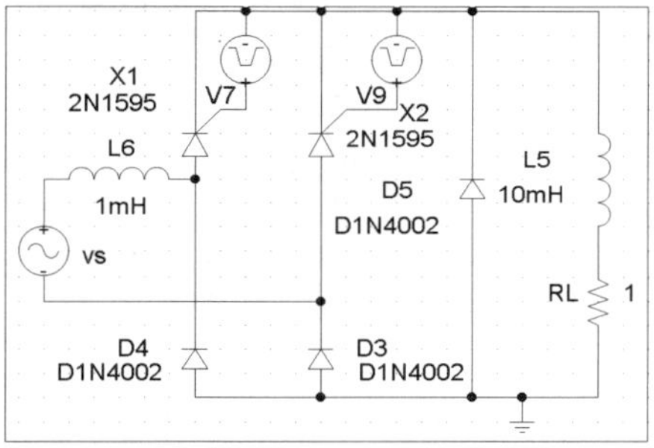

(a)

9.7 SUMMARY

In this chapter, we have seen some of the techniques of conversion and control of ac-dc converters. There are several other techniques mentioned in the references. A particular technique may be more suitable than the others for a specific application. Analysis using circuit modes and numerical methods, such as those available in MATLAB, demonstrated for a few important topologies in this chapter can also be used for other topologies. The important design information gained in this chapter may be summarized as follows:

1. DC voltage from an ac-dc converter can be controlled by the control of firing angle α, the extinction angle β, the pulse-width δ and the number of pulses per half period.
2. All control techniques, except the extinction angle control, have lagging input power factors.

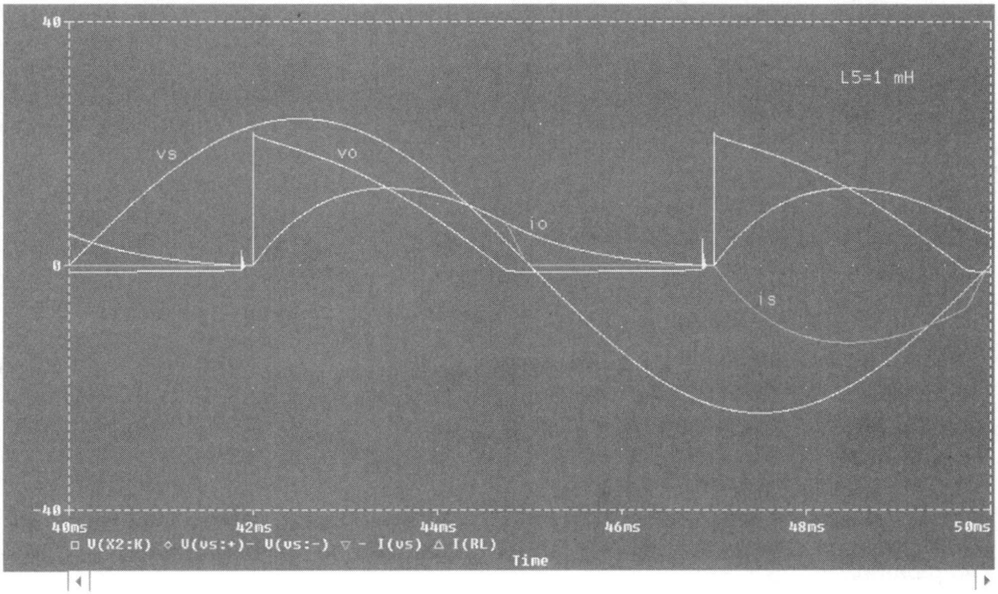

(b)

Figure 9.46
(a) Schematic and (b) waveforms of vs, vo, io, and is for zero source inductance.

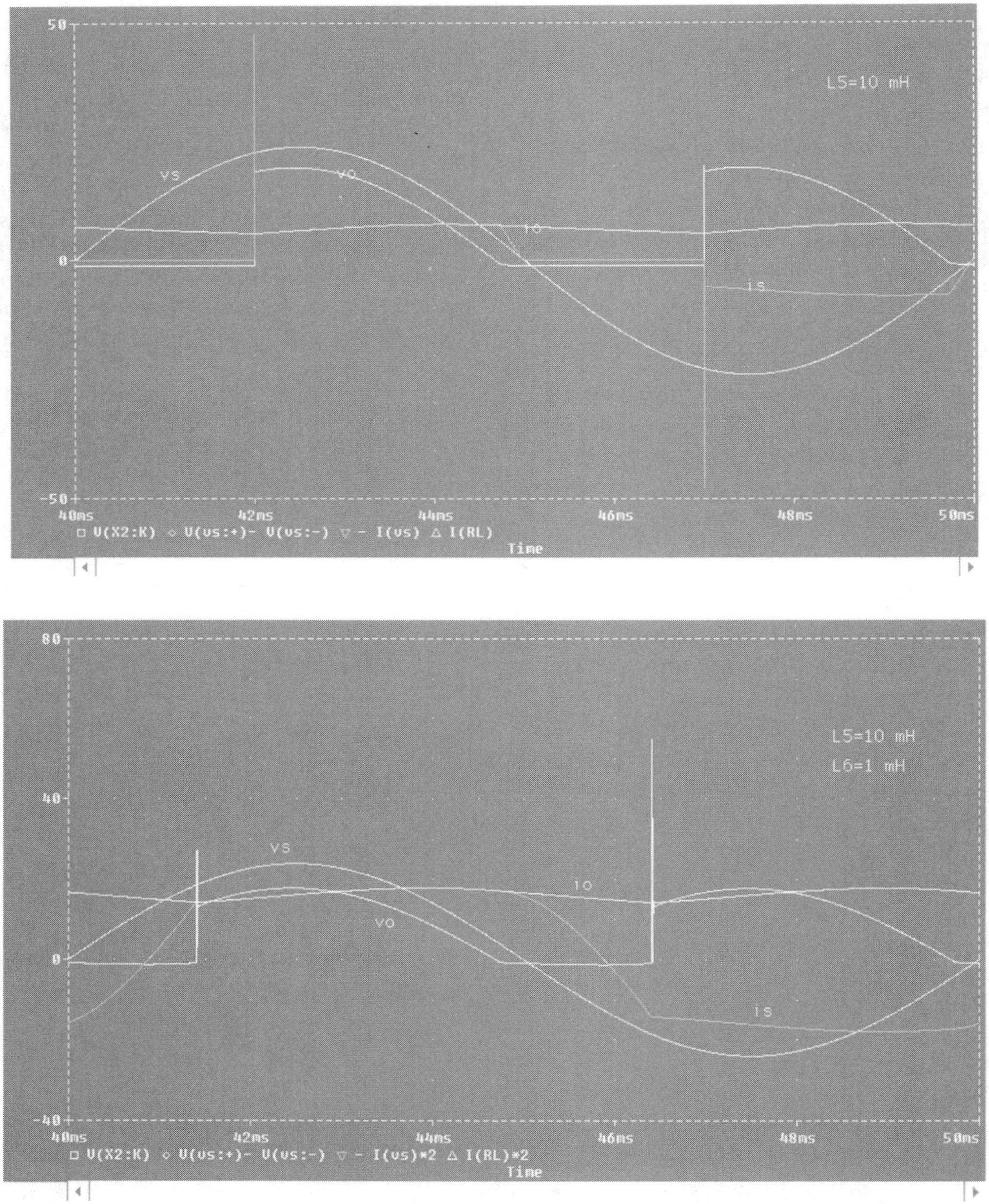

(c)

Figure 9.46 ***(continued)***
(c) Waveforms of vs, vo, io, and is for source inductance L6 = 1 mH.

3. PWM control gives higher input power factor because of quarter-wave symmetry of conduction angle with respect to the input voltage waveform.
4. All control techniques, except the firing angle control, require forced commutation of switching devices.
5. Both the firing angle and the extinction angle control in the full-converter enable two-quadrant operation. PWM control permits single-quadrant operation.
6. The four-quadrant operation can be obtained by connecting two two-quadrant converters in a dual connection.
7. A multiphase ac-dc converter gives high ripple frequency and thus the filter requirements in the output circuit are less constraining.
8. AC-DC converters can be operated both with constant current and constant voltage loads. Constant voltage loads require a continuous current mode of operation at high power levels and a discontinuous current mode at low and medium power levels.
9. Single-phase ac-dc converters are used in low-medium power applications. High-power applications use three-phase converters.
10. The presence of source side inductance introduces the commutation angle constraints and gives rise to an equivalent output source resistance. This resistance is responsible for output voltage drop at a higher load current. On the input side, it also causes line notching.

References

1. Philip T. Krein, *Elements of Power Electronics,* Oxford University Press, New York, 1998.

2. Ned Mohan, Tore M. Undeland, and William P. Robbins, *Power Electronics: Converters, Applications and Design,* 2nd Ed., John Wiley & Sons, Inc , 1995.

3. Joseph Vithayathil, *Power Electronics: Principles and Applications,* McGraw-Hill Inc., New York, 1995.

4. Muhammad H. Rashid, *Power Electronics: Circuits, Devices and Applications,* 2nd Ed., Prentice-Hall Inc., New Jersey, 1993.

5. John G. Kassakian, Martin F. Sclecht, and G. Vergassian, *Principles of Power Electronics,* Addison-Wesley, Boston, 1991.

Problems

9-1 Figure P9.1 shows a converter with two thyristors. Thyristor T1 is fired at α and T2 at $\pi+\alpha$. Analyze and plot the performance characteristics.

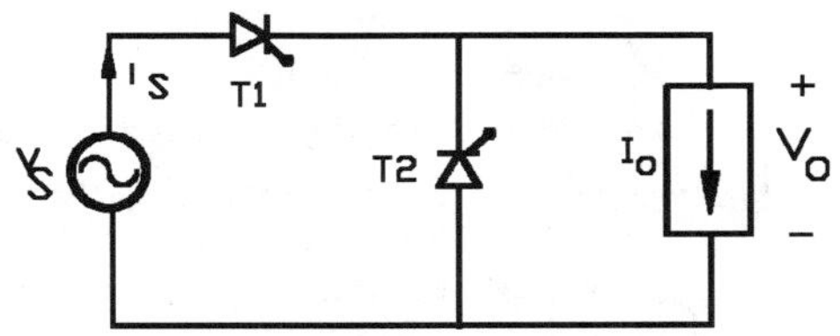

Figure P9.1

9-2 Repeat problem 9-1 when the dc current sink is replaced by an inductance L in series with a dc voltage sink V_o.

9-3 Figure P9.3 shows a semi-converter. Thyristors are fired angle apart. Draw the voltage and current waveforms using MATLAB. Analyze and plot the performance characteristics. Compare its performance with that of a full-bridge converter ac-dc converter.

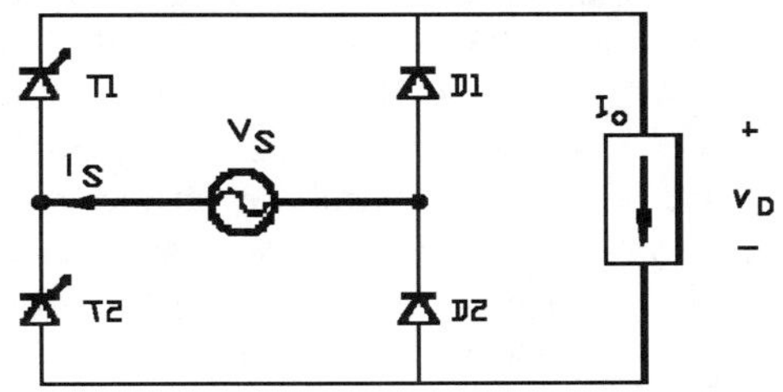

Figure P9.3

9-4 Practical inductors have finite dc resistance. Include a resistance R in series with the large inductor that is represented by a dc current sink in the one-phase full-bridge converter of Figure 9.10. Repeat the analysis, plot the performance characteristics and compare them with those given in Figure 9.15.

9-5 A separately excited 1KW dc motor is driven by a full-bridge converter from a one-phase 220 V 50 Hz supply. The armature inductance of the motor is 250 mH and the back emf at the desired speed is 150 V dc. Find the firing angle α, the rms input current, the current ripple and the input power factor. The current ripple can be reduced by enhancing the armature inductance. Find out the additional inductance to bring the current ripple to 2% level.

9-6 The DC motor of problem 9-5 is running at the rated speed of 1800 rpm. A regenerative brake is applied by operating the converter in the Inverter mode at the maximum negative dc voltage. Find the firing angle and the time for the brakes to bring the motor to a stop.

9-7 Practical dc voltage sinks, such as the armature of a dc motor, have armature resistance, and a battery under charge has a nonzero internal resistance. Include this internal resistance R in the one-phase full-bridge converter of Figure 9.21(a). Repeat the analysis, plot the performance characteristics and compare them with those given in Figure 9.23.

9-8 Replace the dc current sink in the one-phase full-bridge converter with source inductance in Figure 9.25 by a dc voltage sink. Using MATLAB draw the voltage and current waveforms in a switching period. Analyze and plot the performance characteristics. Compare these characteristics with those of the one-phase full-bridge converter with no source inductance as given in Figure 9.23. Extend this idea to a current source driven converter driving a dc voltage sink.

9-9 Consider a three-phase half-bridge converter of Figure P9-9. Draw the voltage and current waveforms using MATLAB. Analyze and plot the performance characteristics. Compare its performance with that of a three-phase full-bridge converter in Section 9.4.1.

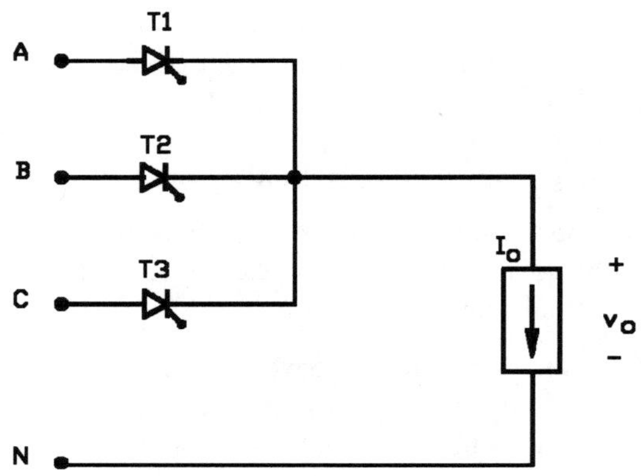

Figure P9.9

9-10 In the three-phase rectifier circuit of Figure 9.34, the rms line-line voltage is given 480 V at 60 Hz. The source side inductance L_S is 40 μH. Calculate the commutation angle u if V_o=480 Vdc and P_o=500 kW.

9-11 Analyze and discuss the three-phase converter circuit of Figure 9.34 by replacing the dc current sink by a dc voltage sink in series with an inductor.

9-12 Calculate the harmonic components, displacement factor and Total Harmonic Distortion in the voltage at the point of common coupling in Figure 9.40.

9-13 In the circuit of Figure 9.40, L_{S1} represents the leakage inductance of a 60 Hz transformer with the following ratings: 3-phase 500 kVA, line-line voltage of 480 volts. L_{S2} is due to a 200 foot-long cable, with a per-phase inductance of 0.1 μH/ft. The converter, at rated input voltage, delivers 25 kW at a voltage of 480 Vdc. Calculate the notch widths and notch depth at the point of common coupling.

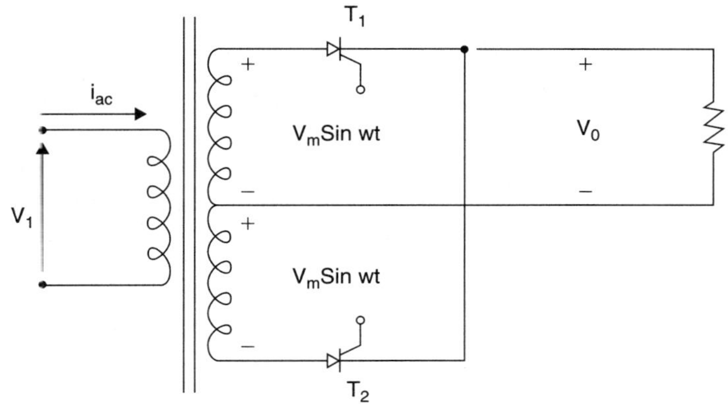

Figure P9.18

9-14 A three-phase full-bridge converter is designed to energize the relay coil of a heavy-duty electromagnetic pickup crane that has an inductance of 1.2 H and needs a constant current of 200 A. The input source is 208 V 60 Hz. Find the firing angle α, the rms input current, the input power factor and the Total Harmonic Distortion THD.

9-15 The full-bridge converter of problem 9-14 is also used to demagnetize the relay coil by operating the converter in inverter mode at maximum negative output voltage for as long as it is necessary to bring down the coil current to zero. Find the firing angle and the time of the inverter operation.

9-16 A three-phase full-bridge converter is connected to a 250 V DC grid. The combined line inductance and the generator inductance is known to be 1 mH. Find out the firing angle α, the minimum dc current outflow to maintain a continuous operation.

9-17 Design a battery charger for the 36-volts battery from a one-phase ac 120 V 60 Hz. It should be capable of charging batteries that are down to 30% terminal voltage. The maximum charging current must be limited to 15 A.

9-18 **a.** Plot the waveforms for i_S, i_O and v_O for $\alpha = 60°$ in the full-wave converter of Figure P9.18 where $v_S = 169 \sin 2\pi 60t$ and the load resistor R = 20 ohms.

b. Calculate the power delivered to the load resistor R and the power factor at the input ac source.

9-19 The load in the ac-dc converter of Figure P9.3 is replaced by the armature of a separately excited dc motor. The ac source is $v_S = 169 \sin 2\pi 60t$. The equivalent circuit of the motor armature consists of R and L that are of sufficient value to enable the source current waveform to be discontinuous. Assume negligible ripple in the motor current. Plot the waveforms for i_O and v_O for

a. $\alpha = 45°$ and the zero speed of the dc motor.

b. $\alpha = 90°$ and the back emf of the motor is 90 Vdc.

9-20 Assume a negligible ripple in the dc current and a delay angle that is $\alpha = 67°$ in the three-phase converter shown on the next page. Find the dc output voltage across the load resistor and plot the waveforms of v_O and I_O.

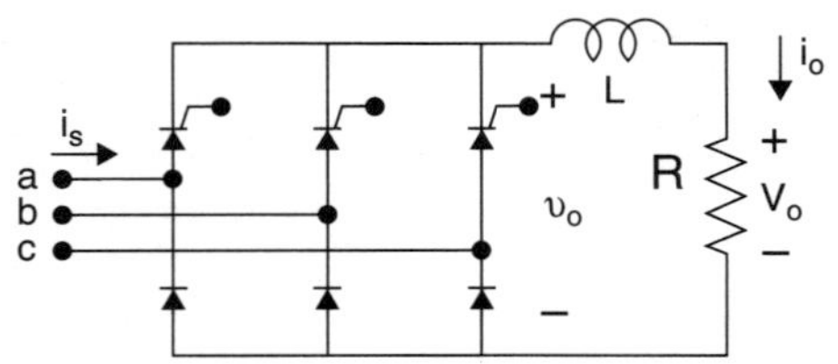

Figure P9.20

9-21 The full-bridge converter of Figure 9.12, with an input voltage 120 Vrms at 60 Hz, is required to provide a dc current of 15 A to the coil of an electromagnet that has an inductance of 0.2 H and a resistance of 2 ohms. Find the phase delay angle α, the rms input current, the input power factor and the Total Harmonic Distortion THD. Use the formulae and MATLAB.

9-22 The coil in problem 9-21 must be de-energized after the intended work is completed. Use the full-bridge converter of Example 9-1 to demagnetize the coil by operating the converter in the inverter mode at maximum negative output voltage for as long as it is necessary to bring down the coil current to zero. Find the firing angle and the time of the inverter operation.

9-23 Repeat problem 9-21 with PWM control.

9-24 Consider the full-bridge converter of Figure 9.20 with L_S=150 mH. Find the maximum output current if the no-load output voltage is desired to be 48 V dc. Select a current lesser than maximum and calculate the performance measures. Draw the waveforms of the output voltage and the source current.

9-25 A single phase 1 hp dc motor is driven by a 220 Vrms 50 Hz supply. The armature inductance of the motor is 100 mH and the back emf at the desired speed is 75 V dc. Use the full converter of Figure 9.21. Draw the voltage and current waveforms using MATLAB. Also find the phase delay angle α, the average output current, the crest factor, the rms input current, the input power factor and the Total Harmonic Distortion THD.

9-26 Find the dc motor current in the problem 9-25 at which the motor current waveform becomes discontinuous at the specified speed (back emf E=75 V). Verify the waveform using MATLAB.

9-27 A 6.5 hp, 2200 rpm self-excited dc motor, driven from a 120 Vrms 60 Hz source, has the following parameters: R_a = 0.02 ohms, L_a = 1 mH, R_f = 150 ohms, k_m = 0.5 N-m/A^2, k_e = 0.608 V/A.rad/sec. The field is separately powered by a full-bridge rectified dc derived from the ac 120 Vrms, 60 Hz voltage source. The motor speed is to be varied from 10% to the rated speed at a constant torque. The maximum peak-to-peak torque ripple should not exceed 5% of the rated torque. Design an ac-dc converter for driving the motor.

9-28 A three-phase, half-wave ac-dc converter is connected to a 380 Vrms (line) supply. The load current is 32 A dc and is independent of the delay angle. The on-state voltage of the thyristor is 1.2 V. (a) Find the average dc output voltage at the delay angles of 0° and 45°, (b) the current and the peak reverse voltage ratings of the thyristor and (c) the average power dissipated in the thyristor.

9-29 A one-phase, full-wave converter is supplied from 220 V, 50 Hz with a source inductance of 0.333 mH. The load current is 4 A dc. Find the overlap angles for (a) transfer of a current from a conducting thyristor to the commutating diode and (b) from the commutation diode to a thyristor when the firing angle is π/3.

9-30 A three-phase, half-wave ac-dc converter is operating in the inversion mode. The ac source is 380 Vrms (line), the extinction angle is 22° and the overlap angle is 4°. Find the average voltage on the load.

9-31 Find the input power factor of a one-phase bridge ac-dc converter at the delay angles of 30° and 60°. Assume constant load current with no ripple, no source inductance and the switches have no voltage drop in the on-state.

CHAPTER 10
AC to Controlled AC

OBJECTIVES

1. Techniques of on-off control, phase delay angle control and pulse width (PWM) control
2. Converter operation with R and L loads
3. Converter operation with source side inductance
4. Operation of the cycloconverter
5. Three-phase to one-phase ac-ac conversion
6. Three-phase ac-ac converters
7. Introduction to dc-link and resonant-link ac-ac conversion

10.1 INTRODUCTION

AC-to-ac converters transfer power from one ac system to another with waveforms of different amplitude, frequency, or phase. These converters are designed for one-quadrant to four-quadrant operation. Typical applications are listed below:

Variable speed drives for appliances and tools,

Four-quadrant PWM drives for traction,

Steel mill roll drives

A general ac-to-ac converter system is shown in Figure 10.1. Input is a sinusoidal voltage source or a current source of frequency ω_s and n phases. Output is required to be the ac voltage or current at the same or a different frequency ω_o and at the same number of phases or m phases. The control signal may be, as shall be seen later, an angle, a pulse width, a voltage or a current signal. Most of the practical sources in power electronics are voltage sources.

Basically two-types of ac-to-ac conversions are in use:

1. Direct conversion, and
2. Indirect conversion.

In the direct conversion, the ac input waveforms are directly converted into the desired output waveforms. Converters that do not involve change of frequency are called the ac controllers. The converters with the change of frequency are called cycloconverters. Cycloconverters can perform only the down conversion of frequency ($\omega_o < \omega_s$) that too in simple multiples (/2, /3, /4, and so on). The cycloconverters convert the input ac waveforms directly into the desired ac waveforms. The output terminals of the

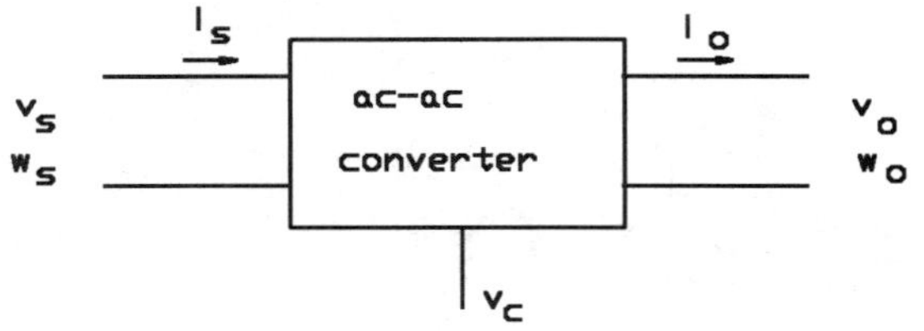

Figure 10.1
An ac-ac converter

cycloconverter are connected to the input terminals through controlled switches in a sequence that generates the desired waveforms. The switches are required to support bipolar voltages and carry bidirectional currents. The design is usually complex. Cycloconversion may not require forced commutation. Hence the cycloconverters may use thyristors. Most of the very high power applications such as traction or heavy industrial drives employ cycloconverters.

Indirect ac-ac conversion involves an intermediate dc stage, called the dc-link or dc bus and the converters are called the dc-link converters. In these converters, the ac is first converted to dc (ac-dc converter) and then inverted to the desired ac form (dc-ac converter). An energy storage component, such as a capacitor, an inductor or a resonant tank, stores the energy difference arising out of power transfer between the two converter stages. The switches require forced commutation. MOSFETs and IGBTs are the preferred devices. Unlike the cycloconverters, the input and output frequencies in the indirect (dc-link) converters can be widely different. This is the only type in which the output frequency can be designed to be higher than the input frequency. These converters are used in medium power applications.

10.2 MODELING AND ANALYSIS

The ac to ac converter system is modeled as an ac voltage source or current source connected to a load through a network of switches. Following sections give the assumptions for a simplified analysis and the criteria for measuring the performance of the converter operation.

10.2.1 Assumptions for analysis

The input source is a one-phase or a three-phase ideal voltage source. A one-phase input voltage in the rest of this chapter will be assumed at a sine wave of frequency f_s ($\omega_s/2\pi$) and a peak amplitude V_S as given below:

$$v_S = V_S \sin\theta \qquad : \theta = \omega_s t \tag{10-1}$$

θ is the conduction angle. Switches including the diodes are assumed to be ideal and unidirectional, that is zero on-state losses, zero off-state losses, zero switching losses and unrestricted voltage or current carrying capacities. Switching devices in a practical design example must be selected with these desired characteristics in mind. Inductors, capacitors and transformers are assumed to be ideal.

10.2.2 Performance characteristics

The steady-state converter performance measures can be put into four categories:

Forward transfer characteristics

Reflective characteristics

Harmonic profile

Component stresses

Voltage Transfer Ratio

The forward voltage transfer ratio is defined as the ratio of the rms output voltage to the rms input voltage:

Voltage transfer ratio

$$T_{vv} = \frac{\tilde{V}_o}{\tilde{V}_s} \tag{10-2}$$

The subscript 'o' refers to the output variable, and the subscript 's' refers to the source side variables.

The plot of $\mathbf{T_{vv}}$ versus the control variable is called the control characteristics of the converter. It is desirable to employ a small control signal for a large variation in the output voltage, that is, a high slope of the control characteristic curve.

Current Reflection Ratio
The reflective characteristics give the effect of load current on the current drawn from the source. The current reflection ratio is defined as the ratio of the rms input current to the rms load current:

Current reflection ratio

$$R_{ii} = \frac{\tilde{I}_s}{\tilde{I}_o} \quad (10\text{-}3)$$

It is desirable to have a $\mathbf{R_{ii}} = 1$.

Harmonic Profile
The output voltage and the source current waveforms are non-sinusoidal. The following measures are used for the undesirable Fourier components of a port variable:

Lowest undesired harmonic frequency in the output voltage

Input power factor

Lowest Undesired Harmonic Frequency is preferable to be as far away as possible from the fundamental frequency.
Input Power Factor is calculated by the following:

$$PF = \frac{\tilde{V}_S \tilde{I}_{S1}}{\tilde{V}_S \tilde{I}_S} \cos \phi_1 \quad (10\text{-}4)$$

The $\boldsymbol{\phi_1}$ is called the phase angle between the source voltage and the fundamental component of the source current.

10.3 AC CONTROLLER

The ac controller controls the amplitude of the ac source waveform. The amplitude can be controlled by several methods:

(i) Turning the converter on and off periodically and controlling the on or off duty cycle,

(ii) Controlling the delay angle, that is, the angle of turning on of switches, or

(iii) Switching at a high frequency by a pulse width modulated waveform.

10.3.1 1-phase on-off control

The ac-ac controller uses a bidirectional switch for controlling ac power to load, as shown in Figure 10.2. Bidirectional refers to the direction of current flow through the switch. The switch is turned on for n cycles and off for m cycles. The average power over n + m cycles is controlled by n as the control parameter. The ratio of n to n + m is called the duty cycle D. The load is considered to be resistive for illustration.

RMS value of the output voltage is calculated from the square root of the ratio of the output and input powers. The voltage transfer ratio T_{vv} is

$$T_{vv} = \frac{V_{Orms}}{V_{Srms}} = \sqrt{\frac{P_O}{P_{in}}} = \sqrt{\frac{n}{n+m}} = \sqrt{D} \quad (10\text{-}5)$$

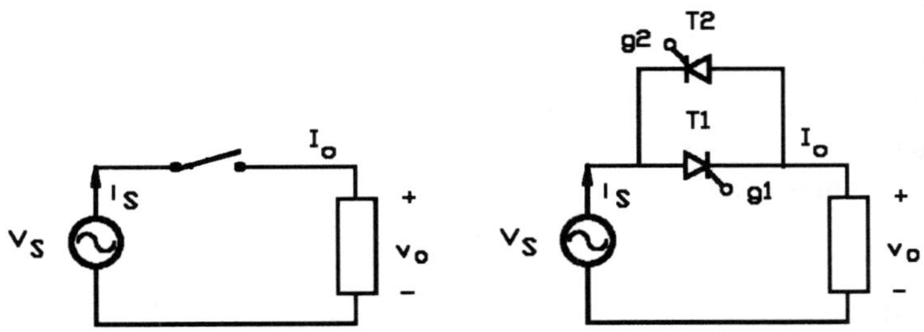

Figure 10.2
On-off ac controller

The power factor is calculated as the ratio of the output power to the apparent input power:

$$pf = \frac{V_{Orms} I_{Srms}}{V_{Srms} I_{Srms}} = \sqrt{D} \qquad (10\text{-}6)$$

The power factor is low for the low duty cycle (small values of D), hence, the on-off control is used for power control only without worrying about the efficiency. The power factor and the harmonic content are highly unsatisfactory. This type of control is used in heavy machines and heating systems that have high thermal time constants.

Example 10-1

A 120 Vrms supply is turned on and off with a 0.4 duty cycle at a frequency of five cycles. Calculate the rms output voltage and the power factor. Determine the harmonic profile of the output voltage waveform and also plot the output voltage waveform as a function of the conduction angle.

```
VS=120*sqrt(2); D=0.4; n=5;
th1=0:pi/100:n*D*2*pi; % angle during on cycles
th2=n*D*2*pi-pi/100:pi/100:n*2*pi;
  % angle during off cycles
th=[th1 th2];
vo=[VS*sin(th1) zeros(size(th2))]; %output voltage
vorms=sqrt(mean(vo.^2)) %rms output voltage
pf=sqrt(D) %power factor
plot(th,vo)
[f,von,phn]=harmonic(vo/VS, 16); % harmonic profile of
  the output voltage waveform upto 16fo
```

The fundamental harmonic f_s (first harmonic in this case) is 13% of the source waveform. The fifth harmonic, which is at the original source frequency, is predominant at 42%. As is evident, the harmonic profile of output voltage waveform is very unsatisfactory.

10.3.2 1-phase delay angle control

A one-phase ac-ac controller with delay control is shown in Figure 10.4. The switch is controlled by any one of the controlling schemes described in Chapter 9, namely the firing angle control, the extinction angle control, the pulse width control, etc. The load may be purely resistive, such as in a light dimmer application, or predominantly inductive, such as in an induction motor drive.

R Load

Consider the controller in Figure 10.4 with a resistive load that uses delay angle α-control. The steady-state operation can be described by four circuit modes, as shown in Figure 10.5.

In mode 1 ($0 \le \theta < \alpha$), thyristor $\mathbf{T_1}$ is in the forward blocking state and thyristor $\mathbf{T_2}$ in the reverse blocking state. Mode 2 begins when thyristor T_1 is turned on at an angle α by the control pulses applied at the gate terminal g_1. During mode 2 ($\alpha \le \theta < \pi$), thyristor T_1 is in the forward on-state and thyristor T_2 in the reverse blocking state. Mode 2 ends when the source voltage reverses in polarity and thyristor $\mathbf{T_1}$ naturally commutates at π conduction angle. In mode 3 ($\pi \le \theta < \pi + \alpha$), thyristor $\mathbf{T_1}$ is in the reverse blocking state and thyristor $\mathbf{T_2}$ is in the forward blocking state. Mode 4 ($\pi + \alpha \le \theta < 2\pi$) begins when thyristor T_2 is turned on at an angle $\pi + \alpha$. Mode 4 ends when the source voltage reverses in polarity again at 2π conduction angle.

Example 10-2

An ac controller with phase delay angle control, shown in Figure 10.4, is supplying power from a 100 V peak, 60 Hz ac source to a resistive load of 10 ohms. For a delay angle of π/4,

a. Plot the output voltage waveform as a function of the conduction angle,
b. Calculate the rms output voltage,
c. Plot the voltage transfer function T_{vv} and
d. Plot the harmonic profile of the output voltage waveform.

The following MATLAB program generates 500-element arrays *theta, vs, vo, io* and *is* that contain respectively the angles θ, the input voltage v_S, the output voltage v_o, the output current i_o and the input current i_S for a firing angle of $\alpha = \pi/4$, $V_S = 100$ V and the load resistance R = 10

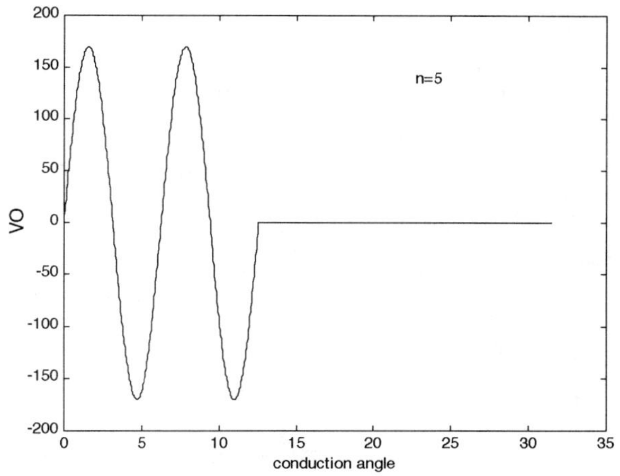

(a)

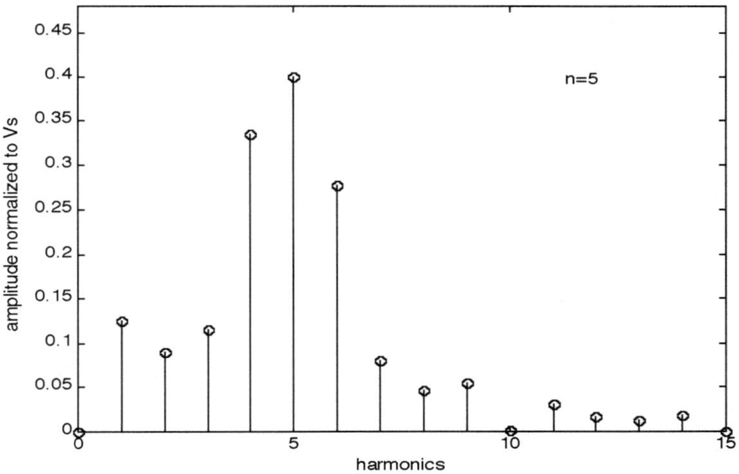

(b)

Figure 10.3
(a) Output voltage waveform in on-off control with duty cycle of 0.4 and (b) harmonics in on-off control

ohms. As a first step, small arrays are constructed for each circuit mode and then combined together to form the total array for each variable. An array spans a 2π angle. Run the program and confirm the waveforms.

```
dtheta=pi/500; %incremental step in the array of the
   conduction angle over a switching period
alpha=pi/4; %delay angle
VS=100; %peak input voltage
RL=10; %load resistance
theta1=[0:dtheta:alpha]; % conduction angles in mode 1
theta2=[alpha:dtheta:pi]; %conduction angles in mode 2
theta3=[pi:dtheta:pi+alpha];
   % conduction angles in mode 3
theta4=[pi+alpha:dtheta:2*pi];
   %conduction angles in mode 4
```

```
theta =[theta1 theta2 theta3 theta4];
  % complete angle array over one period
vo=[ zeros(theta1) (VS * sin(theta2)) zeros(theta3) (VS *
  sin(theta4))]; % output voltage array
vs=VS * sin(theta); %input voltage array
plot(theta,vo)
```

RMS output voltage is calculated as:

$$\tilde{V}_O = \sqrt{\frac{1}{\pi}\int_{\alpha}^{\pi} V_S^2 \sin^2\theta \, d\theta}$$
$$= V_s\sqrt{\frac{2(\pi - \alpha) + \sin 2\alpha}{4\pi}} \tag{10-7}$$

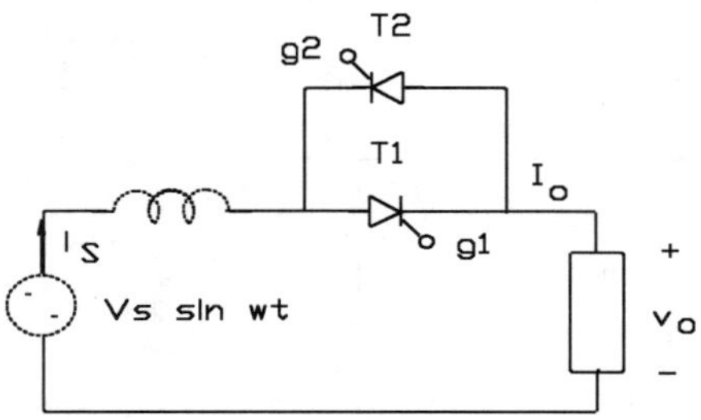

Figure 10.4
A 1-phase ac-ac controller with delay angle control

The voltage transfer ratio is

$$T_{vv} = \frac{\tilde{V}_o}{\tilde{V}_s} = \sqrt{\frac{2(\pi - \alpha) + \sin 2\alpha}{2\pi}} \tag{10-8}$$

Following MATLAB instructions calculate the voltage transfer ratio:

```
Vorms=sqrt{mean(vo .^2)} % rms output voltage
Tvv=vorms/(VS/sqrt{2})) %voltage transfer ratio
```

Vorms = 67.4224   Tvv = .9535

It is left to the readers to write the program for the plots of T_{vv}, the power factor, and the THD versus the delay angle, shown in Figure 10.7.

The slope of the curve $dT_{vv}/d\alpha$ indicates the behavior of a feedback loop involving the delay angle α. A high slope is preferable because a small change in the delay angle α causes a large variation in output voltage. The slope is high for the delay angles in the vicinity of $\pi/2$. Hence, the feedback loop operation is preferable only for the delay angles in the vicinity of $\pi/2$.

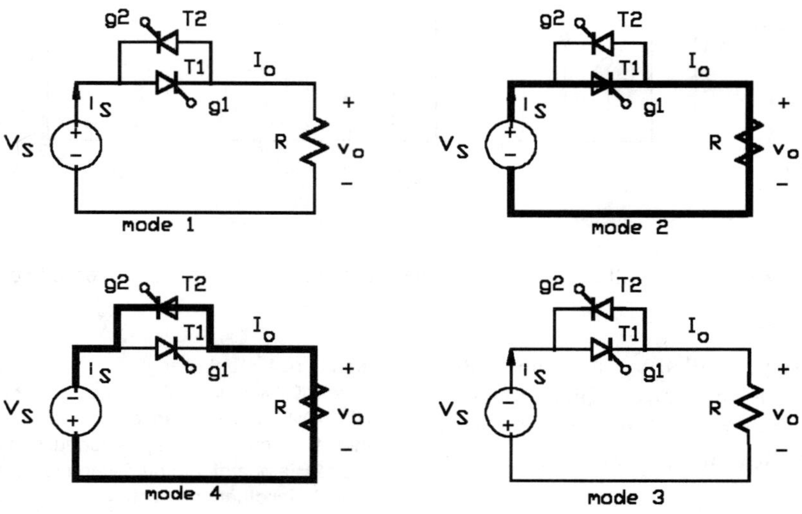

Figure 10.5
Circuit modes

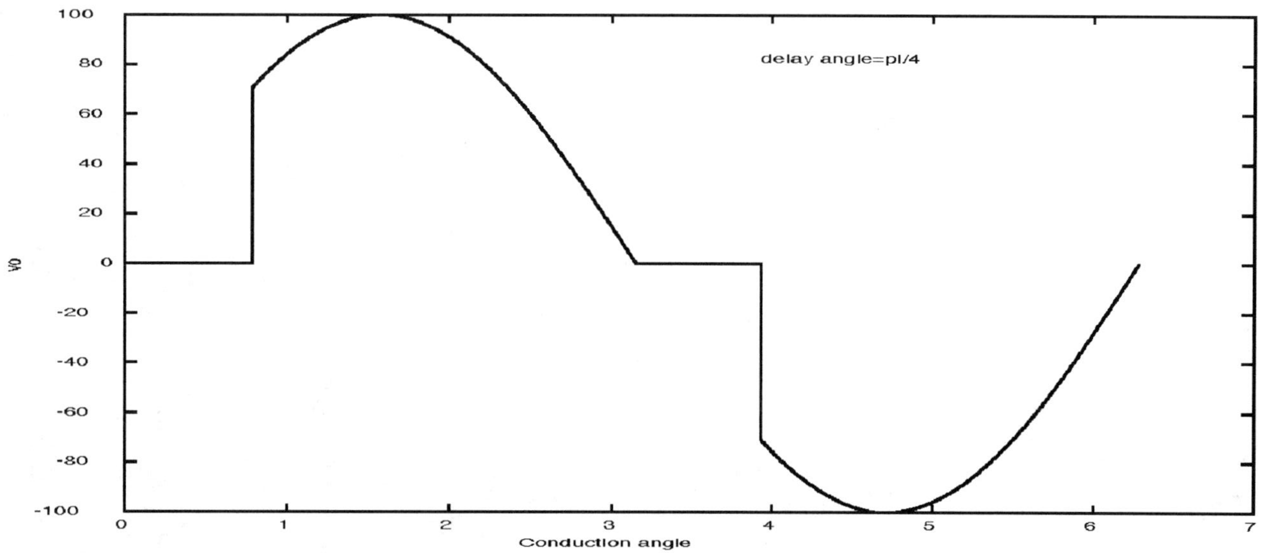

Figure 10.6
MATLAB waveform over a single period

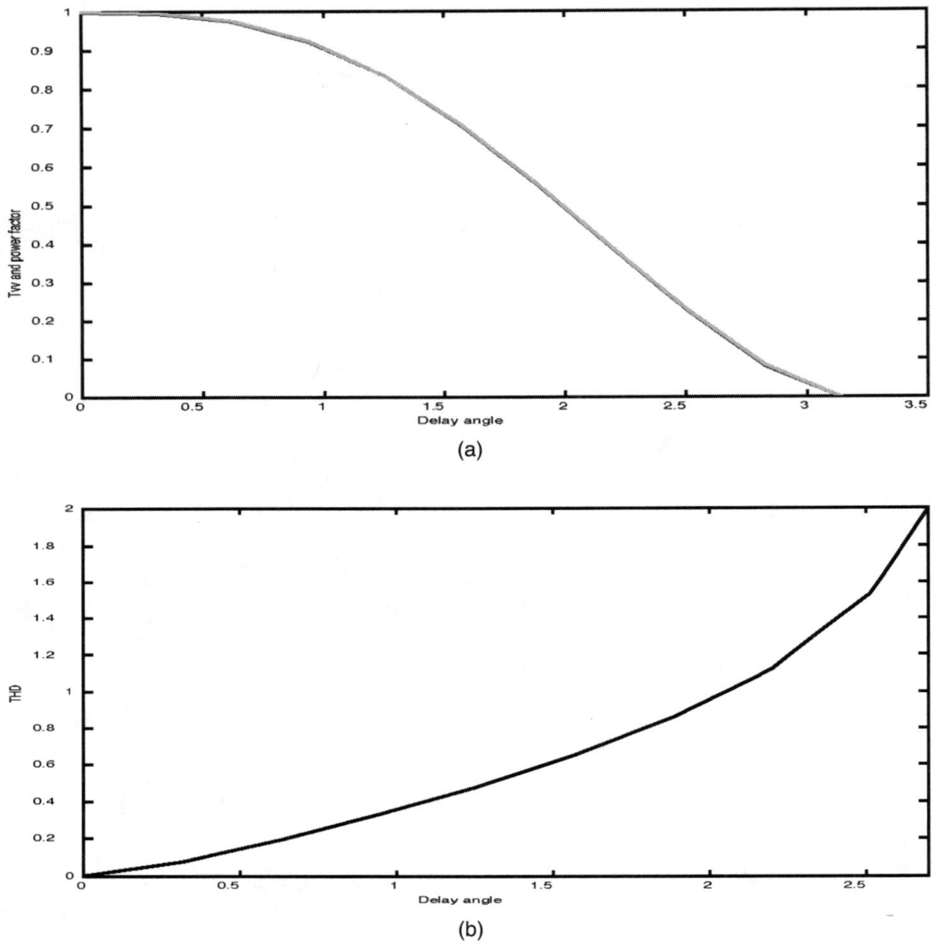

Figure 10.7
(a) The Tvv and power factor versus the delay angle and (b) The THD versus the delay angle

Add the following two MATLAB lines to the program in example 10.2 for finding the harmonic profile of the output voltage waveform:

```
[f, voh,ph] = harmonic(vo/VS, 11);
```

Figure 10.8 shows the harmonic profile of the output voltage.

The phase angle ϕ_1 is negative, therefore, the input power factor is the lagging type. The leading phase angle can be obtained by using the extinction angle control.

1-Phase AC Controller with Source Inductance

The source inductance L_S slows down the stiff change in the output current waveform. The output current becomes zero beyond the conduction

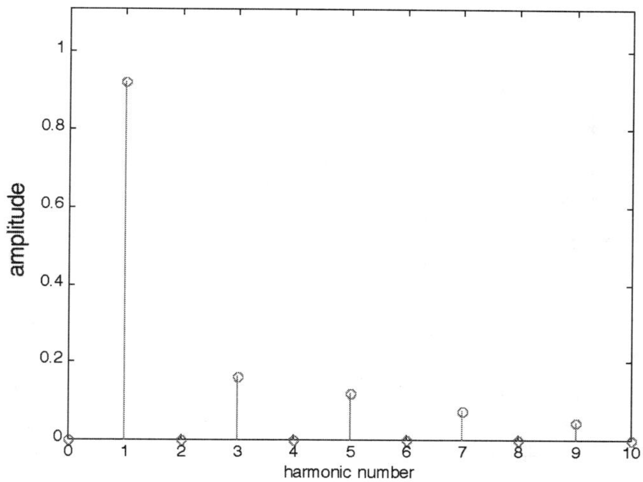

Figure 10.8
Frequency spectrum of output voltage

angle π. During the conduction angle α to π, the current i_o is determined from the following circuit equation:

$$L_S \frac{di_O}{dt} + R\, i_O = V_S \sin \omega t \tag{10-9}$$

$$i_O = \frac{V_S}{Z}\left[\sin(\theta - \phi) - \sin(\alpha - \phi) e^{\frac{R}{\omega L_S}(\alpha - \theta)}\right]$$

$$\text{where } Z = \sqrt{R^2 + (\omega L_S)^2},\ \phi = \tan^{-1}\left[\frac{\omega L_S}{R}\right]$$

The waveforms are shown in Figure 10.9 for $\alpha = 3\pi/8$, R = 2 ohms, L_S = 1 mH and 100 V peak, 60 Hz source.

1-Phase AC Controller with Inductive Load

See Figure 10.10. The thyristors are turned on at zero current and are turned off at zero current by natural commutation. At the turn on of thyristor T_1, a voltage $V_S \sin \alpha$ is applied on the inductor L. The inductor current increases sinusoidally beginning at a positive slope. The inductor current continues to flow beyond the conduction angle π even after the voltage across the inductor becomes negative, see the waveforms in Figure 10.11. The current becomes zero at the angle $2\pi - \alpha$. The turn-on gate pulse to the second thyristor T_2 is applied at the angle $\pi + \alpha$. For delay angle $\alpha < \pi/2$, the turning on angle of the second thyristor T_2 $\pi + \alpha$ is smaller than the extinction angle $2\pi - \alpha$ of thyristor T_1. Shorted by T_1, thyristor T_2 does not turn on. The resulting output current has a dc component. For delay angle $\alpha > \pi/2$, the current in thyristor T_1 becomes zero before the turning on of T_2. The current and voltage waveforms in the two situations are shown in Figure 10.11.

The rms output voltage is calculated from the v_o waveform for α less than or equal to $\pi/2$,

$$\tilde{V}_O = \sqrt{\frac{1}{\pi}\int_{\alpha}^{2\pi - \alpha} (V_S \sin \theta)^2\, d\theta} \quad \theta = \omega_s t \tag{10-10}$$

$$T_{vv} = \frac{\tilde{V}_O}{\tilde{V}_S} = \sqrt{\frac{2(\pi - \alpha) + \sin 2\alpha}{\pi}} \tag{10-11}$$

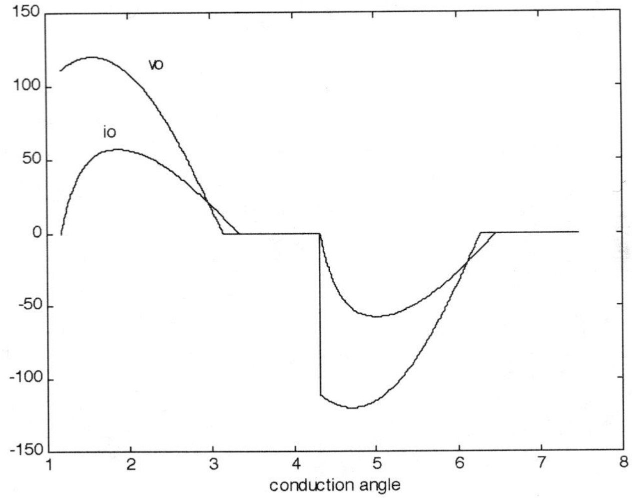

Figure 10.9
Waveforms of the output voltage and output current in the ac controller with source inductance

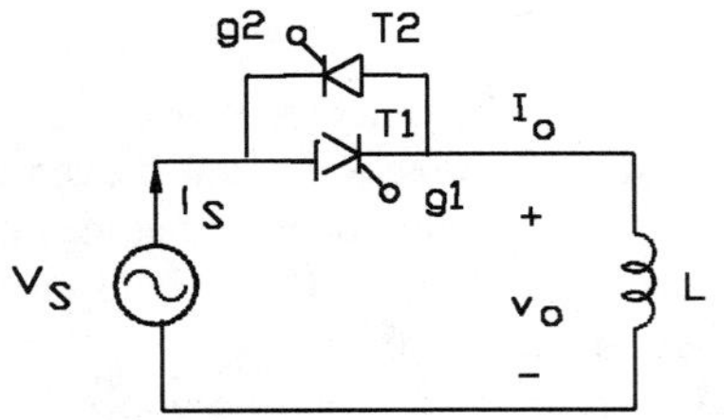

Figure 10.10
A 1-phase controller with inductive load

The output current I_O in the conduction period is obtained from

$$L\frac{di_O}{dt} = V_S \sin \omega_S t \quad \alpha \leq \omega_S t < 2\pi - \alpha \quad (10\text{-}12a)$$

$$i(\theta) = \frac{V_S}{\omega_S L}\int_{\alpha}^{\theta} \sin\theta \, d\theta \quad \theta = \omega_S t$$

$$= \frac{V_S}{\omega_S L}(\cos\alpha - \cos\theta)$$

The rms output current (which is also the input current) is

$$\tilde{I}_O = \sqrt{\frac{1}{\pi}\int_{\alpha}^{2\pi-\alpha} (i(\theta))^2 d\theta}$$

$$\tilde{I}_O = \frac{\tilde{V}_S}{\omega_S L}\sqrt{\frac{2(\pi - \alpha) - \sin 2\alpha}{\pi}} \quad (10\text{-}12b)$$

The plot of the voltage transfer ratio T_{vv} and the rms output current versus the delay angle are shown in Figure 10.12.

A resistor in series with L would render the current waveform non-sinusoidal.

A special case is worth mentioning. The rms value of the fundamental component of the output current is

$$\tilde{I}_{O1} = \frac{1}{\pi}\int_{\alpha}^{2\pi-\alpha} \frac{V_s}{\omega_S L}(\cos\alpha - \cos\theta)\cos\theta \, d\theta$$

$$\tilde{I}_{O1} = \frac{\tilde{V}_S}{2\pi \, \omega_S L}[2(\pi - \alpha) - \sin 2\alpha] \quad (10\text{-}13)$$

The effective value of inductance as seen at the input port is a function of the delay angle,

$$\tilde{I}_{O1} = \frac{\tilde{V}_s}{\omega_S L_{eff}}:$$

$$L_{eff} = L\frac{2\pi}{2(\pi - \alpha) - \sin 2\alpha} \quad (10\text{-}14)$$

Thus a controllable inductance is realized. The controlled inductance is used in practice for inductive phase compensation of unbalanced ac lines.

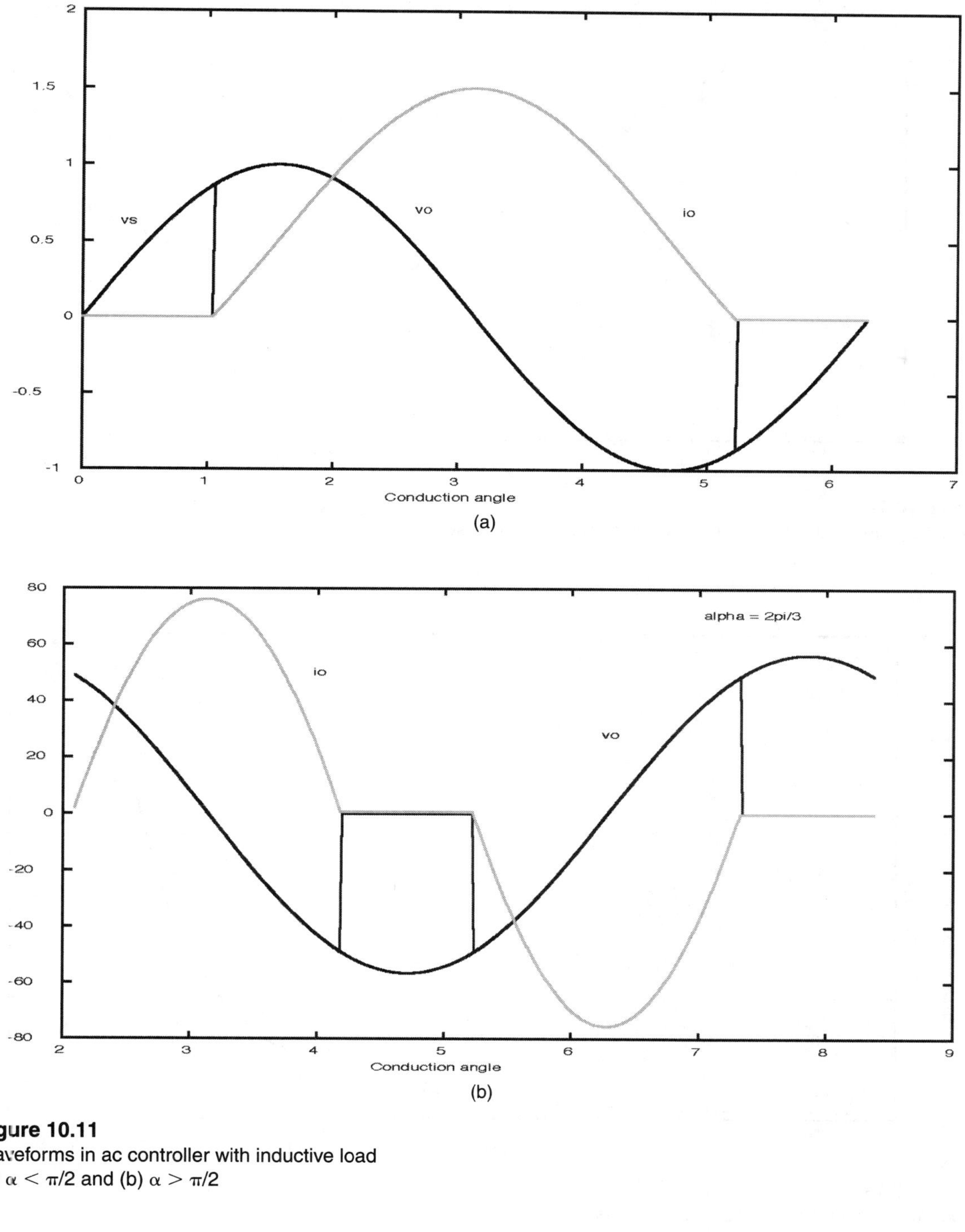

(a)

(b)

Figure 10.11
Waveforms in ac controller with inductive load
(a) $\alpha < \pi/2$ and (b) $\alpha > \pi/2$

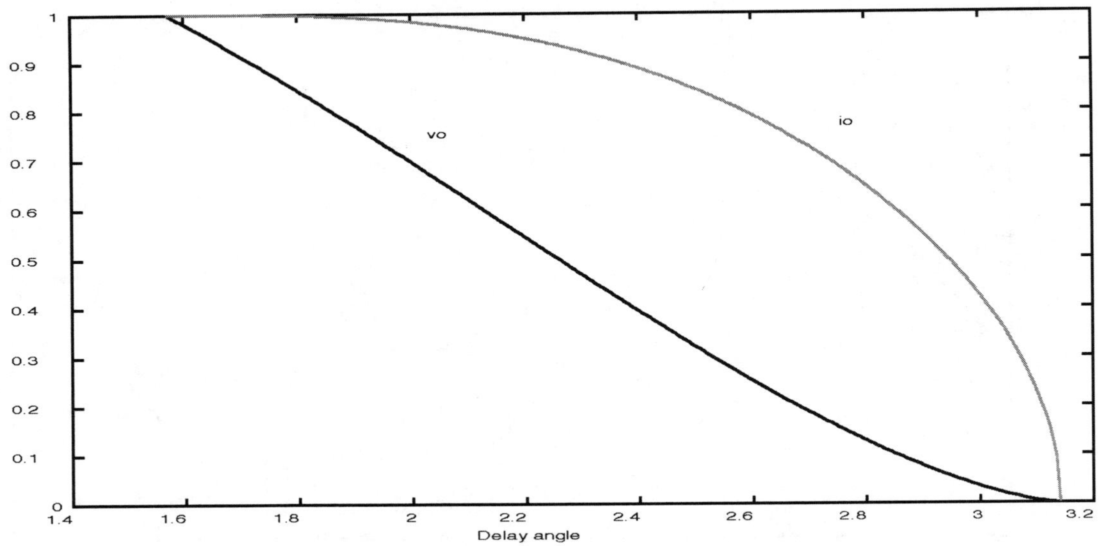

Figure 10.12
The voltage transfer ratio T_{vv} and the rms output current versus the delay angle

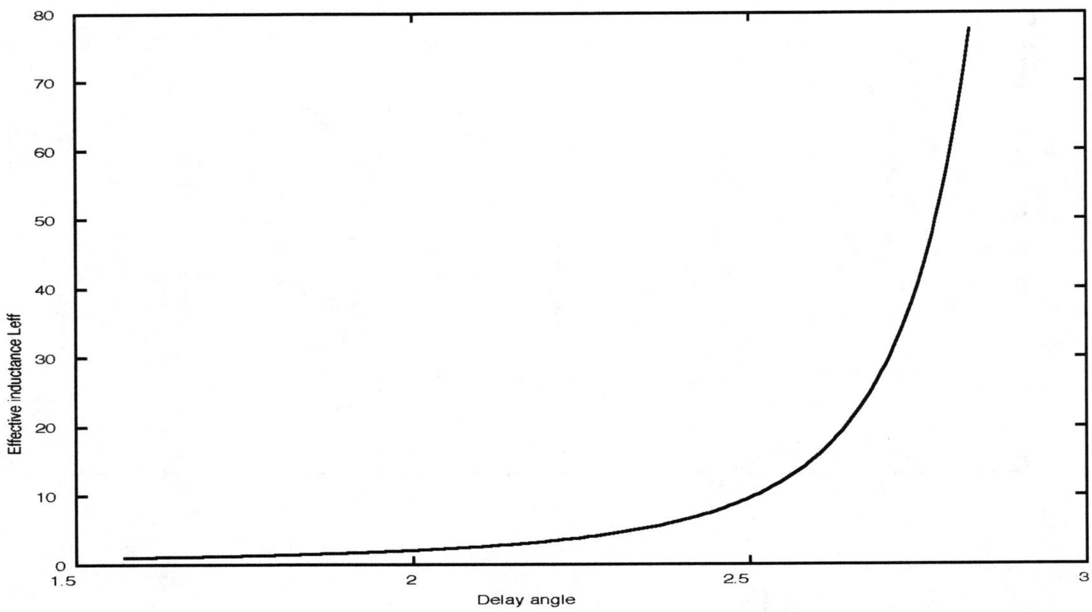

Figure 10.13
Effective inductance L_{eff} versus the delay angle

Example 10-3

Consider the ac controller for a synchronous motor shown in Figure 10.14. The ac source is 120 V rms, 60 Hz. The output voltage v_o is 80 V rms, 60 Hz and is synchronized with respect to the ac source. It is assumed that the voltage emf load is in phase with the input voltage. This is the typical situation in the synchronous motor drive where the torque angle is maintained at p/2. Find the rms output current.

The thyristors are turned on and turned off at zero current by natural commutation. The voltage and current waveforms in the steady state operation are identical to those shown in Figure 10.11(b).

$$v_O = V_O \sin\theta : \theta = \omega_S t \qquad (10\text{-}15)$$

The inductor current lags behind the source voltage by $\pi/2$ radians. The delay angle α is greater than $\pi/2$. The current waveforms in the mode 1 and mode 3 are given by the following:

$$L\frac{di_O}{dt} = (V_S - V_O)\sin\theta$$

$$: \quad \theta = \omega_S t,\ \alpha \geq \pi/2 \qquad (10\text{-}16)$$

$$i_O = \frac{(V_S - V_O)}{\omega_S L}[\cos\alpha - \cos\theta] \qquad (10\text{-}17)$$

The output rms current and the voltage to current transfer ratio are

$$\tilde{I}_O = \sqrt{\frac{1}{\pi}\int_{\alpha}^{2\pi-\alpha} i_O^2\, d\theta} \qquad (10\text{-}18)$$

$$T_{vi} = \frac{\tilde{I}_O}{\tilde{V}_S/\omega_S L} \qquad (10\text{-}19)$$

A plot of $\mathbf{T_{vi}}$ versus α is shown in Figure 10.15.

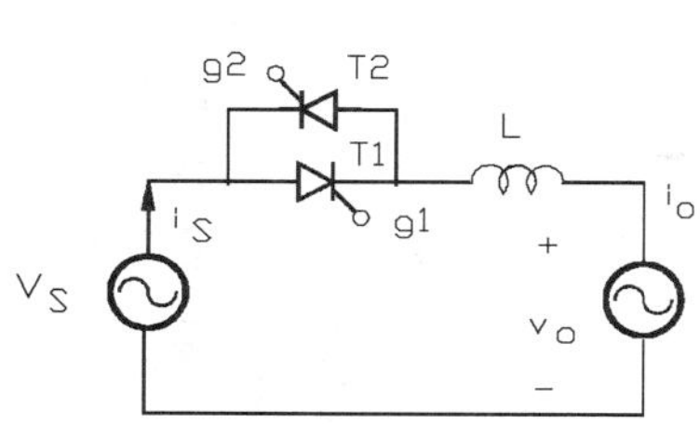

Figure 10.14
The ac controller for synchronous motor

10.3.3 3-phase delay angle control

The three-phase bidirectional ac controller is commonly used in ac motor drives. The following sections discuss the resistive loads with the objective of focusing more on the analysis procedure than the complexity.

Star Connected Load

The three-phase ac controller with a star-connected resistive load is shown in Figure 10.16.

The phase voltages are

$$v_{AN} = V_P \sin\theta \quad : \quad \theta = \omega_S t \qquad (10\text{-}20)$$
$$v_{BN} = V_P \sin(\theta - 2\pi/3)$$
$$v_{CN} = V_P \sin(\theta - 4\pi/3)$$

The line-line voltages are

$$v_{AB} = 3V_P \sin(\theta + \pi/6)$$
$$v_{BC} = 3V_P \sin(\theta - \pi/2) \qquad (10\text{-}21)$$
$$v_{CA} = 3V_P \sin(\theta - 7\pi/6)$$

Converter Operation If the gate currents were continuously applied, the thyristors would behave as diodes and their operation would be similar to that described in the previous chapters. There are two gate control pulses per phase, six in total. The delay angle in all phases is referenced to 0 conduction angle on the phase A voltage.

current commutation	**delay angle**
T5 to T1	α
T6 to T2	$\alpha+\pi/3$
T1 to T3	$\alpha+2\pi/3$
T2 to T4	$\alpha+\pi$
T3 to T5	$\alpha+4\pi/3$
T4 to T6	$\alpha+5\pi/3$

Thyristors T1, T3 and T5 turn on in the positive half cycles of waveforms v_{AN}, v_{BN} and v_{CN} respectively. When turned on, T1 conducts from α to π, T3 conducts from $\alpha+2\pi/3$ to $5\pi/3$ and T5 conducts from $\alpha + 4\pi/3$ to $7\pi/3$ unless forced to commutate earlier.

Thyristors T4, T6 and T2 turn on in the negative half cycles of waveforms v_{AN}, v_{BN} and v_{CN} respectively. Unless forced to commutate earlier, T4 conducts from $\alpha + \pi$ to 2π, T6 conducts from $\alpha + 5\pi/3$ to $8\pi/3$ $(= 2\pi + 2\pi/3)$ and T2 conducts

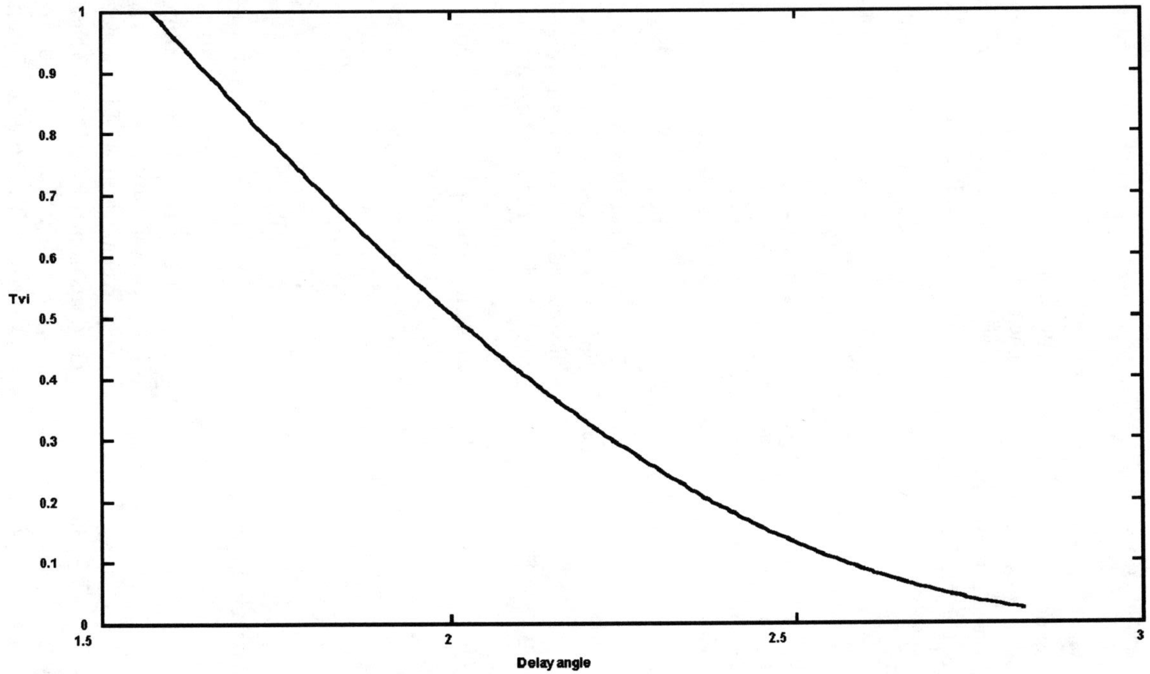

Iorms = 44.5851

Figure 10.15
T_{vi} versus α in Example 10-3

from $\alpha+7\pi/3$ (= $\alpha + 2\pi + \pi/3$) to $10\pi/3$ (= $2\pi + 4\pi/3$).

The delay angle α can be selected in the range 0 to π. The steady-state converter operation differs considerably with the choice of the delay angle α.

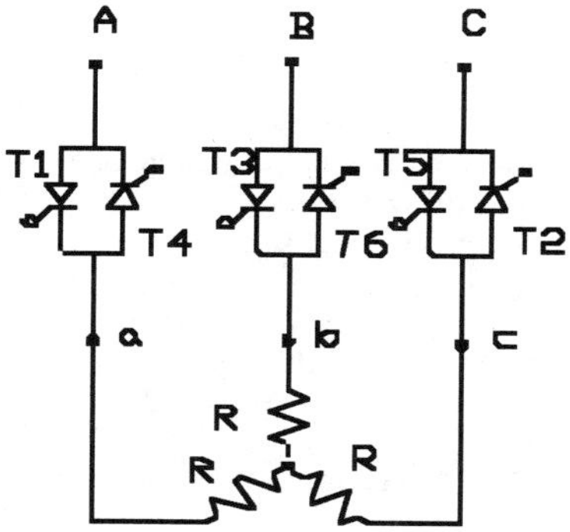

Figure 10.16
The 3-phase ac controller with star-connected load

We consider the case $\alpha > \pi/3$. Other cases are left as an exercise for the reader.

The circuit modes are shown in Figure 10.17. In the first mode, the thyristor T6 was already on. Thyristor T1 is in the forward blocking mode due to the input phase voltage v_{AN}. T1 is turned on at α, referenced to the zero crossing point on the v_{AN} waveform. T5 is reversed biased and the output current i_O flows from phase A through T1, load and T6 to phase B. The phase voltage in the output is half of the line-line voltage v_{AB}. At the angle $\alpha + \pi/3$ in mode 2, thyristor T2 is in the forward blocking mode due to the input phase voltage v_{CN}. T2 is turned on at $\alpha + \pi/3$, α angle referenced to the zero crossing point on the positive half cycle of the v_{CN} waveform, see Figure 10.18. T6 is reversed biased and the output current i_O flows from phase A through T1, load and T2 to phase C. Consequently the phase voltage in the output is half of the line-line voltage v_{AC} ($-v_{CA}$). Likewise from the angle $\alpha+2\pi/3$ to $\alpha+4\pi/3$ in mode

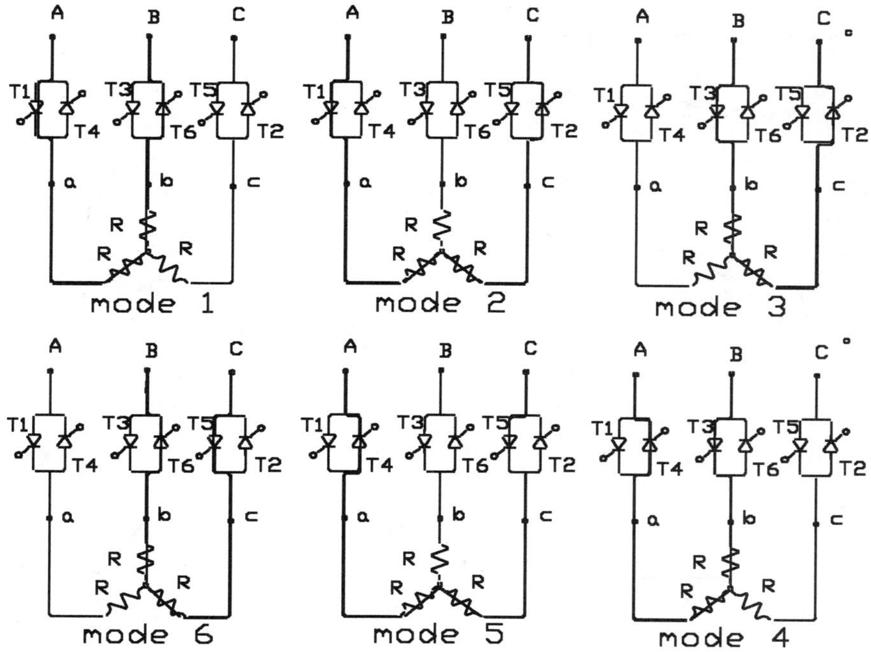

Figure 10.17
Circuit modes

3 and mode 4, the current i_O flows from the input phase B to other phases. From the angle $\alpha+4\pi/3$ to $\alpha+2\pi$, the current i_O flows from the input phase C to other phases in mode 5 and mode 6. The waveforms in mode 1 and mode 2 are shown in Figure 10.18.

The phase output voltage v_{aN} consists of two parts in the positive half cycle:

$$v_{aN} = \frac{v_{AB}}{2} \quad : \quad \alpha \leq \theta < \pi/3 + \alpha$$

$$= \frac{v_{AC}}{2} \quad : \quad \pi/3 + \alpha \leq \theta < \pi \qquad (10\text{-}22)$$

The negative half cycle of v_{aN} is the skew symmetric to the positive half cycle. The transition at $\alpha+\pi/3$, requires natural commutation of thyristor T6. The negative half cycle of v_{aN} is skew-symmetric to the positive half cycle.

Characteristics Output rms phase voltage V_O is calculated as the rms of the v_{aN} waveform over a half period from α to $\alpha+\pi$:

$$\tilde{V}_O = \sqrt{\frac{1}{\pi}\int_{\alpha}^{\alpha+\pi} v_{aN}^2 \, d\theta} \qquad (10\text{-}23)$$

Example 10-4

Plot the waveforms of the phase voltage and v_O for $V_P = 120$ V, $\alpha = 3\pi/8$ in the star connected load of Figure 10.16.

```
th=0:pi/500:pi;% conduction angle
VP=120; % peak voltage of the input phase voltage
alpha=3*pi/8 % delay angle
vAN=VP*sin(th); % input phase A voltage
vBN=VP*sin(th-2*pi/3); vCN=VP*sin(th-4*pi/3);
vAB=sqrt(3)*VP*sin(th+pi/6);% input line-line voltage
vBC=sqrt(3)*VP*sin(th+pi/6-2*pi/3);
  vCA=sqrt(3)*VP*sin(th+pi/6-4*pi/3);
vAC=-vCA;
vanp1=.5*vAB.*(th>alpha);% a part of the output voltage
  in phase A
vanp2=0.5*vAC.*(th>alpha+pi/3);
van=max(vanp1, vanp2);% output voltage in phase A
plot(th,vAN,'—',th, vAB, th, vAC, th,van)
```

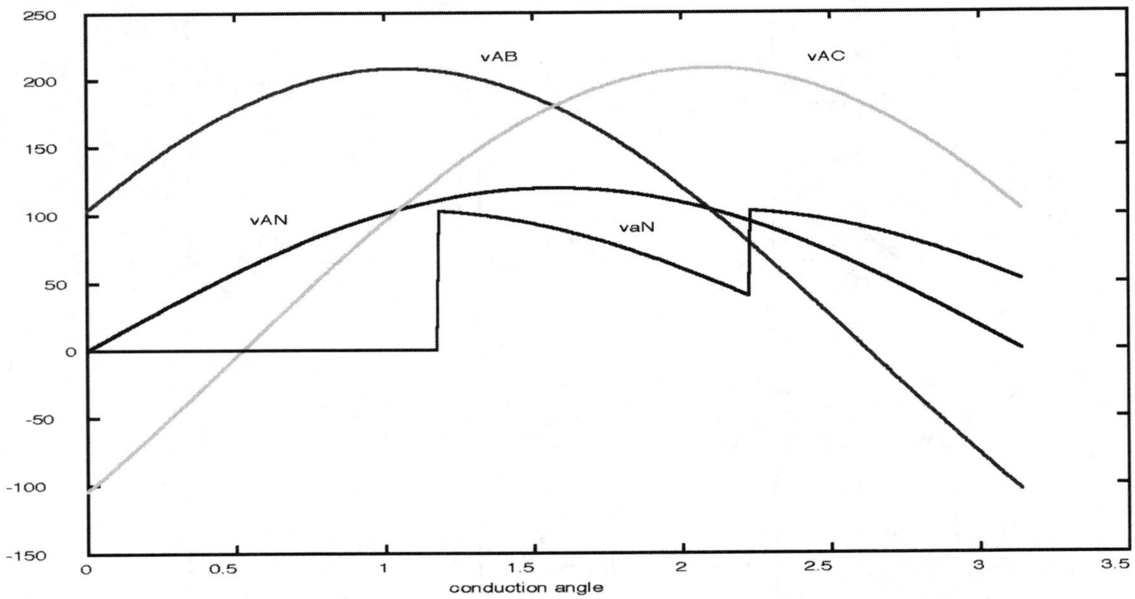

Figure 10.18
Voltage waveforms in mode 1 and mode 2

```
vorms=sqrt(mean(van.^2)) % rms output phase voltage
voa=[van, -van];
  %output voltage waveform over a period
[f,voh,ph]=harmonic(voa/VP);
  % harmonic profile of phase A voltage
axis([0,10,0,1])
pf=cos(ph) %input power factor
vo1rms=voh(2)*VP/sqrt(2);
  % rms voltage of fundamental component
THD=sqrt((vorms ^2-vo1rms^2)/vorms^2) %Total
  harmonic distortion in the output voltage
alpha =1.1781   vorms = 65.3472
cos (alpha) =0.8878   THD =0.3868
```

Readers are encouraged to repeat this example for α in the range 0 to π.

Delta Connected Load

See Figure 10.20. Thyristors are triggered at the delay angle α, referenced to the zero-crossing point at the beginning of the positive voltages in each arm of the Δ-connected load.

The line-line voltages are

$$
\begin{aligned}
v_{AB} &= V_{LL}\sin(\theta)\\
v_{BC} &= V_{LL}\sin(\theta - 2\pi/3)\\
v_{CA} &= V_{LL}\sin(\theta - 4\pi/3)
\end{aligned}
\qquad (10\text{-}24)
$$

Converter Operation

There are two gate control pulses per phase, six in total. The delay angle is referenced to zero conduction angle on the line-to-line voltage v_{AB}.

Thyristors T1, T3 and T5 turn on in the positive half cycles of waveforms v_{AB}, v_{BC} and v_{CA} respectively. When turned on, T1 conducts from α to π, T3 conducts from $\alpha+2\pi/3$ to $5\pi/3$ and T5 conducts from $\alpha + 4\pi/3$ to $7\pi/3$ unless forced to commutate earlier.

Thyristors T4, T6 and T2 turn on in the negative half cycles of waveforms v_{AB}, v_{BC} and v_{CA} respectively. Unless forced to commutate earlier, T4 conducts from $\alpha + \pi$ to 2π, T6 conducts from $\alpha + 5\pi/3$ to $8\pi/3$ $(= 2\pi + 2\pi/3)$ and T2 conducts from $\alpha + 7\pi/3$ $(= \alpha + 2\pi + \pi/3)$ to $10\pi/3$ $(= 2\pi + 4\pi/3)$.

The delay angle α can be selected in the range 0 to π. The line current i_A waveform in the positive half is shown in Figure 10.21.

The line currents are determined from the phase currents:

$$i_a = i_{AB} - i_{CA} = \frac{1}{R}(v_{AB} - v_{CA}) \qquad (10\text{-}25)$$

Due to the delta connection, the triplen harmonics (3rd, 9th, 15th, and so on) are absent in the phase currents.

An alternative form of the delta-connected controller with only three thyristors is given in problem 10-6 at the end of the chapter.

Characteristics

The rms output phase voltage in all resistive load is

$$
\begin{aligned}
\tilde{V}_O &= \sqrt{\frac{1}{\pi}\int_{\alpha}^{\pi}(V_{LL}\sin\theta)^2\,d\theta}\\
&= V_{LL}\sqrt{\frac{1}{2\pi}\left(\pi - \alpha + \frac{1}{2}\sin 2\alpha\right)}
\end{aligned}
\qquad (10\text{-}26)
$$

The forward voltage transfer ratio is

$$T_{VV} = \frac{\tilde{V}_O}{\tilde{V}_{LL}} = \sqrt{\frac{1}{\pi}\left(\pi - \alpha + \frac{1}{2}\sin 2\alpha\right)} \qquad (10\text{-}27)$$

Example 10-5

Plot the waveforms of output phase voltages in a delay angle controlled three-phase ac-ac converter with a delta connected load of 1 ohm, where the ac source is 120 V ac per phase and the delay angle is $2\pi/3$. Plot the harmonic profile and calculate the per-phase rms current and THD of the output current waveform.

```
th=0:pi/500:2*pi; VP=120; alpha=2*pi/3; R=1;
vAB=sqrt(3)*VP*sin(th); vBC=sqrt(3)*VP*sin(th-2*pi/3);
  vCA=sqrt(3)*VP*sin(th-4*pi/3);
vAC=-vCA;
iAB=(vAB/R).*((th>alpha)&(th<pi)|(th>pi+alpha));
iAC=(vAC/R).*(((th<pi/3))|(th>alpha+pi/3)&(th<pi+pi/3)
  |(th>pi+pi/3+alpha));
iA=(iAB+iAC);
```

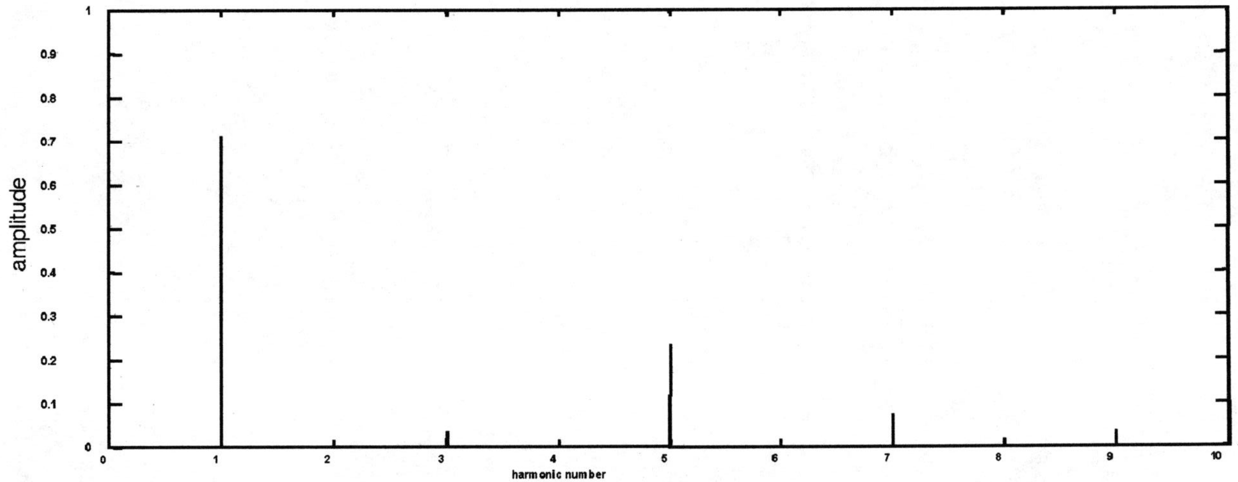

Figure 10.19
Harmonic profile

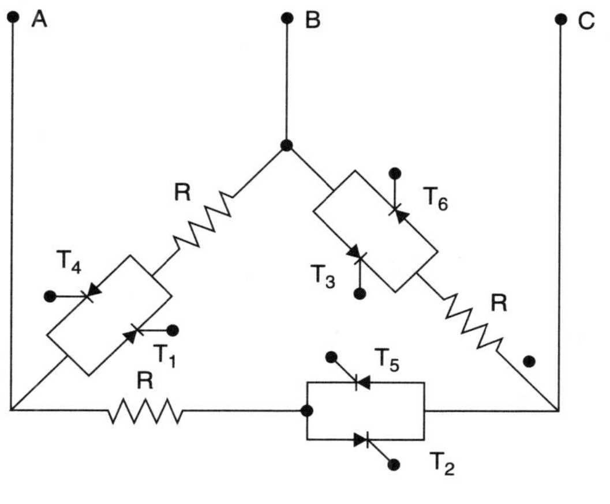

Figure 10.20
The 3-phase ac-ac controller with Δ-connected resistive load

```
plot(th,vAB,'—',th, vAC, th, iAB, th, iA, '+')
iorms=sqrt(mean(iA.^2))
[f,ioh,ph]=harmonic(iA/VP);
axis([0,10,0,1]) cos(ph)
io1rms=ioh(2)*VP/sqrt(2);
vo1rms =ioh(2)*R*VP/sqrt(2);
THD=sqrt((iorms ^2−io1rms^2)/iorms^2)
```

iorms = 91.795 THD =0.5206

10.3.4 PWM control

Phase control of ac-ac converters generates low order harmonics. The unwanted low order harmonics are difficult to reduce. This problem can be overcome by multiplying the ac voltage source waveform with a high frequency pulse train. The amplitude of the controller output voltage is controlled by varying the pulse width. The PWM controller is shown in Figure 10.22. The switching frequency must be an integral multiple of the source frequency.

Example 10-6

A one-phase ac-ac converter is controlled by control of the off time of pulses. Six pulses are used in a half period of the ac source waveform. The off-time duty ratio is 0.6. The load is purely resistive, Δ-connected 1ohm. The ac source is 120 V ac per phase. Plot the harmonic profile and calculate the per-phase rms voltage, the rms voltage of the fundamental component of the converter output voltage and the power factor of the output current waveform at the load. Plot the forward voltage transfer function T_{VV} as a function of the duty ratio.

```
n=6;D=.6; VS=sqrt(2)*120; vop=[ ]; th1=[ ];
thp=0:pi/500:pi/n;
for i=0:n−1
    vop=[vop, (thp<=(D*pi/n) ) ];
    th1=[th1, i*pi/n+thp];
end
vo1=vop.*(VS*sin(th1));
th=[th1, th1+pi]; vo=[vo1 −vo1];
plot(th,vo)
vorms=sqrt(mean(vo.^2))
[f,voh,ph]=harmonic(vo/VS);
axis([0,n*10,0,max(voh)])
vo1rms=sqrt(mean((voh(2)*VS)^2))
pf=voh(2)*VS*cos(ph)
```

vorms=93.5032 vo1rms=103.055
pf=0.6071

10.4 CYCLOCONVERTER

The cycloconverter is used in high power (>100 Kwatts) applications. The thyristor is the device of choice. Natural commutation is the preferred method of switching the thyristors. Output frequency is lower than the source frequency, typical range: 0.2 Hz to 15 Hz. The amplitude and the frequency are both controllable.

10.4.1 1-phase cycloconverter

A one-phase cycloconverter with high filter inductance is shown in Figure 10.23. The cycloconverter converts the source frequency f_s to $f_o = f_s / n$. A period of the desired output waveform, $T_O = n / f_s$, contains n positive half cycles and n negative half cycles of the source waveform. Amplitude of the output voltage waveform is controlled by varying the delay angle in each half cycle. The delay angle profile in T_O is mainly of two types: (a) uniform variation and (b) sinusoidal variation.

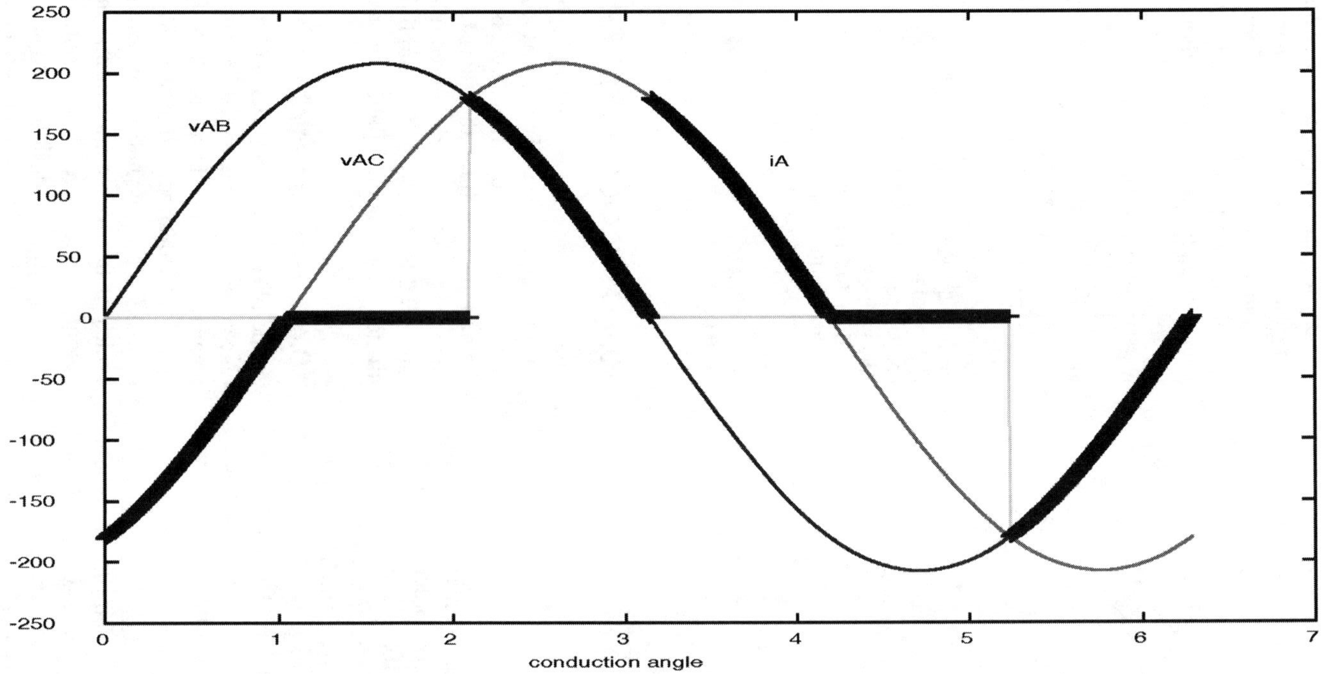

(a)

Figure 10.21
(a) Waveforms

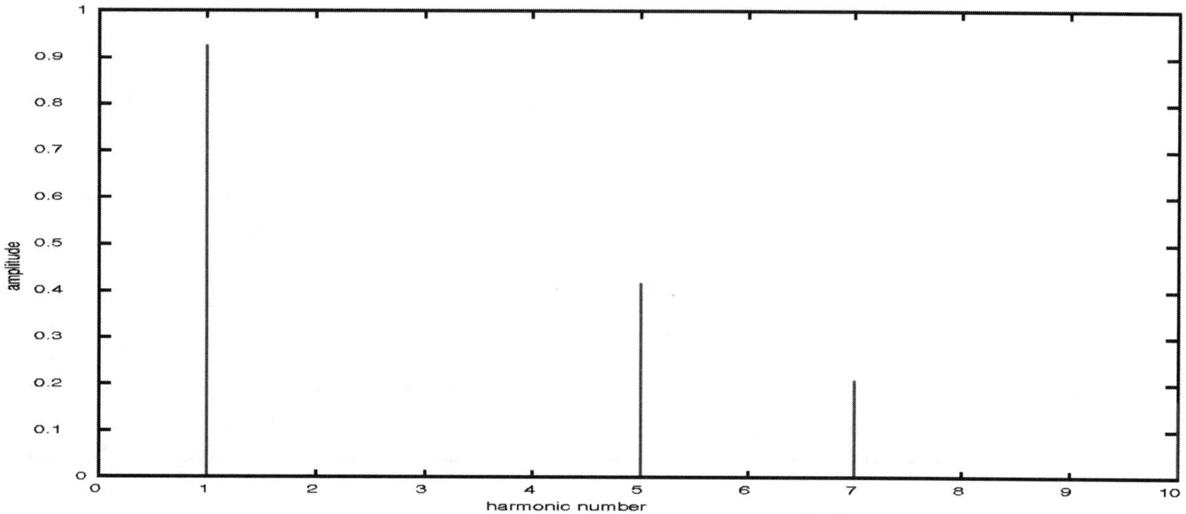

(b)

Figure 10.21 ***(continued)***
(b) Harmonic profile

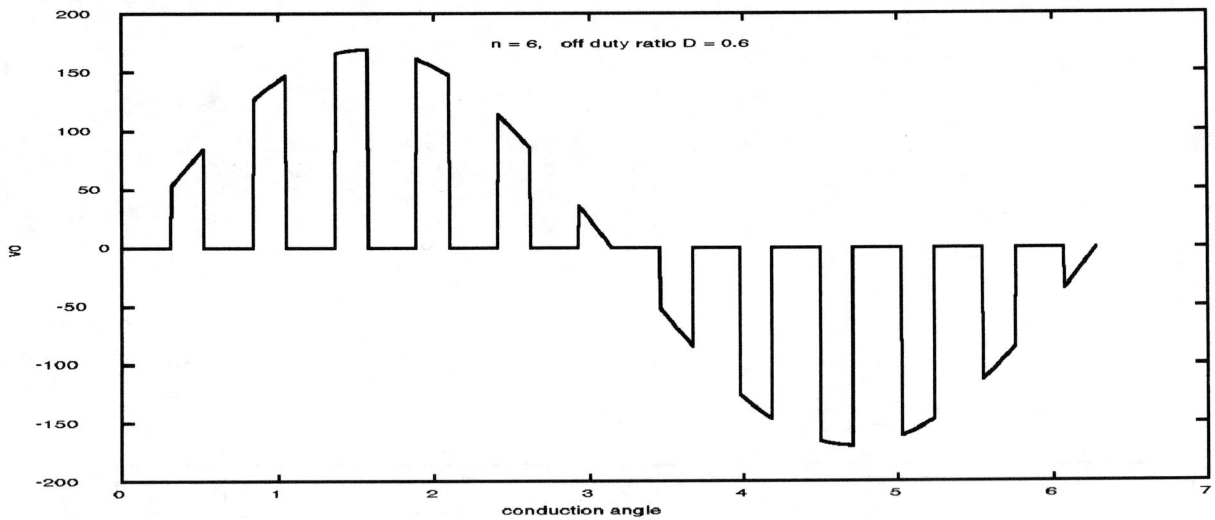

(a)

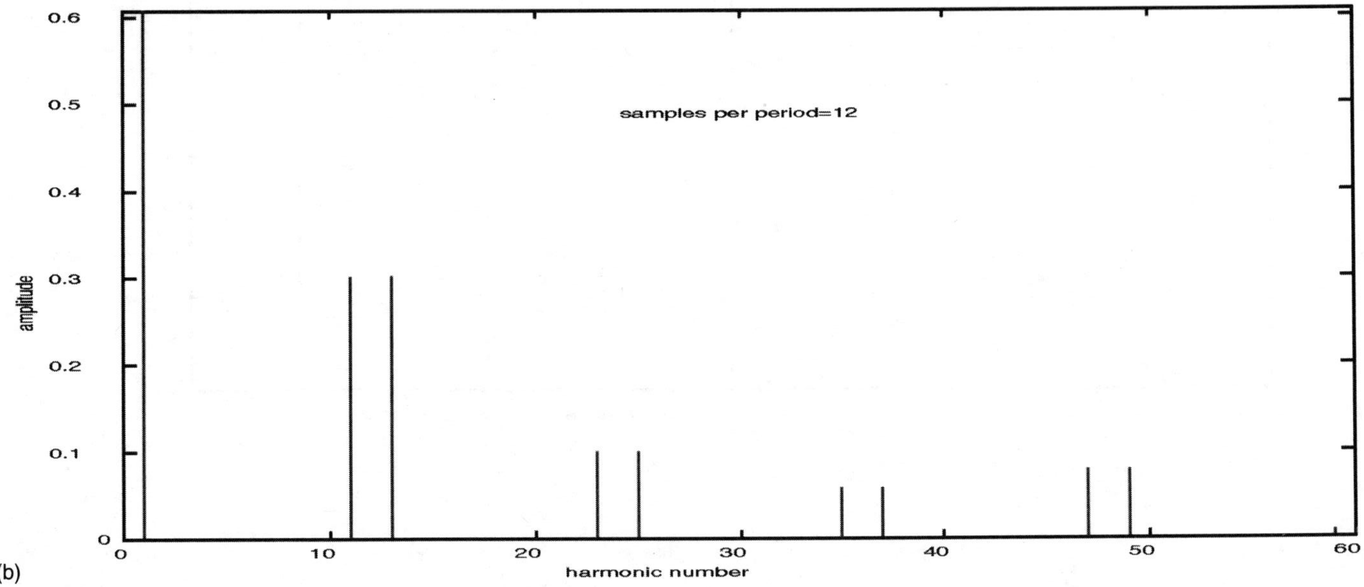

(b)

Figure 10.22
(a) PWM controller with off-duty control (b) Harmonic profile

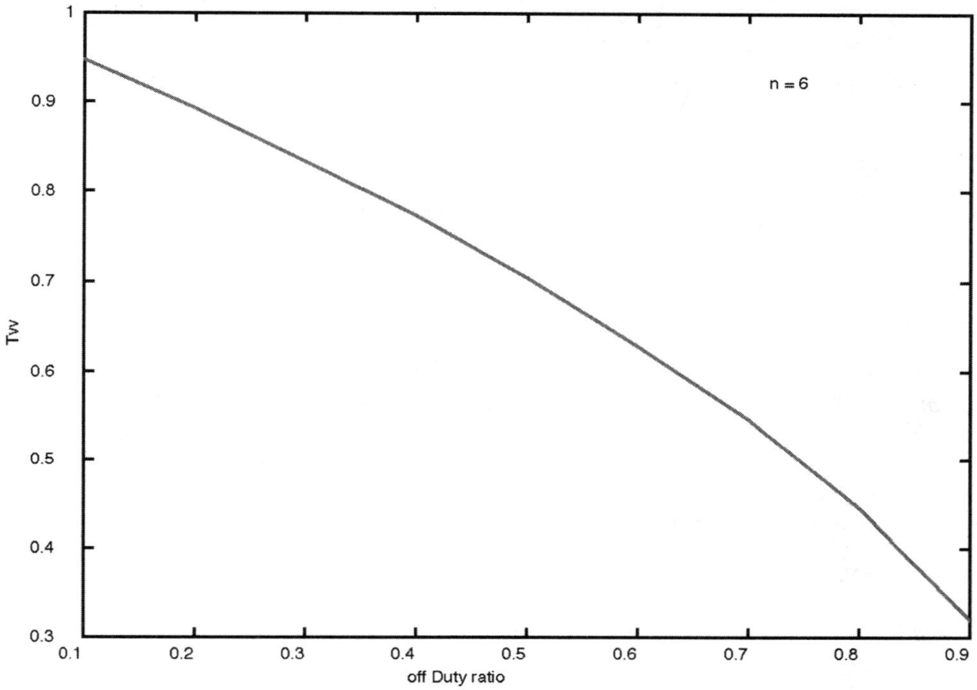

Figure 10.22 ***(continued)***
(c) T_{vv} versus off-duty ratio

Uniform Delay Angle α

The delay angle α is uniform in all half cycle segments of the source waveform. We will illustrate the converter operation using n = 3. In the following discussion, the delay angle is referenced to the point of zero-conduction angle in a cycle of the output voltage waveform (of frequency f_O).

Positive Half Period $T_O/2$

Thyristors T1-T3 are turned on at the delay angle α and $2\pi + \alpha$ in the first and the third positive half cycles of the source waveform. Thyristors T2-T4 are turned on at the delay angle $\pi+\alpha$ in the first negative half period. Thus, the delay angle in the j-th half cycle of the source waveform, is given by $(j-1)\ \pi + \alpha, j = 1, 2, \ldots n$.

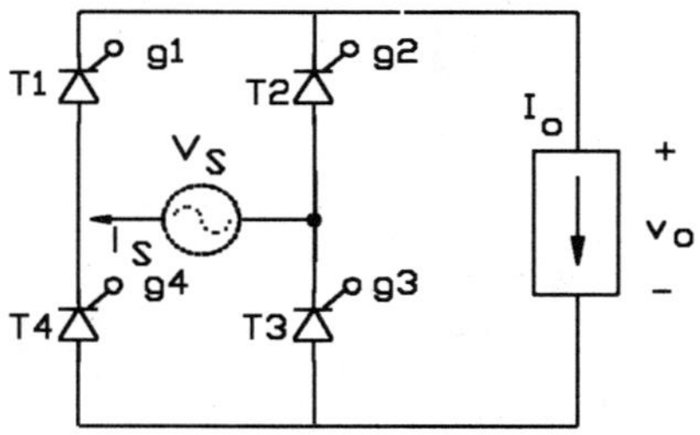

Figure 10.23
The 2-quadrant cycloconverter

Negative Half Period $T_O/2$

Thyristors T2-T4 are turned on at the delay angle $4\pi-\alpha$ and $6\pi-\alpha$. Thyristors T1-T3 are turned on at the delay angle $5\pi-\alpha$ in the first negative half period. All thyristors commutate naturally. See the waveforms in Figure 10.24. Thus, the delay angle in the k-th half cycle of the source waveform is given by $k\,\pi-\alpha$, $k = n + 1, n + 2, \ldots 2n$.

Example 10-7

A one-phase cycloconverter is controlled by the delay angle set at π/3 and the output frequency is desired to be 1/3 of the source frequency (n = 3). The ac source is 120 V rms, 60 Hz. Plot

a. the output voltage waveform,
b. the waveform of the fundamental harmonic of the output voltage,
c. the harmonic profile and
d. the rms output voltage.

```
n=3; VS=sqrt(2)*120; delay=pi/3;
alpha=[pi/3,pi+delay,2*pi+delay,4*pi-delay, 5*pi-delay,
   6*pi-delay,6*pi+delay];
th=alpha(1):pi/512:2*n*pi+alpha(1);
vs=VS*sin(th);
vo=zeros(size(th));
for k=1:2:2*n
      vo1= vs.*((th>=alpha(k))& (th<=alpha(k+1)));
      vo= vo+vo1;
end
for k=2:2:2*n
      vo1= vs.*((th>=alpha(k))& (th<=alpha(k+1)));
      vo= vo-vo1;
end
plot(th,vo,'r')
[f,voh,ph]=harmonic(vo,16);
vo1=voh(2)*sin(th/n);
plot(th,vs,'b:',th,vo,'r*',th,vo1,'r—')
vorms=sqrt(mean(vo.^2))
```

vorms = 119.9512

The fundamental component in the harmonic profile of the output voltage is 44% in amplitude and 1/3 in the frequency of the source.

The above converter produces both positive and negative output voltages but the output current is only positive. Hence this is a two-quadrant topology. The four-quadrant operation is obtained by adding a parallel bridge of inverted thyristors T1′-T4′ as shown in Figure 10.26(a). The T1-T4 bridge is operated during the positive half period $T_O/2$ of the desired output waveform. Thyristors T1-T3 are turned on at α, and 2π+α, and T2-T4 at π+α. All thyristors in the T1′-T4′ bridge are turned off. In the negative half period $T_O/2$, all thyristors in the T1-T4 bridge are turned off. The thyristors T1′-T3′ are turned on at 3π + α and 5π + α. Thyristors T2′-T4′ are turned on at 4π+α. All thyristors commutate naturally. See the waveforms in Figure 10.26(b). Thus the delay angle in the k-th half cycle of the source waveform, is $(k-1)\pi + \alpha$, $k = 1, 2, \ldots 2n-1$.

Sinusoidal Profile of Delay Angle

The delay angle is profiled to obtain a sinusoidal output voltage waveform. To understand the princi-

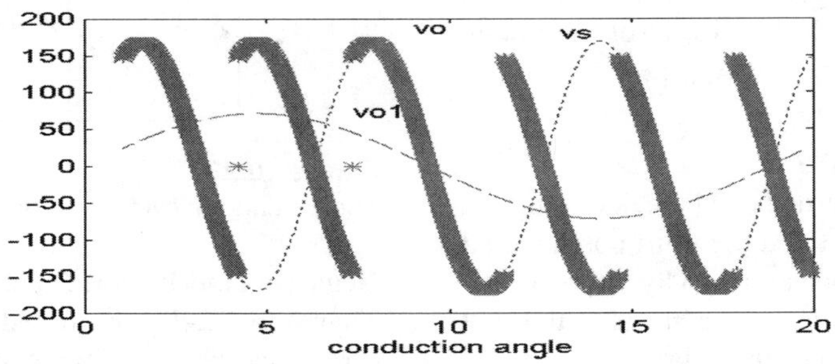

Figure 10.24
The v_O waveform in the 2-quadrant cycloconverter

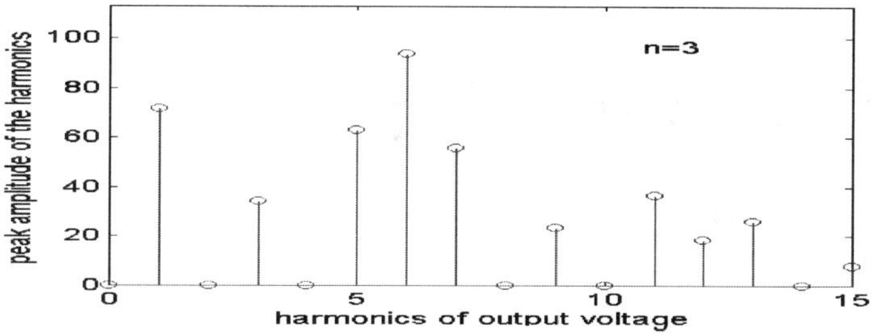

Figure 10.25
Harmonic profile

ple of the methodology, refer to the converter of Figure 10.23. The converter operation over a single period of the source waveform is identical to the operation of an α-controlled full-bridge ac-dc converter with an inductive load. Begin with the assumption that the desired output frequency of the cycloconverter is 1/n of the source frequency. Hence, one cycle of the cycloconverter output voltage consists of 2n half cycles of the source waveform. The average output voltage of the α-controlled ac-dc converter in the j-th half cycle of the source waveform is

$$\overline{V}_{O,j} = \frac{2V_S}{\pi}\cos\alpha_j \qquad j = 1, , \ldots 2n \quad (10\text{-}28)$$

$$\alpha_j = \cos^{-1}\left[\frac{\pi \overline{V}_{O,j}}{2V_S}\right] \quad (10\text{-}29)$$

The average $\overline{V}_{O,j}$ can be shaped over 2n periods into a sinusoidal waveform of the desired frequency ω_o:

$$\overline{V}_{O,j} = V_O \sin\theta_{O,j} \qquad \theta_O = \omega_O t$$

$$\theta_{O,j} = (j-1)\frac{\pi}{n} + \frac{\pi}{2n} \quad (10\text{-}30)$$

V_O is the peak voltage of the desired voltage waveform as the cycloconverter output. Odd integer value of n helps in practical implementation. Furthermore, the term inside the brackets in equation (10-29) should be less than or equal to 1, hence,

$$\frac{\pi\, V_O}{2V_S} \le 1 \Rightarrow V_O \le \frac{2}{\pi}V_S \quad (10\text{-}31)$$

Example 10-8

A one-phase cycloconverter with sinusoidal α-control is required to supply a 120 volts peak voltage at 1/5 of the source frequency (n = 5). The ac source is 120 V rms, 60 Hz. Plot

a. the output voltage waveform,
b. the waveform of the fundamental harmonic of the output voltage, and
c. the harmonic profile.

```
n=5;th[ ]; VS=sqrt(2)*120; VO=100;
for j=1:2*n
    alpha(j)=(j-1)*pi+acos((pi*VO/(2*VS))*sin((j-1)
      *(pi/n)+pi/(2*n)));
end
alpha(2*n+1)=2*n*pi+alpha(1);
th=alpha(1):pi/512:2*n*pi+alpha(1);
vs=VS*sin(th);
vo=zeros(size(th));
for k=1:2:2*n
    vo1= vs.*((th>=alpha(k))& (th<=alpha(k+1)));
    vo= vo+vo1;
end
for k=2:2:2*n
    vo1= vs.*((th>=alpha(k))& (th<=alpha(k+1)));
    vo= vo-vo1;
end
[f,voh,ph]=harmonic(vo,64);
vo1=voh(2)*sin(th/n);
plot(th,vs,'b:',th,vo,'r+',th,vo1,'r—')
vorms=sqrt(mean(vo.^2))
```

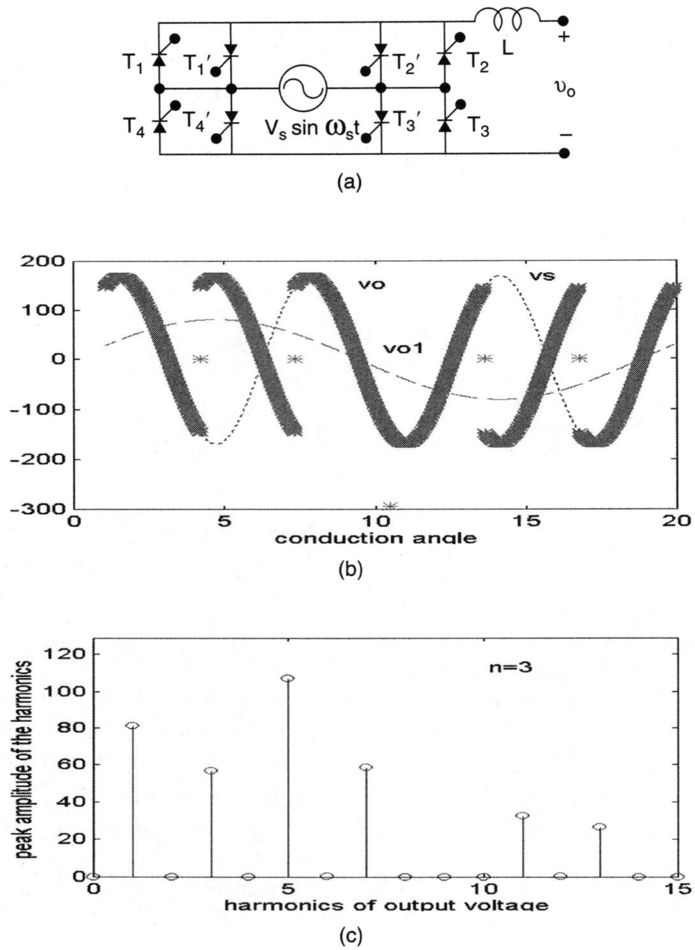

Figure 10.26
(a) The 4-quadrant 1-ϕ cycloconverter (b) the v_O waveform and (c) the harmonic profile

The fundamental component in the harmonic profile of the output voltage is 59% in amplitude of the peak voltage V_S of the source waveform. Note that the fundamental harmonic amplitude is higher than in the corresponding case of the uniform delay angle case. The lowest undesired harmonic frequency can be pushed higher by increasing n. Figure 10.28 shows the harmonic profile for n = 15.

10.4.2 3-phase to 1-phase and 3-phase to 3-phase cycloconverter

The three-phase to one-phase cycloconverter with a large filter inductance is shown in Figure 10.29. The cycloconverter operation is similar to that seen in the previous section except that three phases of source waveforms result in three switchings instead of one

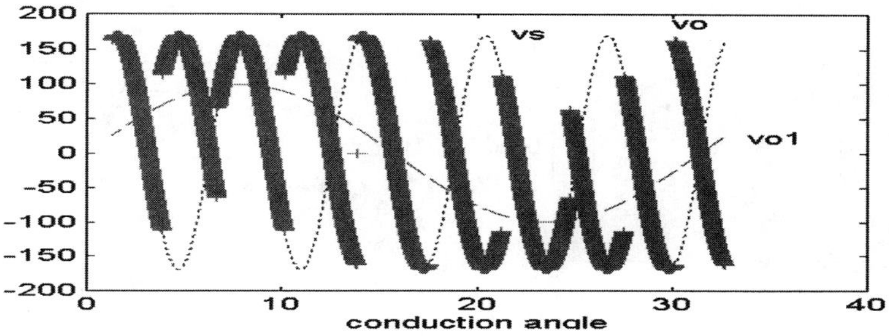

(a)

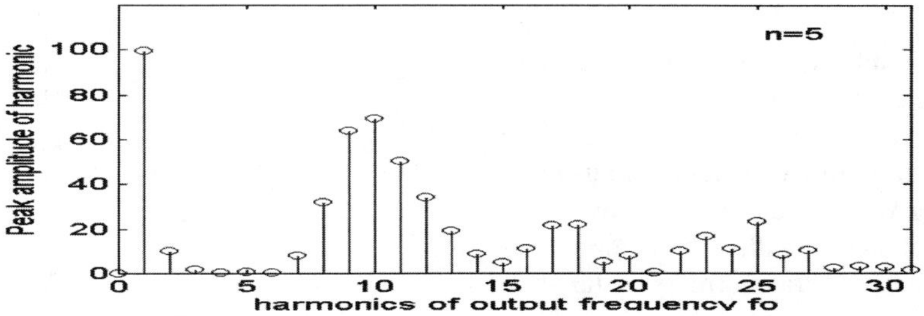

(b)

Figure 10.27
(a) The v_O and v_{o1} waveforms (b) Harmonic profile

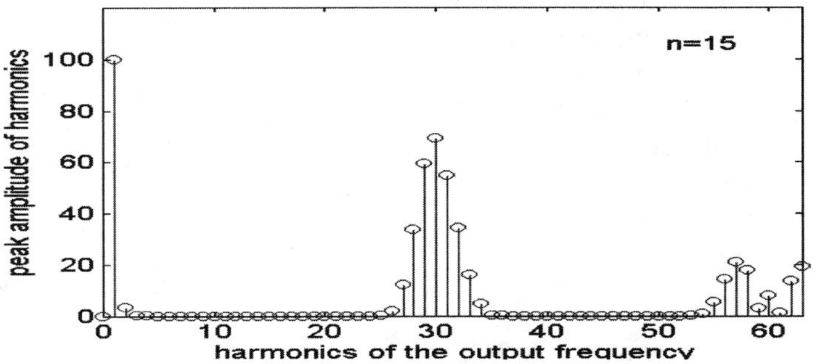

Figure 10.28
The harmonic profile for $n = 15$

per half cycle. This is equivalent to multiplying n by 3. Consequently the converter offers the advantages of high value of 3n at the output frequency of ω_S /n. Sinusoidal profiling of α is most common in these converters. The four-quadrant operation is achieved by adding one reverse thyristor to each of the six thyristors and modifying the delay angle as was demonstrated in the one-phase cycloconverter.

Most of the high power applications of cycloconverters require a three-phase output waveform. One

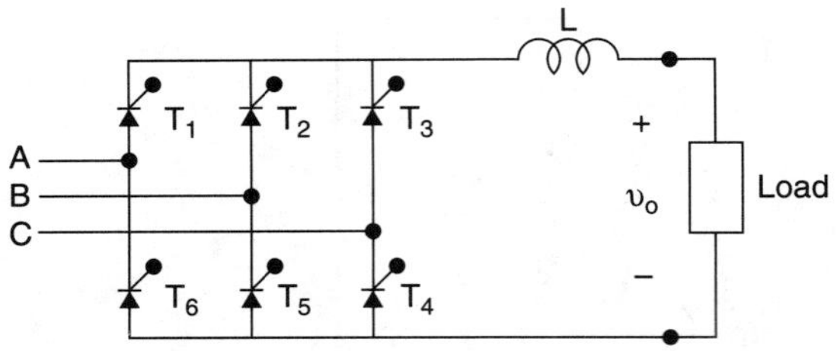

Figure 10.29
The 2-quadrant 3-phase to 1-phase cycloconverter with positive current

example is the adjustable drive for a very large three-phase machine. We could connect three single-phase cycloconverters through isolation transformers into a star or delta circuit at the load side. A four-quadrant cycloconverter would require 12×3 thyristors. An economical alternative with a half number of thyristors and no isolation transformers is shown in Figure 10.30. Each phase is constituted as a four-quadrant full-bridge ac controller.

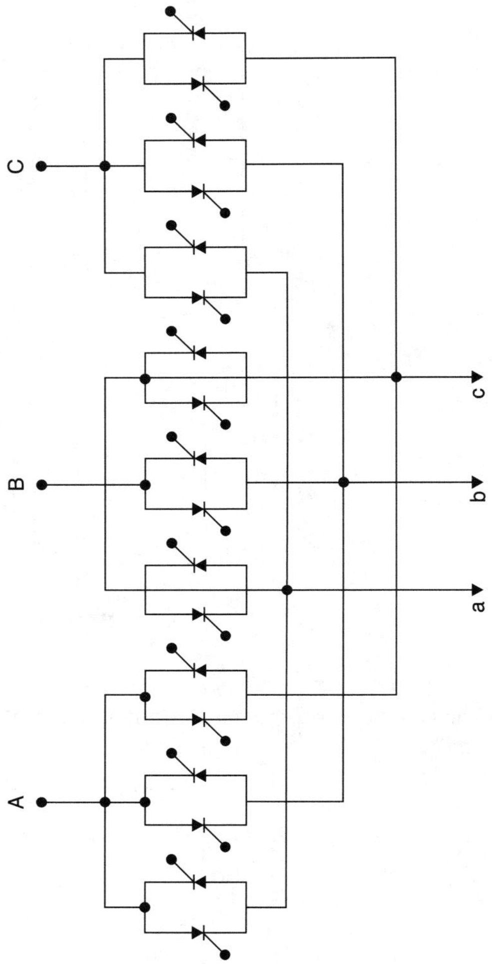

Figure 10.30
A 3-phase to 3-phase cycloconverter

10.5 DC-LINK AC-AC CONVERTER

The ac source is first converted in to a dc voltage or a dc current source. The dc source is then converted into the ac source of the desired specifications: amplitude, frequency and phase. The intermediate dc stage is called the dc link or the dc bus. The output voltage in the dc-link converter may have a different amplitude, a different frequency and a different number of phases than the input voltage source.

$$v_S = V_S \sin \omega_S t,$$
$$v_o = V_o \sin(\omega_o t + \phi),\ \omega_o \neq \omega_S \quad (10\text{-}32)$$

Unlike the cycloconverters, the output frequency need not be an integer multiple or a fraction of the source frequency. The output frequency may be lower or higher than the source frequency. The dc-link converter is expensive but the control is relatively easier than the cycloconverters.

The input energy in the dc-link converter is collected at one rate (frequency) and delivered to the load at a different rate (frequency). Hence, the ac components of power at the input and the output may not be the same. An inductor, a capacitor or their combination (resonant tank), placed in the dc link stores the instantaneous difference between the input and output energies.

In the following sections, we will discuss a few types of dc-link ac converters.

10.5.1 DC current-link ac-ac converter

The dc current-link converter is shown in Figure 10.31. A large inductor is the intermediate storage element. The ac-dc converter at the front end sees a large inductive load. The dc-ac inverter at the load end sees a current source. The load is an ac source or an ac sink v_O, which may represent an induction motor, a synchronous motor or another ac source connected as part of an ac source in a dc transmission bus. The control parameters in the ac-dc converter and the inverter are delay angle α and pulse-width δ respectively. The dc voltage output $\overline{V}_{O1}$ of the ac-dc converter is

$$\overline{V}_{O1} = \frac{2\,V_S}{\pi}\cos\alpha \qquad (10\text{-}33)$$

The voltage sink v_O load and the output current from the dc-ac inverter are

$$v_O = V_O \sin\omega_O t, \qquad (10\text{-}34)$$

$$i_C = \sum_{n=1,3,5,\dots} \frac{4I_{dc}}{n\pi}\sin\frac{n\delta}{2}\sin(n\omega_O t - \theta_n) \qquad (10\text{-}35)$$

I_{dc} is the dc current supplied by the dc-link inductor.

Assume that all the power in the output of the inverter is delivered at the fundamental frequency (about 90%, see Chapter 9) and the output current is a pure sinusoid:

$$i_O = I_{O1}\sin(\omega_O t + \theta_1),$$

$$\text{where } I_{O1} = \frac{4I_{dc}}{\pi}\sin\frac{\delta}{2} \qquad (10\text{-}36)$$

$$I_{dc} = \frac{\pi I_{O1}}{4\sin\delta/2} \qquad (10\text{-}37)$$

The power delivered to the load is equal to the power supplied by the front-end ac-dc converter:

$$P_O = \overline{V}_{O1} I_{dc} \qquad (10\text{-}38)$$

$$= \frac{2V_S}{\pi}\cos\alpha\,\frac{\pi I_{O1}}{4\sin(\delta/2)} = \frac{V_S I_{O1}}{2}\,\frac{\cos\alpha}{\sin(\delta/2)}$$

10.5.2 DC voltage-link AC-AC converter

The dc voltage-link converter is shown in Figure 10.32. A large capacitor is the intermediate storage element. The ac-dc converter sees a large capacitive load. The dc-ac inverter sees a voltage source. The load is an ac source or sink v_O in series with an inductor. The ac source or sink v_O may represent an induction motor or a synchronous motor. The dc voltage output v_{o1} is

$$\overline{V}_{O1} = \frac{2\,V_S}{\pi}\cos\alpha \quad (10\text{-}39)$$

Assume that all power in the output of the inverter is delivered at the fundamental frequency (about 90%, see Chapter 9) and the output current is a pure sinusoid. The rms output voltage of the ac sink for a pulse width δ is obtained as following:

$$v_O = V_O\sin(\omega_O t + \theta_1),$$

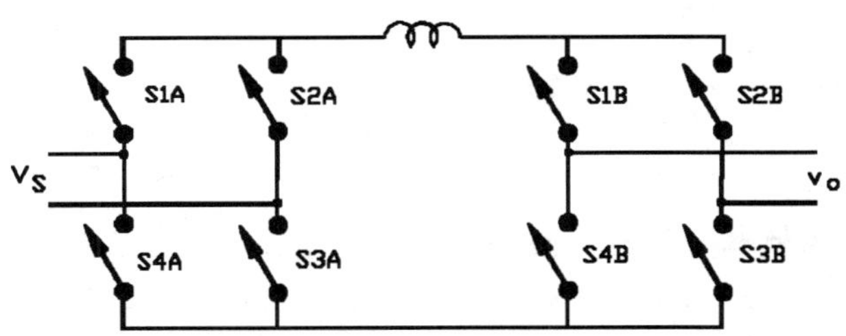

Figure 10.31
DC current-link converter

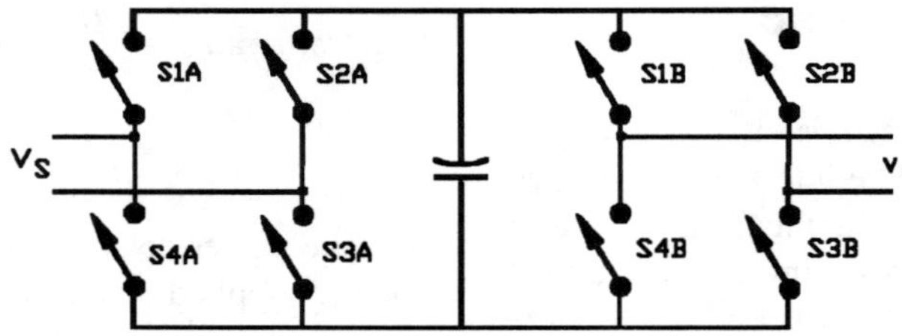

Figure 10.32
DC voltage-link converter

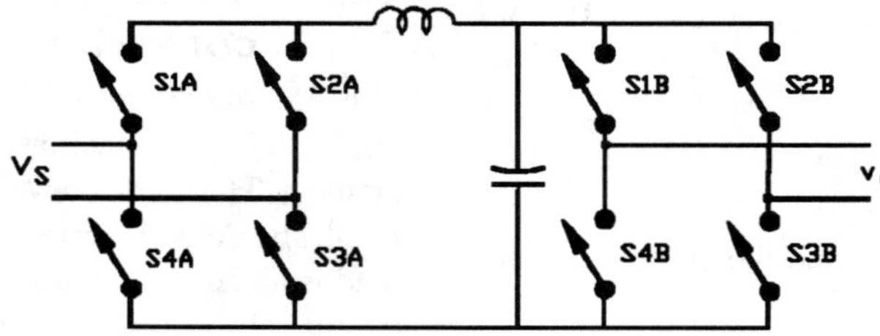

Figure 10.33
LC-link ac-ac converter

$$\text{where} \quad V_O = \frac{4\overline{V}_{O1}}{\pi} \sin\frac{\delta}{2} \tag{10-40}$$

$$V_O = \frac{8V_S}{\pi^2}\cos\alpha \quad \sin\frac{\delta}{2} \tag{10-41}$$

$$\tilde{V}_O = 0.573\ V_S \cos\alpha \sin\frac{\delta}{2} \tag{10-42}$$

The pulse width δ for an ac source as the load is

$$\delta = 2\sin^{-1}\left[\frac{\tilde{V}_O}{0.573V_S\cos\alpha}\right] \tag{10-43}$$

10.5.3 Resonant-link ac-ac converter

The resonant-link ac-ac converter contains a L-C series or parallel or a three-element resonant tank circuit that stores the intermediate energy during the transfer of power from the ac source to ac load. A two-element resonant-link converter is shown in Figure 10.33. The resonant link and the dc-ac inverter part is discussed in the next chapter. Detailed analysis of this converter is beyond the scope of this book.

10.5.4 HVDC transmission

The High Voltage DC (HVDC) transmission is an excellent application of the dc current-link converter. Figure10.34 shows the simplified diagram of a practical HVDC transmission system. Unsynchronized three-phase ac generators of differing frequencies are connected to a dc bus via 12-pulse ac-dc converters. The sinusoidal PWM inverter with a three-phase output is used at the load centers. The HVDC transmission system has inherently better transient stability and ease of 'bussing' than the ac transmission system. The break-even distance of transmission in HVDC systems is several times longer.

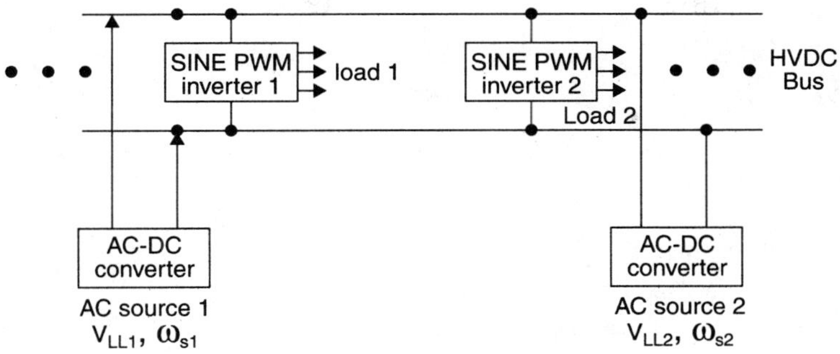

Figure 10.34
HVDC transmission system

10.6 PSPICE SIMULATION

PSPICE Example 10-1 1-Phase AC Controller with R-L Load

Consider the ac controller of Figure 10.35(a) with a source side inductance and the R-C load. Use 1N4002 as the rectifier. The uncontrolled rectifier specifications are

AC source = 24 V peak, 100 Hz RL = 1 ohm
L7 = 750 μH

Perform a transient analysis for at least 10 periods and plot the waveforms in the last two periods of the output voltage and the source current for

a. source voltage vs
b. output voltage vo across the R-L load
c. load current io

The thyristors X1 and X2 are switched by two external pulse sources V7 and V9 respectively as shown in Figure 10.35(a). The pulse sources V7 and V9 are transformer coupled or optically coupled in the practical implementation of the circuit.

Set the pulse source V7 as follows:

Lower level V1	=0 V
Upper level V2	=10 V
Pulse period	=10 ms
Pulse width (on-time)	=100 μs
Delay time TD	=7 ms
Rise time	=0.1 ns
Fall time	=0.1 ns

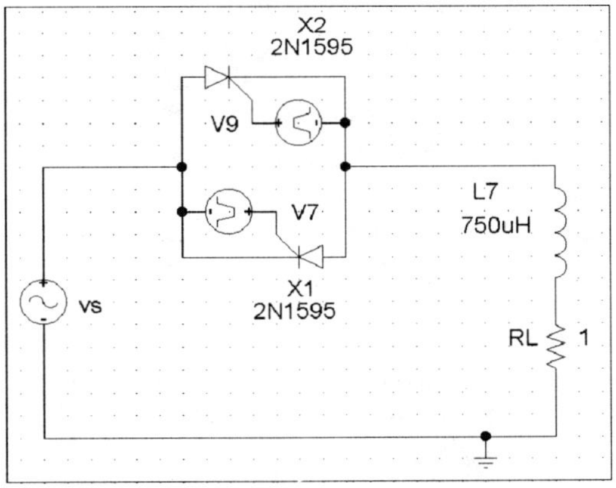

Figure 10.35
(a) Schematic

Set the pulse source V9 with all settings of V7 except the delay time, TD=2 ms. Use the simulation step in the range: 1 ns to 100 ns.

Note that the thyristors continue to conduct beyond π angle because of the inductive load.

10.7 SUMMARY

AC to ac converters transfer power from one ac system to another with waveforms of different amplitude or frequency or a different number of phases. The ac controller refers to the ac-ac converter in which the rms amplitude of the output voltage waveform is

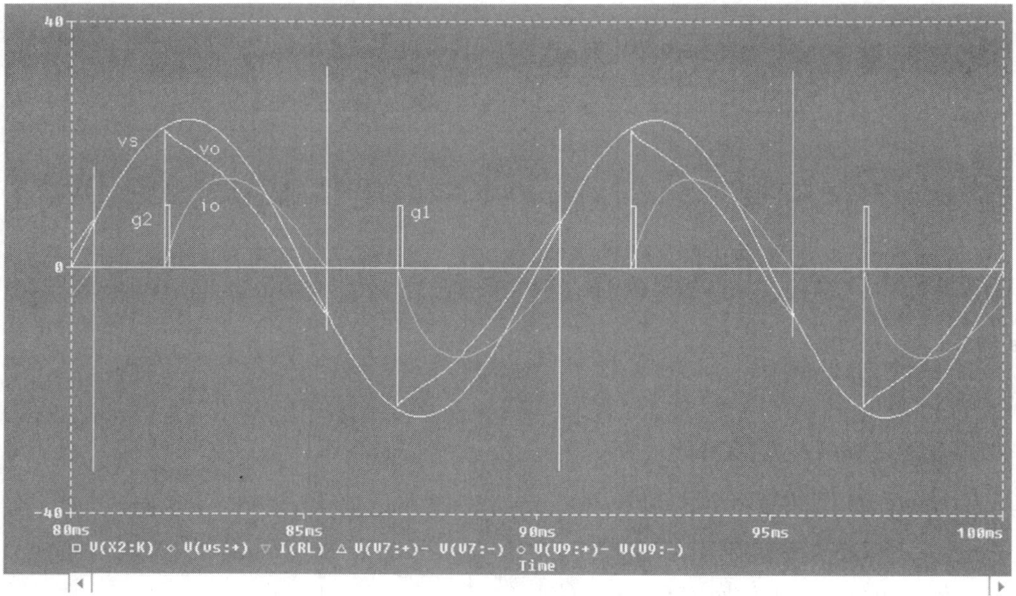

Figure 10.35 (continued)
(b) Waveforms of vs, vo, io, control pulses g1 and g2

controlled through switching without any change in frequency. The cycloconverter transfers power from the ac source to the load at a controlled voltage and frequency that is an integer fraction of the source frequency. Cycloconverters can perform only down conversion of frequency ($\omega_o < \omega_s$) and that too in simple multiples (/2, /3, /4, and so on). The ac controllers and cycloconverters perform direct conversion, where the output terminals are connected to the input terminals through controlled switches in a sequence that generates the desired waveforms. The switches are required to support bipolar voltages and carry bidirectional currents. The design is usually complex. Cyclo-conversion may not require forced commutation. Hence the cycloconverters may use thyristors. Many of the very high power applications, such as traction or heavy industrial drives, employ cycloconverters.

DC-link or resonant-link converters use an intermediate energy storage component or circuit, called link, and the converters are referred to as the link converters or indirect converters. In these converters, the ac is first converted to dc (ac-dc converter) and then inverted to the desired ac form (dc-ac converter). The energy storage component, or circuit, stores the energy difference arising out of the power transfer between the front-end ac-dc converter and the load-end dc-ac inverter stages. The switches require forced commutation. MOSFETs and IGBTs are the preferred devices. The input and output frequencies in indirect converters can be widely different. This is the only type in which the output frequency can be designed to be higher than the input frequency. These converters are used in medium power compact applications.

References

1. Philip T. Krein, *Elements of Power Electronics,* Oxford University Press, New York, 1998.

2. Ned Mohan, Tore M. Undeland, and William P. Robbins, *Power Electronics: Converters, Applica-*

tions and Design, 2nd Ed., John Wiley & Sons, Inc., 1995.

3. Joseph Vithayathil, *Power Electronics: Principles and Applications,* McGraw-Hill Inc., New York, 1995.

4. Muhammad H. Rashid, *Power Electronics: Circuits, Devices and Applications,* 2nd Ed., Prentice-Hall Inc., New Jersey, 1993.

5. John G. Kassakian, Martin F. Sclecht, and G. Vergassian, *Principles of Power Electronics,* Addison-Wesley, Boston, 1991.

6. T.H. Barton, *Rectifiers, Cycloconverters and AC Controllers.* Clarendon Press, Oxford, 1994.

7. L. Gyugyi, B.R. Pelly, *Static Power Frequency Changers.* Wiley, New York, 1976.

8. W. McMurray, *The Theory and Design of Cycloconverter.* MIT Press, Cambridge, MA, 1972.

9. B.R. Pelly, *Thyristor Phase-Controlled Converters and Cycloconverters,* Wiley, New York, 1971.

10. C. Rombaut, G. Seguir, and R. Bausiere, *Power Electronic Converters, Volume 2: AC-AC conversion.* McGraw-Hill, New York, 1987.

Problems

10-1 Replace the thyristors T2 by diode D2 in the converter of Figure 10.4. Analyze the steady-state performance, find the forward voltage transfer ratio and the output phase voltage waveforms for the delay angle $\alpha = \pi/3$.

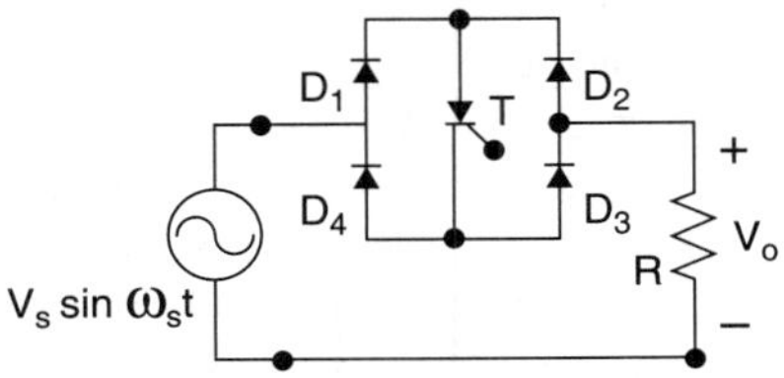

Figure P10.2

10-2 The converter of Figure P10.2 is an economical alternative to a full-bridge ac controller. Draw the output voltage waveforms and prove that the converter's performance is equivalent to that of the full-bridge controller.

10-3 The circuit of Figure P10.2 is used as the on-load tap changer to cope with changing load conditions in resistive heating applications. The output voltage is controlled by the delay angle of thyristors. Find the range of rms output voltage V_O when

a. T1 and T2 are turned off and T3 and T4 are operated as an ac controller and

b. T3 and T4 are turned off and T1 and T2 are operated as an ac controller.

c. Design the turning on sequence and delay angles of thyristors to obtain the rms output voltage in the range V_1 to V_1+V_2 .

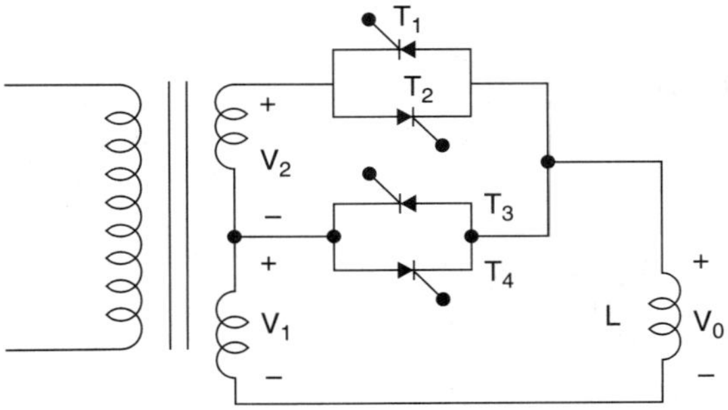

Figure P10.3
Tap changer

10-4 Draw the phase output voltage v_{aN} waveform in three-phase to one-phase ac controller for the delay angle α:
a. $\pi/4$, **b.** $3\pi/2$ and **c.** $11\pi/12$.

10-5 The three-phase converter of Figure P10.5 is an economical alternative to the star-connected converter of Figure 10.16. The added advantage is simpler control circuitry and ease of assembling of thyristors. Draw the waveforms and find the rms output phase voltage.

10-6 The three-phase converter of Figure P10.6 is an economical alternative to the delta-connected converter of Figure 10.20. Draw the waveforms and find the rms output phase voltage.

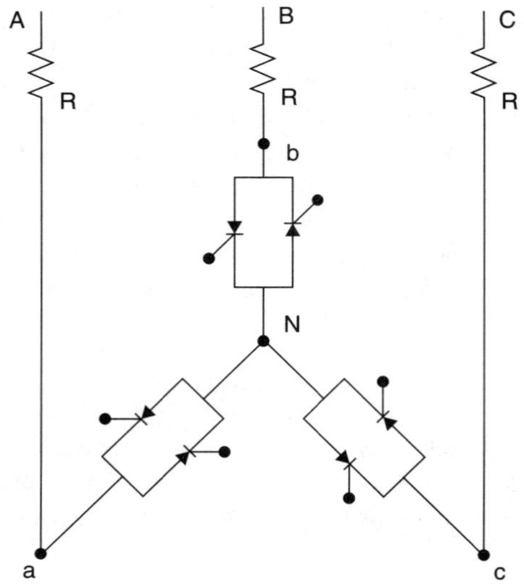

Figure P10.5

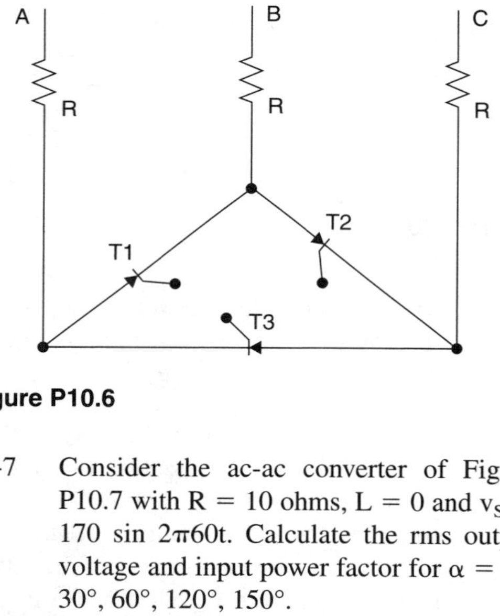

Figure P10.6

10-7 Consider the ac-ac converter of Figure P10.7 with R = 10 ohms, L = 0 and v_S = 170 sin $2\pi 60t$. Calculate the rms output voltage and input power factor for $\alpha = 0°$, 30°, 60°, 120°, 150°.

10-8 Repeat problem 10-7 with L=1 mH.

10-9 A 2 kW resistive heating element is fed from a 120 Vrms, 60 Hz ac bus through an on-off controller. One switching period is equal to 15 periods of the ac bus voltage waveform. Calculate the number of on

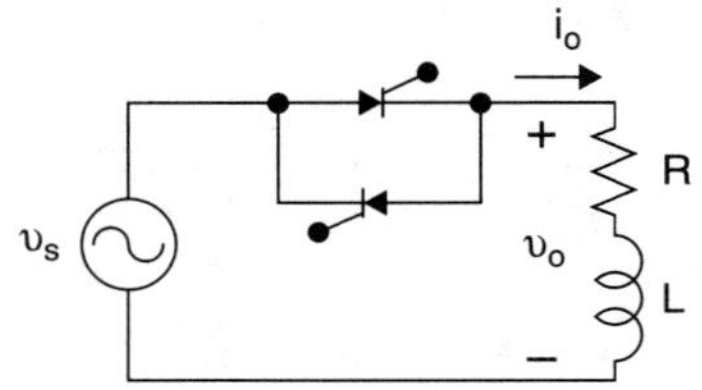

Figure P10.7

periods that would be needed to set the heating power to 60% of the rated power.

10-10 Plot the harmonic profile of the output voltage in problem 10-9. Determine the forward voltage transfer ratio. Plot the ratio of the output power to the maximum power versus the ratio of the number of on periods to the total number of periods in a switching cycle.

10-11 The on-off controller in problem 10-9 is replaced by the phase-controlled ac controller. Find the delay angle to set the heating power to 60% of the maximum.

10-12 Plot the harmonic profile of the output voltage in problem 10-11. Determine the forward voltage transfer ratio. Plot the ratio of the output power to the maximum power versus the ratio of the number of on periods to the total number of periods in a switching cycle.

10-13 A phase controlled thyristor ac controller is loaded by a 10 ohm resistor in series with a 100 mH inductor. The ac source is 120 Vrms, 60 Hz. Determine the rms output voltage for a delay angle of 120°. Plot the waveforms of the output voltage and current.

10-14 Find the delay angle for which the ac controller of problem 10-13 would not respond.

10-15 A thyristor controlled inductor can be used as the compensator of the static VAR (volts amperes reactive) in a 11 kV, 60 Hz ac network. The ac network has an inductance of 5 H. Determine the range of the adjustable VAR.

10-16 A 120 Vrms supply is turned on and off with a 0.75 duty cycle at a frequency of nine cycles.

a. Calculate the rms output voltage and the power factor
b. Determine the harmonic profile of the output voltage waveform and
c. plot the output voltage waveform as a function of the conduction angle.

10-17 An ac controller with phase delay angle control, shown in Figure 10.4, is supplying power from a 350 V peak, 60 Hz ac source to a resistive load of 10 ohms. For a delay angle of $\pi/3$,
a. Plot the output voltage waveform as a function of the conduction angle,
b. Calculate the rms output voltage,
c. Plot voltage transfer function T_{vv} and
d. Plot the harmonic profile of the output voltage waveform.

10-18 Consider the ac controller for a synchronous motor as shown in Figure 10.14. The ac source is 120 V rms, 60 Hz. The output voltage v_O is 90 V rms, 60 Hz and is synchronized with respect to the ac source. It is assumed that the voltage emf load is in phase with the input voltage. The torque angle is maintained at $\pi/2$. Find the rms output current.

10-19 Plot the waveforms of the phase voltage and v_O for $V_P = 120$ V, $\alpha = 2\pi/3$ in the star connected load of Figure 10.16.

10-20 Plot the waveforms of output phase voltages in a delay angle controlled three-phase ac-ac converter with the delta connected load of 2.5 ohms, where the ac source is 120 V ac per phase and the delay angle is $3\pi/4$. Plot the harmonic profile and calculate the per-phase rms current and THD of the output current waveform.

10-21 A 1-phase ac-ac converter is controlled by control of the off time of pulses. Six pulses are used in a half period of the ac source waveform. The off-time duty ratio is 0.75. The load is purely resistive, Δ-connected 1ohm. The ac source is 120 V ac per phase. Plot the harmonic profile and calculate the per-phase rms voltage, the rms voltage of the fundamental component of the converter output voltage and power factor of the output current waveform at the load. Plot the forward voltage transfer function T_{VV} as a function of the duty ratio.

10-22 A one-phase cycloconverter is controlled by the delay angle set at $\pi/3$ and the output frequency is desired to be 1/9 of the source frequency (n=9). The ac source is 120 V rms, 60 Hz. Plot
a. the output voltage waveform,
b. the waveform of the fundamental harmonic of the output voltage,
c. the harmonic profile and
d. the rms output voltage.

CHAPTER 11
Resonant Converters

OBJECTIVES

1. Fundamentals of series, parallel and high-order resonant tank circuits
2. Basics of state-plane analysis of resonant circuits under external excitation
3. Analysis of series resonant dc-dc converter operation with load current continuous and discontinuous in time
4. Analysis of series resonant dc-dc converter in steady-state operation below and above resonance
5. Analysis of parallel resonant dc-dc converter operation with load current continuous and discontinuous in time
6. Introduction to class E resonant conversion
7. Introduction to zero-current and zero-voltage switching topologies
8. Introduction to resonant-link inverters

11.1 INTRODUCTION

Using high switching frequency in all the four types of converters; ac-dc, ac-ac, dc-ac and dc-dc; will be desirable because it will provide tremendous savings in component sizes and weights. As was seen in Chapter 2, the highest switching frequency of converter operation is limited by the losses in the switching devices. Furthermore, the transformers are the primary band-limiting components. Parasitic inductance and capacitance in all components add further limitations.

11.2 RESONANT CONVERTER SYSTEMS

Resonant converter/inverter is defined as a class of converters/inverters in which the topology constitutes of at least one resonant tank circuit as a subcircuit. A resonant tank is a subcircuit consisting of at least one inductor and one capacitor. More specifically, resonant converters have a resonant tank as a part of the circuit during at least one of the circuit modes in the steady-state converter operation. The order of a resonant tank, and therefore of the converter, is determined by the number of independent energy storage elements.

Based on the use of a resonant tank, the resonant converters may be classified as

1. Resonant-link converters
2. Resonant-switch converters

Similarly the resonant inverters may be classified as

1. Resonant-dc link inverters
2. Resonant-ac link inverters

Resonant-Link Converters
These converters consist of a multistate source, resonant tank circuit as the link and the multistate sink. The multistate source is an equivalent representation of a dc voltage/current source and switching network. The state transition is controlled externally or internally such as that in the steering of current through diodes (self-controlled switches).

A multistate sink is an equivalent representation of a resistive load coupled to an inductive or capacitive filter and a diode network for ac-dc conversion. The state transition is controlled either externally or internally.

A resonant-link consists of a resonant tank circuit, such as a series resonant tank or a parallel resonant tank of second order or higher. The multistate source/sink may be connected either to the voltage or the current of the complete resonant tank circuit or to a part of it.

Resonant-Switch Converters
In these converters, the resonant elements L and C are utilized primarily in conjunction with the switch to achieve zero/low loss switching, lossless snubbering and/or soft switching. The converter operation may consist of several modes. At least one of them is a resonant mode. These converters are also referred to in the literature as quasi-resonant, pseudo-resonant and resonant-transition converters.

Resonant-DC-Link Inverters
A resonant tank circuit serves as a buffer between the dc source and an ac load. The dc link is provided by the inductor of the resonant tank circuit. These converters also include zero switching loss topologies.

Resonant-AC Link Inverters
Like in the dc-link inverters, a resonant tank circuit serves as a buffer between the dc source and an ac load. The link between the source and the load does not have a dc path. These converters also include zero switching loss topologies.

11.3 RESONANT TANK CIRCUITS

A resonant tank is a subcircuit consisting of at least one inductor and one capacitor. A second order resonant tank consist of two elements L and C either connected in series or in parallel. A third order resonant tank circuit consists of one L and two C's or one C and two L's. The three elements may be connected in several possible topologies.

11.3.1 Series resonant tank (second order tank)

The input impedance of a series resonant tank circuit, given in Figure 11.1(a), is given by

$$Z_{in} = j\left(\omega L - \frac{1}{\omega C}\right) \tag{11-1}$$

It can be shown after some manipulations (see Problem 11-1) that

$$Z_{in} = -jZ_0 \frac{1-(\omega/\omega_0)^2}{(\omega/\omega_0)} \tag{11-2}$$

where ω_0 and Z_0 are respectively the resonant frequency and the characteristic impedance, given by

$$\omega_0 = \frac{1}{\sqrt{LC}}, \qquad Z_0 = \sqrt{\frac{L}{C}} \tag{11-3}$$

The inductive part of the impedance at the resonant frequency ω_0 is exactly equal to the capacitive part, resulting in an electrical short. The capacitive component $1/\omega C$ dominates at lower frequencies. The inductive component ωL dominates at high frequencies. Hence the series resonant tank behaves like a capacitance at lower frequencies and like an inductance at high frequencies. Let us plot the magnitude and the phase of $\mathbf{Z_{in}}$ as a function of ω/ω_0,

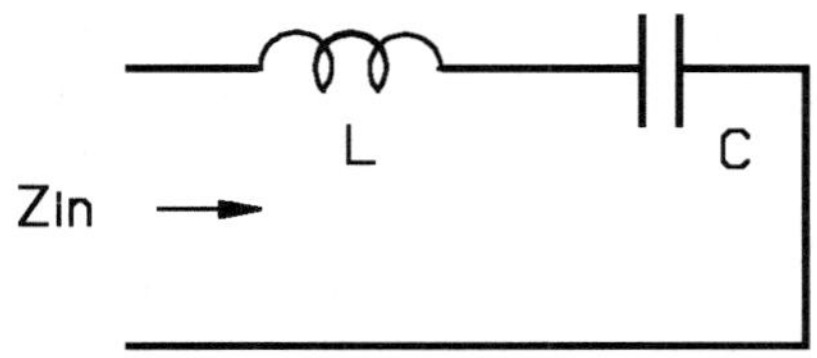

Figure 11.1(a)
Series resonant tank

Example 11-1

Plot the amplitude and phase of the ratio of the input impedance to the characteristic impedance versus normalized frequency for the series resonant tank circuit of Figure 11.1(a).

```
omegan =logspace(-1, 1,100);% normalized frequency
  omegan=omega/ omega_o in a range from 10^-1 to 10^2
Zins=-j*(1-omegan .^2)./ omegan;%Zins=Zin/Z_o
semilogx(omegan, abs(Zins)% plot of amplitude versus
  normalized frequency
semilogx(omegan,angle(Zins))% plot of phase versus
  normalized frequency
```

The phase jumps from $-\pi/2$ to $+\pi/2$ at the resonant frequency implying that the impedance turned from capacitive to inductive.

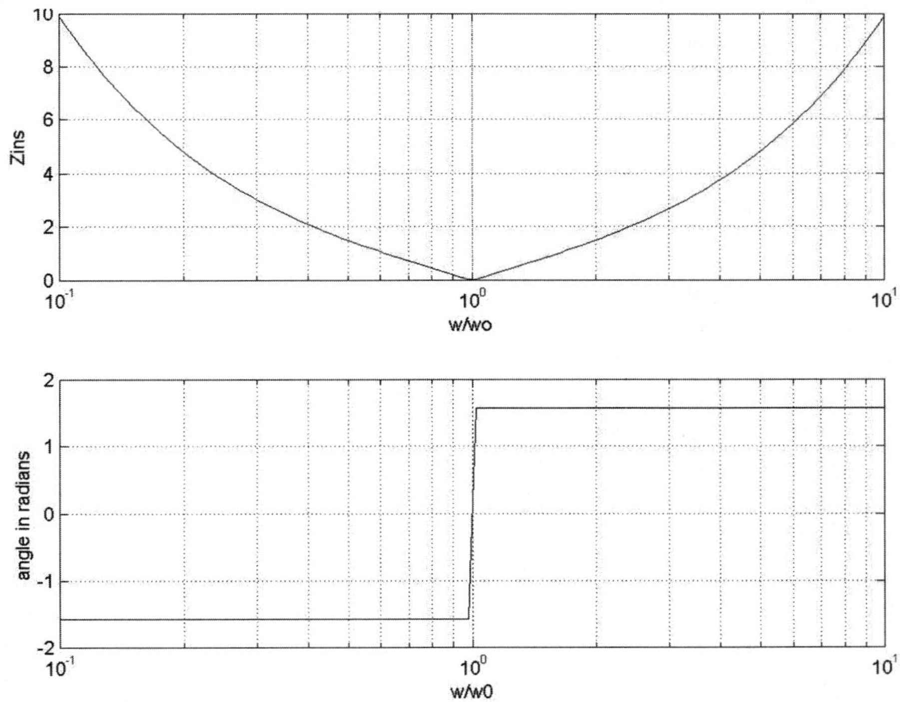

Figure 11.1(b) ***(continued)***
The nomalized input impedance versus normalized frequency of series resonant tank

11.3.2 Parallel resonant tank (second order tank)

Input impedance of a parallel resonant tank circuit can be shown to be

$$Z_{in} = \frac{j\omega L \dfrac{1}{j\omega C}}{j\omega L + \dfrac{1}{j\omega C}} \qquad (11\text{-}4)$$

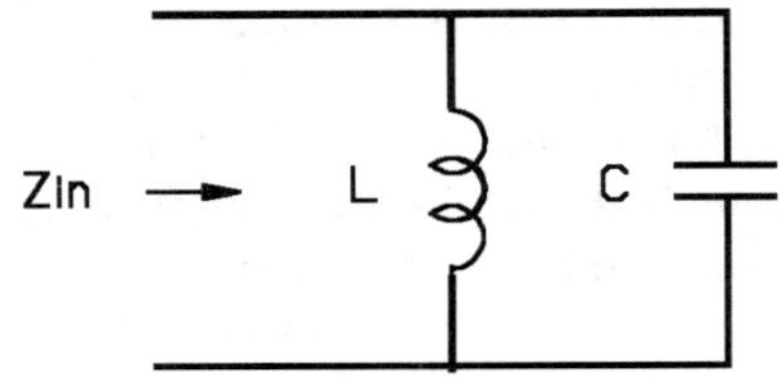

Figure 11.2(a)
Parallel resonant tank

It can be shown after some manipulations (see Problem 11-2) that

$$Z_{in} = j\, Z_0 \frac{(\omega/\omega_0)}{1 - (\omega/\omega_0)^2} \qquad (11\text{-}5)$$

where ω_0 and Z_0 are resonant frequency and characteristic impedance of the tank circuit respectively. At the resonant frequency ω_0, the magnitude of the input impedance becomes infinity, an effectively open-circuit condition. The inductive component ωL dominates at lower frequencies. The capacitive component $1/\omega C$ dominates at high frequencies. Hence the parallel resonant tank behaves like an inductance at lower frequencies and like a capacitance at high frequencies. Let us plot the magnitude and the phase of $\mathbf{Z_{in}}$ as a function of ω/ω_r.

The phase jumps from $+\pi/2$ to $-\pi/2$ at the resonant frequency, implying that the impedance changed from inductive to capacitive.

Example 11-2

Plot the amplitude and phase of the ratio of the input impedance to the characteristic impedance versus normalized frequency for the parallel resonant tank circuit of Figure 11.2(a).

Add the following MATLAB lines to the program of Example 11-1.

```
Zinp=−j*omegan ./(1−omegan .^2); %Zinp=Zin/Zo
semilogx(omegan, abs(Zinp)% plot of amplitude versus
  normalized frequency
semilogx(omegan,angle(Zinp))% plot of phase versus
  normalized frequency
```

11.3.3 Third order resonant tank

Several tank circuits consisting of energy storage elements are shown in Figure 11.3. We will look into the tank of Figure 11.3(c) as an illustration of the impedance characteristics. All third order tank circuits consist of a zero amplitude frequency and an infinite amplitude frequency (resonant frequency).

The input impedance of the tank circuit of part (c) is

$$Z_{in} = \frac{1}{j\omega C_2} + \frac{j\omega L \dfrac{1}{j\omega C_1}}{j\omega L + \dfrac{1}{j\omega C_1}} \qquad (11\text{-}6)$$

$$Z_{in} = -j\, Z_{01}\left[\frac{C_1}{C_2}\frac{1}{(\omega/\omega_0)}\right]\left[\frac{1 - (\omega/\omega_z)^2}{1 - (\omega/\omega_0)^2}\right]$$

where

$$\omega_z = \frac{1}{\sqrt{L(C_1 + C_2)}}, \quad \omega_0 = \frac{1}{\sqrt{LC_1}}, \textit{ and } \omega_0 > \omega_z$$

and

$$Z_{01} = \sqrt{\frac{L}{C_1}}$$

The impedance is capacitive at frequencies below ω_z and above ω_0. Between ω_z and ω_0, the impedance is inductive. The impedance becomes zero at ω_z and infinite at the resonant frequency ω_0. The presence of a zero-amplitude frequency close to the resonant frequency makes it a more sharply tunable circuit than the second order resonant tanks. Plot the magnitude and the phase of $\mathbf{Z_{in}}$ as a function of ω/ω_0 as shown in Figure 11.4.

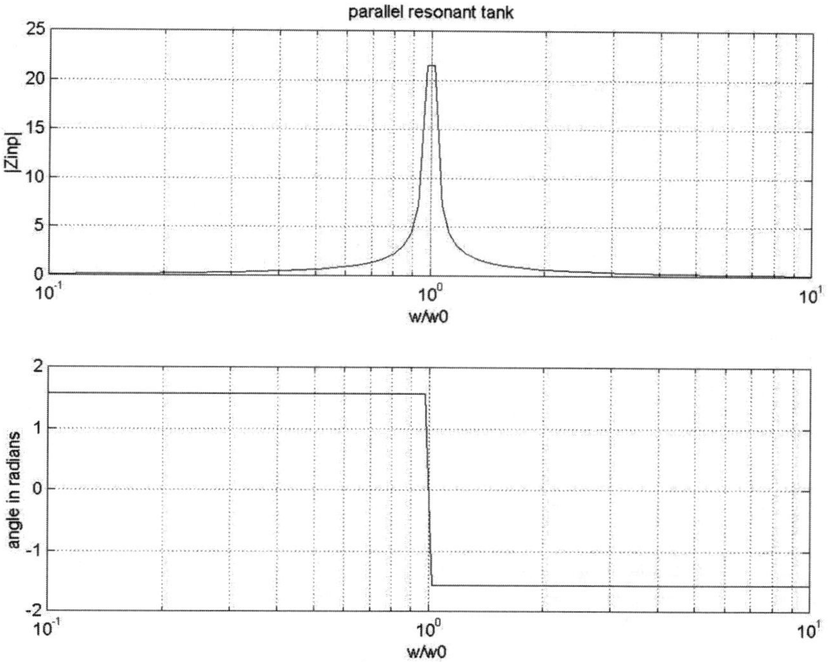

Figure 11.2(b)
Normalized input impedance versus normalized frequency of parallel resonant tank

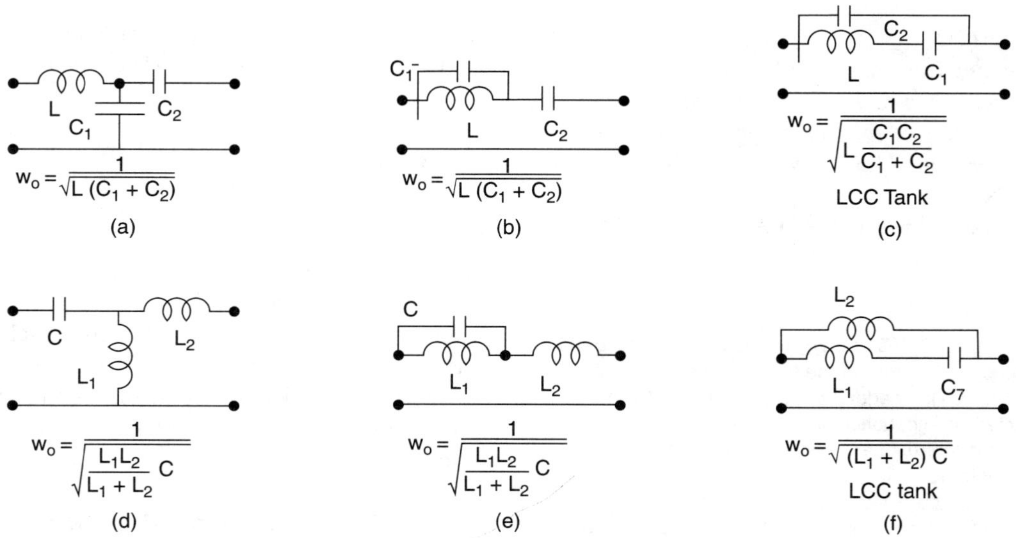

Figure 11.3
Third order tank circuits

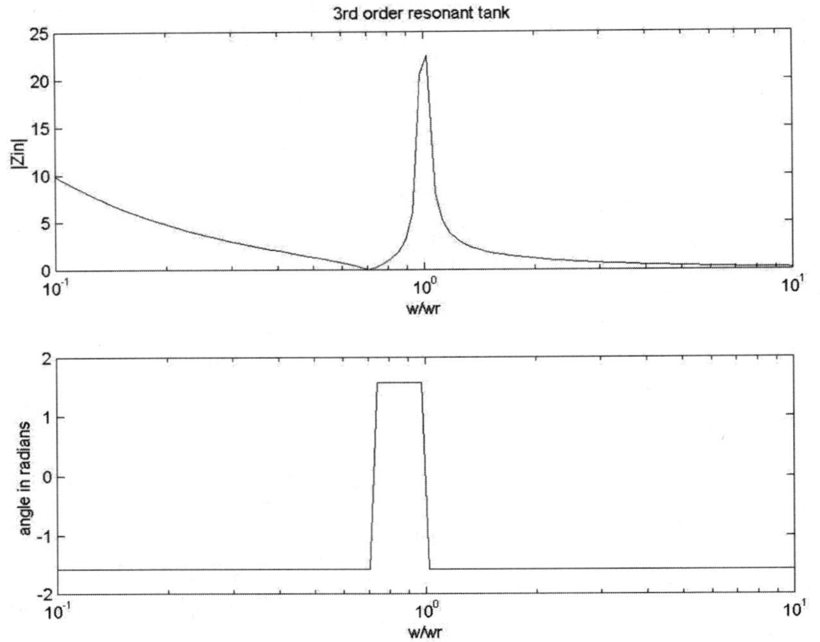

Figure 11.4
Input impedance of a third order resonant tank of Figure 11.3(c)

Example 11-3

Plot the amplitude and the phase of the ratio of the input impedance to the characteristic impedance versus normalized frequency for the resonant tank circuit of Figure 11.3(a) with L = 1 H, C1 = 1 F and C2 = 1 F.

Add the following MATLAB lines to the program of Example 11-2.

```
C1=1; C2=1; L=1;
omegaz =1/sqrt(L *(C1+C2));% omegaz is the resonant
  frequency of the zero of the tank
ZO1=sqrt(L/C1); omegar=1 /sqrt(L*C1);%ZO1 is the
  characteristic impedance and
% omegar is the resonant frequency of the pole of the
  resonant tank
omega=logspace(-1*omegar,1*omegar,100);
% range of frequency from 0.1 to 100 times of omegar
omeganr=omega /omegar; omeganz=omega /omegaz;
  % normalized frequencies
Zin=-j*ZO1*(C1/C2) .*(1 ./omeganr).*(1-omeganz .^2)./
  (1-omeganr .^2);
subplot (2,1,1), semilogx(omeganr,abs(Zin))
subplot(2,1,2), semilogx(omeganr, angle(Zin))
```

11.4 RESONANT TANK IN EXCITATION

To understand the operation of a resonant converter that employs the resonant tank circuit, it will be helpful to investigate a resonant tank circuit under excitation by a dc source without any switching. In the following sections, we will study the behavior of a series resonant tank circuit under dc voltage excitation and a parallel resonant tank circuit under dc current excitation.

11.4.1 Series resonant tank in voltage source excitation

We will study the steady-state response of the series resonant tank under dc voltage excitation, see

Figure 11.5(a). Two equations are required to completely describe the state of the resonant tank circuit:

$$L\frac{di_L}{dt} + v_C = V_g \tag{11-7}$$

$$C\frac{dv_C}{dt} = i_L \tag{11-8}$$

Inductor current i_L and capacitor voltage v_C are the two state variables of the circuit. Differentiating both equations gives

$$L\frac{d^2 i_L}{dt^2} + \frac{dv_C}{dt} = \frac{dV_g}{dt} = 0 \tag{11-9}$$

$$C\frac{d^2 v_C}{dt^2} = \frac{di_L}{dt} \tag{11-10}$$

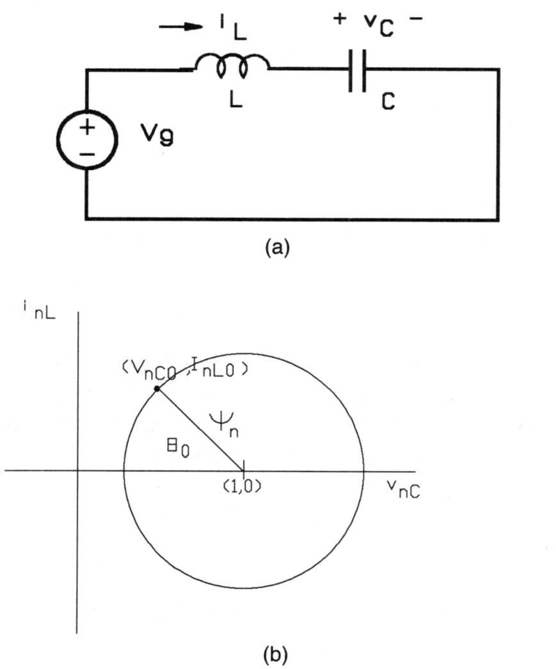

Figure 11.5
(a) Series resonant tank circuit under voltage excitation and (b) state-plane trajectory of the steady-state operation

After some substitutions and manipulations:

$$\frac{d^2 i_L}{dt^2} + \frac{i_L}{LC} = 0 \tag{11-11}$$

$$\frac{d^2 v_C}{dt^2} + \frac{v_C}{LC} = \frac{V_g}{LC} \tag{11-12}$$

Transforming the **t** variable to the conduction angle θ and using the normalization of variables:

$$\omega_0 t = \theta,\ where\ \omega_0 = \frac{1}{\sqrt{LC}},\quad Z_0 = \sqrt{\frac{L}{C}} \tag{11-13}$$

$$v_{nC} = \frac{v_C}{V_g},\quad i_{nL} = \frac{i_L}{V_g/Z_0}$$

$$\frac{d^2}{d\theta^2} i_{nL} + i_{nL} = 0 \tag{11-14}$$

$$\frac{d^2}{d\theta^2} v_{nC} + v_{nC} = 1 \tag{11-15}$$

The solutions for the inductor current and the capacitor voltage are given below:

$$i_{nL} = \Psi \sin\theta \tag{11-16}$$

$$v_{nC} = 1 - \Psi \cos\theta \tag{11-17}$$

where

$$\Psi = \sqrt{I_{nL}0^2 + (1 - V_{nC}0)^2}$$

$$I_{nL}0 = \Psi \sin\theta_0,\ V_{nC}0 = 1 - \Psi\cos\theta_0$$

I_{nL0} and V_{nC0} are the initial values of the inductor current and the capacitor voltage at the conduction angle θ_0. The state variables i_{nL} and v_{nC} are sinusoidal functions of the conduction angle θ. The nonnormalized state variables i_L and v_C are sinusoidal functions of time with a frequency ω_0.

It can be shown (prove it)

$$i_{nL}^2 + (1 - v_{nC})^2 = \Psi^2 1 \tag{11-18}$$

This is the equation of a circle with the center at (1,0) in the v_{nC}-i_{nL} plane, see Figure 11.5(b). This plane is called the state-plane. The state-plane representation of the steady-state response is unique.

1. The trajectory of the normalized state variables is a circular segment on the v_{nC}-i_{nL} plane moving in a clockwise direction. The conduction angle θ is measured with respect to the 180° line on the state-plane.
2. The radius of the circle Ψ is entirely determined by the initial conditions of the state variables.
3. Clockwise movement along the circle corresponds to the increasing time at an angular frequency that is equal to the resonant frequency ω_0.
4. Coordinates of the center of the circle (1,0) correspond to the normalized voltage source V_S and the normalized current source which is zero in our circuit.

The presence of resistance in the circuit will cause the circular trajectory to change to a shrinking helical coil.

11.4.2 Parallel resonant tank in current source excitation

Consider the parallel tank circuit under dc current excitation, as shown in Figure 11.6(a). Two state equations are required to completely describe the steady state of the resonant tank circuit:

$$C\frac{dv_C}{dt} + i_L = I_g \qquad (11\text{-}19)$$

$$L\frac{di_L}{dt} = v_C \qquad (11\text{-}20)$$

These equations are the dual of the equations in the series resonant tank case. Hence, following similar steps, we will get the following solution of the normalized state variables:

$$i_{nL} = 1 + \Psi \sin\theta \qquad v_{nC} = -\Psi\cos\theta \quad (11\text{-}21)$$

where

$$\Psi = \sqrt{(1 - I_{nL}0)^2 + V_{nC}0^2}, \qquad \theta = \omega_0 t,$$

$$\theta_0 = \cos^{-1}\left[\frac{-V_{nC}0}{\Psi}\right]$$

I_{nLO} and V_{nCO} are the initial values of the inductor current and the capacitor voltage. Again, the following relation holds between the normalized current and the voltage (prove it):

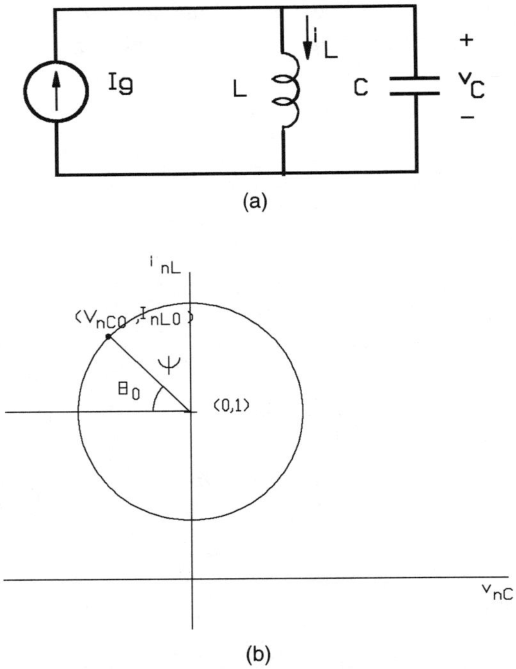

Figure 11.6
(a) Parallel tank circuit under dc current excitation and (b) state-plane representation

$$(1 - i_{nL})^2 + v_{nC}^2 = \Psi^2 \qquad (11\text{-}22)$$

This is the equation of a circle with the center at (0, 1) in the v_{nC}-i_{nL} plane, see Figure 11.6(b). The radius of the circle Ψ is entirely determined by the initial conditions of the state variables. Coordinates of the center of the circle (0, 1) correspond to the normalized voltage source, which is zero, and the normalized current source $I_{ns}=1$ (I_g/I_g).

11.4.3 Resonant tank loaded with a current sink in voltage source excitation

Consider the series tank circuit under dc voltage excitation, the capacitor coupled to a current sink, as shown in Figure 11.7(a). Two state equations are required to completely describe the steady state of the resonant tank circuit:

$$C\frac{dv_C}{dt} + I_o = i_L \qquad (11\text{-}23)$$

$$L\frac{di_L}{dt} + v_C = V_g \qquad (11\text{-}24)$$

The following solution can be obtained for the normalized state variables:

$$i_{nL} = 1 + \Psi \sin\theta \quad v_{nC} = 1 - \Psi\cos\theta \quad (11\text{-}25)$$

where

$$\Psi = \sqrt{(1 - I_{nL0})^2 + (1 - V_{nC0})^2},$$

$$\theta = \omega_0 t, \qquad \theta_0 = \cos^{-1}\left[\frac{1 - V_{nC0}}{\Psi}\right]$$

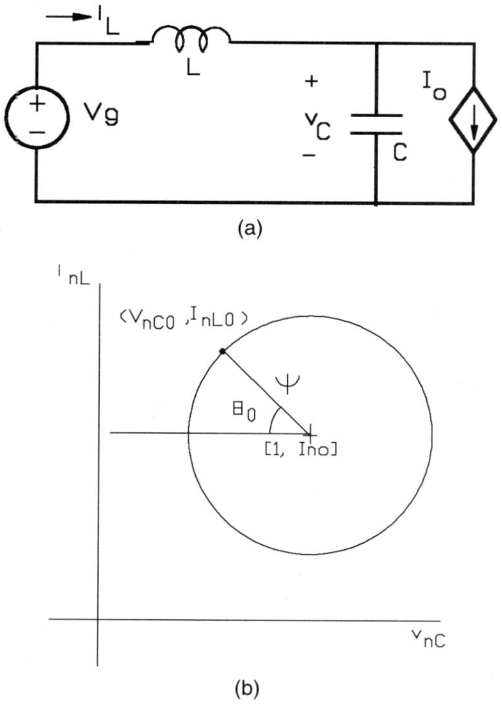

Figure 11.7
(a) Parallel tank circuit under dc current excitation and (b) state-plane representation

I_{nL0} and V_{nC0} are the initial values of the normalized inductor current and the normalized capacitor voltage. Again, the following relation holds between the normalized current and the voltage (prove it):

$$(1 - i_{nL})^2 + (1 - v_{nC})^2 = \Psi^2 \quad (11\text{-}26)$$

This is the equation of a circle with the center at $[1, I_{no}]$ in the v_{nC}-i_{nL} plane, see Figure 11.7(b), where

$$I_{no} = \frac{I_o}{V_g/Z_0}; \qquad Z_0 = \sqrt{\frac{L}{C}}$$

The radius of the circle Ψ is entirely determined by the initial conditions of the state variables. Coordinates of the center of the circle correspond to the normalized voltage source, which is unity, and the normalized current sink I_{no}.

11.5 SERIES RESONANT DC-DC CONVERTER WITH VOLTAGE SINK (CURRENT COUPLED RESONANT CONVERTER)

A series resonant tank circuit serves as the link between the voltage source-switch network and the voltage sink-diode network. The voltage sink is coupled to the current of the resonant tank circuit. We will analyze a half-bridge series resonant converter, shown in Figure 11.8(a) to illustrate the procedure. Two dc voltage sources V_g can be obtained from the capacitor division of a single dc voltage source as shown in Figure 11.8(b). Voltage sink is the equivalent model of a high value capacitor in parallel with load resistance. Switches S_1 and S_2 are turned on and off for an equal amount of time exclusively. The two dc voltage sources and the two switches along with the diodes and the said switching scheme can be modeled by a square wave source V_s as shown in the equivalent circuit of Figure 11.8(c). The voltage sink and the diode network can be modeled by a two-state voltage sink V_k $[+V_o, -V_o]$: $+V_o$ for positive current (entering the top end) and $-V_o$ for negative current (leaving the top end). It can be readily seen from the equivalent circuit that the converter gain $M = V_o/V_g$ equals unity

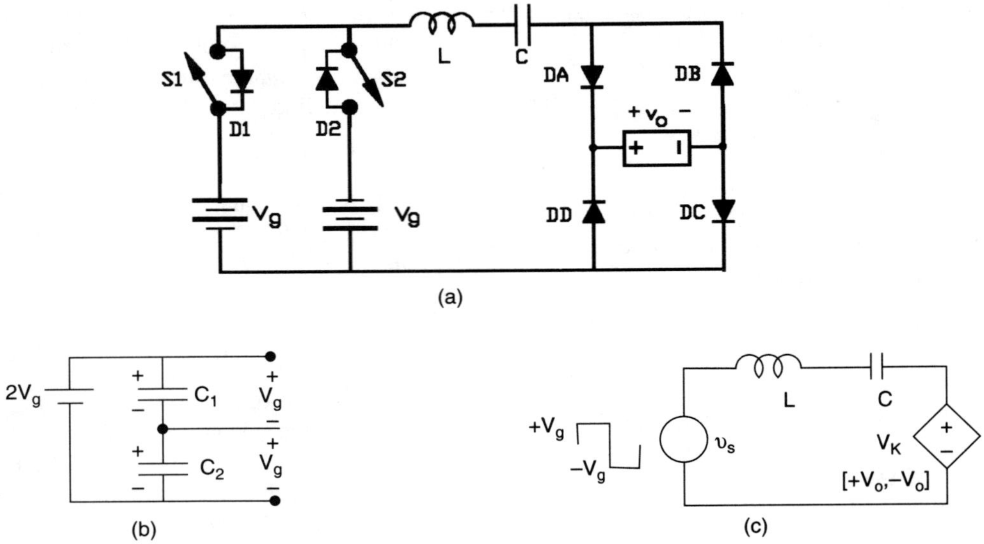

Figure 11.8
(a) Series resonant converter (b) splitting a dc source voltage into two equal half-value dc voltage sources and (c) equivalent circuit model

only when the converter switching frequency is equal to the resonant frequency (at resonance) of the series tank circuit. M otherwise is always less than unity. The current into the voltage sink can be theoretically infinity at the resonance. At other switching frequencies, the current is finite and limited in case of an accidental short in the voltage sink.

We will analyze only the steady-state response of the converter. Based on the wave-shape of the current drawn from the voltage source, there are two possible modes of operation in steady state:

Discontinuous Conduction Mode (DCM) in which the current drawn from the source becomes discontinuous in time (some times zero).

Continuous Conduction Mode (CCM) in which the current drawn from the source is continuous in time.

11.5.1 Discontinuous conduction mode

Switches are turned on and off by a square wave of switching frequency f_s (period=T_s), which is smaller than the resonant frequency f_0. The converter operation consists of six circuit modes: three modes in each half switching period. Inductor current i_L and capacitor voltage v_L are the two state variables. The state variables are continuous in time across the mode boundaries, which means that the final conditions of a circuit mode are also the initial conditions of the next circuit mode. Furthermore, since the switch control waveform is square-wave, the final conditions at the end of the half switching period must be equal in amplitude but opposite in polarity to the initial conditions. Hence, it will suffice to analyze the response in a half switching period.

The current path in each mode is shown by thick lines in the circuits of Figure 11.9(a). Simplified equivalent circuits of each mode are shown in Figure 11.9(b). The final conditions of a circuit mode serve as the initial conditions of the next mode in the sequence.

The first circuit mode m1 begins with zero inductor current and negative capacitor voltage. The

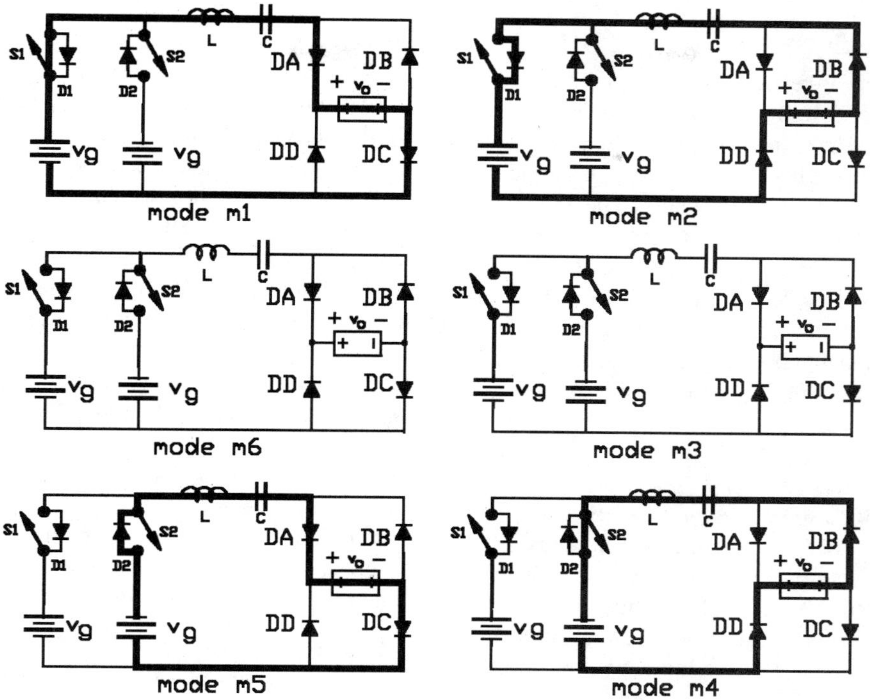

(a)

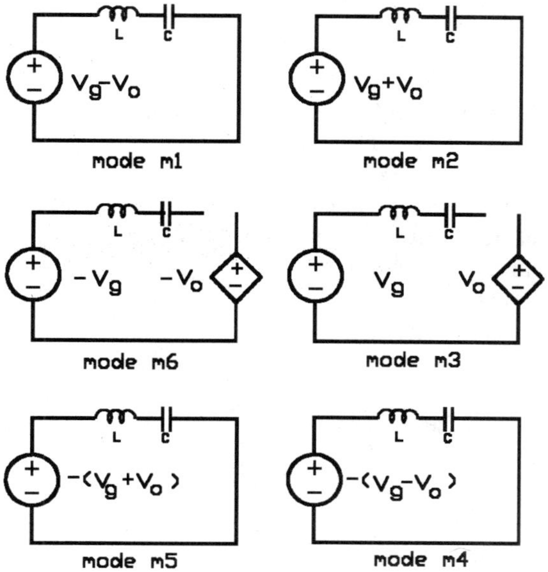

(b)

Figure 11.9
(a) Current path in circuit modes and (b) simplified equivalent circuit modes

capacitor voltage is between $-V_g-V_o$ and $-V_g+V_o$. The net voltage source in the simplified equivalent circuit is V_g-V_o. The inductor current increases to positive values sinusoidally. Negative voltage on the capacitor decreases also sinusoidally. The mode m1 continues until the inductor current becomes zero along with a positive voltage on the capacitor.

In the second mode m2, the inductor current increases with negative polarity. The capacitor voltage decreases although it is still positive. The net voltage source in the simplified equivalent circuit is V_g+V_o. The mode m2 continues until the inductor current becomes zero again. The final capacitor voltage is still positive and is between V_g-V_o and V_g+V_o.

To understand the converter operation in mode 3, hypothesize that the inductor current tends to increase in the positive direction. This action will turn off diodes DB and DD and turn on DA and DC. The net voltage source in the circuit would then become V_g-V_0. Since the capacitor voltage is more than V_g-V_0, the voltage across the inductor would be negative and will discourage the inductor current from increasing. Now hypothesize that the inductor current tends to decrease in the negative direction. This action will maintain diodes DB and DD into the on condition (as they were in mode m2). The net voltage source in the circuit would then become V_g+V_0. Since the capacitor voltage is less than V_g+V_0, the voltage across the inductor would be positive and will discourage the inductor current from decreasing. Consequently, the inductor current will continue to be zero and all diodes DA to DD are off. Thus the inductor current remains at zero and the capacitor voltage remains at the final value it attained in mode m2. The existence of mode m3 is the reason for naming this mode as the discontinuous conduction mode.

The normalization of sources and state variables is very helpful and is done as follows:

$$v_{nC} = \frac{v_C}{V_g}, \quad i_{nL} = \frac{i_L}{V_g/Z_0}$$

$$V_{no} = \frac{V_o}{V_g} = M, \quad Q = \frac{R_L}{Z_0}; \quad Z_0 = \sqrt{\frac{L}{C}}$$

The current and voltage waveforms are sinusoidal functions of time. The trajectory of normalized state variables in each circuit mode is a circular segment on the state-plane. Each circular segment has a radius, which is determined by the initial condition of that mode. The center of the segment is determined by the voltage sources in the simplified equivalent circuits. The steady-state trajectory on the normalized state-plane is given in Figure 11.10. The steady-state trajectory begins at the initial point $[-V_{nc0}, 0]$, at zero angle with respect to the center $[1-V_{no}, 0]$. The radius vector Ψ_1 rotates in the clockwise direction as time elapses at an angular speed ω_0. The circular segment is ended by the changes in circuit conditions that lead to another circuit mode in the mode sequence. The angle subtended by the circular segment is called the conduction angle of that circuit mode.

Mode m1:
initial conditions: $[-V_{nC0}, 0]$, conduction angle: $0 \le \theta < \pi$

center: $[1-M, 0]$

radius

$$\Psi_1 = 1 - M + V_{nC}0\ 2 \tag{11-27}$$

state equations:

$$i_{nL} = \Psi_1 \sin\theta \tag{11-28}$$

$$v_{nC} = (1 - M) - \Psi_1 \cos\theta \tag{11-29}$$

final conditions:

$$v_{nC}(\pi) = 1 - M + \Psi_1, \quad i_{nL}(\pi) = 0$$

Mode m2:
initial conditions: $[1-M+\Psi_1, 0]$, conduction angle: $\pi \le \theta < 2\pi$

center: $[1+M, 0]$

radius

$$\Psi_2 = 1 - M + \Psi_1 - (1 + M) = \Psi_1 - 2M \tag{11-30}$$

state equations:

$$i_{nL} = \Psi_2 \sin\theta \tag{11-31}$$

$$v_{nC} = (1 + M) - \Psi_2 \cos\theta \tag{11-32}$$

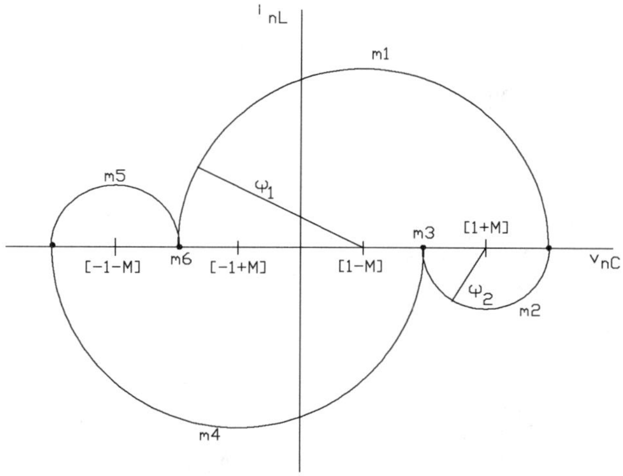

Figure 11.10
State-plane diagram of SRC in DCM operation

final conditions:

$$v_{nC}(2\pi) = V_{nC}0, \quad i_{nL}(2\pi) = 0$$

$$V_{nC}0 = 1 + M - \Psi_2 \tag{11-33}$$

From equations (11-27), (11-30) and (11-31)

$$\Psi_1 + \Psi_2 = 2, \quad and \quad \Psi_1 = 1 + M \tag{11-34}$$

and

$$V_{nC}0 = 2\,M$$

Since V_{nCO} lies between 1+M and 1−M, hence the lower bound and upper bound on the converter gain M are 1/3 and 1 respectively.

Mode m3:
states: $i_{nL} = 0$, $v_{nC} = V_{nCO}$, duration $= \tau$,
equivalent conduction angle $\Delta = \pi(\tau/T_s/2)$

The state plane trajectory for mode 4, mode 5 and mode 6 is the mirror image of the trajectory of mode 1, mode 2 and mode 3. Complete trajectory is determined from the knowledge of the initial condition V_{nCO}. The current and voltage waveforms as functions of time can be deduced from the state-plane and are shown in Figure 11.11.

The half switching period $T_S/2$ is calculated from the conduction angles of the three modes

$$\frac{T_S}{2} = \frac{\pi}{\omega_0} + \frac{\pi}{\omega_0} + \frac{\Delta}{\omega_0} \tag{11-35}$$

The state-plane trajectory for mode 4, mode 5 and mode 6 is the mirror image of the trajectory of mode 1, mode 2 and mode 3. Complete trajectory is determined from the knowledge of the initial condition V_{nCO}.

The switching frequency after normalization with respect to the resonant frequency is

$$f_{ns} = \frac{f_s}{f_0} = \frac{\pi}{2\pi + \Delta} \tag{11-36}$$

The upper bound on f_{ns} is 0.5 corresponding to $\Delta = 0$.

Average output current is

$$\bar{I}_o = \frac{2}{T_s}\left[\int_0^{\pi/\omega_0} i_L\,dt + \int_{\pi/\omega_0}^{(2\pi+\Delta)/\omega_0} |i_L|\,dt\right]$$

$$\bar{I}_{no} = \frac{f_{ns}}{\pi}\left[\int_0^{\pi} i_L\,d\theta + \int_{\pi}^{2\pi} |i_{nL}|\,d\theta\right]$$

$$\bar{I}_{no} = \frac{f_{ns}}{\pi}\left[\int_0^{\pi} \Psi_1 \sin\theta\,d\theta + \int_{\pi}^{2\pi} (-\Psi_2 \sin\theta)\,d\theta)\right]$$

$$\bar{I}_{no} = \frac{2\,f_{ns}}{\pi}(\Psi_1 + \Psi_2) = \frac{4}{\pi} f_{ns} \tag{11-37}$$

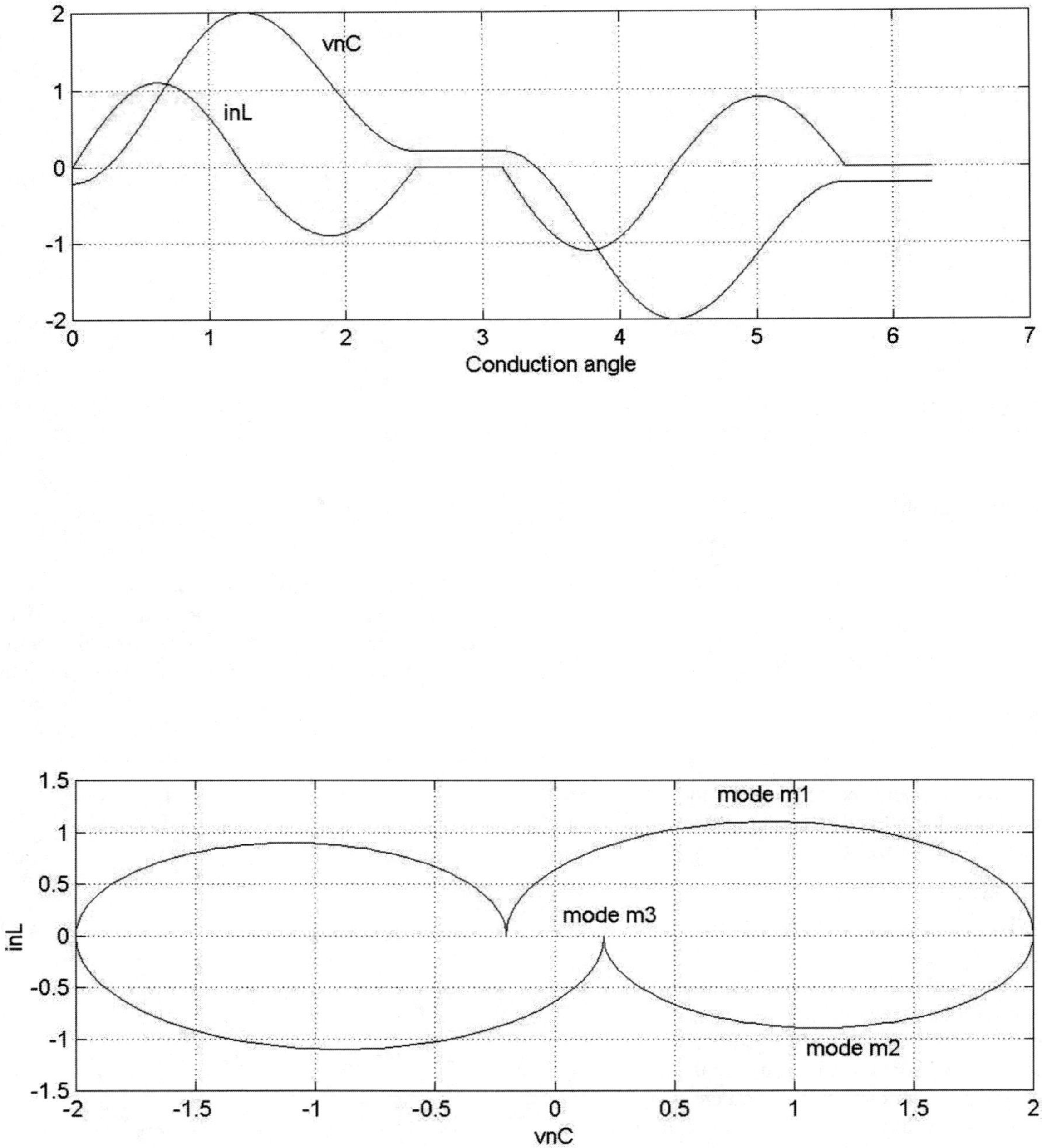

Figure 11.11
Series resonant converter in discontinuous conduction mode

The upper bound on $I_{no}=2\pi$ corresponds to $f_{ns} = 0.5$.

Furthermore, the average output current can also be determined from

$$\bar{I}_o = \frac{V_o}{R_L} \quad hence \quad \bar{I}_{no} = \frac{M}{Q} \qquad (11\text{-}38)$$

$$M = \frac{2}{\pi} Q \; at \; f_{ns} = 0.5 \qquad (11\text{-}39)$$

The average output current is independent of the converter gain (or the output voltage) and varies linearly with switching frequency. Hence the series resonant converter can be thought of as a constant current generator at a fixed switching frequency. Thus it has the inherent protection against a short circuit.

The maximum switching frequency corresponds to $\Delta=0$, therefore, $f_{ns,max}=0.5$.

Example 11-4

Parameters for the series resonant converter of Figure 11.8(a) are

$L = 1$ mH, $C = 1$ nF, $R_L = 200$ ohms, $V_g = 90$ Vdc, $f_{ns} = 0.4$.

a. Find the dc output voltage V_o

b. Draw the waveforms for the current into and the voltage on the voltage sink

c. Find the maximum dc output voltage V_o in DCM and the frequency at which this could be achieved

d. Find the frequency and the value of inductor L for which a minimum 10 W power will be delivered to $R_L = 200$ ohms.

a. The characteristic impedance

$$Z_o = \sqrt{\frac{L}{C}} = \sqrt{\frac{10^{-3}}{10^{-9}}} = 1\; k\Omega$$

The load factor

$$Q = \frac{R_L}{Z_o} = 0.2$$

Converter gain

$$M = \frac{4}{\pi} f_{ns}\, Q = \frac{4}{\pi} \times .4 \times .2 = 0.1018$$

The dc output voltage

$$V_o = M\, V_g = 9.168\; Vdc$$

b.

```
Q=.2; M=0.1018; dth=pi/100; fns=0.4;
Vnco=2*M;
delta=pi/fns -2*pi

thr1=0:dth:pi; thr2=pi:dth:2*pi; thr3=2*pi:dth:2*pi+delta;

thr=[thr1 thr2 thr3];

psi1=1-M+Vnco;
inl1=psi1*sin(thr1);
vnc1=(1-M)-psi1*cos(thr1);

psi2=psi1-2*M;
inl2=psi2*sin(thr2);
vnc2=(1+M)-psi2*cos(thr2);

inl3=zeros(size(thr3));
vnc3=Vnco*ones(size(thr3));

ths=thr*pi/(2*pi+delta);
inl=[inl1 inl2 inl3];
vnc=[vnc1 vnc2 vnc3];

ths_f=[ths pi+ths];
inl_f=[inl -inl];
vnc_f=[vnc -vnc];
plot(ths_f, inl_f, 'r', ths_f, vnc_f, 'b')
```

c. Maximum V_o occurs at $f_{ns}=0.5$,

Max. $V_o = 45$ volts,

Resonant frequency

$$f_o = \frac{\omega_o}{2\pi} = \frac{1}{2\pi\sqrt{LC}} = 159\; KHz$$

Switching frequency

$$f_s = f_{ns} \times f_o = 79.5\; KHz$$

d. The dc output voltage

$$V_o = \sqrt{power \times R_L} \geq \sqrt{10 \times 200} \geq 44.72\; Vdc$$

Converter gain

$$M \geq 0.497$$

Assume that M = 0.5 and it is the maximum M in DCM operation at $f_{ns}=0.5$.

Load factor

$$Q \geq \frac{\pi M}{2} = 0.7854$$

The characteristic impedance

$$Z_o = \frac{R_L}{Q} - Z_o = \frac{200}{.7854} = 254.6 \; ohms$$

The inductor L is

$$L = Z_o^2 C = 64.8 \; \mu H$$

Switching frequency is

$$f_s = f_{ns} \frac{1}{2\pi\sqrt{LC}} = 312.6 \; KHz$$

11.5.2 Continuous conduction mode

The inductor current and, hence, the current supplied by the source is never discontinuous in this mode of operation. The converter operation in this mode can be further subdivided in two types of operations based on the switching frequency below and above the resonant frequency.

Operation Below Resonance

The converter operation consists of four circuit modes: two modes in each half switching period. The converter operation is similar to that in the discontinuous conduction mode except that the mode 3 and mode 6 do not exist. Normalization of sources and state variables is similar to that used in the analysis of the discontinuous conduction mode of operation.

The steady-state trajectory begins at the initial point $[-V_{nLO}, I_{nLO}]$, at an angle θ_0 with respect to the center $[1-M, 0]$. The radius vector Ψ_1 rotates in the clockwise direction as time elapses at an angular speed ω_0. State analysis of all modes is given below. The state-plane trajectory is shown in Figure 11.12 and the time domain waveforms in Figure 11.13.

Mode m1: initial conditions: $[2V_{nCO}, I_{nLO}]$,

$$\Psi_1 \cos\theta_0 = 1 - M + V_{nC}0 \qquad (11\text{-}40)$$

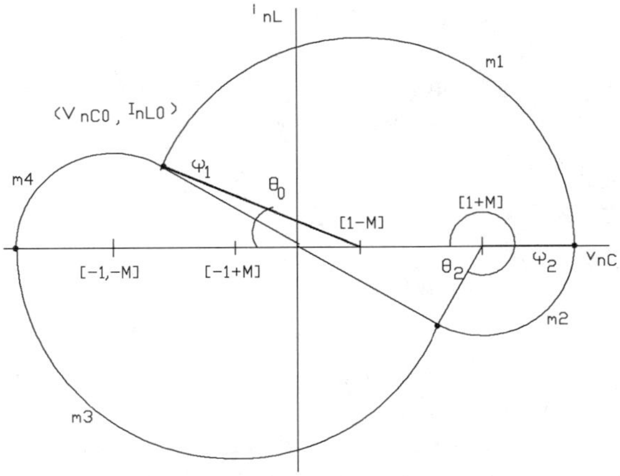

Figure 11.12
State-plane trajectory in continuous conduction mode

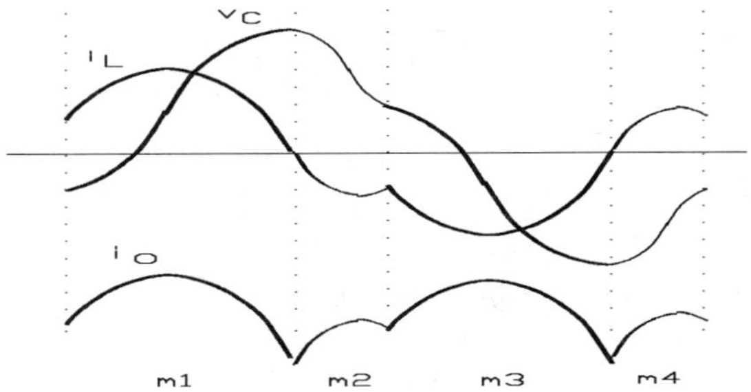

Figure 11.13
Waveforms in CCM operation for $0.5\omega_0 < \omega_s < \omega_0$

center: $[1-M, 0]$

radius

$$\Psi_1 = \sqrt{I_{nL}0^2 + (1 - M + V_{nC}0)^2}$$

conduction angle: $\theta_0 \leq \theta < \pi$

state equations:

$$i_{nL} = \Psi_1 \sin\theta \qquad (11\text{-}41)$$

$$v_{nC} = (1 - M) - \Psi_1 \cos\theta \qquad (11\text{-}42)$$

final conditions:

$$v_{nC}(\pi) = 1 - M + \Psi_1, \quad i_{nL}(\pi) = 0 \qquad (11\text{-}43)$$

Mode m2: initial conditions: $[1-M+\Psi_1, 0]$, conduction angle: $\pi \le \theta < \theta_2$

center: $[1+M, 0]$

radius

$$\Psi_2 = 1 - M + \Psi_1 - (1 + M) = \Psi_1 - 2M \quad (11\text{-}44)$$

state equations:

$$i_{nL} = \Psi_2 \sin\theta \quad (11\text{-}45)$$

$$v_{nC} = (1 + M) - \Psi_2 \cos\theta \quad (11\text{-}46)$$

final conditions:

$$v_{nC}(\theta_2) = V_{nCO}, \quad i_{nL}(\theta_2) = -I_{nLO}$$

$$V_{nCO} = 1 + M - \Psi_2 \cos\theta_2 \quad (11\text{-}47)$$

From equations (11-40) and (11-47)

$$\Psi_1 \cos\theta_0 + \Psi_2 \cos\theta_2 = 2 \quad (11\text{-}48)$$

The state-plane trajectory for mode m3 and mode m4 is the mirror image of the trajectory of mode m1 and mode m2. Complete trajectory is determined from the knowledge of the initial conditions I_{nLO} and V_{nCO}.

The half switching period T_S is calculated from the sum of conduction angles of the two modes:

$$\frac{T_S}{2} = \frac{\pi - \theta_0}{\omega_0} + \frac{\theta_2 - \pi}{\omega_0}$$

The switching frequency after normalization with respect to the resonant frequency is

$$f_{ns} = \frac{f_s}{f_0} = \frac{\pi}{\theta_2 - \theta_0} \quad (11\text{-}49)$$

Average output current is calculated as the following:

$$\bar{I}_{no} = \frac{f_{ns}}{\pi}\left[\int_{\theta_0}^{\pi} |i_{nL}|\, d\theta + \int_{\pi}^{\theta_2} |i_{nL}|\, d\theta\right] \quad (11\text{-}50)$$

$$= \frac{f_{ns}}{\pi}[\Psi_1(1 + \cos\theta_0) + \Psi_2(1 + \cos\theta_2)]$$

$$= \frac{f_{ns}}{\pi}[\Psi_1 + \Psi_2 + \Psi_1\cos\theta_0 + \Psi_2\cos\theta_2]$$

Using equations (11-44) and (11-48)

$$= \frac{f_{ns}}{\pi}[2\Psi_1 - 2M + 2] \quad (11\text{-}51)$$

Hence

$$\Psi_1 = \frac{\pi}{2}\frac{\bar{I}_{no}}{f_{ns}} + M - 1 \quad (11\text{-}52)$$

Average current from the source is calculated as the following:

$$\bar{I}_{ns} = \frac{f_{ns}}{\pi}\left[\int_{\theta_0}^{\pi} i_{nL}\, d\theta + \int_{\pi}^{\theta_2} i_{nL}\, d\theta\right] \quad (11\text{-}53)$$

$$\bar{I}_{ns} = \frac{f_{ns}}{\pi}[\Psi_1(1 + \cos\theta_0) - \Psi_2(1 + \cos\theta_2)]$$

Power drawn from the source equals the power delivered to the load,

$$V_S \bar{I}_S = \bar{V}_O \bar{I}_O$$

$$\bar{I}_{nS} = M \bar{I}_{nO} \quad (11\text{-}54)$$

Hence, the converter gain is obtained from equations (11-50), (11-53) and (11-54)

$$M = \frac{\Psi_1(1 + \cos\theta_0) - \Psi_2(1 + \cos\theta_2)}{\Psi_1(1 + \cos\theta_0) + \Psi_2(1 + \cos\theta_2)} \quad (11\text{-}55)$$

The maximum value of M is unity, which occurs for $\theta_2 = \pi$. Furthermore, it can be shown that by using equations (11-55), (11-44) and (11-48)

$$\theta_0 = \cos^{-1}\left[\frac{1 - M^2 + M\Psi_1}{\Psi_1}\right] \quad (11\text{-}56)$$

Thus, from the given initial conditions, all important parameters, the converter gain, the switching frequency and the normalized output current I_{no} can be determined. Conversely, the initial conditions can be determined from a given switching frequency, an output voltage and an output current. A numerical search technique must be employed to obtain both initial conditions. The MATLAB program to plot is given in the appendix. The program strategy is given below for a specified converter gain M and the switching frequency f_{ns}:

1. Assume the lower limit value on the output current to be the highest value of I_{no} obtained from the discontinuous mode at $f_{ns}=0.5$,

$$\bar{I}_{no_{old}} = \frac{4}{\pi} f_{ns}$$

2. Calculate Ψ_1 from equation (11-51)

$$\Psi_1 = \frac{\pi \bar{I}_{no_{old}}}{2 f_{ns}} + M - 1$$

3. Calculate θ_0 from equation (11-56).
4. Find θ_2 from equation (11-49).
5. Find the new value $I_{no\ new}$ from equations (11-55).
6. Use the new value of I_{no} to be the average of the old and new values and repeat step 1 until the old and new values are close within a small tolerance.

Thus, for a given M and f_{ns}, the MATLAB array is found for I_{no}.

These curves are called the control characteristics of the converter and are shown in Figure 11.14. In a similar manner, one could plot the converter gain M versus the switching frequency with Q as the parameter (see Problem 11-1). These curves are

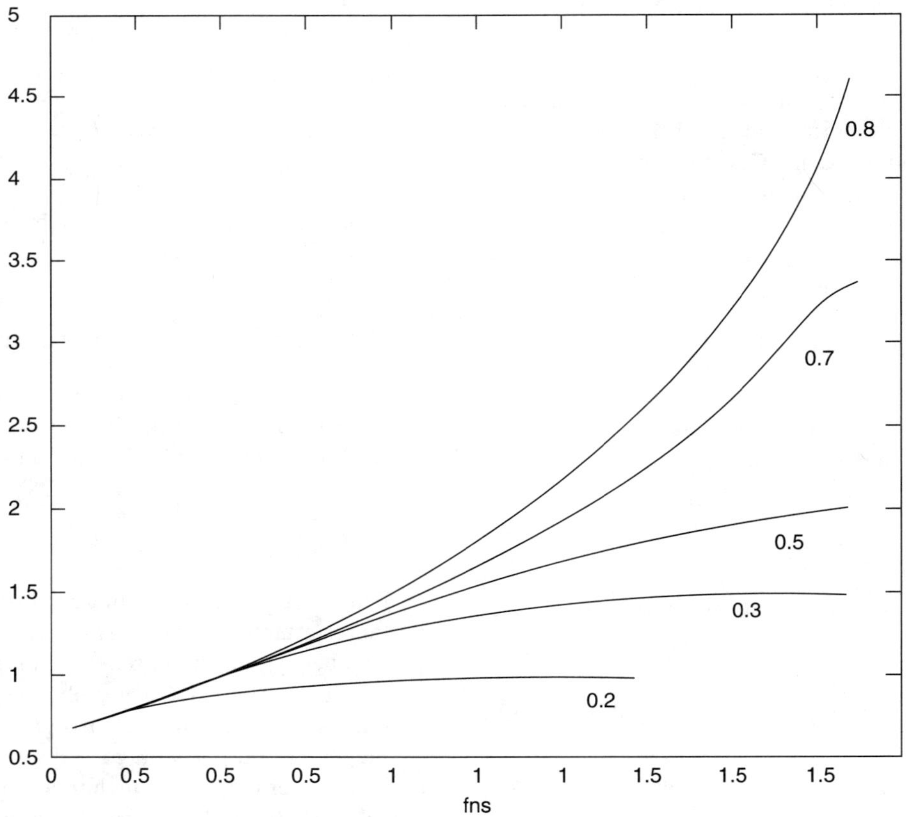

Figure 11.14
Normalized output current versus the normalized switching frequency

shown in Figure 11.15. The maximum value of M is unity, which occurs when the switching frequency is equal to the resonant frequency. This is also evident from the circuit of Figure 11.8 because the series resonant tank circuit (L in series with C) becomes short at the resonant frequency.

Example 11-5

Parameters for the series resonant converter of Figure 11.8(a) are

$L = 1\ mH,\ C = 1\ nF,\ R_L = 200\ ohms,$
$V_g = 90\ Vdc,\ f_{ns} = 0.8.$

a. Find the dc output voltage V_o.

b. Draw the waveforms for the current into and the voltage on the voltage sink.

a. The characteristic impedance

$$Z_o = \sqrt{\frac{L}{C}} = \sqrt{\frac{10^{-3}}{10^{-9}}} = 1\ KHz$$

The load factor

$$Q = \frac{R_L}{Z_o} = 0.2$$

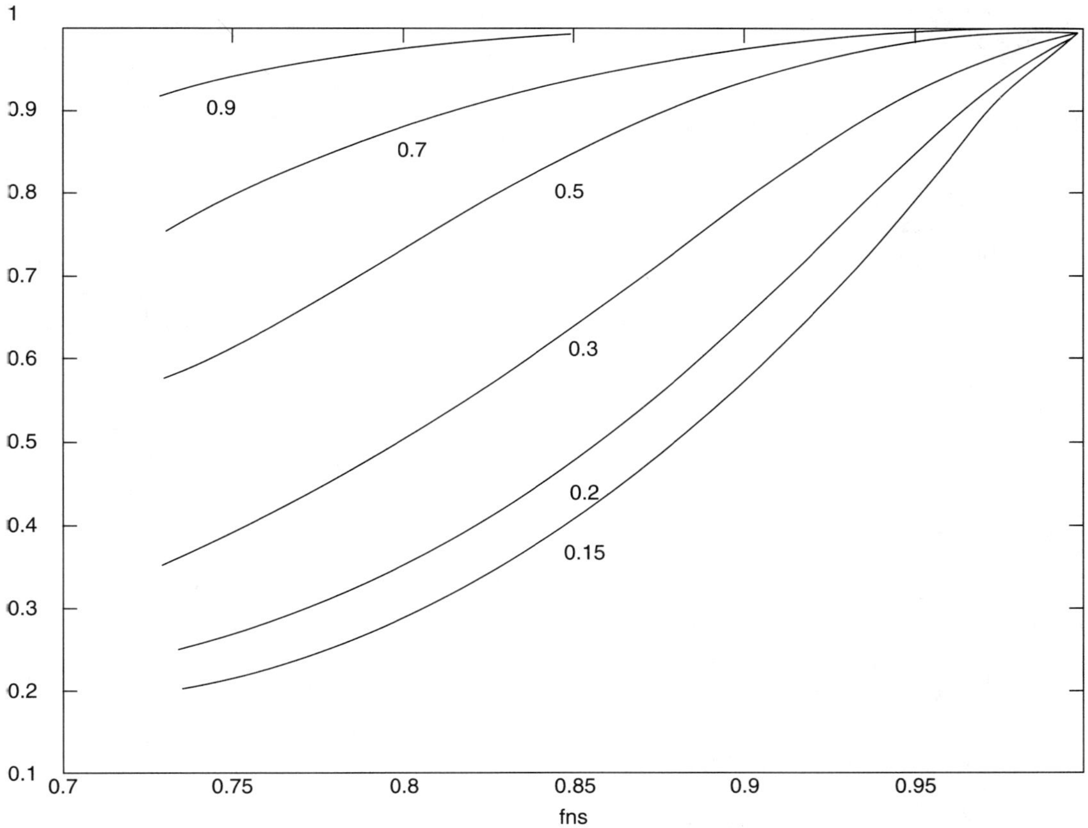

Figure 11.15
Converter gain versus the normalized switching frequency

```
a=0; Q=.2; fns=.8;
Mold=fns*4*Q/pi+.01; Mnew=.3;
while (abs(Mnew-Mold)>= 0.0001)
  psi1=Mold*pi/(2*Q*fns)+Mold-1;
  w1=(1-Mold^2+Mold*psi1)/psi1;
  if (abs(w1)<=1)
    if (w1 >=0)
      theta0=acos(w1);
    else
      theta0=pi-acos(abs(w1));
    end
    theta2=theta0+pi/fns;
    if(theta2>2*pi)
      a=1;
      break;
    end
  else
    a=1;
    break;
  end
  if(a==1)
    a=0;
    M=[ ];
    break;
  else
    psi2=psi1-2*Mold;
      Mnew=(psi1*(1+cos(theta0))-
        psi2*(1+cos(theta2)))/(psi1*
        (1+cos(theta0))+psi2*
        (1+cos(theta2)));
      Mold=(Mnew+Mold)/2;
    end
  end
M=Mold;
if(a==1)
  fns=[ ];
  M=[ ];
end
fns
Ino=M/Q
M
```

b. Add the following lines to the MATLAB program in part (a) for the state-plane diagram and the waveforms.

```
dth=pi/100; Vnc0=1+M-psi2*cos(theta2);
  inl0=psi2*sin(theta2);
thr1=theta0:dth:pi; thr2=pi:dth:theta2;
inl1=psi1*sin(thr1);
vnc1=(1-M)-psi1*cos(thr1);
psi2=psi1-2*M;
inl2=psi2*sin(thr2);
vnc2=(1+M)-psi2*cos(thr2);

thr=[thr1 thr2 ];
ths=thr*pi/(theta2-theta0);
inl=[inl1 inl2 ];
vnc=[vnc1 vnc2 ];

ths_f=[ths pi+ths];
inl_f=[inl - inl];
vnc_f=[vnc - vnc];
plot(ths_f, inl_f, 'r', ths_f, vnc_f, 'b')
plot(vnc_f,inl_f)
```

Operation Above Resonance

The switching frequency f_s is higher than the resonant frequency f_0. The converter operation consists of four circuit modes: two modes in each half switching period. Current flow in each mode is shown by thick lines in the circuit of Figure 11.17(a). Simplified equivalent circuits of each mode are shown in Figure 11.17(b). Normalization of sources and state variables is identical to that in the previous case. The steady-state trajectory begins at the initial point $[-V_{nLO}, -I_{nLO}]$, at an angle θ_0 with respect to the center $[1+M, 0]$. Complete state-plane trajectory is given in Figure 11.17(c).

Mode m1:

initial conditions: $[-V_{nCO}, -I_{nLO}]$

$$-V_{nC}0 = 1 + M - \Psi_1 \cos\theta_0 \qquad (11\text{-}57)$$

center: $[1+M, 0]$

radius

$$\Psi_1 = \sqrt{I_{nL}0^2 + (1 + M + V_{nC}0)^2} \qquad (11\text{-}58)$$

conduction angle: $-\theta_0 \leq \theta < 0$

State equations:

$$i_{nL} = \Psi_1 \sin\theta \qquad (11\text{-}59)$$

$$v_{nC} = (1 + M) - \Psi_1 \cos\theta \qquad (11\text{-}60)$$

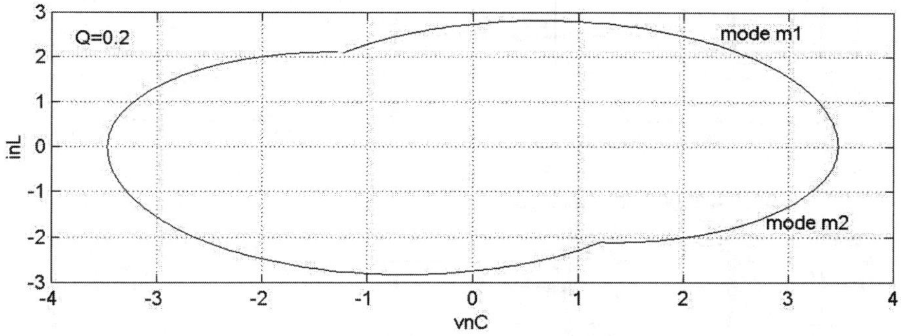

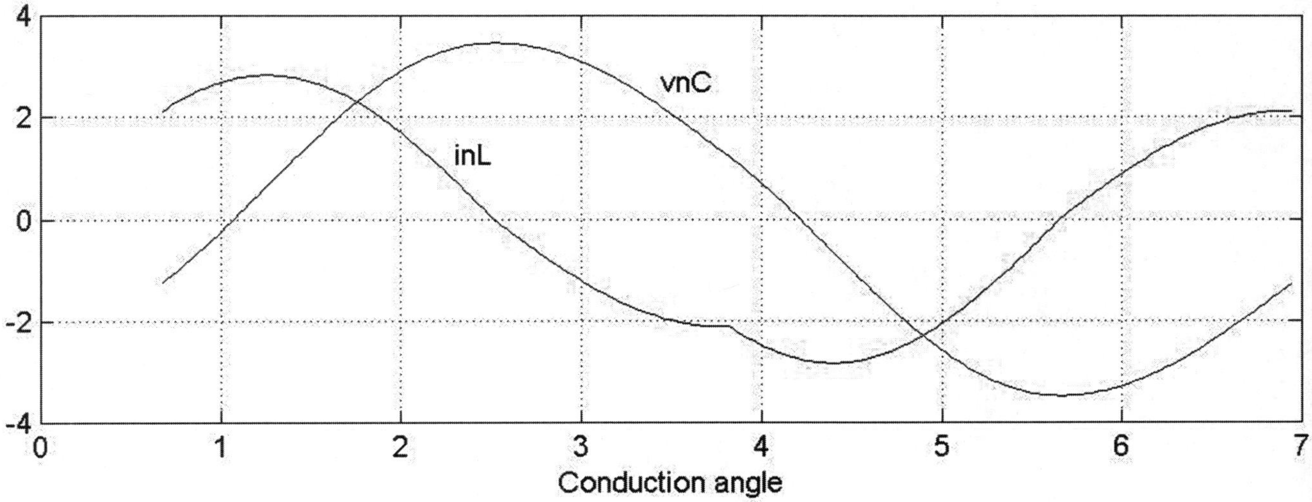

Figure 11.16
Series resonant converter in CCM below-resonance operation

final conditions:

$$v_{nC}(0) = 1 + M - \Psi_1, \; i_{nL}(0) = 0 \quad (11\text{-}61)$$

Mode m2:
initial conditions: $[1+M-\Psi_1, 0]$, conduction angle: $0 \le \theta < \theta_2$

center: $[1-M, 0]$
radius

$$\Psi_2 = \Psi_1 - (1 + M) + (1 - M) = \Psi_1 - 2M \quad (11\text{-}62)$$

$$i_{nL} = \Psi_2 \sin\theta \quad (11\text{-}63)$$

$$v_{nC} = (1 - M) - \Psi_2 \cos\theta \quad (11\text{-}64)$$

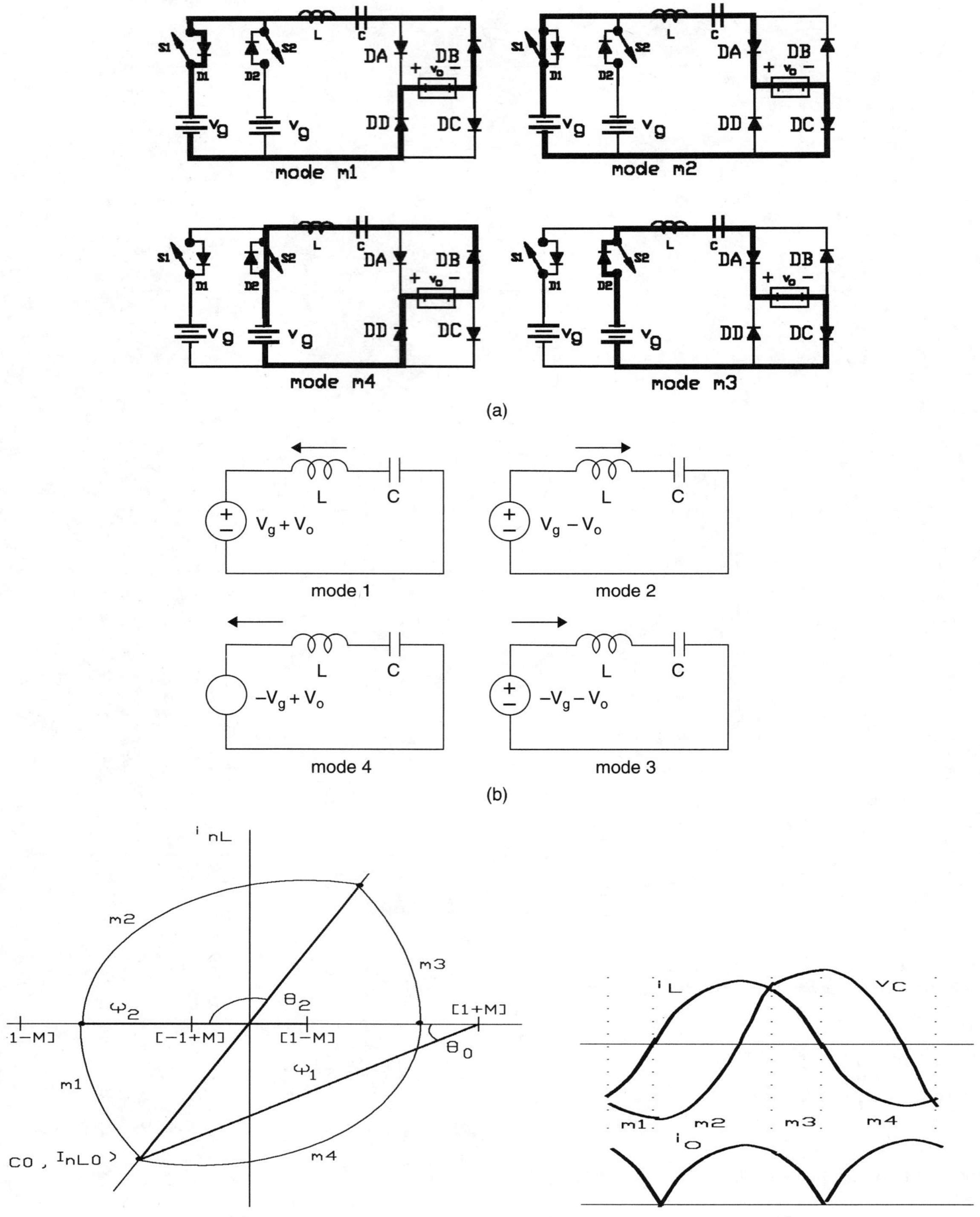

Figure 11.17
CCM operation for $\omega_s > \omega_0$ (a) Circuit modes (b) simplified circuit modes (c) state-plane diagram and (d) waveforms

final conditions:

$v_{nC}(\theta_2) = V_{nCo}, \quad i_{nL}(\theta_2) = I_{nLo}$

$$V_{nC}0 = 1 - M - \Psi_2 \cos\theta_2 \qquad (11\text{-}65)$$

Hence from equations (11-57) and (11-65)

$$\Psi_1 \cos\theta_0 + \Psi_2 \cos\theta_2 = 2 \qquad (11\text{-}66)$$

State-plane trajectory for mode 3 and mode 4 is the mirror image of the trajectory of mode m1 and mode m2.

Half switching period $T_S/2$ is calculated from the conduction angles of two modes

$$\frac{T_S}{2} = \frac{\theta_0}{\omega_0} + \frac{\theta_2}{\omega_0} \qquad (11\text{-}67)$$

Switching frequency after normalization with respect to the resonant frequency is

$$f_{ns} = \frac{f_s}{f_0} = \frac{\pi}{\theta_0 + \theta_2} \qquad (11\text{-}68)$$

Average output current is calculated as the following:

$$\bar{I}_{no} = \frac{f_{ns}}{\pi}\left[\int_{-\theta_0}^{0} |i_{nL}|\, d\theta + \int_{0}^{\theta_2} i_{nL}\, d\theta\right] \qquad (11\text{-}69)$$

$$= \frac{f_{ns}}{\pi}[\Psi_1(1 - \cos\theta_0) + \Psi_2(1 - \cos\theta_2)]$$

Using equations (11-62) and (11-69)

$$\Psi_1 = \frac{\pi\bar{I}_{no}}{2f_{ns}} + M + 1 \qquad (11\text{-}69a)$$

Average current from the source is calculated as the following:

$$\bar{I}_{ns} = \frac{f_{ns}}{\pi}\left[\int_{-\theta_0}^{0} i_{nL}\, d\theta + \int_{0}^{\theta_2} i_{nL}\, d\theta\right] \qquad (11\text{-}70)$$

$$= \frac{f_{ns}}{\pi}[-\Psi_1(1 - \cos\theta_0) + \Psi_2(1 - \cos\theta_2)]$$

Converter gain M is

$$M = \frac{-\Psi_1(1 - \cos\theta_0) + \Psi_2(1 - \cos\theta_2)}{\Psi_1(1 - \cos\theta_0) + \Psi_2(1 - \cos\theta_2)} \qquad (11\text{-}71)$$

The maximum value of M is unity and occurs at $\theta_0 = 0$.

A plot of output dc current I_{no} versus switching frequency f_{ns} with converter gain M as the parameter is given in Figure 11.18.

In a similar manner, one could plot the converter gain M versus the switching frequency with I_{no} as the parameter (see Problem 11-2). These curves are shown in Figure 11.19.

The series resonant converter behaves like a current source. For switching frequencies less than one-half resonant frequency in the discontinuous conduction mode of operation, it appears as an ideal current source, independent of the load resistance and only dependent on the switching frequency. For other switching frequencies, the current source is load dependent. Furthermore, being essentially a current source, this converter is not suitable for multi-output configurations. Voltage regulation is less than desirable.

It is interesting to note that the converter gain M is unity for all load currents when the converter is switched at the resonant frequency. It is also evident from the converter circuit of Figure 11.8 where the series resonant tank (L in series with C) becomes 'short' at the resonant frequency. The average voltage on the output sink must equal the average voltage of the source.

The maximum switch current (which is also the peak value of the inductor current) is a function of the switching frequency and can be several times higher than the dc output current I_o. Hence, the switches are subject to high current stresses and high conducting losses.

The series resonant converter displays very poor load regulation except when the switching frequency is very close to the resonant frequency. At light loads, the converter must be operated in the discontinuous conduction mode to retain control. If both the load and the input voltage vary, the range of switching frequency can be large.

With variable switching frequency control, the sign of the feedback loop gain changes as the switching frequencies pass through the resonant frequency f_0. Hence, in order to keep negative feedback, it is necessary to operate on one side or the other of f_0.

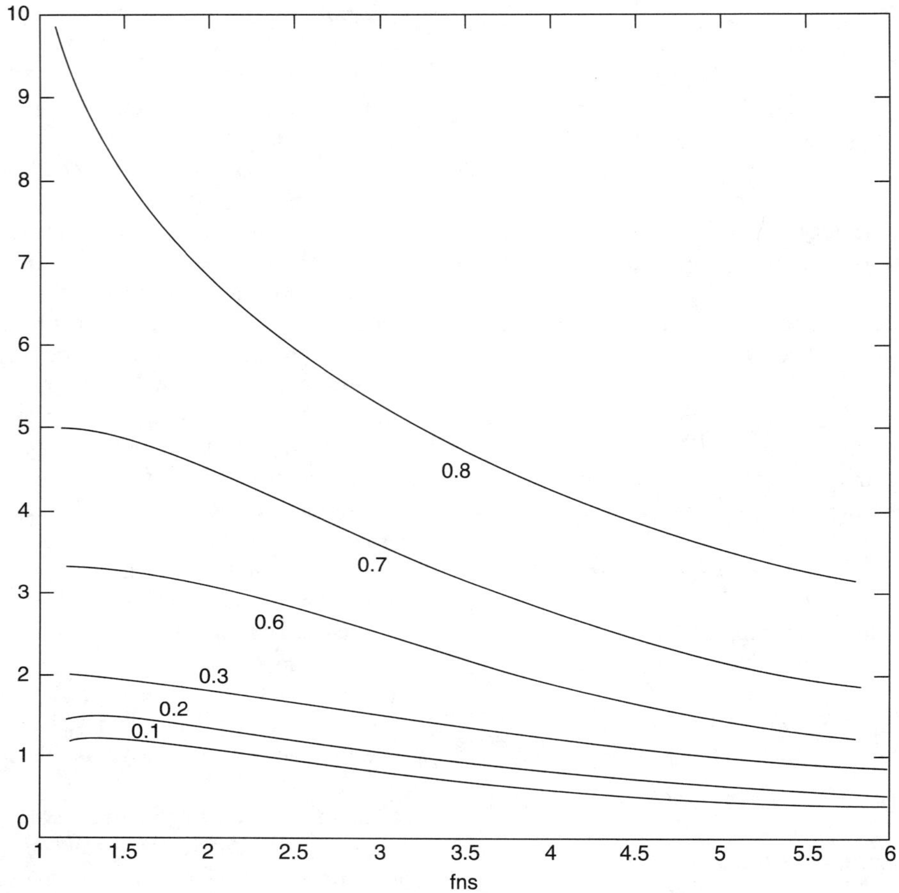

Figure 11.18
Normalized output current versus the normalized switching frequency

11.6 PARALLEL RESONANT DC-DC CONVERTER (PRC) (VOLTAGE COUPLED RESONANT CONVERTER)

Like the series resonant converters, the resonant tank circuit serves as the link between the voltage source-switch network and the current sink-diode network. Current sink is coupled to the voltage across the capacitor or the inductor of the resonant tank circuit. We will analyze a half-bridge series resonant converter, shown in Figure 11.20(a) to illustrate the procedure. Two dc voltage sources V_g can be obtained from the capacitor division of a single dc voltage source as shown in Figure 11.8(b). Current sink is the equivalent model of a high value inductor in series with load resistance. Switches S_1 and S_2 are turned on and off for an equal amount of time exclusively. The two dc voltage sources and two switches along with the diodes and the said switching scheme can be modeled by a square-wave source V_S as shown in the equivalent circuit of Figure 11.20(b). The current sink and the diode network can be mod-

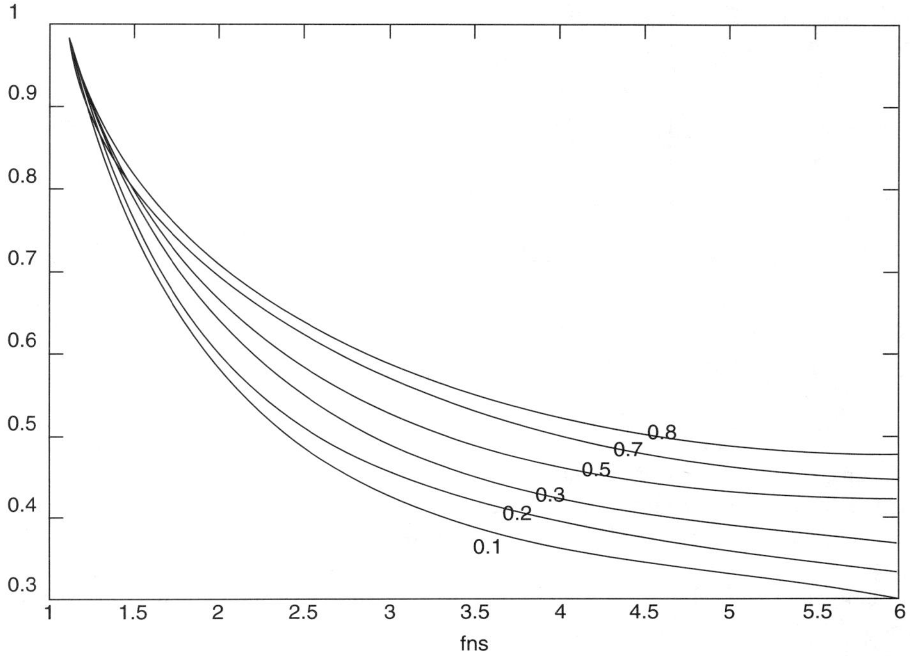

Figure 11.19
Converter gain versus the normalized switching frequency

eled by a three-state current sink I_k [$+I_o$, $-I_o$, short]: $+I_o$ for positive voltage across the current sink and $-I_o$ for negative voltage and 'short' when the external current entering the diode-current sink network is less than I_o in magnitude (I_o freewheels through the diodes presenting a short-circuit condition).

We will analyze only the steady-state response of the converter. Based on the waveshape of the current drawn from the voltage source, there are two possible modes of operation in steady-state:

Discontinuous Conduction Mode (DCM) in which the current drawn from the source becomes discontinuous in time (sometimes zero).

Continuous Conduction Mode (CCM) in which the current drawn from the source is continuous in time.

The state-plane analysis of the steady-state operation of parallel resonant converters are performed by mode-wise analysis on the same lines as was shown in previous sections. Hence, in the interest of space in the following sections, we will present only the mode-wise converter operation without analysis. We suggest that readers analyze the above themselves and check with the analysis given in the literature.

11.6.1 Discontinuous conduction mode

Switches are turned on and off by a square wave of switching frequency f_s (period=T_s), which is smaller than the resonant frequency f_0. Such as in the case of the series resonant converter, the final conditions at the end of the half switching period must be equal in amplitude but opposite in polarity to the initial conditions. Hence, it will suffice to analyze the response in a half switching period. In a

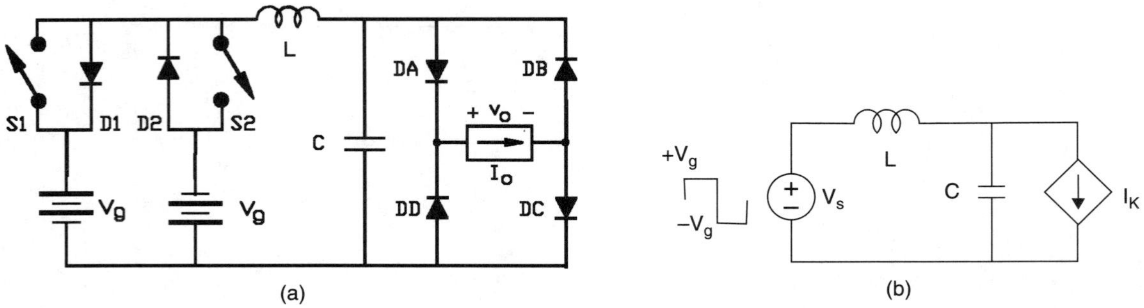

Figure 11.20
(a) Parallel resonant converter (b) Equivalent circuit model

half switching period, the converter operation consists of four circuit modes. Inductor current i_L and capacitor voltage v_L are the two state variables. The current path in each mode is shown by the thick lines in the circuits of Figure 11.21(a). Simplified equivalent circuits of each mode are shown in Figure 11.21(b). The final conditions of one circuit mode serve as the initial conditions of the next mode in the sequence.

The first circuit mode m1 begins with zero inductor current and zero capacitor voltage. The switch S_1 is turned on. The voltage source in the simplified equivalent circuit is V_g. Since the inductor current is less than the current sink amplitude I_o, the sink current freewheels through D_A, D_B, D_C and D_D. This condition maintains zero voltage on the capacitor and the inductor current ramps up linearly until a value of I_o.

In mode m2, the inductor current becomes more than the sink current amplitude, hence the freewheeling is stopped. The clamp on the capacitor is released. The inductor now resonates with the capacitor. The inductor current attains peak value and begins to drop. When the current becomes negative, it is steered from the switch S_1 to diode D_1. Switch S_1 thus commutates naturally (turns off at zero current). Switch S_1 can thus be turned off at any time after instant t_2. The negative inductor current keeps on flowing through diode D_1. Eventually the inductor current becomes zero and it cannot turn back positive because the switch S_1 has been turned off. Hence the current remains zero and the mode m2 ends and m3 begins. In the mode m3, the inductor current remains zero. The sink current flows through the capacitor discharging it linearly to zero voltage. Mode m3 also ends at this point. In the next mode m4, the sink current freewheels through all four diodes D_A-D_D and clamps the capacitor voltage to zero.

The steady-state trajectory on the normalized state-plane is given in Figure 11.21(c).

The current and voltage waveforms as functions of time can be deduced from the state-plane and are shown in Figure 11.21(d). The duration of modes m3 and m4 combined together is the duration for which the current from the voltage source (which is also the inductor current) is discontinuous (equals zero). This duration is also used as the control parameter because it controls the average value of the voltage waveform on the capacitor, which is also the dc output voltage of the converter. The combined duration of modes m1 and m2 approximately equals one period of the resonant frequency. Hence the discontinuous mode of operation is possible for half resonant frequency and lower. Switches are turned on and off at zero current and zero voltage. The maximum current through the switch is $i_{Lmax}=I_o+V_g/Z_0$. The maximum voltage on the capacitor is also the maximum reverse voltage on the rectifier diodes D_A-D_D and equals twice the input source voltage $2V_g$.

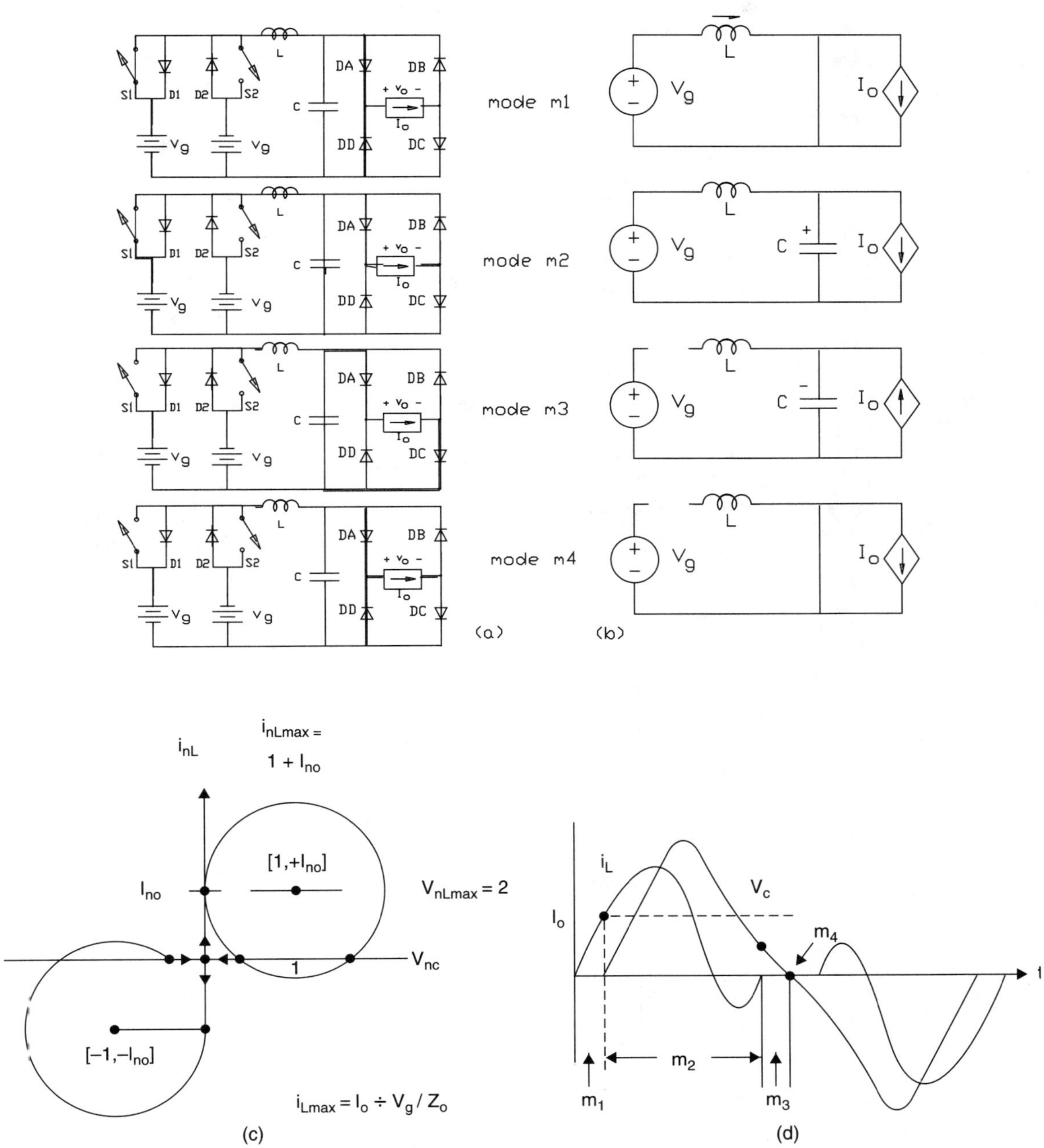

Figure 11.21
Parallel resonant converter in discontinuous conduction mode
(a) Circuit modes (b) Simplified circuit modes (c) State-plane diagram (d) Waveforms in a single switching period

11.6.2 Continuous conduction mode

Operation Below Resonance

Switches are turned on and off by a square wave of switching frequency f_s (period$=T_s$), which is higher than half resonant frequency but smaller than the resonant frequency f_0. The converter operation consists of four circuit modes: two modes in each half switching period. The current flow in each mode is shown by the thick lines in the circuits of Figure 11.22(a). The second and fourth modes are shown as two parts each only to show the steering of the current from the switch to its parallel diode. From a functional viewpoint, modes 2 and 2a represent one circuit mode. Simplified equivalent circuits of each mode are shown in Figure 11.22(b). The steady-state trajectory begins at the initial point $[-V_{nL0}, I_{nL0}]$, at an angle θ_0 with respect to the center $[1, -I_{no}]$. The subscript n refers to normalization similar to the one used in the series resonant converter. The radius vector Ψ_1 rotates in the clockwise direction as time elapses at an angular speed ω_0. The circular segment ends when the changes in the circuit conditions lead to another circuit mode in the mode sequence. The angle subtended by the circular segment is called the conduction angle of the circuit mode.

The steady-state trajectory on the normalized state-plane for a half switching period is shown in Figure 11.22(c). The normalization is similar to the one used in the series resonant converter.

Switching

Switches are turned on at nonzero current and nonzero voltage but turned off at zero current and zero voltage due to natural commutation of current from switches to the parallel diodes. These diodes are, however, turned off at nonzero current, hence, they must have good reverse recovery characteristics. Due to zero current turn-off characteristics,

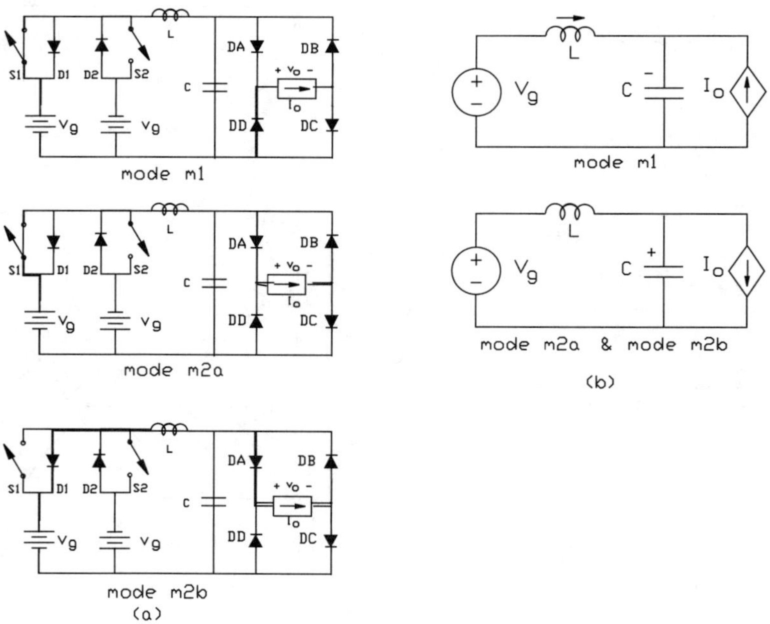

Figure 11.22
CCM operation for $0.5\omega_0<\omega_s<\omega_0$
(a) Circuit modes (b) simplified circuit modes

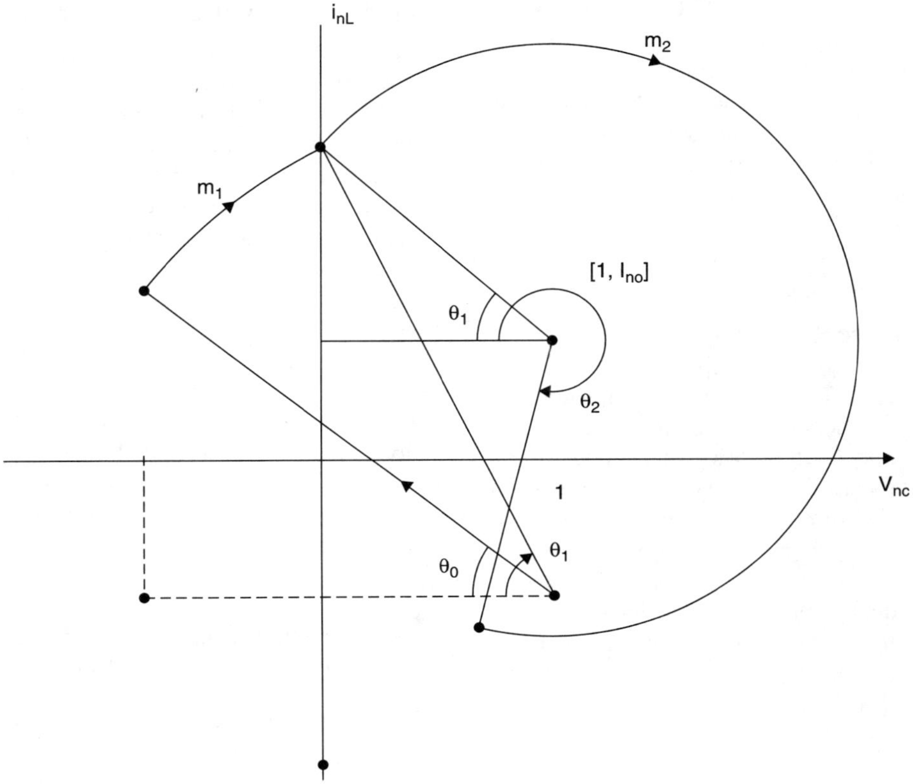

(c)

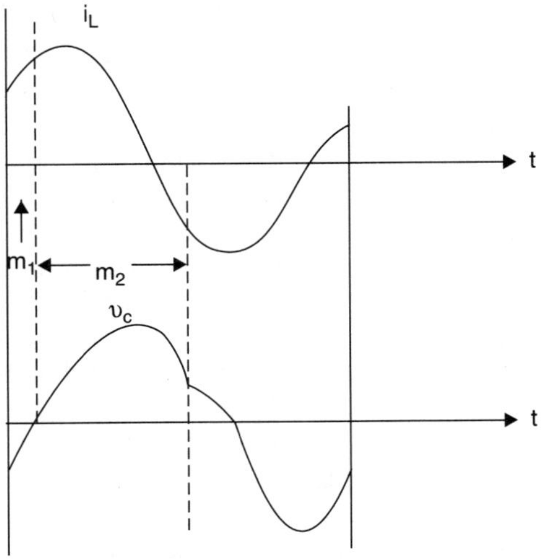

(d)

Figure 11.22 *(continued)*
CCM operation for $0.5\omega_0 < \omega_s < \omega_0$
(c) state-plane diagram and (d) waveforms in a single switching period

bipolar transistors, thyristors and GTOs would serve as the ideal switches. Maximum current through the switch is a function of the switching frequency (initial states in mode m1 and hence the trajectory is unique for a switching frequency). Thus the maximum switch current can be several times higher than the dc output current. Similarly the maximum voltage on the capacitor (also the maximum reverse voltage on the rectifier diodes D_A-D_D) can be several times higher than the input source voltage.

Operation Above Resonance

The switching frequency f_s is higher than the resonant frequency f_0. The converter operation consists of four circuit modes: two modes in each half switching period. Current flow in each mode over a half switching period is shown by the thick lines in the circuits of Figure 11.23(a). Simplified equivalent circuits of each mode are shown in Figure 11.23(b). Normalization of sources and state variables is identical to that in the previous case. The steady-state trajectory begins at the initial point $[-V_{nL0}, -I_{nL0}]$, at an angle θ_0 with respect to the center $[1, -I_{no}]$. State-plane trajectory for a half switching period is given in Figure 11.22(c).

Switching

Switches are turned on at zero current and zero voltage due to natural commutation of the current from the parallel diodes to the switches. The parallel diodes turn off at zero current, hence, they need not have very good reverse recovery characteristics. Switches turn off at nonzero current and nonzero voltage. Turn-off losses can be reduced by using snubbers in parallel with the switches. Since the switch turns on at zero voltage, there is no stored charge on the snubber capacitor, hence no resistor is needed in the snubber. Due to zero voltage turn on characteristics, MOSFETs are ideal switches.

As in the below resonance CCM operation, the maximum current through the switch is a function of the switching frequency and can be several times higher than the dc output current. Similarly the maximum voltage on the capacitor (also the maximum reverse voltage on the rectifier diodes D_A-D_D) can be several times higher than the input source voltage.

11.6.3 Control characteristics

The control characteristics of all three combined modes of operation are given in Figure 11.24. In the DCM operation, the converter output exhibits ideal voltage source characteristics; the output voltage is independent of the load current I_o (ideal load regulation) and is the function of the switching frequency alone. The characteristics also allow multiple outputs with good load regulation.

The CCM operation exhibits high slope of the M versus the switching frequency curves that signify a relatively small frequency change to control a large change of load current.

It is possible to have converter gains less and more than one. Hence, both the step-down and step-up voltages are achievable. Maximum output current is limited to V_g/Z_0 and is achieved only at the resonance.

11.7 HIGH-ORDER RESONANT DC-DC CONVERTER

The resonant link could also be made of more than two energy storage elements. For example a third order resonant tank may be comprised of either two inductors and one capacitor or two capacitors and one inductor. The high-order resonant tank consists of two or more resonant frequencies, thus enabling a close control of linking properties vis. a vis. switching frequency. Inclusion of some important parasitics of electronic components, such as the lead inductance and the junction capacitance of a diode, may turn an otherwise two element resonant topology into a high-order resonant topology.

An example of a third order LCC resonant converter is shown in Figure 11.25. The square-wave voltage source is the model of two dc voltage sources and two switches along with the parallel diodes and the square-wave switching scheme. The multistate voltage sink V_k models the load resistance, the filter capacitor and the rectifier diode network.

The converter control characteristics can be understood from the plot of the impedance of the three-element resonant tank versus frequency given

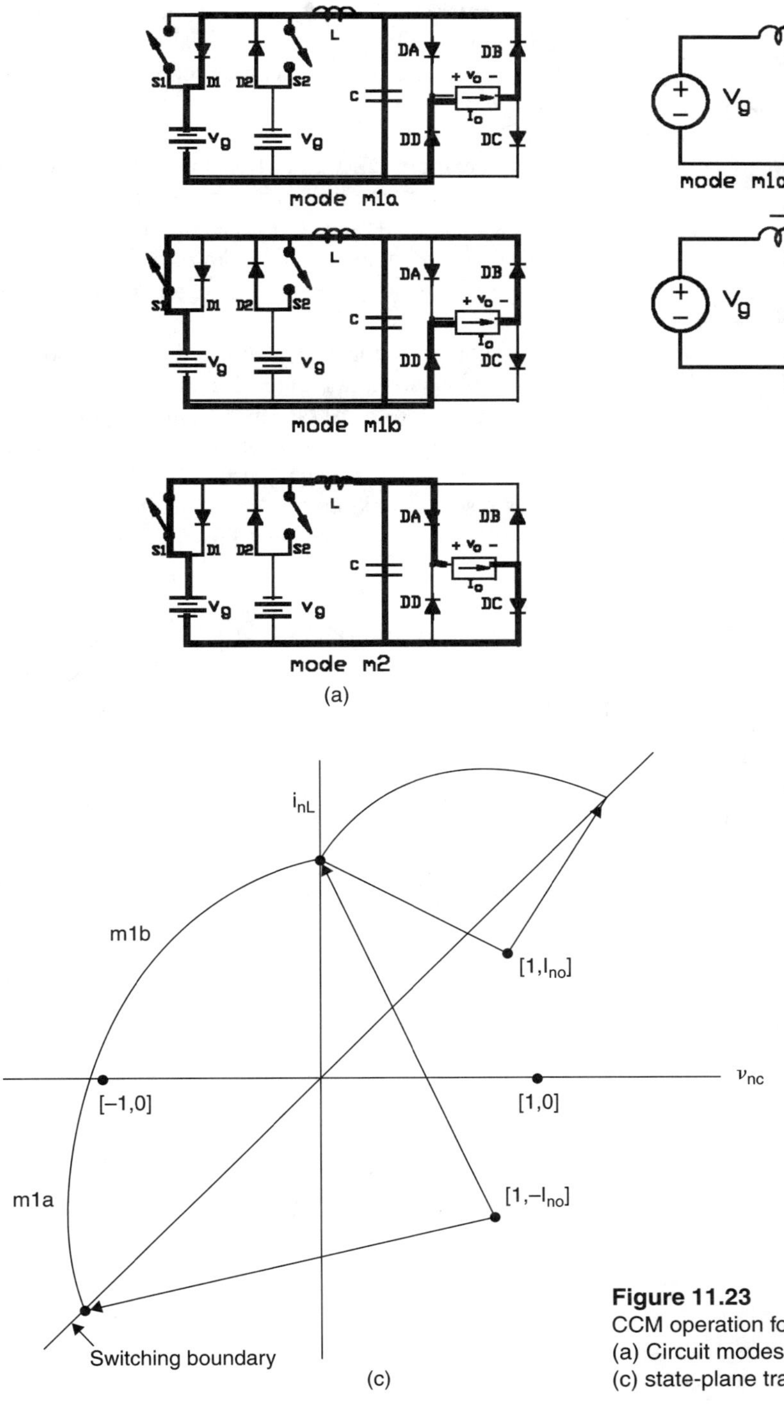

Figure 11.23
CCM operation for $\omega_s > \omega_0$
(a) Circuit modes (b) simplified circuit modes and
(c) state-plane trajectory

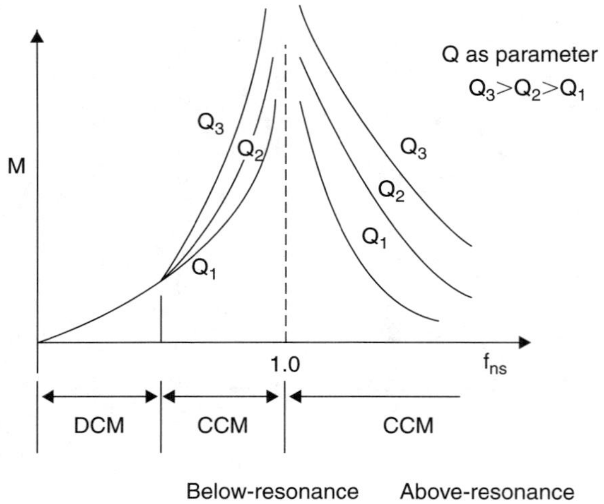

Figure 11.24
Combined control characteristics of parallel resonant converter

in Figure 11.3. It has two resonant frequencies given in Section 11.3.3. If C_1 is large, above topology is reduced to a two-element series-link resonant converter.

The resonant link is short at ω_z, hence the converter exhibits unity gain. The converter gain is zero at ω_0 because the link impedance is infinite. Thus, for a very small range of frequencies from ω_z to ω_0, the converter gain can vary from zero to unity for all values of load currents. These are ideal control characteristics.

Another example of a third order LCC resonant converter, shown in Figure 11.26, that combines advantages of both the current coupling of the series resonant link and the voltage coupling of the parallel resonant link: inherent current limiting of the series resonant link and the excellent voltage regulation of the parallel resonant link. If C_1 is large, the

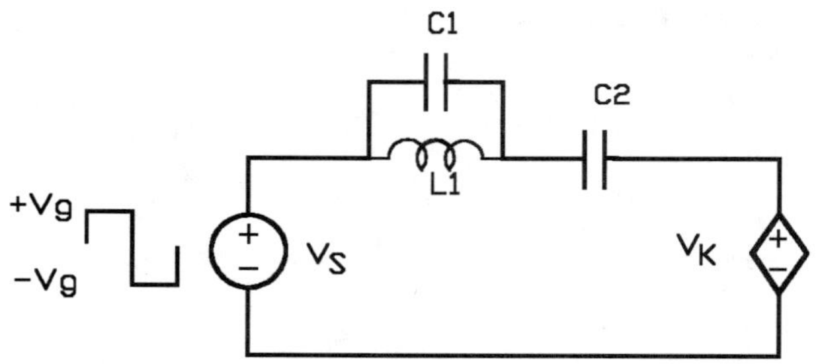

Figure 11.25
A resonant converter with a third order LCC resonant link

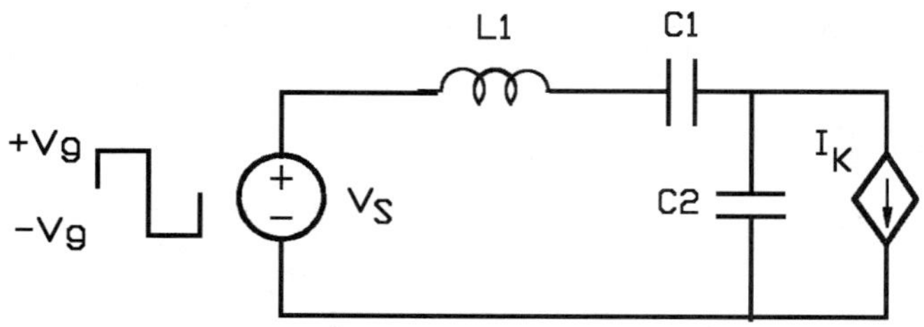

Figure 11.26
A third order LCC resonant converter that combines advantages of both the current coupling of the series resonant link and voltage coupling of the parallel resonant link.

above topology is reduced to a two-element parallel-link resonant converter. Excellent load regulation is obtained at above-resonance operation.

11.8 CLASS E CONVERTERS

The class E converter is a single switch topology that has minimum switching losses. The topology consists of a current source, a single switch with a capacitor in parallel, a series resonant link and a load resistance in case of an inverter or a voltage load in case of a dc-dc converter. This is essentially a third order resonant converter. An example of a class E dc-dc converter is given in Figure 11.27(a). The topology without the diode across the switch is termed as the optimum and with the diode, suboptimum.

The optimum switching scheme is described as follows:

In the time interval during which the switch makes its transition from the on state to the off state, the voltage across the switch remains zero for a time long enough that the current through the switch has been reduced substantially to zero. Then the voltage increases. This assures that high voltage does not appear across the switch while the current through it is nonzero.

During the off state, the voltage across the switch returns to zero prior to the start of the on state, that is, just before the current begins to flow in the switch. This avoids the energy dissipation, which would have occurred if the switch current had begun flowing while the switch voltage was still high. The switch voltage in the off state returns to zero with an approximately zero slope dv/dt. This permits a slight interval prior to turning the switch on in which the voltage across the switch is maintained at zero voltage. All these characteristics lead to nearly zero switching losses.

The optimum operation occurs only for an optimum value of load current. Hence, high efficiency is realized only over a limited range in the vicinity of the optimum value of the load current. Suboptimum converter operation is possible for a larger range of load currents above the optimum value. In this suboptimum operation, the switch voltage returns to zero value (prior to being turned on) at a negative slope dv/dt. This means that the current in the capacitor C_1 (C_1dv/dt) is nonzero and negative. Connecting a diode in antiparallel with the switch, as shown in Figure 11.27(a), permits this negative current to flow, while maintaining a zero voltage across the switch. The switch must be turned on only after the antiparallel diode starts to conduct.

The class E converter is operated at a slightly higher frequency than the resonant frequency of the series resonant link. Steady-state operation consists of four modes as shown in Figure 11.27(b).

Output current is approximately sinusoidal due to the series resonant link. Capacitor C_1 in parallel with the switch allows zero voltage turn off. As the switch is turned off, the capacitor voltage V_{c1} begins to rise at zero slope, builds up to a positive peak and then turns back to zero. The switch is turned on again at the instant v_{c1} comes back to zero. Switching losses are eliminated.

The disadvantages of class E converters are high peak voltage and high peak current through the switch, the inductors and the capacitors.

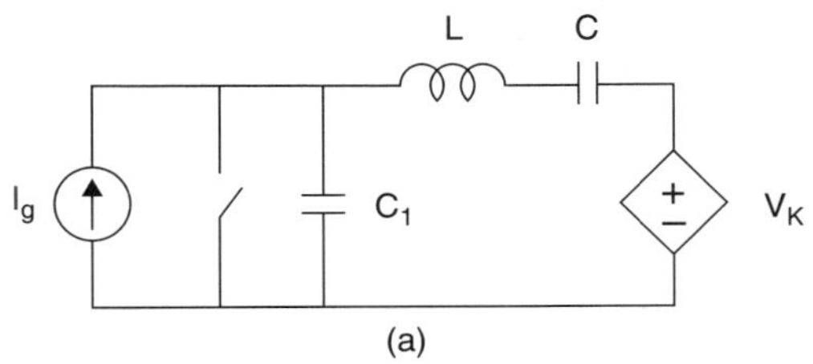

Figure 11.27
(a) Class E converter

11.9 RESONANT SWITCH CONVERTERS

These converters consist of switches with energy storage elements that enable zero-voltage switching and zero-current switching; both turn on and turn off. An inductor in series with the switch will enable

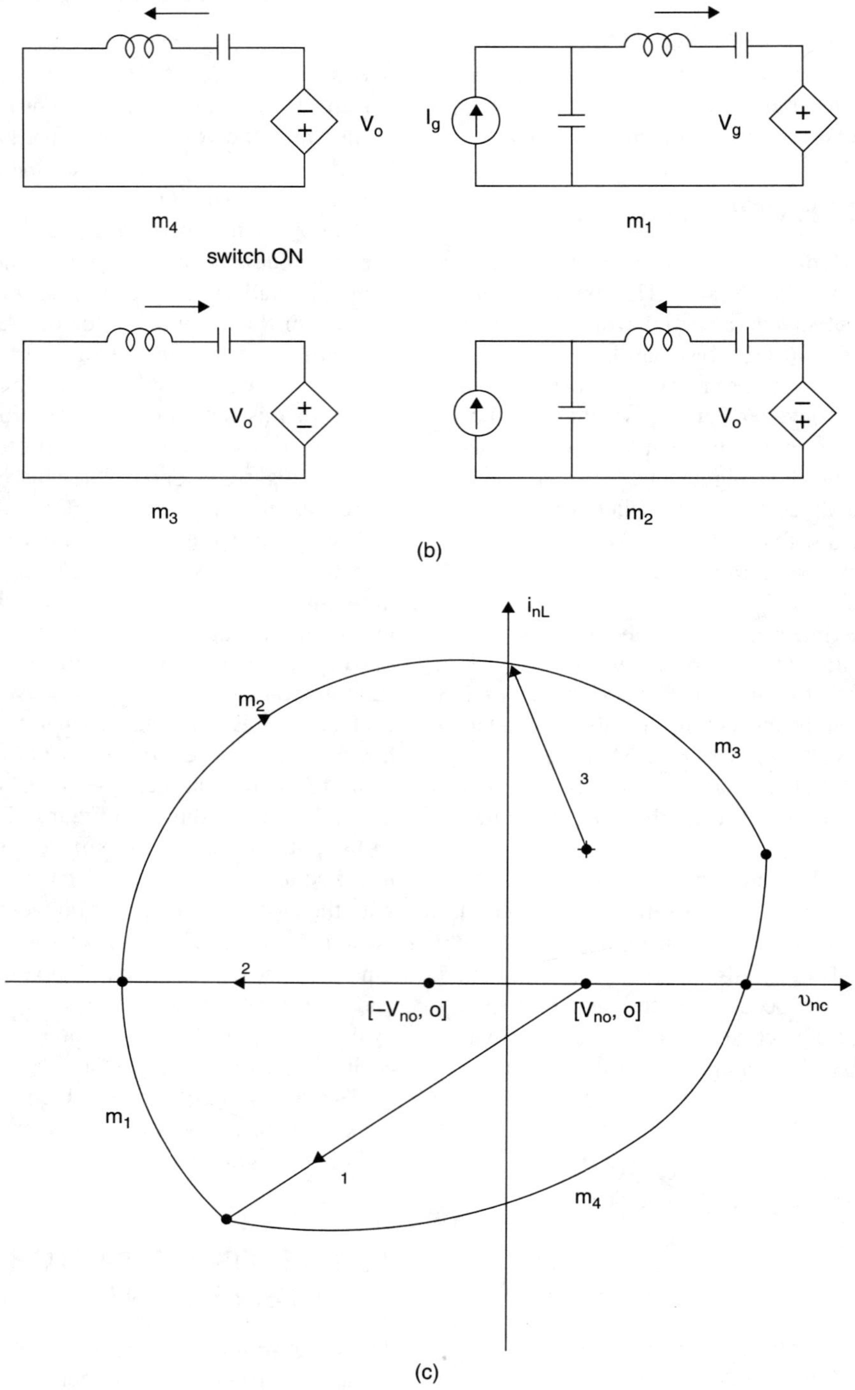

Figure 11.27 ***(continued)***
(b) Circuit modes in the steady-state operation (c) State-plane diagram

zero-current switching. A capacitor in parallel with the switch will enable zero-voltage switching. Figure 11.28 shows how the energy storage elements can be added to a switch to accomplish this objective.

Switching losses can thus be eliminated, which becomes more and more crucial as the converter switching frequencies are pushed higher. Converter control employs either variable frequency control or pulse width/phase control at constant switching frequency operations. In the following section, some illustrative topologies are discussed.

11.9.1 Zero-current switching resonant converters

The dc-dc Buck converter of Chapter 10 can be modified to include the energy storage elements L and C in the topology as shown in Figure 11.29(a). The inductor in series with the switch may represent or include the stray lead inductance. The capacitor in parallel with the switch may represent or include the stray capacitance across the switch. There are four circuit modes in the steady-state converter operation, given in Figure 11.29(b).

Mode m1 begins with the turning on of switch S while the load current is freewheeling through the diode because the current entering into the current sink and the diode combination is less than I_o. The switch turns on at zero current. As a consequence of the switch turning on, the inductor is thus connected directly across the input voltage source. Inductor current, therefore, ramps up while the capacitor is clamped to full source voltage. This mode is represented by a vertical line segment on the state-plane in Figure 11.29(c).

As the inductor current reaches I_o, the diode clamp on the current sink is removed. The inductor resonates with the capacitor as shown in mode m2. The inductor current rises further and then comes back to zero sinusoidally. The surge of inductor current to negative values is arrested by the unidirectional switch. The switch is thus naturally commutated (turns off at zero current). New mode m3 begins in which the current sink gets its current from the input dc source through the capacitor. Mode m3 continues until the capacitor is charged to full input source voltage V_g and stops charging further. Current through the capacitor becomes zero, hence, the current sink is constrained to freewheel through the diode. This is the mode m4 of the converter operation, indicated by a single point (1,0). Mode m4 exists until the switch is turned back on. It should be noted that the switch turns on and off at zero current. Switching losses are eliminated. The peak value of the switch current, which is also the inductor current, is I_o+V_g/Z_0 where Z_0 is the characteristic impedance of the resonant circuit associated with the switch.

11.9.2 Zero-voltage switching resonant converters

An alternative modification of the dc-dc Buck converter is shown in Figure 11.30(a). There are four circuit modes in the steady-state converter operation, given in Figure 11.30(b).

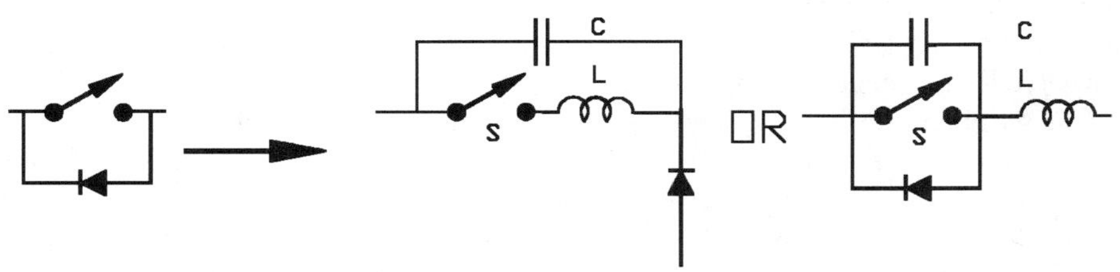

Figure 11.28
Resonant switches

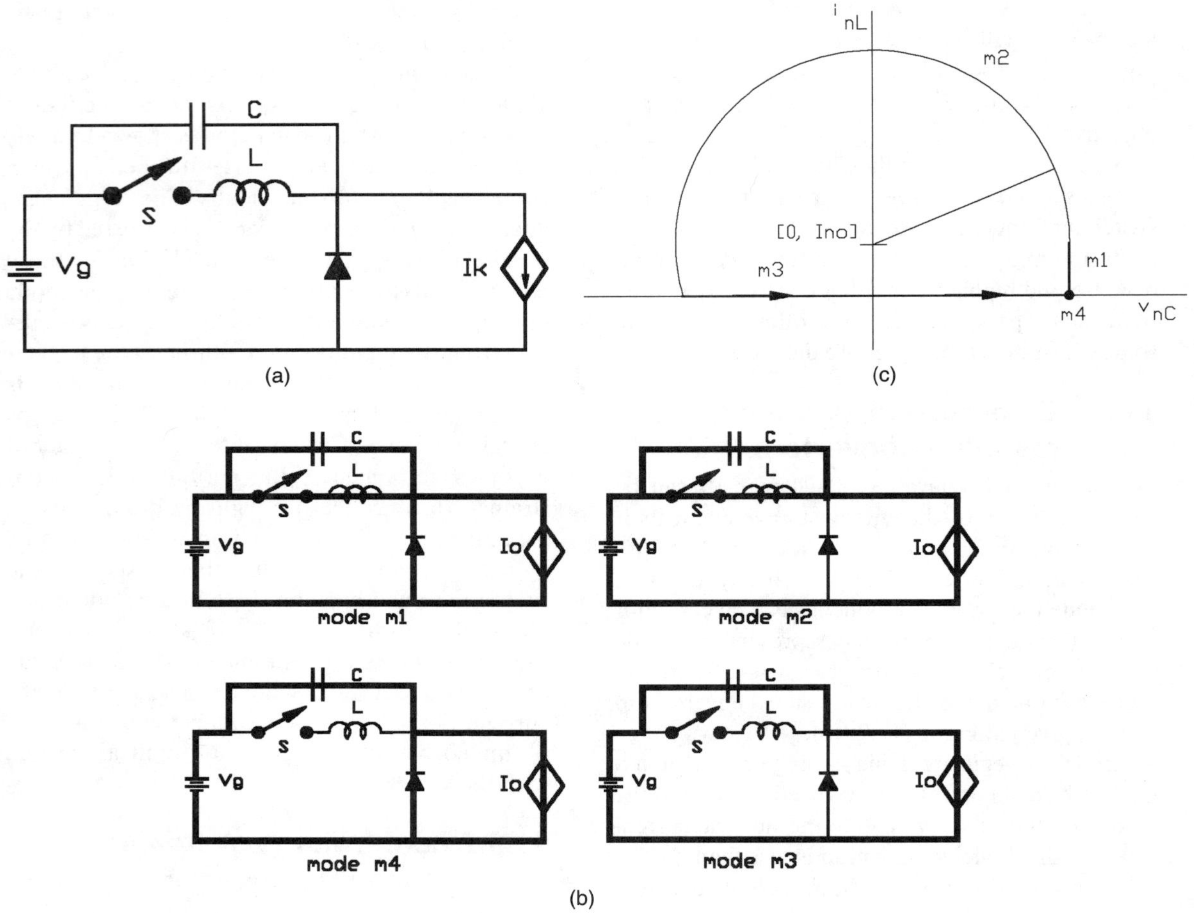

Figure 11.29
(a) DC-DC converter Buck topology with zero-current switching (b) Circuit modes in steady-state operation (c) State-plane trajectory

Mode m1 begins with turning off switch S. The inductor current is constant and equals the current sink I_o. Voltage across the inductor is zero. The capacitor across the switch begins to charge by I_o linearly. The switch turns off at zero voltage and zero slope. This mode is represented by a horizontal line segment on the state-plane in Figure 11.30(c).

As the capacitor reaches V_g, voltage polarity on the inductor reverses. Inductor current begins to decrease, forcing the current sink I_o to freewheel through the parallel diode. The inductor resonates with the capacitor as shown in mode m2. The trajectory on the state-plane in mode m2 is a circular segment with a center [1, 0].

The diode across the switch is forward biased as soon as the capacitor voltage tends to become zero and negative. At that instant, new mode m3 begins. The switch can now be turned on at zero voltage. The inductor current at this instant is negative, hence the current sink I_o continues to freewheel. The inductor

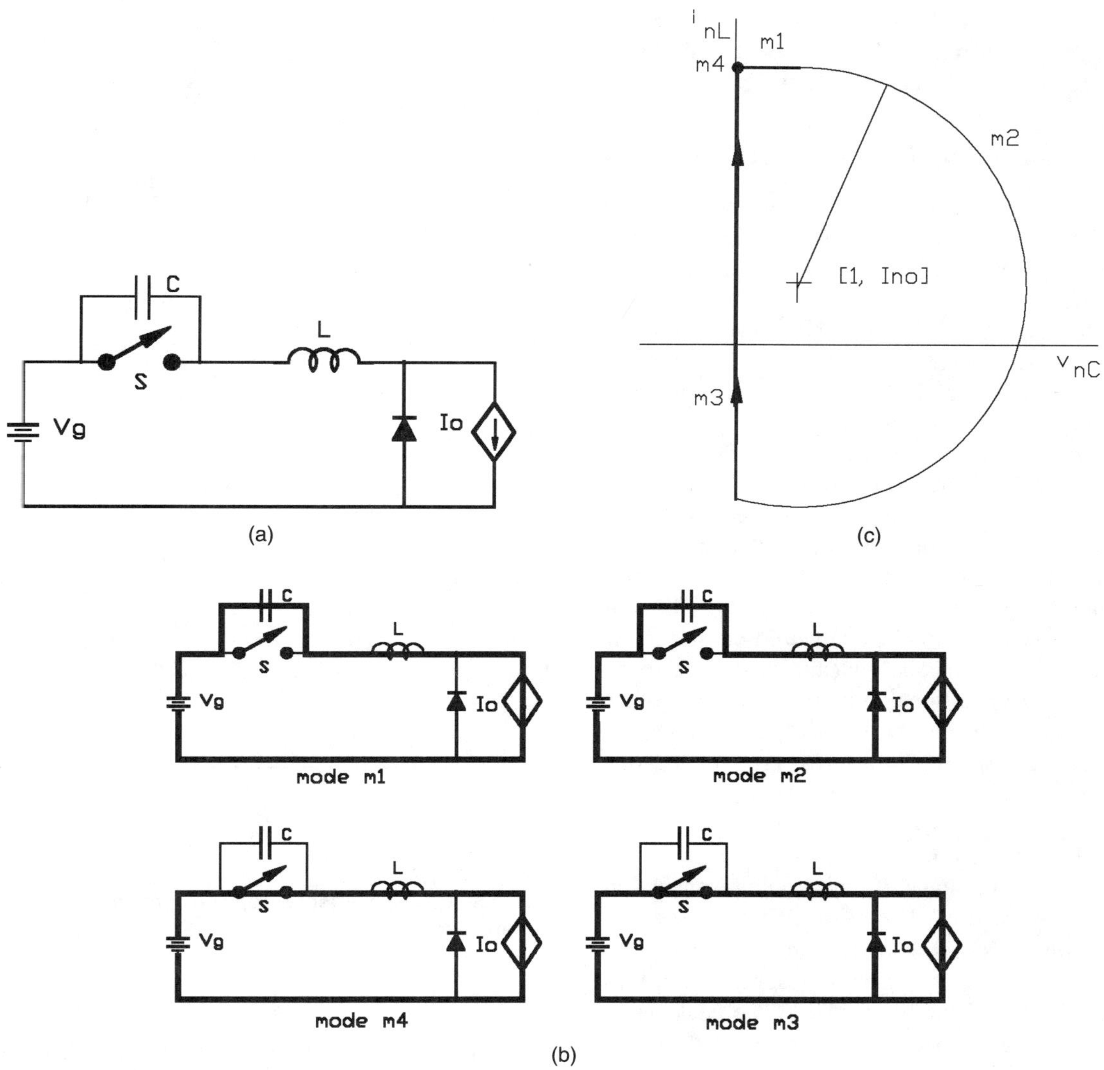

Figure 11.30
(a) DC-DC converter Buck topology with zero-current switching (b) Circuit modes in steady-state operation
(c) State-plane trajectory

experiences across it the dc voltage V_g, causing the inductor current to ramp up. The mode m3 is shown by a vertical line segment on the state-plane in Figure 11.30(c). Mode m3 continues until the inductor current is built up to I_o and stops increasing further because it stops the freewheeling of I_o. In the following mode m4, the circuit is frozen into a state where the inductor current is constant at I_o and the capacitor is clamped to zero by the switch. This state lasts until the switch is turned off. It should be noted that the switch turns on and off at zero voltage. Switching losses are eliminated. The peak value of the voltage across the switch, which is also the capacitor voltage, is $V_g + Z_0 I_o$ where Z_0 is the characteristic impedance of the resonant circuit associated with the switch.

11.10 RESONANT-LINK INVERTERS

The resonant-link converter minus the rectifying network is a resonant-link inverter. The basic concept of inversion is shown by means of the following circuit:

The current sink I_o may be the current to a motor. The state-plane representation of the inverter operation is shown in Figure 11.31. The inductor current is either linearly rising or constant at I_o as long as the switch is turned on. As the switch is turned off, the inductor and capacitor resonate. The resonant operation is represented by the circular trajectory on the normalized state-plane. The center of the trajectory is at $[1, I_{no}]$, the normalized input dc voltage source and the normalized output current. Losses (not shown) would cause the circular trajectory to coil down and shrink to the center point. Hence, for continuous operation of the inverter it is absolutely necessary that the capacitor voltage should return to zero or become slightly negative. The negative voltage would force the antiparallel diode to conduct before the switch is turned on.

The output voltage of this simple resonant converter is a sine wave. Unlike the PWM inverters, the peak voltage is slightly more than twice the input dc voltage source. A three-phase resonant-dc-link inverter is shown in Figure 11.32. The peak output voltage in this inverter is clamped to a slightly higher voltage than the input dc voltage source.

The two-element resonant tank can also be replaced with the three-element or higher order resonant tank.

Converters, in which the input dc voltage source is derived from an ac-dc line rectifier, are called the resonant-ac link converter. Some of the recent advances in the resonant-ac link converter topologies are shown in Figure 11.33.

A resonant inverter based on the concept of the PWM cycloconverter is presented in [ ], where the

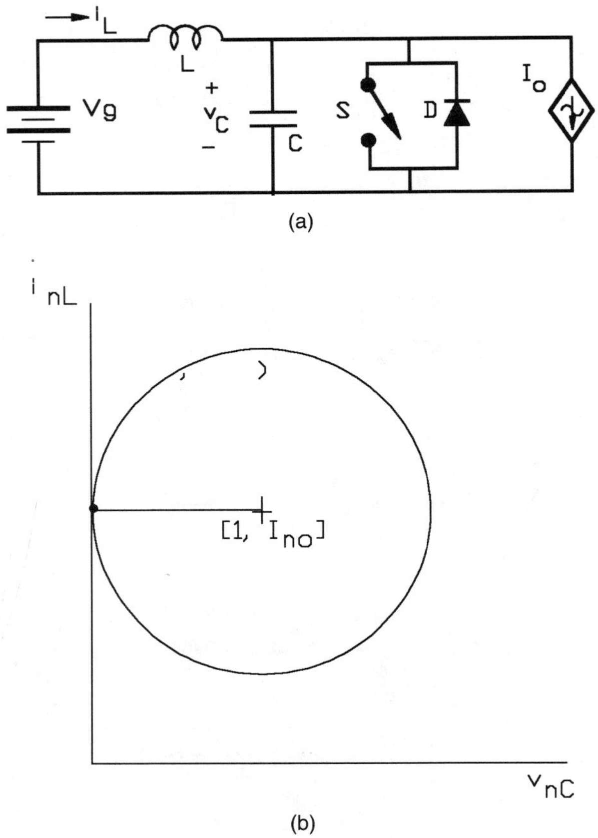

Figure 11.31
Resonant-link inverter (a) Topology and (b) state-plane diagram

low frequency sine wave output is synthesized from switching on and off the operation of the resonant bidirectional inverter for an integral number of half cycles. The concept is illustrated in Figure 11.34.

11.11 BIDIRECTIONAL RESONANT CONVERTER

The bidirectional dc-dc power conversion is of great interest in computer systems and aerospace systems. Figure 11.35 shows a bidirectional resonant con-

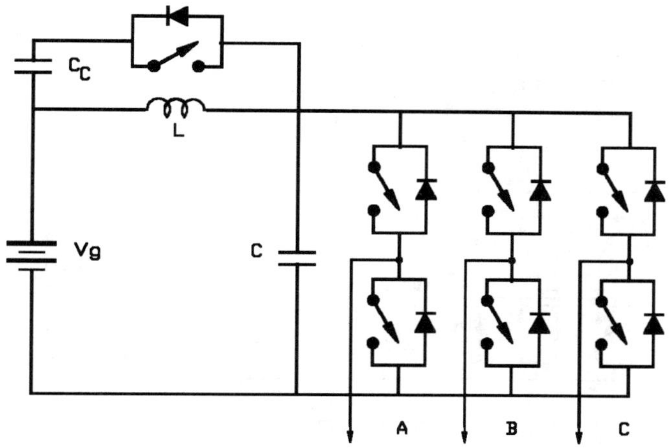

Figure 11.32
A three-phase clamped resonant-link inverter

verter based on the voltage coupled resonant-link topology. This converter is capable of resonant operation from no load to full load with zero-current switching and zero-voltage switching. This converter operates as a Buck-type resonant converter in the forward power flow mode and as the boost type in the reverse power flow mode.

11.12 CONTROL OF RESONANT CONVERTERS

Performance requirements of resonant converters and inverters include the regulation of output characteristics against the variations in the parameters of the input source and the load. The input source parameters may be line ac voltage amplitude and frequency. The load parameters may be the value of the load resistance or reactance. Both together impose very stringent control requirements in the feedback resonant systems. Basically two types of controls are employed in resonant converters:

1. Variable switching frequency control
2. Constant switching frequency control:
   (a) Control of the duration of a single circuit mode
   (b) Phase control
   (c) Resonant frequency control using either switched capacitor or inductor

Variable Switching Frequency Control
Control characteristics of converter gain M, normalized switching frequency f_{ns} and normalized load resistance Q serve as very useful design resources. Ideally, a large variation in load resistance and/or the source voltage should be possible by a minimal change in the switching frequency. From this point of view, three or higher element resonant-link converters perform significantly better than the two-element resonant converters. Variation in the fundamental component of the switching frequency causes the component sizes to be unwieldy and difficult to control.

Constant Switching Frequency Control
Normalized switching frequency f_{ns} in the control characteristics may be replaced by the ratio of the duration of a single circuit mode to the switching period. This duration may be the on-time of a switch (PWM control) or the duration of the discontinuity of the current supplied by the voltage source such as in the DCM operation. Phase difference in the turning on of two switches may also be used as the control parameter.

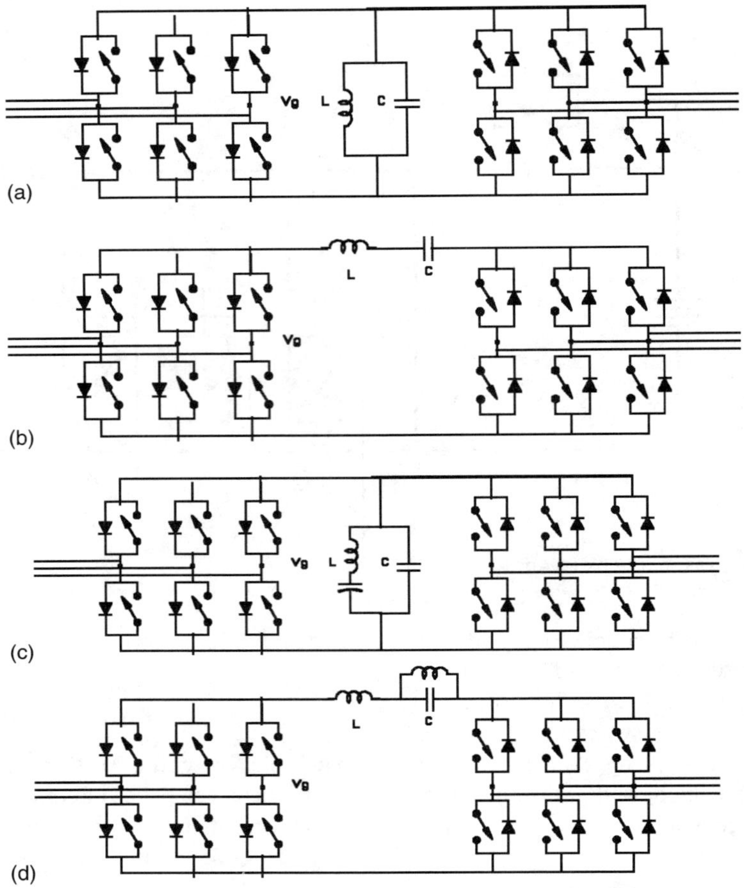

Figure 11.33
Resonant-link power converter with
(a) two-element parallel resonant link (b) two-element series resonant link (c) three-element LCC resonant link and (d) three-element LLC resonant link

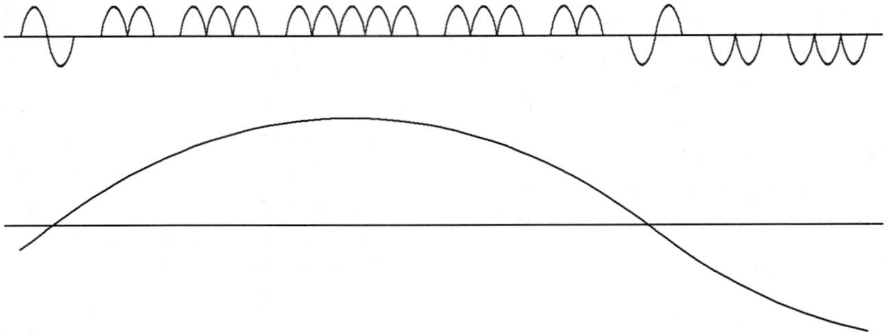

Figure 11.34
High-frequency link resonant cycloconverter

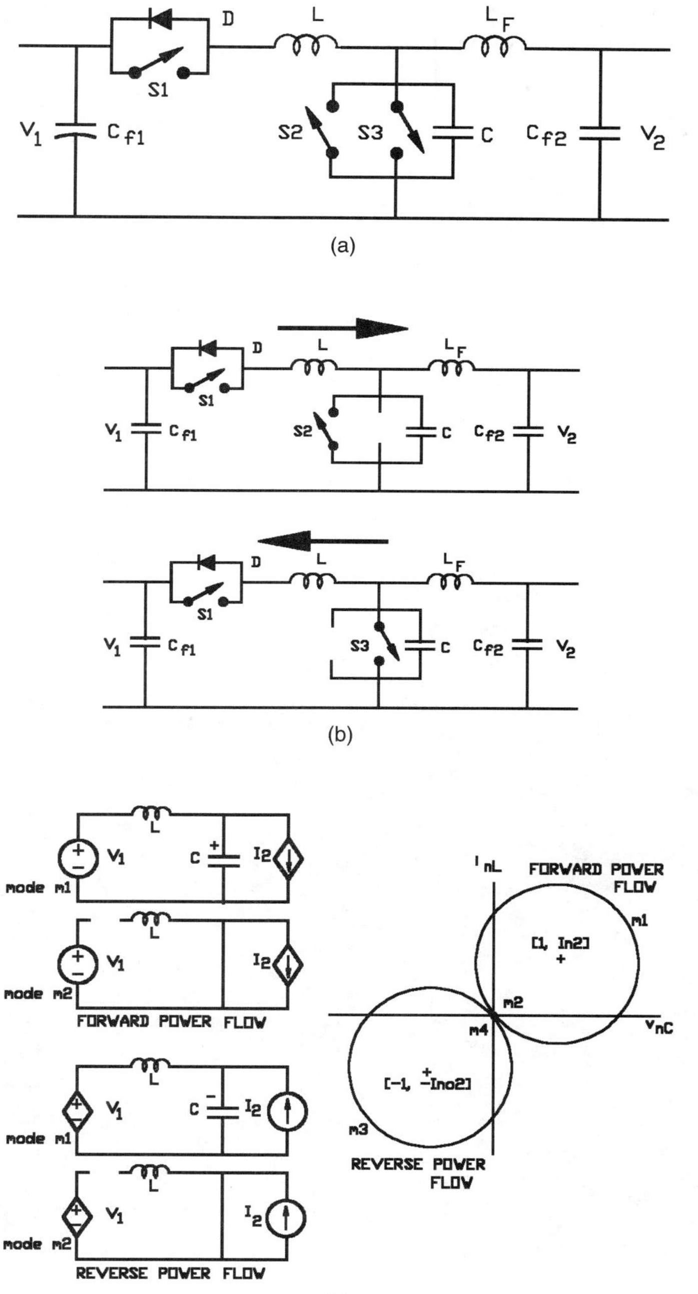

Figure 11.35
(a) Bidirectional resonant converter (b) forward power flow mode (switch S3 off) and (c) circuit modes

Normalized switching frequency may also be used as the control parameter by keeping switching frequency constant and varying the resonant frequency within a single switching period. The resonant frequency may be altered by switching capacitance or inductance in or out of the tank circuits. Usually this mode of control involves complicated switching and many circuit modes. The control operation, however, is easier due to the fixed and high frequency of the fundamental component of the switching frequency. Component selection is also easier.

11.13 PSPICE SIMULATION

PSPICE Example 11-1 Half-Bridge Series Resonant Converter in Continuous Conduction Mode Below Resonance

Consider the half-bridge series resonant converter of Figure 11.36(a) with a voltage sink in the load circuit. The voltage sink is the model of a diode bridge rectifier followed by a high value capacitor in parallel with the load resistor. Use a MOSFET IRF150 with an antiparallel connected 1N4148

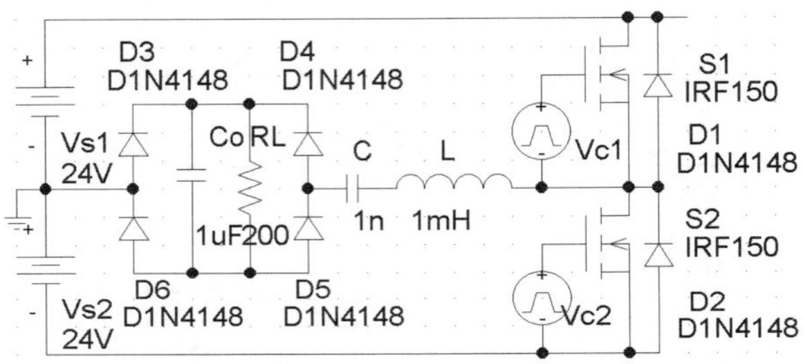

(a)

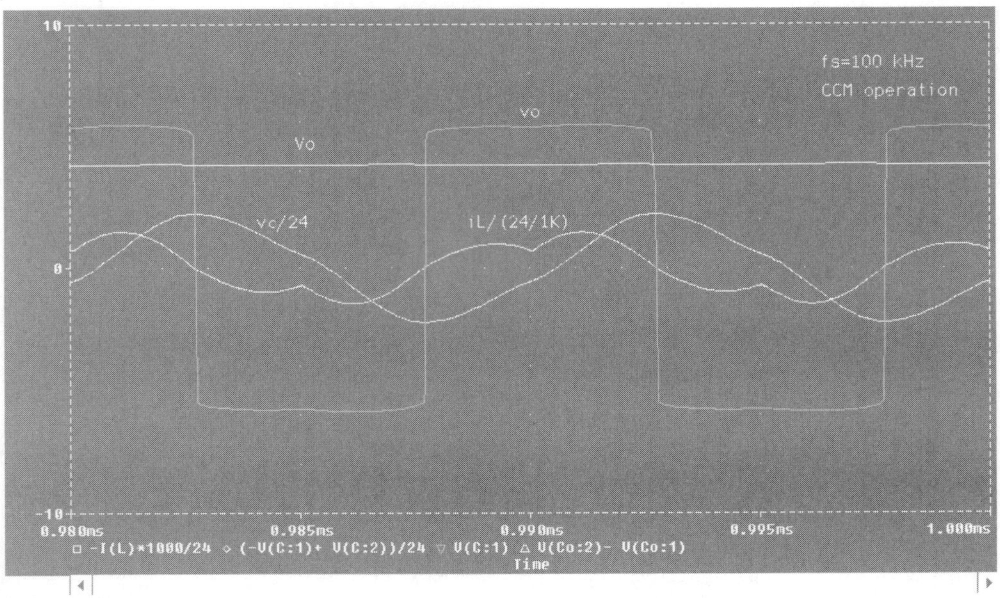

(b)

Figure 11.36
(a) Schematic (b) Waveforms of the inductor current i_L and the capacitor voltage v_C

diode for permitting reverse flow of current for the switches. A high-speed diode will be better. The dc-dc resonant converter specifications are

Two DC sources each of 24 V,
L = 1 mH   C = 1 nF
C_O =1 μF   R = 200 ohms
Switching frequency = 100 kHz

Perform a transient analysis for at least 100 switching periods and plot the steady-state waveforms for the following in the last two periods:

a. the current iL in the resonant inductor
b. the voltage vc across the resonant capacitor C
c. the output voltage waveform vo across the voltage sink
d. the dc output voltage V_O on the load resistor RL

The MOSFETs S1 and S2 are switched by two external pulse sources Vc1 and Vc2 respectively as shown in Figure 11.36(a). Both pulse sources are transformer coupled or optically coupled in the practical implementation of the circuit.

Set the pulse source Vc1 as follows:

Lower level V1	=0 V
Upper level V2	=10 V
Pulse period	=10 μs
Pulse width (on-time)	=4.95 μs
Delay time TD	=0
Rise time	=0.1 ns
Fall time	=0.1 ns

Set the pulse source Vc2 with all settings of Vc1 except the delay time, TD=5 μs.

The resonant frequency f_0 is 159 kHz. Hence, the operation at the switching frequency of 100 kHz is a below-resonance operation.

Use the simulation step in the range: .1 ns to 1 ns. Plot the normalized variables for the inductor current and the capacitor voltage as follows:

$$v_{nC} = \frac{v_C}{V_g} \quad where \quad V_g = 24\ V$$

$$i_{nL} = \frac{i_L}{(V_g/Z_0)} \quad where \quad Z_0 = \sqrt{\frac{L}{C}} = 1000\ ohms$$

The current and voltage on the resonant components are as per the expected waveforms in the continuous conduction mode of operation.

11.14 SUMMARY

The resonant converter is defined as a class of converters in which the resonant tank is a part of the circuit in at least one of the circuit modes in steady-state converter operation. The order of a resonant tank, and therefore of the converter, is determined by the number of independent energy storage elements. The resonant converters are classified as resonant-link and resonant-switch converters. Similarly, resonant inverters are classified as resonant-dc link and resonant-ac link inverters.

This chapter has demonstrated state-plane analysis for a series resonant dc-dc converter in discontinuous conduction mode (DCM) and continuous conduction mode (CCM). In series resonant converters, the output is taken in series with C or L of the resonant tank. In DCM, the series resonant converter exhibits ideal current source characteristics. In CCM, the dc-dc converter operates as a constant voltage source with unity converter gain when the switching frequency is equal to the resonant frequency. At below and above resonant frequency, the converter gain M is less than unity, and the converter characteristics are in between the characteristics of a constant output current and a constant output voltage converter. Essentially, the series resonant converter acts as the voltage step-down converter.

In parallel resonant converters, the output is taken in parallel with C or L of the resonant tank. In DCM operation, the parallel converter exhibits ideal voltage source characteristics. In CCM, the converter gain can be more than unity and theoretically infinite at the resonant frequency. Essentially, the parallel resonant converter acts as the voltage step-up converter.

The resonant converters show sinusoidal waveforms of the currents and voltages, in any case, the waveforms have much softer transitions than the

PWM converters. The maximum repetitive voltage and/or current through the switching devices are, however, several times higher than their counterparts in PWM converters.

The class E converter was discussed and is essentially a three-element resonant converter used in low-power, very high frequency applications. Zero-voltage switching and zero-current switching using resonant techniques, which were discussed, help in minimizing switching losses.

References

1. R. Severns, "Topologies for three element resonant converters," *Proceedings of IEEE Annual Power Electronics Conference 1990,* pp. 712–722, Los Angeles.
2. B. Ray, "Single cycle Resonant Bi-directional DC/DC Power Conversion," *IEEE Annual Power Electronics Conference 1993,* pp. 44–50, San Diego, Calif.
3. A. K. S. Bhat, "A resonant converter suitable for 650 V bus operation," *IEEE Trans. Power Electronics,* Vol. 6, No. 4, 739–748, October, 1991.
4. S. Freeland and R.D. Middlebrook, "A unified analysis of converters with resonant switches," *IEEE Power Electronics Specialists Conference Record,* 1987.
5. M. K. Kazimiercuzk, "Steady-state analysis and design of a buck zero-current-switching resonant dc/dc converter," *IEEE Trans. Power Electronics,* Vol. 3, No. 3, 286–296, July, 1988.
6. M.K. Kazimiercuzk and D. Czarkowski, *Resonant Power Converters.* Wiley, New York, 1995.
7. W. A. Tabiscz and F. C. Lee, "Zero-voltage-switching multiresonant technique—A novel approach to improve performance of high-frequency quasi-resonant converters," *IEEE Trans. Power Electronics,* Vol. 4, No. 4, October, 1989.

Problems

11-1 Analyze the soft-switching type resonant Buck converter as shown in Figure P11.1 with L = 10 μH, C = 0.1 μF and V_S 12 V. Draw the waveforms for the inductor current and the capacitor voltage for below-resonance operation in continuous conduction mode.

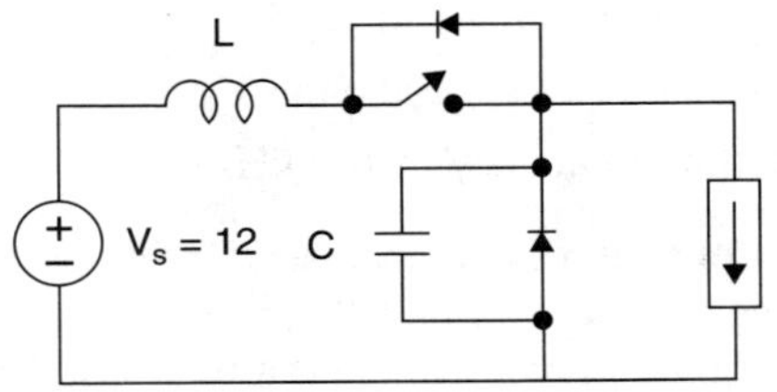

Figure P11.1

11-2 Repeat Problem 11-1 for the input voltage $v_S = 5$ volts.

11-3 Analyze the soft-switching type resonant Buck converter with zero-voltage-switching capability as shown in Figure P11.3 with L = 20 μH, C = 0.1 μF and $V_S = 12$ V. Draw the waveforms for the inductor current and the capacitor voltage for below-resonance operation in continuous conduction mode.

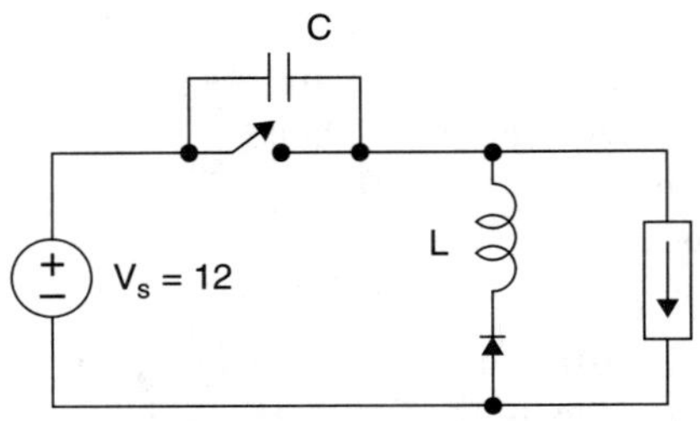

Figure P11.3

11-4 Design the ZVS soft-switching resonant Buck converter as shown in Figure P11.3 to convert 175 Vdc to 24 Vdc at a power level of 50 watts. Assume the load to be a constant 2 amp dc current sink. Find the period of switching, the gate pulse width and the ratings of the switch and the diodes.

11-5 Analyze the steady-state operation of the resonant converter, shown in Figure P11.5 with L = 20 μH, C = 0.1 μF and V_S = 12 V. Draw the waveforms for the inductor current and the capacitor voltage for below-resonance operation in continuous conduction mode.

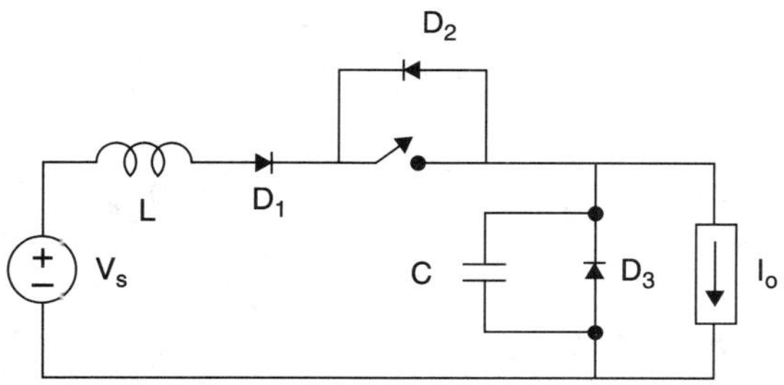

Figure P11.5

11-6 Analyze the steady-state operation of the resonant boost converter, shown in Figure P11.6 with L = 10 μH, C = 1 μF, V_S = 12 V and V_O =48 V. Draw the waveforms for the inductor current and the capacitor voltage for below-resonance operation in continuous conduction mode.

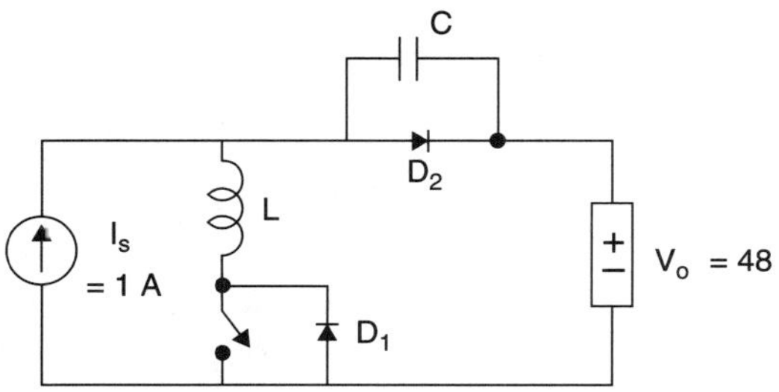

Figure P11.6

11-7 Analyze the steady-state operation of the resonant Buck converter, shown in Figure P11.7 with L = 10 μH, C = 0.1 μF and V_S = 175 V. Draw the waveforms for the inductor current and the capacitor voltage for below-resonance operation in continuous conduction mode.

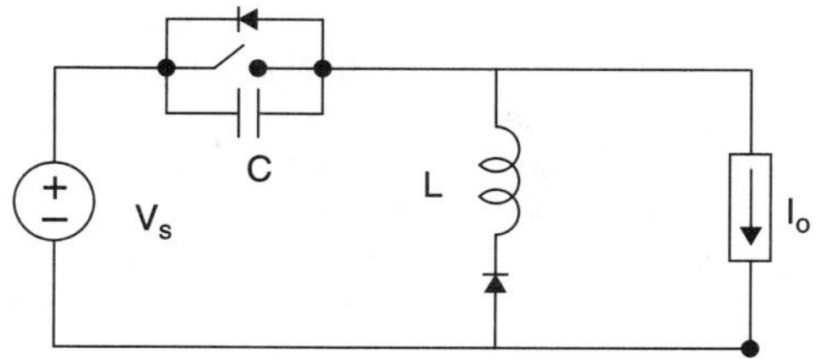

Figure P11.7

11-8 Analyze the steady-state operation of the resonant flyback converter, shown in Figure P11.8 with L = 10 μH, C = 1 μF, V_S = 12 V and V_O = −48 V. Draw the waveforms for the inductor current and the capacitor voltage for below-resonance operation in continuous conduction mode.

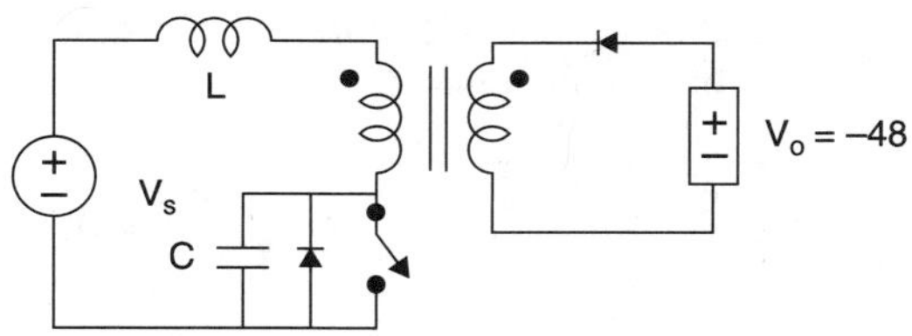

Figure P11.8

11-9 Plot the amplitude and the phase of the input impedance versus frequency for a fourth order resonant tank circuit of Figure P11.9 with L1=1 H, L2 = 2 H, C1=1 F and C2=1 F.

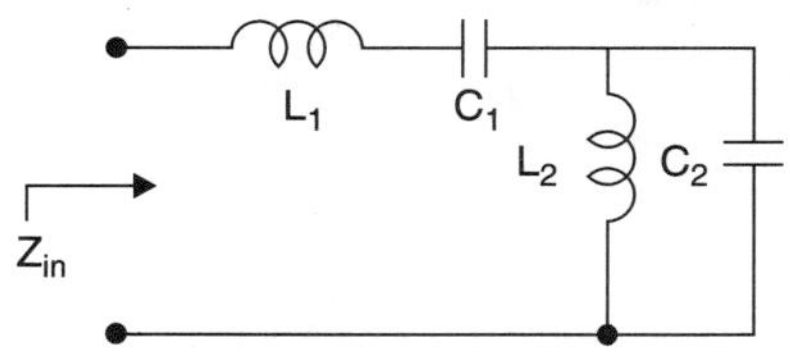

Figure P11.9

11-10 The parameters for the series resonant converter of Figure 11.8(a) are L = 1 mH, C = 2 nF, R_L = 50 ohms, V_g = 90 Vdc, f_{ns} = 0.4.

a. Find the dc output voltage V_o,

b. Draw the waveforms for the current into and the voltage on the voltage sink,

c. Find the maximum dc output voltage V_o in DCM and the frequency at which this could be achieved,

d. Find the frequency and the value of Inductor L for which a minimum 10 watts power will be delivered to R_L.

11-11 The parameters for the series resonant converter of Figure 11.8(a) are; $L=1$ mH, $C=2$ nF, $R_L=100$ ohms, $V_g=90$ Vdc, $f_{ns}=0.8$.

a. Find the dc output voltage V_o,

b. Draw the waveforms for the current into and the voltage on the voltage sink.

11-12 Find the circuit modes in the steady-state below-resonance operation of the converter with the switching waveforms as shown in Figure P11.12. The resonant inductors L1 and L2 represent the turn-on snubber inductors.

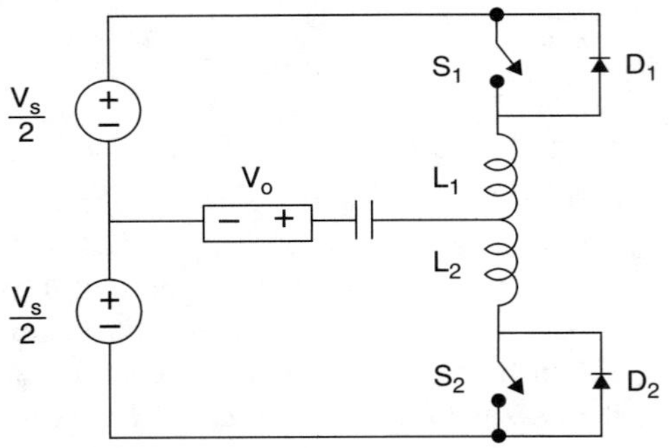

Figure P11.12

11-13 Find the circuit modes in the steady-state below-resonance operation of the converter with the switching waveforms as shown in Figure P11.13. The resonant capacitor represents the drain to source capacitance of the MOSFET switches.

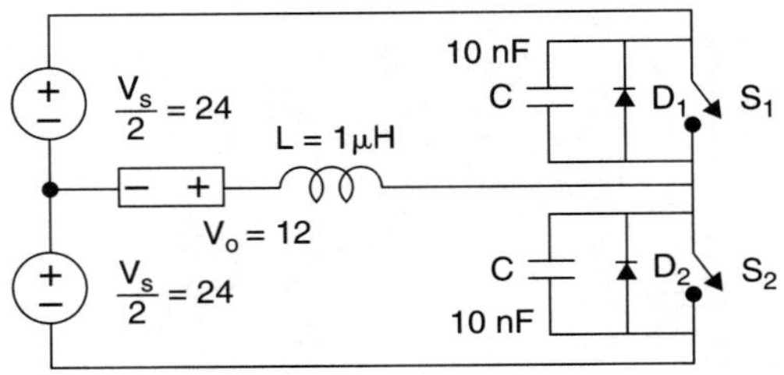

Figure P11.13

PART FOUR

POWER ELECTRONIC APPLICATION SYSTEMS

CHAPTER 12

Electric Utility Interface: Power Factor Correction and Static Var Control

OBJECTIVES

1. Need for power factor correction in electric utility
2. Limits of passive filtering
3. Power factor correction using active current shaping
4. Interface for bidirectional interface
5. Three-phase utility interface
6. Introduction to static VAR compensators

12.1 INTRODUCTION

There is a growing concern regarding harmonic pollution of the power distribution system. Adoption of the International Electrical Commission standard IEC-555-3 and IEEE-519-1992 has helped greatly in the awareness for clean ac line currents and a power factor close to unity.

12.2 ELECTRIC UTILITY DISTRIBUTION SYSTEM

The electric utility distribution system can be represented by a Thevenin equivalent voltage source v_S in series with a source inductance L_s, see Figure 12.1.

The line voltage at the point of common coupling is dependent on the load current and the source inductance,

$$\tilde{V}_{1t} = \tilde{V}_S - j\omega\, L_S\, \tilde{I}_1 \qquad (12\text{-}1)$$

where subscript 1 refers to the fundamental harmonic (n=1). I_1 is the fundamental harmonic of the line current. Actually, the line current drawn from the electric utility by a power electronic system consists of several components: real and imaginary components at the line frequency and high frequency harmonics:

$$\tilde{I} = \tilde{I}_{1P} + j\tilde{I}_{1Q} + \sum_{n>1}\tilde{I}_n \qquad (12\text{-}2)$$

The imaginary (reactive) component, I_{1Q}, as shall be seen later in this chapter, is the main contributer to the line voltage regulation. Minimizing I_{1Q} is referred to as the static VAR control. The real component of the fundamental harmonic, I_{1P}, contributes little to the line voltage regulation. The phase angle of the fundamental harmonic current with respect to the line voltage is a very important parameter that determines the power factor. Recall

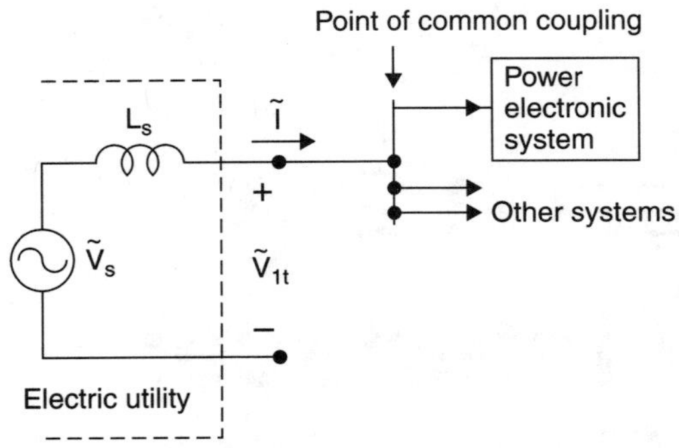

Figure 12.1
Electric utility system

the definition of the power factor from Chapter 2, which is repeated here

$$PF = \frac{\tilde{I}_1}{\tilde{I}_S} \cos \theta_1 \tag{12-3}$$

The subscript s refers to the line source. The subscript 1 refers to the fundamental component. The power factor is also the ratio of the real power used by the load to the power supplied by the line source. A poor power factor is translated in heavy expenses to the user; hence, considerable effort has been directed in improving the power factor.

Suppression of high frequency harmonics in the electric utility is another very important requirement by electric utility vendors. The harmonics in a typical unfiltered single-phase line current are given below:

TABLE 12.1 TYPICAL HARMONIC DISTORTION IN AN UNFILTERED SINGLE PHASE LINE CURRENT

N=	3	5	7	9	11
I_n/I_1 (%)	73.2	36.6	8.1	5.7	4.1

Compare these numbers with the percent amplitudes of high frequency harmonics as specified by IEEE519-1992 standards, given in Table 12-2.

Obviously, there is a great need for reduction of the harmonics. The interface design must aim at following two-fold goals:

TABLE 12.2 HARMONIC DISTORTION OF LINE CURRENT (I_N/I_1, FOR ALL HARMONICS LESS THAN 11TH)

As per IEEE-519-1992		
For $I1/I_{SC}$	I_n/I_1	THD (%)
>0.05	4%	5%
0.02–0.05	7%	8%
0.01–0.02	10%	12%

Note: $I_{SC} = V_S/\Omega_e I_{sh}$ is the short circuit current at the point of common coupling.

(a) to prevent all high frequency harmonics from being coupled to other systems connected to the point of common coupling, and

(b) to shield the power electronic system under consideration from the transients and high frequency harmonics present in the electric utility.

The utility interface may be a passive filter or an active current-shaping system.

12.3 PASSIVE FILTERING

The simplest technique is to use an inductor in series with the line source to shape the current waveform. This technique was discussed in the chapter on the ac-to-dc converter, and it is not repeated here.

The ac line current can also be smoothened by using L-C filters in the line just before connecting it to a power electronic converter system. One such filter is shown in Figure 12.2. Inductors L_1 and L_2 basically add a source side inductance. It reduces high order harmonics in the input current, reduces output voltage ripple and reduces voltage and current stresses on the rectifier diodes. The capacitor suppresses the high frequency harmonics and transients from being coupled to and from the source. The power factor improves slightly but the ac line voltage, available to the converter, becomes dependent on the load.

The disadvantage of the passive filtering is the size of the filter components at low line frequency. Look to switch-mode techniques for reducing the size of the filter components. The active current shaping is one such technique.

12.4 ACTIVE CURRENT SHAPING: POWER FACTOR CORRECTION

AC line-operated, switch-mode power supplies and converters for motor drives all require an ac-dc converter as interface to the ac line. However, the uncontrolled version, the bridge rectifier, is more popular. A large capacitor is connected to the output of this rectifier for reducing the voltage ripple. Since the capacitor draws line current only when the output voltage is below the line voltage, the line current consists of a highly non-sinusoidal waveform, narrow and high peaks as shown in Figure 12.3.

The power factor is ideal (=1) when the bridge rectifier is loaded by a current sink and poorest when loaded by a voltage sink such as the one in Figure 12.3. In the case of current sink loading, the rectifier diodes conduct for π line angle and the power stress on the diodes is uniformly distributed over a half line period. In the case of voltage sink loading, the rectifier diodes conduct for a much shorter line angle and the power stress on the diodes is highly concentrated in a short duration.

Similarly, the power factor is ideal when a voltage sink load is connected to a switched constant current waveform source and poorest when supplied by a switched constant dc voltage waveform source. A power factor correction device should act like an interface between the bridge rectifier and the voltage sink load. The interface appears as the current sink load to the bridge rectifier and a switched constant current waveform source to the voltage sink load.

The power factor for the current waveform in Figure 12.3 is less than 0.7, which is a very poor

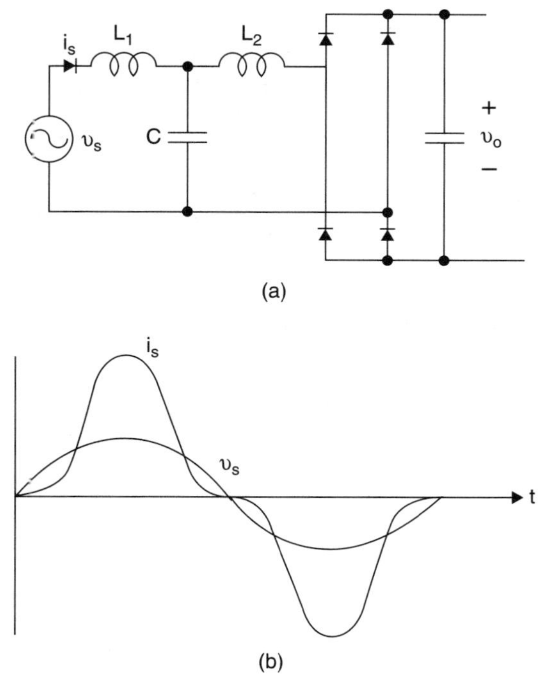

Figure 12.2
Passive filter for improving the power factor

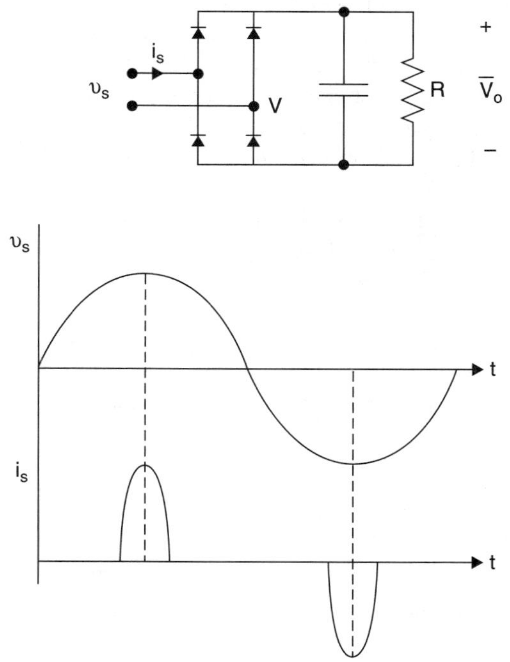

Figure 12.3
The uncontrolled bridge rectifier, the line voltage and the line current waveforms

power factor. Two consequences happen due to a poor power factor:

1. Diodes in the bridge rectifier must be rated for higher currents, which are several times more than the peak value of the fundamental component.
2. The user pays for high reactive power from the ac line.

Improving a power factor in the ac-dc rectifier involves

(a) shaping the line current to a sinusoidal waveform and

(b) reducing the phase difference between the line voltage and the line current.

In essence, the ac-dc converter in the power factor correction *emulates a resistor.*

The *resistor emulator,* also called the *Power Factor Preregulator (PFP),* is basically a dc-dc converter. The PFP topology is given in Figure 12.4. The input to PFP is a full-rectified sinusoidal voltage waveform. A voltage sink V_O represents the load in parallel with a large C. The dc-dc converter is switched at a frequency f_S that is several times higher than the line frequency f_L. A large value of inductor would be needed to appear as a current sink load to the bridge rectifier. The current through the diode in PFP switches between zero and a constant value I_O. The PFP acts as an ideal interface that appears as the current sink load to the bridge rectifier and a switched constant current waveform source to the voltage sink load.

For the practical values of the inductor, the input current of PFP, as we know from Chapter 6, possesses a triangular wave shape. The input current waveform before the diode bridge is modified to contain a strong fundamental sinusoid at the line frequency but with harmonics at several times higher frequency than the line frequency.

Since the switching frequency is very high in comparison to the line frequency, the input and output voltages of the PFP converter may be considered to be constant throughout a switching period. Thus, the PFP converter can be analyzed like a regular dc-dc converter. The line voltage and the input voltage to the PFP are given by

$$v_S = V_S \sin\theta$$
$$v_1 = V_S\left|\sin\theta\right|, \qquad \theta = 2\pi f_L t \qquad (12\text{-}4)$$

The θ is the line phase angle.

The voltage transfer ratio of PFP is required to vary with the angle θ in a half line period. The voltage transfer ratio of the dc-dc PFP is

$$T_{vv}(\theta) = \frac{\overline{V}_O(\theta)}{v_S(\theta)} = \frac{\overline{V}_O}{V_S|\sin\theta|}, \quad \textit{where } f_S >> f_L \qquad (12\text{-}5)$$

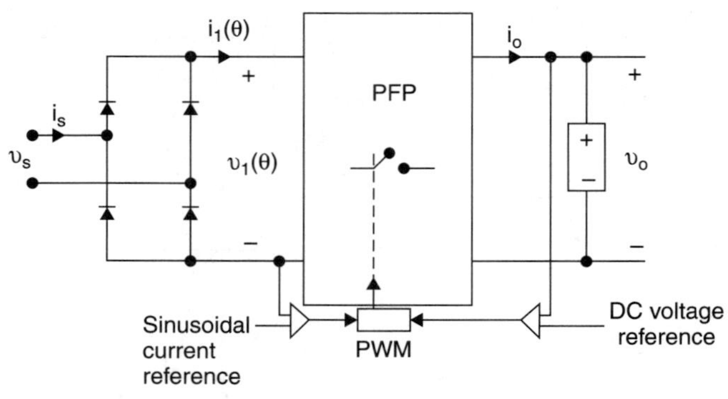

Figure 12.4
A dc-dc converter as the Power Factor Preregulator

where V_O is the local average dc output voltage from the PFP.

The T_{vv} in a line period is shown in Figure 12.5. The high voltage transfer ratio at the zero and π angles can be achieved by using Boost, Buck-Boost or flyback topologies. Buck topology cannot provide high voltage transfer ratios.

For Unity Power Factor

The current from the diode bridge must be identical in shape and in phase with the voltage waveform, hence,

$$i_1 = I_S |\sin\theta| \tag{12-6}$$

The input and output powers, averaged over a switching period, are given by:

$$p_i = v_1\, i_1 = V_S I_S \sin^2\theta, \quad p_O = \overline{V}_O i_O \tag{12-7}$$

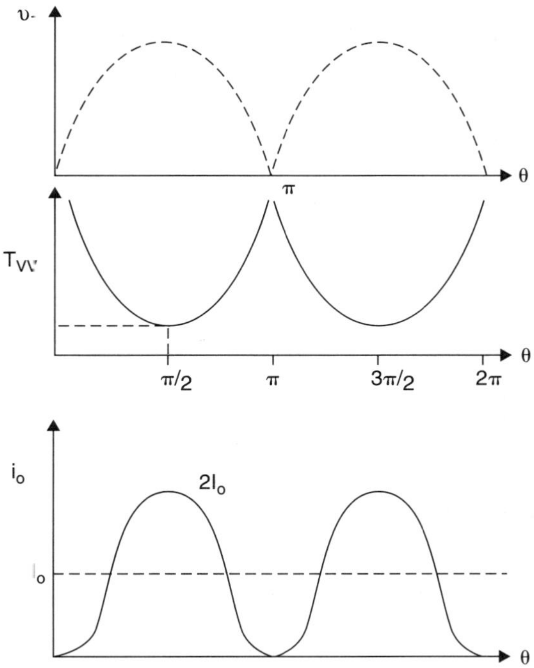

Figure 12.5
The voltage transfer ratio and the output current waveform of PFP as required for the unity power factor (zero phase and sinusoidal current)

Assuming the conversion efficiency to be 1 ($p_i = p_O$), the output current requirement is determined as

$$i_O = \frac{V_S I_S \sin^2\theta}{\overline{V}_O} \tag{12-8}$$

The input and output powers, as averaged over the line period, are:

$$P_i = \frac{V_S I_S}{2}, \qquad P_O = \overline{V}_O I_O \tag{12-9}$$

where I_O is the average dc output current from the PFP.

The output current of the PFP is then

$$i_O = 2I_O \sin^2\theta = I_O(1 - \cos 2\theta) \tag{12-10}$$

The PFP output current required for a unity power factor, as a function of the angle θ is shown in Figure 12.5.

The above discussion points out two important requirements for PFP:

1. The voltage transfer ratio T_{vv} must be varied over a half line period.
2. The current output of the PFP should be varied simultaneously over the half line period according to the profile shown in Figure 12.5.

Thus, two separate control loops are required. The control system is given in Figure 12.6. In one of the loops, T_{vv} is controlled by the pulse width as the control parameter. The output voltage is sensed, compared with a dc reference, and the error signal is used to modify the pulse width of the switching device in the PWM converter. In the other loop, the output current is sensed, compared with a sinusoidal reference, and the error is used to control the pulse width. These two loops are merged into one and called the current-mode control. In this method, the reference current waveform is made a function of the input voltage v_1 waveform and the error voltage,

$$i_{ref} = e\, v_1 = (V_O - V_{ref}) V_S |\sin\theta| \tag{12-11}$$

The error voltage is sampled only once in a switching period T_S and held at the sampled value during the feedback to regulate the output voltage

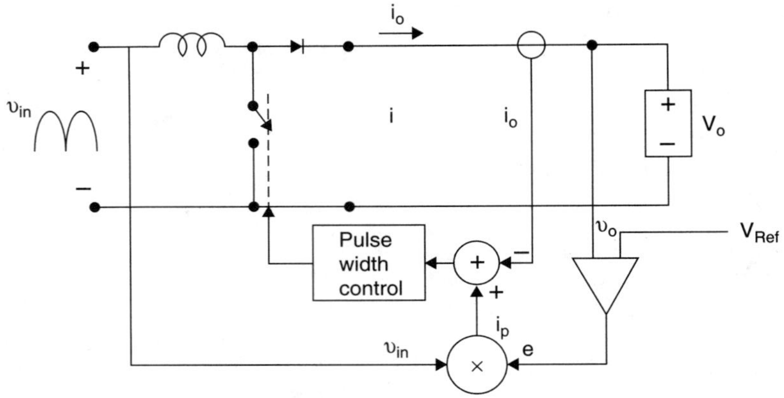

Figure 12.6
Control loops in power factor correction

around the nominal value V_O. This method can yield power factors in the range of 0.95 to 0.99 and reduces the THD of the line current to within 3%.

12.4.1 PWM control of power factor

Consider the ac-dc rectifier with a Boost type PFP incorporated into it, as shown in Figure 12.7. The voltage transfer ratio in the Boost-PFP is obtained by combining equation (12-11) with equation (6-22) in Chapter 6,

$$T_{vv} = \frac{V_O}{V_S|\sin\theta|} = \frac{1}{1-D}, \quad \textit{where} \quad \textit{D: Duty Ratio}$$

$$D = 1 - \frac{V_S}{V_O}|\sin\theta| \qquad (12\text{-}12)$$

Example 12-1

A full-bridge rectifier from 60 Hz, 120 volts rms ac source is designed to provide 500 volts dc to a resistive load of 10 ohms. Use an ideal Boost type Power Factor Preregulator for achieving unity power factor. The PFP converter in Figure 12.7 is switched at 10 times the line frequency. Find the inductance L and plot the duty ratio, the current waveform at the input of the PFP and the average input current over a half line period.

From Chapter 6, the triangular current in the Boost converter in a switching period, is given by

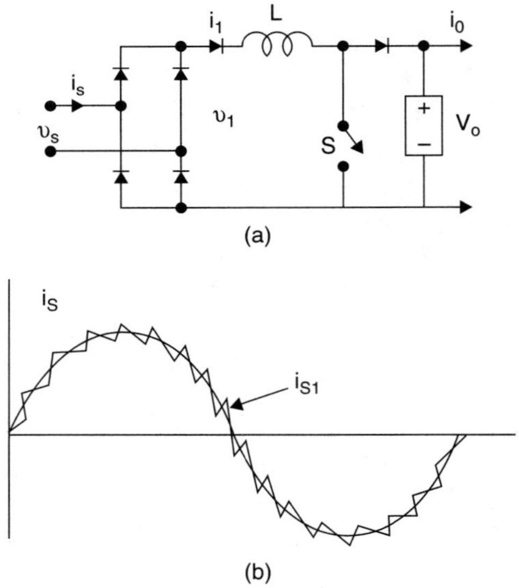

Figure 12.7
Boost type Power Factor Preregulator

$$i_L = \frac{v_1(\theta)}{L}t \qquad 0 \le t < DT_S$$

$$= \frac{v_1(\theta)}{L}DT_S - \frac{v_1(\theta) - V_O}{L}(t - DT_S)$$

$$DT_S \le t < T_S$$

The line angle θ corresponds to the t=0 instant of the switching period.

The average value of the inductor current, which is also the input current to the PFP, is

$$i_{1,avg} = \frac{1}{2}\frac{v_1(\theta)}{L}DT_S$$

```
Tl=1/60; N=64; VS=120*sqrt(2); VO=200; R=10;
  IO=VO/R; Ts=Tl/N; L=0.001;
theta=pi/(N):2*pi/N:pi;% the line angle
D=1-(VS/VO)*abs(sin(theta));% duty ratio as a function
  of the line angle th=[ ]; i1=[ ]; i1avg=[ ]; iL=[ ];% null
  arrays, th: the line angle array, i1: the input current
  array %i1avg: the mean input current to PFP, and iL:
  the inductor current array in a switching period
[m1,m2]=size(theta);
ILO=2*IO*(sin(theta)).^2;% initial current at the beginning
  of a switching period
for =1:m2
     t1=linspace(0,D(i)*Ts,128);% switch on-time
     t2=linspace(D(i)*Ts,Ts,128);% switch off-time
     t=[t1,t2];% a switching period
     th=[th,(t+(i-1)*Ts)*360/Tl];
     iL1=ILO(i)+VS*abs(sin(2*pi*(i-1)/N)).*t1/L; % the
       inductor current during on-time
     iL2=max((max(iL1)+(VS*abs(sin(2*pi*(i-1)/N))
       -VO)*(t2-D(i)*Ts)/L),0);
     % inductor current during off-time
     iL=[iL1,iL2];% inductor current in a switching period
     Vo=[Vo,VS/(1-D(i))];
     i1=[i1,iL];
     i1avg=[i1avg,(mean(iL))];
end
plot(th,i1/(2*IO),theta*(max(th))/pi,i1avg/(2*IO),‘*’,theta*
  (max(th))/(pi),D, ‘−−’)
```

Example 12-2

Find the frequency spectrum of the line current in Example 12-1.

To perform the FFT analysis, determine the line current i_S waveform over a full line period from the i_1 waveform.

Add the following lines at the end of the MATLAB program of Example 12-1:

m12_1dd.m

```
th_l=[th, 180+th];
is=[i1, -i1];
```

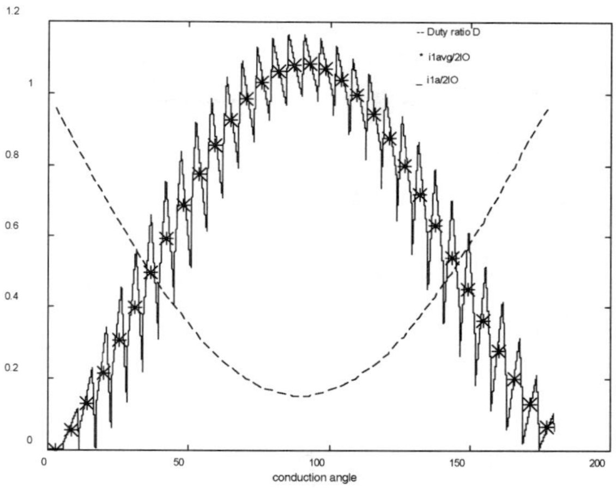

Figure 12.8
Line current in Boost PFP for 500 volts at R=10 ohms

```
plot(th,is/(2*IO))
y=fft(is);
bar(abs(y)/(IO*2*length(y)))
axis([0,100, 0,1])
```

Note that the line current contains a high fundamental harmonic component. The next significant harmonic is 16 times the line frequency.

The literature may be referred to for several other types of power factor correction strategies. An interesting strategy employs the constant current ripple method.

12.4.2 Constant current ripple control of power factor

The input current to the PFP, i_1 is maintained within a band of tolerance of (+/−) ΔI with respect to the desired unity PF current waveform as given in Figure 12.5. The circuit is illustrated in Figure 12.10. If the actual current increases more than the nominal value as given by equation (12-9), plus the half of the tolerance band, the switch S1 is turned off and S2 is turned on. Opposite switching occurs when the actual current decreases below the nominal value minus half of the tolerance band. The

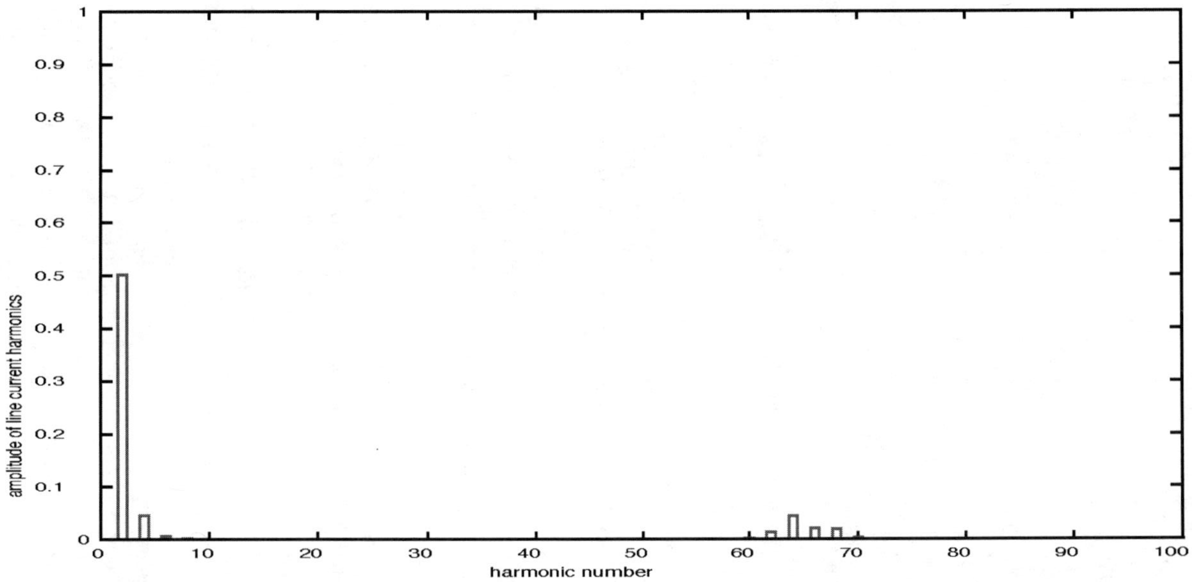

Figure 12.9
Harmonics in the line current waveform

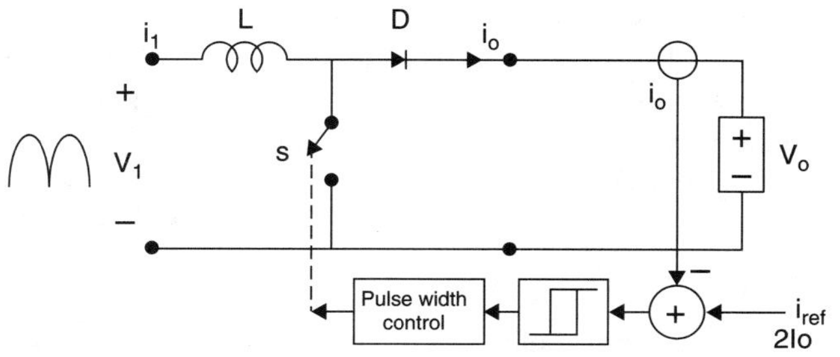

Figure 12.10
The current ripple control of power factor correction

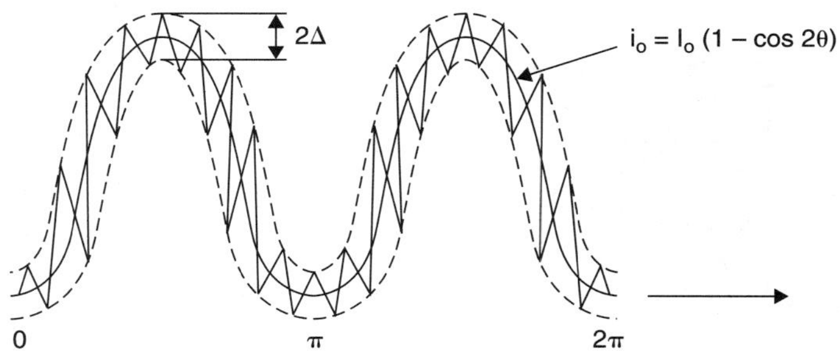

Figure 12.11
Waveforms in the constant current ripple method

switching period is variable along the current waveform, see Figure 12.11.

12.5 INTERFACE FOR BIDIRECTIONAL POWER FLOW

In motor drive applications with regenerative braking, the power flow from the ac line is required to be bidirectional. A bidirectional converter can be designed using the phase delay angle control, but at the penalty of poor input power factor. We will use a Sine-PWM based topology that will enable bidirectional power flow at unity power factor.

Figure 12.12 shows the switch-mode converter as the interface utilizing Sine-PWM switching. The sources side inductance smoothes the line current.

DC sink V_O represents the dc motor in forward motoring or braking condition. The ac utility is modeled by the ac source v_S in series with an inductor L. To start with, assume that the ac line source v_S and the series inductor L are removed and the converter is operated in the inverter mode using unipolar Sine-PWM switching. Refer section 7.8 in chapter 7 for the inverter operation. The unipolar switching is not necessary; it is selected for illustration only. The inverter is operated to produce a sine wave ac voltage waveform of frequency ω_S, which is the frequency of the ac line source. Peak voltage of the fundamental harmonic in the inverter output is derived from equation (7-55) as

$$V'_{OPeak} = aV_O$$

Where a is the PWM constant, see section 7.8.1, and it is less than or equal to unity. Please note that above equation is valid when the number of pulses per half period of the ac output waveform is large. The complete ac waveform of the fundamental component is given by

$$v'_O = aV_O \sin \omega_s t$$

This is the desired waveform, which should be applied to the switch-mode bridge in Figure 12.12. Now let us bring the ac line source and L back into the circuit. It is possible to obtain the desired ac waveform from the line source at unity power factor by introducing a delay angle δ, see the phasors in Figure 12.13. $[]_{S1}$ is the fundamental component current from the line source v_S.

The voltage and current waveforms are shown in Figure 12.14. The average waveform v'_O, shown by the dotted line, is applied as input to the switch-diode bridge and it is related to the ac line source as

$$v'_O(t) = V_s \sin(\omega_S t + \delta)$$

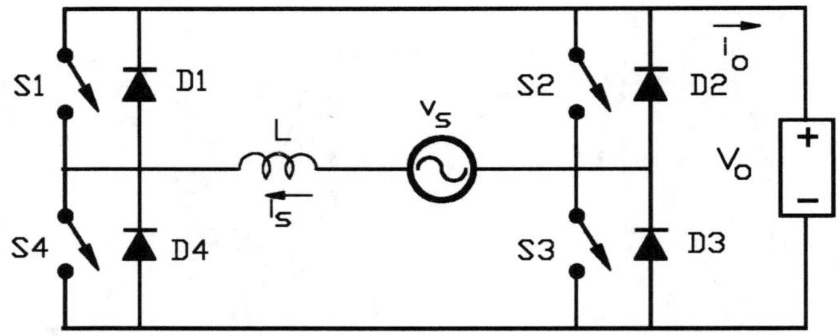

Figure 12.12
A switch-mode converter for the ac-line interface

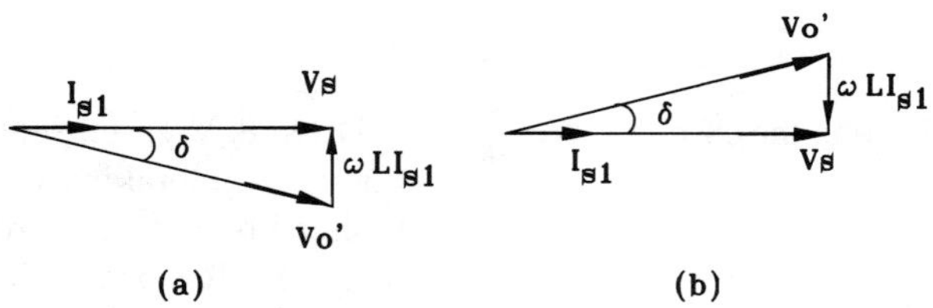

Figure 12.13
(a) Power flow from ac line to dc sink V_O and
(b) power flow from dc sink V_O to ac line

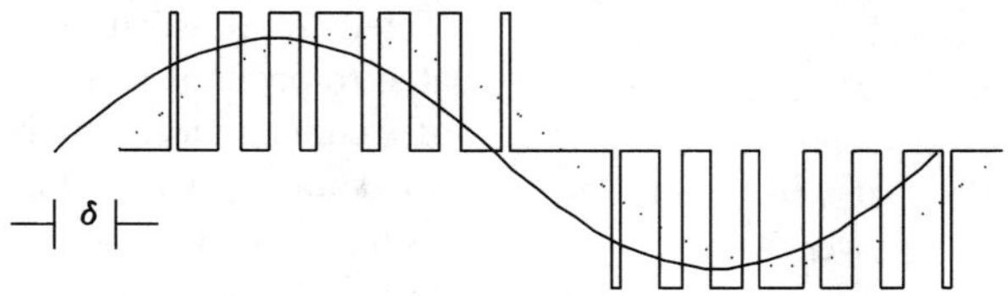

Figure 12.14
Voltage and current waveforms in the bi-directional interface circuit

V_S is the peak voltage and ω_S is the frequency of the ac line source.

This interface circuit has been analyzed in detail in reference [2]. We will present the results only. The real power flow from the line source is given by

$$P = a\frac{V_s V_O}{2\omega_S L}\sin\delta$$

The factor a controls the amount of power flow and δ controls the direction of flow. For a positive δ, the power flows from the utility to the motor (sink V_O) in the forward motoring mode. For a negative δ, the power flows back to the utility from the motor in the braking mode. In both operations, it is required that the voltage sink V_O (ensured by $a < 1$) must be greater than the peak voltage source V_S. The converter always operates in Continuous Conduction Mode.

12.6 3-PHASE UTILITY INTERFACE

The active input line current shaping in 3-ϕ can be performed using the bidirectional switch-mode converter of Figure 12.12 in each phase. The dc output voltage V_O should be greater than the peak voltage in each phase, or 1.633 times the rms line-line voltage:

$$V_O \geq 2\sqrt{2}\,\tilde{V}_{\phi 1}$$
$$\geq \frac{2\sqrt{2}}{\sqrt{3}}\,\tilde{V}_{LL}$$

The dc output current for a unity power factor is obtained from equating the input and output powers:

$$I_O = \frac{3\,\tilde{V}_S\,\tilde{I}_S}{V_O} \qquad (12\text{-}13)$$

12.7 STATIC VAR COMPENSATORS

The equivalent circuit of an electrical utility for the fundamental harmonic components is given in Figure 12.1. L_s is the source inductance due to long wires in the distribution system.

I_L is the current drawn by a user and, in general, it has real and imaginary components. The voltage available to a user is given by

$$\tilde{V}_{1t} = \tilde{V}_S - j\omega\, L_S \tilde{I}_1$$

$$\textit{where}\ \ \tilde{I}_1 = \tilde{I}_{1P} + j\tilde{I}_{1Q} \qquad (12\text{-}14)$$

$$\tilde{V}_{1t} = \tilde{V}_S + \omega\, L_S \tilde{I}_{1Q} - j\omega\, L_S \tilde{I}_{1P} \qquad (12\text{-}15)$$

The reactive component of the load current, I_Q, directly affects the real part of V_U. On the other hand, the real part of the load current, I_P, is in quadrature with V_U, hence does not contribute significantly to its magnitude. The line voltage regulation is highly dependent on the reactive current drawn from the electric utility. The reactive current is responsible for reactive power drawn from the source. The line voltage control by the controlling I_Q is referred to as the static

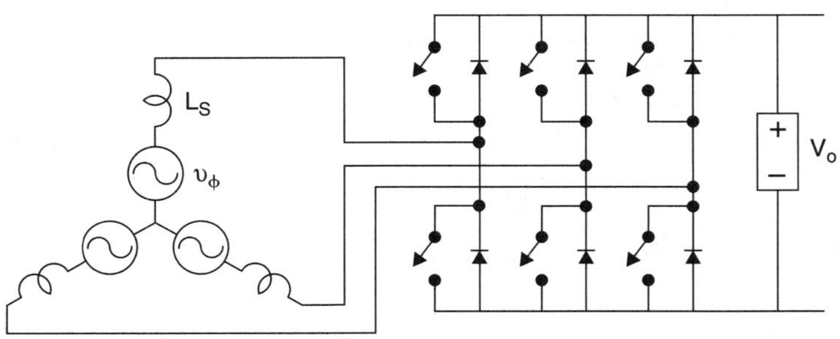

Figure 12.15
Three-phase switch-mode interface

VAR control. In electric utility network, it is desired to regulate the voltage within a narrow range, +5% to −10%, and have balanced loads on all phases. The static VAR controllers are used to prevent voltage flickers caused by the industrial loads that cause very rapid changes in the reactive power. Primarily, there are three types of static VAR controllers:

1. Switched-inductor compensator
2. Switched-capacitor compensator
3. Switch-mode compensator

12.7.1 Switched-inductor compensator

This compensator is used for compensating negative VAR drawn by the load. Connecting a variable inductor across the utility terminals controls the reactive current drawn from the electric utility. The variable inductor can be realized by controlled phase-delay switching of the inductor, as shown in Figure 12.16. Switch S1 is turned on at an angle α with respect to the line voltage. Nominally, the current in an ideal inductor lags the inductor voltage by $\pi/2$, hence, the phase-delay angle should be greater than $\pi/2$. After turning on of S1, the inductor current increases sinusoidally and commutates naturally at an angle $3\pi/2-\alpha$. Switch S2 is turned on at $\pi+\alpha$ and commutates at angle $5\pi/2-\alpha$. The thyristor is the preferred switching device. Figure 12.16 also shows the waveforms of the inductor current I_L and its fundamental component i_{L1}at delay angle α.

The fundamental harmonic of the inductor current waveform is given by, the proof is left for the readers,

$$I_Q = I_{L1} = \frac{V_S}{\pi\omega L}(2\pi - 2\alpha + \sin 2\alpha) \quad (12\text{-}16)$$

where $\pi/2 \leq \alpha \leq \pi$

The variable equivalent inductance is given by

$$L_{eq} = \frac{V_S}{\omega\,|I_{L1}|} = L\frac{\pi}{(2\pi - 2\alpha + \sin 2\alpha)} \quad (12\text{-}17)$$

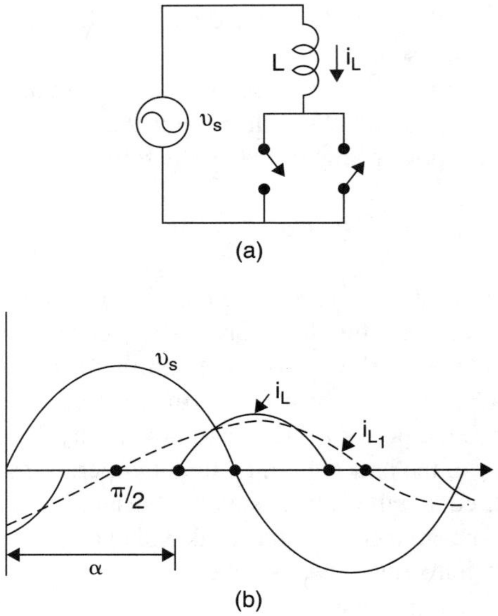

Figure 12.16
A switched-inductor compensator (a) Compensator and (b) Waveforms

Example 12-3

A single-phase electric utility, 120 Vac, 60 Hz, is to be regulated within (±) 5% at 10 kW power. The VAR load varies within (±) 25% of a nominal value of −4.5 KVAR. Design an appropriate conventional VAR compensator. Assume the line inductance to be 1.0 mH.

Negative VAR is compensated by a switched-inductor type compensator. The resistive and reactive components of the current are calculated as follows by assuming the terminal voltage at the load point being approximately equal to the utility voltage:

$$\tilde{I}_{1P} = \frac{P}{\tilde{V}_{1t}} \approx \frac{10{,}000}{120} = 83.3\,A$$

$$\tilde{I}_{1Q} = \frac{Q}{\tilde{V}_{1t}} \approx \frac{4{,}500}{120} = 37.5\,A$$

Subscript 1 refers to a fundamental harmonic. Use the circuit of Figure 12.16 for the switched-inductor type compensation. The delay angle α for the nominal value of I_Q is biased in the linear region

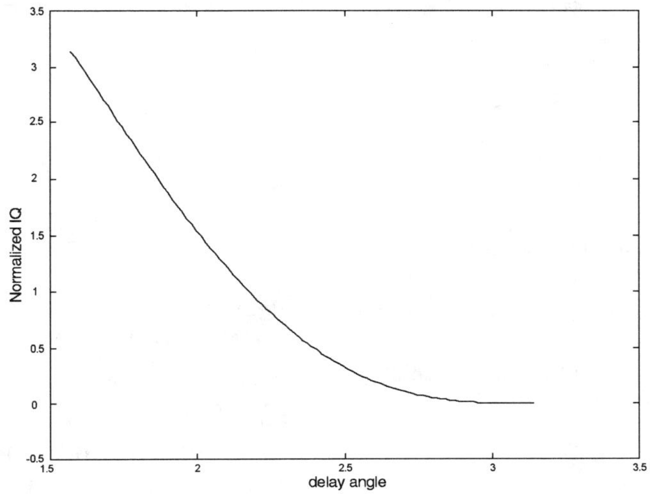

(c)

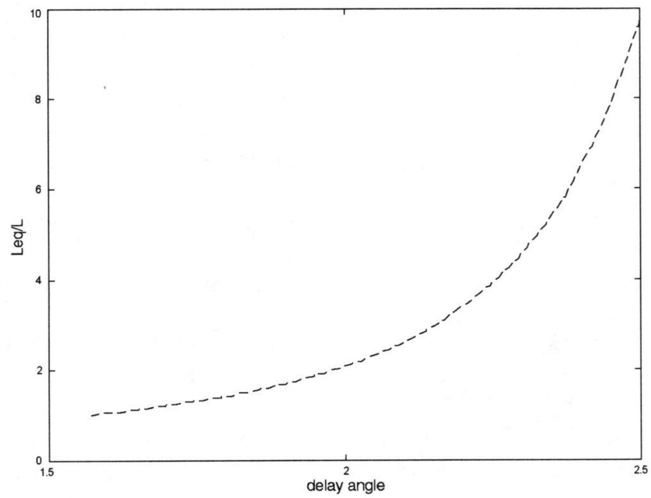

(d)

Figure 12.16 *(continued)*
(c) $I_Q/(V_S/\omega L)$ as a function of α (d) L_{eq}/L as a function of α

with the highest slope, as shown in Figure 12.16(c). The bias angle is selected to be $\alpha_0 = 1.8$ rad. The inductor L for this nominal operation is calculated by

$$\tilde{I}_Q = 37.5 = \frac{\tilde{V}_S}{\pi\omega L}[2\pi - 2 \times 1.8 + \sin(2 \times 1.8)]$$

$$L = 6.1\ mH$$

The reactive component current for a maximum 5% variation in the terminal voltage at the load point is

$$\tilde{I}_{1Q} = \frac{0.05*120}{\omega\ L_S} \approx \frac{6}{2\pi 6010^{-3}} = 15.9\ A$$

The reactive components of current for KVAR of 4500+25% and 4500−25% are given by:

$$\tilde{I}_{1Q'} \approx \frac{1.25*4500}{120} = 46.875\ A$$

$$\tilde{I}_{1Q''} \approx \frac{0.75*4{,}500}{120} = 28.125\ A$$

VAR compensation requires correction in I_Q = 46.875 − 15.9 = 30.975 A at one end to I_Q = 28.875 − 15.9 = 12.225 A at the other end of the VAR compensation range.

The delay angles for I'_Q and I''_Q are calculated from the following:

$$30.975 = \frac{\tilde{V}_S}{\omega L}[2\pi - 2\alpha' + \sin 2\alpha']$$

$$\approx \frac{120}{2\pi 60 \times 4.24 \times 10^{-3}}[2\pi - 2\alpha' + \sin 2\alpha']$$

$$12.225 = \frac{\tilde{V}_S}{\omega L}[2\pi - 2\alpha'' + \sin 2\alpha'']$$

$$\approx \frac{120}{2\pi 60 \times 4.24 \times 10^{-3}}[2\pi - 2\alpha'' + \sin 2\alpha'']$$

The delay angles $\alpha' - \alpha''$ are found by numerical search in MATLAB:

```
L=0.0061;omega=2*pi*60; Vsrms=120;IQ=46.875;
alpha1=pi/2; alpha2=pi-pi/10;f1=1;fm=-1;
while(abs(f1-fm)>=0.001)
  alpham=(alpha1+alpha2)/2;
  f1=-IQ+(Vsrms/(pi*omega*L))*(2*pi-2*alpha1+sin
    (2*alpha1));
  %f2=-IQ+(Vsrms/(pi*omega*L))*(2*pi-2*alpha2+sin
    (2*alpha2));
  fm=-IQ+(Vsrms/(pi*omega*L))*(2*pi-2*alpham+sin
    (2*alpham));
  if(f1*fm>0)
    alpha1=alpham;
  else
    alpha2=alpham;
  end
end
alpha=alpham
```

The range of delay is found to be
$\alpha' = 1.6508$ radians to $\alpha'' = 1.9506$ radians

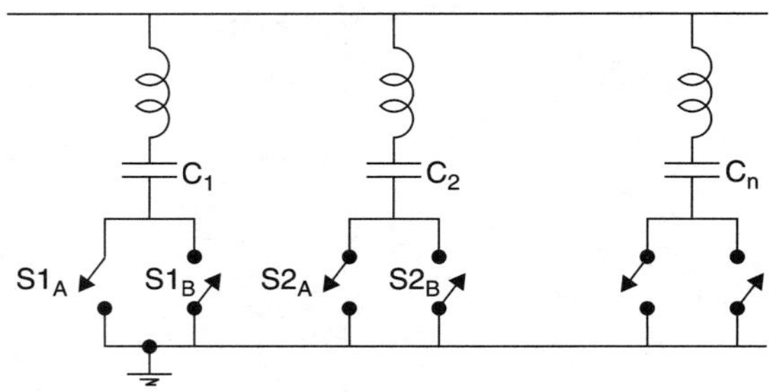

Figure 12.17
A switched-capacitor compensator

12.7.2 Switched-capacitor compensator

This compensator is used for compensating positive VARs in the load. Figure 12.17 shows a switched-capacitor compensator. Each switch is bidirectional, such as two thyristors back to back. These compensators use integral half-cycle control, that is, the capacitors are fully in or fully out of the circuit in each half cycle of the line frequency. The inductors in series with each capacitor branch are for limiting the current. Unlike the continuous control of the static VAR in the switched inductor compensator, the VAR in the switched-capacitor method is compensated in discrete steps.

12.7.3 Switch-mode compensator

The drawback of the above methods is that the compensation is done at line frequency. Hence, the value and the size of the inductor and capacitor banks are large and expensive. Switch-mode control of the reactive current is easier and less expensive. The switch-mode PFP of Figure 12.12 can be used as the per-phase compensator at high switching frequencies. Current-mode control is the preferred mode of control. Both the magnitude and the phase of line ac current can be controlled by this compensator.

12.8 SUMMARY

1. Harmonic standards and recommended practices have been established by international

agencies to limit the harmonics injected by power electronic components.

2. The source side inductance reduces the input current harmonics.
3. The power factor correction circuits are dc-dc converters that perform active current shaping on the input side.
4. Bidirectional switch-mode power factor correction circuits are introduced.
5. Static VAR compensators are used to control both the magnitude and the phase of the ac line currents from the utility.

References

1. Philip T. Krein, *Elements of Power Electronics,* Oxford University Press, New York, 1998.

2. Ned Mohan, Tore M. Undeland, and William P. Robbins, *Power Electronics: Converters, Applications and Design,* 2nd Ed., John Wiley & Sons, Inc., 1995.

3. Joseph Vithayathil, *Power Electronics: Principles and Applications,* McGraw-Hill Inc., New York, 1995.

4. Muhammad H. Rashid, *Power Electronics: Circuits, Devices and Applications,* 2nd Ed., Prentice-Hall Inc., New Jersey, 1993.

5. John G. Kassakian, Martin F. Sclecht, and G. Vergassian, *Principles of Power Electronics,* Addison-Wesley, Boston, 1991.

6. Kjeld Thorborg, *Power Electronics,* Prentice-Hall Inc., *1988.*

7. Daniel M. Mitchell, *Switching Regulator Analysis,* McGraw-Hill Book Company, New York, 1988.

Problems

12-1 Connect a capacitor across the output of the Boost type Power Factor Preregulator of Figure 12.3. Find the expression for the voltage ripple.

12-2 Find the expression for the fundamental harmonic of the inductor current in the switched-inductor VAR compensator.

12-3 A switching power supply with a high Power Factor Preregulator (PFP) is shown in Figure P12.3. The output voltage of the PFP is required to be 150 Vdc. Determine the range of variation of duty cycle D when the ac bus

a. varies in the range 90–130 Vrms with the nominal at 120 Vrms,

b. varies in the range 190–260 Vrms with the nominal at 120 Vrms,

c. may be either of the two ac buses.

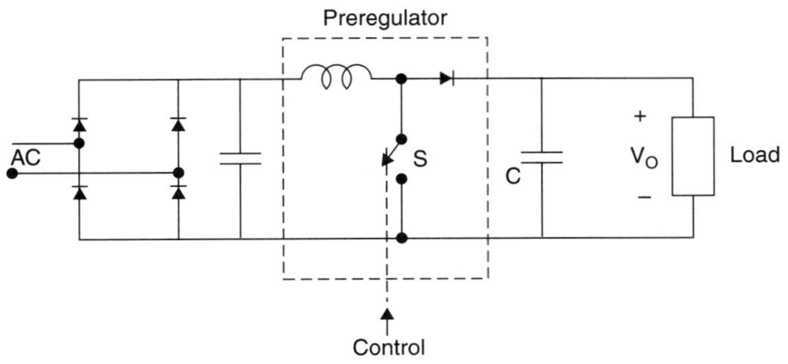

Figure P12.3

12-4 A full-bridge rectifier from 60 Hz, 120 V rms ac source is designed to provide 350 V dc to a resistive load of 10 ohms. Use an ideal Boost type Power Factor Preregulator of Figure 12.7, which is switched at 10 times the line frequency.

a. Find the inductance L and plot the duty ratio, the current waveform at the input of the PFP and the average input current over a half line period.

b. Find the frequency spectrum of the line current in Problem 12-4.

12-5 A single-phase electric utility, 120 V, 60 Hz, is to be regulated within (+/−) 5% at 5 kW power. The VAR load varies within (+/−) 30% of a nominal value of −3.5 KVAR. Design an appropriate conventional VAR compensator. Assume the line inductance to be 0.8 mH.

CHAPTER 13
Converter Control

OBJECTIVES

1. Averaging a variable and a circuit
2. Linearizing a nonlinear variable and a circuit
3. State-space averaging
4. Developing the transfer function of converter systems
5. Feedback control of converters

13.1 INTRODUCTION

Converter control refers to specifying the desired nominal operating condition and then regulating the converter so that it stays close to the nominal performance in the face of disturbances, noise and modeling errors.

Controller action:

1. to easily determine the operating point on the control characteristics
2. to regulate the converter performance in the vicinity of the operating point.

So far we have focused on the analysis of the steady-state performance of power electronic converters. The converter operation in steady state represents a single point, called the operating point, on the static control characteristics curves. The transient response of the power electronic converter can be explained using a dynamic model in the vicinity of the operating point.

The dynamic model can be one of several:

1. Averaged circuit model
2. Linearized model
3. State-space model and averaged state-space model
4. Discrete time-sampled data model

The first three categories will be discussed in this chapter. The fourth category needs a much greater background in discrete control systems, which is beyond the scope of this book. If readers are interested, they can research this topic on their own.

13.2 AVERAGED MODEL

We can analyze the dynamic behavior of the power electronic converter from an operational model in the vicinity (or local) of the steady-state operating point. The converter operation involves switching modes, therefore an overall converter model is a nonlinear model. The averaged model, localized to the steady-state operating point, is a linear model. The analytical techniques for linear systems are very well known.

13.2.1 Averaging

Local average of a variable x taken over the interval T:

$$\bar{x} = \frac{1}{T}\int_{t-T}^{t} x(\tau)d\tau \qquad (13\text{-}1)$$

$\bar{x}$ is smoother than x and it is a continuous function of time unless x has impulses in it. The derivative of the local average variable $\bar{x}$ is also the average of the derivative of x:

$$\frac{d\bar{x}}{dt} = \frac{1}{T}\frac{d}{dt}\int_{t-T}^{t} x(\tau)d\tau = \frac{1}{T}\int_{t-T}^{t}\left(\frac{dx}{d\tau}\right)d\tau \qquad (13\text{-}2)$$

This leads to the following in reference to the voltage-current relationship of an inductor:

$$\overline{v_L} = L\frac{d\bar{i}}{dt} = \frac{1}{T}\int_{t-T}^{t}\left(L\frac{di}{d\tau}\right)d\tau \qquad (13\text{-}3)$$

Extending the above reasoning, we could find the local average of current and voltage in a capacitor:

$$\overline{i_C} = C\frac{d\bar{v}}{dt} \qquad (13\text{-}4)$$

The local averaging concept can be extended to the circuit laws: the Kirchoff's voltage law and Kirchoff's current law:

$$\sum_{m=1,2,\ldots M}\overline{v_m} = 0 \quad , \quad \sum_{n=1,2,\ldots N}\overline{i_n} = 0 \qquad (13\text{-}5)$$

M is the number of elements in a loop (mesh) and N is the number of branches forming a node.

A switch in series with a voltage source could also be easily accounted for in the averaging process, as shown in Figure 13.1. The average Thevenin voltage at the terminals a-a′ is

$$\overline{v_{aa'}} = \bar{D}\,V_S \quad where \quad \bar{D} = \frac{\tau}{T_S} \qquad (13\text{-}6)$$

D is the duty ratio.

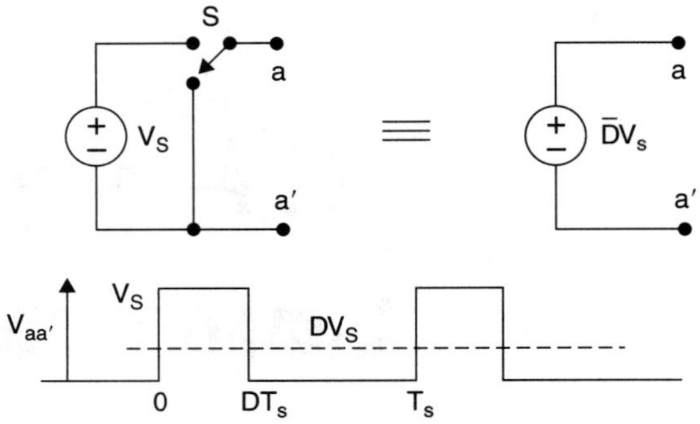

Figure 13.1
Averaging a source connected to a two-position switch that is turned on and off periodically

The average model of a bridge network of switches connected to a voltage source is shown in Figure 13.2.

$$\begin{aligned}\overline{v_{aa'}} &= \frac{1}{T_S}[\bar{D}T_S V_S - (1 - \bar{D})T_S V_S] \\ &= (2\bar{D} - 1)V_S \end{aligned} \qquad (13\text{-}7)$$

13.2.2 Averaged model

The averaged model of a converter circuit can be derived as follows:

1. Replace the instantaneous voltages and currents by their averages.
2. Replace the switches by elements that have nonzero average voltage and nonzero average current. The switch control function determines the average value.
3. Components R, L and C are unchanged.
4. The steady-state models like the voltage sink and the current sink are replaced by actual components as they would be required for determining the transient response.

Example 13-1 Averaged Model of Buck Converter

Develop the averaged model for a dc-dc Buck converter in which the source voltage is 24 Vdc and the output voltage is 9 Vdc operating in the continuous conduction mode. Other components are L = 1 mH and C = 100 μF.

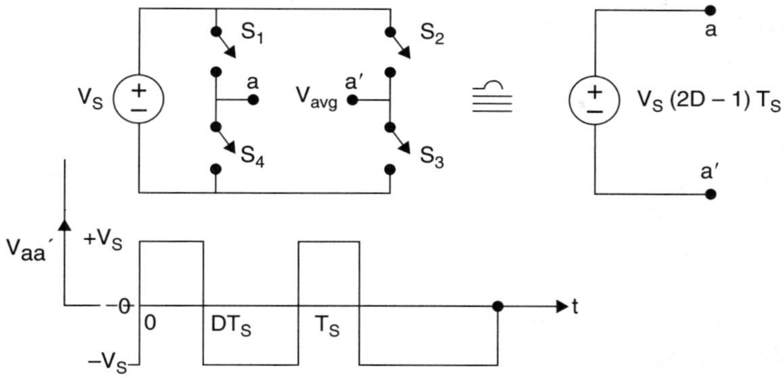

$$V_{aa'} = \frac{1}{T_S}[V_S DT_S - (1 - D)T_S V_S]$$

$$= V_S(2D - 1)$$

Figure 13.2
Averaging a source connected to a bridge network of switches

Figure 13.3
(a) Buck Converter and (b) averaged model

The Buck converter circuit is repeated in Figure 13.3(a) for convenience. The average duty ratio is calculated as

$$\overline{D} = \frac{\overline{V}_O}{V_S} = \frac{9}{24} = 0.375$$

The averaged model is shown in Figure 13.3(b).

Dynamic and Steady-State Response from the Averaged Model

The steady-state response of the converter is obtained from the averaged model by 'shorting' the inductors and 'opening' the capacitors in the circuit. A step change in the average duty ratio D from 0.375 to 0.5 produces a step change in the averaged dc voltage source $\overline{V}_{aa'}$ from 9 to 12 V.

The transient response of the output voltage (hence the output current) is obtained using the Laplace Transform:

$$V_O(s) = \frac{dV_S}{s} \frac{R // \frac{1}{sC}}{\left(R // \frac{1}{sC}\right) + sL} \tag{13-8}$$

$$= \frac{d\,V_S}{LC} \frac{1}{s\left(s^2 + \frac{1}{RC}s + \frac{1}{LC}\right)}$$

The transient response $\overline{V}_O(t)$ would have an over-damped response if

$$\left(\frac{1}{RC}\right)^2 - \frac{4}{LC} \geq 0 \Rightarrow R \leq \frac{1}{2}\sqrt{\frac{L}{C}} \quad (13\text{-}9)$$

Example 13-2 Dynamic Response from the Averaged Model

In the Buck converter of Example 13-1,

a. *for output dc current $I_O = 20$ A, determine the transient response of the output voltage V_O when the duty ratio D undergoes a step change to 0.5 (in the next switching period).*
b. *Repeat (a) for $I_O = 2$ A. Compare the nature of the two responses.*
c. *Find L for the critically damped response at $I_O = 20$ A.*

a. The step change in the duty ratio d is (0.5−0.375) = 0.125.

The load resistance is calculated from 9 volts: $R = \dfrac{V_O}{I_O} = \dfrac{9}{20}$

The time-domain responses of the average dc output voltage for 20 A and 2 A load are given respectively in Figure 13.4 and Figure 13.5.

b. The load resistance at 2 A is: $R = \dfrac{V_O}{I_O} = \dfrac{9}{2}$

Note that the response is oscillatory, like the second order system response at low current (high R).

c. L for the critically damped response at $I_O = 2$ A is obtained from equation (13-9)

$$L = 4R^2C = 4.(9/2)^2.10^{-4} = 8.1.10^{-3} \quad or \quad 8.1\ mH$$

The time-domain responses of the average dc output voltage is shown in Figure 13.6.

13.2.3 Local average at the switching frequency

The averaging in the previous section can be described as the averaging at dc, the zero frequency. A generalized average, which is local to a frequency ω, may be defined as

$$\overline{x}^{\omega} = \frac{1}{T}\int_{t-T}^{t} e^{-j\omega\tau}\, x(\tau)d\tau \quad (13\text{-}10)$$

The ω may be the switching frequency of the converter. The ω-average is particularly useful for analyzing the dynamic behavior of resonant converters. Details of this topic are beyond the scope of

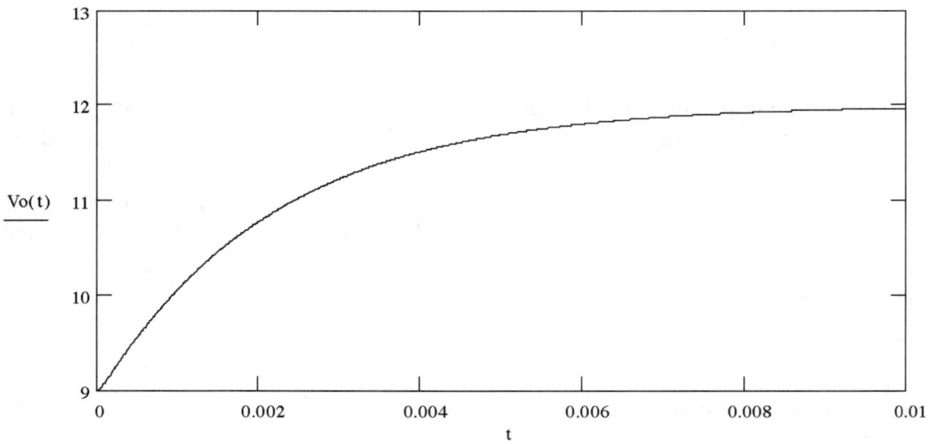

Figure 13.4
Average output voltage transient response at $I_O = 20$ amps

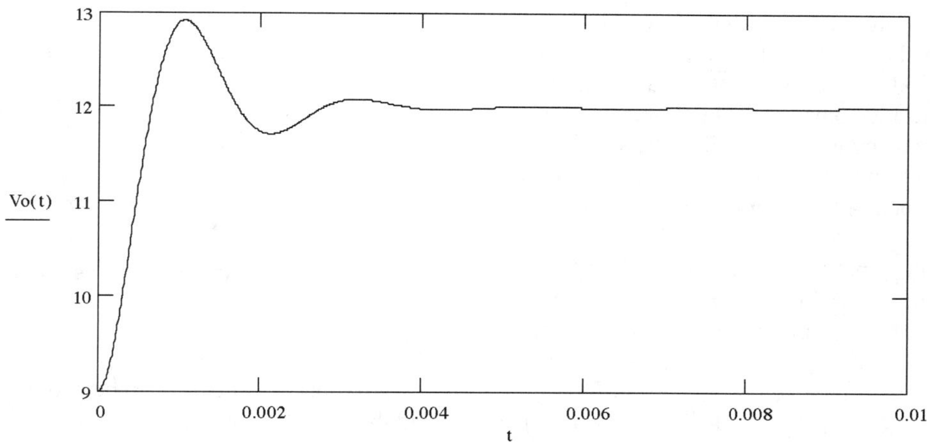

Figure 13.5
Transient response of the average output voltage at I_O = 2 amps

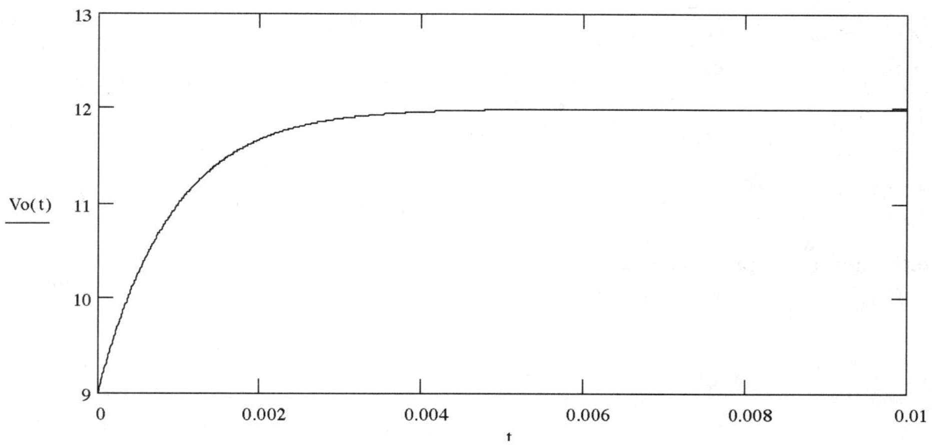

Figure 13.6
Critically damped transient response of the average output voltage at I_O = 20 A for L = 8.1 mH

this book. Readers are recommended to see details in the literature.

13.3 LINEARIZED MODEL

Rigorous determination of the stability and the control characteristics of power electronic converters requires solving the nonlinear model. Linearization of the nonlinear model greatly simplifies analysis. Linearized models approximately describe small deviations or perturbations from nominal operation of a system.

13.3.1 Linearization

Linearization of a variable or a circuit involves expanding nonlinear terms and expressions into a

Taylor series around a nominal value and then retaining only the first order terms. The nominal value is selected to be the steady-state operating point in power electronic converters. Thus achieved the linear model approximately describes the small deviations in the vicinity of the steady-state operating point. The linearized model is also called the *small signal model.*

The small signal variable using the Taylor series expansion of a variable x(a, b) around a nominal value $\overline{X}$ and then retaining the first order term, is given by:

$$\hat{x}(a,b) \approx \frac{\partial x}{\partial a}\hat{a} + \frac{\partial x}{\partial b}\hat{b} \qquad (13\text{-}11)$$

Following the reasons as given in averaging, the derivative of the perturbation variable x is also the perturbation of the derivative of x.

$$\begin{aligned}\frac{d\hat{x}}{dt} &\approx \frac{d}{dt}\left(\frac{\partial x}{\partial a}\right)\hat{a} + \frac{\partial x}{\partial a}\frac{d\hat{a}}{dt} + \frac{d}{dt}\left(\frac{\partial x}{\partial b}\right)\hat{b} + \frac{\partial x}{\partial b}\frac{d\hat{b}}{dt}\\ &\approx \frac{\partial}{\partial a}\left(\frac{dx}{dt}\right)\hat{a} + \frac{\partial x}{\partial a}\frac{d\hat{a}}{dt} + \frac{\partial}{\partial b}\left(\frac{dx}{dt}\right)\hat{b} + \frac{\partial x}{\partial b}\frac{d\hat{b}}{dt}\\ &\approx \frac{\partial}{\partial a}\left(\frac{dx}{dt}\right)\hat{a} + \frac{\partial}{\partial b}\left(\frac{dx}{dt}\right)\hat{b} \qquad (13\text{-}12)\end{aligned}$$

The perturbations concept can be extended to the circuit laws: Kirchoff's voltage law and Kirchoff's current law:

$$\sum_{m=1,2,\ldots M} \hat{v}_m = 0, \qquad \sum_{n=1,2,\ldots N} \hat{\imath}_n = 0 \qquad (13\text{-}13)$$

M is the number of elements in a loop (mesh) and N is the number of branches forming a node.

Example 13-3 Linearized Model of Buck Converter in CCM and the Transient Response

Develop the linearized model of the dc-dc Buck converter of Example 13-1.

The converter is operating in the Continuous Conduction Mode. The average dc current in a switching period in steady state is

$$\overline{I}_O = \frac{V_S}{R}D$$

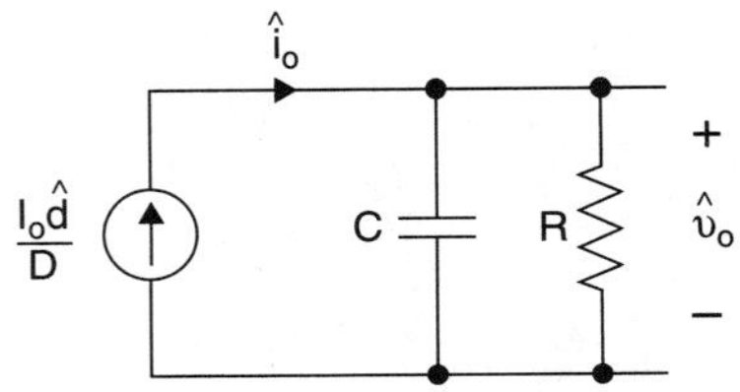

Figure 13.7
Linearized model of the Buck Converter in CCM

Assuming the duty ratio to be perturbed to D + $\hat{d}$ and the output current to I_O + i ^$_O$,

$$\hat{\imath}_O \approx \frac{\partial i_O}{\partial D}\hat{d} = \frac{V_S}{R}\hat{d} = \frac{I_O}{D}\hat{d}$$

The linearized model is shown in Figure 13.7. Compare it with the averaged model in Figure 13.2.

The transient response of the output voltage due to a step change in the duty ratio by d is obtained using the Laplace Transform:

$$V_O(s) = \frac{I_O}{D}\frac{d}{s}\left[R\,//\frac{1}{sC}\right] = \frac{d\,I_O}{D\,C}\frac{1}{s\left(s+\frac{1}{RC}\right)}$$

The time domain of the change in the output voltage is given in Figure 13.8.

Example 13-4 Linearized Model of Buck Converter in DCM

The converter is operating in the Discontinuous Conduction Mode. The output voltage and the average dc current in a switching period in steady state is

$$\frac{V_O}{V_S} = \frac{D}{D+\Delta} \Rightarrow \Delta = D\left(\frac{V_S}{V_O} - 1\right)$$

$$\overline{I}_O = \frac{V_S T_S}{2L}D\Delta \Rightarrow \overline{I}_O = \frac{V_S T_S}{2L}D^2\left(\frac{V_S}{V_O} - 1\right)$$

Assuming the duty ratio to be perturbed to D + d^ and the output current to I_O + i^$_O$,

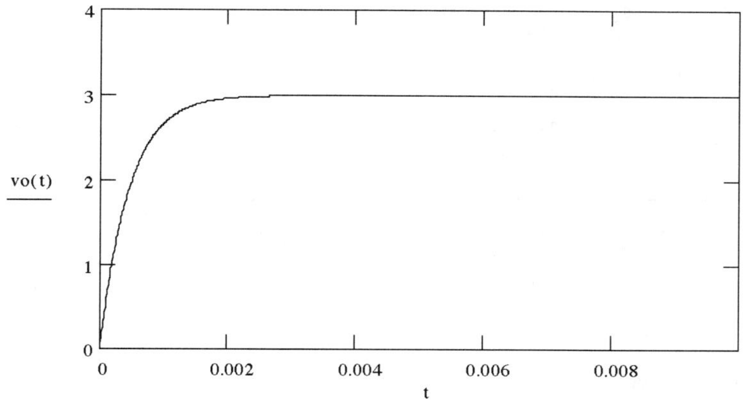

Figure 13.8
Transient response of the output voltage of the linearized Buck converter

$$\hat{i}_O \approx \frac{\partial \bar{I}_O}{\partial D}\hat{d} + \frac{\partial \bar{I}_O}{\partial V_O}\hat{v}_O$$
$$\approx \frac{V_S T_S}{2L} 2D\left(\frac{V_S}{V_O} - 1\right)\hat{d} + \frac{V_S T_S}{2L} D^2\left(-\frac{V_S}{V_O^2}\right)\hat{v}_O$$
$$\approx \frac{2I_O}{D}\hat{d} - \frac{D + \Delta}{R}\hat{v}_O$$

The linearized model is shown in Figure 13.9.

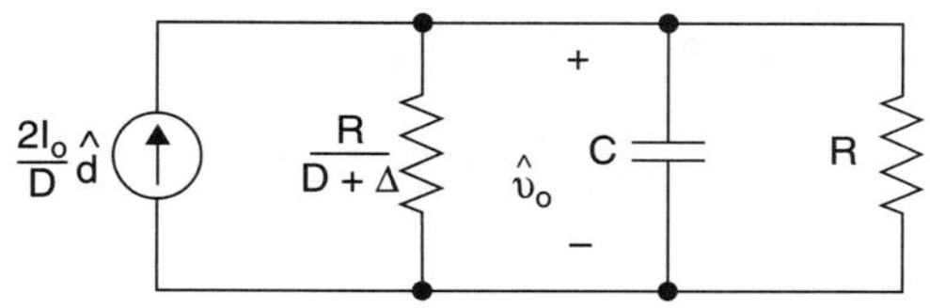

Figure 13.9
Linearized model of the Buck Converter in DCM

13.4 STATE-SPACE AVERAGED MODEL

We will use the Buck converter to illustrate the small signal transfer function using the state-space averaging method.

Continuous Conduction Mode of Operation

D: duty ratio at the operating point (in steady state)
V_O: dc output voltage at the operating point
$\hat{d}$: small perturbation in the duty ratio
$\hat{v}_o$: small perturbation in the output voltage
V_S: input dc voltage source

The steady-state operation consists of two circuit modes, as shown in Figure 13.10 :

(i) mode m1: switch on, duration $D.T_S$ and

(ii) mode m2: switch off, duration $(1-D)T_S$

The procedural steps of developing the desired model are listed below:

1. A state variable vector is constructed of inductor current i_L and capacitor voltage v_C.

$$\mathbf{x} = \begin{bmatrix} i_L \\ v_C \end{bmatrix} \qquad (13\text{-}14)$$

Small perturbations in the state variables, the duty ratio, the source voltage and consequent perturbation in the output voltage are:

$$\mathbf{x} = \mathbf{X} + \hat{\mathbf{x}}, \quad d = D + \hat{d}, \quad v_S = V_S + \hat{v}_S, \quad v_O = V_O + \hat{v}_O \qquad (13\text{-}15)$$

In the following analysis, we will assume the source voltage to be constant, hence, $v_S = V_S$.

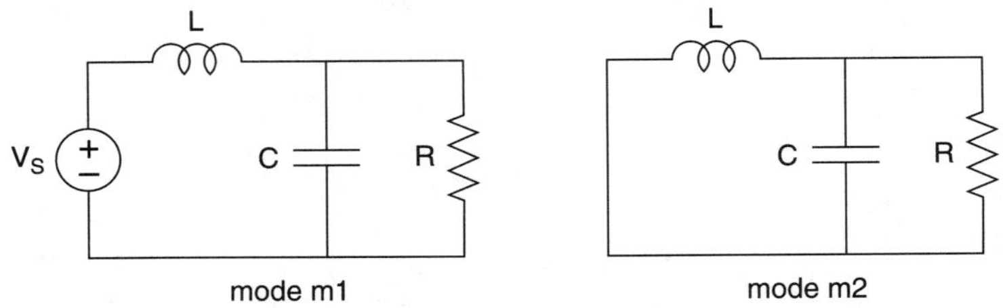

Figure 13.10
Circuit modes in Buck converter

The state-space model is given by

*mode m*1:

$$\dot{\mathbf{x}} = \mathbf{A}_1\mathbf{x} + \mathbf{B}_1 V_S \quad during \quad d.T_S$$
$$v_O = \mathbf{C}_1\mathbf{x}$$

*mode m*2: (13-16)

$$\dot{\mathbf{x}} = \mathbf{A}_2\mathbf{x} + \mathbf{B}_2 V_S \quad during \quad (1-d).T_S$$
$$v_O = \mathbf{C}_2\mathbf{x}$$

where

$$\mathbf{A}_1 = \begin{bmatrix} 0 & -\dfrac{1}{L} \\ \dfrac{1}{C} & -\dfrac{1}{RC} \end{bmatrix}, \quad \mathbf{A}_2 = \begin{bmatrix} 0 & -\dfrac{1}{L} \\ \dfrac{1}{C} & -\dfrac{1}{RC} \end{bmatrix},$$

$$\mathbf{B}_1 = \begin{bmatrix} \dfrac{1}{L} \\ 0 \end{bmatrix}, \quad \mathbf{B}_2 = \begin{bmatrix} 0 \\ 0 \end{bmatrix}$$

$$\mathbf{C}_1 = [0 \quad 1], \quad \mathbf{C}_2 = [0 \quad 1],$$

A_1, and A_2 are the state matrices and B_1 and B_2 are excitation vectors.

2. **Averaging:** The state equations of two circuit modes are *time weighted* and *averaged* over the switching period T_S:

$$\dot{\mathbf{x}} = [\mathbf{A}_1 d + \mathbf{A}_2(1-d)]\mathbf{x} + [\mathbf{B}_1 d + \mathbf{B}_2(1-d)]V_S$$
$$v_O = [\mathbf{C}_1 d + \mathbf{C}_2(1-d)]\mathbf{x} \qquad (13\text{-}17)$$

3. **Steady-state:** The steady state values of the state variables, the sources and the duty ratio are assumed as

$$\mathbf{x} = \mathbf{X} \quad , \quad v_O = V_O \quad , \quad d = D$$
$$\dot{\mathbf{x}} = \mathbf{A}\,\mathbf{X} + \mathbf{B}\,V_S = \mathbf{0}$$
$$V_O = \mathbf{C}\,\mathbf{X}$$

$$where \quad \mathbf{A} = \mathbf{A}_1 D + \mathbf{A}_2(1-D)$$
$$\mathbf{B} = \mathbf{B}_1 D + \mathbf{B}_2(1-D) \qquad (13\text{-}18)$$
$$\mathbf{C} = \mathbf{C}_1 D + \mathbf{C}_2(1-D)$$

$$\mathbf{A} = \begin{bmatrix} 0 & -\dfrac{1}{L} \\ \dfrac{1}{C} & -\dfrac{1}{RC} \end{bmatrix}, \quad \mathbf{B} = \begin{bmatrix} \dfrac{D}{L} \\ 0 \end{bmatrix}, \quad C = [0 \quad 1]$$

Hence, the *steady-state transfer function* is

$$\mathbf{X} = -\mathbf{A}^{-1}\mathbf{B}V_S$$
$$\frac{V_O}{V_S} = -\mathbf{C}\mathbf{A}^{-1}\mathbf{B} \qquad (13\text{-}19)$$

$$\frac{V_O}{V_S} = [0 \quad 1]\begin{bmatrix} 0 & -\dfrac{1}{L} \\ \dfrac{1}{C} & -\dfrac{1}{RC} \end{bmatrix}^{-1}\begin{bmatrix} \dfrac{D}{L} \\ 0 \end{bmatrix}$$

$$= [0 \quad 1]\frac{1}{1/LC}\begin{bmatrix} -\dfrac{1}{RC} & \dfrac{1}{L} \\ -\dfrac{1}{C} & 0 \end{bmatrix}\begin{bmatrix} \dfrac{D}{L} \\ 0 \end{bmatrix} \qquad (13\text{-}20)$$

$$= D$$

4. The state equation during the perturbation is (from equations (13-15) and (13-16))

$$\dot{\mathbf{x}} + \dot{\mathbf{x}} = [\mathbf{A}_1(D + \hat{d}) + \mathbf{A}_2(1 - D - \hat{d})]$$
$$(\mathbf{X} + \hat{\mathbf{x}}) + [\mathbf{B}_1 + \mathbf{B}_2(1 - D - \hat{d})]V_S \qquad (13\text{-}21)$$

$$= \mathbf{A}\,\mathbf{X} + \mathbf{B}\,V_S + \mathbf{A}\,\mathbf{x} + [(\mathbf{A}_1 - \mathbf{A}_2)\mathbf{X} + (\mathbf{B}_1 - \mathbf{B}_2)V_S]d + (\mathbf{A}_1 - \mathbf{A}_2)\,\mathbf{x}d$$

Using equation (13-17) and ignoring term $\hat{\mathbf{x}}\,\hat{\mathrm{d}}$,

$$\dot{\mathbf{x}} = \mathbf{A}\,\mathbf{x} + [(\mathbf{A}_1 - \mathbf{A}_2)\mathbf{X} + (\mathbf{B}_1 - \mathbf{B}_2)V_S]d \quad (13\text{-}22)$$

Similarly,

$$V_O + v_O = [\mathbf{C} + (\mathbf{C}_1 - \mathbf{C}_2)\,d]\,(\mathbf{X} + \mathbf{x}) \quad (13\text{-}23)$$

Using the steady-state output $V_O = \mathbf{C}\,\mathbf{X}$ and ignoring the term $\hat{\mathbf{x}}\,\hat{\mathrm{d}}$,

$$v_O = \mathbf{C}\,\mathbf{x} + [(\mathbf{C}_1 - \mathbf{C}_2)\mathbf{X}]d \quad (13\text{-}24)$$

5. Transforming to s-domain

$$s\mathbf{x}(s) = \mathbf{A}\,\mathbf{x}(s) + [(\mathbf{A}_1 - \mathbf{A}_2)\mathbf{X} + (\mathbf{B}_1 - \mathbf{B}_2)V_S]d(s)$$
$$v_O(s) = \mathbf{C}\,\mathbf{x}(s) + [(\mathbf{C}_1 - \mathbf{C}_2)\mathbf{X}]d(s) \quad (13\text{-}25)$$

The small signal transfer function can be shown to be

$$\frac{v_O(s)}{d(s)} = \mathbf{C}[s\mathbf{I} - \mathbf{A}]^{-1}[(\mathbf{A}_1 - \mathbf{A}_2)\mathbf{X} + (\mathbf{B}_1 - \mathbf{B}_2)V_S] + (\mathbf{C}_1 - \mathbf{C}_2)\mathbf{X} \quad (13\text{-}26)$$

I is the identity matrix. In our example, $A_1 - A_2 = 0$ and $C_1 - C_2 = 0$. Thus, the transfer function of the Buck converter in the CCM operation is obtained as

$$G_S(s) = \frac{\hat{v}_O(s)}{\hat{d}(s)} = [0 \quad 1]\begin{bmatrix} s & \frac{1}{L} \\ -\frac{1}{C} & s + \frac{1}{RC} \end{bmatrix}^{-1} \begin{bmatrix} \frac{1}{L} \\ 0 \end{bmatrix} V_S = \frac{(V_S/LC)}{s^2 + \frac{1}{RC}s + \frac{1}{LC}} \quad (13\text{-}27)$$

Example 13-5 Transfer Function of Buck Converter Using State-Space Averaging, Pole Zero Location and Step Response

Find the small signal transfer function for a Buck converter of Figure 13.3 at (a) I_O=2 A and (b) I_O=20 A. Find and plot in both cases the poles and zeros of the system and the step response for a change in the duty ratio from 0.375 to 0.5.

Step change in duty ratio d=0.125.

a. I_O=2 A: R = 9/2 ohms

Substituting the values of L, C and V_s in equation (13-27), the transfer function is obtained as

$$G_S(s) = \frac{2.4 \times 10^8}{s^2 + 2.22 \times 10^3 s + 10^7}$$

The transfer function does not have any zero but has two poles.

The output voltage for a step change in the duty ratio is obtained from

$$\hat{v}_O(s) = G_S(s)\,\hat{d}(s) = G_S(s)\,\frac{d}{s}$$

Use the MATLAB program to find the location of poles and the step response of the small signal transfer function:

```
Vs=24; L=1e-3; C=1e-4; R=9/2; d=0.125;
num=Vs/(L*C); % numerator contains a constant term
den=[1 1/(R*C) 1/(L*C)]; % the coefficients of the
  denominator polynomial in s
  [z,p,k]=zpkdata(num, den) %the zeros, poles and the
    DC gain of the transfer function
  pzmap(num, den) % the pole-zero map in the s-plane
  vo=d*step(num, den) % response of the output voltage
    for a step change in d
```

b. I_O=20 amps: R = 9/20 ohms

Substituting the values of L, C and V_s in equation (13-27), the transfer function is obtained as

$$G_S(s) = \frac{2.4 \times 10^8}{s^2 + 2.22 \times 10^4 s + 10^7}$$

Change the value of R to 9/20 in the MATLAB program and run it again. The pole zero locations and the step response are shown in Figure 13.11.

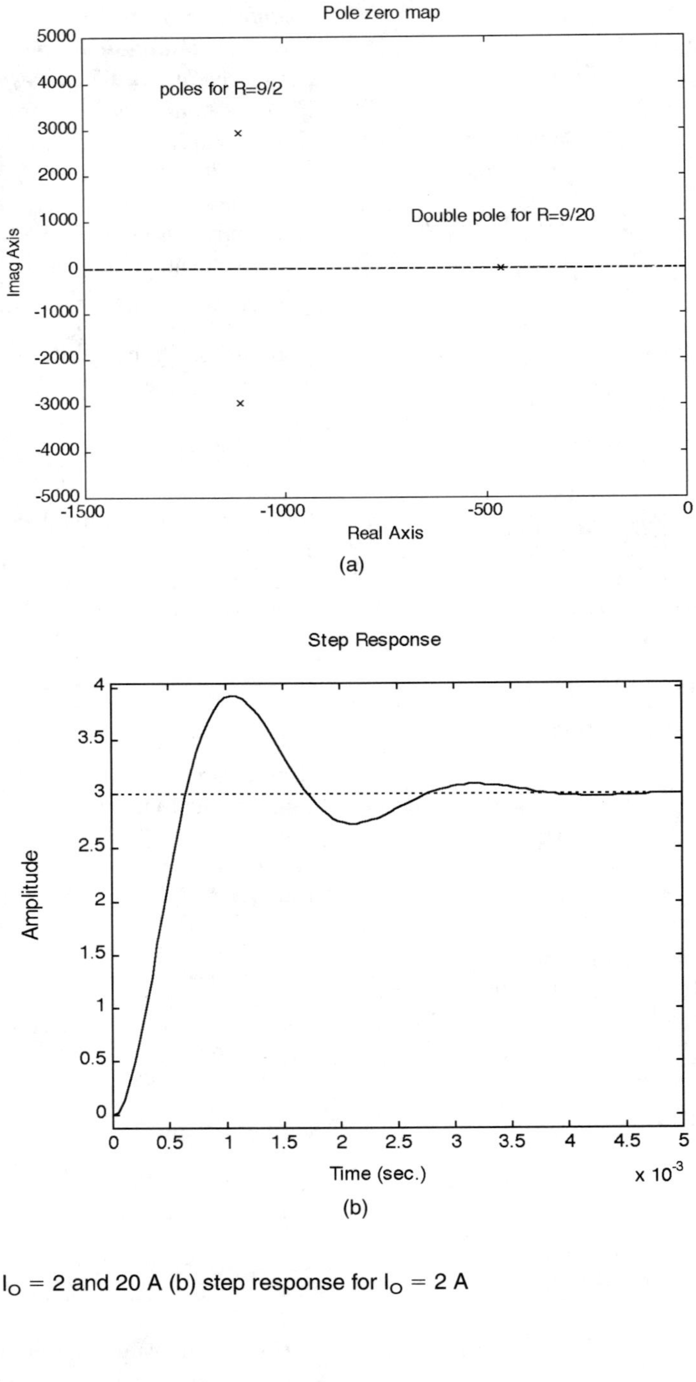

Figure 13.11
(a) Pole-zero plots for $I_O = 2$ and 20 A (b) step response for $I_O = 2$ A

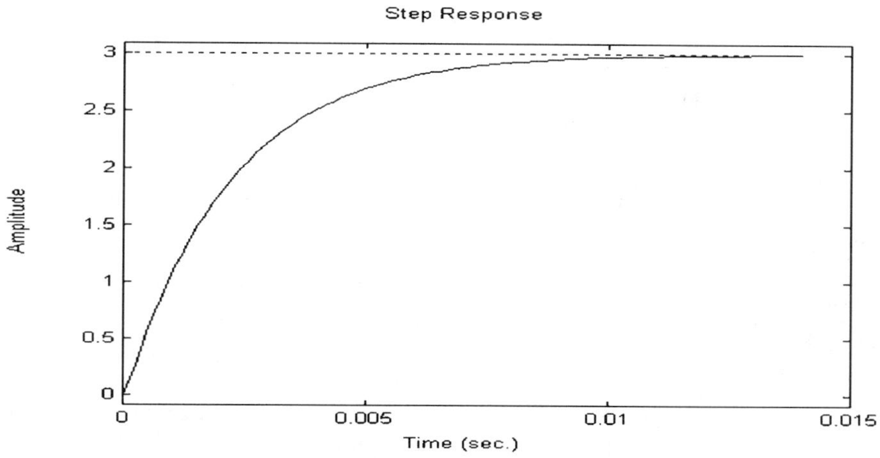

(c)

Figure 13.11 ***(continued)***
(c) step response for $I_O = 20$ A

Note that at low value of I_O (light load), the step response is faster and has an overshoot, and the poles are closer to the origin in the s-plane. At the high value of I_O (heavy load), the step response is sluggish and the poles are farther to the left of the origin in the s-plane.

13.5 FEEDBACK CONTROL

Converter response to a change in the input source, a load or a parameter, can be modified to exhibit the desired characteristics by employing feedback control techniques. We will illustrate feedback control using the example of the Buck converter in CCM, as given in Figure 13.12.

The output voltage is subtracted from the reference voltage and the error voltage is passed through a controller block and then used to control the pulse width of the Buck converter. The open-loop transfer function without the controller is

$$G_{OL}(s) = G_P(s)G_S(s) \qquad (13\text{-}28)$$

where $G_P(s)$ and $G_s(s)$ are the PWM transfer function and the transfer function of the Buck converter respectively. We would use the $G_S(s)$ of equation (13-27) that was developed using the state-plane averaging method. Next, we will determine the $G_P(s)$ in the following section.

13.5.1 Transfer function of pulse width modulator

The waveforms of the control voltage and the reference voltage, the two inputs to the pulse width modulator, are shown in Figure 13.13. The control voltage v_c is compared with a reference voltage v_R. The reference voltage has the saw-tooth waveform of the switching frequency. The duty ratio is given by

$$d = \frac{v_c}{V_R} \qquad (13\text{-}29)$$

where

$$d = D + \hat{d} \quad v_C = V_C + \hat{v}_C \quad and \quad D = \frac{V_c}{V_R}$$

d^ and v^$_C$ are small perturbations in the nominal duty ratio D and the nominal control voltage V_C respectively. The small perturbations are assumed to be at much smaller frequencies than the switching

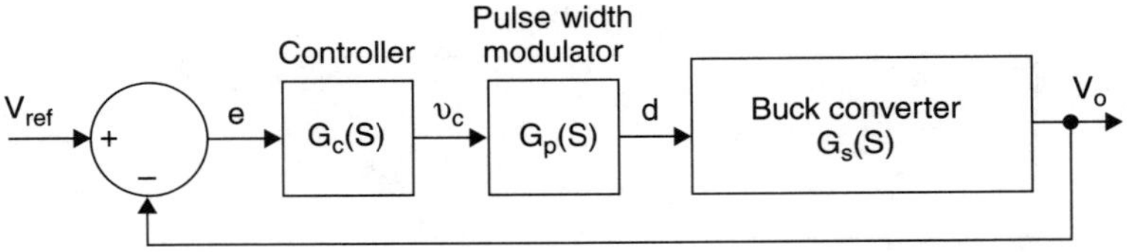

Figure 13.12
Feedback control system for the Buck Converter

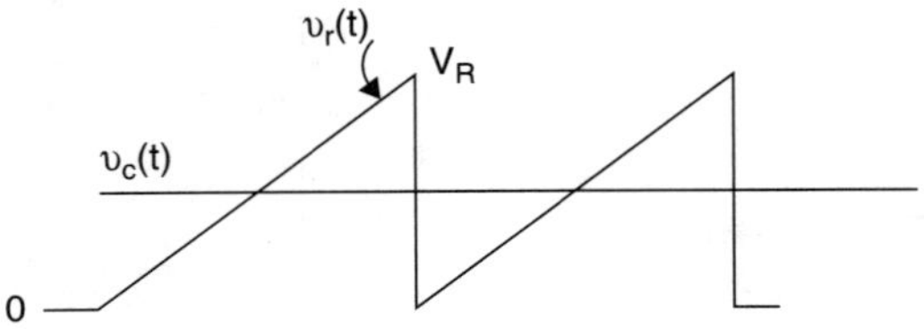

Figure 13.13
Waveforms in the pulse-width modulator

frequency. V_R is the peak voltage of the reference saw-tooth waveform.

As the result of perturbations in the control voltage, the duty ratio is modified as:

$$d = D + \hat{d} = \frac{V_c + \hat{v}_c}{V_R} = \frac{V_c}{V_R} + \frac{\hat{v}_c}{V_R} \tag{13-30}$$

$$\hat{d} = \frac{\hat{v}_c}{V_R}$$

The transfer function of the pulse-width modulator is

$$G_P(s) = \frac{\hat{d}(s)}{\hat{v}_c(s)} = \frac{1}{V_R} \tag{13-31}$$

13.5.2 Transient response of open-loop system

The open-loop transfer function for the Buck converter in CCM without the controller is obtained from equations (13-28) and (13-31):

$$G_{OL}(s) = \frac{V_S}{V_R} \frac{(1/LC)}{s^2 + \frac{1}{RC}s + \frac{1}{LC}} \tag{13-32}$$

The highest power of s in the denominator polynomial denotes the order of the open-loop transfer function. The Buck converter is a second-order system.

The close-loop transfer function in the control system of Figure 13.12 with $G_C(s)=1$ is

$$G_{CL}(s) = \frac{G_P(s)G_S(s)}{1 + G_P(s)G_S(s)} \tag{13-33}$$

The converter response of the closed-loop system consists of two parts: (i) transient response and (ii) the steady-state response. The steady-state response is the response of the converter when all transient behavior has subsided. It is determined by the input excitation, the reference input parameter in the control system. The error between the reference and the actual output parameter (output voltage in our example) is determined by the location of zeros of the open-loop transfer function $G_{OL}(s)$. The transient response is solely determined by the pole positions of the open-loop transfer function.

We now state important parameters of the time-domain response of a control system:

Response Time t_r: The time the output variable of the control system takes from the application of the input step change to 90% of the steady-state value.

Settling Time t_S: The time the output variable takes to settle within a small percentage (2%–5%) of the steady-state value.

Overshoot: The overshoot in transient response expressed as the percentage of the steady-state value of the output variable.

Steady-state error: The error between the reference input and the actual steady-state value of the output variable.

Readers are encouraged to find the above parameters from the plots of Figure 13.11.

The ideal characteristics of the time-domain response of the feedback control system can be listed as follows:

1. Low steady-state error
2. Low response time to a step change in the input source voltage, the load current or the value of a controlling parameter
3. No or low overshoot in the transient response
4. Low settling time to the steady-state condition

The frequency-domain criteria for ideal control are based on the Bode diagrams of the open-loop transfer function. Compare the open-loop system transfer function of equation (13-32) with the following standard second-order control system:

$$G_{OL}(s) = \frac{A}{s^2 + 2\xi\,\omega_n\,s + \omega_n^2} \qquad (13\text{-}34)$$

$$\omega_n = \frac{1}{\sqrt{LC}} \quad \xi = \frac{1}{2\omega_n RC} = \frac{Z_0}{2R} = \frac{1}{2Q}$$

where

$$A = \frac{V_S}{V_R LC}$$

The ξ and ω_n are called the damping constant and the natural frequency (the resonant frequency in our case) respectively.

We now state important definitions that concern the frequency domain design of the controller.

Gain crossover frequency: The frequency $\omega_{crossover}$ at which the gain of the open-loop transfer function is 0 dB on the Bode diagram.

Phase Margin: The amount by which the phase of the transfer function is above $-180°$ at the gain crossover frequency.

Phase margin

$$PM = \Phi_{OL} + 180° \qquad (13\text{-}35)$$

Phase crossover frequency: The frequency at which the phase of the transfer function is 180° on the Bode diagram.

Gain Margin: The amount by which the gain of the transfer function is above 0 dB at the phase crossover frequency.

Example 13-6 Gain Crossover Frequency and Phase Margin of the Buck Converter

Plot the Bode diagram and determine the gain crossover frequency and the phase margin for the Buck converter of Example 13-5 with the peak voltage of the saw-tooth reference waveform of 10 V.

```
Vs=24; L=1e-3; C=1e-4; R=9/2; d=.125; VR=10;
num=Vs/(VR*L*C); den=[1 1/(R*C) 1/(L*C)];
[mag,phase,w]=bode(num, den,{1000,100000});
  % Bode diagram of the transfer function
[mingain, I]=min(abs(mag-1))
PM=180+phase(I)
wcrossover=w(I)
```

From the control theory, we know that for the damping constant $\zeta > 0.7$ the transient response has less than 5% overshoot. Hence, the damping constant of the Buck converter in CCM becomes smaller at the high value of Q or high values of R (light load), causing unacceptable overshoot. At high load current, that is, the damping constant $\zeta > 1$, the response is very sluggish as was evident in the response in Example 13-5 at $I_O = 20$ A.

The objective of low steady-state error requires the gain of the open-loop transfer function to be high at low frequencies. It can be achieved by adding a pole at the origin 1/s (a pole at zero frequency). However, it adds a $-90°$ phase all across the Bode diagram of the open-loop transfer function. This will render the phase margin to be negative and a larger amount of phase correction would be needed.

The objective of improving the transient response is met by designing for

(a) a high gain-crossover frequency. However, the gain-crossover frequency must be an order of

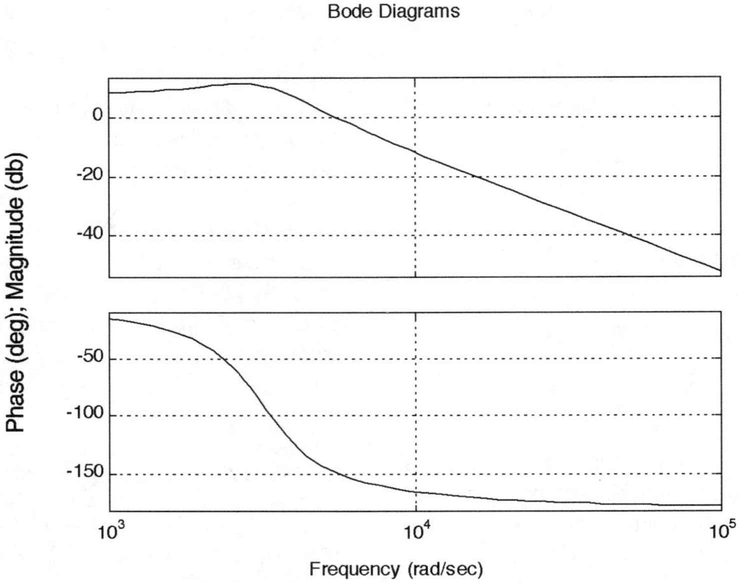

Figure 13.14
The root-locus plot of the unity feedback Buck converter system at I_O =2 A

magnitude less than the converter switching frequency, and

(b) a phase margin in the range of 45° to 60°.

The modifications are performed by adding a controller $G_C(s)$ and feedback as shown in Figure 13.12. The open-loop transfer function with controller included is

$$G_{OL}(s) = G_C(s)G_P(s)G_S(s) \qquad (13\text{-}36)$$

The controller design is a very important subject but is beyond the scope of this book. Readers are encouraged to read texts in the area of control system design for further knowledge.

13.6 SUMMARY

Converter control refers to regulating the converter operation so that it stays close to the nominal performance in the face of disturbances and noise. The steady-state converter operation represents a single point, called the operating point, on the static control characteristics curves. The converter systems involve switching; therefore, they are usually nonlinear and difficult to analyze. The transient response of the power electronic converter is derived from the analysis of a dynamic model in the vicinity of the operating point. This chapter described the development of the dynamic model by using the methods of averaging, linearizing and/or state-space averaging.

In the averaging method, the converter circuit is averaged over a switching period. Thus the obtained averaged model can be solved using conventional familiar techniques. The steady-state waveforms of averaged voltages and averaged currents in the converter are obtained as the solution of the averaged circuit model.

The linearized model of a power electronic converter is obtained from expanding the nonlinear terms in the steady-state solution (may also be from the averaged model) by Taylor's series around the nominal operating point. The linearized model is also called the small signal model. The linearized models can be analyzed and then stabilized by using well-known linear control system analytical techniques.

The averaged state-space modeling is a powerful technique of dynamic modeling. This method focuses on those variables that are central to describing the dynamic response of power converter systems. The last item discussed in this chapter was about the feedback control of power converters.

References

1. Philip T. Krein, *Elements of Power Electronics,* Oxford University Press, New York, 1998.

2. Ned Mohan, Tore M. Undeland, and William P. Robbins, *Power Electronics: Converters, Applications and Design,* 2nd Ed. , John Wiley & Sons, Inc., 1995.

3. Joseph Vithayathil, *Power Electronics: Principles and Applications,* McGraw-Hill Inc., New York, 1995.

4. John G. Kassakian, Martin F. Sclecht, and G. Vergassian, *Principles of Power Electronics,* Addison-Wesley, Boston, 1991.

5. Keith Billings, *Switch Mode Power Supply Handbook,* McGraw-Hill Inc., New York, 1989.

6. Kjeld Thorborg, *Power Electronics,* Prentice-Hall Inc., 1988.

7. Daniel M. Mitchell, *Switching Regulator Analysis,* McGraw-Hill Book Company, New York, 1988.

Problems

13-1 Determine the averaged model of the Buck converter in the discontinuous conduction mode of operation.

13-2 Include the ESR (series resistance) of the filter capacitor in the circuit of the Buck converter. Modify the averaged model in the discontinuous conduction mode of operation.

13-3 Determine the averaged model of the Boost converter in the discontinuous conduction mode of operation.

13-4 Determine the averaged model for the full-bridge phase controlled ac-dc converter.

13-5 Find the small signal transfer function for a series resonant converter as shown in Figure 11.8 in Chapter 11 with the following specifications: $V_g = 48$ V, L = 1 μH, C = 1 μF and the switching frequency f_S = 40 kHz. The converter is operating in discontinuous mode below resonance. Plot the unit step response, the Bode diagrams and the Nyquist diagrams.

13-6 Consider the Buck regulator with $V_s = 125$ V, L = 125 μH, R = 47 ohms, C = 1 μF, D = 0.6 and f_S = 20 kHz.

a. Determine the averaged model and plot the unit step response using MATLAB.

b. Determine the small signal model and plot the unit step response using MATLAB.

13-7 The converter shown below is called the Sepic converter. Develop the averaged model and the small signal model.

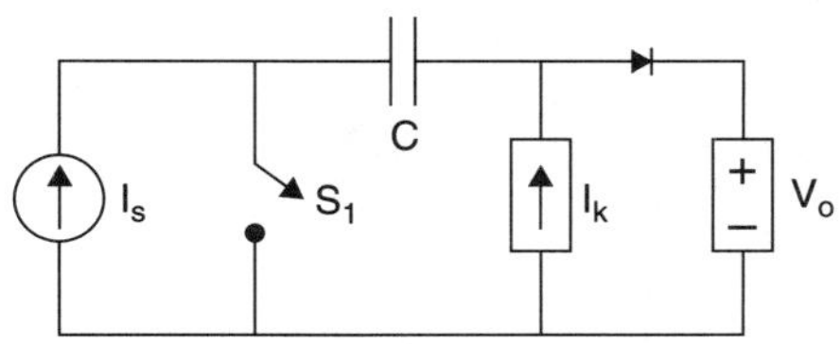

Figure P13.7

13-8 Develop the averaged model and the small signal model for the flyback converter.

13-9 Include the leakage inductance on the primary side of the transformer in the flyback converter. Modify the averaged model and the small signal model.

13-10 Design the closed-loop control system for an inverter that gives 120 Vrms, 60 Hz ac from a 12-V battery. The load varies from 10 to 200 W.

13-11 Design the closed-loop control system for a switch-mode power supply using the Buck converter that gives 5 V from a 12-V battery with a 1% voltage regulation for a variation of load in the range from 10 to 200 W.

13-12 Develop the averaged model for a dc-dc Buck converter in which the source voltage is 48 Vdc and the output voltage is 12 Vdc operating in the continuous conduction mode. Other components are L = 0.5 mH and C = 100 μF.

13-13 In the Buck converter of Problem 13-12,

a. For output dc current I_O = 10 amps, determine the transient response of the output voltage V_O when the duty ratio D undergoes a step change to 0.5 (in the next switching period).

b. Repeat (a) for I_O = 2 amp. Compare the nature of the two responses.

c. Find L for the critically damped response at I_O =10 amps.

13-14 Develop the linearized model of the dc-dc Buck converter of Problem 13-12.

13-15 Find the small signal transfer function for a Buck converter of Problem 13-12 at (a) I_O=2 amps and (b) I_O =20 amps. Find and plot in both cases the poles and zeros of the system and the step response for a change in the duty ratio from 0.375 to 0.5.

13-16 Plot the Bode diagram and determine the gain-crossover frequency and the phase margin for the Buck converter of Problem 13-12 with peak voltage of the saw-tooth reference waveform of 10 V.

13-17 Develop the averaged and small-signal model for the Buck-Boost converter, shown in Figure P13.16.

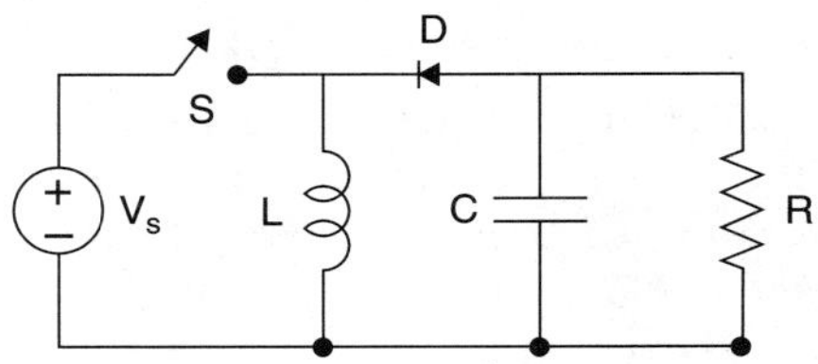

Figure P13.16

14
Applications I: Power Supply and . . .

OBJECTIVES

1. The requirements of a dc power supply
2. Control techniques in dc regulated power supply
3. Protection of dc power supply
4. Cross regulation in multioutput power supply
5. Electronic lamp ballast
6. Induction heating
7. Electromagnetic interference considerations

14.1 INTRODUCTION

Switch-mode design makes it possible to build efficient and smart power electronic converters. This chapter will describe the low-power to medium-power applications of power electronic converters, such as the following:

DC regulated power supply
Battery charger
Uninterruptible (ac) power supply
Electronic ballast
Induction heating
Switch-mode welding

14.2 DC POWER SUPPLY SYSTEM

The dc power supplies are used extensively in electronic and electric applications. The general requirements of the power supplies are:

1. **Regulated Output:** The output parameter, be it voltage, current or frequency, must be held constant within a specified tolerance in face of variations in the input voltage, load variations, interference and noise.
2. **Isolation:** The load may be required to be electrically isolated from the input.
3. **Multiple Outputs:** More than one output with different ratings, different polarities and different isolation requirements are often required in electronic applications, especially in instrumentation.
4. **High Conversion Efficiency**
5. **High Power Density:** for small size and weight.
6. **Precise and Wide-Ranging and Fast Control**
7. **High Quality Output:** Low harmonic distortions in the output and input power.

The block diagram of a dc power supply is given in Figure 14.1. The input source is the ac line voltage. Instead of using a direct controlled ac-dc converter, the dc power supply uses an uncontrolled ac-dc rectifier followed by a dc-dc converter. The former operation must be performed at the low line frequency (50-60 Hz). There are two disadvantages in this method: 1. the filters use high value L and C components that have large size and more weight and 2. the power supply response to variations in the line and load conditions and interference is sluggish. The dc-to-dc converter, on the other hand, can be operated at very high frequencies (1 MHz), which would overcome the above disadvantages.

The dc-dc converter may be one of two types: 1. PWM dc-dc converters or 2. resonant dc-dc converters. The dc-dc converter may or may not have isolation. The output voltage is regulated by using a feedback control loop. The control variable may either be the voltage (voltage-mode control) or the output current (current-mode control). There may be one or more output ports. The ports may have different voltage and current ratings. Usually only one output is regulated. The variation of an unregulated output due to the load variation in the regulated output is called cross regulation. Certain topologies are more suited for minimizing cross regulation and some techniques are available for minimizing it. An EMI filter at the input prevents electromagnetic interference from being fed back to the ac source.

The dc-dc converter can be based on any one of the topologies or their derivatives described in Chapter 6 and Chapter 11. They are listed as follows:

PWM converters

1. Buck converter
2. Boost converter
3. Buck-Boost converter
4. Cŭk converter (derived from the Buck-Boost converter)
5. Flyback converter (derived from the Buck-Boost converter)
6. Forward converter (derived from the Buck converter)
7. Push-pull converter

Resonant converters

8. Series resonant dc-dc converter
9. Parallel resonant dc-dc converter
10. Pseudo-resonant converter
11. Class D and E resonant converter
12. ZVS and ZCS and clamped resonant variations

Most types of dc-dc converters are driven by either a voltage source or a current source (a large inductor in series with a voltage source). The current-source converters exhibit a low power-to-weight ratio.

Medium to high-power (>250 W) power supplies today use low switching frequencies, in the range of 10 to 40 kHz. The low-wattage dc power supplies in portable applications use high switching frequencies in the range of 100kHz to 1MHz.

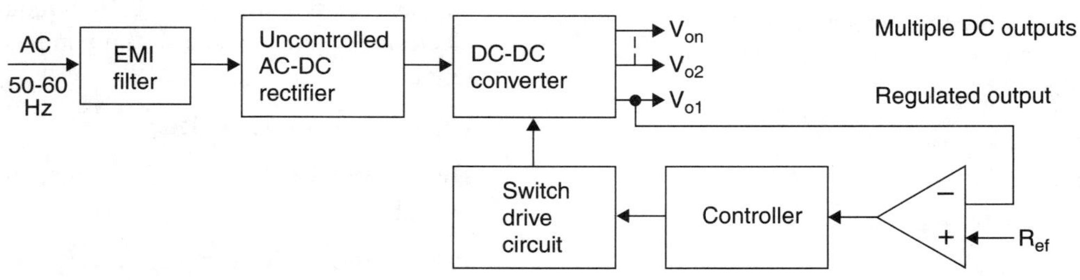

Figure 14.1
DC power supply system

Frequencies higher than 1MHz do not offer significant advantages in terms of the size and power-weight ratio because of the difficulties in the increased switching losses, core losses and hysteresis losses in the transformers. Furthermore, the high switching frequency gives rise to more electromagnetic interference (EMI). Packaging of components and magnetics becomes a very critical issue.

14.3 CONTROL OF SWITCH-MODE DC POWER SUPPLIES

The switch-mode dc power supplies are controlled by both hard-wired analog controllers and software-driven digital controllers. The latter is preferable for the following reasons: 1. less sensitivity to the aging of components and electrical noise and 2. adaptibility to digital control methods such as the proportional-integro-differential control (PID) of the power supply's response to the source and load variations, transients, and interference. The following section will discuss two types of control strategies: (a) the voltage-mode control and (b) the current-mode control.

14.3.1 Voltage-mode control

The output voltage V_o is compared with a reference voltage V_R. The error voltage is amplified and used to control, for example, the pulse width of the switch in the PWM converter. The controller block in Figure 14.1 in this case is a pulse-width controller. The controller, see Figure 14.2(a), is a comparator that compares the amplified error voltage (the control voltage v_c with a reference saw-tooth waveform of the desired switching frequency. The output of the comparator is the pulse wave of desired pulse width, see the waveforms in Figure 14.2(b).

14.3.2 Current-mode control

In the current-mode control, the reference saw-tooth waveform is replaced by a voltage proportional to the current output of the power supply. This action is equivalent to adding an inner control loop in the control system. See Figure 14.3. There are basically two types of current-mode controls:

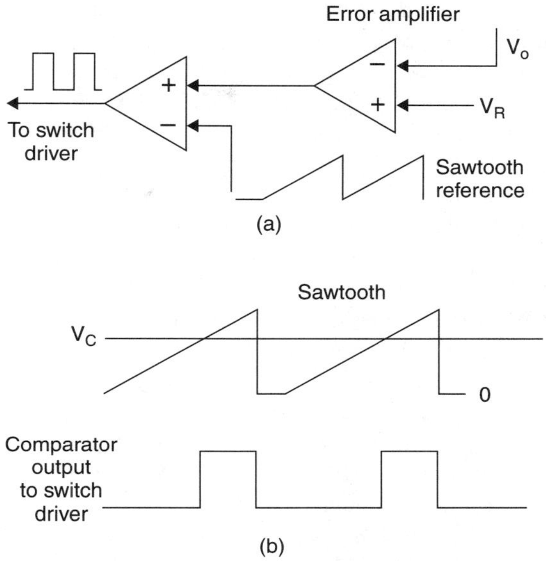

Figure 14.2
(a) Pulse-width controller in PWM controlled power supplies using voltage-mode control and
(b) waveforms

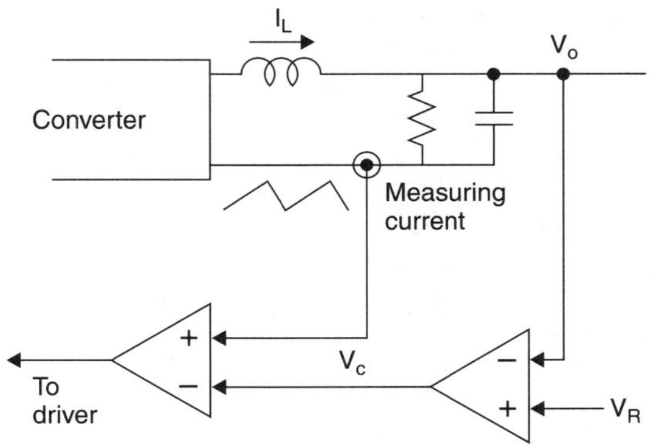

Figure 14.3
The current-mode control of PWM dc power supply

1. Variable frequency control:
   (a) Window control
   (b) Constant off-time control
2. Constant switching frequency control.

In the ***window control,*** the control voltage determines the average value of the output current $I_{o,avg}$. The switch in PWM converter is turned on when the sensed output current falls a preset half window level below the average value, $I_o - \Delta I_o/2$. The switch is turned off when the sensed output current becomes more than $I_o + \Delta I_o/2$. Obviously this scheme can be used only in the continuous conduction mode of operation.

In the ***constant off-time control,*** the control voltage determines the peak value of the output current I_o. The switch in the PWM converter is turned off for a fixed time. The output current in PWM power supply falls linearly. The switch is turned on after the off-time has elapsed, whereupon the output current rises linearly. The switch is kept on until I_o reaches the peak value.

In the ***constant switching frequency control,*** turn-on of the switch in the PWM converter is synchronized to a clock pulse-train of the desired switching frequency. The output current I_o rises linearly. The switch is kept on until I_o reaches the peak value, which is determined by the control voltage v_c. The switch is turned off. I_o falls linearly. The switch is turned on again at the arrival of the next clock pulse.

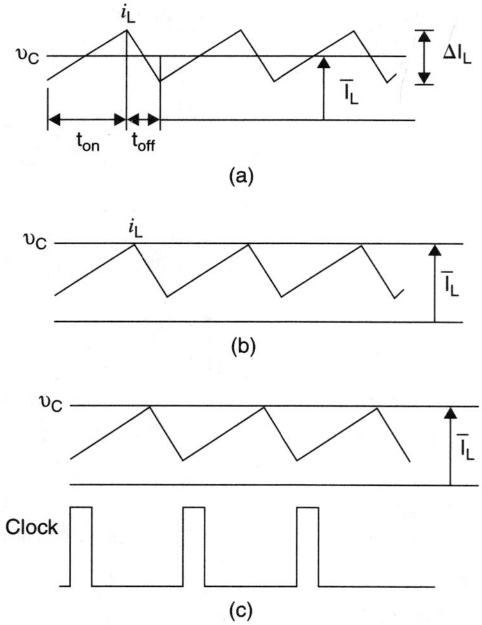

Figure 14.4
Current mode control using (a) Window control, (b) Constant off-time control and (c) Constant switching frequency control

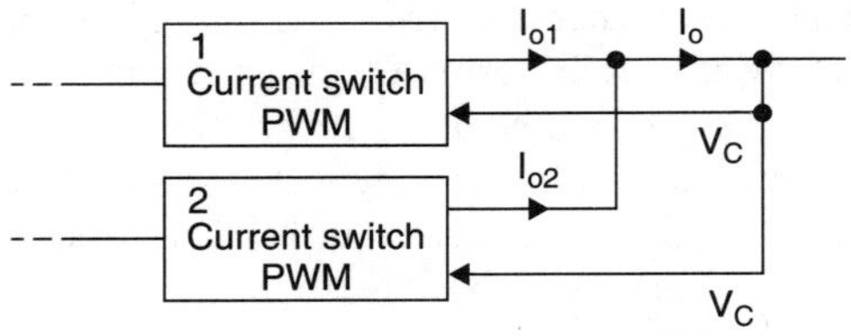

Figure 14.5
Paralleling of current controlled PWM power supply modules

The current-mode control has more advantages than the conventional voltage-mode control because of the automatic limiting of the peak value of I_o. Furthermore, several power supply modules can be paralleled to increase the current capacity relatively easily, see Figure 14.5. Each module can have an independent current-mode control. The control voltage v_c to all modules is derived from a single point.

14.4 PROTECTION OF DC POWER SUPPLIES

The protection features in dc power supplies are as important as the control features themselves. Present day integrated circuits have most of these provisions. These features can also be provided in software.

14.4.1 Soft-start

The soft start in a dc power supply is provided by increasing the pulse width slowly. The consequence of slowly increasing the output voltage does not cause severe stresses on the filter components and the switching devices.

14.4.2 Current limiting

The output current can be sensed by measuring the voltage across a small resistance in series with the load. Thus sensed current (actually voltage) is compared with a reference. The error is amplified and used to decrease the pulse width so as to limit the current but decrease the output voltage. The load characteristics are shown in Figure 14.6(a).

The constant current in the short-circuit condition at the output causes stresses on the power supply components. This problem can be avoided by reducing the limiting current as the output voltage falls, see Figure 14.6(b). This is called the ***fold-back current limiting.*** A part of the decreasing output voltage is subtracted from the sensed voltage across the sensing resistance, before it is compared with the reference.

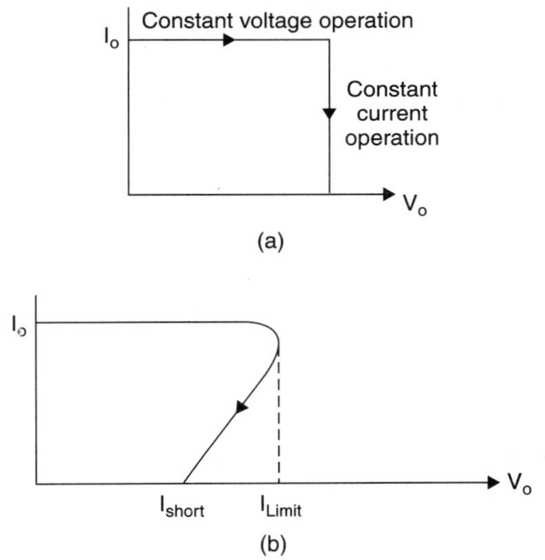

Figure 14.6
Current limiting: (a) Constant current limiting and (b) fold-back current limiting

14.4.3 Over-voltage and under-voltage protection

The control of the power supply is shut down when the output voltage is beyond the window of the under-voltage and over-voltage limit settings. Thyristors and special integrated circuits can be used directly for this purpose.

14.4.4 In-rush current protection

At turn-on, the filter capacitor at the output of the uncontrolled rectification is uncharged; therefore, it draws heavy in-rush current from the ac source. An inductance in series before or after the uncontrolled rectifier serves to limit the in-rush current. An alternative is a series resistance R_S after the uncontrolled rectifier that is shunted by a thyristor T, as shown in Figure 14.7. Initially the capacitor C has zero voltage, whereupon the transistor Q goes into the saturation state. The consequent voltage between the gate-to-cathode voltage of the thyristor is in the forward blocking state. The in-rush current through the uncontrolled rectifier is limited by the series resistance R_S. The transistor comes out of sat-

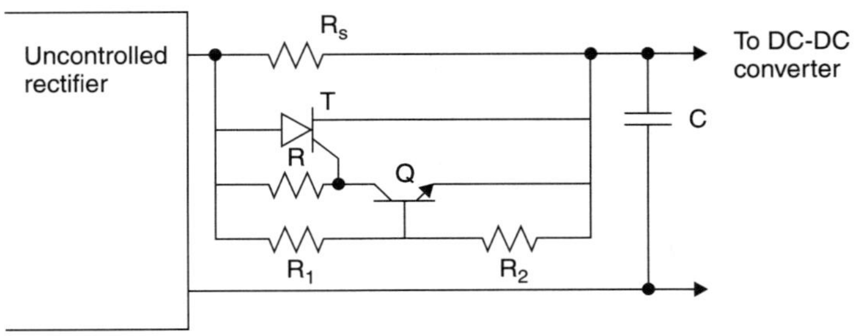

Figure 14.7
In-rush current protection in dc switch-mode power supply

uration when a sufficient voltage is developed on the filter capacitor C. The gate-to-cathode voltage rises on the thyristor. It turns on and shunts the series resistance R_S and remains on thereafter.

14.5 ELECTRICAL ISOLATION

The electrical isolation in switch-mode supplies is usually provided in the high frequency signal part of the dc converter because of the small size of the high frequency isolation transformer. In the feedback controlled system, the control parameter, such as the pulse width of the switches in PWM converters, is on the primary side of the isolation transformer. The output voltage is sensed on the secondary side of the isolation transformer. The feedback loop in these power supplies also requires electrical isolation. See Figure 14.8 for an electrically isolated power supply. The PWM controller in the feedback loop can be placed either before or after the feedback isolation transformer. The controller placement after the isolation transformer simplifies its interfacing with the switch-mode converter in the forward path.

The isolation transformer in the forward path conducts a large amount of power, whereas in the feedback path it conducts only a low power signal. The feedback isolation transformer may be replaced by an opto-coupler. The state-of-the-art opto-coupler characteristics are not very stable with respect to temperature and time, hence, the transformer isolation is preferable.

14.6 EQUIVALENT SERIES RESISTANCE (ESR)

The equivalent series resistance of the output filter contributes very significantly to the peak-peak ripple output voltage, see Figure 14.9:

$$V_{r,p-p} = ESR \times \Delta I_o \tag{14-1}$$

The filter capacitor in the output must be selected to be as low an ESR as possible.

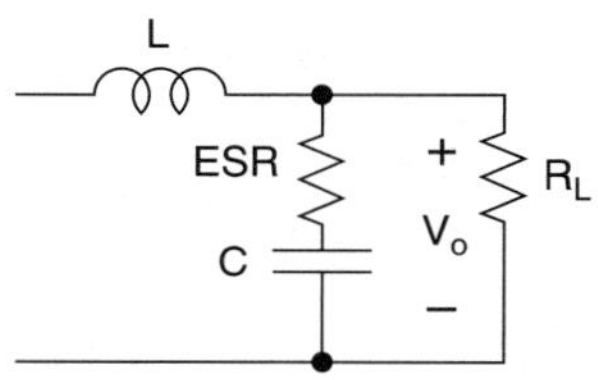

Figure 14.9
ESR in dc power supply

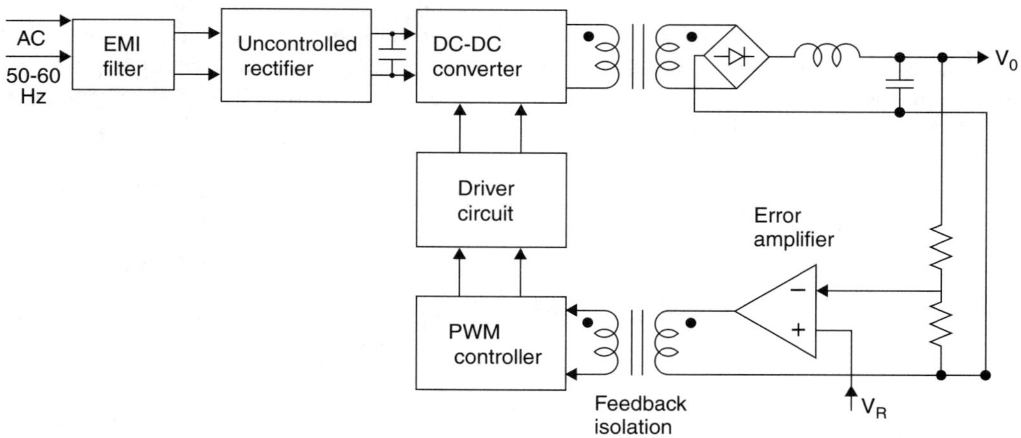

Figure 14.8
Electrical isolation in PWM dc power supplies

14.7 SYNCHRONOUS RECTIFIERS

In low voltage power supplies, called voltage regulator modules, such as the 48 V to 1.2 V, 1.8 V or 3 V dc supplies, the on-state voltage of rectifier diodes is the major source of power inefficiency and the consequent heat dissipation problems. Even the Schottky rectifier diodes have a large on-state voltage drop. MOSFETs with low on-state resistance have low on-state voltage drop and, thus, have low on-state power losses. These rectifiers are called the synchronous rectifiers.

14.8 CROSS-REGULATION IN MULTIPLE OUTPUTS

Multiple output power supplies are very commonly used in instrumentation and computer applications. Usually one of the outputs is regulated by feedback control. The cross regulation in these power supplies refers to the effect of the load and the line variation in the feedback regulated output on the voltage of the unregulated outputs and vice-versa.

Consider a multioutput power supply, as shown in Figure 14.1, with the voltage transfer ratios, T_{vv1}, T_{vv2}, . . ., with loads, R_{L1}, R_{L2}, ... The cross regulation parameters are then defined as,

Load cross regulation:

$$X_{2,L} = \frac{dT_{vv2}}{dR_{L1}}\Big|_{\Delta V_S=0,\ \Delta R_{L2}=\Delta R_{L3}=\ldots=0}$$

Line cross regulation:

$$X_{2,S} = \frac{dT_{vv2}}{dV_S}\Big|_{\Delta R_{L1}=\Delta R_{L2}=\Delta R_{L3}=\ldots=0}$$

A poor cross regulation would seriously limit the range of control in the regulated supply. Both types of cross regulation parameters must be designed to be as small as possible.

Most often, the multiple outputs are taken from a multi-winding transformer in the output circuit, see Figure 14.10. The leakage inductances in the secondary windings are the main source of cross regulation. The minimization of leakage inductances and some other topological method (refer to the literature) are used for reducing cross regulation.

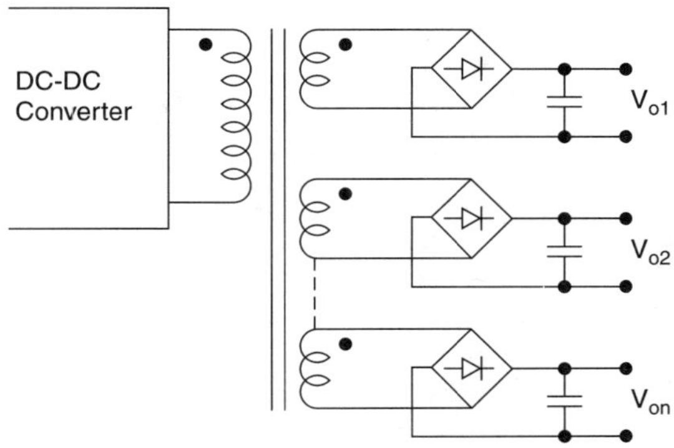

Figure 14.10
Transformer coupled multioutput power supplies

14.9 BATTERY CHARGING SYSTEMS

The battery charging systems constitute a substantial part of power converter applications. The increase in the usage of portable appliances and communication equipment is responsible for a spurt in the demand of the low-power battery charger systems.

The battery voltage in discharge should not be allowed to fall below a voltage level, which is called the final discharge voltage; otherwise, the battery life is shortened. A 400 amp-hour battery can be discharged at a constant current of 400 amperes for 1 hour or 10 amperes for 40 hours before its voltage falls to the final discharge level.

The battery can be brought back to its fully charged state by following the charge path shown in Figure 14.11. The initial charging takes place at a constant current, raising the battery voltage to the trickle charge level. After that a constant voltage is applied to the battery bank through the inductance. The charging current then continuously decreases until it reaches to the trickle level.

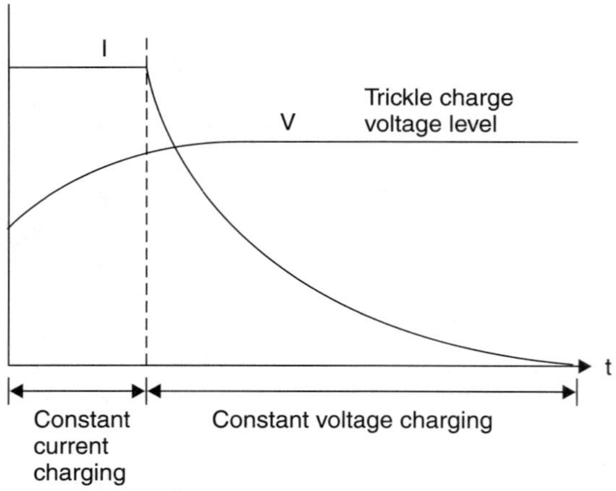

Figure 14.11
Charging of a battery

14.10 UNINTERRUPTIBLE (AC) POWER SUPPLY (UPS)

Medical and emergency facilities require uninterrupted ac power supplies as protection against power outages and over-voltage and under-voltage situations from the electric utility. The UPS also protects the load from line transients and harmonic disturbances. They are, in general, not long-term sources. They are available to provide power during power shutdown or as an interim source until the backbone supply is restored.

Three configurations of UPS are given in Figure 14.12. In the first configuration, the **standby power supply** shown in Figure 14.12(a), the load is supplied normally by the ac line directly. The ac line also simultaneously charges a battery bank. The inverter is also normally inactive. In case of outage or under-voltage or even in over-voltage in the ac line, the battery bank takes over and the inverter is activated. A static switch turns off the direct ac line path and connects the load to the output of the inverter.

In the second configuration, the **true UPS** shown in Figure 14.12(b), the load is supplied normally by the battery or ac line through the ac-dc rectifier and the inverter. The ac line simultaneously charges the battery bank in normal-voltage and over-voltage ac line conditions. There is no need to switch the power supply to the load in case of outage or under-voltage in the ac line. In case of inverter failure, the static switch turns off the inverter path and connects the load to the ac mains directly. The inverter output should be synchronized to the line voltage source.

In the third configuration, the **standby UPS** shown in Figure 14.12(c), the load is supplied either by the battery through an inverter or the ac line through the ac-ac converter or tap-changer and the inverter. When the ac line is supplying power, the battery bank acts as the standby and is maintained in the trickle charge condition. The power flow in the inverter is bi-directional. It can act as the battery charger in the trickle charge condition.

14.10.1 Rectifier

The ac-dc rectifier performs two functions simultaneously: 1. to provide a dc voltage source to the inverter and 2. to charge the battery bank. The rectifier could be selected to be any one of the uncontrolled (Chapter 8) or controlled (Chapter 9) ac-dc converter topologies. Alternately we could use the uncontrolled ac-dc converter followed by a high frequency isolated or nonisolated dc-dc converter as described in Chapter 6. Figure 14.13 shows a three-phase uncontrolled ac-dc rectifier with a separate one-phase controlled dc-dc converter (battery charger) for the battery bank. The rectifier supplies power directly to the inverter.

The dc-dc converter may also be based on the resonant converter topologies of Chapter 11.

14.10.2 Inverter

The UPS inverters have generally stringent harmonic distortion requirements; typically, the Total Harmonic Distortion (THD) is specified to be less than 5%, and the amplitude of individual harmonics is less than 3% of the fundamental amplitude. An added advantage of low THD is the improved dynamic response of the UPS in load variations. The feedback control loop with fast response is also required for good dynamic behavior.

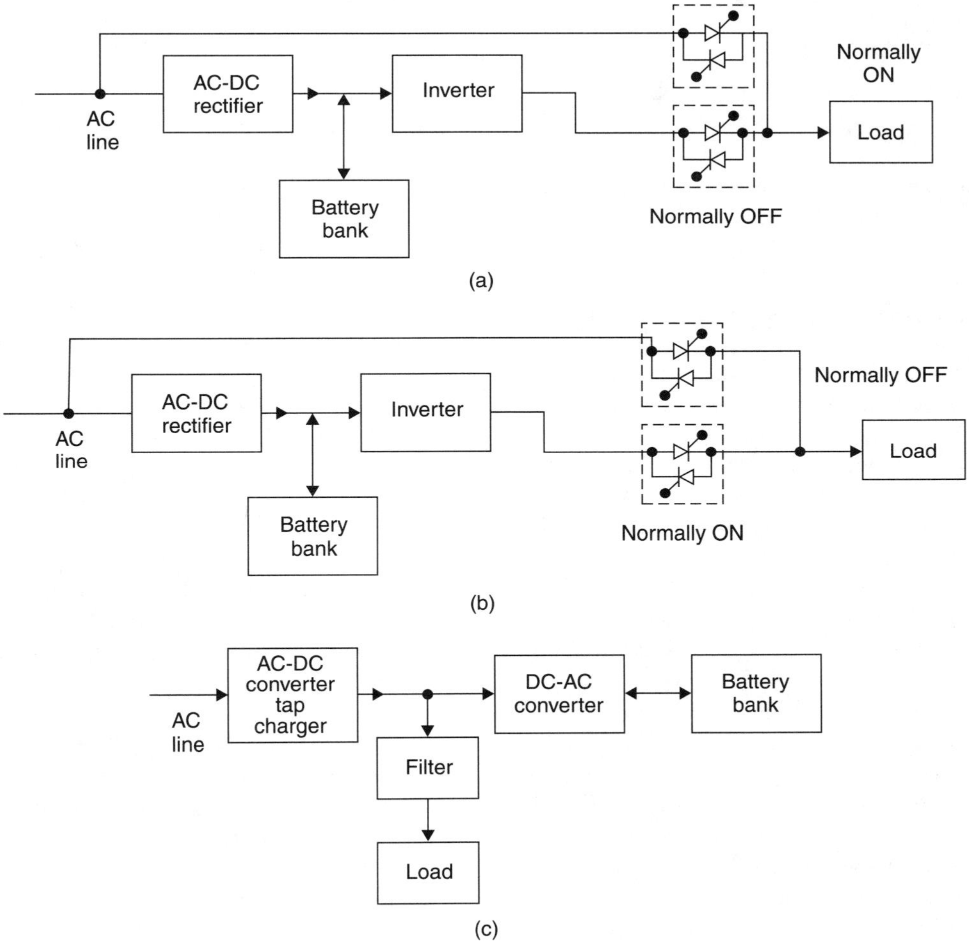

Figure 14.12
Uninterruptible ac power supply (UPS) configurations: (a) Standby power supply, (b) True UPS and (c) Alternate UPS

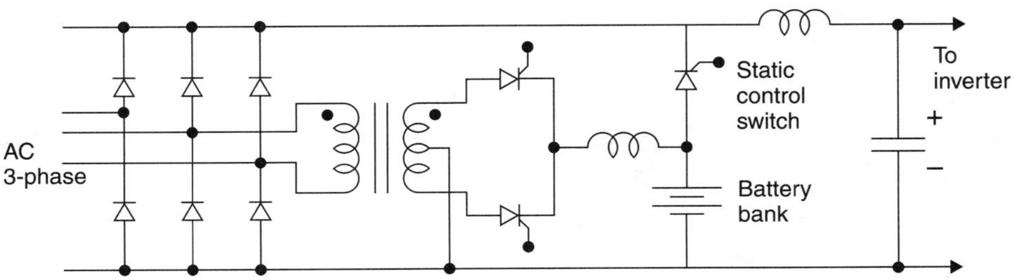

Figure 14.13
A 3-phase ac-dc rectifier with separate 1-ϕ controlled battery charger

The Sine-PWM inverter is the most commonly used inverter. Others are:

1. Resonant-link inverter
2. Integral half-cycle inverter

The inverters are connected in parallel in the applications involving more than a few kilowatts of power.

14.11 ELECTRONIC LAMP BALLAST

Lighting consumes approximately 15% of energy in residential buildings and 30% in commercial buildings. Fluorescent lamps are three to four times more energy efficient than incandescent lamps. The energy efficiency of fluorescent lamps can be further boosted by 20% to 30% by operating at frequencies higher than 25 kHz.

Conventional fluorescent lamps use an inductive ballast (also called the choke) in series for stable operation with a 60-Hz ac source, see Figure 14.14(a). A modern high frequency fluorescent system is shown in a block diagram in Figure 14.14(b). The uncontrolled ac-dc converter with filter is followed by a high frequency dc-ac inverter. The inverter is designed from the several topologies discussed in Chapter 7. One of the prominent topologies is the Class-E resonant inverter: half-bridge or full-bridge.

The switching frequency of the inverter lies in the range of 25 to less than 40 kHz and 60 kHz or higher. The current from the source has a very poor power factor, hence, the power factor correction is very necessary. The EMI filter prevents the conducted electromagnetic interference signal from being fed back to the ac line.

14.12 INDUCTION HEATING

Induction heating consists of a high power transformer whose secondary winding is the metal workpiece itself. The metal workpiece acts as a shorted secondary winding. The induced current in a shorted secondary winding generates intense heat. The induction current is confined to a small thickness deep into the metal surface owing to the skin effect. The skin depth at which the current falls to the 1/e of the surface current is

$$\delta = k\sqrt{\frac{\rho}{f}} \tag{14-2}$$

The k is a constant, the ρ is the resistivity of the metal workpiece, and the f is the frequency of the induction current. The line frequency is used for heating large workpieces to greater depths. High frequencies up to 1 MHz are used for hardening and annealing of metal gears.

The metal load is modeled by a resistance in series or in parallel with inductance, both referred to as the primary side of the induction transformer. Figure 14.15 shows two prominent induction heating circuits.

In the voltage-source series resonant inverter, the output current is nearly sinusoidal at the switching frequency slightly below the resonance. The power is controlled by a variable switching frequency control.

In the current-source parallel resonant inverter, the output current is nearly sinusoidal at the switch-

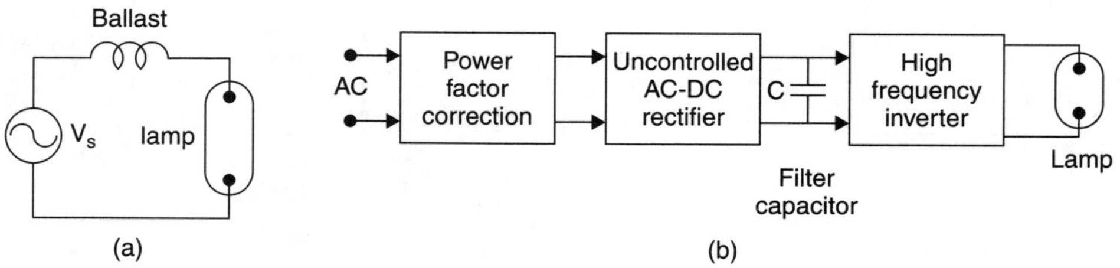

Figure 14.14
(a) Conventional fluorescent lamp with inductive ballast and (b) Block diagram of the electronic ballast

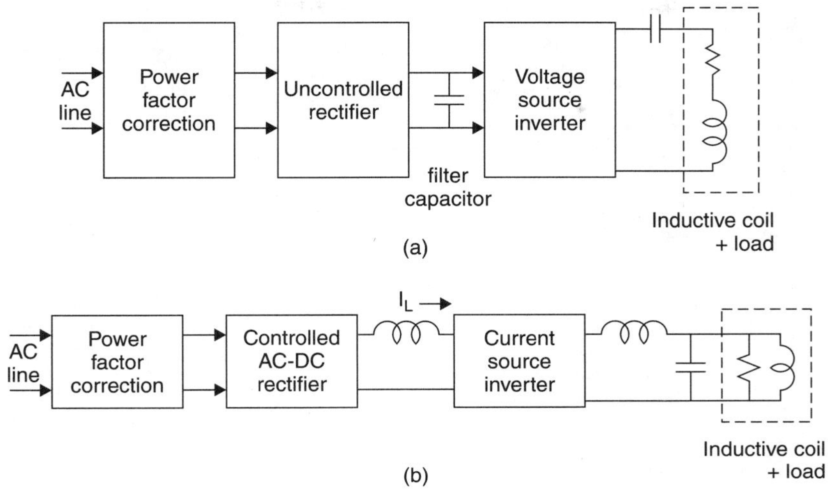

Figure 14.15
Induction heating: (a) Voltage-source series resonant inverter and (b) Current-source parallel resonant inverter

ing frequency slightly above the resonance. The power is controlled by variable switching frequency control. Alternately, the current I_L can be used as the control parameter while keeping the switching frequency constant.

14.13 SWITCH-MODE WELDING

In arc welding, an arc is established between two electrodes, one of the electrodes being the metal workpiece. The electrical isolation of the workpiece is the most important requirement. Another requirement is to maintain a low-ripple current after the arc is established. The block diagram of the switch-mode welder is given in Figure 14.16. Isolation is provided by a high frequency transformer. A small inductance at the output limits the welding current.

14.14 ELECTROMAGNETIC INTERFERENCE CONSIDERATIONS

The fast switching of large currents and voltages in power electronic circuits is the source of electromagnetic noise. The fast rise and fall times of the switched waveforms contain significant energy at the harmonic frequencies in the radio frequency (RF) band. These harmonics cause interference in nearby electric systems and even in the ac line source. The electromagnetic interference can be classified into two types:

1. ***Conducted EMI:*** EMI that is carried by the wires to the nearby systems including the ac power lines, and
2. ***Radiated EMI:*** EMI that is radiated through space.

The conducted EMI is several orders of magnitude higher than the radiated EMI. The radiated EMI can be suppressed by using ground planes in the printed circuit boards liberally and shielding and packaging of the power electronic converters in aluminum.

The conducted EMI can be further classified into two types:

1. Differential mode EMI between the input or output port terminals,
2. Common mode EMI at all port terminals with respect to a common point such as the ground.

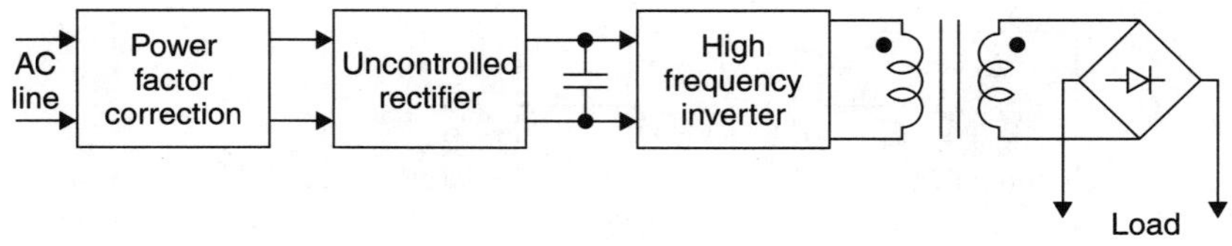

Figure 14.16
Block diagram of the switch-mode welder

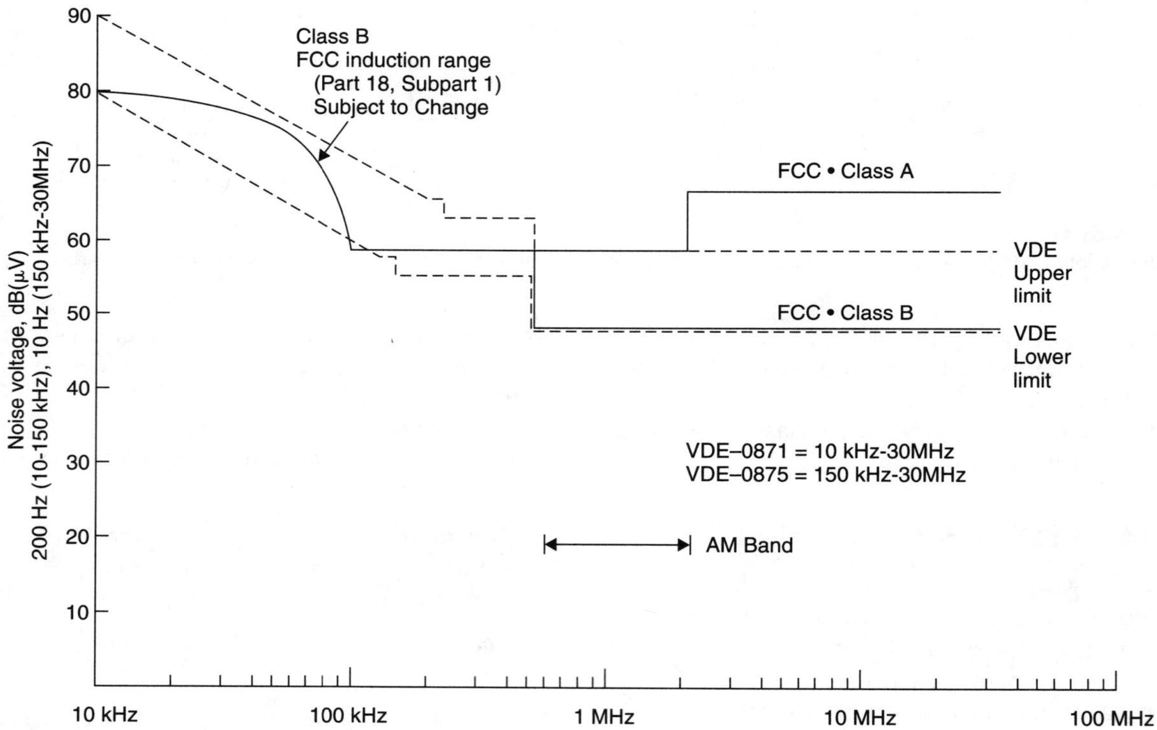

Figure 14.17
The FCC and VDE standards for conducted EMI

There are several standards, such as the CISPR, IEC, VDE, FCC and military standards, that specify the maximum limit on the conducted EMI. Of these the VDE and FCC standards are used in industrial, commercial and residential applications. Figure 14.17 shows these standards.

The best way to reduce EMI is to prevent them at the source itself. The dV/dt and di/dt snubbers help greatly in achieving this objective. Other preventive measures are as follows:

1. use a twisted pair of leads in place of simple wires
2. use ground-planes liberally in the printed circuit board design
3. reduce stray capacitance by minimizing exposed metal surfaces

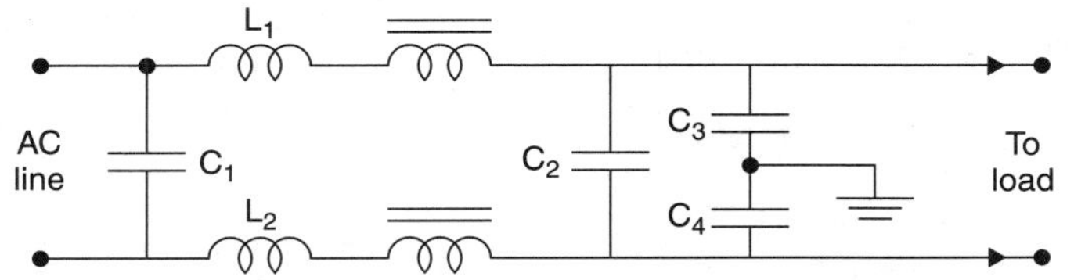

Figure 14.18
Filter for conducted EMI

In addition to preventive measures, filters are used to reduce the conducted EMI to ac lines. One such filter is shown in Figure 14.18.

14.15 SUMMARY

Switch-mode power electronic converters have brought about the efficient and smart use of electrical energy. The low-power to medium-power applications of power electronic converters include dc regulated power supply, battery charger, uninterruptible power supply, electronic ballast, induction heating, switch-mode welding, etc.

Some of the common requirements of all power electronic applications are high conversion efficiency, high power density, precise, wide-ranging and fast control, and high quality output with low harmonic distortions, generation of low internal EMI and immunity to outside EMI. Other requirements may differ with applications. The dc power supply requires (a) regulation of output voltage or current even when the input source or load conditions vary over a wide range, (b) electrical isolation between the source and the load, and (c) low cross regulation error in multiple outputs, etc. Due to the great expansion in portable appliances and communication equipment, there is the requirement for high voltage to low voltage regulator modules (VRMs).

The applications of battery charging systems have also grown very rapidly with portable designing. Another very important low-power application is the electronic ballast design for the lighting industry. Electronic ballasts allow for energy-efficient and smart lighting systems. Use of switch-mode designs have made it possible to design smart and energy-efficient welding and induction heating appliances and industrial machines.

This chapter described the use of synchronous rectifiers for improving the power efficiency of voltage regulator modules. The chapter also discussed two types of control strategies in switching power supply design: (a) the voltage-mode control and (b) the current-mode control.

Low-power applications use high switching frequencies, from 100 KHz to 1 MHz, primarily for reducing component size and weight. A further push for high frequency is blocked due to the yet unsolved and under-solved problems of packaging, EMI, heat dissipation, and limitations of magnetics design. Medium-power applications (>250 W) use low switching frequencies in the range of 10 to 40 kHz.

References

1. Philip T. Krein, *Elements of Power Electronics,* Oxford University Press, New York, 1998.

2. Ned Mohan, Tore M. Undeland, and William P. Robbins, *Power Electronics: Converters, Applications and Design,* 2nd Ed., John Wiley & Sons, Inc., 1995.

3. Joseph Vithayathil, *Power Electronics: Principles and Applications,* McGraw-Hill Inc., New York, 1995.

4. Muhammad H. Rashid, *Power Electronics: Circuits, Devices and Applications,* 2nd Ed., Prentice-Hall Inc., New Jersey, 1993.

5. John G. Kassakian, Martin F. Sclecht, and G. Vergassian, *Principles of Power Electronics,* Addison-Wesley, Boston, 1991.

6. Keith Billings, *Switch Mode Power Supply Handbook,* McGraw-Hill Inc., New York, 1989.

7. Kjeld Thorborg, *Power Electronics,* Prentice-Hall Inc., 1988.

8. Daniel M. Mitchell, *Switching Regulator Analysis,* McGraw-Hill Book Company, New York, 1988.

Problems

14-1 Consider the step-down battery charger circuit shown below.

$L = 100\ \mu H$ $R = 2$ ohms $V_S = 169$ volts
$V_K = 48$ volts $f_S = 20$ kHz

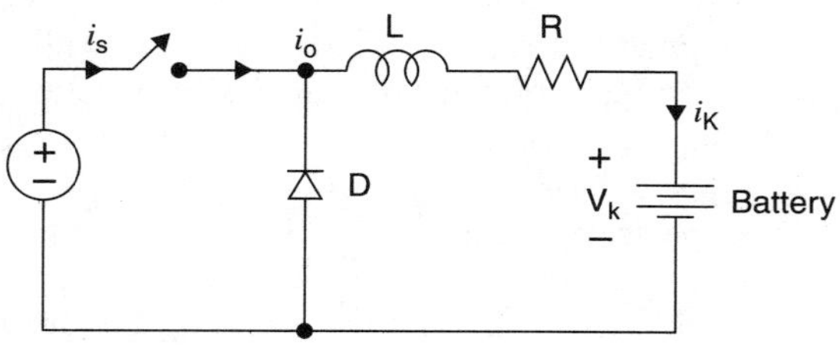

Figure P14.1

a. Determine the duty ratio D for steady-state operation.
b. Plot the waveforms for I_S, I_O, and I_K, assuming zero energy on the inductor at the time of applying power to the circuit.
c. Calculate the power delivered by the source V_S.

14-2 The duty ratio is abruptly changed to D=0.4 for only one period during the steady-state operation of the battery charger of Problem 14-1.
a. Plot the waveforms for I_S, I_O, and I_K during this one period.
b. Plot the waveforms for I_S, I_O, and I_K at the end of the period of change.

14-3 A thyristor controlled inductor can be used as the compensator of the static VAR (volts amperes reactive) in an 11 KV, 60 Hz ac network. The ac network has an inductance of 5 H. Determine the range of the adjustable VAR.

14-4 A switching power supply using the forward converter is shown in Figure P14.4 with
$V_S = 170$ Vdc $D = 0.4$ $f_S = 40$ kHz
The transformer consists of three windings. The primary and the feedback windings have each 100 turns. Determine
a. waveform of the voltage across the primary winding,
b. waveform across the feedback winding and
c. maximum forward and reverse blocking voltages on the switch.

14-5 A switching power supply is shown in Figure P14.5 with $V_S = 170$ Vdc, $f_S = 40$ kHz, and the number of turns in primary = 100. Determine
a. the maximum permissible duty cycle D
b. the number of turns in the secondary winding to give $V_O = 12$ V.

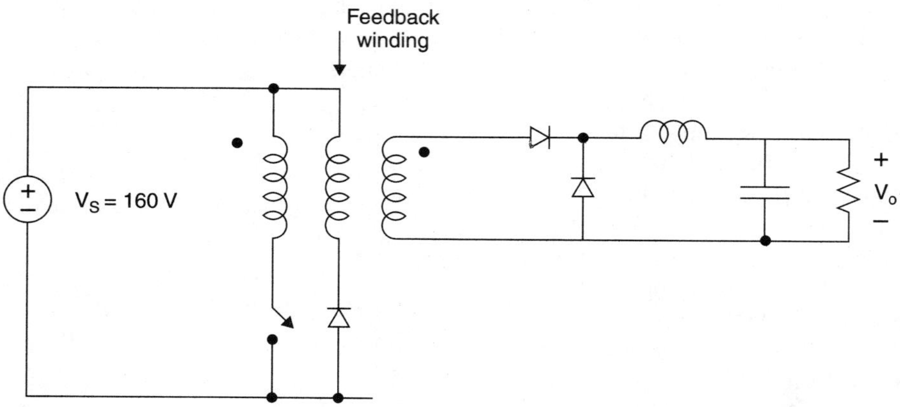

Figure P14.4

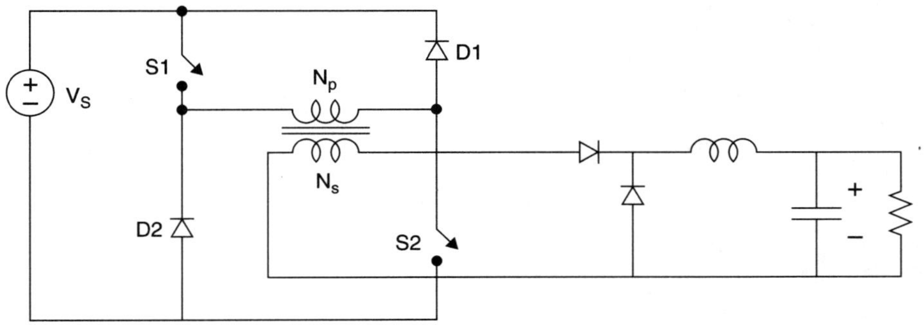

Figure P14.5

14-6 Design a 10 W, 5 V-3.3 V converter. The switching frequency should be 500 kHz or higher. Suggest the converter topology with the types of the switches and diodes.

14-7 Design a 100 W, electronic ballast for 200 Vrms, 20 kHz application. The input source is a 120 Vrms, 60 Hz. Suggest the converter topology with the types of switches and diodes.

CHAPTER 15
Applications II: Motor Drives . . .

OBJECTIVES

1. DC motor drives
2. Adjustable speed drives (ASD)
3. Induction motor drives
4. Variable frequency drives
5. Synchronous motor drives
6. Servo motor drives

15.1 INTRODUCTION

Modern motor drives are designed with switch-mode principles. They are more efficient and smart in comparison to the nonswitch-mode versions. Motor drive applications range from low wattage in appliances and robotics to very high wattage (in kilowatts) in industrial drives. The most common motor drives of today can be classified in four categories that are based on the type of motor; dc motor drives, induction and reluctance motor drives, synchronous ac motor drives, and servo motor drives.

A typical motor drive system is shown in Figure 15.1.

Motor drives can also be classified in two categories based on the type of application:

1. servo motor drives
2. adjustable speed drives

Both types require feedback control. The servo drives have relatively stringent requirements for the precision in the angular position or speed and the response time and response characteristics, such as the overshoot and settling time.

15.2 DC MOTOR DRIVES

The dc motor drives are still the workhorses for some applications in the industrial world of today. These are excellent drives with high precision and low initial costs. The equivalent circuit and dynamic equations of a dc motor were introduced in Chapter 2, hence they will not be repeated here.

15.2.1 Terminology in DC drives

1. ***Forward Motoring:*** The normal mode of operation of the dc motor is when it rotates in the forward direction. The forward and reverse directions are preselected and only depend on the polarity of dc source

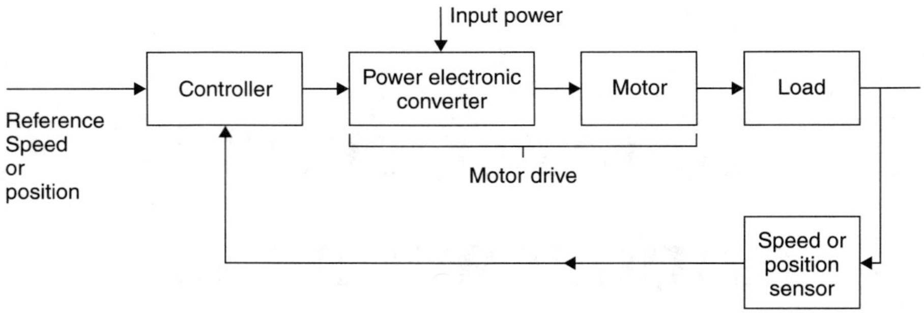

Figure 15.1
A typical motor drive system

vis. a vis. to the armature winding. The configuration is given in Figure 15.2(a). Back emf, E_m is less than dc source voltage V_S. Energy flows from the dc source to the load. Both armature and field currents are positive. It is the first quadrant operation in the speed versus the torque plane.

2. ***Regenerative Braking (Plugging):*** The dc motor operates in this mode when, during the forward motion of the motor, the polarity of the dc supply across the armature is reversed for stopping or reversing the direction of rotation. The configuration is given in Figure 15.2(b). The motor acts as the generator. The kinetic energy of the motor is converted into electrical energy and returned to the source. The armature and field currents are opposite in polarity.

3. ***Dynamic Braking:*** The dynamic braking refers to removing the dc source and connecting a resistor R_b across the armature winding for stopping the motor. The configuration is given in Figure 15.2(c). The kinetic energy of the motor is dissipated in R_b. The armature and field currents are opposite in polarity.

4. ***Four-Quadrant Operation:*** The four-quadrant operation refers to the operation of the dc motor with regenerative braking in both forward and reverse directions of rotation. The configuration is given in Figure 15.2(d). In the first and third quadrants, the energy flows from the dc source to the load. The second and the fourth quadrant operation indicate regenerative braking where the kinetic energy of the motor is returned to the dc source.

5. ***Form Factor:*** Form factor is defined as the ratio of the rms value of the ripple to the average value of the armature current in dc motor:

$$F_a = \frac{\tilde{I}_a}{\bar{I}_a} \tag{15-1}$$

The rms value of armature current signifies the amount of ripple in armature current and the consequent power losses, including the heating losses, in the dc motor. The average value of armature current signifies the dc electrical power applied to the motor and the consequently developed mechanical torque.

The form factor is zero for a pure dc armature current. A high value of the form factor signifies more losses, hence, high motor temperature and poor efficiency. The high value also indicates the high value of peak armature current. High peak current may result in excessive arcing in commutator and brushes. High peak current and high ripple in the armature current also signifies **pulsation** in the developed mechanical torque, which is not desirable. Designers must always aim for the lowest value of the form factor.

15.2.2 Low-power dc motor drives

The low power dc motors (up to 5 kW) are often permanent-magnet dc motors. An example is the dc motor in the cassette driver in tape recorders. The

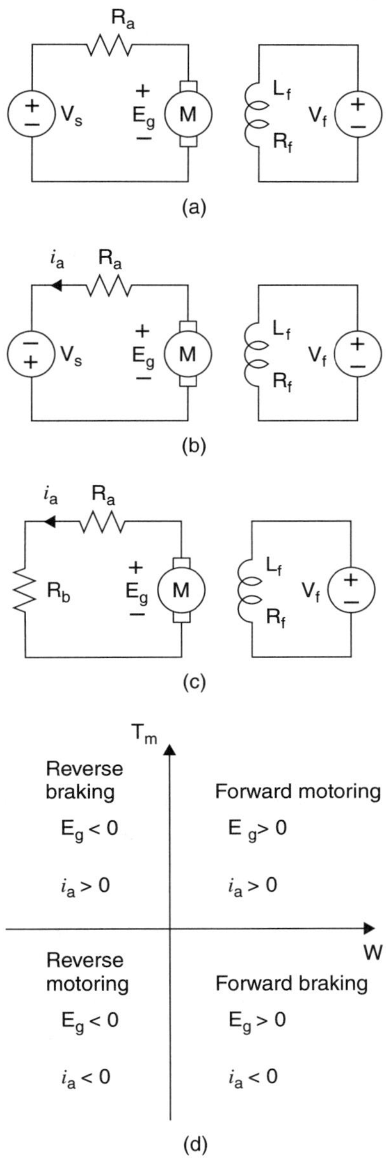

Figure 15.2
(a) Forward motoring, (b) Regenerative braking, (c) Dynamic braking and (d) Four quadrant operation

field flux in these motors being constant, the motor equations of Chapter 2 are modified as follows:

$$\begin{aligned} T_m &= k_T I_a \\ E_m &= k_e \omega \\ V_S &= I_a R_a + E_m \end{aligned} \tag{15-2}$$

The steady-state speed-torque relation is

$$\omega = \frac{1}{k_e}\left(V_S - \frac{R_a}{k_T} T_m\right)$$

$$\frac{\omega}{\omega_0} = 1 - AT_m \tag{15-3}$$

where $\quad A = \dfrac{R_a}{k_T V_S} \quad \textit{and} \quad \omega_0 = \dfrac{V_S}{k_e}$

The speed-torque characteristics are shown in Figure 15.3.

Since the armature resistance R_a is usually small, the motor speed does not vary significantly for a large variation in the torque. This is a desirable feature in the applications of the tape recorder and the analog recorder. The motor speed is varied by controlling the dc source voltage V_S.

15.2.3 DC servo drives

DC Motor with Separately Excited Field Winding

Medium to high power dc motors generally have a separately excited field winding on the stator. The speed-torque relation and characteristics are:

$$\omega = \frac{1}{k_e I_f}\left(V_S - \frac{R_a}{k_T I_f} T_m\right)$$

$$\frac{\omega}{\omega_0} = 1 - AT_m$$

where $\quad A = \dfrac{R_a}{k_T I_f} \quad \textit{and} \quad \omega_0 = \dfrac{V_S}{k_e I_f} \qquad (15\text{-}4)$

In practice the dc motor is operated at constant torque, hence, the field current I_f is constant for speeds less than the rated speed. The motor speed is controlled by varying source voltage V_S as is done in permanent-magnet dc motors.

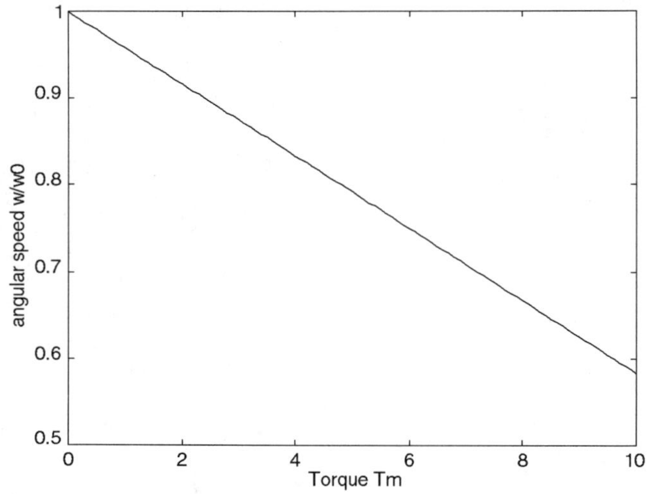

Figure 15.3
The speed-torque relation of permanent-magnet dc motor

For speeds more than the rated speed, the source voltage V_S is held constant and the field current I_f is decreased. In this region of operation, the power, the product $\omega_m T$, is held constant, hence the torque capability decreases as speed is increased. The motor is said to be operating in the field-weakening region. The speed can be increased up to 50 to 100% of the rated value.

Desired Features of the Power Electronic Converter

1. The converter should allow the reversal of both the voltage and current polarities for four-quadrant operation.
2. The converter should be able to operate in current-controlled mode during fast acceleration and deacceleration.
3. The average (dc) output voltage should be a linear function of the control parameter for accurate control of the position or speed in servo applications.
4. The form factor of the armature current should be low in order to achieve minimum fluctuations in the torque and speed of the motor.
5. The control response should be fast with low overshoot and low settling time.

The switch-mode medium to high frequency dc-dc converter is one choice. The other choice is the line-frequency dc-dc converter, also called the dc chopper.

The full-bridge (also called the H-bridge) dc-dc converter produces a four-quadrant operation. A detailed discussion of the converter is given in Chapter 6.

The peak-peak ripple in armature current is rewritten as follows:

$$\Delta I_{p-p,\max} = \frac{V_S}{2L_a f_s} \quad \textit{for bipolar voltage switching}$$
$$\Delta I_{p-p,\max} = \frac{V_S}{8L_a f_s} \quad \textit{for unipolar voltage switching} \tag{15-5}$$

The current ripple, hence the torque pulsation, can be reduced by enhancing the armature inductance by adding an external inductor and by selecting a higher switching frequency of the converter.

Control in DC Servo Motor Drives

A servo control system with and without an internal current control loop is shown in Figure 15.4. In the

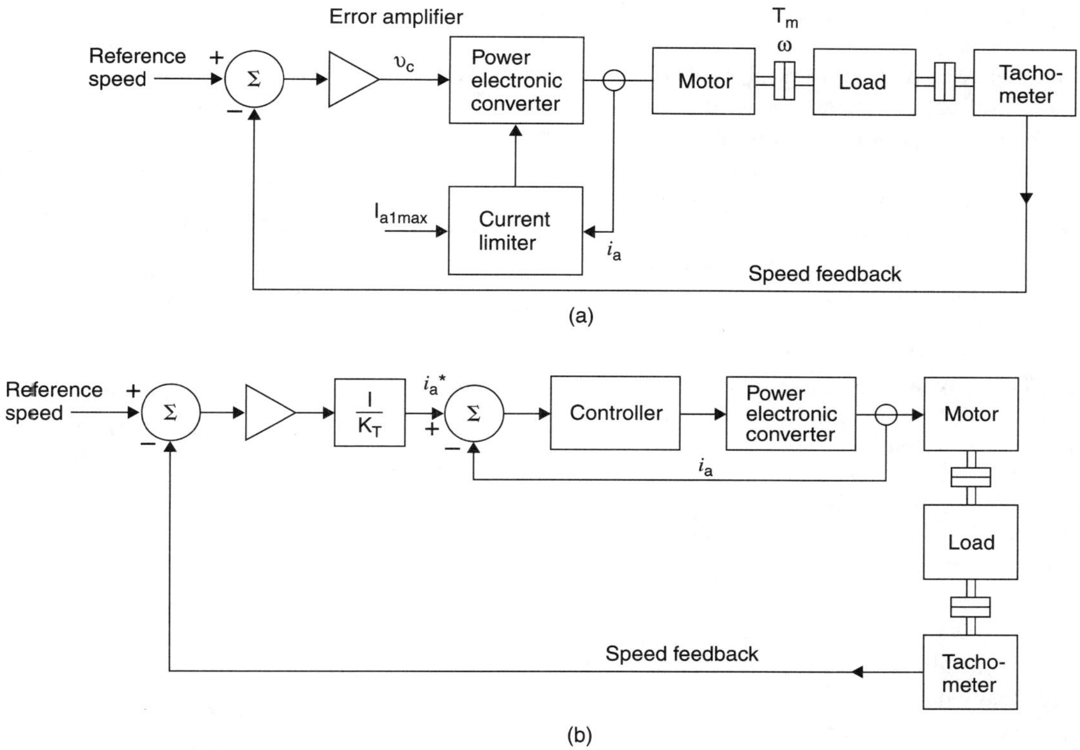

Figure 15.4
DC servo drives (a) Without internal current loop and (b) with internal current loop

first method, a current limiting circuit prevents the armature current from exceeding an acceptable limit during the acceleration and deacceleration of the dc motor. In the second method, the internal current control loop directly controls the armature current and, hence, the mechanical torque.

In practical designs, a blanking time is introduced between the turning off of one pair of the switches and the turning on of the other pair of switches so as to avoid short-circuiting the dc source. A nonlinearity creeps in the performance. The current-mode control minimizes the nonlinearity.

15.2.4 Adjustable speed drives

The dynamic response requirements in the adjustable speed drive systems are relaxed in comparison to those in the servo drives. The adjustable speed drives involve both the switch-mode dc-dc converter and the line-frequency converter.

Using Switch-Mode DC-DC Converter

The modes of operation of dc-dc converters in the adjustable speed drive systems are

1. The H-bridge dc-dc converter for four-quadrant operation.
2. The two-quadrant operation when the speed reversal is not demanded. A half-bridge converter would be most appropriate.
3. The single-quadrant converter is used in the applications where no braking is required. Only speed control is needed in these applications.

Using Line-Frequency Controlled AC-DC Converter
The line-frequency ac-dc converters, also called the phase (α) controlled converters, are used in high power drives. These converters are described in detail in Chapter 9. Generally an additional inductor is added in series with the armature winding to reduce the ripple in armature current. The line current is unidirectional, but the output voltage can reverse polarity, hence, the two-quadrant operation. The thyristor is the most preferred switching device.

Effect of Discontinuous Conduction Mode (DCM) Operation The armature current can become discontinuous in light load conditions and the output voltage rises in the discontinuous conduction mode, which leads to higher speeds, see Figure 15.5. In open-loop and discontinuous conduction mode operation, the speed is poorly regulated in the variable torque situation.

15.2.5 Traction motor drives

The traction motor drives, also called the chopper drives, are heavy-duty dc motor drives at low switching frequencies. These are four-quadrant drives with a regenerative braking arrangement, which is particularly ideally suited for transit systems with frequent stops.

Although the dc series motors are traditionally used in traction applications due to their high starting torque capability, regenerative braking is a difficult and unstable operation. Furthermore, the dc series motors are sensitive to dc source fluctuations. The separately excited dc motors are very stable in regenerative braking.

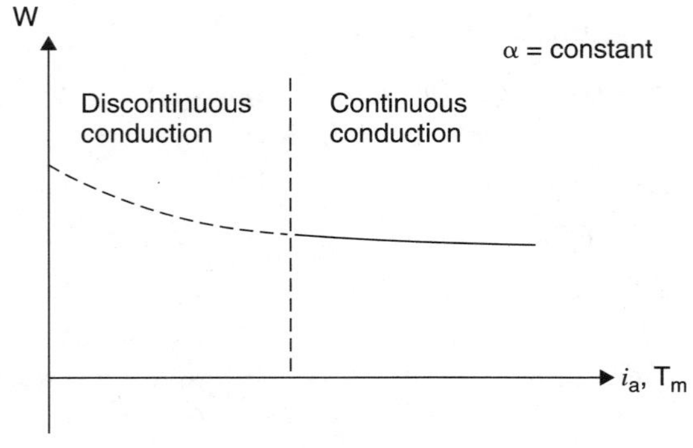

Figure 15.5
Effect of discontinuous armature current on the motor speed

15.3 INDUCTION MOTOR DRIVES

Low cost and ruggedness are the chief reasons for the popularity of the induction motor drives in industry. The induction motor drives can be categorized as:

1. ***Servo drives***

Use precise control scheme. Applications include computer peripherals, machine tools and robotic tools and equipment.

2. ***Adjustable drives***

Use speed control with braking. Applications include fans, compressors, pumps, blowers and process control systems.

Induction Motor Characteristics
The induction motor was introduced briefly in Chapter 2. The important terms are recollected here as follows:

ω_s: synchronous speed
ω_m: induction motor speed
ω_{slip}: slip speed $= \omega_s - \omega_m$
s: slip, the ratio of the slip speed to the synchronous speed
f: frequency of the ac voltage source
f_{sl}: slip frequency
ϕ_{ag}: air-gap flux
T_{em}: electromagnetic torque developed
I_r: rotor current
I_m: magnetizing current to establish the air-gap flux
I_s: stator current

The important relations are:

$$\omega_s \propto f,\ f_{sl} = sf,$$
$$\omega_r = (1 - s)\omega_s \quad \propto f \quad \textit{for low } s$$
$$V_S \approx k_s f \phi_{ag},\quad I_r \approx k_r f \phi_{ag} \qquad (15\text{-}6)$$

$$T_m \propto s f \phi_{ag}^2 \quad \propto s f \left[\frac{V_S}{f}\right]^2$$

$$\%P_r = \frac{s}{1 - s}$$

The air-gap flux is proportional to the voltage-to-frequency ratio of the ac source:

$$\phi_{ag} \propto \frac{V_S}{f} \tag{15-7}$$

The torque is proportional to the product of the source frequency and the square of the air-gap flux.

The equivalent circuit of an induction motor is shown in Figure 15.6.

The induction motor characteristics can be summarized as follows:

1. The motor speed at low slip is a linear function of frequency f of the ac source.
2. The air-gap flux must be kept constant for constant torque operation. This is equivalent to keeping the ratio of the voltage to frequency constant.
3. The rotor current I_r is proportional to the slip frequency f_{sl}.
4. Power losses in the induction motor are low at low slip values.

Thus, the best strategy of the induction motor control is to vary

(a) source frequency f to control the motor speed, and

(b) source voltage in proportion to f.

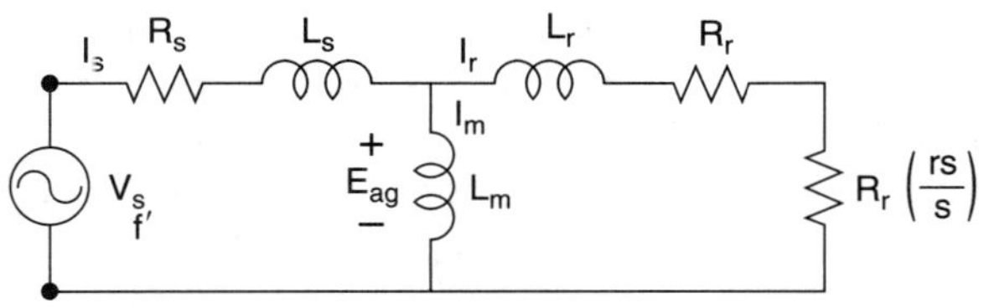

Figure 15.6
Equivalent circuit of the induction motor

The above strategy and the linear relations are valid only for low values of slip. Complete induction motor control characteristics are plotted as the motor torque, normalized to the rated torque, versus the motor speed, normalized to the synchronous speed. See Figure 15.7.

The linear relation is shown by solid lines. The maximum torque produced by the motor is called the pull-out torque. Commonly used induction motor drives use only the solid line portion of these characteristic curves. The starting rotor current, at slip $=1.0$, is approximately six times the rated current.

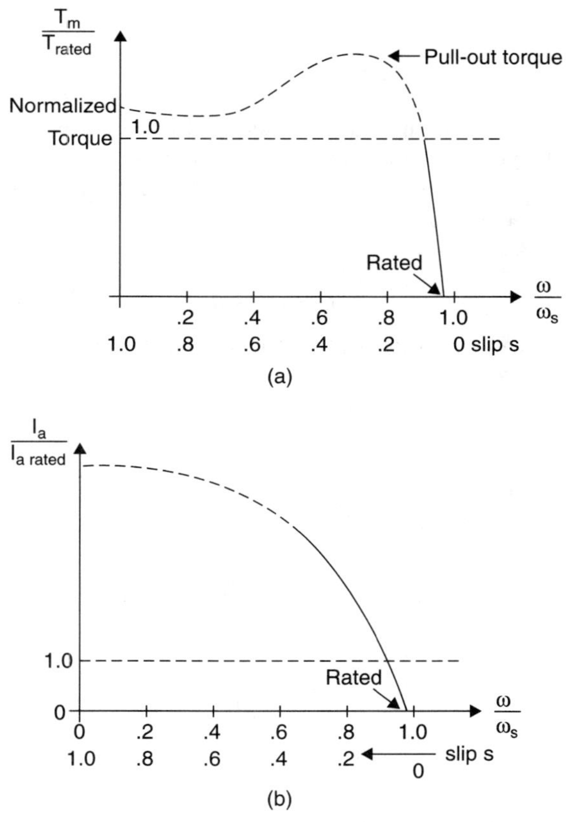

Figure 15.7
Induction motor control characteristics: (a) Normalized torque and (b) normalized rotor current versus normalized motor speed.

15.3.1 V-F control in induction motor drives

The synchronous speed of the induction motor and, hence, the motor (rotor) speed can be controlled by varying the frequency f of the stator ac supply. The back emf E_{ag} of the induction motor is directly proportional to the product of frequency f and air-gap flux ϕ_{ag}:

$$E_{ag} = k f \phi_{ag} \qquad (15\text{-}8)$$

E_{ag} can be made linearly related to f at constant air-gap flux ϕ_{ag}. For low impedance of the stator winding ($R_S + j2\pi f L_S$), the E_{ag} can be approximated by the source voltage V_S. Hence, the constant air-gap flux means

$$\phi_{ag} \propto \frac{E_{ag}}{f} \propto \frac{V_S}{f} \qquad (15\text{-}9)$$

The air-gap flux is maintained constant by the varying source voltage in proportion to f.

Torque in terms of the motor (rotor) speed is derived as follows:

$$T_{em} = k f_{sl}; \quad \phi_{ag} = constant$$
$$T_{em} = k\omega_{sl} = k(\omega_s - \omega_r) = k\omega_s - k\omega_r \quad (15\text{-}10)$$

The characteristics of the induction motor are given in Figure 15.8.

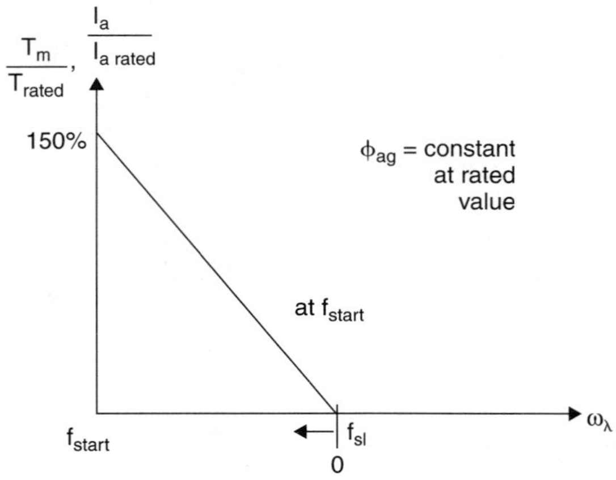

Figure 15.8
Start-up motor characteristics

Starting Torque

The slip frequency at zero speed equals the starting stator frequency f_{sl} itself. The starting torque and the starting rotor current are

$$I_{r,start} = k_1 f_{start},$$
$$T_{em,start} = k_2 f_{start} : at\ constant\ \phi_{ag} \qquad (15\text{-}11)$$

Knowing the starting torque requirement, the starting rotor current could be limited by selecting a proper starting frequency:

$$f_{start} = \frac{T_{em,start}}{T_{em,rated}} f_{sl,rated} = \frac{I_{r,start}}{I_{r,rated}} f_{sl,rated} \qquad (15\text{-}12)$$

Constant-Torque Operation

In the region of speed below the rated value, the solid curves in Figure 15.7 show the torque-speed characteristics at the low values of slip frequency f_{sl}. The source voltage is varied in proportion to the source frequency for holding the air-gap flux ϕ_{ag} constant. The motor can deliver a constant torque at the rated value by drawing the rated current $I_{r,\ rated}$.

The slip frequency remains constant in the constant-torque operation. The power loss $P_r = 3I_r^2 R_r$ is also constant. This power is dissipated in the rotor. Heat removal from the rotor at low speeds poses problems, hence, cooling fans may be required.

Voltage Boosting at Low Speed

The induction motor control by maintaining constant air-gap flux ϕ_{ag} is in reality making the ratio of back emf E_{ag} to frequency f constant. At high speeds (high E_{ag}), the source voltage V_S is approximately equal to the back emf E_{ag}. At low speeds, the back emf E_{ag} is low, hence, the voltage drop across the stator resistance R_S and the leakage inductance of the stator winding L_S cannot be ignored anymore. The impedance $2\pi f L_r$ is negligible in comparison to the resistance R_r/s in the equivalent circuit of the induction motor. Hence, the rotor current I_r and E_{ag} would be in phase. The magnetizing current I_m lags E_{ag} by 90 degrees, hence,

$$I_S = I_r - jI_m \quad and$$
$$V_S = E_{ag} + (R_S + j2\pi f L_S) I_S \qquad (15\text{-}13)$$

Using the above two equations and ignoring the imaginary part due to low frequencies,

$$V_S \cong E_{ag} + j2\pi f L_S I_m + R_S I_r \quad and$$
$$E_{ag} \propto f, \quad I_m \propto f \quad at\ constant\ \phi_{ag} \qquad (15\text{-}14)$$
$$V_S \cong kf + R_S I_r$$

At low speeds (low f), the $R_S I_r$ term cannot be ignored. That is why the V_S-f curve as shown in Figure 15.9 does not pass through the origin. Hence, at low f the source voltage V_S must be boosted to compensate for the $R_S I_r$ drop so that E_{ag} can be made truly proportional to f and the air-gap flux ϕ_{ag} can be maintained constant.

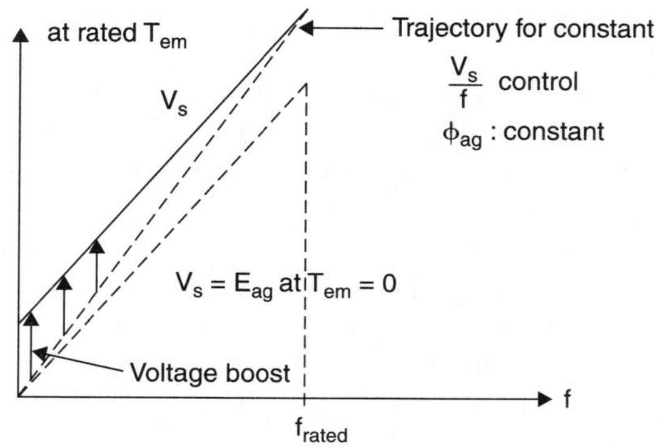

Figure 15.9
Voltage boosting at low speeds (low f) to hold the air-gap flux constant

Constant-Power Operation
The motor speed can be increased beyond the rated value by increasing the frequency of the ac source connected to the stator winding. In this region the source voltage is held constant at the rated value. The ratio V_S/f decreases, consequently the air-gap flux ϕ_{ag} decreases. The rotor current is constant, hence, the power loss in the rotor is also constant. The torque is given as

$$T_{em} \propto \frac{\omega_{sl}}{f^2} \qquad (15\text{-}15a)$$

Maximum torque is given by

$$T_{em,\max} \propto \frac{f_{rated}}{f} T_{rated} \qquad (15\text{-}15b)$$

The slip s is constant in this region. Hence, the rotor current I_r, which is proportional to the slip s, is also constant. The motor speed is given by

$$\omega_r = (1 - s)\omega_s \propto f \qquad (15\text{-}16)$$

The maximum output power of the motor $P_{max} = \omega_r T_{em,\,max}$ is also constant. Hence, this region of operation is called the constant-power region.

Constant Slip-Frequency Operation
Beyond a speed in the range 1.5-2 times the rated value, the air-gap flux ϕ_{ag} reduces so much that the motor torque approaches the pull-out torque value. At still higher speeds, the motor can deliver only a fixed fraction of the pull-out torque and the slip frequency ω_{sl} becomes constant. The maximum torque declines as

$$T_{em,\max} \propto \frac{1}{f^2} \qquad (15\text{-}17)$$

Both the torque and the motor current I_r decline with increasing speed.

High Voltage Operation
In most induction motors, the insulation is rated for much higher voltages than the rated voltage of the motor. Hence the motor can be run at high speeds by applying higher voltage to the stator. The V_S/f ratio is held constant, hence, the air-gap flux is also constant. Consequently the motor current I_r is also held constant at the rated value. Thus, the motor delivers the rated torque at higher power and higher speeds.

15.3.2 Braking in induction motors

Regenerative braking can be applied in the induction motor by reducing the frequency of the stator supply. The torque-speed characteristics beyond the synchronous speed are shown in Figure 15.10.

The slip is negative for rotor speed beyond the synchronous speed ω_s. As a result the developed torque is also negative, and it acts in the direction opposite to the direction of the rotation of the motor.

This is regenerative braking. The induction motor operates in the generator mode.

Braking is applied by reducing the frequency of the ac source to the stator. The synchronous speed at the reduced f is lower than the motor speed ω_r. Hence, the slip is turned negative. Then the motor goes in the generator mode and an opposing torque develops that applies the brake. The frequency f must either be decreased by a large amount or reduced continuously so that the corresponding synchronous speed is continuously lower than the motor speed.

15.3.3 Variable-frequency drive

The block diagram of a variable-frequency converter (non-cycloconverter type) for the induction motor control is given in Figure 15.11. The ac-dc converter may be either the uncontrolled or the controlled rectifier. The inverter output is a single phase or 3-ϕ ac voltage with variable amplitude and frequency, recognizing that v/f ratio is maintained constant.

The variable-frequency inverter for induction motor control must have the following capabilities to:

(a) adjust the frequency according to the desired output speed,

(b) maintain a constant air-gap flux in the constant torque region

(c) supply the rated current to the motor.

Three basic types of inverters will be discussed in the following sections:

1. PWM voltage source inverter (PWM-VSI)
2. Square-wave voltage source inverter (square-wave VSI)
3. Current source inverter (CSI)

These inverter categories have been discussed in detail in Chapter 7. The PWM-VSI uses an uncontrolled ac-dc converter at the input. The square-wave VSI and CSI use the controlled ac-dc converter.

Variable-Frequency PWM-VSI Drives

Figure 15.12 shows the block diagram of variable-frequency PWM-VSI drive and Figure 15.13 shows the schematic and waveforms in a three-phase drive. The output current waveform is approximately sinusoidal. The frequency of the ripple current from the ac-dc converter is equal to the switching frequency of the inverter, which is several times

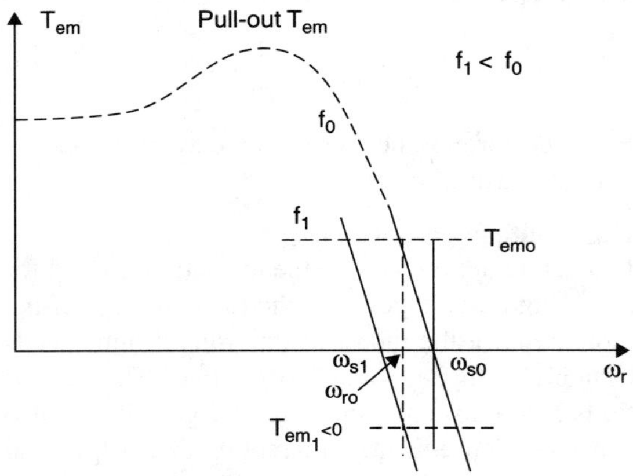

Figure 15.10
Induction motor characteristics in braking

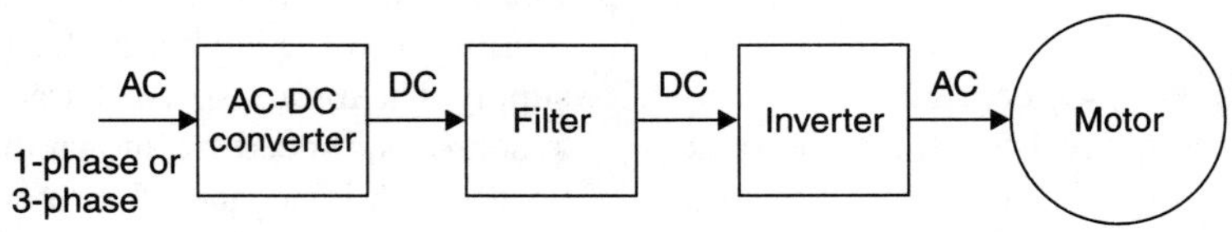

Figure 15.11
Variable-frequency converter block diagram

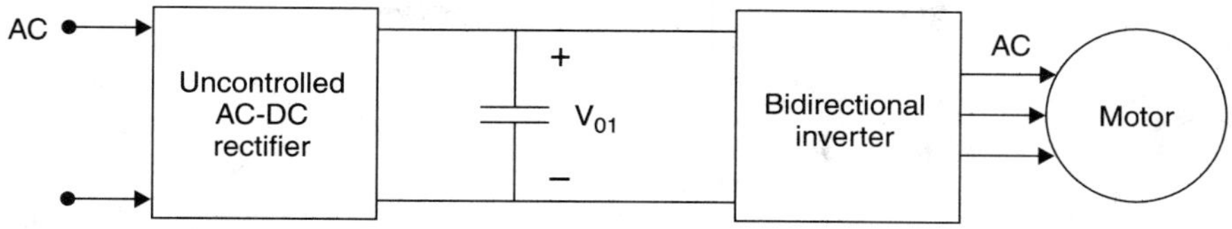

Figure 15.12
Block diagram of PWM-VSI drive for induction motor control

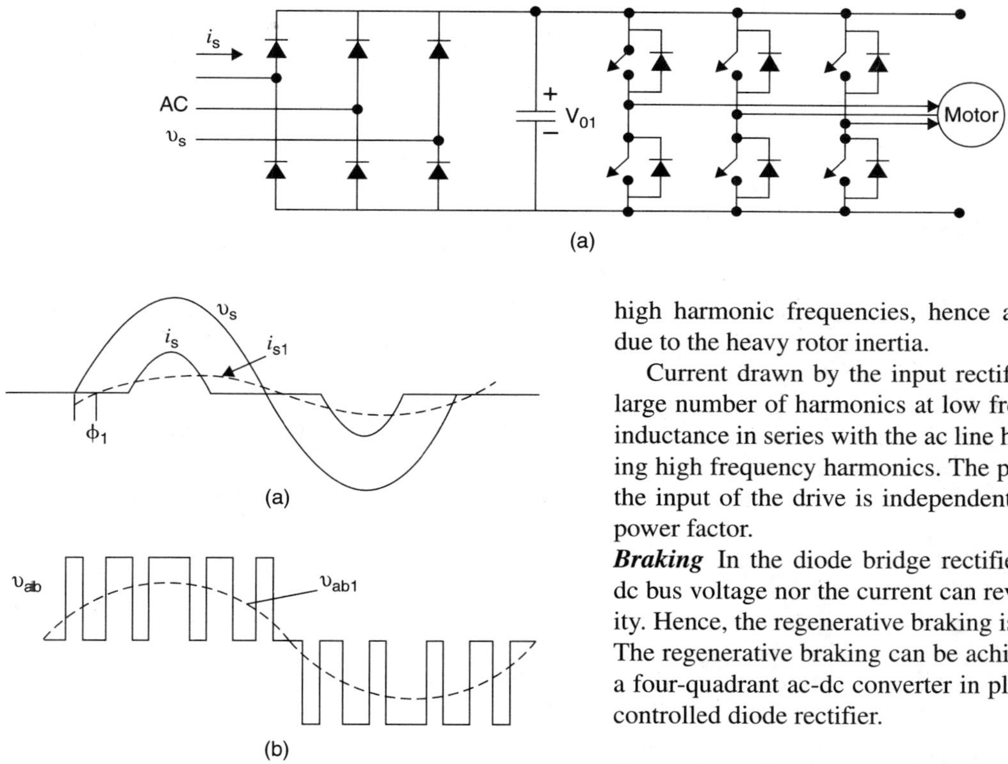

Figure 15.13
A 3-phase PWM-VSI for induction motor control: (a) schematic and (b) waveforms

higher than the desired output frequency. Hence, a low value capacitor is sufficient for filtering. The harmonics, however, may have higher amplitudes than the fundamental. As a result the power losses in the PWM-VSI inverter is significantly higher than the square-wave VSI. The pulsating torque is at high harmonic frequencies, hence are ineffective due to the heavy rotor inertia.

Current drawn by the input rectifier contains a large number of harmonics at low frequencies. An inductance in series with the ac line helps in reducing high frequency harmonics. The power factor at the input of the drive is independent of the motor power factor.

Braking In the diode bridge rectifier, neither the dc bus voltage nor the current can reverse in polarity. Hence, the regenerative braking is not possible. The regenerative braking can be achieved by using a four-quadrant ac-dc converter in place of the uncontrolled diode rectifier.

Servo Control in PWM-VSI Drives

Precise control such as required in the servo drives is achieved by using field-oriented space vector control. The block diagram is shown in Figure 15.14. The controller uses high-power Digital Signal Processor (DSP) integrated circuits for modeling of the induction motor, for calculating the three-phase reference currents for error detection and actual control. The theory of the controller is beyond the scope of this book and, therefore, not discussed.

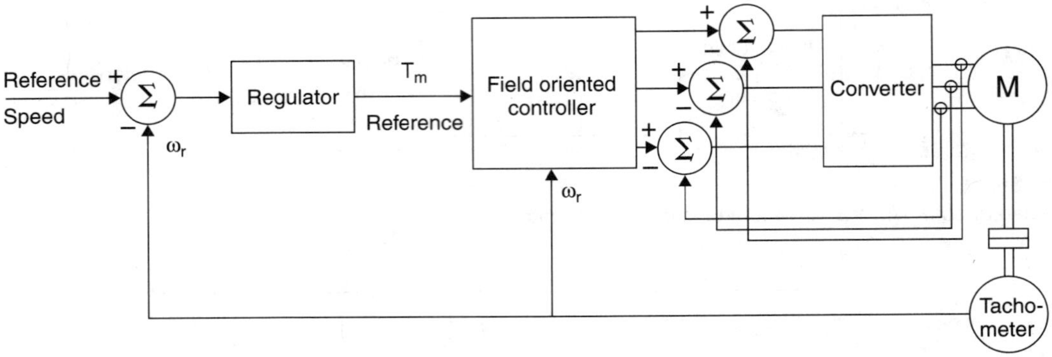

Figure 15.14
Field-oriented space vector control for servo control using induction motor

Adjustable Speed Control in PWM-VSI Drives

Figure 15.15 shows an adjustable speed control for the induction motor. No feedback control is required if the accuracy is not the premium requirement. The reference $\omega_{r,\,ref,}$ the control inputs ω_s, f and V_S are calculated in place of the actual values. The complete control circuit can be realized using the special purpose power-integrated circuits or the Digital Signal Processor (DSP) integrated circuits. The switching frequency is varied in proportion to the desired value of the fundamental frequency of the inverter output.

Speed Control Circuit Through the ramp limiting at the input of the controller, the designer can specify the maximum acceleration (rising slope in speed-up) and deacceleration (falling slope in the speed-down) rates.

Current Limiting The motor current, in the forward motoring mode, increases if the reference ω_s is increased faster than the actual motor speed can follow. In the braking mode, if the reference ω_s is decreased too fast, the negative slip would become large. As a result, the braking current through the inverter would become large. The current limiting circuit limits the current in both conditions.

Slip Compensation The slip compensation is required to keep the rotor speed constant. The compensator estimates the developed electromagnetic torque and adds a proportional term to the reference speed $\omega_{r,\,ref}$.

Voltage Booster The booster adds a term in proportion to the estimated torque to the calculated value of V_S in order to keep the air-gap flux constant.

Accuracy in the controller can be increased by using the measured values in place of the calculated values of the input control parameters.

Variable Frequency Square-Wave VSI Drives

The block diagram of the square-wave PWM-VSI drive is given in Figure 15.16. The ac-dc converter at the input is designed around a controlled topology given in Chapter 9. The inverter with square-wave control is discussed in detail in Chapter 7. Because of the square-wave control, the inverter output performs only frequency control. The amplitude is controlled by varying the dc bus voltage at the output of the ac-dc converter. In a three-phase drive, three switches are on at any time and each switch conducts for 180 degrees.

The dc bus voltage from a phase-controlled ac-dc converter and the line-line rms inverter output voltage are given as

$$\begin{aligned}\overline{V}_{O1} &= 1.35\,\tilde{V}_{LL}\cos\alpha \\ \tilde{V}_{O2,LL} &= 0.78\,\overline{V}_{O1} = 1.05\,\tilde{V}_{LL}\cos\alpha\end{aligned} \tag{15-18}$$

The maximum inverter output voltage is approximately equal in both the square-wave drive ($\alpha = 0$) and PWM drive (in over-modulation). This allows the use of standard induction motors.

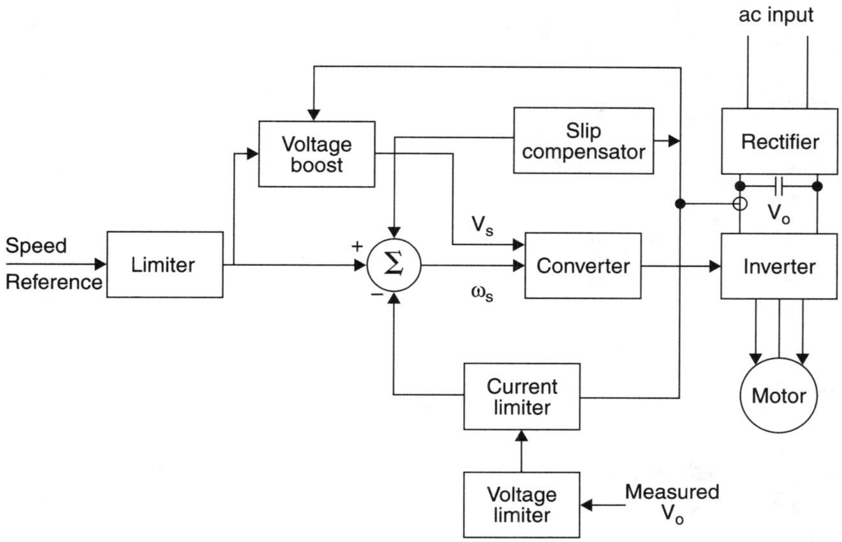

Figure 15.15
An adjustable speed control for the induction motor

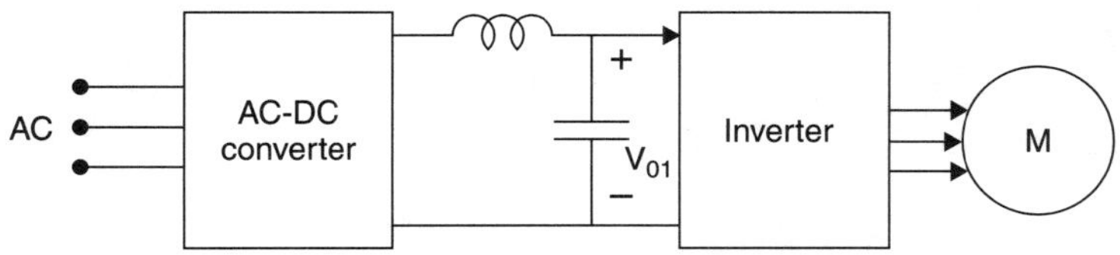

Figure 15.16
Block diagram of the square-wave VSI drive

The motor speed in the drive with constant V_S/f, is given as

$$\omega_r = \omega_{rated} \cos \alpha \tag{15-19}$$

Input power factor of the square-wave drive is

$$PF_{in} \approx 0.955 \cos \alpha \approx 0.955 \frac{\omega_r}{\omega_{rated}} \tag{15-20}$$

The input power factor is poor at low speeds. An alternate design that will overcome this problem is to have an uncontrolled ac-dc rectifier followed by a step-down dc-dc converter.

The inverter output has low order harmonics of substantial amplitudes. As a result the square-wave drive contains high torque ripple and high speed ripple.

Variable-Frequency CSI Drive

CSI drives are mostly used in large horsepower applications. A variable-frequency CSI drive is given in Figure 15.17. The ac-dc converter at the input is a controlled converter. The CSI inverter is discussed in detail in Chapter 7. A large inductor is used in the dc link that makes the input to the inverter appear as a current source.

The inverter uses forced commutation of switches. The induction motor in this drive has non-standard specifications. The regenerative braking can be easily performed in CSI drives.

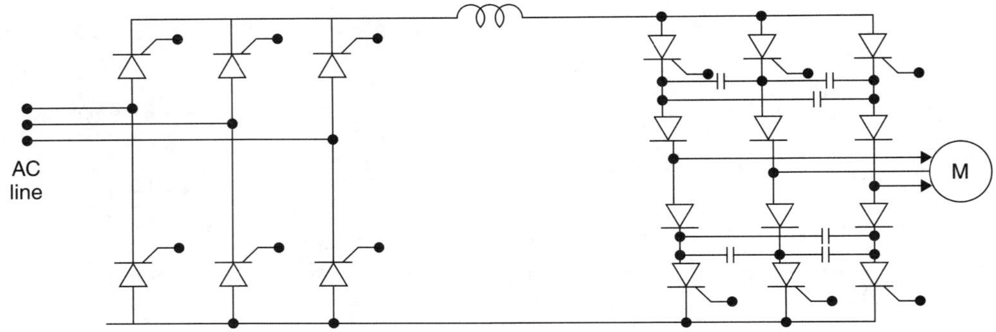

Figure 15.17
Variable-frequency CSI drive

15.3.4 Line-frequency variable-voltage drives

The line-frequency variable-voltage drive is used for induction motors that have large rotor resistance and hence pull-out torque at large slip values. The drive permits speed control over a wide range.

The electromagnetic torque in terms of the ac source voltage is given by

$$T_{em} \propto f\phi_{ag}^{\ 2} \propto \frac{V_S^2}{f} \tag{15-21}$$

For constant-torque operation the speed is controlled by varying the source voltage. Speed control in this manner results in poor efficiency at low speeds (large slip) because of dissipation in large rotor resistance. This technique is widely used in fractional horsepower fan or pump drives, cranes, and hoist drives that have intermittent-load operation.

15.3.5 Soft start in induction motor drive

Starting current in the induction motors may be approximately six times the rated current. The soft start of the induction motor, reduced current starting, is actually the reduced voltage starting. The circuit in Figure 15.18 can also be used for soft starting in constant speed control.

Each thyristor conducts for an entire half cycle. The back-to-back thyristor pairs are shorted out of the circuit by mechanical contactors in the steady-state condition.

15.3.6 Speed control by static slip power recovery

The electromagnetic torque in terms of the rotor current is given by

$$T_{em} \approx k_T \phi_{ag} I_r \tag{15-22}$$

The rotor current of the induction motor can be approximated by, (refer to the equivalent circuit of the induction motor)

$$I_r \approx \frac{E_{ag}}{R_r\left(1 + \dfrac{1-s}{s}\right)} \approx \frac{V_S}{R_r/s}; \quad L_r \text{ is ignored}$$

$$I_r \propto \frac{f\phi_{ag}}{R_r/s} \tag{15-23}$$

R_r is the rotor resistance.

The rotor current must be kept constant, for constant torque operation at the rated air-gap flux. It is equivalent to maintaining the rotor resistance to the slip ratio constant. Hence, the motor speed can be controlled by varying an external resistance in the rotor circuit in proportion to the slip s.

An alternate scheme is to recover the power in the rotor circuit rather than it dissipating in the external rotor resistor. A slip power recovery circuit is shown in Figure 15.19. The induction mo-

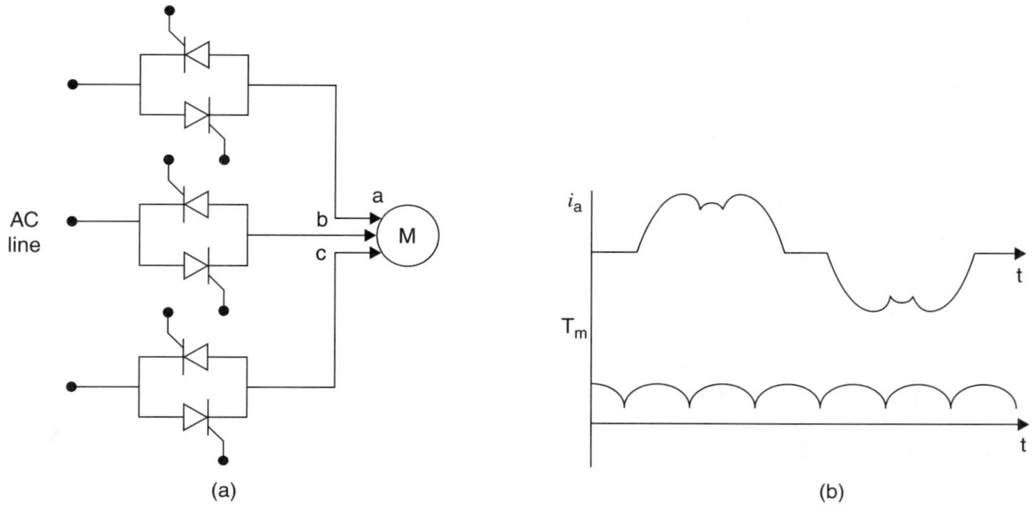

Figure 15.18
Soft start of induction motor (a) Circuit and (b) Waveforms

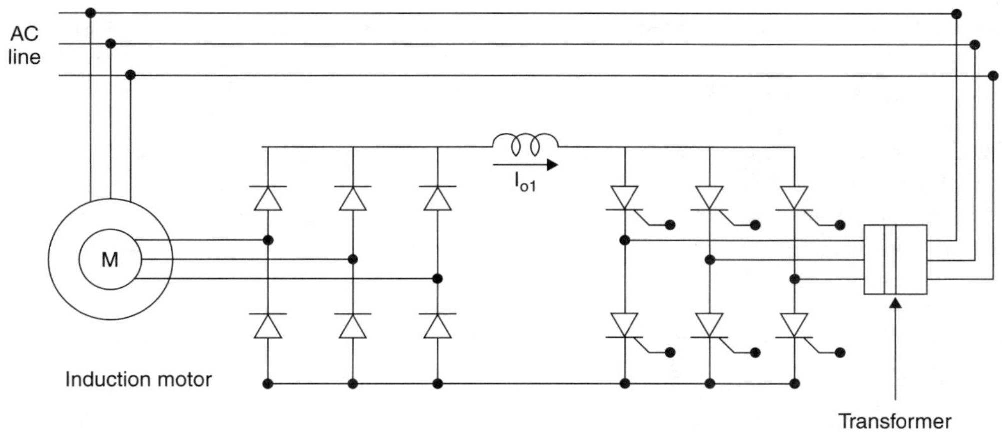

Figure 15.19
Slip power recovery for speed control

tor in this circuit should be a wound rotor motor with slip rings. This motor is more expensive than the squirrel-cage type. However, the use of the slip power recovery is competitive in large power rating adjustable speed drives over a small range.

15.4 SYNCHRONOUS MOTOR DRIVE

Synchronous motors of low wattage (a few kWs) use a permanent magnet in the stator and they are referred to as ***brushless dc motors.*** Large power applications use large synchronous motors with wound rotor field.

The synchronous motor drives can be categorized as:

1. ***Servo drives***

Use precise control scheme. Applications include computer peripherals, machine tools and robotic tools and equipment.

2. ***Adjustable drives***

Use speed control with braking. Applications include large fans, heat pumps, and compressors.

Synchronous Motor Characteristics

The synchronous motor was introduced briefly in Chapter 2. The equivalent circuit of the synchronous motor is shown in Figure 15.20.

The important terms are recollected here as follows:

ω_s: synchronous speed
f: frequency of the ac voltage source
ϕ_{ag}: air-gap flux
T_{em}: electromagnetic torque developed
I_s: stator current
E_f: induced voltage on the stator winding, also called the excitation voltage
E_s: induced voltage on the phase

The important relations are:

$$\omega_s \propto f,$$
$$V_S \approx k_s f\phi_{ag}, \quad I_r \approx k_r f\phi_{ag}$$

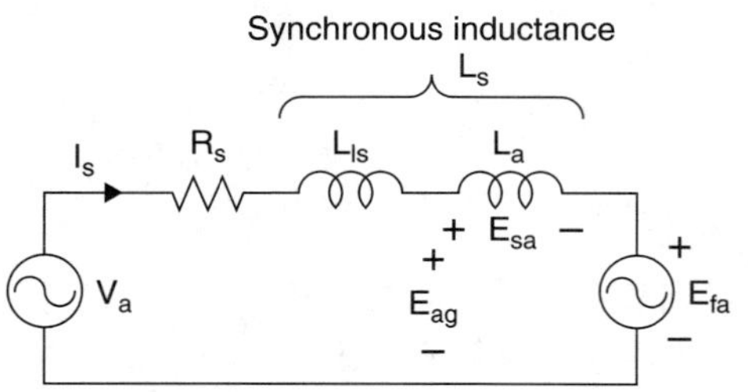

Figure 15.20
Equivalent circuit of a single-phase synchronous motor

$$T_{em} \propto s f\phi_{ag}^2 \quad \propto s f\left[\frac{V_S}{f}\right]^2 \tag{15-24}$$

$$\% P_r = \frac{s}{1 - s}$$

The air-gap flux is proportional to the voltage-to-frequency ratio of the ac source:

$$\phi_{ag} \propto \frac{V_S}{f} \tag{15-25}$$

The torque is proportional to the product of the source frequency and the square of the air-gap flux.

15.4.1 Servo-motor drives

The synchronous motor is ideally suited for servo applications that require fast and accurate response in the control system. The servo drives can be classified in two categories based on the profile of the air-gap flux density throughout the circular periphery of the motor: (a) sinusoidal profile and (b) trapezoidal profile.

Air-Gap Flux with Sinusoidal Profile

The air-gap flux density distribution in the circular air-gap is sinusoidal, see Figure 15.21. The induced emfs in the three phases in the motor are also sinusoidal. The torque angle δ is held at 90 degrees. The rotor field position is sensed by mechanical angle resolvers (sensors). Currents in the three phases are

$$\begin{aligned} i_a &= I_S \cos[\theta_e(t)] \\ i_b &= I_S \cos[\theta_e(t) - 120°] \\ i_c &= I_S \cos[\theta_e(t) - 240°] \end{aligned} \tag{15-26}$$

where, $\theta_e(t)$ is electrical angle, which is equal to the mechanical angle in the two-pole rotor.

The drive block diagram is shown in Figure 15.22.

Air-Gap Flux with Trapezoidal Profile

The motor windings are made of concentrated coils in place of the evenly distributed coils. Thus the flux density distribution in the circular air-gap is trapezoidal. See Figure 15.23. The induced emfs are also trapezoidal. The flat top or bottom of the emf waveform occurs for at least 120 electrical degrees during each half cycle.

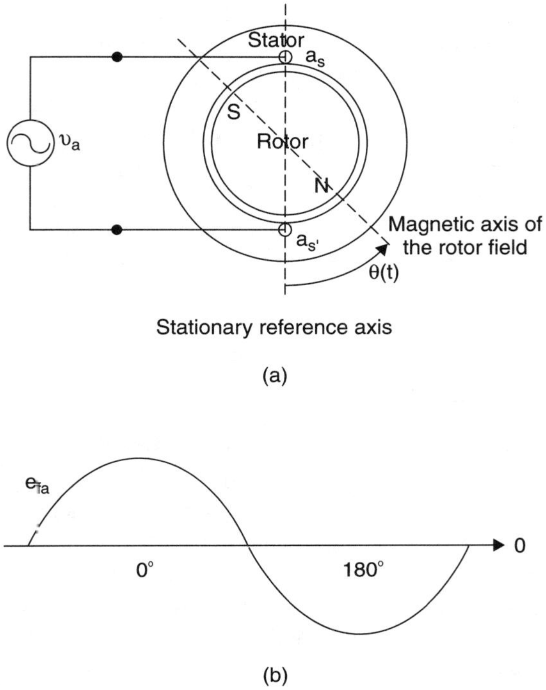

Figure 15.21
(a) Rotor position sensing reference for a two-pole synchronous motor and (b) sinusoidal profile of the flux density in the air-gap

The torque is constant and depends only on the amplitude of the stator current. The servo motor drive for trapezoidal air-gap flux has an identical block diagram shown in Figure 15.22. The reference currents have trapezoidal shape.

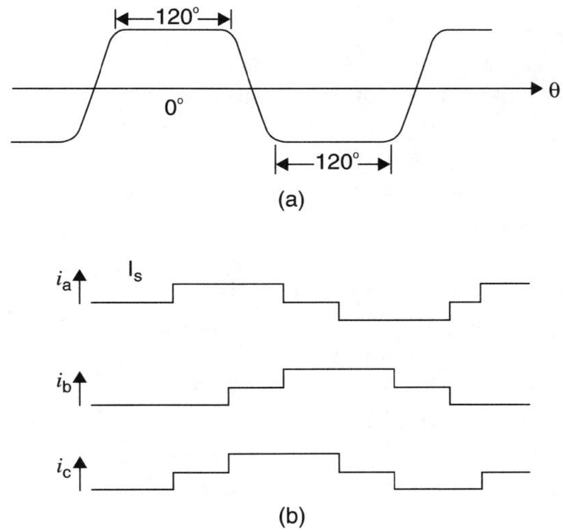

Figure 15.23
(a) Trapezoidal profile of the flux density in the air-gap and (b) waveforms

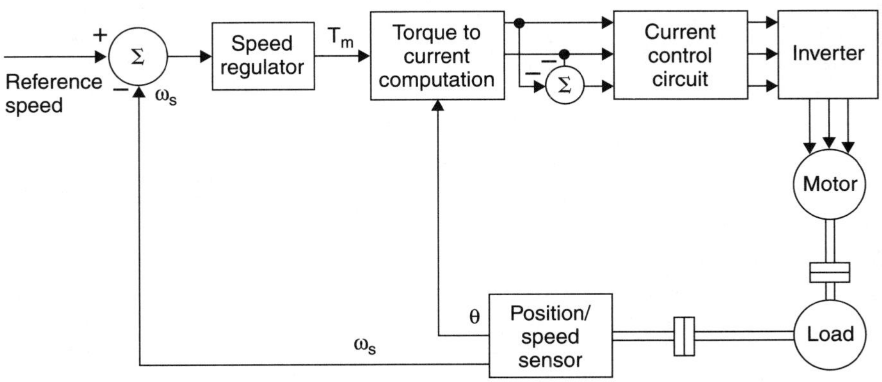

Figure 15.22
Synchro-motor servo drive with sinusoidal air-gap flux

15.4.2 Load commutated inverter (LCI) drives

It is an adjustable speed drive for very large horsepower synchronous motors (>1000 hp), typically of nonsalient pole construction. The schematic and the waveforms are shown in Figure 15.24.

The LCI drive output is basically a voltage-to-current converter. It supplies a three-ϕ current to the synchronous motor. The frequency and the phase of the current is synchronized to the rotor position. The input is a controlled ac-dc current converter. The inverter produces the three-phase current for the motor. The amplitude of the output current is controlled by the ac-dc converter through phase control. The thyristor commutation in the inverter is performed by the internally induced emfs in the motor.

At start-up and at speeds lower than 10% of the rated speed, the induced emf in the motor is not large enough to effect current commutation. The input ac-dc converter operates in the inverter mode to force the dc bus current I_{ob} to become zero, thus providing current commutation. The synchronous motor can be started as an induction motor. As the speed approaches the synchronous speed, the rotor field is excited to change to the synchronous motor mode.

The LCI drive is more efficient (>95%) than the CSI induction motor drives in very large power applications. The LCI converter is simpler and has lower losses.

The LCI drive has an inherent capability to provide regenerative braking by converting the synchronous motor into a synchronous generator. The inverter acts as the rectifier and the input ac-dc converter operates in the inverter mode to return the braking power to the line.

15.4.3 Cycloconverter drives

Cycloconverter drives are described in detail in Chapter 7. The cycloconverter output is derived from the ac line input without going through a dc link. The output frequency is lower than the input frequency (usually less than one-third). Cycloconverter drives are used in low speed and very large horsepower.

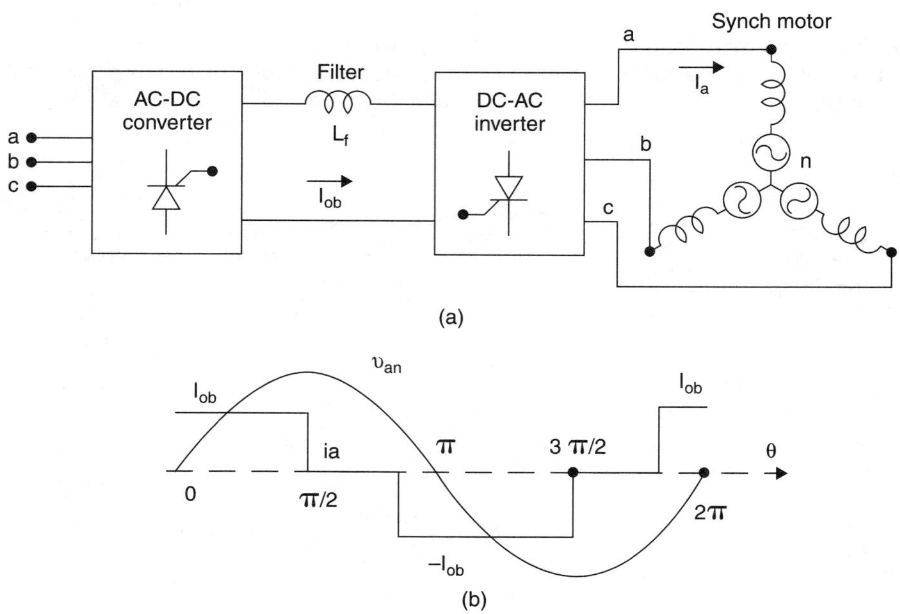

Figure 15.24
LCI drive: (a) Schematic and (b) Waveforms

15.5 SUMMARY

This chapter discussed the switch-mode design of motor drives that are more efficient and smart compared to the nonswitch-mode varieties. The system requirements and the control issues are discussed for dc motor drives, induction motor drives and synchronous ac motor drives in servo and adjustable-speed applications. The first part contains the discussion on the design of dc motor drives in low power and medium-to-high power applications. The second part discusses the constant-torque, constant-power and constant slip-frequency operation of induction motors. Issues related to braking are also discussed. The PWM drives of induction motors with voltage source and current source are introduced. The last topic covered the synchronous drives.

References

1. Philip T. Krein, *Elements of Power Electronics,* Oxford University Press, New York, 1998.

2. Ned Mohan, Tore M. Undeland, and William P. Robbins, *Power Electronics: Converters, Applications and Design,* 2nd Ed., John Wiley & Sons, Inc., 1995.

3. Joseph Vithayathil, *Power Electronics: Principles and Applications,* McGraw-Hill Inc., New York, 1995.

4. Muhammad H. Rashid, *Power Electronics: Circuits, Devices and Applications,* 2nd Ed., Prentice-Hall Inc., New Jersey, 1993.

5. John G. Kassakian, Martin F. Sclecht, and G. Vergassian, *Principles of Power Electronics,* Addison-Wesley, Boston, 1991.

6. Kjeld Thorborg, *Power Electronics,* Prentice-Hall Inc., 1988.

Problems

15-1 An electric car uses a dc motor that is supplied power from a battery bank of 200 Vdc through a dc-dc converter. The converter operates at 2 KHz. The armature circuit of the motor has a resistance of 0.2 ohms and an inductance of 100 μH. The back emf at a speed of 40 miles per hour is 60 volts. The duty cycle D= 33.3%. Determine

a. the peak-peak ripple current and

b. the average dc current in the armature of the motor.

15-2 The electric car of Problem 15-1 uses a regenerative braking system. The duty cycle D at the time of braking is 0.25. The dc motor operates as the dc generator and supplies 20 A. Find the average dc current returned (to charge) to the battery.

15-3 A dc motor with a voltage rating of 400 V for both armature and field is used as a separately excited motor. The field is supplied at the rated voltage. The armature is supplied by a thyristor bridge from a three-phase, 460 V, 60 Hz ac bus. The source inductance per phase of the ac bus is 100 μH. The armature is rated for 120 A and has a resistance of 0.12 ohm. The no-load speed is 1200 RPM. Determine the delay angle required to achieve a speed of 800 RPM at the rated current in the armature.

15-4 The dc motor of Problem 15-3 is supplied by a fully-controlled thyristor bridge from a three-phase, 440 V, 60 Hz ac utility supply. The source inductance per phase of the ac bus is 100 μH. The armature is rated for 120 A and has a resistance of 0.12 ohm. The no-load speed is 1200 RPM. Determine

a. the minimum permissible delay angle

b. the delay angle required to achieve a speed of 800 RPM at the rated current in the armature.

15-5 The dc-link voltage of a three-phase voltage source inverter is provided by a one-phase diode bridge rectifier from a 120 V one-phase supply. The input side capacitor is 2000 μF. The ac motor is regeneratively braked, causing a reverse dc current of 20 A. Determine

a. the rate at which the dc-link voltage will rise due to braking

b. the value of the braking resistor needed on the dc side.

15-6 Design a 100 kW inverter for an electric vehicle to be operated from a 400-V battery bank. The desired switching frequency is 10 kHz. Suggest the converter topology with the types of switches and diodes.

15-7 A 6.5 hp, 2200 rpm self-excited dc motor, driven from a 120 Vrms 60 Hz source, has the following parameters: $R_a = 0.02$ ohms, $L_a = 1$ mH, $R_f = 150$ ohms, $k_m = 0.5$ N-m/A^2, $k_e = 0.608$ V/A.rad/sec. The field is separately powered by a full-bridge rectified dc derived from the ac 120 Vrms, 60 Hz volts source. The motor speed is to be varied from 10% to the rated speed at a constant torque. The maximum peak-to-peak torque ripple should not exceed 5% of the rated torque. Design an ac-dc converter for driving the motor.

CHAPTER 16
Temperature Control, Protection and Packaging

OBJECTIVES

1. Basics of heat transfer and concept of thermal resistance
2. Modeling of heat transfer systems
3. Surge voltage protection of power semiconductor devices
4. Fault current protection of power semiconductor devices
5. Fuse protection
6. Techniques of circuit layout and shielding

16.1 INTRODUCTION

Power dissipation in electrical components raises the internal temperature and affects performance and reliability. A high internal temperature may be detrimental to the physical structure of the component. Removal of heat and control of internal temperature profile are very important considerations in the designing of the small-size lightweight power electronic systems of today. Packaging and layout of components play very important roles in the reliability of design.

This chapter will concentrate on thermal issues such as the temperature control of components and the design of a heat transfer system. Fundamental concepts of heat transfer processes are explained with primary focus on the static model of heat transfer. The topic of transient heat transfer is beyond the scope of this book, and it is left to the readers to refer to literature on the subject. The discussion of thermal issues is followed by the specification and the selection of heat sinks.

16.2 TEMPERATURE CONTROL IN SEMICONDUCTOR DEVICES

The theoretical upper limit on the internal temperature of a semiconductor device is the temperature at which the intrinsic carrier density of the most lightly doped region is equal to the majority carrier doping density in that region. The p-n junction would loose rectifying property at this limit temperature. The maximum allowed temperature is far below this theoretical limit. Vendors specify a maximum temperature, which is traditionally referred to as the maximum junction temperature T_{jmax}. This is also used as the worst-case temperature in the design procedure. Typical semiconductor packages are designed for T_{jmax} of 125°C.

16.3 HEAT TRANSFER BASICS

Power losses in electrical components appear in the form of heat. The accumulation of heat energy increases the temperature of the internal structure of components. The heat must be removed from the body of the component by transferring it to the ambient environment. The heat transfer takes place in three ways:

1. Conduction
2. Convection and/or
3. Radiation

Conduction is the heat transfer among stationary interfaces by the vibratory motion of atoms or molecules.

Convection is the mechanical transport of heat by a moving fluid or gas. The fluid flow may be natural or forced. Fluid flow in natural convection is caused by temperature gradients. The forced convection is created by pumps or fans.

In heat transfer by radiation, heat energy is converted into electromagnetic radiation, which is absorbed by other components in the vicinity. In enclosed packages, the radiation from hot components may be absorbed by neighboring components at the lower temperatures. As a consequence, the neighboring components are required to function at elevated temperatures, something they were not planned to do. Shiny sheet metal partitions serve as radiation shields.

Fortunately, the heat transfer in power electronic systems is dominated by conduction and convection processes.

16.4 HEAT TRANSFER SYSTEM

A typical heat transfer system is shown in Figure 16.1. In this illustration the heat is transferred from the semiconductor switching device to the ambient. The electrical power dissipated in the switching device is transported out of the system in the form of heat. Interfaces 1 and 2 are constituted of materials that provide an easy transit for heat without absorbing any part of the heat. The heat is transferred via interface 1 from the semiconductor chip to the metal casing of the device. Interface 2 provides the easy transit to heat from the metal casing of the device to the aluminum heat sink. Interface 2 might have to simultaneously provide electrical isolation. Heat transfer through interfaces 1 and 2 is predominantly by conduction. The heat transfer from the heat sink to the ambient is by convection.

The thermal management problem is to design a heat transfer system that constrains the T_J to the limiting value at interface 1.

As all materials offer electrical resistance to the flow of current, all materials offer resistance to the flow of heat. The later is called *Thermal Resistance* R_θ. The thermal resistance of a section of the material of cross-section A and length l is given by

$$R_\theta = \frac{\rho_\theta l}{A} \qquad (16\text{-}1)$$

Parameter ρ_θ is the thermal resistivity of the material, the thermal analog of electrical resistivity ρ.

Heat flows from a high temperature interface to a low temperature interface. Rate of flow of heat is termed the *thermal power* Q,

$$Q = \frac{dH}{dt} \qquad (16\text{-}2)$$

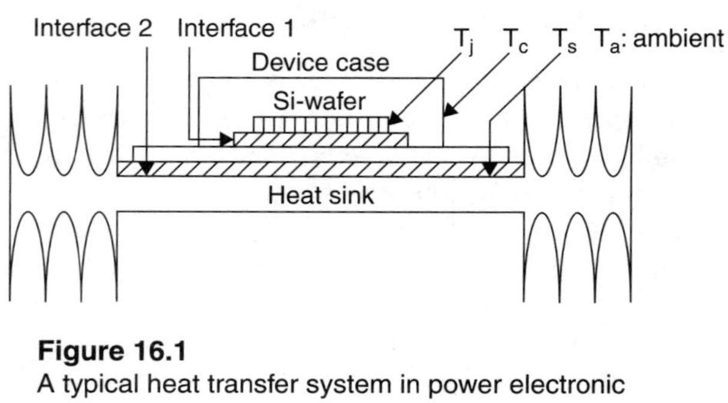

Figure 16.1
A typical heat transfer system in power electronic systems

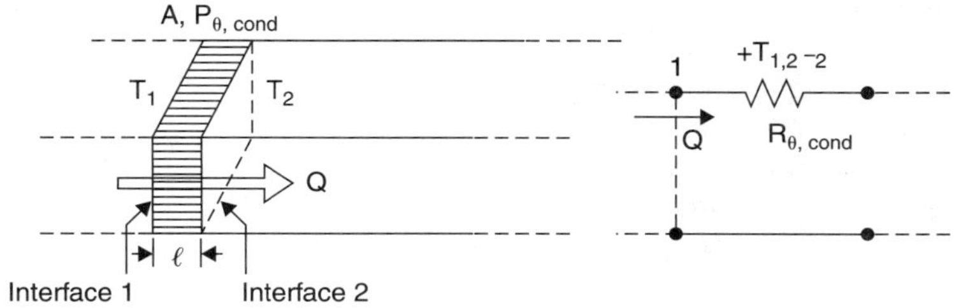

Figure 16.2
Heat transfer between interfaces

The amount of the rise in temperature of interface 1 with respect to the temperature of interface 2 for a thermal power Q is, see Figure 16.2,

$$T_{1,2} = R_\theta Q \quad ; \quad T_{1,2} = T_1 - T_2 \qquad (16\text{-}3)$$

The thermal interfaces are of three kinds owing to the three types of heat transfer processes: the conductive, convective and radiative interfaces.

16.4.1 Thermal resistance in conduction

The conductive thermal interface may be modeled by a set of two parallel plates separated by a material of specific thermal resistivity. Plate surfaces have several imperfections such as nonparallel surfaces and scratches. Thermal grease is generally used to fill up the scratches and small imperfections. Too much grease increases the thermal resistance of the interface.

The heat transfer equation in thermal conduction

$$T_{1,2} = R_{\theta,cond}\, Q \quad ; \quad T_{1,2} = T_1 - T_2 \quad , \quad Q = P_{diss} \qquad (16\text{-}4)$$

In a simple case, interface 1 would be the semiconductor junction in a switching device and interface 2 would be the ambient. Thermal power Q is equal to the dissipated electrical power in interface 1.

Conduction thermal resistivities of some commonly used materials are given in Table 16-1.

Mylar, mica, Beryllia and AIN are used for electrical isolation between the metal body of a semiconductor device and the aluminum heat sink.

TABLE 16-1 THERMAL RESISTIVITY OF MATERIALS USED IN POWER ELECTRONIC SYSTEMS

Material	Thermal Resistivity (°C-cm/W)
Air	3050
Mylar	635
Silicon grease	520
Mica	150
Alumina (Al_2O_3)	6.0
Silicon	1.2
Beryllia (BeO)	1.0
Aluminum Nitride (AIN)	0.64
Aluminum	0.48
Copper	0.25

Silicone grease, impregnated with metallic oxides such as ZnO_2, is used to fill scratches and imperfections on the mating surfaces in the heat transfer path. The silicone grease is also called the *thermal compound.*

The aluminum surfaces oxidize over time and accumulate a layer of aluminum oxide. The aluminum oxide acts as the insulating cover and offers greater thermal resistance to heat transfer. Black anodization is frequently used for the protection of aluminum surfaces from oxidation. The anodizing produces an intentional layer of oxide on the aluminum surface, but its thickness is small and its contribution to the thermal resistance can be ignored.

Example 16-1 Calculation of T_j

The current through an IRF 1010 MOSFET switch in a dc-dc Buck converter is shown in Figure 16.3. The on-state resistance $R_{DS,on}$ of the MOSFET is known to be 14 milli-ohms. Find out the junction temperature of the MOSFET in steady state if the junction to case thermal resistance $R_{\theta jc}$ is 1.0 °C/Watt and the case temperature is maintained constant at 25°C.

$$i = 50 + \frac{50}{4 \times 10^{-4}} t \qquad 0 \le t \le 0.4\ ms$$

$$= 0 \qquad 0.4 \le t \le 1\ ms$$

$$Q = \frac{1}{1ms} \int_0^{0.4ms} i^2 R_{DSon} dt$$

$$T_j = T_C + R_{\theta jC} Q$$

```
T = 0.001; % time period
Rthjc = 1; Rdson = 14e − 3; Tc = 25;
t1 = 0:1e − 5:0.0004; t2 = 0.0004:1e − 5:.001;
t = [t1, t2,t1 + T, t2 + T];% two periods
i1 = 50 + (50/4e − 4)*t1;
i2 = zeros(size(t2));
i = [i1, i2, i1, i2];
Pdiss = (i. ^ 2)*Rdson; % instantaneous power
pmean = mean(Pdiss) % average power
Tj = Tc + Rthjc*pmean % junction temperature
plot(t,i, 'r',t,Pdiss, 'r + ')
```
pmean = 32.8854 Tj = 57.8854

Example 16-2

A power resistor has the aluminum casing, as shown in Figure 16.4. The maximum temperature at the hot spot of the resistor is 125°C. Find the maximum permissible power that the resistor can dissipate at 25°C ambient. Plot the maximum permissible dissipation power as a function of the ambient temperature in the range of 25 to 70 °C.

The thermal equivalent circuit is shown in Figure 16.4(b). The Q is the amount of heat generated at the hot spot of the resistor.

The thermal resistivity of aluminum is $\rho_\theta = 0.48$ °C−cm/W (from Table 16-1). Thermal resistance on the left side and the right side of the hot spot are calculated using equation (16-1),

$$R_{\theta L} = R_{\theta R} = \frac{0.48 \times 2.5}{0.75 \times 0.75} = 0.2133$$

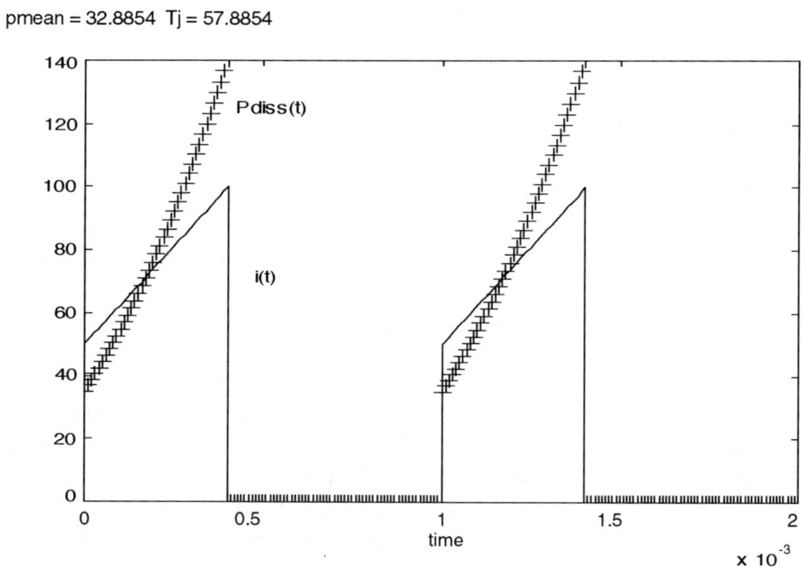

Figure 16.3
Current and instantaneous waveforms

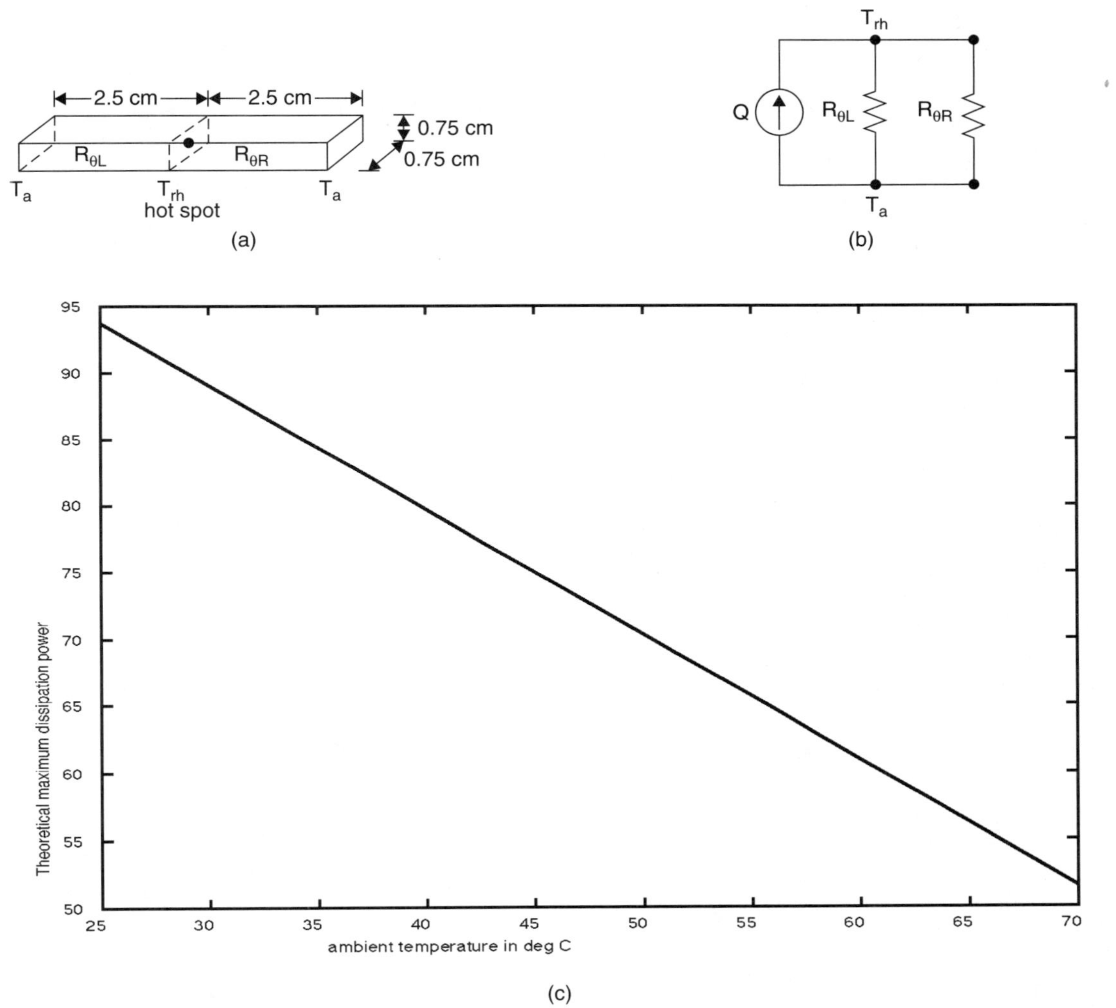

Figure 16.4
(a) Power resistor (b) thermal equivalent circuit and (c) power versus the ambient temperature

The parallel combination of two thermal resistors in parallel is

$$R_{\theta T} = R_{\theta l} / 2 = 0.1067$$

The temperature difference between the ambient and the hot spot is $T_{rh} - T_a = 125 - 25 = 100°C$. The maximum power that can be dissipated in the resistor is

$$Q = \frac{(T_{rh} - T_a)}{R_{\theta T}} = \frac{100}{0.1067} = 937.207 \ watts$$

The average dissipated power as a function of the ambient temperature is

$$Q = \frac{T_{rh}}{R_{\theta T}} - \frac{T_a}{R_{\theta T}}$$

16.4.2 Thermal resistance in convection

Convection is the mechanical transport of heat by a moving fluid such as air. The fluid moves from a high density area to a low density area by either natural convection or by a forced convection such as by a fan. The following equation describes the static relationship in convection:

$$T_{1,2} = R_{\theta,conv}\,Q \quad ; \quad R_{\theta,conv} = \frac{1}{h(T_{1,2},\nu)A} \qquad (16\text{-}5)$$

Parameter h is called the *film coefficient of heat transfer.* "A" is the area of the interface through which the heat is being convected. Besides the fluid temperature and the fluid velocity, the parameter h is dependent on the nature of flow, the turbulence. The turbulence is highly dependent on the geometry of the enclosure or heat sink, especially the space between the fins on the heat sink.

All heat transferred through conduction ends finally at a convective interface. Fins on an aluminum heat sink are the final convective interface through which the heat is removed by the flow of air. In a heat exchanger, the heat is removed by a moving liquid.

The air flow through the fins of a heat sink is designed to cause a turbulent flow. The turbulent flow causes a significant amount of mixing and, therefore, better heat transfer; that is, it offers a low thermal resistance. The relationship among fin spacing, flow rate and turbulence is given by the *Reynold's number:*

$$\mathrm{Re} = \frac{\rho\,\nu\,\omega}{\eta} \qquad (16\text{-}6)$$

where ρ is the fluid density, ν is the fluid velocity, ω is the channel width and η is the coefficient of viscosity of the fluid. A high Reynolds number means high turbulence. The fluid, flowing in a wide channel, will acquire turbulence at a lower velocity ($\nu\omega$ product) and flowing in a narrow channel at a high velocity. Fins on the heat sink are designed using this concept. Closely spaced fins are avoided. Likewise, close spacing between the components in a package, for example on a printed circuit board, impede convection and, hence, must be avoided.

16.4.3 Thermal resistance in radiation

Heat transfer through radiation follows the Stefan-Boltzmann law:

$$Q = 5.7 \times 10^{-8}\,EA(T_1^{\,4} - T_2^{\,4}) \qquad (16\text{-}7)$$

where thermal power Q is in watts, E is emissivity of the surface, A is the outer surface area including fins in square meters, and T_1 and T_2 are the surface temperature and the ambient temperature in Kelvin respectively. The emissivity of a heat sink surface can be obtained from the design handbooks. E for the black anodized aluminum surface is 0.9 in contrast with that of a shiny polished aluminum surface as low as 0.05.

The static relationship for heat transfer through radiation and the radiation thermal resistance is given by:

$$T_{1,2} = R_{\theta,rad}\,Q \quad ;$$

$$R_{\theta,rad} = \frac{[T_1 - T_2]}{5.7 \times 10^{-8}\,E\,A\,[T_1^{\,4} - T_2^{\,4}]} \qquad (16\text{-}8)$$

16.5 STATIC THERMAL MODEL OF HEAT TRANSFER SYSTEMS

Drawing from the circuit theory, we could make the following analogies:

$$T_{1,2} \Leftrightarrow V \quad Q \Leftrightarrow i \quad R_\theta \Leftrightarrow R$$

Like voltage, the temperature is measured between two interfaces, hence, two subscripts are used to indicate the difference. A single subscript in T indicates the measurement with respect to the ambient temperature T_a. The ambient temperature resembles the earth reference in an electrical system; that is to say the ambient temperature does not alter for any amount of heat transfer to or from it.

The heat transfer system of Figure 16.1 could be modeled by an equivalent analogous thermal circuit as shown in Figure 16.5:

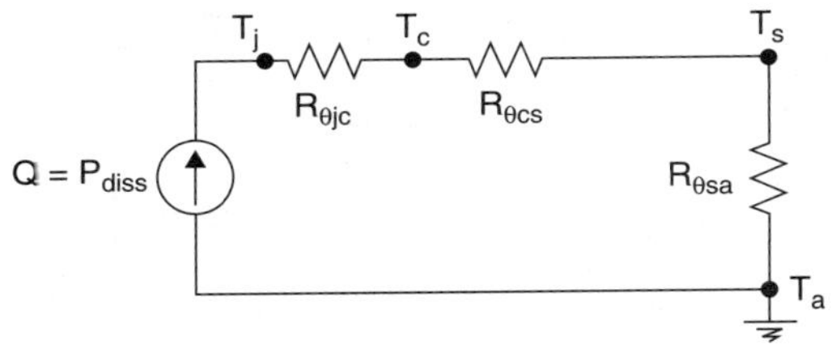

Figure 16.5
Analogous thermal circuit model

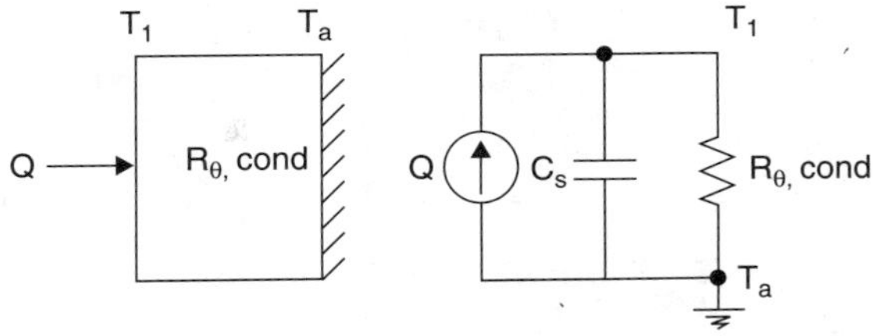

Figure 16.6
The thermal equivalent circuit in transient heat transfer

16.6 TRANSIENT THERMAL IMPEDANCE

At start-up, at shut-down or during steep load variation, the instantaneous dissipation in the semiconductor devices may greatly exceed the average steady-state dissipation limits. The permissible surge in the maximum junction temperature is determined by the magnitude and the duration of the transient condition.

The thermal resistance of the static model changes to the thermal impedance, which includes *heat capacity.* The heat capacity C_s is defined as the amount of heat (energy) stored in a unit mass that would raise its temperature by a unit degree:

$$C_s = \frac{dH}{dt} \quad \text{where} \quad dH = Q\,dt$$

$$\text{hence,} \quad Q = C_s \frac{dT}{dt} \tag{16-9}$$

Here, the temperature T is referenced to ambient temperature. This equation is analogous to i-v relation of a capacitor in electrical systems:

$$i = C\frac{dv}{dt} \tag{16-10}$$

Heat capacity C_s is analogous to the electrical capacitance. The heat capacity C_s is reference to the ambient. Masses in a thermal system constitute the heat storage devices. Figure 16.6 shows a thermal mass and its equivalent thermal circuit with heat capacity included.

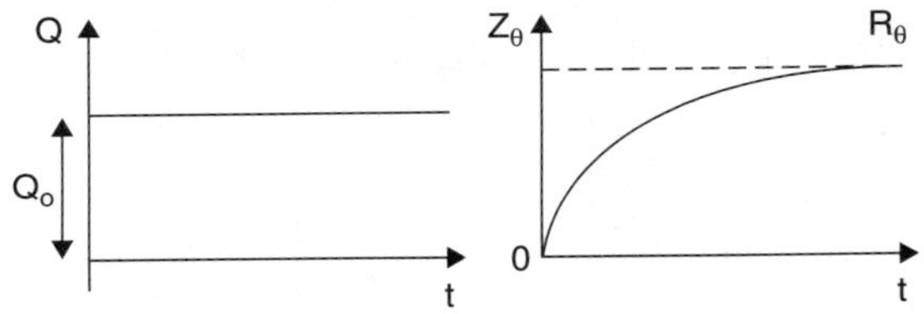

Figure 16.7
Transient thermal impedance

Thermal impedance in the Figure 16.6 circuit using the Laplace transform is

$$T_1 = Z_\theta\, Q \ : \ \text{where} \ \ Z_\theta = \frac{R_\theta}{1 + sR_\theta C_S} \tag{16-11}$$

For a step change in the amount of heat imparted, that is to say, upon application of a step power input Q_0, the steady-state rise in the temperature at interface 1 is given by

$$T_{1,ss} = T_{1,0} + R_\theta\, Q_0 \tag{16-12}$$

The instantaneous temperature follows an exponential path,

$$T_1 = T_{1,0} + R_\theta Q_0 (1 - e^{t/\tau_\theta}) \tag{16-13}$$

where $\tau_\theta = R_\theta C_s$

Rearranging variables in the above equation,

$$T_1 = T_{1,0} + Z_\theta(t)\, Q_0; \ Z_\theta = R_\theta(1 - e^{t/\tau_\theta}) \tag{16-14}$$

Impedance Z_θ can be said to be an exponential function of time, as shown in Figure 16.7.

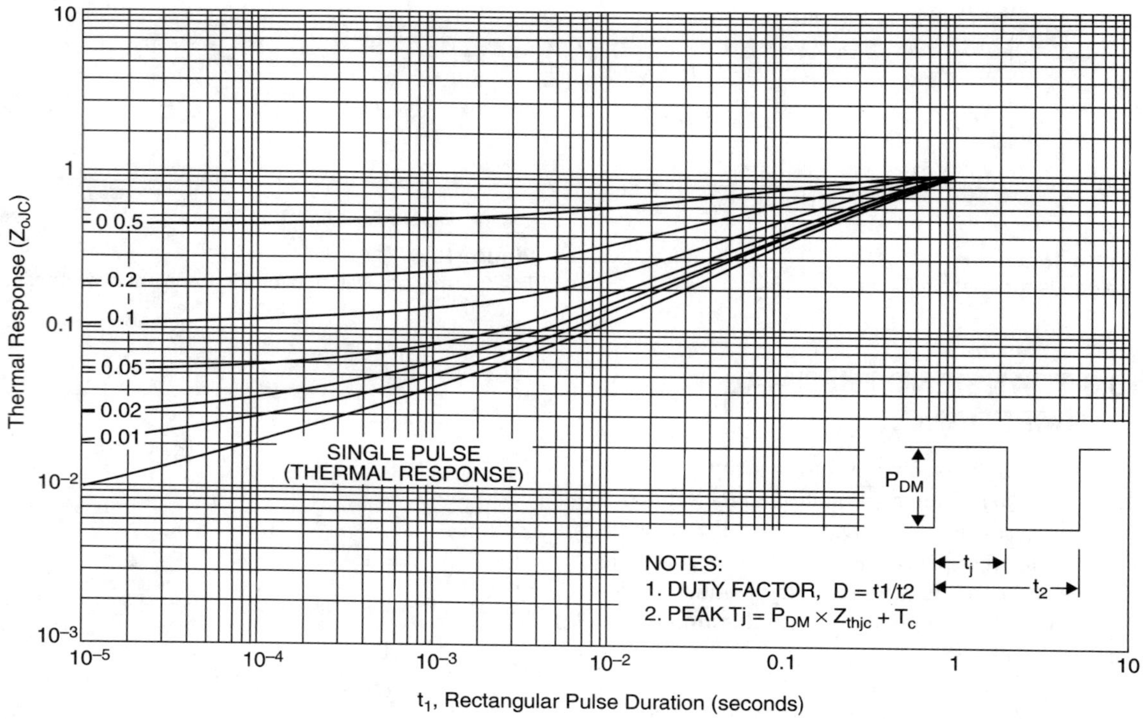

Figure 16.8
Transient thermal impedance for MOSFET IRF 1010 (Courtesy, International Rectifier)

Thermal systems in practice are nonlinear high order systems. Manufacturers provide plots of the transient thermal impedance versus time for the semiconductor device packages. Figure 16.8 shows these characteristics for MOSFET IRF 1010.

The thermal equivalent circuit of the heat transfer system of Figure 16.1 with heat capacity included, is shown in Figure 16.9:

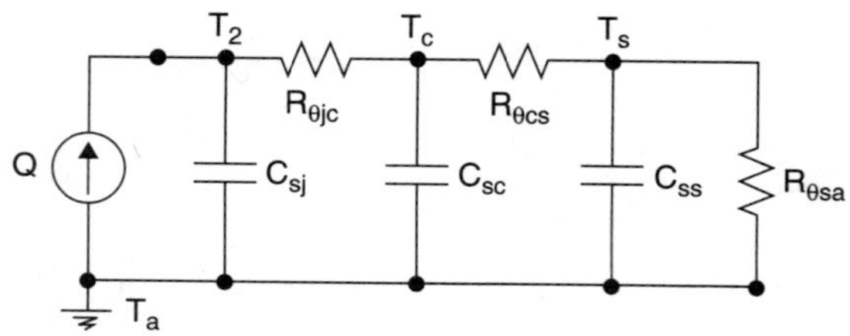

Figure 16.9
Transient thermal equivalent circuit of the heat transfer system of Example 16-1

Example 16-3 Junction Temperature Profile of a Heat Transfer System

Include the heat capacity in the thermal equivalent circuit of a MOSFET switch that is subjected to a step change in the current. The case of MOSFET is maintained at 25°C. Plot the junction temperature T_j. The thermal parameters of MOSFET are $R_\theta = 1°C/W$ and $C_S = 0.01$ W-sec/°C.

The thermal time constant is calculated from equation (16-13) $\tau_\theta = R_\theta C_S = 0.01$.

The step change in power for the given step change of current is $Q_0 = I^2 R_{Dson}$. The junction temperature profile is then determined from equation (16-14), using MATLAB.

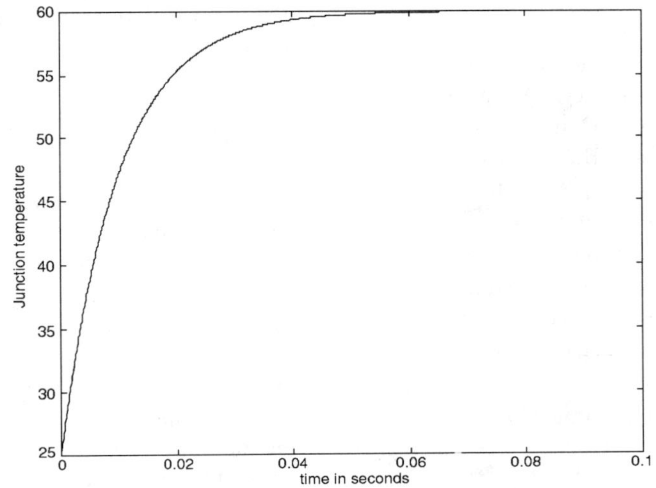

Figure 16.10
Transient response of the junction temperature in MOSFET

```
T = 0.1; % time of observation
Rthjc = 1; Rdson = 14e − 3; Tc = 25; I = 50; Cs = 0.01;
tau = Rthjc*Cs;
t = 0:1e − 4:T;
Q0 = (I.^2)*Rdson;% dissipated power
Zth = Rthjc*(1− exp(− t/tau));
Tj = Tc + Zth*Q0; % junction temperature
plot(t,Tj, 'r')
```

Example 16-4 Transient Thermal Design for a MOSFET

Use the device characteristics of MOSFET IRF 1010 in the Buck converter of Example 16-1. Find the peak device current, the peak dissipation and the case to ambient thermal resistance $R_{\theta ca}$. Use a maximum junction temperature of 150°C and the ambient temperature to be at 40°C and the case temperature should not exceed 60°C.

$$\text{Duty ratio} = 40\%$$

The thermal impedance at this duty ratio and at the time 400 mS is obtained from the graphs of Figure 16.8:

$$Z_{qjc} = 0.4°C/W.$$

The maximum permissible power dissipation in the device is

$$P_{\max} = Q = \frac{T_{j,\max} - T_C}{Z_{\theta jc}} = 225\ W$$

The maximum current through the MOSFET is determined from the knowledge of the on-state resistance R_{DSon} at $T_{j,\max}$, see the device characteristics:

$$R_{DSon} = 1.75 \times 0.014 = 0.0255 \text{ ohms}$$

The maximum current through the MOSFET is

$$I_{\max} = \sqrt{\frac{P_{\max}}{R_{DSon}}} = \sqrt{\frac{225}{.0255}} = 95\ A$$

The heat sink attached to the device case would quickly stabilize at the steady-state temperature because of a large mass (high heat capacity), and hence it exhibits low-pass filter characteristics. The required case to ambient thermal resistance is

$$R_{\theta ca} = \frac{T_c - T_a}{P_{avg}} = \frac{60 - 40}{225 \times 0.4} = 0.22 \text{ deg. } C$$

16.7 HEAT SINK

Refer to the heat transfer system using a heat sink, as given in Figure 16.1. The thermal equivalent system is given in Figure 16.5. Consider that this system is part of a PWM dc-dc Buck converter.

The worst-case design in power electronic systems requires the following knowledge:

$T_{j,\,max}$: maximum junction temperature
$T_{a,max}$: maximum ambient temperature
$V_{on,\,max}$: maximum on-state voltage and
$I_{on,\,max}$: maximum on-state current.

The maximum allowable junction-to-ambient thermal resistance $R_{\theta ja}$ can be estimated as

$$R_{\theta\,ja} = (T_{j,\max} - T_{a,\max}) \,/\, P_{diss} \quad (16\text{-}15)$$

where

$$P_{diss} = P_{on,\max} + P_{sw,avg}$$

P_{diss} is the total power loss in the switching device, the $P_{on,\,max}$ is the maximum on-state power loss and $P_{sw,\,avg}$ is the average switching power loss over a switching period.

$$P_{on,\,\max} = \frac{\tau\max}{T} I^2_{on,\,\max} R_{on,\,\max} \quad (16\text{-}16)$$

$$P_{sw,\,avg} = \frac{1}{T}\int_0^T v_{sw}\, i_{sw}\, dt \quad (16\text{-}17)$$

τ_{max} is the maximum pulse width in a PWM converter.

The junction-to-ambient thermal resistance $R_{\theta ja}$ is a sum of three components, see the equivalent circuit of Figure 16.6:

$$R_{\theta} ja = R_{\theta jc} + R_{\theta cs} + R_{\theta sa} \quad (16\text{-}18)$$

where

$R_{\theta jc}$: junction-to-case thermal resistance
$R_{\theta cs}$: case-to-sink thermal resistance
$R_{\theta sa}$: sink-to-ambient thermal resistance

The junction-to-case resistance $R_{\theta jc}$ is obtained from the data sheet of the device. The case-to-sink resistance $R_{\theta cs}$ can be obtained from the handbook of design by knowing the type of the case of the semiconductor device and the type of thermal compound, silicon grease and insulator used for electrical isolation and thermal conductance. The sink-to-ambient resistance $R_{\theta sa}$ is determined from equations (16-15) to (16-18). The heat sink is then selected from the vendor provided data sheets by using the value of the just determined $R_{\theta sa}$.

The contact area between the device and the heat sink is extremely important for minimizing thermal resistance between the device case and the sink. The surfaces should be flat, smooth, and free of dirt, corrosion and surface oxides. Silicone grease and thermal compounds are normally applied to improve the heat transfer capability.

The semiconductor device should be mounted on the heat sink by maintaining an appropriate amount of mounting pressure. In the case of stud mounted devices, excessive mounting torque may cause mechanical damage to the silicon wafers. The use of lubricants in the stud mountings must be avoided as they increase undue tension.

The semiconductor device may also be cooled by heat pipes mounted in between the device and the finned heat sinks. The heat pipe is partially filled with low-vapor-pressure liquid. When the heat from the device is imparted to the liquid in the pipe, the liquid flows to the condensing end where it condenses and dumps the heat to the finned heat sinks. The condensed liquid returns to the device end for a fresh cycle of heat transfer.

16.8 SURGE VOLTAGE PROTECTION

AC-line voltages may experience large surge transient voltages. This happens more often now because both industrial and commercial users are computerized heavily. Modern electrical equipment, both portable and fixed, uses more and more switching power supplies. When they are turned on and off frequently, they may cause large voltage transients on the ac line utility. Hence, surge voltage protection devices are becoming increasingly essential for electrical equipment.

Selenium diodes and metal oxide varistors (MOV) are commonly used for the protection of power electronic systems from the input voltage surge transients.

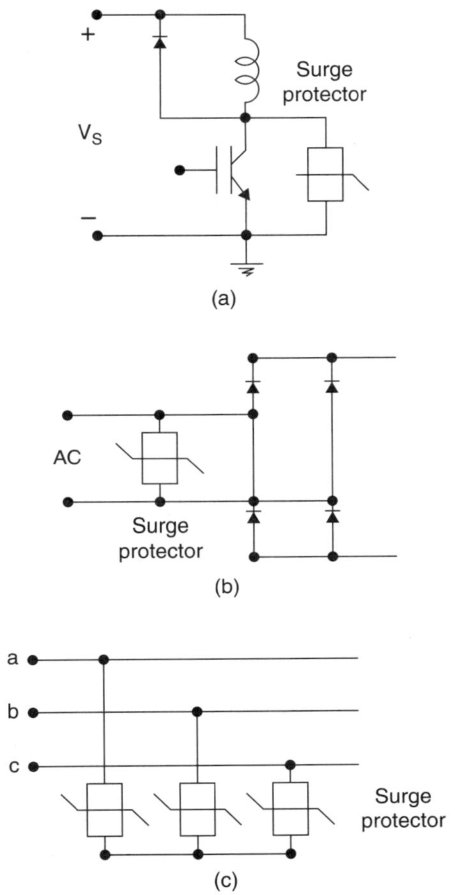

Figure 16.11
Surge voltage protection

Figure 16.11 shows the application of the surge protection devices.

16.8.1 Selenium diode

Selenium diodes are basically zener diodes with sharply defined reverse breakdown voltage on the characteristic curves, see Figure 16.12. Normally the selenium diode is biased in the reverse bias region at nearly half of the desired limiting surge voltage. At this operating point it draws very small current from the source. In the event of surge voltage occurring, the selenium diode breaks down

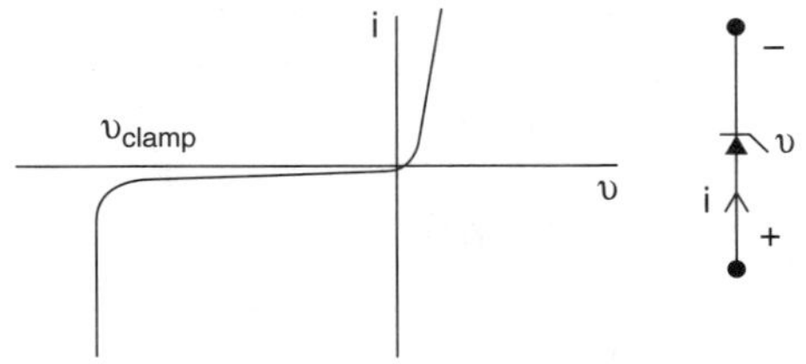

Figure 16.12
Selenium diode characteristics

and conducts heavy current to limit the voltage across it. The selenium diode must be capable of dissipating the surge energy, the surge voltage multiplied by the current, without exceeding the maximum limiting temperature rating. A typical selenium cell withstands a continuous operating dc voltage of 25 volts with a clamping voltage of 72 volts. For ac line applications, two selenium diodes are connected back-to-back. The selenium cells are connected in series for enhancing the voltage capability.

The selenium diodes do not help in limiting high dV/dt because of high parasitic capacitance.

16.8.2 Metal oxide varistor (MOV)

A metal oxide varister is a nonlinear resistance device. It is constituted of metal oxide particles separated by insulation film. The i-v characteristics of a MOV is given by

$$i = K\, v^{\alpha} \tag{16-19}$$

The k is a constant and the α ranges between 30 to 40. The insulation film in the MOV becomes conductive at high applied voltages and acts to limit the voltage across it.

16.9 FAULT CURRENT PROTECTION

Fusing is generally used for protection from short-circuit or excessive-current conditions. The fuses may be used in series with both the wires of a single-phase utility and on all three lines of the three-phase utility. The equipment manufacturers, however, recommend connecting fuses in series with each device separately. The choice is generally guided by cost constraints.

16.9.1 Fuse terminology

$i_{f,\,let}$: maximum current that is let through the fuse

t_C: fusing time or the fault clearing time

t_{melt}: duration from the instant the fault current exceeds the let-through level to the instant the fuse begins to melt

t_{arc}: The duration of the arc across the fuse ends

$V_{arc,\,max}$: maximum arc voltage across the fuse

i^2-t: The *let-through energy,* the energy required to melt the fuse. The current through the fuse is i.

16.9.2 Fuse operation

The typical fuse current profile is shown in Figure 16.13. The fusing operation takes a definite time to complete. The fusing time, also called the fault clearing time t_c, is measured from the time the fault current increases beyond a let-through current level. This time consists of two segments:

$$t_C = t_{melt} + t_{arc} \tag{16-20}$$

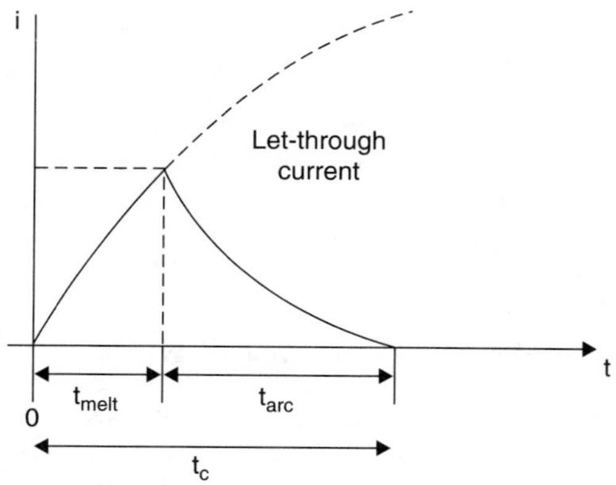

Figure 16.13
Fuse current profile

The melting time t_{melt} is the duration from the instant the fault current exceeds the let-through level to the instant the fuse begins to melt due to the rising temperature of the fuse. After time t_{melt}, an arc erupts across the fuse ends. The fuse resistance increases and the current reduces. The arc generates a large amount of heat, which vaporizes the fuse element. The arc length increases and the fuse current further reduces. Finally after a time t_{arc}, the current extinguishes and the fuse opens. The fault is said to clear. The time t_{melt} is dependent on the through current amplitude. But, the time t_{arc} is independent of the current amplitude and depends on the power factor.

Complete Fuse Protection: The fuse current decay profile with time is always lower than the device maximum current-time characteristics.

Short-term Fuse Protection: The fuse current decay profile is lower than the device maximum current-time characteristics only for a short duration from the onset of the fault condition.

16.9.3 Fuse selection

Estimate the fault current and apply the following criteria:

1. The fuse must carry continuously the rated current of the device it is protecting,
2. The i^2-*t let-through energy* of the fuse must be less than the value of the device to be protected,
3. The fuse must withstand the voltage after fault clearing,
4. The peak arc voltage must be less than the maximum voltage rating of the device.

Thyristors have better over-current capability than transistors. Fast-acting fuses are sufficient for thyristor protection.

IGBTs and bipolar power transistors are protected by a crowbar circuit in addition to a fast-acting fuse. Figure 16.14 shows a crowbar circuit that basically converts the transistor circuit to a thyristor circuit. The triggering circuit of the thyris-

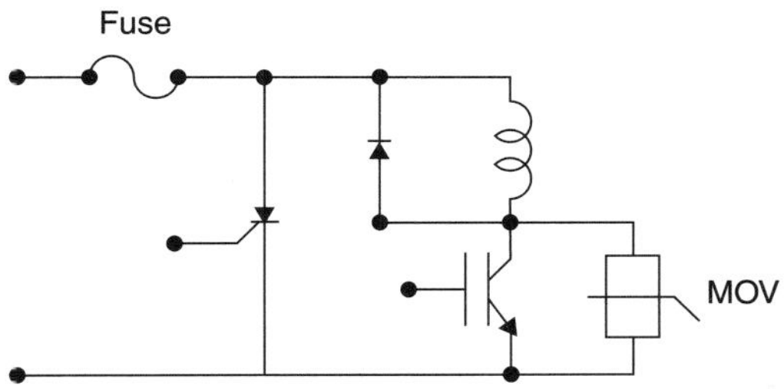

Figure 16.14
Crowbar protection

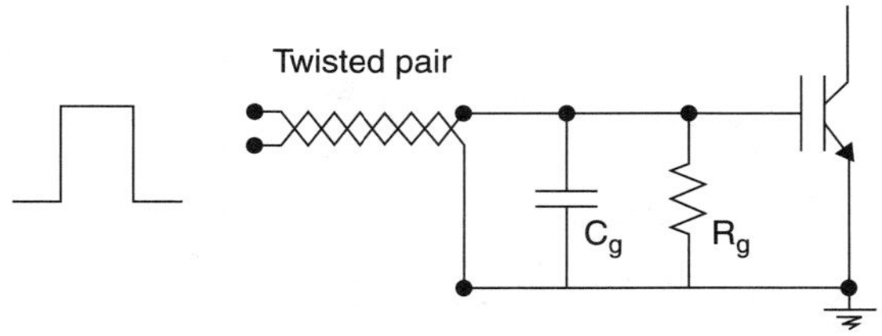

Figure 16.15
Twisted pair in control paths

tor is either voltage sensed or current sensed. The thyristor produces a short circuit and helps blow the series fuse.

16.10 CIRCUIT LAYOUT TECHNIQUES

The physical layout of power electronic circuits plays a very crucial role in the reliability of operation. This is particularly very crucial in the gate-drive circuits and control circuits.

16.10.1 Minimizing stray inductance

Stray inductance includes the self-inductance of the conductors and the leads of the electric components. The leads and the conductor lengths should be kept at the minimum needed. This is particularly important on the terminals and the conductors that carry heavy currents. The device vendors separate the control path conductors from the power path conductors. The device manufacturers call the terminals on the control path the sense terminals. The conductors to the sense terminals are required to carry small signal currents only, hence the stray inductance poses no problem. Furthermore, the effect of stray inductance on the control paths is nullified by using twisted pair conductors in place of the plain parallel line conductors. Figure 16.15 shows the use of the twisted pair in the gate-drive circuit.

The power path conductors, such as the bus bars, are required to be wide and with a large area of cross section to minimize the self-inductance. Copper strips, with a thin insulator sandwiched between them as shown in Figure 16.16(a), comprises a transmission line (or x-line) and provides an excellent means of reducing the stray inductance. The actual design can be adapted to meet the manufacturing and cost constraints. Figure 16.16(c) shows the use of transmission lines in the layout of the dc-dc Buck converter of Figure 16.16(b).

16.10.2 Shielding and partitioning

The large body components, specially the ungrounded heat sinks, serve as the radiator of both the electromagnetic energy and the thermal energy in heat form. Furthermore, the high input impedance input terminals, such as the input of opamp in the control circuits, may pick up the electromagnetic signal and noise from these sources. The picked up signals may cause undesired and very detrimental feedback leading to oscillatory responses and instability in the control systems.

Physical separation and partitioning of circuits as blocks and shielding using metal partitions helps in picking up these noises and preventing undesired feedback. The metal partitions must be connected to ground. This is applicable to both the packaging in enclosures and printed circuit layouts. It is highly recommended to liberally use ground plains, to guard high impedance terminals by use of ground tracks and to use grounded metal plains such as are used in the double-sided and multilevel printed circuit boards.

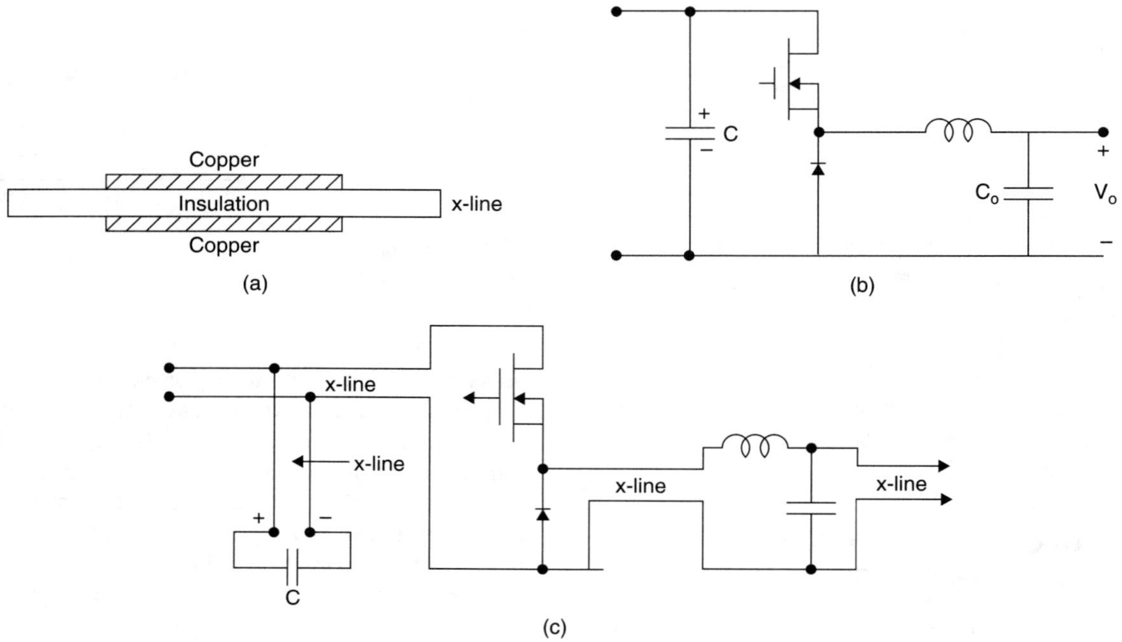

Figure 16.16
(a) A low-inductance transmission line copper strip (b) a dc-dc Buck converter circuit and (c) use of transmission line strips in the circuit layout

16.11 SUMMARY

This chapter has introduced readers to concepts and issues involved in the packaging issues, such as the temperature control, fault protection and lay-out techniques in power electronic converter systems.

The important points are as follows:

1. Concerted design efforts are required to not exceed the vendor specified maximum junction temperature of semiconductor devices.
2. The heat transfer between the thermal interfaces takes place through conduction, convection and radiation. Of these, the conduction and convection processes are most popularly used.
3. The heat transfer system can be modeled by a thermal equivalent circuit, which includes the thermal resistances, thermal power sources and heat storage elements. The thermal response of the system can be obtained by time-tested electrical circuit techniques.
4. The vendor specified thermal impedance characteristics can be used to estimate the transient thermal response.
5. Metal oxide varistors and selenium diodes are used for protection against over voltage and surge voltages in line-operated power electronic converters.
6. Fuse selection and placement must be done very carefully for reliable operation of power electronic converters.
7. Careful attention must be given to the physical layout of components in the converter design to avoid and minimize stray inductance and to avoid detrimental feedback and noise pick up.

References

1. Philip T. Krein, *Elements of Power Electronics*, Oxford University Press, New York, 1998.
2. Ned Mohan, Tore M. Undeland, and William P. Robbins, *Power Electronics: Converters, Applications and Design,* 2nd Ed., John Wiley & Sons, Inc., 1995.
3. Joseph Vithayathil, *Power Electronics: Principles and Applications,* McGraw-Hill Inc., New York, 1995.
4. John G. Kassakian, Martin F. Sclecht, and G. Vergassian, *Principles of Power Electronics,* Addison-Wesley, Boston, 1991.
5. Kjeld Thorborg, *Power Electronics,* Prentice-Hall Inc., 1988.

Problems

16-1 A thyristor has a thermal resistance of 0.8°C/W between its inner semiconductor junction and the heat sink and 1.7°C/W between the heat sink to the ambient. What will be the power loss in the thyristor if the maximum junction temperature is 150°C and the ambient temperature is (a) 25°C and (b) 41°C?

16-2 Specify the thermal resistance of the heat sink in Problem 16-1 when the junction temperature is not allowed to increase beyond 125°C at the ambient temperature of 25°C. Find the maximum temperature of the case of the thyristor in this design.

16-3 The thermal impedance characteristics of a semiconductor switch and the pulsed power load through the switch are shown below. The switch is rated for P_{max} = 50 W. Estimate the maximum junction temperature of the semiconductor device at the ambient temperature of 25°C.

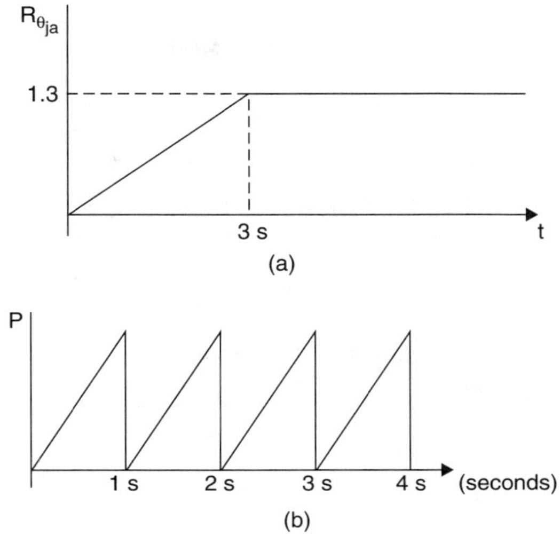

Figure P16.3

16-4 Refer to Figure 16.3. The on-state resistance $R_{DS,\,on}$ of the MOSFET is known to be 14 milli-ohms. Find the junction temperature of the MOSFET in steady state if the junction-to-case thermal resistance R_{jc} is 1.0°C/W and the case temperature is maintained constant at 25°C.

16-5 A power resistor has the aluminum casing, as shown in Figure 16.4. The maximum temperature at the hot spot of the resistor is 125°C. Find the maximum permissible power that the resistor can dissipate at 25°C ambient. Plot the maximum permissible dissipation power as a function of the ambient temperature in the range of 25 to 70°C.

16-6 Include the heat capacity in the thermal equivalent circuit of a MOSFET switch which is subjected to a step change in the current. The case of MOSFET is maintained at 25°C. Plot the junction temperature T_j. The thermal parameters of MOSFET are R_θ = 1 °C/W and C_S = 0.01 W-s/°C .

16-7 Use the device characteristics of MOSFET IRF 1010 in the Buck converter of Example 16-1. Find the peak device current, the peak dissipation and the case to ambient thermal resistance R_{ca}. Use a maximum junction temperature of 150°C and the ambient temperature to be 40°C and the case temperature not to exceed 60°C.

16-8 Two identical MOSFETs, that share current equally, are mounted on a common heat sink. The device parameters are:

Thermal resistance from the junction to the device case $R_{\theta JC} = 1.2°C/W$

Thermal resistance from the device case to the heat sink $R_{\theta CS} = 0.3°C/W$

Thermal resistance from the heat sink to the ambient $R_{\theta SA} = 0.75°C/W$

Maximum junction temperature $T_{Jmax} = 125°C$.

a. Draw the thermal equivalent circuit for the combination.
b. Find the total power dissipated at the ambient temperature $T_A = 25°C$.
c. Find the maximum power dissipated if one of the transistors is disconnected.

16-9 Repeat Problem 16-8 when the two MOSFETs are mounted on the heat sink with two identical mica washers per transistor, each with a thermal resistance of 0.15°C/W.

APPENDIX A

A.1 REVIEW OF BASIC PRINCIPLES

Before we discuss power electronic circuits and systems, it is worthwhile reviewing the basic principles of electrical engineering that we use throughout this book.

Three main laws that govern power electronic systems are:

$$v = Ri \tag{A-1}$$

$$\lambda = Li \tag{A-2}$$

$$q = Cv \tag{A-3}$$

The above three laws are applicable to resistance, inductance and capacitance respectively.

The **v** and **i** are instantaneous voltage and current respectively. The λ and q are flux linkage and charge respectively.

A coil with self-inductance **L,** carrying current **i,** produces a magnetic flux ϕ. The flux ϕ is a vector and it points in the direction of the thumb if the fingers of the right hand curl in the direction of current flow. The magnetic field density **B** is given by

$$B = \frac{\phi}{A} \tag{A-4}$$

A is the area of the coil normal to the flux vector. Flux is measured in **Webers** or **Lines.** The flux linkage λ is the magnetic flux ϕ multiplied by **N,** the number of times the current coils around it.

A voltage **v,** when applied to a capacitor, two metal surfaces facing each other, having a capacitance **C,** will dump a charge q on it.

Magnetic Flux and the Induced Voltage

When a conductor, while moving in a magnetic field, sweeps through a magnetic flux ϕ, causing a flux linkage λ, a voltage is induced between its terminals which is given by

$$v = \frac{d\lambda}{dt} = L\frac{di}{dt} \tag{A-5}$$

Polarity of the induced voltage depends on the flux vector direction. The same relation is true when the conductor is fixed and the magnetic flux is varying with time. It is the relative variation with time that matters.

Charge and the Current

Rate of flow of charge through a material, a conductor or the dielectric of a capacitor, gives rise to a current **i** that is given by

$$i = \frac{dq}{dt} = C\frac{dv}{dt} \tag{A-6}$$

or 1 coulomb of charge /second past a point is equal to 1 amp.

Kirchoff's Voltage Law

$$\sum_{l=1}^{L} v_{ml} = 0 \quad for \quad all \quad m \in M$$

M is the number of meshes (loops) in the circuit. **L** is the number of elements in series in the m-th mesh (loop). The $\mathbf{v_{ml}}$ is the voltage across the l-th element in the m-th mesh. Voltage on any element except the voltage source or sink, is regarded positive if the current enters the element terminal when the mesh is traversed in (say) the clockwise direction.

Kirchoff's Current Law

$$\sum_{j=1}^{J} i_{nj} = 0 \quad for \quad all \quad nodes \quad n \in N \tag{A-7}$$

N is the number of nodes in the circuit. **J** is the number of branches connected to the n-th node. The $\mathbf{i_{nj}}$ is the current entering the n-th node from the j-th branch. The current entering a node is regarded as positive.

Conservation of Energy

When a current **i** is incident upon or drawn from an element (including the source and sink) which has voltage **v** across its terminals, for a time segment **dt,** the energy transaction is given by

$$dE = vi\, dt \tag{A-8}$$

The law of conservation of energy states that the net energy transaction over the segment **dt** in a circuit consisting of N elements is zero:

$$dE_T = \sum_{l}^{N} v_j i_j \, dt = 0, \quad hence \sum_{l}^{N} v_j i_j = 0 \tag{A-9}$$

The rate of flow of energy through a node is called power. Energy imparted to an element is considered as positive. Since the net energy transaction is zero, its rate (net power) is also zero. Alternately saying, the sum of the incoming powers from the sources to a system (or circuit) equals the sum of the outgoing powers including the power losses in the circuit elements.

A.2 BASIC MATHEMATICAL METHODS

A.2.1 A periodic variable

A periodic signal with period $\mathbf{T_S}$ is defined by

$$v(t + T_S) = v(t) \tag{A-10}$$

The MATLAB instruction for specifying a periodic variable (with period $\mathbf{T_S}$) such as a sinusoidal voltage is given in the following example:

m_1A1.m

```
Ts=1; dt=Ts/512; ; % period=1 second, dt is the time
  step in the time array
t=0:dt:Ts; %time array over a period
v=10*sin(2*pi*t/Ts); %array of sinusoidal voltage over the
  period
plot(t,v) ) %plot of voltage versus time
grid
xlabel('time in seconds')
ylabel('amplitude')
```

We have selected 512 elements in the array with a purpose. This number is a simple power of 2 and is very helpful, as we shall see later, in speeding the determination of the harmonic profile of waveform represented by the array. Readers could select any power of 2.

An example of a non-sinusoidal voltage waveform is given below:

m_1A_2.m

```
Ts=1; dt=Ts/512;
t1=0:dt:Ts/4-dt;
t2=Ts/4:dt:Ts/2-dt;
t3=Ts/2:dt:Ts;
t=[t1, t2, t3];
v=[10*ones(size(t1)), -5*ones(size(t2)), zeros(size(t3))];
plot(t,v)
axis([0,Ts,-12, 12])
grid
xlabel('time in seconds')
```

Another example of a non-sinusoidal ramping current waveform is given below. Add the following instruction to m_1A_2.m.

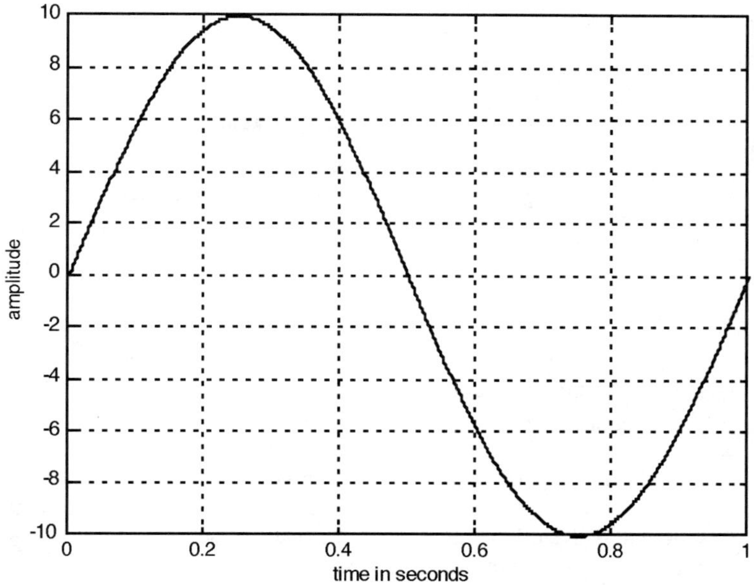

Figure A.1
A sinusoidal waveform

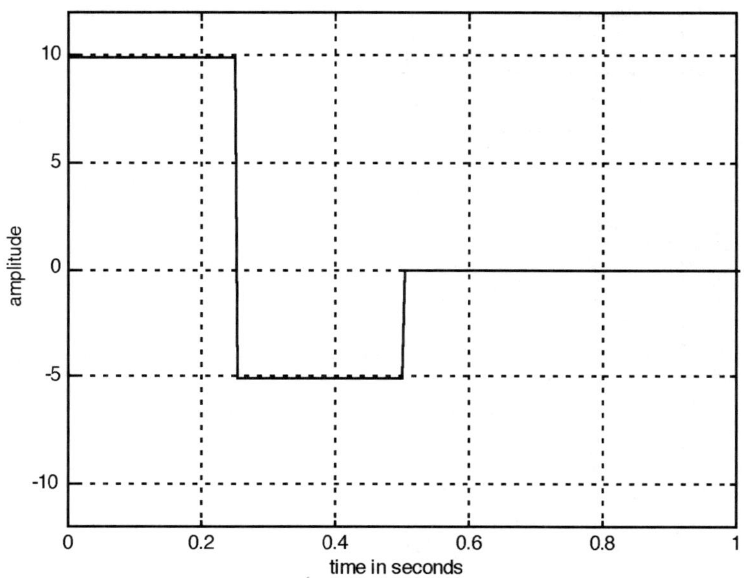

Figure A.2
A pulse waveform

m_1A_3.m

```
Ts=1; dt=Ts/512;
t1=0:dt:Ts/4-dt; t2=Ts/4:dt:Ts/2-dt;
t3=Ts/2:dt:Ts;
t=[t1, t2, t3];
slope1=5/(Ts/4);slope3=5/(Ts/2);
i1=slope1*t1;
i1max=max(i1);
i=[i1, i1max*ones(size(t2)), i1max-slope3*(t3-Ts/2 )];
plot(t,i)
axis([0,Ts,-2, 7])
grid
xlabel('time in seconds')
ylabel('amplitude')
```

A periodic signal has the half-wave odd symmetry if it satisfies the following condition:

$$v(t + \frac{T_S}{2}) = -v(t) \text{ or } v(\theta + \pi) = -v(\theta) \quad \text{(A-11)}$$

A periodic signal has the half-wave odd symmetry and quarter-wave even symmetry simultaneously if it satisfies the following condition:

$$v\left(\frac{T_S}{4} - t\right) = v\left(\frac{T_S}{4} + t\right): \quad t_0 \leq t < \frac{T_S}{4} \quad \text{(A-12)}$$

$$\text{or, } v\left(\frac{\pi}{2} - \theta\right) = v\left(\frac{\pi}{2} + \theta\right): \quad \theta_0 \leq \theta < \frac{\pi}{2} \quad \text{(A-13)}$$

A.2.2 Averaging a periodic variable

The average value of a periodic variable (signal) with period $\mathbf{T_S}$ is calculated by

$$\overline{V} = \frac{1}{T_S}\int_{t_0}^{t_0 - T_S} v dt \quad \text{(A-14)}$$

$$\text{or } \overline{V} = \frac{1}{2\pi}\int_{\theta_0}^{\theta_0 + 2\pi} v d\theta: \ \theta = \omega_S t, \ \omega_S = \frac{2\pi}{T_S} \quad \text{(A-15)}$$

The t_0 and θ_0 are the instant and the angle respectively at the beginning of the period under consideration. The average value of a variable is the average area under the waveform in time domain over one period. Average value is also the dc component of the periodic variable. The MATLAB program is given below:

```
vdc=mean(v) % average value of waveform in m_1A_2.m
vdc = 1.2476
```

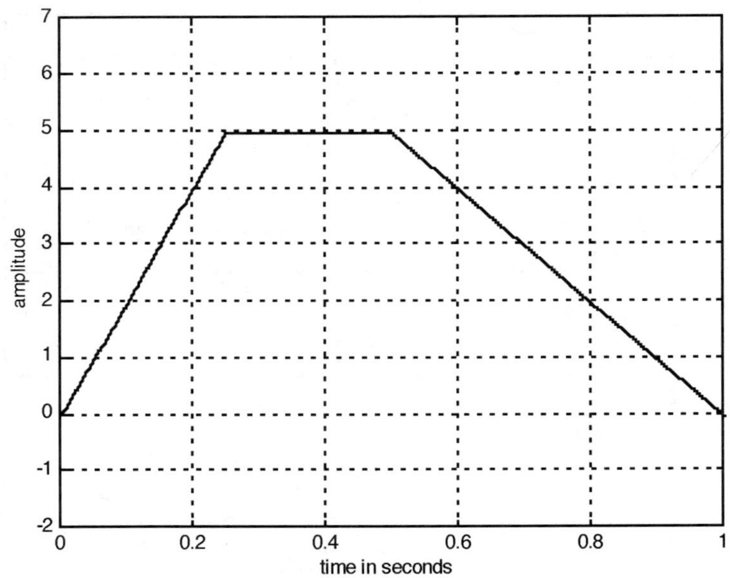

Figure A.3
A trapezoidal waveform

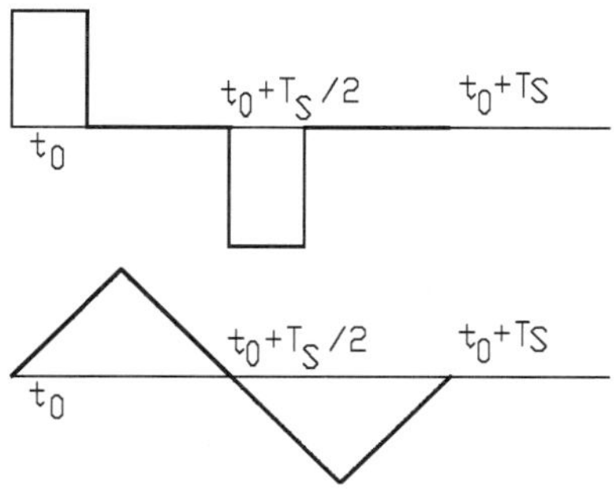

Figure A.4
Periodic waveforms and symmetry

The dc component of a nonsinusoidal periodic variable (voltage or current) is its average amplitude over a period.

Periodic average of a sinusoidal waveform (sine or cosine with zero dc component) is zero.

$$\frac{1}{T_S}\int_{t_0}^{t_0+T_S} V\sin(\omega_S t+\phi)\,dt = 0$$

$$\frac{1}{T_S}\int_{t_0}^{t_0+T_S} V\cos(\omega_S t+\phi)\,dt = 0 \qquad \text{(A-16)}$$

Periodic average of the product of sinusoids of two different frequencies:

$$= \frac{1}{T_S}\int_{t_0}^{t_0+T_S} V_1\sin(\omega_{S1}t+\phi_1).\; V_2\sin(\omega_{S2}t+\phi_2)dt \qquad \text{(A-17)}$$

$$= \frac{V_1V_2}{T_S}\int_{t_0}^{t_0+T_S}\frac{1}{2}[\cos((\omega_S 1-\omega_S 2)\, t+\phi_1-\phi_2)-\cos((\omega_S 1+\omega_S 2)\, t+\phi_1+\phi_2)]\,dt \qquad \text{(A-18)}$$

When $\omega_{S1} \neq \omega_{S2}$, the two terms inside the integral are themselves sinusoids of $\omega_{S1} - \omega_{S2}$ and $\omega_{S1} + \omega_{S2}$ frequencies, hence, their averages are zero. Therefore,

$$= 0 \qquad if\ \omega_{S1} \neq \omega_{S2} \qquad \text{(A-19)}$$

$$= \frac{V_1V_2}{2}\cos(\phi_1-\phi_2), \quad if \quad \omega_{S1} = \omega_{S2} \qquad \text{(A-20)}$$

It is worth remembering that the ***product of two sinusoids of differing frequencies is zero.*** The above result also means that the periodic average of the square of a sinusoid equals $V_1^2/2$.

In general, the voltage and current on both the source side and the sink side of a power electronic converter are either pure dc or periodic ac signals.

A.2.3 RMS value of a periodic variable

The rms stands for root mean square. The rms value of a periodic variable with period $\mathbf{T_S}$ is calculated by

$$\tilde{V} = \sqrt{\frac{1}{T_S}\int_t^{t+T_S} v^2\,dt} \qquad \text{(A-21)}$$

or

$$\tilde{V} = \sqrt{\frac{1}{2\pi}\int_\theta^{\theta+2\pi} v^2\,d\theta} \qquad \text{(A-22)}$$

The rms value of a waveform is found by squaring the amplitude, averaging the area under the squared waveform over one period, and then finding the square root in that order.

The MATLAB instruction is given below:

```
vrms=sqrt(mean(v .^2)) % rms value of waveform in
  m_1A_2.m
vrms = 5.5847
```

A.2.4 Harmonic components of a periodic signal

The harmonic profile of a periodic signal is obtained from the Fourier series expansion:

$$v = \overline{V} + \sum_1^\infty V_n\sin(n\theta+\phi_n) \;:\; \theta = \omega_S t \qquad \text{(A-23a)}$$

where

$$\overline{V} = \frac{1}{T_S}\int_{t_0}^{t_0+T_S} v\,dt = \frac{1}{2\pi}\int_{\theta_0}^{\theta_0+2\pi} v\,d\theta$$

$$V_n = \sqrt{a_n^2+b_n^2}\ and\ \phi_n = \tan^{-1}\left[\frac{a_n}{b_n}\right]$$

$$a_n = \frac{2}{T_S}\int_t^{t+T_s} v\cos(n\omega_S t)dt = \frac{1}{\pi}\int_\theta^{\theta+2\pi} v\cos(n\theta)d\theta$$

$$b_n = \frac{2}{T_S}\int_t^{t+T_s} v\sin(n\omega_S t)dt = \frac{1}{\pi}\int_\theta^{\theta+2\pi} v\sin(n\theta)d\theta$$

The harmonic profile can also be expressed in the exponential form:

$$v = A_0 + \sum_{n=-\infty}^{\infty} A_n e^{n\omega_s t+\phi_n} \quad \text{(A-23b)}$$

where

$$\overline{V} = A_0 \quad V_n = 2\,|A_n| \quad \phi_n = angle(A_n) + \pi/2$$

The coefficients $\mathbf{A_n}$ are determined from the Discrete Fourier Transform of the MATLAB array of the signal over one period. Finding coefficients a_n and b_n involves lengthy computing time. Fast Fourier Transform algorithm provides a quick method of determining the harmonic components A_n and ϕ_n. The MATLAB program for determining the harmonic profile along with the phase angle of the fundamental component is given below. Readers are recommended to type in this program and store as a function **harmonic.m.**

```
harmonic.m
function [f,y,ph]=harmonic(x,n)
%f is the array of harmonics
%y(k) is the amplitude of the k-th harmonic
%ph is the phase of the fundamental harmonic
%n is the number of harmonics to be displayed on the
  plot
N=size(x,2);%number of elements in the x array
f=0:1:N-1;
y1=fft(x);
y=2*(abs(y1))/N;%the n-th harmonic
y(1)=y1(1)/N;%the dc component
%ph=angle(y1(2))+pi/2;
ph=angle(y1(2))+pi/2;% phase angle of fundamental
  harmonic
stem(f,y,'r')
axis([0,n-1,0, 1.2*max(y)])
```

The fft algorithm works fastest when the number of elements is a simple power of 2, such as 128, 256, 512, and 1024. This explains why we have selected 512 elements in the arrays in the previous MATLAB examples.

The harmonic profiles of the voltage and the current waveforms in MATLAB programs **m_1A_2.m** and **m_1A_3.m** are shown below:

```
% in file m_1A_2.m
[f,y,ph]=harmonic(v,16);
title('harmonic profile of voltage waveform ')
xlabel('harmonic number')
ylabel('amplitude')
ph*180/pi
ans = 72.0387 degrees
% in file m_1A_3.m
[f,y,ph]=harmonic(I,16);
title('harmonic profile of current waveform')
xlabel('harmonic number')
ylabel('amplitude')
ph*180/pi
ans = -62.7683 degrees
```

Harmonic profile of a signal with half-wave odd symmetry:

$$v = \sum_{odd\ n}^{\infty} V_n \sin(n\theta + \phi_n) \quad \text{(A-24)}$$

The dc term is zero and only the odd harmonics exist.

Harmonic profile with half-wave odd symmetry and quarter-wave even symmetry:

$$v = \overline{V} + \sum_{odd\ n}^{\infty} V_n \sin(n\theta) \quad \text{(A-25)}$$

The dc term is zero, the odd harmonics exist and the phase terms are missing.

A.2.5 Integration and differentiation of an ac signal

The MATLAB function for integrating an ac signal over an interval 0 to t is given below:

```
function y=integ(x,th)
% y array is the integration of array x with respect to
  variable th
dth=th(2)-th(1); %step in th array
y=cumsum(x)*dth; %cumulative sum at each element of
  the x array
end
```

The MATLAB function for differentiating an ac signal over an interval 0 to t is given below:

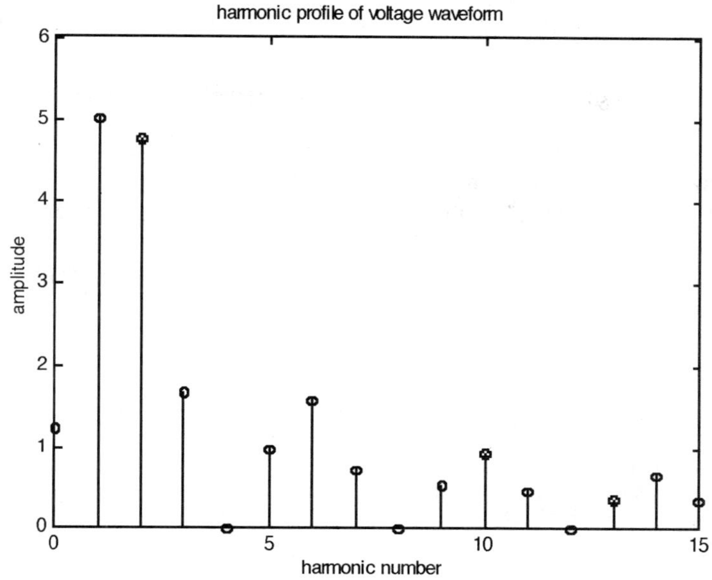

Figure A.5
A harmonic profile of voltage in m_1A_2.m

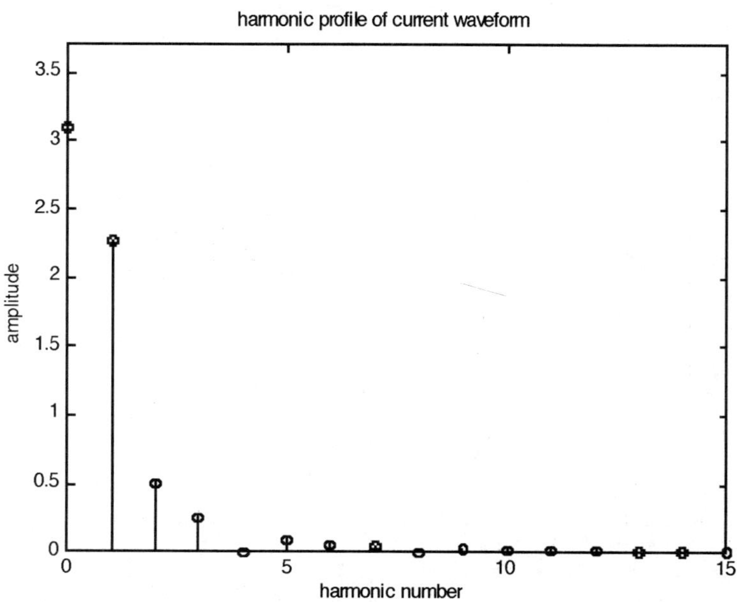

Figure A.6
Harmonic profile of current in m_1A_3.m

```
function y=dxdt(x,t)
[m,n]=size(x);
y=diff(x) ./ diff(t);
y=[y, y(n-1)];
end
```

Readers are encouraged to type and store these programs as they will be used extensively in this text.

Following the MATLAB program calculates and plots the integration and differentiation of a sinusoidal signal over a single period:

m_1A_4.m

```
th=0:pi/512:2*pi; %angle over a period
v=sin(th);
vi=integ(v,th); %integration of v with respect to angle th
vd=dxdt(v,th); %differentiation of v with respect to angle th
plot(th, v, th, vi, th, vd)
xlabel('angle in radians')
gtext('v')
gtext('vi')
gtext('vd')
```

A.3 ENERGY AND POWER

Energy flow through the pair of terminals of a source, over an elemental time is given by

$$dW = v_S(t)i_S(t)dt = p(t)dt \qquad \text{(A-26)}$$

Instantaneous power **p(t)** is the rate of change of energy flowing through the terminals. The average power flowing through the terminals over a time interval τ is calculated as following:

$$P = \frac{1}{\tau}\int_{t}^{t+\tau} v_S(t)\, i_S(t)dt \qquad \text{(A-27)}$$

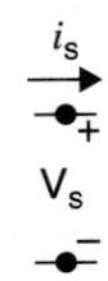

Figure A.8
Power flow through the source terminals

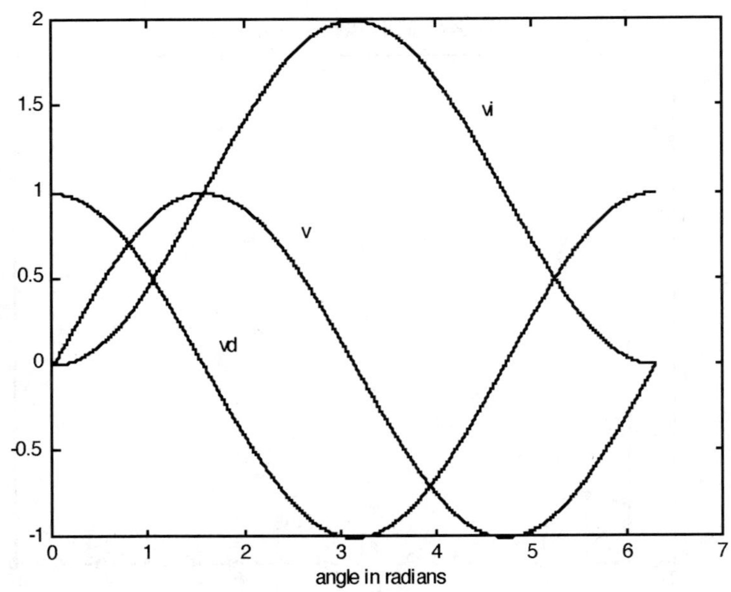

Figure A.7
Plots of voltage, differentiated voltage, and the integrated voltage

Consider the voltage and current waveforms of MATLAB program **m_1A_2.m** and **m_1A_3.m.** The MATLAB instructions for power calculations are given below:

```
Pinst=v.*i; % the instantaneous power. v and i are the
  voltage and current arrays over a period
plot(th,Pinst) %instantaneous power over a period
Pavg=mean(Pinst) % average power
Pavg = 0
```

Average power from an ac sinusoidal voltage source to resistance R:

$$P_S = \frac{1}{T_S}\int_t^{t+T_S} \frac{(V_S \sin \omega_S t)^2}{R}\, dt \tag{A-28}$$

$$= \frac{V_S^2}{2RT_S}\int_t^{t+T_S} [1 - \cos 2\omega_S t]dt = \frac{V_S^2}{2R} \tag{A-29}$$

Average power from a non-sinusoidal voltage source to resistance R:

$$P_S = \frac{1}{T_S}\int_t^{t\_T_S} \frac{\left[\overline{V}_S + \sum_n V_n \sin(n\omega_S t)\right]^2}{R}\, dt$$

The expression inside the integral contains terms with zero exponent, single exponent and square of sinusoids of single frequency and products of sinusoids of unequal frequencies. The average of the product terms, which have the single exponent of sine terms or the products of sinusoids of unequal frequencies, are zero (see equations (A-16) and (A-17)):

$$\int_t^{t+T_S} \sin n\omega_S t\, dt = 0 \tag{A-30}$$

and $$\int_t^{t+T_S} \sin n\omega_S t \sin m\omega_S t\, dt = 0, \quad n \neq m \tag{A-31}$$

$$P_S = \frac{1}{RT_S}\int_t^{t\_T_S} \left[\overline{V}_S^2 + \sum_n V_n^2 \sin^2 (n\omega_S t)\right] dt$$

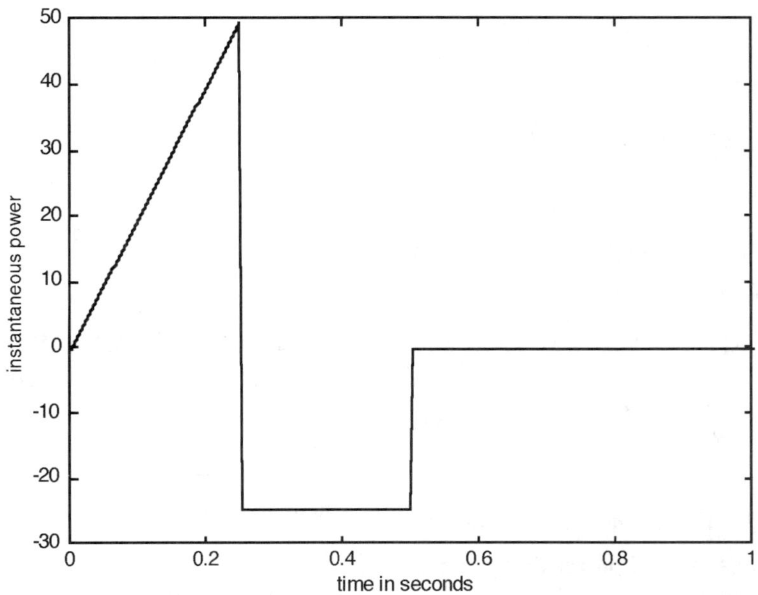

Figure A.9
Instantaneous power

$$\text{Hence, } P_S = \frac{V_S^2}{R} + \sum_n \frac{V_{sn}^2}{2R} \tag{A-32}$$

Therefore, the power is supplied individually by each harmonic component of the source voltage waveform.

A.3.1 Power from a sinusoidal voltage source to an impedance

The current waveform from a sinusoidal source to a R-L combination is a sine wave with a phase difference, as given by

$$i_S = I_S\, e^{j\theta} \sin \omega_S t \qquad I_S = \frac{V_S}{2\sqrt{R^2 + \omega_S^2 L^2}}$$

$$\phi = \tan^{-1}\left[\frac{\omega_S L}{R}\right] \tag{A-33}$$

Power delivered by the sinusoidal source is given by

$$S = \frac{1}{T_S}\int_t^{t+T_S} vs\, i_s\, dt \tag{A-34}$$

$$S = \frac{1}{T_S}\int_t^{t+T_S} V_S I_S\, e^{j\phi} \sin^2 \omega_S t\, dt \tag{A-35}$$

$$S = \frac{1}{2T_S}\left[\int_t^{t+T_S} V_S I_S\,(\cos\phi + j\sin\phi)\,(1 - \cos 2\omega_S t)\, dt\right] \tag{A-36}$$

Recognizing that the integral of cos ($2\omega_S t+\phi$) term over a single period is zero,

$$S = \tilde{V}_S \tilde{I}_S\,(\cos\phi + j\sin\phi) \;:\; \tilde{V}_S = \frac{V_S}{\sqrt{2}} \;\; \tilde{I}_S = \frac{I_S}{\sqrt{2}}$$

$$\phi = \tan^{-1}\frac{\omega_S L}{R} \tag{A-37}$$

$$S = P + jQ$$

$$P = \tilde{V}_S \tilde{I}_S \cos\phi, \quad Q = \tilde{V}_S \tilde{I}_S \sin\phi \tag{A-38}$$

The real part of S is the **real power** dissipated in the resistance, the imaginary part oscillates between the inductor and the source, called the **reactive power.** S is called the **apparent power (also called the Volt-Ampere).**

Power Factor is defined the ratio of the real power to the apparent power:

$$pf = \frac{P}{S} = \cos\phi \tag{A-39}$$

A.3.2 Power from a non-sinusoidal voltage source to an impedance

A non-sinusoidal voltage source is expressed as

$$v_S(t) = \overline{V_S} + \Sigma V_{Sn} \sin n\omega_S t \tag{A-40a}$$

V_{Sn} is the peak amplitude of the n-th harmonic of the voltage source waveform. Current to an impedance load is given by

$$i_S(t) = \overline{I_S} + \Sigma I_{sn} \sin(n\omega_S t - \phi_n) \tag{A-40b}$$

where

$$\overline{I_S} = \frac{\overline{V_S}}{R}, \quad I_{Sn} = \frac{V_{Sn}}{\sqrt{R^2 + X_n^2}}, \quad \phi_n = \tan^{-1}\frac{X_n}{R}$$

and $\mathbf{X_n}$ is the reactance of the load at the n-th harmonic. Average power from the source in this case is given by

$$P_S = \left[\frac{1}{T_S}\int_{t_0}^{t_0+T_S} [\overline{V_S} + \Sigma V_{Sn} \sin n\omega_S t]\,[\overline{I_S} + \Sigma I_{Sn} \sin(n\omega_S t - \phi_n)]dt\right] \tag{A-41}$$

$$P_S = \frac{\overline{V}_S^2}{R} + \sum_n \frac{V_{Sn}^2}{2\sqrt{R^2 + X_n^2}} \cos\phi_n \tag{A-42}$$

A.3.3 Power from a dc voltage source that supplies dc current

If the load is purely resistive, the current drawn from the dc source is a pure dc. Average power drawn from the dc source over a time interval t is calculated from

$$P_S = \overline{V_S}\,\overline{I_S} \tag{A-43}$$

Where $\mathbf{V_S}$ and $\mathbf{I_S}$ are respectively the dc source voltage and dc current from the source. The same equation is also valid for a dc voltage sink, which is sinking a dc current.

A.3.4 Power from a dc voltage source that supplies ac current

If the load consists of inductors, capacitors, and/or ac sinks, the current drawn from a dc voltage source usually has a periodic ac component in addition to the dc component. Consider that a periodic ac current from the source is expressed by a Fourier series:

$$i_S = \bar{I}_S + \sum_{1}^{\infty} I_{Sn}\, e^{-j\phi_n} \sin(n\omega_S t) \;:\; \omega_S = \frac{2\pi}{T_S} \quad \text{(A-44)}$$

$$i_S = \bar{I}_S + \sum_{1}^{\infty} I_{Sn}\, e^{-j\phi_n} \sin(n\theta) \;:\; \theta = \omega_S t \quad \text{(A-45)}$$

where the ω_S is the frequency of the fundamental component of the ac current waveform. The average power out of the dc voltage source is calculated by

$$P_S = \frac{1}{2\pi}\int_{\theta_0}^{\theta_0+2\pi} V_S\left[\bar{I}_S + \sum_{1}^{\infty} I_{Sn} e^{-j\phi_n} \sin(n\theta + \phi_n)\right] d\theta \quad \text{(A-46)}$$

Recognizing that the integral of sin $(n\theta+\phi)$ term over one period is zero,

$$P_S = V_S \bar{I}_S \quad \text{(A-47)}$$

The average power delivered by a dc voltage source in such a case is zero at all harmonic frequencies. The same is also true for a dc voltage sink, which is sinking an ac current.

The above conclusion is also true if the voltage source is ac and the current supplied is dc or the source is a current source in place of a voltage source. The average power transported across a pair of terminals involves only the dc components of the voltage and current.

A.4 PSPICE SIMULATION

In this section we will look at some examples of simulation of power electronic converter circuits using PSPICE for WINDOWS applications. The readers are expected to learn the basics of PSPICE simulation elsewhere.

PSPICE Example 1-1 Switching a Battery to R-L Circuit

Draw the circuit as shown in Figure A.10(a). Perform a transient analysis for 10 mS. The inductor current response is shown in Figure A.10.

The time constant of the circuit is L/R=1 ms. Expectedly, the inductor current rises exponentially from zero initial current to 1.2 amps. The steady state reaches in approximately 5 time constants. The inductor voltage jumps from zero initial value to 12 volts instantly and then decays exponentially to zero in the steady state. Readers are encouraged to solve the circuit analytically and verify with the simulation results.

PSPICE Example1-2 Full-Wave Rectification

The circuit in Figure 4.11(a) is a full-bridge rectifier. The converter operation will be analyzed in Chapter 8; here we simulate the circuit and note the response.

The ac voltage source is a 10 Hz, 24 volts peak sine wave. Perform the transient analysis for 4 periods, that is 40 ms.

Note that the output voltage has a dc component of approximately 21.5 volts and a 3 volts peak-peak ripple. The source current and the output current have non-sinusoidal wave shape, in fact, short bursts of current. This shows that the rectifier diodes conduct for only a short duration in a period. The source current has a zero dc component, but the output current has a dc component of approximately 215 mA and a 700 mA peak-peak ripple. The peak current through any diode in the steady state is nearly 3.5 times the dc output current. The initial source current, which is also called the inrush current, and which is also conducted by the rectifier diodes, is even higher, 1400 mA, approximately 7 times.

Figure A.11(c) shows the harmonic profile of the source current and the output current waveforms. The profile is determined by performing the FFT analysis of the waveforms. The source current has zero dc component and harmonics at 10 Hz, 30 Hz,

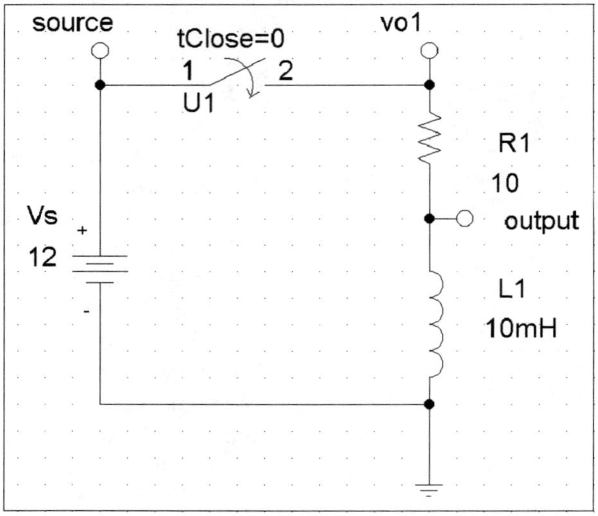

(a)

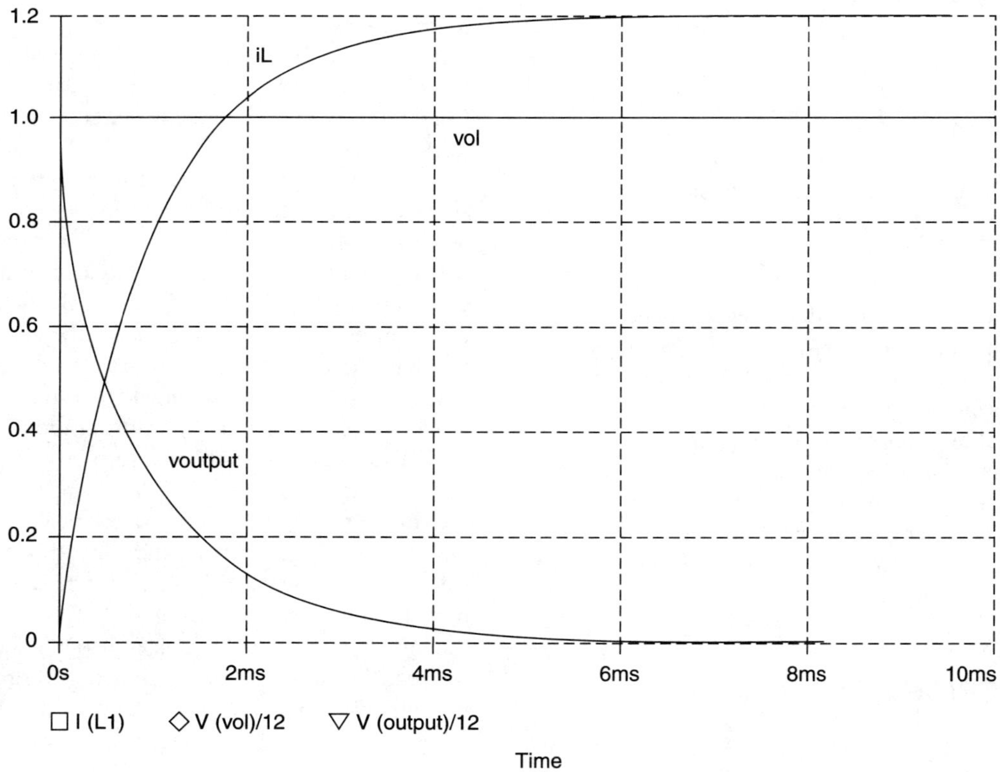

(b)

Figure A.10
(a) Circuit and (b) the inductor current waveform

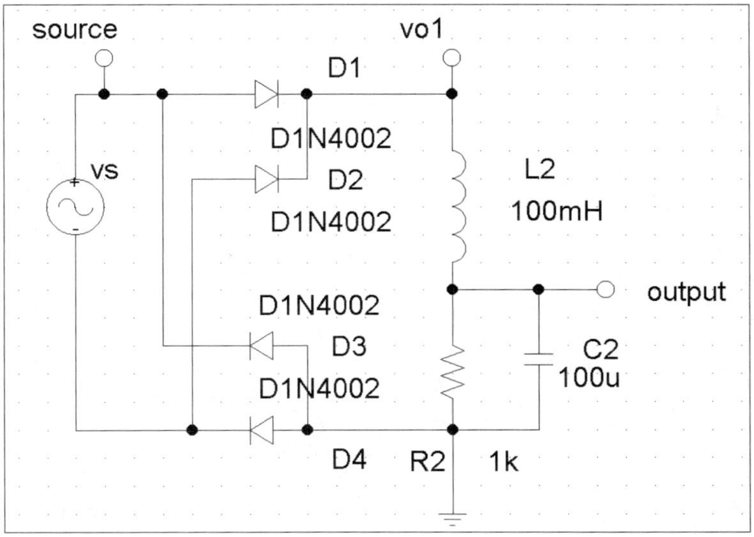

(a)

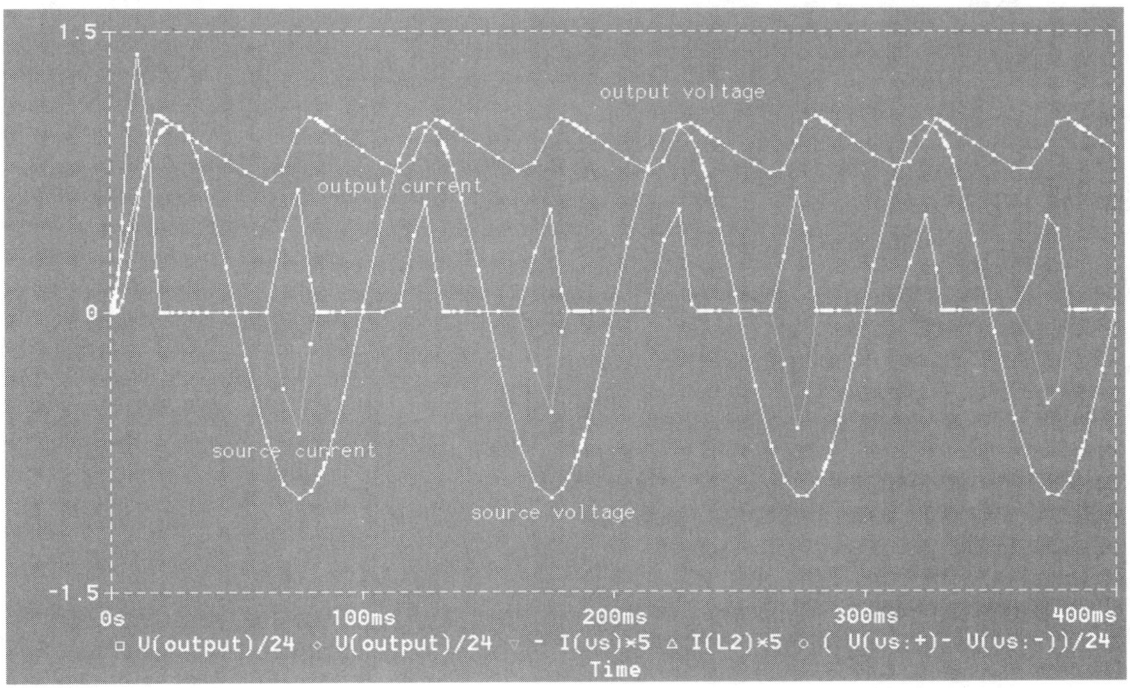

(b)

Figure A.11
The simulation of a full-bridge rectifier (a) The schematic (b) the waveforms for the source voltage, the output voltage, the source current and the output current, which also flows through the inductor and

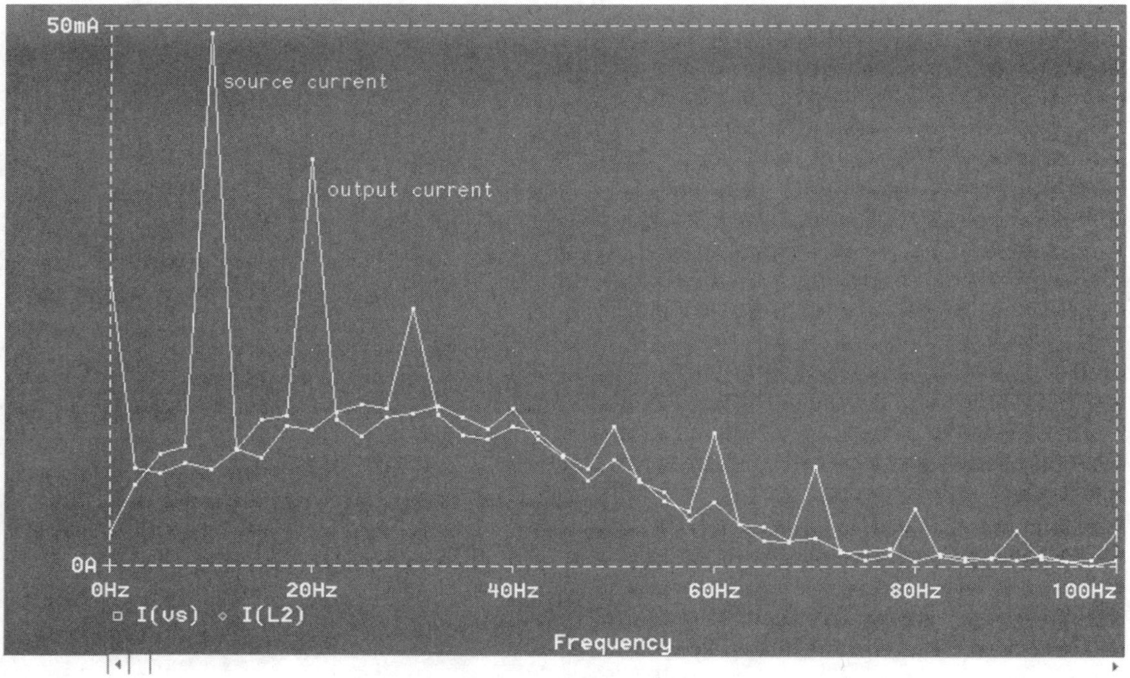

(c)

Figure A.11 ***(continued)***
(c) Harmonic profiles of the source current and the output current

50 Hz, and so on. The output current has a dc component and harmonics at 20 Hz, 40 Hz, 60 Hz, and so on. The readers should note that there are nonzero harmonics at unexpected frequencies. This error may be attributed to the accuracy of simulation procedure. Later on we would do more rigorous simulation.

APPENDIX B

B.1 ELECTROMAGNETICS

Study of electromagnetics is an indispensable necessity. Motors, transformers, microphones, and magnetically levitated high-speed vehicles, all involve electromagnetic phenomena. In the following sections, we will study electrostatic and magnetic fields and the interrelationships between them.

Electrostatic Fields

A stationary charge **q** in a medium establishes an electric field in the space around it, see Figure B.1. The space may consist of a free space, liquid or solid material. Electric field intensity **E** decreases as the square of the distance **r** from the charge and is given by

$$E = \frac{q}{4\pi\epsilon_r^2} \qquad \text{(Coulomb's Law)} \qquad \text{(B-1)}$$

Epsilon ϵ is called permittivity, also called dielectric constant, of the medium in which the field is established.

Electric field is a vector pointing radially outward from the center of the field if the charge is positive and inward if negative. The radial lines emanating from the charge, which show the direction of electric field vectors, are called the electrostatic lines of force or simply the electric flux lines. In case of an electric dipole, a positive charge and a negative charge separated by a distance d, the radial flux lines adjust as shown in Figure B.2. Electric field vectors originate at the positive charge and terminate on the negative charge.

Another charge of similar polarity, when brought in the field, is dispelled away from the field producing charge. The potential (energy) of

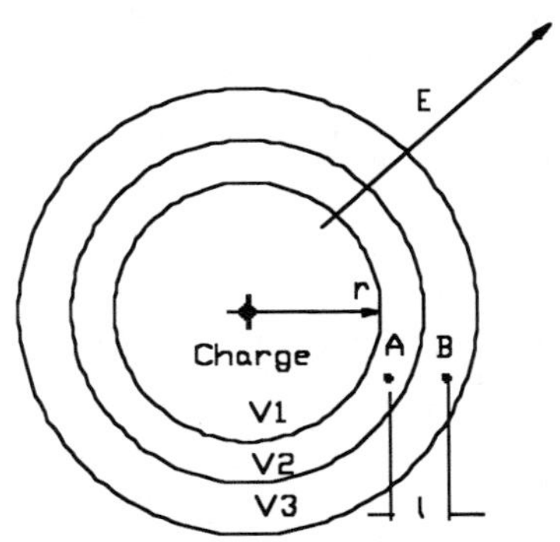

Figure B.1
Electric field due to a stationary charge

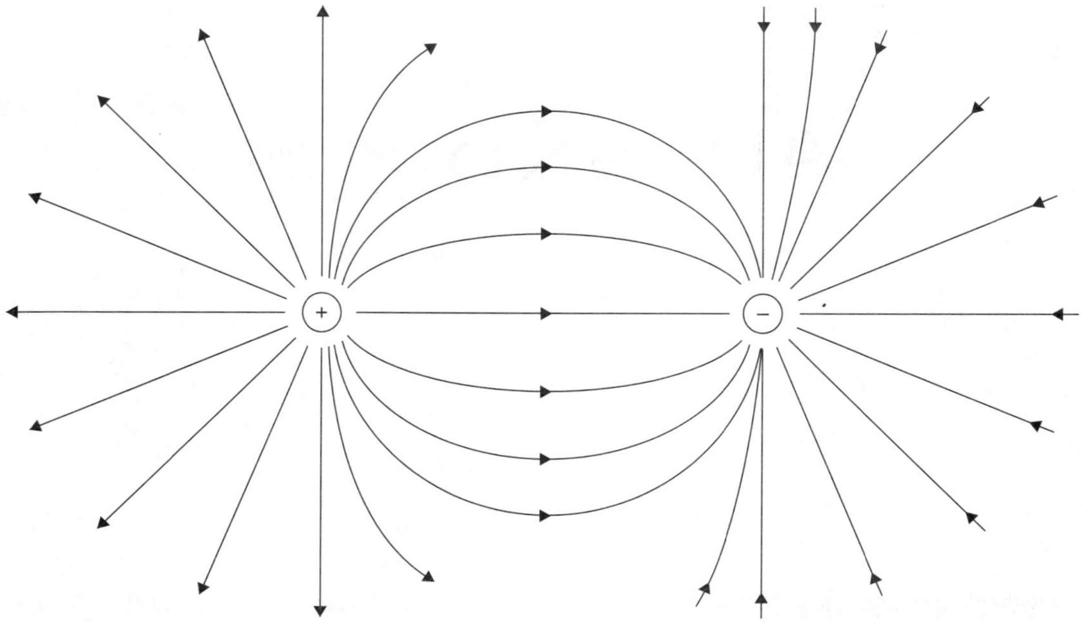

Figure B.2
Electric flux lines of an electric dipole, two opposite charges separated by a distance

the other charge is said to be higher near the center of the field where **E** is higher. Thus, there are equipotential contours around the field producing charge. **E** points in the direction of increasing potential. If left alone, the other charge (of similar polarity) will move in the opposite direction of the arrow. The potential difference between any two points A and B, separated by a distance **l** is

$$V_{BA} = -\int_A^B \vec{E}.\,\vec{dl} \tag{B-2}$$

Work done (energy expended) to move an infinitesimal charge **dq** through a potential difference of **v** volts in the direction of arrow of the electric field, is expressed by

$$dw = v\,dq \tag{B-3}$$

and the power expended is

$$p = \frac{dw}{dt} = v\frac{dq}{dt} = v\,i \tag{B-4}$$

E field produces electric flux. The electric flux inside a bounded surface of area **A** and the flux density are denoted by ψ and **D** respectively and are defined as

$$= \int_A \vec{D}.\,d\vec{A}, \quad \vec{D} = \varepsilon\,\vec{E} \tag{B-5}$$

More on permittivity of materials in later sections of Chapter 2.

Total electric flux through the surface is equal to the total charge enclosed by that surface. Hence,

$$q = \int d\psi = \int \vec{D}.\vec{dA} \tag{B-6}$$

Potential energy density (per unit volume) stored in an electrostatic field (E field alone) is given by

$$w_E = \frac{1}{2}\,\vec{D}.\,\vec{E} \tag{B-7}$$

APPENDIX C

C.1 SEMICONDUCTOR BASICS

Electrical current is defined as the rate of flow of charge. The charge can flow in a material if it (1) contains charge carriers and (2) they are free to move. Semiconductor materials fall in the middle of the metals and the insulators in both respects. Semiconductors are the four valency materials, meaning the atoms have four electrons in the outermost orbit and have a vacancy for four more.

Some of the electrons in the outer orbit of atoms in a semiconductor acquire enough energy to break loose from the atomic structure, called lattice. They leave behind the holes (the absence of electrons). The free electron has a negative charge and the hole a positive charge. The free electron moves like a particle in the lattice. The hole moves by attracting an electron from a neighboring atom. Hence the electrons and holes may be considered as the negative and positive charge carriers respectively.

In the semiconductor material, two phenomena are occurring simultaneously:

1. *generation* of free electrons due to thermal energy, and
2. loss of free electrons due to collision with a hole, called *recombination.*

The two processes are in equilibrium at a specific temperature.

The number of free electrons and holes are equal at any time and thus the charge neutrality is maintained. The thermal equilibrium densities of the electrons and holes, $\mathbf{n_i}$ and $\mathbf{p_i}$ respectively, in a pure (intrinsic) semiconductor are

$$n_i = p_i = C\, e\,\frac{qE_g}{2kT} \tag{C-1}$$

C is a constant, **q** is the electron charge, $\mathbf{E_g}$ is the band gap energy of the semiconductor material, **k** is the Boltzmann's constant and **T** is the temperature in Kelvin. For silicon, $\mathbf{n_i \approx 1.4 \times 10^{10}\ /cm^3}$ **at 25°C.** The charge carrier density and hence, the conductivity, increases with the increase in the temperature. This is in contrast with most of the metals in which the conductivity decreases with increasing temperature due to increased collision occurrences.

The rate of recombination of electrons (or holes) is proportional to the electron (or hole) density:

$$\frac{dn}{dt} = -\frac{n}{\tau_n}\ ,\quad \frac{dp}{dt} = -\frac{p}{\tau_p} \tag{C-2}$$

The $\boldsymbol{\tau_n}$ and $\boldsymbol{\tau_p}$ are the mean lifetimes of electrons and holes respectively. The lifetime is in the range from nanoseconds to microseconds and is dependent on the manufacturing process and the geometry of the diode.

The charge carriers move randomly on a microscale and diffuse from a high density area to a low density area on a macroscale. The mean lengths traveled by charge carriers before recombination are

$$L_n = \sqrt{D_n \tau_n}, \qquad L_p = \sqrt{D_p \tau_p} \qquad \text{(C-3)}$$

$\mathbf{D_n}$ and $\mathbf{D_p}$ are the diffusion constants of electrons and holes respectively.

The intrinsic semiconductor can be doped metallurgically by a third valency (such as boron) or a fifth valency (such as phosphorous) material. The doping of an intrinsic semiconductor material is also referred to as the addition of impurity atoms. An atom of the third valency material can absorb an electron from the bulk of the semiconductor material, thus creating an excess hole. The hole population increases and the semiconductor is said to be *p-doped* or *p-semiconductor.* The accepting of an extra electron renders the *acceptor* atom to be a negatively charged ion.

An atom of the fifth valency material can give an electron to the bulk of the semiconductor material, thus creating an excess electron. The electron population increases and the semiconductor is said to be *n-doped* or *n-semiconductor.* The donation of an electron renders the *donor* atom to be a positively charged ion. The total charge neutrality is maintained at all times.

The number of impure atoms is a very small fraction of the total atoms in a volume of the semiconductor (10^{16}-10^{19} in 10^{23} cm^{-3}). Hence, the addition of impure atoms does not affect the thermal density of charge carriers in the bulk of the material that is still intrinsic. The following relation holds good among the electron and hole densities in a doped semiconductor:

$$n \times p = n_i^2 \qquad \text{(C-4)}$$

In the n-semiconductor, the electrons are in the majority, hence called the *majority carriers.* Majority carriers in the p-semiconductor are holes. Both types of carriers exist nonetheless in both types of semiconductors. Both contribute to the flow of current, see Figure C.1.

The number of impure atoms is considerably greater than the number of free electrons in the intrinsic semiconductor (10^{16}-10^{19} versus 10^{10} cm^{-3}). The majority carriers far outweigh the intrinsic carri-

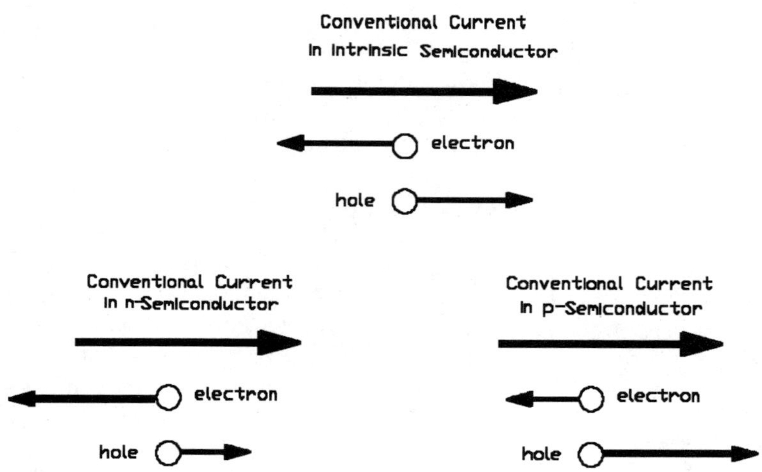

Figure C.1
Electron and hole currents

ers. The majority carrier density in n-semiconductor approximately equals the density of donor atoms $\mathbf{N_D}$. The majority carrier density in p-semiconductor approximately equals the density of acceptor atoms $\mathbf{N_A}$.

$$n_n \approx N_D, \quad p_p \approx N_A$$

The minority carrier densities are

$$n_p \approx \frac{n_i^2}{N_D}, \qquad p_n \approx \frac{n_i^2}{N_A} \tag{C-5}$$

C.2 CHARGE TRANSPORT IN HOMOGENOUS-STRUCTURE SEMICONDUCTOR DEVICES

The homogenous structure devices are fabricated from an intrinsic or doped semiconductor material. Electrons and holes in such a device differ in mobility. Electrons move like particles where as the holes move by pushing away an electron from the atom. The electron mobility μ_n is higher than the hole mobility μ_p: for silicon μ_n = **1500 cm²/V-s** and μ_p = **500 cm²/V-s.** Table C.1 shows some properties of important semiconductors. Both mobilities decrease with temperature.

Charge carrier transport in solid materials is the result of two phenomena:

1. **electrical drift** (under the influence of an electric field)
2. **density diffusion** (under the gradient of carrier density)

In metals the first phenomenon is very predominant. The charge carrier (electron) density is very high, hence their displacement from one part to another does not make an appreciable difference. Thus, the diffusion component of current is negligible. In semiconductors, however, the charge carrier densities are low, hence any differential displacement causes density gradients. Thus, the charge transport is constituted of both the drift and diffusion processes. The drift current density in semiconductors is given by

$$J_{drift} = (\mu_n n + \mu_p p)qE \tag{C-6}$$

where **nq** : electron density × electron charge
pq : hole density × electron charge

TABLE C.1 PARAMETERS OF VARIOUS SEMICONDUCTORS AT 25°C

Parameter	units	Si	Ge	GaAs	3C-SiC	6H-SiC	Diamond
E_g	eV	1.12	0.66	1.42	2.2	2.9	5.5
n_I	/cm³	1.4×10^{10}	3×10^{13}	1.8×10^6	—	—	—
μ_n	cm²/V-s	1360	3900	8500	1000	600	2200–
μ_p	cm²/V-s	490	1900	400	—	—	—
D_n	cm²/s	34	100	213	—	—	—
D_p	cm²/s	12	50	10	—	—	—
τ_n	s	10^{-6}	10^{-3}	10^{-8}	—	—	—
τ_p	s	10^{-6}	10^{-3}	10^{-8}	—	—	—
ϵ_p	F/m	11.8	16	13.1	9.7	10	5.5
σ_θ	W/cm- C	1.5	—	0.5	5.0	5.0	20
E_{BR}	V/cm	3×10^5	—	4×10^5	4×10^6	4×10^6	1×10^7
T_{opt}	K	300	—	460	873	1240	1100

σ_θ : thermal conductivity, T_{opt}: operating temperature, E_{BR}: breakdown electric field

Conductivity of the homogenous semiconductor is given by

$$\sigma = \mu_n nq + \mu_p pq \tag{C-7}$$

Since both **n** and **p** increase with temperature, the conductivity has a positive temperature coefficient. Mobility's, however, decrease with increasing temperature. Increase in the charge carrier densities **n** and **p** overshadows the decrease in mobilities, hence, the homogenous-structure devices have overall positive temperature coefficient. On the other hand, in the metals the number of conducting electrons also increases with temperature, but they do not alter the charge carrier density profile appreciably. Marginal increase in charge carrier densities is overshadowed by the reduction in the mobility due to increased collisions.

Spatial concentration of charge carriers in the homogenous-structure semiconductor gives rise to density gradients. Charge carriers drift to equalize the concentration densities, thus giving rise to diffusion current. The diffusion current density is

$$J_{diff} = (D_n \frac{\partial n}{\partial x} + D_p \frac{\partial p}{\partial x})q \tag{C-8}$$

where $\partial \mathbf{n}/\partial \mathbf{x}$: electron density gradient, and $\partial \mathbf{p}/\partial \mathbf{x}$: hole density gradient, assuming that the carrier densities differ in x-dimension only.

Diffusion constants and mobilities are related by the **Einstein relation:**

$$\frac{D_p}{\mu_p} = \frac{D_n}{\mu_n} = \frac{kT}{q} \tag{C-9}$$

Both kinds of charge transport may exist in semiconductor devices in general. One or the other may be dominant depending upon the device structure or the application.

C.3 HETEROGENEOUS-STRUCTURE DEVICES

Power electronic semiconductor devices are fabricated by metallurgical joining of the metals and semiconductor materials in several physical structures. Characteristics of these devices are determined by the geometry and the physics of the junctions of the component materials. There are three basic types of junctions:

Metal-semiconductor junction

p-n semiconductor junction

n^--n^+ or p^--p^+ semiconductor junction

The n^- and n^+ are symbols for the lightly doped and the heavily doped n-semiconductors respectively.

Depending on the dopant density and the type (n or p type) of the semiconductor, the metal-semiconductor junction can result in several useful devices or an ohmic junction that provides an electrical connection to the semiconductor region of the device. These devices are discussed in several semiconductor books. One of them is the Schottky Barrier diode, a very useful power device, which is discussed in Chapter 3.

The remaining two types of junctions give rise to all semiconductor heterogeneous structures that are known today. These structures are fabricated by a controlled diffusion of an impurity of one type (n or p type) into the surface of another type by a process called epitaxial deposition. The two types of semiconductors may also be just one type with different impurity doping densities, for example, n^--n^+ junction consists of a lightly doped n^- joined to a heavily doped n^+ type semiconductor. Both types of heterogeneous structures have similar characteristics because both the p-n and n^--n^+ junctions have similar density gradients of the charge carriers across the junction. Charge carrier densities in both types of junctions are identically prone to drift effect due to the application of electric field. In order to understand the characteristics of the semiconductor devices we need to study the p-n junction in detail.

HARRIS Semiconductor

RURP30120

April 1995

30A, 1200V Ultrafast Diode

Features

- Ultrafast with Soft Recovery . . . . <110ns
- Operating Temperature . . . . +175°C
- Reverse Voltage 1200V
- Avalanche Energy Rated
- Planar Construction

Applications

- Switching Power Supplies
- Power Switching Circuits
- General Purpose

Description

The RURP30120 (49031) is an ultrafast diode with soft recovery characteristic (t_{RR} < 110ns). It has low forward voltage drop and is silicon nitride passivated ion-implanted epitaxial planar construction.

This device is intended for use as a freewheeling/clamping diode and rectifier in a variety of switching power supplies and other power switching applications. Its low stored charge and ultrafast recovery with soft recovery characteristic minimize ringing and electrical noise in many power switching circuits, reducing power loss in the switching transistors.

PACKAGING AVAILABILITY

PART NUMBER	PACKAGE	BRAND
RURP30120	TO-220AC	RUR30120

NOTE: When ordering, use the entire part number.

Package

JEDEC TO-220AC

Symbol

Absolute Maximum Ratings T_C = +25°C, Unless Otherwise Specified

		RURP30120	UNITS
Peak Repetitive Reverse Voltage	V_{RRM}	1200	V
Working Peak Reverse Voltage	V_{RWM}	1200	V
DC Blocking Voltage	V_R	1200	V
Average Rectified Forward Current (T_C = +110°C)	$I_{F(AV)}$	30	A
Repetitive Peak Surge Current (Square Wave, 20kHz)	I_{FSM}	60	A
Nonrepetitive Peak Surge Current (Halfwave, 1 Phase, 60Hz)	I_{FSM}	300	A
Maximum Power Dissipation	P_D	125	W
Avalanche Energy (L = 40mH)	E_{AVL}	30	mj
Operating and Storage Temperature	T_{STG}, T_J	-65 to +175	°C

File Number **3397.2**

Typical Performance Curves

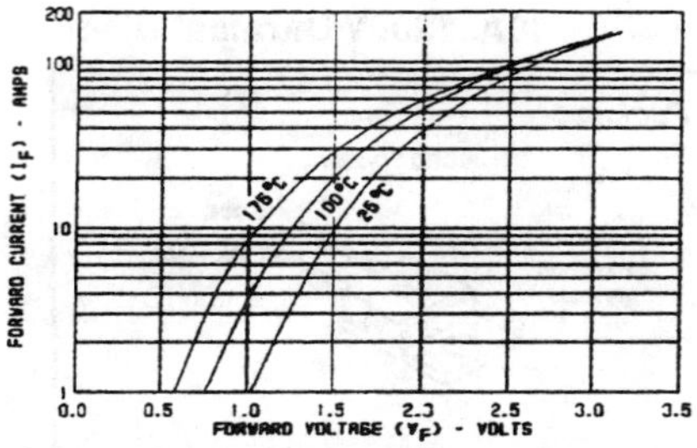

FIGURE 3. TYPICAL FORWARD CURRENT vs FORWARD VOLTAGE DROP

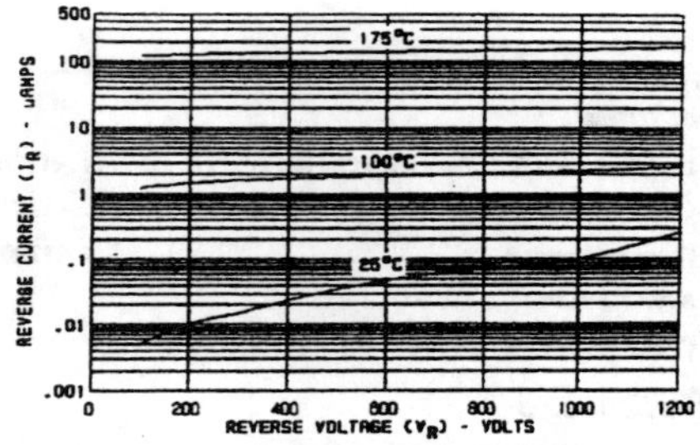

FIGURE 4. TYPICAL REVERSE CURRENT vs VOLTAGE

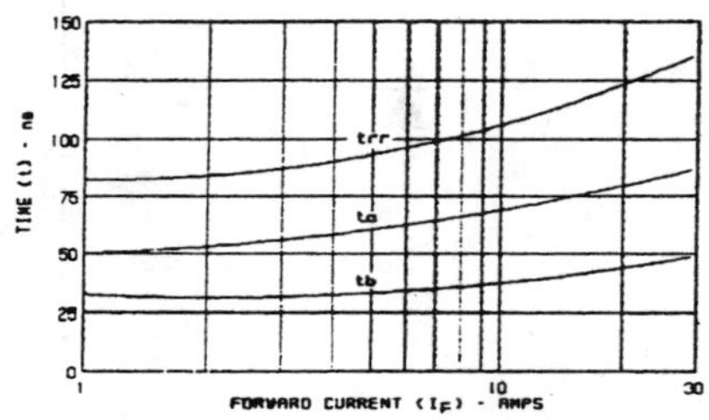

FIGURE 5. TYPICAL t_{RR}, t_A AND t_B CURVES vs FORWARD CURRENT

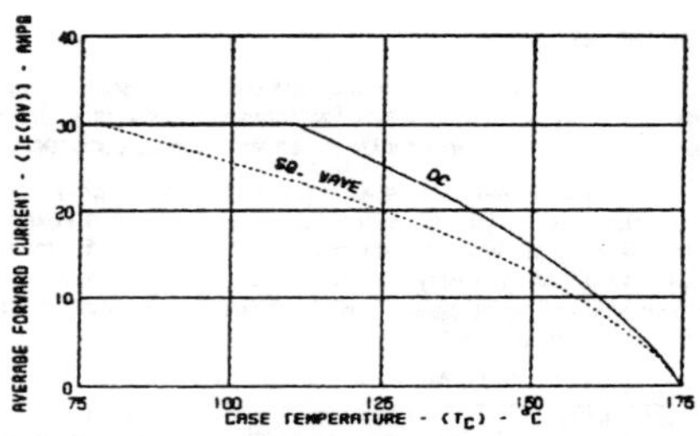

FIGURE 6. CURRENT DERATING CURVE FOR ALL TYPES

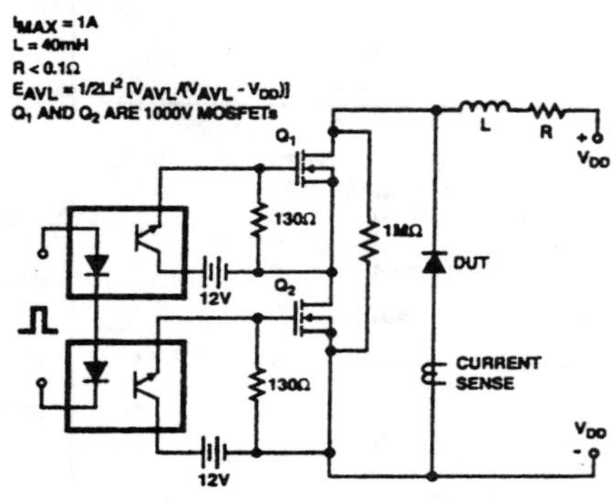

FIGURE 7. AVALANCHE ENERGY TEST CIRCUIT

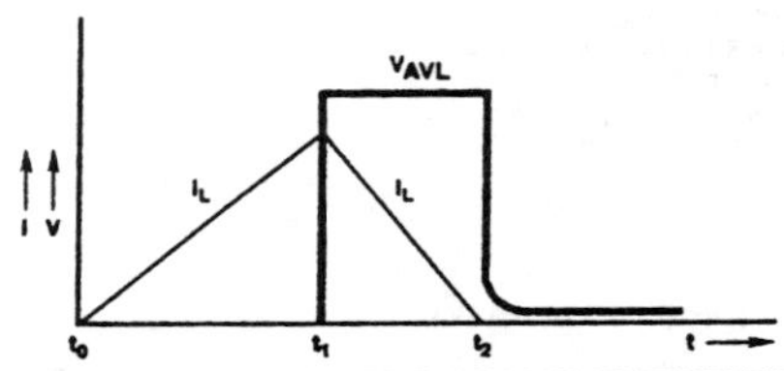

FIGURE 8. AVALANCHE CURRENT AND VOLTAGE WAVEFORMS

Electrical Specifications T_C = +25°C, Unless Otherwise Specified

SYMBOL	TEST CONDITION	LIMITS MIN	TYP	MAX	UNITS
V_F	I_F = 30A, T_C = +25°C	-	-	2.1	V
V_F	I_F = 30A, T_C = +150°C	-	-	1.9	V
I_R	V_R = 1200V, T_C = +25°C	-	-	500	µA
I_R	V_R = 1200V, T_C = +150°C	-	-	1	mA
t_{RR}	I_F = 1A, dI_F/dt = 100A/µs	-	-	110	ns
	I_F = 30A, dI_F/dt = 100A/µs	-	-	150	ns
t_A	I_F = 30A, dI_F/dt = 100A/µs	-	90	-	ns
t_B	I_F = 30A, dI_F/dt = 100A/µs	-	45	-	ns
$R_{θJC}$		-	-	1.2	°C/W

DEFINITIONS

V_F = Instantaneous forward voltage (pw = 300µs, D = 2%).
I_R = Instantaneous reverse current.
t_{RR} = Reverse recovery time (See Figure 2), summation of $t_A + t_B$.
t_A = Time to reach peak reverse current (See Figure 2).
t_B = Time from peak I_{RM} to projected zero crossing of I_{RM} based on a straight line from peak I_{RM} through 25% of I_{RM} (See Figure 2).
$R_{θJC}$ = Thermal resistance junction to case.
E_{AVL} = Controlled avalanche energy (See Figures 7 and 8).
pw = pulse width.
D = duty cycle.

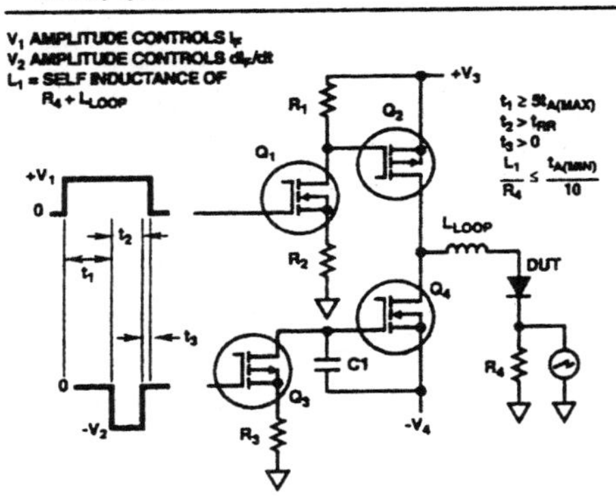

FIGURE 1. t_{RR} TEST CIRCUIT

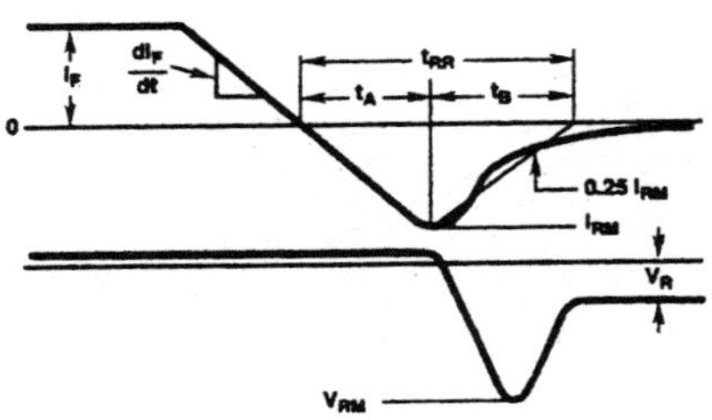

FIGURE 2. t_{RR} WAVEFORMS AND DEFINITIONS

APPENDIX D
MATLAB Progams

MATLAB Program for discontinuous mode operation, Figure 8.30

```
VS=120*sqrt(2); VO=120;omegaL=2*pi*60*0.1;
dth=pi/512;
alpha=.566:dth:pi/2-dth;
Iosc=120/omegaL; Vodc=2*VS/pi;
Iob=2*VS*sin(alpha)/(pi*omegaL);
VO=[]; IO=[]; pf=[];td=[];
n=size(alpha,2);

for i=1:n
    VOi=VS*sin(alpha(i));
    th1=alpha(i):dth:pi+alpha(i)-dth;
    th2=pi+alpha(i):dth:2*pi+alpha(i)-dth;
    th=[th1 th2 ];
    vs1=VS*sin(th1);
    vs2=VS*sin(th2);
    vs=[vs1 vs2];
    vl1=vs1-VOi;
    io1=integ(vl1/omegaL,th);
    io1=io1.*(io1>=0);
    io2=io1;
    io=[io1 io2];
    IOi=mean(io);
    is=[io1 -io1];
    [f,ish,ph]=harmonic(is);
    isrms=sqrt(mean(is.^2));
    is1rms=ish(2)/sqrt(2);
    phase1=ph-alpha(i);
    df=cos(phase1);
    pfi=df*is1rms/isrms;
    tdi=sqrt((isrms/is1rms)^2-1);
    VO=[VO VOi]; IO=[IO IOi]; pf=[pf pfi]; td=[td tdi];
end
io=[io1 io2];
is=[io1 -io2]; %input current array
isrms=sqrt(mean(is .^2));
iodc=mean(io);
    IO=[IO iodc];
    Rii=[Rii iodc/isrms];
    [f, ish,ph] =harmonic(is);
    phi=[phi ph(1)];
    df=cos(ph);
    pf=[ pf (((ish(1)/sqrt(2))/isrms) * df)];
    td = [ td (sqrt ( (isrms/(ish(1)/sqrt(2))) ^ 2 -1 ))];
ripple1=VS/(iodc*omegal)*(cos(alpha0)-(2/pi *(alpha0-
  alphan)+1).*cos(alphan));
ripple=[ripple ripple1];
    if(iodc >.001)
CF1= (VS/omegal*(cos(alphan)+cos(alpha0))
  +VO/omegal*(alphan+alpha0-pi))/iodc;
    CF=[CF CF1]
    end
end
plot(alpha,Tvv, alpha,Rii,'*', alpha,pf,'+', alpha, Iob)
plot(alpha,td)
plot(alpha, CF)
```

MATLAB Program for Section 8.5.2

```
%notching in constant current case
VS=120*sqrt(2); InO=.2; Ls1=1; Ls2=1;
u=acos(1-2*InO);
dth=pi/512;
th1=0:dth:u-dth;
th2=u:dth:pi-dth;

th=[th1 th2];
th=[th pi+th];
vs=VS*sin(th);
vs1=(Ls2/(Ls1+Ls2))*VS*sin(th1);
vs2=VS*sin(th2);
vpcc=[vs1 vs2 ];
vpcc=[vpcc -vpcc];
plot(th, vs,'—',th, vpcc)
```

MATLAB Program for Section 8.5.2

```
%notching in constant voltage case
VS=120*sqrt(2); VO=120; Ls1=1; Ls2=1;
alpha=.7854; beta=pi-.1;
dth=pi/512;
th1=0:dth:alpha-dth;
th2=alpha:dth:beta-dth;
th3=beta:pi/100:pi-dth;
th=[th1 th2 th3];
th=[th pi+th];
vs=VS*sin(th);
vs1=VS*sin(th1);
vs2=(Ls2/(Ls1+Ls2))*(VS*sin(th2) +VO*Ls1/Ls2 );
vs3=VS*sin(th3);
vpcc=[vs1 vs2 vs3];
vpcc=[vpcc -vpcc];
plot(th, vs,'—',th, vpcc)
```

APPENDIX E

The MATLAB functions used in Chapter 1 are given below:

```
function [beta]=betam(a,w)
b=pi-a:pi/1000:pi+a;
f=cos(a)-cos(b)-w*(b-a);
[x,n]=min(f.^2);
beta=b(n);
end
```

```
function [f,y,ph]=harmonic(x,n)
%f is the array of harmonics
%y(k) is the amplitude of the k-th harmonic
%ph is the phase of the fundamental harmonic
%n is the number of harmonics to be displayed on the plot
N=size(x,2);%number of elements in the x array
f=0:1:N-1;
y1=fft(x);
y=2*(abs(y1))/N;%the n-th harmonic
y(1)=y1(1)/N;%the dc term
%ph=angle(y1(2))+pi/2;
ph=angle(y1(2))+pi/2;% phase angle of fundamental
  harmonic
stem(f,y,'r')
axis([0,n-1,0, 1.2*max(y)])
%end
```

```
function y=integ(x,th)
dth=th(2)-th(1);
y=cumsum(x)*dth;
```

APPENDIX F

Use the MATLAB function to find the intersection of two curves or the solution of two simultaneous equations with two variables.

```
function y=intsect(x1,x2,t,ll,ul)
% to find intersection of two arrays x1 and x2 , both
  functions of independent variable t (an array), within
  the segment from lower limit (t=ll) to the upper limit
  (t=ul)
tp=((t>=ll)&(t<=ul)); % the t array with 1's in the
  intended segment and 0's elsewhere
xd=abs(x1-x2); % absolute value of the difference
  x1-x2
xd1=xd.*(tp)+max(xd)*(~tp); % the difference array in
  the intended segment is raised over the pedestal of the
  maximum value of the difference array outside the
  intended segment. This is done to eliminate ambiguity
  in the determination of the minimum of the difference
  array
[a,b]=min(xd1); % a is the minimum value and b is the
  index of the segment of the difference array at which
  the minimum occurs
y=t(b); % the time at which the minimum occurs
end
```

INDEX